CADMIUM CONTAMINATION IN SOILS AND PLANTS

CADMIUM CONTAMINATION IN SOILS AND PLANTS

Mechanisms and Mitigation

Edited by

FANG-JIE ZHAO

College of Resources and Environmental Sciences, Nanjing Agricultural University, Nanjing, China

PENG WANG

College of Resources and Environmental Sciences, Nanjing Agricultural University, Nanjing, China; College of Sciences, Nanjing Agricultural University, Nanjing, China

STEVE MCGRATH

Sustainable Soils and Crops, Rothamsted Research, Harpenden, United Kingdom

ELSEVIER

ACADEMIC PRESS

An imprint of Elsevier

elsevier.com/books-and-journals

Academic Press is an imprint of Elsevier

125 London Wall, London EC2Y 5AS, United Kingdom
50 Hampshire Street, 5th Floor, Cambridge, MA 02139, United States

ISBN: 978-0-443-34104-5

For Information on all Academic Press publications visit our website at https://www.elsevier.com/books-and-journals

Publisher: Andre Wolff
Acquisitions Editor: Chiara Giglio
Editorial Project Manager: Sudip Saha
Production Project Manager: Aishwarya R
Cover Designer: Mark Rogers

Typeset by Aptara, New Delhi, India

Contents

7. Transformation of cadmium in paddy soils

Peng Wang, Hui Huang, and Fang-Jie Zhao

8. Impacts of cadmium on soil microorganisms and fauna

Steve McGrath and Fang-Jie Zhao

9. Assessment of cadmium bioavailability in soils

Taige Li, Hao Zhang, Xing Liu, and Jun Luo

10. Molecular mechanisms of cadmium uptake and translocation in plants

Fang-Jie Zhao and Xin-Yuan Huang

11. Toxicity and detoxification of cadmium in plants

Zhong Tang and Fang-Jie Zhao

12. Cadmium hyperaccumulation

Hendrik Küpper, Qi Tao, and Filis Morina

13. Role of mycorrhiza in cadmium uptake and detoxification

Xin Zhang and Baodong Chen

14. Modeling cadmium transfer from soil to crops

Meie Wang, Chi Peng, Rufus L. Chaney, and Weiping Chen

15. Dietary intake of cadmium and impacts on human health

Di Zhao and Fang-Jie Zhao

16. Ecological risk assessment of cadmium contamination in soils

Koen Oorts and Steve McGrath

17. Soil cadmium standards for agricultural land

Sha Zhang, Jing Song, Yongming Luo, and Roger Brewer

18. Amendments and agronomic practices to reduce cadmium uptake by crops

Zhong-Rui Xu, Fang-Jie Zhao, and Peng Wang

19. Breeding and biotechnology to limit cadmium accumulation in crops

Xin-Yuan Huang and Fang-Jie Zhao

20. Phytoremediation of cadmium-contaminated soils

Longhua Wu, Pengjie Hu, Tong Zhou, Jiawen Zhou, Zhu Li, and Peter Christie

Contributors

Lu Bai State Environmental Protection Key Laboratory of Eco-Industry, Chinese Research Academy of Environmental Sciences, Beijing, China

Roger Brewer Hawai'i Department of Health, Honolulu, HI, United States

Rufus L. Chaney Chaney Environmental, Beltsville, MD, United States

Baodong Chen State Key Laboratory of Regional and Urban Ecology, Research Centre for Eco-Environmental Sciences, Chinese Academy of Sciences, Beijing, China; University of Chinese Academy of Sciences, Beijing, China

Mingming Chen Academy for Advanced Interdisciplinary Studies, Nanjing Agricultural University, Nanjing, China

Weiping Chen Research Center for Eco-Environmental Sciences, Chinese Academy of Sciences, Beijing, China

Peter Christie State Key Laboratory of Soil and Sustainable Agriculture, Institute of Soil Science, Chinese Academy of Sciences, Nanjing, China

Yingxuan Deng School of Environment, Henan Normal University, Xinxiang, China

Hui Gao Soil Chemistry Group, Wageningen University & Research, Wageningen, The Netherlands

Deyi Hou School of Environment, Tsinghua University, Beijing, China

Pengjie Hu State Key Laboratory of Soil and Sustainable Agriculture, Institute of Soil Science, Chinese Academy of Sciences, Nanjing, China

Hui Huang College of Ecology and Environment, Nanjing Forestry University, Nanjing, China

Xin-Yuan Huang College of Resources and Environmental Sciences, Nanjing Agricultural University, Nanjing, China

Yifan Huang Soil Chemistry Group, Wageningen University & Research, Wageningen, The Netherlands

Xiyue Jia School of Environment, Tsinghua University, Beijing, China

Hendrik Küpper Department of Plant Biophysics and Biochemistry, Biology Centre, Institute of Plant Molecular Biology, Czech Academy of Sciences, České Budějovice, Czech Republic; Department of Experimental Plant Biology, Faculty of Science, University of South Bohemia, České Budějovice, Czech Republic

Wenkai Lei The Analysis and Test Center, Capital Normal University, Beijing, China; State Key Laboratory of Atmospheric Environment and Extreme Meteorology, Institute of Atmospheric Physics, Chinese Academy of Sciences, Beijing, China

Jiawei Li State Key Laboratory of Earth System Numerical Modeling and Application, Institute of Atmospheric Physics, Chinese Academy of Sciences, Beijing, China

Jinbo Li School of Grassland Science, Beijing Forestry University, Beijing, China

Taige Li State Key Laboratory of Water Pollution Control and Green Resource Recycling, School of Environment, Nanjing University, Nanjing, China

Zhu Li State Key Laboratory of Soil and Sustainable Agriculture, Institute of Soil Science, Chinese Academy of Sciences, Nanjing, China

Wanli Lian School of Environment, Henan Normal University, Xinxiang, China

Xing Liu State Key Laboratory of Water Pollution Control and Green Resource Recycling, School of Environment, Nanjing University, Nanjing, China

Jun Luo State Key Laboratory of Water Pollution Control and Green Resource Recycling, School of Environment, Nanjing University, Nanjing, China

Yongming Luo State Key Laboratory of Soil and Sustainable Agriculture, Institute of Soil Science, Chinese Academy of Sciences, Nanjing, China

Steve McGrath Sustainable Soils and Crops, Rothamsted Research, Harpenden, United Kingdom

Filis Morina Department of Plant Biophysics and Biochemistry, Biology Centre, Institute of Plant Molecular Biology, Czech Academy of Sciences, České Budějovice, Czech Republic

Koen Oorts ARCHE Consulting, Leuven, Belgium

Yuepeng Pan State Key Laboratory of Atmospheric Environment and Extreme Meteorology, Institute of Atmospheric Physics, Chinese Academy of Sciences, Beijing, China

Chi Peng Institute of Environmental Engineering, School of Metallurgy and Environment, Central South University, Changsha, China

Jing Song State Key Laboratory of Soil and Sustainable Agriculture, Institute of Soil Science, Chinese Academy of Sciences, Nanjing, China; University of Chinese Academy of Sciences, Nanjing, China

Zaijin Sun Technical Centre for Soil, Agriculture and Rural Ecology and Environment, Ministry of Ecology and Environment, Beijing, China

Zhong Tang College of Resources and Environmental Sciences, Nanjing Agricultural University, Nanjing, China

Qi Tao College of Resources, Sichuan Agricultural University, Chengdu, China; New Cornerstone Science Laboratory, State Key Laboratory of Crop Gene Exploration and Utilisation in Southwest China, Sichuan Agricultural University at Wenjiang, Chengdu, Sichuan, China

Long Wang School of Resources and Environment, Henan Agricultural University, Zhengzhou, China

Meie Wang Research Center for Eco-Environmental Sciences, Chinese Academy of Sciences, Beijing, China

Peng Wang College of Resources and Environmental Sciences, Nanjing Agricultural University, Nanjing, China; College of Sciences, Nanjing Agricultural University, Nanjing, China

Liping Weng Soil Chemistry Group, Wageningen University & Research, Wageningen, The Netherlands

Longhua Wu State Key Laboratory of Soil and Sustainable Agriculture, Institute of Soil Science, Chinese Academy of Sciences, Nanjing, China

Zhong-Rui Xu College of Resources and Environmental Sciences, Nanjing Agricultural University, Nanjing, China

Xinping Yang College of Resources and Environmental Sciences, Nanjing Agricultural University, Nanjing, China

Hao Zhang Lancaster Environment Centre, Lancaster University, Lancaster, United Kingdom

Sha Zhang State Key Laboratory of Regional and Urban Ecology, Ningbo Observation and Research Station, Institute of Urban Environment, Chinese Academy of Sciences, Xiamen, China; Zhejiang Key Laboratory of Pollution Control for Port-Petrochemical Industry, CAS Haixi Industrial Technology Innovation Center in Beilun, Ningbo, China

Xin Zhang State Key Laboratory of Regional and Urban Ecology, Research Centre for Eco-Environmental Sciences, Chinese Academy of Sciences, Beijing, China; University of Chinese Academy of Sciences, Beijing, China

Di Zhao College of Resources and Environmental Sciences, Nanjing Agricultural University, Nanjing, China

Fang-Jie Zhao College of Resources and Environmental Sciences, Nanjing Agricultural University, Nanjing, China

Jiawen Zhou State Key Laboratory of Soil and Sustainable Agriculture, Institute of Soil Science, Chinese Academy of Sciences, Nanjing, China

Tong Zhou State Key Laboratory of Soil and Sustainable Agriculture, Institute of Soil Science, Chinese Academy of Sciences, Nanjing, China

About the Editors

Fang-Jie Zhao is a Professor at Nanjing Agricultural University, China. He obtained his PhD from the University of Newcastle Upon Tyne, United Kingdom, in 1992. His research focuses on the biogeochemistry of trace elements, molecular mechanisms of trace element uptake and detoxification in plants, bioremediation of contaminated soils, and biofortification of micronutrients in food crops. He has published over 450 peer-reviewed articles and has been a Clarivate Web of Science highly cited researcher since 2017. He is the recipient of the 2024 UNESCO The World Academy of Sciences Award in Agricultural Science and the 2022 IFA Norman Borlaug Plant Nutrition Award by the International Fertilizer Association.

Peng Wang is a Professor at Nanjing Agricultural University, China. He obtained his PhD from the Institute of Soil Science, Chinese Academy of Sciences, in 2011. He researches soil biogeochemistry, environmental remediation, and human health, specializing in cadmium and arsenic risk control within soil–crop–human systems. His discoveries of mechanisms, such as the voltaic cell effect and pH-dependent adsorption governing cadmium activity, have advanced practical remediation. He has developed novel high-throughput characterization techniques for soil–plant systems. He has published over 160 articles in journals such as *PNAS, Nature Health*, and *ES&T*, with over 11,000 citations and an h-index of 58. He is a Highly Cited Researcher (Clarivate and Elsevier) and serves as a Section Editor for *Plant and Soil (Springer Nature)* and an Associate Editor for *Applied Geochemistry (Elsevier)*.

Steve McGrath of Rothamsted Research obtained his PhD from the University of Sheffield, United Kingdom, in 1980. His research focuses on understanding the biogeochemical processes that occur in soils and their impacts on crop productivity, food quality, and the environment. He has contributed significantly to the development of sustainable soil management practices and the development of guidelines for soil protection and to reduce the potential risks associated with heavy metal contamination in the food chain. He has published over 375 peer-reviewed articles and has been a Clarivate Web of Science highly cited researcher since 2000. His research has been recognized with numerous awards, including election as a Fellow of the Royal Society in 2024 and as the President of the International Society of Trace Element Biogeochemistry in 2013.

Preface

Cadmium (Cd) is a so-called trace metal, yet its consequences for agriculture, ecosystems, and human health are profound. Highly mobile within soil–plant–food systems and acutely toxic even at low concentrations, Cd presents one of the most persistent and challenging problems in modern agri-environmental management. Across the globe, Cd contamination of soils threatens soil ecological functioning, compromises food safety, and poses serious risks to public health.

This book brings together the current state of knowledge on Cd in soil–plant systems, offering a comprehensive and authoritative synthesis of research from across disciplines. Comprising 20 chapters, it spans analytical methods for Cd determination; the global extent and drivers of soil Cd contamination; inputs, outputs, and fluxes within soil–plant systems; and the biogeochemical processes governing Cd transformation and mobility in soils. The volume further explores the mechanisms of Cd uptake, transport, and detoxification in plants; the impacts of Cd on soil microorganisms, fauna, and crops; ecosystem and human health risk assessment; and emerging remediation strategies to mitigate Cd contamination.

The authors have long-standing expertise in Cd research, ensuring both depth and rigor. Collectively, this book provides an essential reference for researchers, postgraduate students, environmental professionals, and policymakers concerned with soil and crop contamination, agricultural sustainability, food safety, public health, and environmental remediation.

Fang-Jie Zhao
Peng Wang
Steve McGrath

CHAPTER

1

Introduction

Fang-Jie Zhao[a], Peng Wang[a,b], Steve McGrath[c]

[a]College of Resources and Environmental Sciences, Nanjing Agricultural University, Nanjing, China [b]College of Sciences, Nanjing Agricultural University, Nanjing, China [c]Sustainable Soils and Crops, Rothamsted Research, Harpenden, United Kingdom

Cadmium (Cd) is a metal belonging to Group IIB in the periodic table with a relatively high density (8.65 g cm^{-3}). The element was discovered as a contaminant present in zinc (Zn) compounds by the German chemist Friedrich Stromeyer in 1817. Cd is a relatively rare element, comprising about 0.1–0.5 mg kg^{-1} of the Earth's crust, and is ranked 64th in abundance among all elements (Agency for Toxic Substances and Disease Registry, 2012; Krauskopf & Bird, 1995). Cd occurs naturally in all rocks and is enriched in certain ores, such as zinc/lead/copper sulfides, and sedimentary shales or phosphorites (McLaughlin et al., 2021). Cd is used primarily for the production of rechargeable nickel–Cd batteries, as well as pigments, coatings, plating, and photovoltaic cells. Global refinery production of Cd is about 24,000 MT annually (U.S. Geological Survey, 2025).

During pedogenesis, weathering of rocks releases Cd from mineral lattices to more soluble forms in soil. Soil Cd that is derived from its parent materials is termed geogenic Cd. Geogenic soil Cd concentrations can vary from <0.01 mg kg^{-1} up to tens or hundreds of mg kg^{-1} in soils that develop on carbonates, shales, or phosphorites (McLaughlin et al., 2021). For example, soils that developed from phosphorite concretions in Jamaica contain unusually high concentrations of Cd, giving rise to the median and 90 percentile Cd concentrations for the country as high as 4.0 and 78 mg Cd kg^{-1}, respectively (Smolders & Mertens, 2013). The original source of Cd in the phosphorite concretions in Jamaica is guano deposits (seabird droppings) formed in the Late Miocene or Pliocene, reflecting the highly efficient transfer of Cd from ocean water into pelagic birds. As another example the terra rossa (red) soils that developed from Cd-enriched carbonate rocks in the karst region of southwestern China have elevated Cd concentrations (Wen et al., 2020). During pedogenesis, much of the calcium carbonate minerals have been weathered and leached, leaving Cd in the residual fractions largely associated with amorphous iron/manganese oxides.

In addition to geogenic Cd, human activities, such as metal mining and smelting, combustion of coal, incineration of wastes, applications of Cd-containing fertilizers, especially phosphate

Cadmium Contamination in Soils and Plants: Mechanisms and Mitigation.
DOI: https://doi.org/10.1016/B978-0-443-34104-5.00017-7

fertilizers, and irrigation with contaminated water, have also added Cd to soils; this portion of Cd can be classed as anthropogenic Cd and reflects the level of man-made pollution. Anthropogenic Cd pollution is of great concern to human and ecological health, although it is difficult to separate anthropogenic and geogenic Cd in soil. Because of long-distance transport of Cd via the atmosphere and water, it is likely that all the soils on our planet have received anthropogenic Cd inputs to some extent. Combining both geogenic and anthropogenic sources, soil Cd concentrations typically vary from 0.1 to a few mg kg^{-1} with the median values of large-scale surveys settling in the range of 0.1–0.3 mg Cd kg^{-1} (Smolders & Mertens, 2013). Much higher concentrations can be found in soils with high geogenic or anthropogenic Cd sources. It has been estimated that about 7% of the global cropland soils contain Cd at levels above the thresholds for ecological or human health (Hou et al., 2025).

Cd has no known biological functions in organisms except in some marine diatoms, such as *Thalassiosira weissflogii*, which can use Cd, instead of Zn, as a catalytic metal atom in carbonic anhydrase, an enzyme required for CO_2 fixation (Lane & Morel, 2000). The evolution of the Cd-enzyme may be a result of low Zn bioavailability in the ocean surface waters. In the terrestrial environment, Cd is known as a widespread toxin, which can cause toxicity to organisms living in soil and human diseases such as renal dysfunction, osteoporosis, and cancers (Åkesson et al., 2014; International Agency for Research on Cancer, 1993; Nordberg, 2009). The itai-itai disease (renal tubular osteomalacia) is the best known Cd-induced disease, which is caused by chronic exposure to Cd via consumption of rice grown in Cd-contaminated paddy soil (Aoshima & Horiguchi, 2019). This endemic disease started to occur in the Jinzu River basin in Toyama Prefecture, Japan, in the 1910s, because of upstream lead–zinc mining and smelting, although it was not until 1960s that Cd poisoning was finally identified as the cause of the disease. Cd accumulates in the liver and kidneys and has a long half-life (10–20 years) in human body (International Agency for Research on Cancer, 1993). Thus chronic exposure to Cd affects human health, mainly in older people.

Consumption of food is the main route of Cd exposure for the general population, accounting for approximately 90% of the total Cd intake of nonsmokers (Clemens et al., 2002). Compared with other metals or metalloids, Cd has relatively high bioavailability in soil and is readily taken up by crop plants and accumulated in edible parts. There are three barriers that restrict the entry of toxic elements from soil to the food chain: (1) solubility. Some elements (e.g., Pb, Hg, Cr^{3+}, Ag, Au, Ti, Sn, and Zr) have very low solubility in soil so that plants absorb only very small amounts of them; (2) translocation. After absorption by the roots, some elements (e.g., Pb, Hg, and Al) can be strongly adsorbed on the root surfaces or strongly sequestered in the root cells, limiting their translocation to plant shoots; (3) phytotoxicity. Some elements (e.g., Cu, Zn, Ni, Co, and Mn) can cause toxicity to plants and reduce yields, and the affected crops are less likely to be consumed by humans or domestic animals (Chaney, 1980) (Fig. 1.1). Cd is also phytotoxic. However, the soil–plant barrier does not protect humans and animals from the toxicity of Cd, because Cd has a relatively high solubility in soil and a high translocation in plants and is toxic to animals at levels that do not cause phytotoxicity (Chaney, 1980). The ability of Cd to bypass these three barriers underlies its problematic nature in the soil–plant–animal/human nexus.

The high transfer potential of Cd is demonstrated in Fig. 1.2, showing that the median transfer factor (i.e., grain metal concentration/soil metal concentration) of Cd is approximately 10, 100, and 1000 times those of copper (Cu), arsenic (As), and lead (Pb), respectively, among 200 paired soil–rice grain samples collected in the study of Chen et al. (2018). The median transfer factor of Cd is approximately 1, meaning that a soil containing 0.2 mg kg^{-1} total Cd, that is, within the

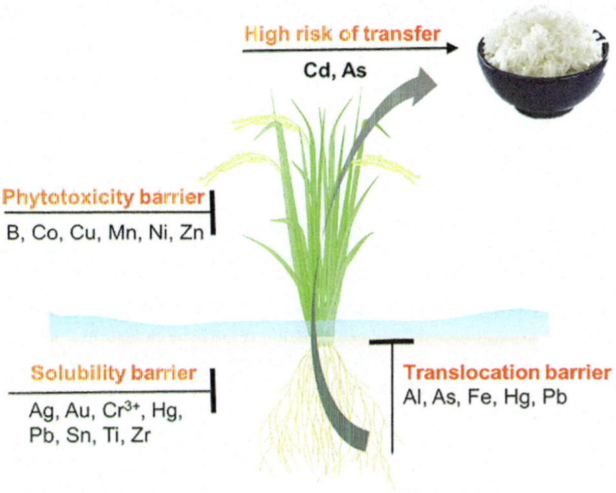

FIGURE 1.1 Soil–plant barriers and the risk of toxic element transfer to the food chain, based on the concept proposed by Chaney (1980) with modifications. *Cd*, Cadmium. *From Zhao and Wang (2020) with copyright permission.*

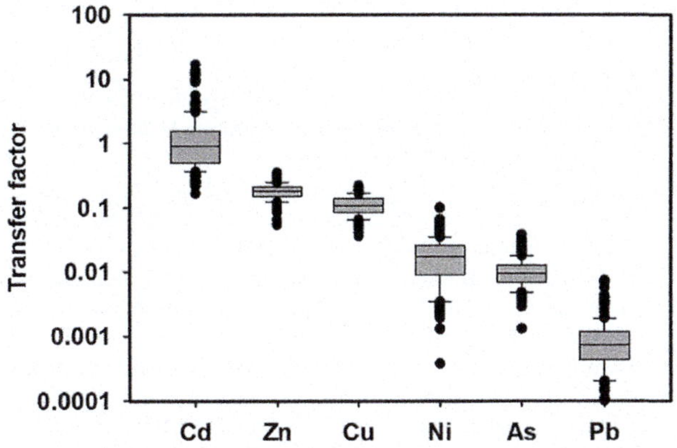

FIGURE 1.2 Boxplot of the transfer factors (i.e., grain metal concentration/soil metal concentration) for cadmium (Cd), zinc (Zn), copper (Cu), nickel (Ni), arsenic (As), and lead (Pb) in 200 paired soil–rice grain samples collected from southern China. *Data from Chen et al. (2018) with copyright permission.*

normal background range, could produce rice grain that exceeds the Chinese Cd limit for rice of 0.2 mg kg^{-1}. Another feature shown in this diagram is the large variation (approximately 2 orders of magnitude) in the Cd transfer factor among the 200 paired samples. This variation is caused by soil factors and crop management practices that affect Cd bioavailability, as well as crop cultivars that differ in Cd uptake and translocation (McLaughlin et al., 2021; Zhao et al., 2022). This example illustrates not only the high potential of Cd in the transfer from soil to the food chain but also the need to understand its behavior in soil–plant systems to mitigate its potential impact on human health.

The last quarter of century has seen a large increase in publications on Cd in soil and plants, as well as assessments on the risk of Cd to ecological functions and human health. Much progress has been made in understanding the mechanisms of Cd transformation in soil and Cd uptake and translocation in plants, paving the foundations for developing technological solutions to remediate Cd-contaminated soils or to limit Cd uptake by food crops. This book aims to synthesize and present the state-of-the-art knowledge of Cd in soil–plant systems comprehensively, focusing on the mechanisms and mitigation strategies.

References

Agency for Toxic Substances and Disease Registry. (2012). *Toxicological profile for cadmium.* U.S. Department of Health and Human Services, Public Health Service.

Åkesson, A., Barregard, L., Bergdahl, I. A., Nordberg, G. F., Nordberg, M., & Skerfving, S. (2014). Non-renal effects and the risk assessment of environmental cadmium exposure. *Environmental Health Perspectives, 122*(5), 431–438. http://ehp.niehs.nih.gov/wp-content/uploads/122/5/ehp.1307110.pdf. doi:10.1289/ehp.1307110.

Aoshima, K., & Horiguchi, H. (2019). Chapter 1: Cadmium toxicity, current topics in environmental health and preventive medicine. In S. Himeno, & K. Aoshima (Eds.), *Historical lessons on cadmium environmental pollution problems in japan and current cadmium exposure situation*. Springer Nature Singapore Pte Ltd. doi:10.1007/978-981-13-3630-0.

Chaney, R. L. (1980). Health risks associated with toxic metals in municipal sludge. In G. Bitton (Ed.), *Sludge: Health risks of land applications* (pp. 59–83). Ann Arbor Science Publisher.

Chen, H., Yang, X., Wang, P., Wang, Z., Li, M., & Zhao, F. J. (2018). Dietary cadmium intake from rice and vegetables and potential health risk: A case study in Xiangtan, southern China. *Science of the Total Environment, 639*, 271–277. www.elsevier.com/locate/scitotenv. doi:10.1016/j.scitotenv.2018.05.050.

Clemens, S., Palmgren, M. G., & Krämer, U. (2002). A long way ahead: Understanding and engineering plant metal accumulation. *Trends in Plant Science, 7*(7), 309–315. doi:10.1016/S1360-1385(02)02295-1.

Hou, D., Jia, X., Wang, L., McGrath, S. P., Zhu, Y. G., Hu, Q., Zhao, F. J., Bank, M. S., O'Connor, D., & Nriagu, J. (2025). Global soil pollution by toxic metals threatens agriculture and human health. *Science, 388*(6744), 316–321. https://www.science.org/toc/science/388/6744. doi:10.1126/science.adr5214.

International Agency for Research on Cancer. (1993). Beryllium, cadmium, mercury and exposures in the glass manufacturing industry. In *IARC monographs on the evaluation of carcinogenic risk of chemicals to humans no. 58* (p. 444). IARC Working Group.

Krauskopf, K. B., & Bird, D. K. (1995). *Introduction to geochemistry*. McGraw-Hill.

Lane, T. W., & Morel, F. M. M. (2000). A biological function for cadmium in marine diatoms. *Proceedings of the National Academy of Sciences of the United States of America, 97*(9), 4627–4631. doi:10.1073/pnas.090091397.

McLaughlin, M. J., Smolders, E., Zhao, F. J., Grant, C., & Montalvo, D. (2021). Managing cadmium in agricultural systems. *Advances in Agronomy, 166*, 1–129. http://www.elsevier.com/wps/find/bookdescription.cws_home/675816/description#description. doi:10.1016/bs.agron.2020.10.004.

Nordberg, G. F. (2009). Historical perspectives on cadmium toxicology. *Toxicology and Applied Pharmacology, 238*(3), 192–200. doi:10.1016/j.taap.2009.03.015.

Smolders, E., & Mertens, J. (2013). Chapter 10: Cadmium. In *Heavy metals in soils: Trace metals and metalloids in soils and their bioavailability* (pp. 283–311). *Springer Science + Business Media.* doi:10.1007/978-94-007-4470-7_1.

U.S. Geological Survey. (2025). Cadmium statistics and information. https://www.usgs.gov/centers/national-minerals-information-center/cadmium-statistics-and-information

Wen, Y., Li, W., Yang, Z. F., Zhang, Q., & Ji, J. F. (2020). Enrichment and source identification of Cd and other heavy metals in soils with high geochemical background in the karst region, Southwestern China. *Chemosphere, 245*, 125620. doi:10.1016/j.chemosphere.2019.125620.

Zhao, F. J., Tang, Z., Song, J. J., Huang, X. Y., & Wang, P. (2022). Toxic metals and metalloids: Uptake, transport, detoxification, phytoremediation, and crop improvement for safer food. *Molecular Plant, 15*(1), 27–44. https://www.cell.com/molecular-plant/home. doi:10.1016/j.molp.2021.09.016.

Zhao, F. J., & Wang, P. (2020). Arsenic and cadmium accumulation in rice and mitigation strategies. *Plant and Soil, 446*(1–2), 1–21. doi:10.1007/s11104-019-04374-6.

CHAPTER

2

Analytical methods for cadmium in soil–plant systems

Mingming Chen[a], *Peng Wang*[b,c]

[a] Academy for Advanced Interdisciplinary Studies, Nanjing Agricultural University, Nanjing, China [b] College of Resources and Environmental Sciences, Nanjing Agricultural University, Nanjing, China [c] College of Sciences, Nanjing Agricultural University, Nanjing, China

2.1 Introduction

Cadmium (Cd) is a heavy metal element that is highly toxic to living organisms. It is widely present in soil, industrial wastewater, and municipal waste and can enter the food chain through the soil–plant system, thus posing a serious threat to human and ecosystem health. The concentrations of Cd in soils and plants are usually in the parts per billion (ppb) to parts per million (ppm) range, thus necessitating careful sample preparation and sensitive and accurate analytical methods for quantification. This chapter reviews the sample pretreatment methods and current major methods for analyzing Cd in soil and plants, compares the advantages and limitations of different analytical techniques, and aims to provide a reference for the study of Cd-related environmental and health research.

The measurement and analysis of Cd in the soil and plant samples typically involve the following steps:

1. **Sampling**: the operation of collecting representative samples to maintain quantitative representation of the soils and plants.
2. **Sample preparation**: the samples are conducted by drying, homogenization, grinding, etc. while avoiding contamination.
3. **Digestion**: the samples are extracted using microwave digestion or acid digestion to obtain soluble and exchangeable Cd. Samples for X-ray fluorescence spectroscopy (XRF) analysis do not need digestion.
4. **Analytical measurement**: the main analytical techniques used to quantify Cd in soil–plant systems are classical spectrometric analysis techniques [inductively coupled plasma mass spectrometry (ICP-MS), inductively coupled plasma optical emission spectrometry

Cadmium Contamination in Soils and Plants: Mechanisms and Mitigation.
DOI: https://doi.org/10.1016/B978-0-443-34104-5.00015-3

(ICP-OES), graphite furnace atomic absorption spectrometry (GFAAS), XRF, etc.], optical sensing, electrochemical, and portable or benchtop XRF.

5. **Analytical validation**: the calibration curve, recovery, precision, and the accuracy tests must be rigorously performed in accordance with International Union of Pure and Applied Chemistry (IUPAC) recommendations.

2.2 Sampling and sample preparation

In the analysis of Cd in soil and plants, sample collection and pretreatment are the first steps for accurately assessing the Cd concentration in the actual samples (de la Guardia & Garrigues, 2014; Mohd Rosdi et al., 2024). These steps must be operated carefully to be representative and to minimize the errors associated with sampling and extraction. Typically, different sample types and analytical methods require different sampling and pretreatment techniques.

2.2.1 Sampling and extraction of Cd in soils

Soil sampling equipment varies in capacity and material, depending on the volume of sample required and the species to be analyzed. Sampling is usually carried out using augers, spatulas, and shovels made of stainless steel or polymerized materials, and the equipment must be thoroughly cleaned before each use to avoid contamination of the sample, which could lead to inaccurate results. Typically the sampling area is divided into strata, considering the soil type, topography, degree of erosion, vegetation, and drainage conditions. A widely used soil sampling method is to collect samples at several points and then mix them to form a composite sample (Dean, 2013). Depth of collection should be decided according to the purpose of the study. According to the stratification of the soil, soil samples are collected separately at different depth levels using soil augers or digging tools. Generally, they can be categorized as surface layer (0–20 cm), subsurface layer (20–40 cm), deep layer (below 40 cm), etc. Shallow sampling and sampling at different depths can be used to assess compositional variability, which is critical for identifying nutrients or contaminants (Guo, 2009; Pan et al., 2020). Stratified sampling at each sampling point should be carried out continuously, minimizing the time interval, avoiding mutual contamination of soil samples from different depth layers, and ensuring the representativeness and comparability of soil samples from different depths.

After collection, samples should be stored in clean, dry containers, or plastic bags to avoid the effects of humidity and temperature on the samples (Rhymes et al., 2021). A composite sample can be obtained by meticulous homogenizing samples to remove stone and plant residues (Campos-M & Campos-C, 2017). Soil samples usually need to be dried for the soil chemical analysis (Terzi & Kalkan, 2024). In 2006 the International Organization for Standardization (ISO) recommended drying in an oven at a maximum temperature of 40°C, which ensures a shorter drying time and prevents microbial propagation (Koopmans & Groenenberg, 2011). Particle sizes in soil vary widely, from submicron to several centimeters. The air-dried soil samples are ground with a mortar or grinder to break them up sufficiently. For better analysis of the elemental content of the soil the soil samples should be ground to a certain particle size (usually through a 100 or 200 mesh sieve) after drying, and contamination should be avoided during grinding (Buyang et al., 2019). This process, known as quartering, involves thoroughly mixing the individual simple samples to ensure that the composite sample is representative of the entire sample area.

The soil samples are digested in order to convert all forms of Cd (free Cd^{2+}, organically bound Cd, mineral bound Cd, etc.) to Cd^{2+} ions for the detection of the total Cd content. The commonly used soil digestion methods are acid digestion and microwave digestion (Da Silva et al., 2014; Park et al., 2020). For a complete quantitative analysis of total Cd content, hydrofluoric acid is required to solubilize the silicates, to destroy the soil structure, and to make Cd more soluble. Generally, soil Cd acid digestion is carried out in a heating block using a 3:1 ratio of nitric and hydrochloric acids (i.e., aqua regia) and a certain amount of Hydrogen Fluoride acid (HF) (Krishnamurti et al., 1994). Moreover, the use of nitric and hydrofluoric acids with a microwave digestion system is also common for the determination of Cd.

2.2.2 Sampling and extraction of Cd in plants

Plant sample collection requires the selection of representative samples from the same plant population. Different plant organs, for example, roots, stems, leaves, and grains, may be collected separately. Fresh samples should be kept in cool boxes and brought back to the laboratory quickly. Attached soil particles, dust, and impurities should be removed thoroughly with deionized water to avoid data distortion because of extraneous particle contamination. After cleaning the samples are dried to constant weight in a drying oven (usually 60°C) or freeze-dryer. Dried plant parts are ground into powder for homogenization and improved extraction efficiency (Juranović Cindrić et al., 2012).

Effective sample digestion is essential for accurate analysis of Cd in plants. It is important to ensure that the element Cd in the plant is completely solubilized and released in a form that is compatible with the chosen analytical instrument. Common oxidizing acids such as nitric acid, hydrochloric acid, perchloric acid, or various mixtures (with hydrochloric and hydrofluoric acids) are combined with heat to dissolve the matrix. In general, closed microwave-assisted digestion procedures with various acids/acid mixtures are considered to be relatively fast and environment-friendly methods for plant materials compared to conventional open vessel digestion procedures (Krushevska et al., 1993; Lamble & Hill, 1998; Zeiner & Juranović Cindrić, 2017).

Table 2.1 presents the pretreatment methods used in various soil and plant samples for subsequent quantification of Cd. For different types of soil and plant samples the most suitable extraction technique can be selected according to the experimental purpose and equipment conditions. Sample preparation is still considered to be the weak link in sample analysis, and therefore it is an area that needs appropriate optimization and improvement.

Total Cd means the sum of all forms of Cd (including free, bound, mineral, etc.) in samples, representing the total concentration of Cd in soil or plants. Commonly used microwave digestion and acid digestion methods completely release the total Cd from the sample and convert it into a detectable ionic form (e.g., Cd^{2+}), which can be used to detect total Cd by ICP-OES, ICP-MS, GFAAS, or XRF. Cd ions usually refer to the free or extractable state Cd^{2+}, reflecting the bioavailability of Cd and the risk of environmental transport (Viala et al., 2021; Zhang et al., 2024a). After extraction by leaching solutions (e.g., $CaCl_2$, DTPA solution) a number of optical sensing methods (fluorescence or colorimetric) or electrochemical methods allow direct detection of Cd^{2+}. Free Cd ions are present in soil and plant samples at very low levels, possibly as low as ppb or even lower (Degryse et al., 2006; Lofts et al., 2004). If

TABLE 2.1 Pretreatment procedures for soil and plant samples.

Samples	Analyte	Drying	Digestion	References
Soils	Cd, Pb	Air-dried	MW, HNO_3 + HCl (3:1), 180°C	da Silva et al. (2016)
	As, Sb, Cd, Pb	Air-dried	HNO_3 + HCl + HF, 150°C	Taheri et al. (2022)
	As, Cd, Pb	Dried 80°C	HNO_3 + HCl + HF, 100°C	Nyachoti et al. (2023)
	As, Cd, Hg, Pb	Dried 105°C	MW, HNO_3 + HCl (3:1), 260°C	Saleem et al. (2024)
	Cd, Zn	Dried 40°C	HNO_3, 105°C	Brekken and Steinnes (2004)
Plants	Cu, Cd, Cr, Pb, Zn	Dried 60°C	Dry ashing, HNO_3 + HCl (4:1), 200°C	Terfi and Sadi (2015)
	Cd, Pb, Cu, Zn	Dried 65°C	HNO_3 + $HClO_4$, 150°C	Zhang et al. (2011)
	Cd, Co, Pb, Zn	Dried 60°C	MW, HCl + HNO_3 + HF (3:3:2), 185°C	Ge and Zhang (2015)

MW, microwave oven.

combined with sample digestion, total Cd can also be measured indirectly by optical sensing and electrochemical methods. The detection target (total Cd or Cd^{2+}) of the different methods depends mainly on the sample preparation and the sensitivity of the instrument.

2.3 Classical spectrometric analysis techniques

2.3.1 Inductively coupled plasma optical emission spectrometry

As an advanced atomic emission spectroscopy, ICP-OES is widely used for major, minor, and trace analysis of soil and plant samples (Cheng et al., 2013; Demırezen & Aksoy, 2004; Hamurcu et al., 2010). This technique employs an inductively coupled plasma to produce excited atoms and ions, which emit characteristic electromagnetic radiation. The sample solution is converted to an aerosol and directed into the central channel of the plasma (Hossain et al., 2021; Salamanca et al., 2024; Yan et al., 2021). For example, Ferreira et al. (2025) determined Pb, Cd, As, Cr, Cu, and Zn in 38 soil samples collected from Santo Amaro, Brazil, using the ICP-OES technique to assess the current environmental impacts of the lead smelter that operated in the city from 1960 to 1993.

The advantages of this technique include a wide linear dynamic range, good repeatability, and the ability to determine multiple elements simultaneously with sufficient accuracy. However, the limit of detection (LOD) of Cd detection in soil and plant samples by ICP-OES is relatively high, which is suitable for high-concentration samples. In particular, LOD, which represents the lowest concentration of analytes that can be reliably detected under specific analytical methods and experimental conditions (Ershadi & Shayanfar, 2018; Ikonen et al., 2020), is an important

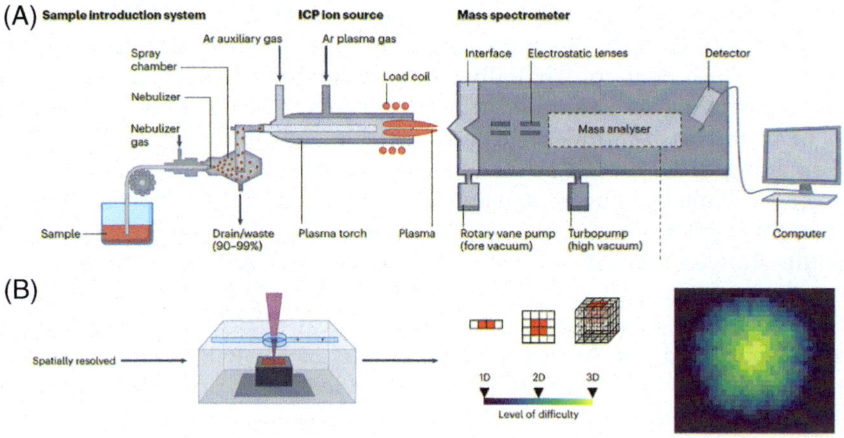

FIGURE 2.1 Graphical outline of (A) ICP-MS and (B) LA-ICP-MS. An inductively coupled plasma (ICP) mass spectrometer comprises four main parts: a sample introduction system, an ICP ion source, a mass spectrometer, and a detection system. Spatially resolved analysis achieved by laser ablation in combination with ICP-MS detection by scanning a laser beam across a sample surface. *MS*, Mass spectrometry. *Adapted from Van Acker et al. (2023) with copyright permission.*

indicator of the detection capability of the ICP-OES method. In ICP-OES analysis, samples are usually diluted before measurement to ensure that the results are within the linear range. The dilution of the sample affects the LOD of the analytical method and the accuracy of the results, and an appropriate dilution needs to be selected. In general, it is important to minimize the number of dilutions while ensuring that the sample does not exceed the linear range of the instrument in order to increase the sensitivity of the assay and reduce the LOD.

2.3.2 Inductively coupled plasma mass spectrometry

ICP-MS is the most suitable analytical technique for determining trace metal elements in soil–plant systems, enabling a fast and simultaneous quantification with multielement capability and wide linear range (Beauchemin, 2008; Brown & Milton, 2005; Thompson & Walsh, 2012). The sample solutions are converted to aerosols in a nebulizer. Argon is used as a carrier gas to bring the aerosol to the center of the ICP torch. As soon as the sample enters the plasma, it is rapidly atomized and ionized, and the ions are introduced into the mass spectrometer, where they are separated according to the mass-to-charge ratio (Fig. 2.1A). There are eight stable isotopes of Cd, including ^{112}Cd, ^{111}Cd, and ^{114}Cd. In ICP-MS analysis the high abundance isotope ^{111}Cd with least interference is usually selected for detection (Cuello-Nuñez et al., 2018). The ICP-MS method with ultralow LOD (ppt level), good precision, and sensitivity is particularly suitable for trace Cd analysis in soil and plants. Its main disadvantages are high cost and polyatomic interferences. Various reactions and collision systems have been frequently used for correcting polyatomic interferences in ICP-MS, such as dynamic reaction cell, collisional-reaction cell, and collision-reaction interface (Balaram, 2021). In these cases the interference of polyatomic ions is suppressed by introducing different collision gases into the system (You et al., 2022).

With the development of science and technology, combining ICP-MS with laser ablation sampling (LA-ICP-MS) enables direct elemental analysis of solid materials without the need for prior digestion, also enabling spatially resolved analysis with micrometer resolution. By scanning a laser beam across a sample surface the LA-ICP-MS mapping provides rich spatial information of metals in soil and plant systems (Fig. 2.1B). Pessôa et al. (2017) performed LA-ICP-MS imaging of Cd in the seeds of sunflower grown in Cd-contaminated soil. They optimized the LA parameters (laser intensity, frequency, and spot size), which greatly improved the accuracy of the signals (Pessôa et al., 2017). Regarding the accumulation of Cd in sunflower seeds the results showed that Cd is translocated to the seeds, and the cotyledons contained the highest concentration (Cd-high group), ranging from 10 to 20 ppm. Wu et al. (2024a) utilized LA-ICP-MS to study the mechanisms and physiological responses of rice leaves coexposed to submicron plastics and Cd. In situ imaging results of LA-ICP-MS confirmed a significant reduction (29.3%–77.9%) in Cd accumulation in the leaves of rice plants during coexposure (Wu et al., 2024b).

2.3.3 Atomic absorption spectrometry

Atomic absorption spectrometry (AAS) is a simple, widely used, and quantitative technique of metal analysis by which approximately 70 metals can be measured (Morrison & Talmi, 1970). Typical AAS includes a light source, a sample input section (atomizer), a monochromator, a detector, and a data processing system (Fig. 2.2A). This technique is used to measure the concentration of the elements through passing the light of a specific wavelength, emitted by a source of radiation, fixed for a particular element in a sample (Beaty & Kerber, 1978; Massas et al., 2013). Over the decades, researchers have developed various kinds of atomizers, such as flames, graphite furnaces, and quartz tube, for the detection of Cd (Báez et al., 2007; García et al., 2009; Piccolo & O'Connor, 1968). The AAS methods commonly used for analyzing Cd in soil and plants are flame atomic absorption spectrometry (FAAS) and graphite furnance AAS (GFAAS). FAAS is a fast and simple-to-operate technique with good accuracy, but the sensitivity of Cd detection is relatively low, and the LOD is usually at the level of mg/L (ppm), which is suitable for samples with high Cd content. In addition, GFAAS has high sensitivity and great precision at ppb levels, which is suitable for trace Cd analysis. Because of the long and complicated graphite furnace heating process the analytical speed of this method is relatively slow, and the equipment cost is relatively high. Ribeiro Menezes et al. (2021) used the simple and robust GFAAS methods for determination of As, Cd, and Pb in hemp seeds. By optimizing the sample preparation and instrumental conditions the samples could be previously treated in either an acid or alkaline medium, achieving limits of quantification of 0.04 ppm (Cd) with good accuracy and precision.

2.3.4 X-ray fluorescence spectroscopy

XRF is a nondestructive elemental analytical technique requiring minimal sample preparation. It is suitable for both liquid and solid materials (Tsuji et al., 2012). The soil or plant sample is excited by an X-ray source of sufficient energy (greater than the binding energy of the K or L shells of the atoms), which causes the metal atoms to become unstable. Immediately afterwards, another electron from the outer orbital with a higher energy fills the gap in the inner orbital.

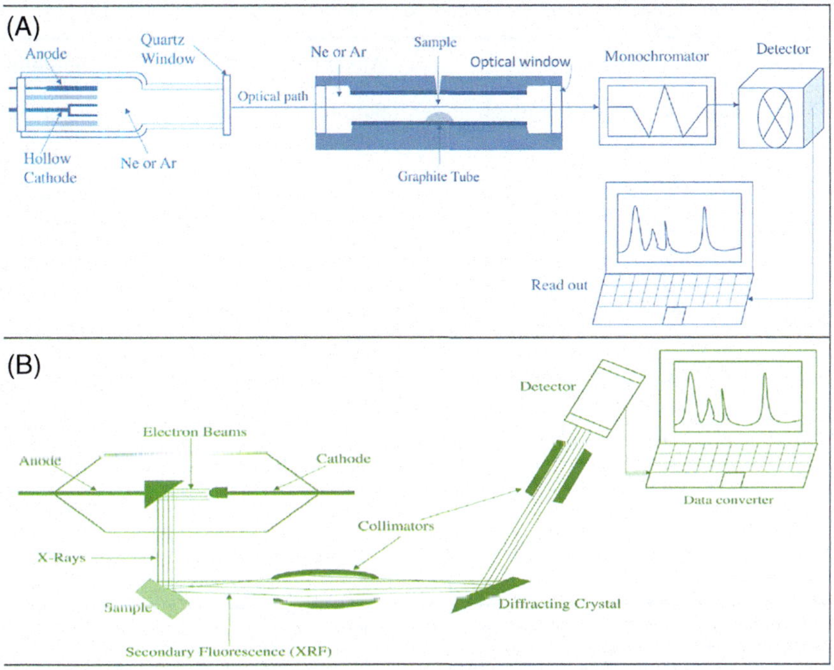

FIGURE 2.2 Graphical outline of typical GFAAS technique (A) and XRF analyzer (B). *GFAAS*, Graphite furnace atomic absorption spectrometry; *XRF*, X-ray fluorescence spectroscopy. *Adapted from Hossain et al. (2021) with copyright permission.*

As the electron drops to a lower energy state, it emits fluorescent X-rays characteristic of the element (Beckhoff et al., 2007). In general an X-ray spectrometer is composed of the following parts: a radiation (X-ray) source, a sample chamber, collimators, and a detector (Zhang, 2006) (Fig. 2.2B). The major advantage of XRF is the ability to perform quantitative analysis over a wide range of elements without digestion. Dekeyrel et al. (2024) developed and optimized the monochromatic energy dispersive x-ray fluorescence (MEDXRF) method for Cd analysis. Within 200 s measurement time the average limit of quantification was 0.178 ppm for biological samples and 0.205 ppm for soil samples. The optimized MEDXRF technique has advantages in terms of cost-effectiveness and efficiency of cadmium detection. Without sample digestion, XRF is particularly suitable for on-site rapid detection and analysis of high-concentration Cd samples (e.g., Cd-contaminated soils and plant samples).

Table 2.2 provides a comparative overview of traditional spectrometric methods (ICP-OES, ICP-MS, GFAAS, and XRF) used for analyzing total Cd in soil and plant samples. Each method exhibits distinct characteristics in terms of sensitivity, LOD, linear response range, advantages, and limitations. In terms of sensitivity and LOD, ICP-MS stands out with its ultralow LOD (ppt-level) and exceptional sensitivity, making it the ideal choice for trace Cd analysis in plants and soils. In contrast, XRF with low sensitivity is well-suited for rapid screening of high-concentration samples. When considering cost, the techniques rank as follows: ICP-MS

TABLE 2.2 Comparison of spectrometric techniques for Cd analysis in soil and plant samples.

Methods	LOD	Sensitivity	Advantages	Limitation
ICP-OES	0.1-1 (ppb)	Medium-high	Simultaneous multielement analysis; fast analysis; high throughput	Higher LOD than ICP-MS; requires sample digestion; spectral interference possible
ICP-MS	0.001–0.01 (ppt)	Extremely high	Ultralow detection limits; isotopic analysis; high precision for trace elements	High instrument cost; polyatomic/mass interference; complex sample preparation
GFAAS	0.005–0.02 (ppb)	High	Low cost; high sensitivity for single elements; minimal sample volume required	Single-element analysis; limited linear range; short graphite tube lifespan
XRF	1–10 (ppm)	Low	Nondestructive; rapid in situ analysis; no sample digestion needed	Poor sensitivity; homogeneous samples; semiquantitative results

ICP-MS, inductively coupled plasma mass spectrometry; *ICP-OES*, inductively coupled plasma optical emission spectrometry; *LOD*, limit of detection; *XRF*, X-ray fluorescence spectroscopy.

> ICP-OES > GFAAS ≈ XRF. ICP-MS is particularly advantageous for high-precision trace Cd analysis, making it a preferred method for research and environmental monitoring. ICP-OES, on the other hand, is cost-effective and well-suited for high-throughput polymetallic analysis. GFAAS offers a practical solution for trace Cd single-element analysis, especially for laboratories with limited budgets. XRF is especially suitable for rapid on-site detection or for scenarios where destructive analysis is not possible. Given the diversity of soil and plant samples, the selection of the most appropriate analytical technique should be guided by the specific experimental objectives and available equipment conditions.

2.4 Optical sensing methods

2.4.1 Fluorescence methods

Fluorescent sensors for metals contain two essential features: a metal chelating or binding moiety and at least one fluorophore capable of absorbing and emitting light (Fig. 2.3A). Once the metal binds specifically to the sensor, the electronic or molecular structure of the sensor must be altered to induce changes in fluorescence properties (Carter et al., 2014). Numerous fluorescent probes (small-molecule fluorophores, polymers, quantum dots, carbon nanotubes, metal nanoclusters, etc.) have been developed for Cd detection (Sharma et al., 2024). Moreover, there have been several reports of Cd^{2+} ions fluorescence detectors using various fluorescence detection mechanisms, such as intramolecular charge transfer, photoinduced electron transfer

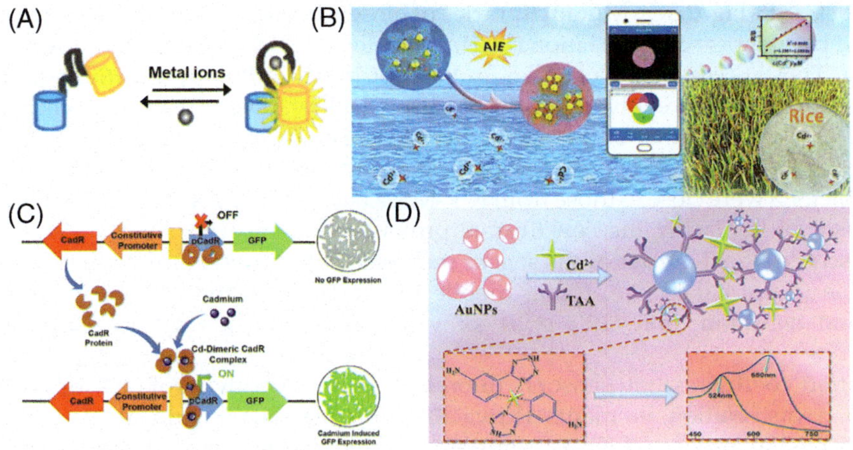

FIGURE 2.3 Optical sensing methods for Cd analysis: (A) schematic diagram of the principle of metal fluorescence sensor. (B) An AIE fluorescent sensor with smartphone platform for quantitative detection of Cd^{2+} in rice samples. (C) Strategies for construction of wild-type cadmium-induced bacteria. (D) Highly sensitive colorimetric detection of Cd in rice based on the TAA@AuNPs sensor. *AIE*, Aggregation-induced emission; *TAA*, (1*H*-tetrazol-5-yl) aniline. *Adapted from Cai et al. (2022), Li et al. (2024), Wang et al. (2020) with copyright permission.*

(PET), chelation-enhanced fluorescence (CHEF), and fluorescence resonance energy transfer (Aich et al., 2015; Liu et al., 2019a; Ma et al., 2013; Yu et al., 2012). Tümay et al. (2020) designed pyrene-modified nanocrystalline cellulose (NP-1) as a "turn-on" fluorescence sensor for measurement of Cd^{2+} in rice samples. Once the Cd^{2+} ions were introduced, the synergistic effect of PET and CHEF led to significant "turn-on" fluorescence response obtained at 453 nm for pyrene moieties. With high applicability and stability, the linear range of the sensor was from 11.2 to 6744.6 ppb. The LOD of this method for the detection of Cd^{2+} was 10.70 ppb, which was lower than the WHO's permissible limit of Cd in plants. Wang et al. (2020) established an aggregation-induced emission–based fluorescent sensor for detecting Cd^{2+} (Fig. 2.3B). The colorimetric fluorescent sensor was obtained by mixing orange emission glutathione-stabilized gold nanoclusters (AuNCs) with blue emission ethylenediamine-functionalized graphene oxide (EDA-GO) and introduced Cu^{2+} to quench the orange emission of AuNCs. The addition of Cd^{2+} can induce the aggregation of Cu^{2+}-GSH (Glutathione)-AuNCs and emit orange fluorescence, which changed the fluorescence color of the sensor from blue to red with the LOD as low as 3.66 ppb in solution. By integrating fluorescent paper strips with a smartphone the visual on-site detection platform has a sensitive detection of Cd^{2+} with the LOD of 11.2 ppb. After microwave digestion with HNO_3 and H_2O_2, this sensor successfully detected the Cd content in rice, which was below the specified safety threshold.

The Cd whole-cell biosensors constructed by genetically encoded fluorescent sensors have attracted widespread attention (Wu et al., 2024b). In general, Cd-responsive whole-cell fluorescent biosensors are constructed using cadmium resistance transcriptional regulators (CadR, CadC, and CmtR) as sensing elements and reporter genes (e.g. Green fluorescent protein (GFP), Red fluorescent protein (RFP) or luciferase) as signal generators (Hui et al., 2022; Jia et al., 2021; Wei et al., 2024). Cai et al. (2022) improved the performance of a cadmium-responsive biosensor by directed evolution of CadR (Fig. 2.3C). Mutant libraries of biosensors in *Escherichia coli* were produced by error-prone polymerase chain reaction (PCR) and selected by five-round fluorescence-activated cell sorting. They successfully isolated a mutant epCadR5 with high sensitivity and specificity in detecting trace Cd. The evolved biosensor showed a low LOD of 0.45 ppb, which is 6.8 times more specific to Cd than that of the wild type. Moreover, the epCadR5 bacteria could produce fluorescence response in a Cd-contaminated soil matrix, and the quantitative values of Cd from epCadR5 bacteria were close to those from ICP-MS. Zhang et al. (2024b) optimized the Cd whole-cell biosensor by replacing the constitutive promoter and incorporating a 5′-UTR sequence. The fluorescence signal output of the sensor increased significantly and was sufficiently robust to be visible to the naked eye. Using mobile phone cameras the biosensor presented good linearity and detection limit (1.81 ppb) for visual detection of Cd^{2+} in real environmental samples.

In addition to fluorescent sensors, fluorescence microscopy provides opportunities for mapping Cd distribution in plants (Kováčik & Babula, 2017). Luo et al. (2023) reported a Cd^{2+} fluorescent probe, which is composed of *p*-dimethylaminobenzaldehyde linking *p*-nitrophenylacetonitrile and di-2-picolylamine (the recognition group of Cd^{2+}). This new probe showed good linear relationship with different concentrations of Cd^{2+} with a low LOD of 3.52 ppb. By fluorescent imaging, it was observed that Cd^{2+} was absorbed by rice root cells and accumulated in roots and then was transported to shoots through the xylem for further dispersal to different organs of rice. Chen et al. (2025) designed a Cd^{2+}-dependent DNAzyme fluorescent probe for in situ imaging of Cd^{2+} in rice cell walls and vacuoles. Introduction of Cd^{2+} into the sensing system activated the Cd^{2+}-dependent DNAzyme, leading to the separation of black hole quencher (BHQ2) and X-rhodamine, allowing a fluorescence signal to be recovered. The probe presented not only high specificity for Cd^{2+} but also a detection sensitivity of less than 0.01 ppb. They successfully observed the adsorption, transportation, and storage of Cd^{2+} in the rice cells. The combination of fluorescence microscopy and specific Cd^{2+} fluorescent probe can be used for the rapid in situ identification of crops with low Cd accumulation.

2.4.2 Colorimetric methods

Colorimetric methods for Cd^{2+} analysis have attracted attention because of their ease of use, low cost, fast response time, and portability (Litluechai et al., 2024; Tekuri & Trivedi, 2017). In particular the results obtained with colorimetric sensors can be easily judged by naked-eye observation without complicated instruments (Gai et al., 2023). The colorimetric process usually consists of two types, one in which the color of the material itself changes upon reaction with the target, and the other in which a special color-generating substrate, such as 3,3′,5,5′-tetramethylbenzidine, is used to produce a colorimetric substance during a catalytic reaction that results in a color change.

Li et al. (2024) developed a high-efficiency colorimetric sensor based on gold nanoparticles (AuNPs) modified with 3-(1*H*-tetrazol-5-yl) aniline (TAA) for the detection of Cd^{2+} (Fig. 2.3D). The direct interaction between Cd^{2+} and TAA@AuNPs leads to agglomeration of the AuNPs solution, which changes the color of the system from wine-red to gray-blue. The proposed sensor has a low LOD of 3.5 ppb for Cd^{2+} by UV–vis spectroscopy with an excellent linear relationship. Moreover, the colorimetric sensor was successfully used to monitor Cd^{2+} in soil and rice samples. Liu et al. (2024) prepared the zirconium-based metal–organic framework (MOF) (PCN-222) as a colorimetric nanoplatform for the determination of Cd^{2+}. In the presence of PO_4^{3-} the robust bonding between PO_4^{3-} and Zr nodes triggers the collapse of PCN-222, leading to the release of ligands (TCPP) with visible alteration in solution color to light red. The introduction of Cd^{2+} ions changes the color to green because of metal ions' coordination with the porphyrin ring and carboxyl group. The colorimetric method using UV spectroscopy and a smartphone platform offers great potential for the rapid and portable screening of Cd^{2+} in actual environment analysis.

2.5 Electrochemical methods

2.5.1 Principles of detection

Over the past few years, electrochemical techniques have received much attention for Cd^{2+} analysis because of their high sensitivity, rapidity, low cost, and reliability (Yi et al., 2022). Electrochemical techniques, including square wave voltammetry (SWV), differential pulse voltammetry (DPV), anodic stripping voltammetry (ASV), and linear sweep voltammetry, record the response current generated by scanning potential (Liu et al., 2019b; Lu et al., 2019; Pizarro et al., 2020). ASV is the most commonly used electrochemical technique for Cd^{2+} detection. The specificity of ASV is attributed to the requirement of a specific potential for Cd oxidation. The peak position and peak current are introduced for Cd^{2+} identification and Cd^{2+} quantification, respectively. Owing to better sensitivity, low signal-to-noise ratio, and good sensing performances than other technologies, SWV and DPV are widely used in combination with ASV techniques. Two types of ASV have been developed for Cd^{2+} detection: differential pulse stripping voltammetry (DP–ASV) and square wave adsorptive stripping voltammetry (SW–ASV).

In general, electrochemical determination of Cd^{2+} is carried out in a cell consisting of a 3-electrode system: working electrode, auxiliary electrode, and reference electrode. Specifically the electroanalysis of Cd^{2+} is carried out in two steps: self-reduction and reoxidation. First, Cd^{2+} is efficiently accumulated on the electrode surface by an adsorptive or faradaic procedure and then electrochemically reduced and deposited as elemental Cd via applying a constant potential for a certain period of time. In the reoxidation process, Cd is reoxidized to Cd^{2+} by utilizing a voltage scan in the anodic direction, which induces dissolution of the Cd analyte presented on the electrode surface, leading to a rapid electrochemical reaction and the generation of a strong oxidation peak current, which is proportional to the Cd^{2+} level (Fig. 2.4).

2.5.2 Applications

As electrochemical reactions are fundamental procedures carried out on the surface of the working electrode, the characteristics of the working electrode determine the detection performance. The perfect working electrode for electrochemical detection of Cd^{2+} should have ideal electrical characteristics and preenrichment of Cd^{2+}. Researchers are striving to design and construct various modified working electrodes based on different materials (polymers, MOF, organic groups, and oxides) for ASV detection of Cd^{2+} (Hassan et al., 2019; Magesa et al., 2019;

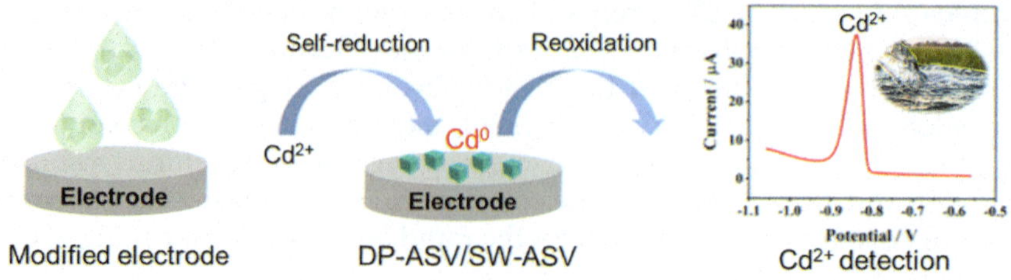

FIGURE 2.4 Schematic diagram of electrochemical detection of Cd^{2+} in soil and plants. Different materials (polymers, metal–organic framework, organic groups, and oxides) were modified on working electrodes. The DP–ASV and SW–ASV techniques, including self-reduction and reoxidation, were carried out for the electroanalysis of Cd^{2+}. The resulting characteristic peak oxidation current is proportional to the Cd^{2+} concentration in the actual samples. *DP–ASV*, Differential pulse stripping voltammetry; *SW–ASV*, square wave adsorptive stripping voltammetry.

Wan et al., 2021). Chamjangali et al. (2017) developed the uniform composite film of multiwalled carbon nanotubes (MWCNTs) and poly alizarin red S (MWCNTs-PARS) on the surface of glassy carbon electrodes (GCEs) (MWCNTs-PARS/GCE) for the determination of Cd in soil acid extracts. The prepared MWCNTs/PARS can combine the synergetic advantages of MWCNTs and ARS in the ASV determination of Cd^{2+}, achieving the sensing performance with a linear response of 5–160 ppb and a detection limit (LOD) of 0.43 ppb by using DP–ASV. Lu et al. (2019) reported the graphene aerogel (GA) and MOF composites for simultaneous DP–ASV detection of Cd^{2+}. In the GAs–MOF nanocomposite the GA enhances the conductivity of the composites, and UiO-66-NH_2 provides rich binding site for heavy-metal ions because of the interaction between hydrophilic groups and metal cations. The method can be used for the detection of Cd^{2+} in the leaching solutions of soil and vegetables with high accuracy and reliability.

Electron-rich organic molecule receptors generally contain numerous functional groups (e.g., N, O, and S) that show superior affinity toward Cd^{2+} (Kokab et al., 2020; Priya et al., 2019). The special binding property and abundant adsorption sites make them superior candidates as receptors with good detection performance when introduced in sensors for Cd^{2+}. A *N*-terminal carboxybenzyl-blocked tripeptide (Cbz-APL)-modified GCE-based sensor was constructed for sensitive and selective determination of Cd ions in soil and tobacco plants (Kokab et al., 2020). The adsorbed tripeptide molecules act as stepping stones between the electrode surface (host) and Cd^{2+} (guest) in the analyte solution. They increase charge transfer process via decrease in resistance, thus underlining the mediator role of the modifier. SW–ASV was utilized to further investigate the sensing performance of the developed sensor, with a low detection limit of 5×10^{-4} ppt and linearity range of 1.7 ppb to 0.01 ppt. Many metallic oxides can exhibit low toxicity, excellent biocompatibility, and good catalytic and adsorption capacity toward heavy metal ions, giving significant promise for electrochemical detection of Cd^{2+}. Huang et al. (2020) prepared a composite of CeO_2 nanomaterials supported on graphite as sensitive electrode materials. The morphology and structure of CeO_2 nanomaterials determine the electrocatalytic ability of the interface and the sensitivity for detecting Cd^{2+} ions. An electrochemical sensor with the rod-CeO_2/graphite composite modified on electrode surfaces achieved a detection limit of 0.39 ppb for Cd^{2+}. In the last few years the electrochemical detection of Cd^{2+} in the environment has remained a very active area of research. There is still an urgent need to develop more efficient

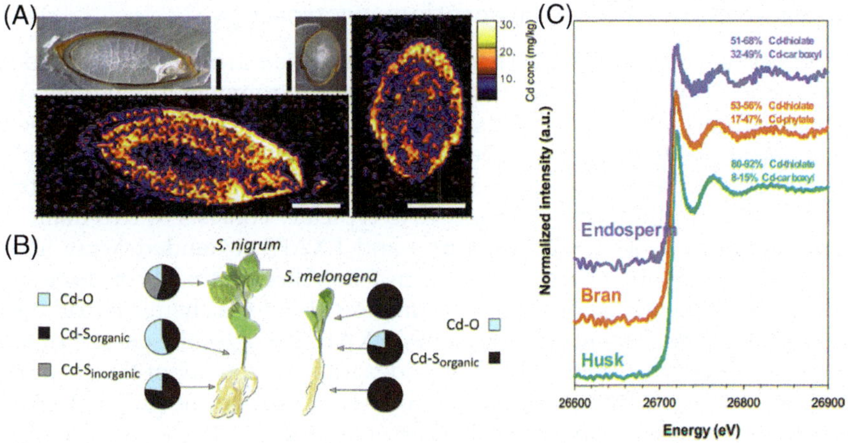

FIGURE 2.5 Synchrotron-based methods for cadmium analysis in soil–plant systems: (A) representative synchrotron-based XFM images of the Cd distribution in both longitudinal and transverse sections of rice grain with husk. (B) The chemical Cd species in *Solanum nigrum* and *Solanum melongena* (roots, stem, and leaves). (C) Cd XANES spectra of various fractions of the rice grain. *XANES*, X-ray absorption near edge structure; *XFM*, X-ray fluorescence imaging. *Adapted from Gu et al. (2020), Pons et al. (2021) with copyright permission.*

electrode materials and to improve the performance and reliability of electrochemical detection and analysis.

2.6 Synchrotron-based methods

2.6.1 Synchrotron-based X-ray fluorescence imaging

X-ray fluorescence imaging (XFM) is a powerful technique for quantitatively determining the concentration and distribution of elements in various types of samples (Silva et al., 2017). A beam of X-rays of a specific energy is focused on the sample, which is scanned through the beam in a raster pattern, generating an elemental profile based on the characteristic fluorescent X-rays emitted from each part of the sample. In particular, synchrotron-based XFM analysis is typically orders of magnitude faster than benchtop systems with low detection limits and high resolution, which is critical for the analysis of plant tissues (Kopittke et al., 2018; van der Ent et al., 2018).

Using synchrotron-based XFM, Tananonchai et al. (2023) investigated the accumulation and formation of Cd in elephant grass (*Pennisetum purpureum*) treated with EDTA to enhance metal uptake. Imaging results indicated Cd accumulation and distribution in the internal tissues of root: endodermis, xylem, and phloem, whereas in stems and leaves it was found in epidermis and phloem. Gu et al. (2020) utilized synchrotron-based X-ray fluorescence microscopy (μ-XRF) to reveal the distribution patterns of Cd in rice grain. In both longitudinal and transverse sections (Fig. 2.5A). The Cd concentration was significantly higher at the periphery of the starchy endosperm and lower in the middle of the endosperm. The pattern of Cd distribution was variable among different samples, which may influence the bioavailability of seed Cd in the human diet (Gu et al., 2020).

2.6.2 Synchrotron-based X-ray absorption spectroscopy

X-ray absorption spectroscopy (XAS) has been widely used to obtain information on the local chemical environment of trace metals in a variety of environmental materials. When X-rays strike a material surface, they are absorbed by the material. The absorption coefficient does not remain constant throughout the wavelength band but rather jumps at specific energies, forming the absorption edge. Discrete peaks or fluctuations appear in the spectrum near the absorption edge or in the high energy range. XAS typically consists of two main components: XANES (X-ray absorption near edge structure) and EXAFS (extended X-ray absorption fine structure). XANES covers the energy range from 50 eV before the absorption edge to 50 eV after the absorption edge, which provides information about the chemical oxidation stage and coordination geometry of elements in complexes. EXAFS involves the energy range from 50 to 800 eV behind the absorption edge, which provides information about the structure of atomic clusters in a small range (i.e., the coordination number of the near-neighbor atoms, the spacing between atoms, and the atomic species) (Chantler et al., 2024).

Several studies have investigated the binding of Cd to organic matter in soils using XANES and EXAFS. Collins et al. (1999) reported the involvement of 7–8 O atoms in the complexation of Cd by sodium humate. Using EXAFS, Karlsson et al. (2005) investigated the coordination chemistry of Cd in soil organic matter (SOM) from different types of organic soils. They found that Cd complexes in SOM are a mixture of 4-coordination with S (thiols) and 4- and 6-coordinations with O/N ligands. Wiggenhauser et al. (2021) employed XAS to identify inorganic and organic forms of Cd bound to both S- and O-functional groups in soils. They found that Cd–S was the dominant binding form in flooded soils, contributing in part to the immobilization of Cd in anoxic paddy soil.

In studies of plants, Küpper et al. (2004) reported that 0.3–3.1 S atoms, along with O/N ligands, were involved in a 6-coordination of Cd in the tissue of the Cd/Zn hyperaccumulator *Thlaspi caerulescens* (now called *Noccaea caerulescens*). Pons et al. (2021) utilized XAS to study the chemical Cd species in plants. In both *Solanum nigrum* (*S. nigrum*) and *Solanum melongena* (*S. melongena*), Cd in roots and new leaves is mainly bound to thiol ligands, with only a small fraction of inorganic S ligands in *S. nigrum* leaves (Fig. 2.5B). They suggested binding of Cd to S ligands as a detoxification mechanism, possibly involving the sequestration of Cd complexed with glutathione or phytochelatins in plant vacuoles. Gu et al. used synchrotron-based XAS for analyses of the speciation of Cd in rice grain. Most of the Cd in rice grains was complexed with S (66%–92%), likely in the form of Cd bound with thiol-rich proteins (Gu et al., 2020) (Fig. 2.5C). The rest was in the form of cadmium-carboxylate complexes and cadmium–histidine.

2.7 Comparison of different methods

A survey of the literature of Cd analysis published globally over the past decade shows that the ICP-MS technique was the most widely used technique for the quantification of Cd, which was used in 49% of analytical methods, followed by ICP-OES. The main reasons for the ICP-MS being the most widely used technique are (1) good detection limits (~ppt for many elements), (2) simple sample preparation and injection method, (3) small volumes of sample required (<1 mL), (4) high analytical throughput (~40 samples per hour), (5) the capability of simultaneous multielement (50 or more elements) analyses, (6) high accuracy and sensitivity, and (7) high precision (relative standard deviation <1%). ICP-MS is also capable of determining isotopes and has the capability to quantify almost all the elements of the periodic table. These techniques (ICP-OES and ICP-MS) have a wide dynamic detection limit, ranging from extreme trace to high concentrations. Despite the wide range of advantages of these techniques, there are some limitations, such as matrix interference and spectral problems in the determination of trace elements. Moreover, there is a chance of error because of different sample digestion procedures. Other drawbacks include high cost of the instruments and the need for specialized operations.

Various AAS techniques are also recognized as robust, economical, and reliable analytical techniques for Cd detection in soil and plant sample matrices. Their main drawback is that all of them are typically single-element techniques. Sample preparation, introduction, specifications, and performance differ substantially for each technique. XRF is a user friendly and nondestructive analytical method with the attractive benefits of no sample digestion and multielement analysis capability. Nevertheless, it is not suitable for analyzing Cd in soil and plant samples from uncontaminated environment. It also faces some limitations such as radiation exposure hazards, strong matrix effects, and the need for specialized training to operate the instrument. In addition, synchrotron-based techniques have lower detection limits and high resolution, but they are often limited by high cost, accessibility, energy instability, short target life, and safety issues.

Up to now, conventional methods for Cd detection, including ICP-MS, ICP-OES, and GFAAS, have been mainly used in laboratories. However, these methods are not suitable for real-time and routine in-field monitoring. As a result, various optical biosensors and bioassays (fluorescence or colorimetric) have been developed and applied to allow rapid measurements and field applications for Cd^{2+} analysis without the limitations of conventional methods. If combined with microwave or acid digestion, total Cd in soil and plant samples is converted to free Cd^{2+}. Depending on the sample extraction method, these optical and electrochemical analytical methods can measure both free Cd ions and, indirectly, total Cd in the sample. Integration of fluorescence or colorimetric sensors with smartphone platforms enables compact, intelligent, and portable on-site detection of Cd in environmental samples. They still exhibit some disadvantages, for example, cross-sensitivity with other metal ions, synthesis of new materials, and stability of fluorescent and colorimetric signals. Electrochemical techniques for Cd have many unique advantages such as high sensitivity and selectivity, time-saving, impressive cost-effectiveness, and miniaturization. The present electrochemical detection methods still face some challenges, such as electrode material instability, poor signal reproducibility, and time-consuming preconcentration steps.

2.8 Conclusion and outlook

Sampling is one of the most critical steps in soil–plant analysis. The accuracy of results depends fundamentally on the quality of this operation. Most of the analytical procedures for Cd analysis require liquid-form samples, which means that samples must be processed or digested based on the complexity of the sample matrix. Microwave and acid digestion methods are commonly used to dissolve Cd in the sample for further analysis. Although in some analytical techniques (EDX, XRF, etc.), soil and plant samples can be analyzed directly without the need for digestion.

The choice of the best technique is largely dependent on the research objectives. The first and the most important consideration before selecting a technique is the detection limit requirement.

Other issues that need to be considered include (1) numbers of samples to be measured at a time; (2) budget and cost of measurements; and (3) the presence of problematic components, for example, high levels of dissolved solids and volatile organics in the sample matrices.

For analysis of total or extractable fractions of Cd in soil and plant samples, ICP-MS is the dominant technique because of its high sensitivity and simultaneous detection of multiple elements in the sample. Its main limitations are high cost, the need for professional operations, and lack of real-time applicability. In recent years, various optical sensors, including fluorescence, colorimetric, and whole-cell biosensors, have been developed to allow rapid measurements for Cd^{2+} analysis. Electrochemical techniques have also received considerable attention for Cd^{2+} detection because of high sensitivity and selectivity, time-saving, cost-effectiveness, and miniaturization. Therefore, in future testing and analysis, it is important to focus not only on the accuracy and sensitivity of the measurement method but also on the cost of instrument testing. Future measurement and analysis should be developed in the direction of accuracy, sensitivity, high throughput, controllable cost, and intelligence.

2.9 Summary

- Soil and plant samples require careful pretreatment procedures—including drying, grinding, and digestion (e.g., microwave or acid) to prevent contamination and ensure reliable analysis.
- Classical spectrometric techniques offer distinct advantages and limitations. ICP-MS (highest sensitivity, costly, prone to interference); ICP-OES (multielement, wide linear range, limited detection limits); GFAAS (sensitive, cost-effective for single-element, limited linear range); and XRF (nondestructive, no digestion needed, lower sensitivity, requires homogeneous samples).
- Fluorescence and colorimetric sensing methods provide high sensitivity and can be integrated with smartphone platforms for on-site visual monitoring of Cd^{2+} or total Cd.
- Electrochemical techniques achieve highly sensitive detection of Cd^{2+} by employing modified electrodes, making them well-suited for analyzing soil and plant extracts.
- Synchrotron-based methods provide spatially resolved imaging and chemical-speciation data, thereby elucidating the distribution and behavior of Cd within soil–plant systems.

References

Aich, K., Goswami, S., Das, S., Mukhopadhyay, C. D., Quah, C. K., & Fun, H. K. (2015). Cd^{2+} triggered the FRET "ON": A new molecular switch for the ratiometric detection of Cd^{2+} with live-cell imaging and bound X-ray structure. *Inorganic Chemistry, 54*(15), 7309–7315. http://pubs.acs.org/journal/inocaj. doi:10.1021/acs.inorgchem.5b00784.

Báez, A., Belmont, R., García, R., Padilla, H., & Torres, M. C. (2007). Chemical composition of rainwater collected at a southwest site of Mexico City, Mexico. *Atmospheric Research, 86*(1), 61–75. doi:10.1016/j.atmosres.2007.03.005.

Balaram, V. (2021). Strategies to overcome interferences in elemental and isotopic geochemical analysis by quadrupole inductively coupled plasma mass spectrometry: A critical evaluation of the recent developments. *Rapid Communications in Mass Spectrometry, 35*(10). doi:10.1002/rcm.9065.

Beaty, R., & Kerber, J. (1978). *Concepts, instrumentation and techniques in atomic absorption spectrophotometry*. The Perkin-Elmer Corporation.

Beauchemin, D. (2008). Inductively coupled plasma mass spectrometry. *Analytical Chemistry, 80*(12), 4455–4486. doi:10.1021/ac8006945.

Beckhoff, B., Kanngießer, B., Langhoff, N., Wedell, R., & Wolff, H. (2007). *Handbook of practical X-ray fluorescence analysis*. Springer Science & Business Media.

Brekken, A., & Steinnes, E. (2004). Seasonal concentrations of cadmium and zinc in native pasture plants: Consequences for grazing animals. *Science of the Total Environment, 326*(1–3), 181–195. doi:10.1016/j.scitotenv.2003.11.023.

Brown, R. J. C., & Milton, M. J. T. (2005). Analytical techniques for trace element analysis: An overview. *TrAC Trends in Analytical Chemistry, 24*(3), 266–274. doi:10.1016/j.trac.2004.11.010.

Buyang, S., Yi, Q., Cui, H., Wan, K., & Zhang, S. (2019). Distribution and adsorption of metals on different particle size fractions of sediments in a hydrodynamically disturbed canal. *Science of the Total Environment, 670*, 654–661. doi:10.1016/j.scitotenv.2019.03.276.

Cai, Y., Zhu, K., Shen, L., Ma, J., Bao, L., Chen, D., Wei, L., Wei, N., Liu, B., Wu, Y., & Chen, S. (2022). Evolved biosensor with high sensitivity and specificity for measuring cadmium in actual environmental samples. *Environmental Science & Technology, 56*(14), 10062–10071. doi:10.1021/acs.est.2c00627.

Campos-M, M., & Campos-C, R. (2017). Applications of quartering method in soils and foods. *International Journal of Engineering Research and Applications, 7*(1), 35–39. doi:10.9790/9622-0701023539.

Carter, K. P., Young, A. M., & Palmer, A. E. (2014). Fluorescent sensors for measuring metal ions in living systems. *Chemical Reviews, 114*(8), 4564–4601. http://pubs.acs.org/journal/chreay. doi:10.1021/cr400546e.

Chamjangali, M. A., Boroumand, S., Bagherian, G., & Goudarzi, N. (2017). Construction and characterization a non-amalgamation voltammetric flow sensor for online simultaneous determination of lead and cadmium ions. *Sensors and Actuators, B: Chemical, 253*, 124–136. doi:10.1016/j.snb.2017.06.098.

Chantler, C. T., Bunker, G., D'Angelo, P., & Diaz-Moreno, S. (2024). X-ray absorption spectroscopy. *Nature Reviews Methods Primers, 4*(1). https://www.nature.com/nrmp/. doi:10.1038/s43586-024-00366-8.

Chen, J., Pan, J., Duan, M., Fan, F., Liu, J., Hu, W., Yu, D., Sun, Z., Sheng, X., Tan, Y., Tan, Y., Sun, X., Tang, N., Wang, W., Tang, W., Ye, N., Chen, J., Liu, Z., & Yuan, D. (2025). In situ images of Cd^{2+} in rice reveal Cd^{2+} protective mechanism using DNAzyme fluorescent probe. *Journal of Hazardous Materials, 483*. https://www.sciencedirect.com/science/journal/03043894. doi:10.1016/j.jhazmat.2024.136650.

Cheng, B.-Y., Fang, W.-T., Shyu, G.-S., & Chang, T.-K. (2013). Distribution of heavy metals in the sediments of agricultural fields adjacent to urban areas in Central Taiwan. *Paddy and Water Environment, 11*(1–4), 343–351. doi:10.1007/s10333-012-0325-3.

Collins, C. R., Ragnarsdottir, K. V., & Sherman, D. M. (1999). Effect of inorganic and organic ligands on the mechanism of cadmium sorption to goethite. *Geochimica et Cosmochimica Acta, 63*(19–20), 2989–3002. doi:10.1016/S0016-7037(99)00226-4.

Cuello-Nuñez, S., Benning, J., Liu, C., Branton, P., Hu, J., McAdam, K. G., Coburn, S., Braybrook, J., & Goenaga-Infante, H. (2018). Fractionation of cadmium in tobacco and cigarette smoke condensate using XANES and sequential leaching with ICP-MS/MS. *Analytical and Bioanalytical Chemistry, 410*(26), 6795–6806. link.springer.de/link/service/journals/00216/index.htm. doi:10.1007/s00216-018-1265-6.

da Silva, F. B. V., do Nascimento, C. W. A., Araújo, P. R. M., da Silva, L. H. V., & da Silva, R. F. (2016). Assessing heavy metal sources in sugarcane Brazilian soils: an approach using multivariate analysis. *Environmental Monitoring and Assessment, 188*(8). www.wkap.nl/journalhome.htm/0167-6369. doi:10.1007/s10661-016-5409-x.

Da Silva, Y. J. A. B., Do Nascimento, C. W. A., & Biondi, C. M. (2014). Comparison of USEPA digestion methods to heavy metals in soil samples. *Environmental Monitoring and Assessment, 186*(1), 47–53. doi:10.1007/s10661-013-3354-5.

Dean, J. R. (2013). *Environmental trace analysis: Techniques and applications environmental trace analysis: Techniques and applications* (pp. 1–254). Wiley. http://onlinelibrary.wiley.com/book/10.1002/9781118683422. doi:10.1002/9781118683422.

Degryse, F., Smolders, E., & Merckx, R. (2006). Labile Cd complexes increase Cd availability to plants. *Environmental Science & Technology, 40*(3), 830–836. doi:10.1021/es050894t.

Dekeyrel, J., Atkinson, R., Chavez, E., da Silva, M., Idarraga-Castaño, O., Pulleman, M., & Smolders, E. (2024). Using optimized monochromatic energy dispersive X-ray fluorescence to determine the cadmium concentration in cacao and soil samples. *Heliyon, 10*(20), e39034. doi:10.1016/j.heliyon.2024.e39034.

de la Guardia, M., & Garrigues, S. (2014). The social responsibility of environmental analysis. *Trends in Environmental Analytical Chemistry, 3–4*, 7–13. doi:10.1016/j.teac.2014.09.001.

Demırezen, D., & Aksoy, A. (2004). Accumulation of heavy metals in *Typha angustifolia* (L.) and *Potamogeton pectinatus* (L.) living in Sultan Marsh (Kayseri, Turkey). *Chemosphere, 56*(7), 685–696. doi:10.1016/j.chemosphere.2004.04.011.

Ershadi, S., & Shayanfar, A. (2018). Are LOD and LOQ reliable parameters for sensitivity evaluation of spectroscopic methods? *Journal of AOAC International*, *101*(4), 1212–1213. http://www.ingentaconnect.com/content/aoac/jaoac. doi:10.5740/jaoacint.17-0363.

Ferreira, S. L. C., Lima, C. C., Garcia, R. J. L., da Silva Júnior, J. B., Coutinho, J. J., Garcia, K. S., Rocha Soares, S. A., & Oliveira dos Santos, L. (2025). Risk assessment of toxic elements in surface soils collected near and far from a deactivated lead smelter in Bahia, Brazil. *Analytical Letters*, *58*(2), 285–298. www.tandf.co.uk/journals/titles/00032719.asp. doi:10.1080/00032719.2024.2323676.

Gai, P., Pu, L. i, Wang, C., Zhu, D., & Li, F. (2023). CeO_2@NC nanozyme with robust dephosphorylation ability of phosphotriester: A simple colorimetric assay for rapid and selective detection of paraoxon. *Biosensors and Bioelectronics*, *220*, 114841. doi:10.1016/j.bios.2022.114841.

García, R., Belmont, R., Padilla, H., Torres, M. C., & Báez, A. (2009). Trace metals and inorganic ion measurements in rain from Mexico City and a nearby rural area. *Chemistry and Ecology*, *25*(2), 71–86. doi:10.1080/02757540902758776.

Ge, J., & Zhang, J. (2015). Heavy metal contamination and accumulation in soil and plant species from the Xinqiao Copper Deposit, Anhui Province, China. *Analytical Letters*, *48*(3), 541–552. www.tandf.co.uk/journals/titles/00032719.asp. doi:10.1080/00032719.2014.946039.

Gu, Y., Wang, P., Zhang, S., Dai, J., Chen, H. P., Lombi, E., Howard, D. L., Van Der Ent, A., Zhao, F. J., & Kopittke, P. M. (2020). Chemical speciation and distribution of cadmium in rice grain and implications for bioavailability to humans. *Environmental Science and Technology*, *54*(19), 12072–12080. http://pubs.acs.org/journal/esthag. doi:10.1021/acs.est.0c03001.

Guo, M. (2009). Soil sampling and methods of analysis. *Journal of Environmental Quality*, *38*(1), 375. doi:10.2134/jeq2008.0018br.

Hamurcu, M., Özcan, M. M., Dursun, N., & Gezgin, S. (2010). Mineral and heavy metal levels of some fruits grown at the roadsides. *Food and Chemical Toxicology*, *48*(6), 1767–1770. doi:10.1016/j.fct.2010.03.031.

Hassan, K. M., Elhaddad, G. M., & AbdelAzzem, M. (2019). Voltammetric determination of cadmium(II), lead(II) and copper(II) with a glassy carbon electrode modified with silver nanoparticles deposited on poly(1,8-diaminonaphthalene). *Microchimica Acta*, *186*(7). doi:10.1007/s00604-019-3552-0.

Hossain, M., Karmakar, D., Begum, S. N., Ali, S. Y., & Patra, P. K. (2021). Recent trends in the analysis of trace elements in the field of environmental research: A review. *Microchemical Journal*, *165*. www.elsevier.com/inca/publications/store/6/2/0/3/9/1. doi:10.1016/j.microc.2021.106086.

Huang, W., Zhang, Y., Li, Y., Zeng, T., Wan, Q., & Yang, N. (2020). Morphology-controlled electrochemical sensing of environmental Cd^{2+} and Pb^{2+} ions on expanded graphite supported CeO_2 nanomaterials. *Analytica Chimica Acta*, *1126*, 63–71. doi:10.1016/j.aca.2020.06.010.

Hui, C. y., Guo, Y., Gao, C. x., Li, H., Lin, Y. r., Yun, J. p., Chen, Y. t., & Yi, J. (2022). A tailored indigoidine-based whole-cell biosensor for detecting toxic cadmium in environmental water samples. *Environmental Technology and Innovation*, *27*. http://www.journals.elsevier.com/environmental-technology-and-innovation/. doi:10.1016/j.eti.2022.102511.

Ikonen, K. E., Wehde, K. E., Khalida, H., & Kenttämaa, H. I. (2020). Bias, limit of detection, and limit of quantitation for the ASTM D2425 method updated in 2019. *Journal of Chromatography A*, *1614*. www.elsevier.com/locate/chroma. doi:10.1016/j.chroma.2019.460705.

Jia, X., Liu, T., Ma, Y., & Wu, K. (2021). Construction of cadmium whole-cell biosensors and circuit amplification. *Applied Microbiology and Biotechnology*, *105*(13), 5689–5699. doi:10.1007/s00253-021-11403-x.

Juranović Cindrić, I., Zeiner, M., Krpetić, M., & Stingeder, G. (2012). ICP-AES determination of minor and major elements in Cornelian cherry (*Cornus mas* L.) after microwave assisted digestion. *Microchemical Journal*, *105*, 72–76. doi:10.1016/j.microc.2012.03.011.

Karlsson, T. ö, Persson, P., & Skyllberg, U. (2005). Extended X-ray absorption fine structure spectroscopy evidence for the complexation of cadmium by reduced sulfur groups in natural organic matter. *Environmental Science & Technology*, *39*(9), 3048–3055. doi:10.1021/es048585a.

Kokab, T., Shah, A., Nisar, J., Khan, A. M., Khan, S. B., & Shah, A. H. (2020). Tripeptide derivative-modified glassy carbon electrode: A novel electrochemical sensor for sensitive and selective detection of Cd^{2+} ions. *ACS Omega*, *5*(17), 10123–10132. pubs.acs.org/journal/acsodf. doi:10.1021/acsomega.0c00760.

Koopmans, G. F., & Groenenberg, J. E. (2011). Effects of soil oven-drying on concentrations and speciation of trace metals and dissolved organic matter in soil solution extracts of sandy soils. *Geoderma*, *161*(3-4), 147–158. doi:10.1016/j.geoderma.2010.12.012.

Kopittke, P. M., Punshon, T., Paterson, D. J., Tappero, R. V., Wang, P., Blamey, F. P. C., Ent, A., & Lombi, E. (2018). Synchrotron-based X-ray fluorescence microscopy as a technique for imaging of elements in plants. *Plant Physiology, 178*(2), 507–523. doi:10.1104/pp.18.00759.

Kováčik, J., & Babula, P. (2017). Fluorescence microscopy as a tool for visualization of metal-induced oxidative stress in plants. *Acta Physiologiae Plantarum, 39*(8). doi:10.1007/s11738-017-2455-0.

Krishnamurti, G. S. R., Huang, P. M., Van Rees, K. C. J., Kozak, L. M., & Rostad, H. P. W. (1994). Microwave digestion technique for the determination of total cadmium in soils. *Communications in Soil Science and Plant Analysis, 25*(5–6), 615–625. doi:10.1080/00103629409369067.

Krushevska, A., Barnes, R. M., & Amarasiriwaradena, C. (1993). Decomposition of biological samples for inductively coupled plasma atomic emission spectrometry using an open focused microwave digestion system. *The Analyst, 118*(9), 1175–1181. doi:10.1039/AN9931801175.

Küpper, H., Mijovilovich, A., Meyer-Klaucke, W., & Kroneck, P. M. H. (2004). Tissue- and age-dependent differences in the complexation of cadmium and zinc in the cadmium/zinc hyperaccumulator *Thlaspi caerulescens* (ganges ecotype) revealed by X-ray absorption spectroscopy. *Plant Physiology, 134*(2), 748–757. http://www.plantphysiol.org/. doi:10.1104/pp.103.032953.

Lamble, K. J., & Hill, S. J. (1998). Microwave digestion procedures for environmental matrices. *Analyst, 123*(7). http://pubs.rsc.org/en/journals/journal/an. doi:10.1039/a800776d.

Li, G., Yang, Q., Liu, X., Liu, L., Han, J., & Li, X. (2024). Highly sensitive colorimetric detection of the toxic metal cadmium in rice using the 3-(1*H*-tetrazol-5-yl) aniline@AuNPs sensor. *Journal of Photochemistry and Photobiology A: Chemistry, 447*, 115228. doi:10.1016/j.jphotochem.2023.115228.

Litluechai, A., Prompa, A., Uppachai, P., Jarernboon, W., Butwong, N., & Mukdasai, S. (2024). Highly sensitive colorimetric detection of Cd(II) based on silica sol modified with dithizone and cationic surfactant. *RSC Advances, 14*(44), 32101–32108. doi:10.1039/d4ra03983a.

Liu, X., Chen, H., Hu, X., Liu, J., Zhong, Y., Li, P., Zhu, X., Zhang, Y., Liu, M., & Yao, S. (2024). Traffic lights inspired colorimetric sensing platform for cadmium/lead ions: PO43- induced structural change of porphyrin metal-organic framework. *Sensors and Actuators B: Chemical, 420*, 136455. doi:10.1016/j.snb.2024.136455.

Liu, X., Li, N., Xu, M. M., Jiang, C., Wang, J., Song, G., & Wang, Y. (2019a). Turn on fluorescent detection for Cd^{2+} based on surfactant controlled squaraine aggregation. *Spectrochimica Acta—Part A: Molecular and Biomolecular Spectroscopy, 208*, 236–242. doi:10.1016/j.saa.2018.10.012.

Liu, Y., Li, T., Ling, C., Chen, Z., Deng, Y., & He, N. (2019b). Electrochemical sensor for Cd^{2+} and Pb^{2+} detection based on nano-porous pseudo carbon paste electrode. *Chinese Chemical Letters, 30*(12), 2211–2215. doi:10.1016/j.cclet.2019.05.020.

Lofts, S., Spurgeon, D. J., Svendsen, C., & Tipping, E. (2004). Deriving soil critical limits for Cu, Zn, Cd, and Ph: A method based on free ion concentrations. *Environmental Science and Technology, 38*(13), 3623–3631. doi:10.1021/es030155h.

Lu, M., Deng, Y., Luo, Y. i, Lv, J., Li, T., Xu, J., Chen, S.-W., & Wang, J. (2019). Graphene aerogel–metal–organic framework-based electrochemical method for simultaneous detection of multiple heavy-metal ions. *Analytical Chemistry, 91*(1), 888–895. doi:10.1021/acs.analchem.8b03764.

Luo, K., Zhou, L., Xie, C., Yang, Q., Tan, L., & Lin, Q. (2023). High-fidelity fluorescent probes for visualizing the inhibitory behavior of selenium on cadmium uptake in rice. *Journal of Hazardous Materials, 457*, 131748. doi:10.1016/j.jhazmat.2023.131748.

Ma, Y., Wang, F., Kambam, S., & Chen, X. (2013). A quinoline-based fluorescent chemosensor for distinguishing cadmium from zinc ions using cysteine as an auxiliary reagent. *Sensors and Actuators B: Chemical, 188*, 1116–1122. doi:10.1016/j.snb.2013.08.008.

Magesa, F., Wu, Y., Tian, Y., Vianney, J.-M., Buza, J., He, Q., & Tan, Y. (2019). Graphene and graphene like 2D graphitic carbon nitride: Electrochemical detection of food colorants and toxic substances in environment. *Trends in Environmental Analytical Chemistry, 23*, e00064. doi:10.1016/j.teac.2019.e00064.

Massas, I., Kalivas, D., Ehaliotis, C., & Gasparatos, D. (2013). Total and available heavy metal concentrations in soils of the Thriassio plain (Greece) and assessment of soil pollution indexes. *Environmental Monitoring and Assessment, 185*(8), 6751–6766. doi:10.1007/s10661-013-3062-1.

Mohd Rosdi, N. A. N. B., Abd Hamid, N., Mohd Ali, S. F., Sino, H., & Lee, L. C. (2024). A critical review of soil sampling and data analysis strategies for source tracing of soil in forensic investigations. *Critical Reviews in Analytical Chemistry, 54*(8), 3520–3558. www.tandf.co.uk/journals/titles/10408347.asp. doi:10.1080/10408347.2023.2253473.

Morrison, G. H., & Talmi, Y. (1970). Microanalysis of solids by atomic absorption and emission spectroscopy using an RF [radiofrequency] furnace. *Analytical Chemistry, 42*(7), 809–811. doi:10.1021/ac60289a012.

Nyachoti, S., Godebo, T. R., Okwori, O. F., Kodsup, P., & TatahMentah, M. K. (2023). Occurrence and spatial distribution of lead, arsenic, cadmium, and uranium in soils of southern Louisiana. *Water, Air, and Soil Pollution, 234*(11). https://www.springer.com/journal/11270. doi:10.1007/s11270-023-06716-8.

Oğuz Tümay, S. ü, Şanko, V., Demirbas, E., & Şenocak, A. (2020). Fluorescence determination of trace level of cadmium with pyrene modified nanocrystalline cellulose in food and soil samples. *Food and Chemical Toxicology, 146*, 111847. doi:10.1016/j.fct.2020.111847.

Pan, F., Yu, Y., Yu, L., Lin, H., Wang, Y., Zhang, L., Pan, D., & Zhu, R. (2020). Quantitative assessment on soil concentration of heavy metal–contaminated soil with various sample pretreatment techniques and detection methods. *Environmental Monitoring and Assessment, 192*(12). doi:10.1007/s10661-020-08775-4.

Park, J., Kim, J. Y., Lee, K., Kim, M. S., Kim, M. J., & Choi, J. W. (2020). Comparison of acid extraction and total digestion methods for measuring Cd isotope ratios of environmental samples. *Environmental Monitoring and Assessment, 192*(1). https://link.springer.com/journal/10661. doi:10.1007/s10661-019-8017-8.

Pessôa, G. d. S., Lopes Júnior, C. A., Madrid, K. C., & Arruda, M. A. Z. (2017). A quantitative approach for Cd, Cu, Fe and Mn through laser ablation imaging for evaluating the translocation and accumulation of metals in sunflower seeds. *Talanta, 167*, 317–324. https://www.journals.elsevier.com/talanta. doi:10.1016/j.talanta.2017.02.029.

Piccolo, B., & O'Connor, R. T. (1968). Atomic absorption spectroscopy. *Journal of the American Oil Chemists' Society, 45*(11), 789–792. doi:10.1007/BF02631956.

Pizarro, J., Segura, R., Tapia, D., Navarro, F., Fuenzalida, F., & Aguirre, M. J. (2020). Inexpensive and green electrochemical sensor for the determination of Cd(II) and Pb(II) by square wave anodic stripping voltammetry in bivalve mollusks. *Food Chemistry, 321*, 126682. doi:10.1016/j.foodchem.2020.126682.

Pons, M.-L., Collin, B., Doelsch, E., Chaurand, P., Fehlauer, T., Levard, C., Keller, C., & Rose, J. (2021). X-ray absorption spectroscopy evidence of sulfur-bound cadmium in the Cd-hyperaccumulator *Solanum nigrum* and the non-accumulator *Solanum melongena*. *Environmental Pollution, 279*, 116897. doi:10.1016/j.envpol.2021.116897.

Priya, T., Dhanalakshmi, N., Thennarasu, S., Karthikeyan, V., & Thinakaran, N. (2019). Ultra sensitive electrochemical detection of Cd^{2+} and Pb^{2+} using penetrable nature of graphene/gold nanoparticles/modified L-cysteine nanocomposite. *Chemical Physics Letters, 731*, 136621. doi:10.1016/j.cplett.2019.136621.

Rhymes, J. M., Cordero, I., Chomel, M., Lavallee, J. M., Straathof, A. L., Ashworth, D., Langridge, H., Semchenko, M., De Vries, F. T., Johnson, D., & Bardgett, R. D. (2021). Are researchers following best storage practices for measuring soil biochemical properties? *SOIL, 7*(1), 95–106. www.soil-journal.net. doi:10.5194/soil-7-95-2021.

Ribeiro Menezes, I. M. N., Nascimento, P. d. A., Gonzalez, M. H., & Oliveira, A. (2021). Simple and robust GFAAS methods for determination of As, Cd, and Pb in hemp products using different sample preparation strategies. *Food Analytical Methods, 14*(5), 1043–1053. http://www.springer.com/life+sci/food+science/journal/12161. doi:10.1007/s12161-020-01958-2.

Salamanca, A. MC., Mogollon, O. LM., Chaali, N., Ariza-Nieto, C., Beltran-Medina, J. éI., Cuadros, R. EO., & Cruz, E. ND. (2024). ICP-OES analysis of total As and Cd in Columbian *Oryza sativa* L. rice. *Food Additives & Contaminants: Part B, 17*(1), 16–27. doi:10.1080/19393210.2023.2278805.

Saleem, M., Pierce, D., Wang, Y., Sens, D. A., Somji, S., & Garrett, S. H. (2024). Heavy metal(oid)s contamination and potential ecological risk assessment in agricultural soils. *Journal of Xenobiotics, 14*(2), 634–650. https://www.mdpi.com/2039-4713/11. doi:10.3390/jox14020037.

Sharma, A., Verwilst, P., Li, M., Ma, D., Singh, N., Yoo, J., Kim, Y., Yang, Y., Zhu, J. H., Huang, H., Hu, X. L., He, X. P., Zeng, L., James, T. D., Peng, X., Sessler, J. L., & Kim, J. S. (2024). Theranostic fluorescent probes. *Chemical Reviews, 124*(5), 2699–2804. http://pubs.acs.org/journal/chreay. doi:10.1021/acs.chemrev.3c00778.

Silva, A. L. M., Cirino, S., Carvalho, M. L., Manso, M., Pessanha, S., Azevedo, C. D. R., Carramate, L. F. N. D., Santos, J. P., Guerra, M., & Veloso, J. F. C. A. (2017). Elemental mapping in a contemporary miniature by full-field X-ray fluorescence imaging with gaseous detector vs. scanning X-ray fluorescence imaging with polycapillary optics. *Spectrochimica Acta Part B: Atomic Spectroscopy, 129*, 1–7. doi:10.1016/j.sab.2016.12.006.

Taheri, M., Mahmudy Gharaie, M. H., Mehrzad, J., Stone, M., & Afshari, R. (2022). Profiles and potential health risks of heavy metals in polluted soils in NE-Iran. *Toxin Reviews, 41*(2), 523–535. http://www.tandfonline.com/loi/itxr20. doi:10.1080/15569543.2021.1906706.

Tananonchai, A., Tang, M. T., Pao, C. W., Sampanpanish, P., Tanthanuch, W., & Tancharakorn, S. (2023). The study of EDTA enhanced Cd accumulation and formation in Napier grass using synchrotron μX-ray fluorescence imaging and X-ray absorption spectroscopy. *Radiation Physics and Chemistry, 207*. www.elsevier.com/locate/radphyschem. doi:10.1016/j.radphyschem.2023.110851.

Tekuri, V., & Trivedi, D. R. (2017). A new colorimetric chemosensors for Cu^{2+} and Cd^{2+} ions detection: Application in environmental water samples and analytical method validation. *Analytica Chimica Acta, 972*, 81–93. http://www.journals.elsevier.com/analytica-chimica-acta/. doi:10.1016/j.aca.2017.03.035.

Terfi, S., & Sadi, F. (2015). Optimization of extraction of toxic metals from medicinal plants, *Malva sylvestris* L., and *Pistacia lentiscus*. *Analytical Letters, 48*(7), 1190–1197. doi:10.1080/00032719.2014.976869.

Terzi, M. H., & Kalkan, M. (2024). Evaluation of soil pollution by heavy metal using index calculations and multivariate statistical analysis. *Environmental Monitoring and Assessment, 196*(5). https://www.springer.com/journal/10661. doi:10.1007/s10661-024-12569-3.

Thompson, M., & Walsh, N. (2012). *Handbook of inductively coupled plasma spectrometry*. Springer Science & Business Media.

Tsuji, K., Nakano, K., Takahashi, Y., Hayashi, K., & Ro, C. U. (2012). X-ray spectrometry. *Analytical Chemistry, 84*(2), 636–668. doi:10.1021/ac202871b.

Van Acker, T., Theiner, S., Bolea-Fernandez, E., Vanhaecke, F., & Koellensperger, G. (2023). Inductively coupled plasma mass spectrometry. *Nature Reviews Methods Primers, 3*(1). doi:10.1038/s43586-023-00235-w.

van der Ent, A., Przybyłowicz, W. J., de Jonge, M. D., Harris, H. H., Ryan, C. G., Tylko, G., Paterson, D. J., Barnabas, A. D., Kopittke, P. M., & Mesjasz-Przybyłowicz, J. (2018). X-ray elemental mapping techniques for elucidating the ecophysiology of hyperaccumulator plants. *New Phytologist, 218*(2), 432–452. http://www.wiley.com/bw/editors.asp?ref=0028-646X&site=1. doi:10.1111/nph.14810.

Viala, Y., Sappin-Didier, V., Bussière, S., Coriou, C., & Nguyen, C. (2021). Simple models efficiently predict free cadmium Cd^{2+} in the solutions of low-contaminated agricultural soils. *Science of the Total Environment, 778*, 146428. doi:10.1016/j.scitotenv.2021.146428.

Wan, J., Shen, Y., Xu, L., Xu, R., Zhang, J., Sun, H., Zhang, C., Yin, C., & Wang, X. (2021). Ferrocene-functionalized Ni(II)-based metal-organic framework as electrochemical sensing interface for ratiometric analysis of Cu^{2+}, Pb^{2+} and Cd^{2+}. *Journal of Electroanalytical Chemistry, 895*. doi:10.1016/j.jelechem.2021.115374.

Wang, H., Da, L., Yang, L., Chu, S., Yang, F., Yu, S., & Jiang, C. (2020). Colorimetric fluorescent paper strip with smartphone platform for quantitative detection of cadmium ions in real samples. *Journal of Hazardous Materials, 392*, 122506. doi:10.1016/j.jhazmat.2020.122506.

Wei, Y., Shi, D., Chen, T., Zhou, S., Yang, Z., Li, H., Yang, D., Li, J., & Jin, M. (2024). CRISPR/Cas9-based engineered *Escherichia coli* biosensor for sensitive and specific detection of Cd(II) in drinking water. *Chemosphere, 362*, 142607. doi:10.1016/j.chemosphere.2024.142607.

Wiggenhauser, M., Aucour, A. M., Bureau, S., Campillo, S., Telouk, P., Romani, M., Ma, J. F., Landrot, G., & Sarret, G. (2021). Cadmium transfer in contaminated soil-rice systems: Insights from solid-state speciation analysis and stable isotope fractionation. *Environmental Pollution, 269*. https://www.journals.elsevier.com/environmental-pollution. doi:10.1016/j.envpol.2020.115934.

Wu, X., Yin, S., Liu, Y., Zhu, Y., Jiang, T., Liang, S., Bian, S., Cao, Y., Wang, G., & Yang, J. (2024a). Molecular mechanisms and physiological responses of rice leaves co-exposed to submicron-plastics and cadmium: Implication for food quality and security. *Journal of Hazardous Materials, 463*, 132957. doi:10.1016/j.jhazmat.2023.132957.

Wu, X. J., Dong, D. W., Qu, J. Y., Han, Y., You, Z. Y., Li, P. P., & Han, J. G. (2024b). An enzyme-fused phycobiliprotein synthesis system developed for visual whole-cell biosensors for the detection of cadmium during wastewater treatment. *Environmental Technology and Innovation, 36*. https://www.sciencedirect.com/science/journal/23521864. doi:10.1016/j.eti.2024.103882.

Yan, C., Yang, X., Li, Z., Liu, Y., Yang, S., Deng, Q., & Wen, X. (2021). Switchable hydrophilicity solvent-based preconcentration for ICP-OES determination of trace lead in environmental samples. *Microchemical Journal, 168*, 106529. doi:10.1016/j.microc.2021.106529.

Yi, Y., Zhao, Y., Zhang, Z., Wu, Y., & Zhu, G. (2022). Recent developments in electrochemical detection of cadmium. *Trends in Environmental Analytical Chemistry, 33*, e00152. doi:10.1016/j.teac.2021.e00152.

You, Z., Akkuş, A. ı, Weisheit, W., Giray, T., Penk, S., Buttler, S., Recknagel, S., & Abad, C. (2022). Multielement analysis in soils using nitrogen microwave inductively coupled atmospheric-pressure plasma mass spectrometry. *Journal of Analytical Atomic Spectrometry, 37*(12), 2556–2562. doi:10.1039/d2ja00244b.

Yu, F., Li, P., Song, P., Wang, B., Zhao, J., & Han, K. (2012). An ICT-based strategy to a colorimetric and ratiometric fluorescence probe for hydrogen sulfide in living cells. *Chemical Communications, 48*(23), 2852–2854. doi:10.1039/c2cc17658k.

Zeiner, M., & Juranović Cindrić, I. (2017). Review – trace determination of potentially toxic elements in (medicinal) plant materials. *Analytical Methods*, *9*(10), 251. doi:10.1039/C7AY00016B.

Zhang, C. C (2006). *Fundamentals of environmental sampling and analysis* (pp. 1–436). Wiley. doi:10.1002/0470120681.

Zhang, J., Wang, Y., Zhang, J., Ding, Y., Yu, H., & Jin, H. (2011). Evaluation of mineral element contents in *Paris polyphylla* var. *yunnanensis* from Southwest China. *African Journal of Pharmacy and Pharmacology*, *5*(15), 1792–1796. http://www.academicjournals.org/AJPP/PDF/pdf2011/22%20Oct/Zhang%20et%20al.pdf. doi:10.5897/AJPP11.433.

Zhang, M., Chen, X., Wang, N., Guan, L., Wang, L., Chen, X., Yang, Z., Sun, Y., Fan, Y., Meng, Y., Liu, M., Chen, W., Wu, F., Song, R., Wang, S., Lu, X., Wang, J., Guo, L., Zhao, L., … Ye, W. (2024a). GhCOMT33D modulates melatonin synthesis, impacting plant response to Cd^{2+} in cotton via ROS. *Physiologia Plantarum*, *176*(6). doi:10.1111/ppl.14647.

Zhang, T., Zhu, K. Li, Zhang, X., Yu, X., Shen, L., Gao, D., Chen, Y., Wang, Q., Chen, S., & Bao, L. (2024b). Development of CadR-based cadmium whole cell biosensor for visual detection of environmental Cd^{2+}. *Analytica Chimica Acta*, *1330*, 343299. doi:10.1016/j.aca.2024.343299.

CHAPTER

3

Global patterns of cadmium contamination in soils

Deyi Hou, Xiyue Jia

School of Environment, Tsinghua University, Beijing, China

3.1 Introduction

Anthropogenic activities have significantly influenced the cycling of trace metals in the geologic environment (Cloquet et al., 2006; Lee et al., 2019; Li et al., 2025; Zhang et al., 2025). Soil is the major sink of heavy metals released from industrial and agricultural processes (Hu & Cheng, 2013). Soil pollution by cadmium (Cd) is particularly of concern because of its high toxicity, high mobility, and easy uptake by major food crops such as rice and wheat and thus threat to global food security (Harris & Taylor, 2013; Hou, 2023; Larison et al., 2000; Li et al., 2017). The World Health Organization (WHO) lists Cd as one of the top 10 chemicals of major public health concern (WHO, 2020). The occurrence of Cd pollution in paddy soil and rice grain has been reported in many countries, for example, China, Japan, Bangladesh, Thailand, and South Korea (Arao et al., 2011; Khaokaew et al., 2011(Zhao et al., 2015)). In the United States, Cd has been identified as a contaminant of concern, that is, above a health guidance level, at 1011 out of 1867 sites/events on the Agency for Toxic Substances and Disease Registry (ATSDR) Substance Priority List (ATSDR, 2020). The excessive anthropogenic input of Cd relative to other trace metals may also be attributed to its low boiling point (765°C) and high solubility in acidic environments (Cloquet et al., 2006; Kubier et al., 2019).

Globally, urban waste, fly ash, and atmospheric deposition were reported to be the most important sources of Cd pollution in soil (Nriagu & Pacyna, 1988). The total emission of Cd from anthropogenic activities into the soil environment has been estimated at 9900–45,000 t/year, with the above processes accounting for the majority of this input (Nriagu & Pacyna, 1988). Worldwide atmospheric emissions of Cd have been estimated at 3200–15,000 t/year, with anthropogenic sources contributing 3100–12,000 t/year, whereas natural emissions occur at one order of magnitude lower level (Nriagu, 1989). However, these estimates are outdated, representing a major research gap that needs to be addressed.

The anthropogenic input of Cd has increased as industrial output has grown, but it has also decreased in some regions of the world as environmental stewardship has strengthened.

Cadmium Contamination in Soils and Plants: Mechanisms and Mitigation.
DOI: https://doi.org/10.1016/B978-0-443-34104-5.00010-4

In Europe the use of phosphorus fertilizers declined by approximately 70% between 1980 and 2010, and atmospheric deposition of Cd fell from nearly 3 g Cd/ha/year in the 1990s to 0.35 g Cd/ha/year in the 2010s (Six & Smolders, 2014). From the mid-1960s to the early 1990s, Cd concentrations in Greenland ice decreased by nearly 60% because of efforts to reduce industrial emissions in various developed countries in the Northern Hemisphere (Boutron et al., 1991). In contrast, atmospheric emissions of Cd increased significantly in China, the new "world factory," where Cd emissions grew from approximately 300 t in 2000 to 800 t in 2010 (Cheng et al., 2015a). However, emission control measures have been strengthened in China, leading to reduced Cd emissions in various industrial sectors. For instance, Cd emissions from coal-fired power plants in China peaked in 2005 and decreased by 58% from 2005 to 2010 (Tian et al., 2014).

The spatial distribution of soil pollution by Cd is governed by a complex interplay of natural and anthropogenic processes. In agricultural soil, mineral fertilizers and manure represent the main influx, whereas crop harvest represents the main efflux of Cd on a regional scale (Imseng et al., 2018) (see also Chapter 5). Phosphate rocks tend to contain elevated Cd levels; for instance, phosphate fertilizers in New Zealand prior to 1997 mainly originated from guano-derived phosphorite, which contained about 450 mg Cd/kg of phosphorus (Salmanzadeh et al., 2017). In mining regions, mine tailings and smelter emissions often result in large-scale agricultural soil pollution, with more severe contamination in close proximity (Schaider et al., 2007; Shi et al., 2022; Velleux et al., 2006). In urban areas, Cd pollution may originate from point sources, including manufacturing plants, waste disposal facilities, incinerators, and fossil fuel combustors (Cui et al., 2019; Hellström et al., 2007; Ramos-Ruiz et al., 2017). Diffusive sources, such as vehicular emissions, can also contribute to Cd pollution in urban topsoil (Hjortenkrans et al., 2007; Murphy & Hutchinson, 2015; Szwalec et al., 2020). Soil parent materials and soil types also play a key role in soil Cd distribution, as the geologic weathering of bedrock is a main source of Cd in natural soil. For instance, in the Swiss and French Jura Mountains, Oolitic carbonates were found to be highly enriched in Cd, with the associated soils containing an average Cd concentration of 0.8 mg/kg and reaching a maximum of 2 mg/kg (Quezada-Hinojosa et al., 2009). In California, United States, soils developed from shale parent materials contained an average Cd concentration of 7.5 mg/kg (Lund et al., 1981). The complex interplay of environmental and social processes has resulted in a highly heterogeneous distribution of Cd on a global scale. In this study, we intend to develop spatially explicit, continuous prediction of soil pollution by Cd on a global scale against a uniform threshold.

3.2 Materials and methods

3.2.1 Data compilation

The following keyword combination was used to synthesize a global database of soil Cd concentrations: TOPIC: ("soil" OR "land" OR "geochem*") AND TOPIC: ("Spatio temporal" OR "regional scale" OR "provincial" OR "province" OR "county" OR "mapping" OR "map" OR "spatial distribution" OR "spatial variability" OR "spatial variation" OR "spatial interpolation" OR "hectare" OR "acre" OR "km" OR "principal component analysis" OR "kriging" OR "GIS" OR "multisite" OR "multiple sites" OR "forest sites") AND TOPIC: ("metal*" OR "cadmium" OR "soil pollut*" OR "soil contam*" OR "trace element" OR "toxic element") NOT TOPIC: ("marine" OR "ocean"). The search was conducted using the Web of Science tool, covering the following databases: Web of Science Core Collection, MEDLINE, Data Citation Index, Biosis Previews, Inspec, SciELO Citation Index, Chinese Science Citation Database, and KCI-Korean Journal Database. Articles and reviews published in English were retrieved for further screening. Based on peer-reviewed studies a snowball method was used to identify additional databases pertaining to government agencies through references from relevant papers. More details about the database synthesis method can be found in this study (Hou et al., 2025).

3.2.2 Regulatory thresholds

Different countries have varying types of regulatory thresholds for Cd in soil (Table 3.1). These thresholds include screening values, trigger values, intervention values, and clean values. In general, they can be broadly categorized into two groups: the first group of thresholds is used as a screening value, above which soil pollution concentration becomes a concern, and usually, risk assessment is required to determine further action. The second group of thresholds is used as an intervention value, above which remediation or risk management is often required. Currently there is no consensus on threshold values on a global scale. For the purpose of the present study, we selected 1 mg/kg as the uniform threshold for the following reasons: (1) this threshold has been used by various countries; (2) it is modestly conservative based on all the thresholds listed in Table 3.1.

3.2.3 Data analysis

The Cd concentrations from various studies were synthesized using 10 km × 10 km equal-area grids (approximately 2-decimal-degree equivalent) in EPSG: 4326 (WGS84). The 10-km resolution was chosen because it is large enough to evaluate regional-scale spatial distribution rather than site-specific pollution events, and it is also small enough to reasonably capture small-scale variation because of natural background, for example, associated with differing parent materials and weathering conditions during geologic time spans, as well as different levels of anthropogenic pollution from atmospheric deposition and agricultural practices (Hou et al., 2017, 2020). Statistical inference was used to derive exceedance for each 10 km × 10 km grid, and more details can be found in (Hou et al., 2025). A series of covariates were used to construct extremely randomized trees models for predicting the distribution of Cd exceedance, including lithological types, climatic variables, soil texture variables, basic physicochemical properties, elevation, and slope, as well as socioeconomic variables (Hou et al., 2025). It should be noted that in our previous study (Hou et al., 2025), the predictive model for cropland, whose threshold was also 1 mg/kg, was reported. In this study, we are reporting the predictive model for all soils based on the 1 mg/kg threshold. Because of this difference, in regions where cropland only accounts for a small portion of all soils, the reported exceedance rates could be largely different.

3.3 Results and discussion

3.3.1 Overview of global soil pollution by cadmium

The global exceedance rate of Cd was estimated to be 8% using a probability cut-off value of 0.5. The rate dropped to 4%, 2%, 1.1%, and 0.4% at cut-off values of 0.6, 0.7, 0.8, and 0.9,

TABLE 3.1 National regulatory thresholds for cadmium in soil (mg/kg).

Country	Threshold type	Thresholds		
Land use type		Residential	Industrial	Agricultural[a]
Austria	Trigger value	2		1
Austria	Intervention value	10		
Belgium	Risk investigation level	6	30	2
Belgium	Cleanup level	6	30	2
Canada	Soil quality guidelines	10	22	1.4
China	Screening level[b]	20	65	0.45
China	Intervention level	47	172	2.625
Czech Republic	Maximum admissible values	20	30	0.4
Czech Republic	Maximum admissible values for agricultural, other than sandy and loamy sand soils			1
Denmark	Ecotoxicological soil quality criteria			0.3
Finland	Threshold value	1		
Finland	Lower guideline value	10		
Finland	Upper guideline value	20		
France	VDSS	10		
France	VDI usage is sensible	20		
France	VDI usage is nonsensible	60		
Germany	Triggering level	20	60	
Germany	Action level[c]			0.07
Italy	Limit values	2	15	
Slovakia	Limit values: sand, loamy sand			0.4
Slovakia	Limit values: sandy loam, loam			0.7
Slovakia	Limit values: clay-loam, clay			1
Sweden	Guideline values	0.4	12	
Netherlands	Target value	0.8		
Netherlands	Intervention value	12		
United States	US-RSL-residential	71	980	

[a] Some of the values are not specifically designed for, but only broadly applicable to agricultural soil.
[b] Agricultural soil represents average for paddy field and nonpaddy field under various pH ranges.
[c] Agricultural soil represents average for different types of crops.

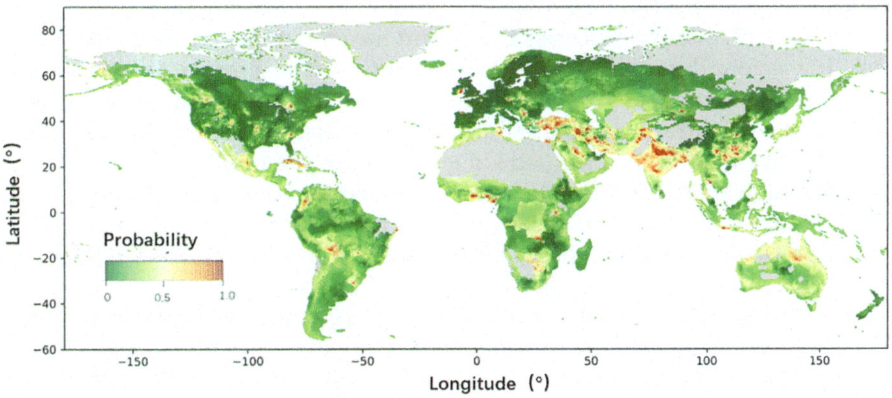

FIGURE 3.1 Probability of Cd exceedance above the 1 mg/kg threshold. *Red* indicates high probability, whereas *green* indicates low probability. High probability is predicted for South Asia, the Middle East, and Central America.

respectively. Cd exceedance was the most notable in northern and central India; Pakistan; Bangladesh; southern China; southern parts of Thailand; Iran; Turkey; Nigeria; South Africa; Mexico; and Cuba. Both anthropogenic sources and geogenic enrichment likely contributed to the elevated Cd concentrations in these regions (6, 8, 27, and 28). In the following sections, we discuss the distribution of Cd exceedance in various regions (Fig. 3.1).

3.3.2 Distribution pattern by continents

3.3.2.1 Asia

The Cd exceedance rate in Asia was estimated to be 15%, which was the second highest among all continents. Moreover, there was a relatively high confidence in the predicted exceedance. When the probability cut-off value was increased from 0.5 to 0.6, the Cd exceedance rate in Asia dropped from 15% to 9%, making it the highest among all continents. The majority of the exceedances were located in the so-called metal-enriched corridor in lower Eurasia, which extended from southern Europe to the Middle East, South Asia, and southern China (Hou et al., 2025). Our previous study showed that this metal-enriched corridor may have been formed because of a combination of factors, including strong weathering of metal-enriched parent rocks (Alloway, 2013; Moghadam et al., 2017); plant-pumping effects (Imseng et al., 2018; Meharg & Meharg, 2021); a lower degree of leaching associated with precipitation and terrain (Alloway, 2013); and a long history of mining and smelting activities dating back to ancient civilizations (Fao, 2021).

The elevated Cd levels in southern China are well-documented. Our model predicted that 5.5% of soil in China exceeded the 1 mg/kg threshold at a cut-off value of 0.6. Much of this soil was distributed in southern China, particularly in Guangxi Province and Guizhou Province. A previous study estimated the average Cd concentration in Guangxi and Guizhou provinces to be 1.0 and 0.8 mg/kg, respectively (Chen et al., 2015). Mining activities in southern China is a major source of Cd. For example, in Chenzhou, a region with nearly 20,000 km^2 of area in southern China, which is also famous for its mining activities, the geometric mean of Cd concentrations was estimated to be 1.4 mg/kg (Zhai et al., 2008). In another large-scale study, researchers selected 29 mountains across China and measured the Cd concentration in various soil horizons. It was found that the top O-horizon contained an average Cd concentration of 1.2 mg/kg, which was consistent with our previous finding that mountainous areas tend to have higher toxic metal concentrations (Hou et al., 2025). The large mountainous area in southern China may have attributed to the elevated Cd levels in this region.

In South Asia, high soil Cd exceedance rates were widely distributed. The exceedance rate in India was predicted to be 60% at a cut-off value 0.5, 20% at a cut-off value of 0.7, and 5% at a cut-off value 0.9. In Pakistan the predicted soil Cd exceedance rate reached 68%, 19%, and 4% at cut-off values 0.5, 0.7, and 0.9, respectively. In Bangladesh the exceedance rate was predicted to be 82%, 47%, and 7% at cut-off values 0.5, 0.7, and 0.9, respectively. The predicted probability distribution suggested high confidence in the existence of elevated Cd levels in the South Asia region, which was consistent with many local and regional studies that identified high Cd concentrations in soil. For instance, in the Indian Himalayan state of Himachal Pradesh, India, soil Cd concentration was estimated to be 2.3 mg/kg. The possible sources of soil pollution were mostly anthropogenic, because the mean Cd concentration in the background forest area was only 0.03 mg/kg, while that in urban, peri-urban, and rural areas reached 5.0, 3.2, and 1.2 mg/kg, respectively (Kashyap et al., 2019). Soils in agricultural fields along a national highway in North India were found to contain Cd at a mean concentration of 2.7 mg/kg (Gupta et al., 2021). Sampling of soils along the Sutlej River in India estimated the mean Cd concentration to be 2.4 mg/kg (Setia et al., 2023). In northern Pakistan, sampling in 75 towns and villages estimated an average Cd concentration of 1.8 mg/kg (Abbasi et al., 2013). In Chitral, a city of Pakistan, Cd concentration was estimated to be 2.0 mg/kg (Rehman et al., 2020). In the basin of River Swat, Pakistan, mean Cd concentration was estimated to be 3.2 mg/kg (Khan et al., 2022). In industrial areas of Jhenaidah and Kushtia districts, Bangladesh, soils were found to contain a mean Cd concentration of 1.2 mg/kg (Kormoker et al., 2021). These local studies suggested that many of the elevated Cd concentrations in South Asian countries may have been because of a combination of industrial activities, transportation, and agricultural activities.

In the Middle East, high exceedance rates were predicted for many countries, especially Turkey and Iran. In Turkey the Cd exceedance rate was estimated to be 72% at a probability cut-off value of 0.5. When the cut-off value was increased to 0.6 and 0.7, the exceedance rate dropped to 50% and 27%, respectively. In Iran the Cd exceedance rates corresponding to cut-off values of 0.5, 0.6, and 0.7 were estimated to be 48%, 28%, and 12%, respectively. The predicted exceedance rates were generally consistent with many local-to-regional studies conducted in these countries. In the Çarsamba plain, located in the central Black Sea region of Turkey, mean Cd concentration was measured to be 1.3 mg/kg (Saglam et al., 2011). In Sultan Marsh, Turkey, the largest wetland in the Middle East and Europe, the average soil Cd concentration was measured to be 4.1 mg/kg (Yalcin et al., 2007). In another study conducted in the central Black Sea region of Turkey the mean Cd concentration was estimated to be 1.5 mg/kg (Özkaraova Güngör & Orkun, 2014). In a heavily industrialized region the average Cd concentration reached 7.4 mg/kg (Kara et al., 2014). At a major potato production region in central Anatolia of Turkey, average Cd concentration in agricultural soil reached 1.7 mg/kg (Kara et al., 2004). In Isfahan, central Iran the mean Cd concentration was estimated to be 1.8 mg/kg (Amini et al., 2005). In Hamadan Province of western Iran, an area with intensive agricultural production, the mean Cd concentration reached 1.7 mg/kg. In Zarrinshahr and Mobarakeh, central Iran, the mean Cd concentration was estimated to be 1.4 mg/kg (Fig. 3.2).

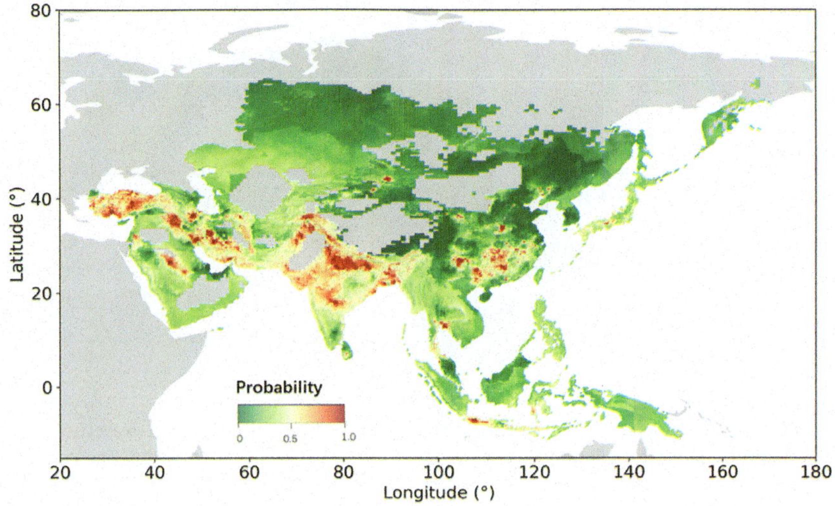

FIGURE 3.2 Soil pollution by cadmium in Asia. *Red* indicates a high probability of exceeding 1 mg/kg, whereas *green* indicates a low probability of exceedance.

3.3.2.2 Europe

The overall exceedance rate in Europe was estimated to be 1.4%, the lowest among all six continents. A major hotspot was located in Ireland. We estimated that 19% of the areas in Ireland had an average Cd level exceeding 1 mg/kg. These results were generally consistent with recent continental-scale studies in the European Union (EU). Tóth et al. examined 23,000 topsoil samples from EU countries to propose priority areas for detailed assessment. They found that the average Cd concentration was 0.09 mg/kg, and the largest area of elevated Cd levels was located in the limestone shale belt of Ireland (Tóth et al., 2016). Another continental-scale survey of agricultural soil in Europe showed that Ap soil had a median Cd concentration of 0.18 mg/kg, whereas Gr soil had a median Cd concentration of 0.20 mg/kg (Reimann et al., 2014). Elevated Cd levels in large parts of Ireland and sporadic locations of Spain were attributed to sulfide deposits (Reimann et al., 2014).

Our modeling results predicted elevated Cd levels in the Balkan Peninsula, a major part of which had not been included in the previous EU study. This was generally consistent with some previous local studies. In Novi Sad, Serbia, and Banja Luka, the average Cd concentration was estimated to be 1.66 mg/kg (Škrbić & Đurišić-Mladenović, 2010). Another soil investigation in Southern Serbia and Northern Kosovo found an average Cd concentration of 1.9 mg/kg. Moreover, the Cd concentration increased to 4.2 mg/kg after a flooding event, suggesting that Pb/Zn industrial activities in this region may have polluted the Ibar River, subsequently causing Cd pollution in soil via sediment deposition (Barać et al., 2016). An earlier nationwide soil quality survey from industrial areas estimated an average Cd concentration of 1.3 mg/kg in Serbia (Slavković et al., 2004). In an intensively cultivated area in Greece the average Cd concentration was found to reach 3.3 kg (Golia et al., 2009). The sources of Cd enrichment were mainly attributed to phosphate fertilizer usage (Golia et al., 2015). In northern Greece a local study showed that the average Cd concentration reached 1.5 mg/kg, and in areas

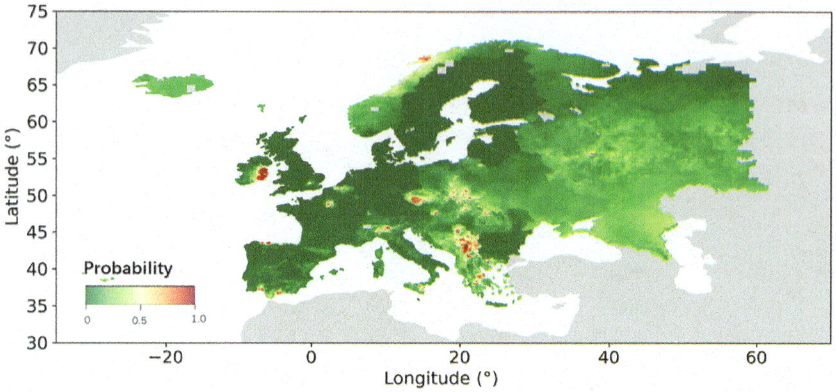

FIGURE 3.3 Soil pollution by cadmium in Europe. *Red* indicates a high probability of exceeding 1 mg/kg, whereas *green* indicates a low probability of exceedance.

impacted by a river passing through mining sites, the average soil Cd concentration reached 8.9 mg/kg (Sofianska & Michailidis, 2015). In the Karditsa region of central Greece the average Cd concentration was found to be 3.2 mg/kg (Antoniadis & Golia, 2021).

There were some less extensive hot spots in several other European countries. In the Czech Republic a large area of over 600 km^2 was affected by a lead smelter since the 18th century and had an average Cd concentration of 2.5 mg/kg (Suchara & Sucharová, 2004). In Italy, we estimated the Cd exceedance rate to be 2.6%, nearly twice the European average. A study in Pavia Province, northern Italy, showed an average Cd concentration of 1.22 mg/kg, which was attributed to anthropogenic pollution sources in the northeastern part of the province (Vives et al., 2008). In Vallecamonica, Italy, the median Cd concentration reached 1.9 mg/kg. The area was impacted by ferroalloy plant emissions for over a century, but the elevated Cd levels were attributed to both anthropogenic sources and geogenic origins (Borgese et al., 2013). Some types of rock material may also contribute to elevated Cd levels at some parts of Europe. For instance, black shale may contain 1 mg/kg of Cd, whereas most common rock types contain low concentration of Cd, ranging between 0.05 and 0.25 mg/kg. In some areas of Scandinavia, black shale caused elevated Cd concentrations (Reimann et al., 2014). Karst development in Croatia may have also resulted in elevated Cd levels (Reimann et al., 2014) (Fig. 3.3).

3.3.2.3 North America

North America generally had low Cd levels, with the exception of a few hotspots in the United States, Mexico, and Cuba (Fig. 3.4). The overall Cd exceedance rate in North America was predicted to be 4%, which was the second lowest among all continents. The United States Geological Survey conducted a nationwide geochemical and mineralogical survey of soils, in which they found that approximately 5% of surficial soils contained Cd levels above 0.7 mg/kg (Smith et al., 2013). Most existing reports of Cd pollution were located in urban areas. In Anniston, Alabama, a heavy metal source-tracking study found that average Cd level reached 1.7 mg/kg in the industrial zone (within 500 m of former industrial sites) and the vicinity of the industrial zone, whereas it reached 0.7 mg/kg in other areas (Ha et al., 2014). In New York City a survey of 904 urban gardens showed that the average and mean Cd levels reached 1.6

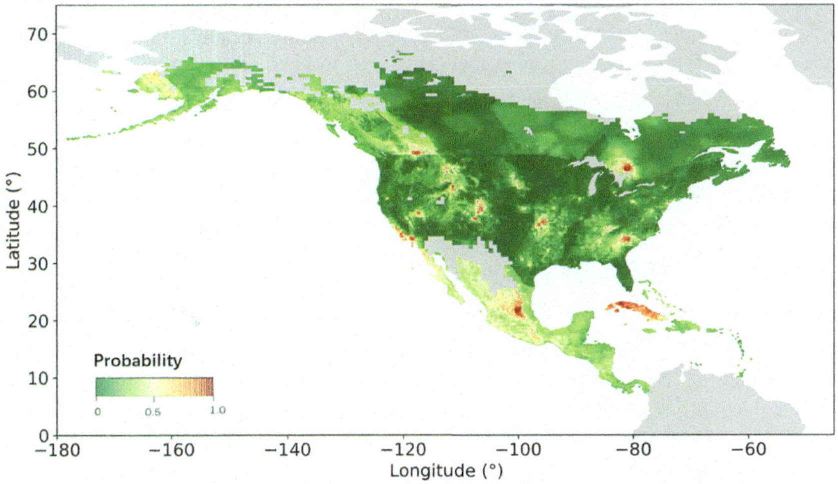

FIGURE 3.4 Soil pollution by cadmium in North America. *Red* indicates a high probability of exceeding 1 mg/kg, whereas *green* indicates a low probability of exceedance.

and 1.2 mg/kg, respectively (Cheng et al., 2015b). In Los Angeles, California, 270 soil samples were collected from residential areas, schools, and parks, and the average Cd concentration was estimated to be 9.6 mg/kg (Hung et al., 2023). Soil pollution by Cd in these urban areas has been attributed to industrial combustion processes as well as garden usage of phosphate fertilizers (Cheng et al., 2015b; Ha et al., 2014). Cd exceedance was more prevalent in South Carolina. The geometric mean of Cd concentration was estimated to be 2.2 mg/kg, based on a survey of 50 sites (Franklin et al., 2003). Besides these sporadic hotspots of Cd pollution in the United States the most prominent region was in Cuba. In Mayabeque Province, Cuba, the average and median Cd concentrations reached approximately 2.4 and 1.8 mg/kg, respectively, well above the Cuban quality reference value of 0.6 mg/kg (Sosa et al., 2022). The elevated Cd concentrations were attributed to both pedogenical and anthropogenic sources. In Mexico, four areas of San Luis Potosi were investigated for heavy metal contamination. The average Cd concentration ranged from 3.7 to 7.5 mg/kg, suggesting elevated Cd levels in the region (Ngo et al., 2021).

3.3.2.4 *South America*

In South America the Cd exceedance rate was predicted to be 5%, which was lower than those in Asia, Africa, and Oceania but higher than in North America and Europe. Hotspots of Cd pollution were mainly present in Brazil, Bolivia, and Colombia (Fig. 3.5). The elevated Cd levels were because of both geogenic enrichment and anthropogenic sources. The surface soil of native forest areas in the Amazonian Craton was found to be enriched in Cd, with an average concentration of 10 mg/kg and a maximum concentration of 27 mg/kg (de Lima et al., 2020). The high background concentrations were attributed to parent materials rich in ferruginous deposits. Moreover, the soils were highly acidic, with a median pH of 4.8, suggesting relatively strong Cd mobility in this region. A large coal-fired power plant caused the average soil Cd level in surrounding areas to reach 65 mg/kg (da Silva Júnior et al., 2019). Brazil is the largest sugarcane producer. A survey of agricultural fields with a long sugarcane cultivation history showed an

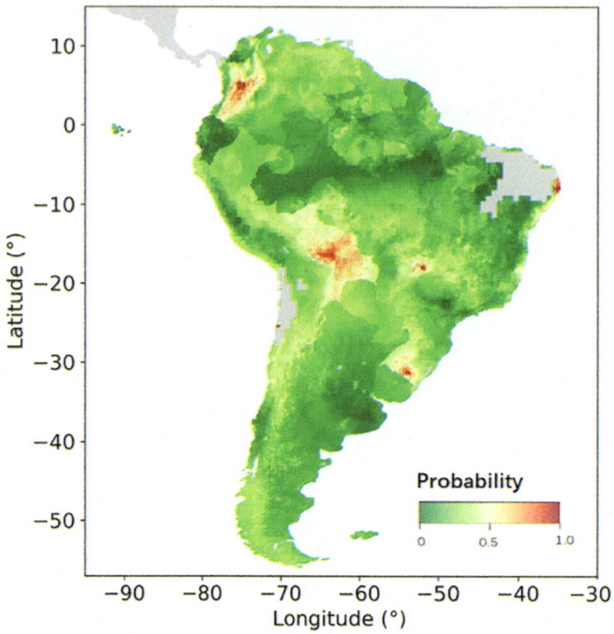

FIGURE 3.5 Soil pollution by cadmium in South America. *Red* indicates a high probability of exceeding 1 mg/kg, whereas *green* indicates a low probability of exceedance.

average Cd concentration of 1.9 mg/kg, which was attributed to the intensive use of phosphorus fertilizers (da Silva et al., 2016). In the Recife metropolitan region, the fourth-largest Brazilian metropolis, the average Cd concentration was found to be 1.5 mg/kg, and potential sources included automotive exhaust emissions, metallurgical and cement industries, and phosphate–based fertilizers used for over 44 years (da Silva et al., 2017). In the Pilcomayo basin of Bolivia, mining of the Potosi deposits began nearly five centuries ago and contaminated river water and sediment for over 200 km downstream of the mines (Miller et al., 2004). Irrigation with this contaminated water may have resulted in Cd levels typically above 1 mg/kg in the Mondragon region. In Colombia a nationwide survey of cacao-growing soil indicated that average Cd concentrations in over half of the regions were above 1 mg/kg, whereas most of the other regions had an average Cd concentration near 1 mg/kg (0.7–0.98 mg/kg) (Bravo et al., 2021). Despite the overall low exceedance rate, these localized hotspots still highlight the importance of Cd pollution in South America, especially in agricultural soils. More frequent and wide-range soil investigations are needed to better understand the situation and to develop policy instruments to protect soil health and food safety.

3.3.2.5 *Oceania*

Oceania was predicted to have a high Cd exceedance rate, reaching 15%. This was mainly because of high predicted Cd exceedance rate in Australia, which reached 16%. However, it should be noted that this high exceedance rate was based on a 0.5 probability cut-off value. If the cut-off value was set at 0.6, the Cd exceedance rate in Australia dropped to 3%, and if the cut-off value was set at 0.7, the exceedance rate further dropped to 0.6%. These results suggested that the machine learning model had relatively low confidence in the predicted Cd exceedance rate in Australia. To put it in perspective, we compared the exceedance rates in Asia and Oceania. At a cut-off value of 0.5 the exceedance rate in Oceania (15.3%) was slightly higher than in Asia (14.9%). However, at a cut-off value of 0.6, the exceedance rate in Oceania decreased to 3%, whereas in Asia, it remained higher at 9%. At an even higher confidence level of 0.8, the exceedance rate in Oceania further dropped to 0.1%, while in Asia, it remained at 2.5%.

Most of the predicted Cd exceedance in Australia was located in rangelands in the northern part of the country. The elevated exceedance rate may have been because of the strong weathering conditions in this area (Hou et al., 2025). A previous study showed that these areas had a much higher degree of weathering in comparison with the southern coastal areas (Reimann et al., 2012). The predicted high Cd exceedance contradicted the results from a national geochemical survey, in which 1300 samples were collected from 1186 catchments at a sampling density of approximately 1/5200 km^2. The median geochemical background concentration of Cd was estimated to be 0.04 mg/kg (Reimann & de Caritat, 2017). The authors concluded that there was no diffusive Cd pollution on a continental scale. However, it should be noted that these samples were collected from floodplains or similar landforms located at the outlet/lowest point of large catchments. As our recent studies suggest, flat terrain tends to render lower toxic metal concentration (Hou et al., 2025). This sampling design may have underestimated the background concentration. Moreover, most local-scale studies in Australia have suggested relatively high Cd levels in industrial and mining-affected urban area. In Mount Isa city, Queensland, the geometric mean of Cd concentrations in <180 μm grain size fraction was estimated to be 1.8 mg/kg, and in >180 μm grain size fraction, it was estimated to be 2.6 mg/kg (Taylor et al., 2010). The elevated Cd levels were attributed to mining and smelting. Within 2 km of smelter stack the average Cd concentration was 2.8 mg/kg. In contrast the average Cd concentration was 1.9 mg/kg for soil samples located 2–4 km away from smelter stack. In the small city of Port Pirie, South Australia, the mean Cd concentration reached 5 mg/kg, likely because of historical and current emissions from smelting and refining (Maliki et al., 2015). In Iron Cove, Sydney, the mean Cd level was estimated to be 1 mg/kg (Snowdon & Birch, 2004). In Lake Macquarie, NSW, the vicinity of smelting activities was investigated, and the mean soil Cd concentration was estimated to be 4.6 mg/kg. These results suggested that Cd pollution may well exist in areas impacted by intensive urbanization and mining and smelting, as well as strong weathering (Fig. 3.6).

3.3.2.6 *Africa*

There is a general lack of soil pollution investigation data in African countries. Nevertheless, several studies have been conducted on a local scale, providing some basis for model prediction. Furthermore, using the relationship built upon covariates and Cd exceedances in other areas, the machine learning model could provide a continuous prediction of Cd exceedance probability (Fig. 3.7). The overall Cd exceedance rate was estimated to be 6%, which is much higher than those of Europe, North America, and South America but lower than that of Asia. Moreover, the exceedance rate dropped to 3% when a probability cut-off value of 0.6 was used, and it further decreased to 1.5% when the probability cut-off value was set at 0.7. The most prevalent Cd exceedances were predicted to occur in Nigeria, Ghana, the Democratic Republic of Congo, and Botswana. In Akure City, Nigeria, the average Cd concentration was estimated to be 1.9 mg/kg, and the sources may have included smelting, fuel combustion, waste disposal, cement manufacturing, and emission from chemical plants (Adewumi, 2022). In the Jebbu-Ode city, Nigeria, the average and median Cd concentrations were estimated to be 3.1 and 2.2 mg/kg, respectively. Elevated Cd concentrations were detected not only in industrial areas but also in schools, residential areas, and agricultural fields (Adedeji et al., 2019). In the Ibadan metropolis

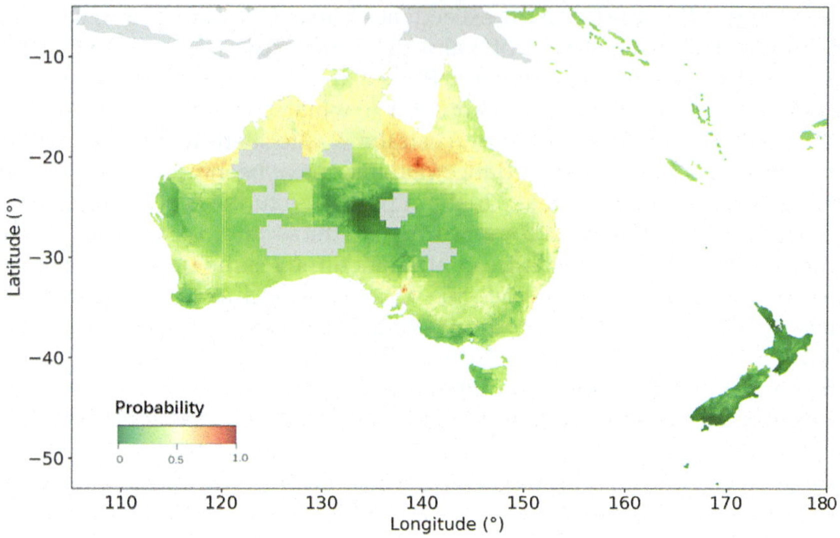

FIGURE 3.6 Soil pollution by cadmium in Oceania. *Red* indicates high probability of exceeding 1 mg/kg, whereas *green* indicates low probability of exceedance.

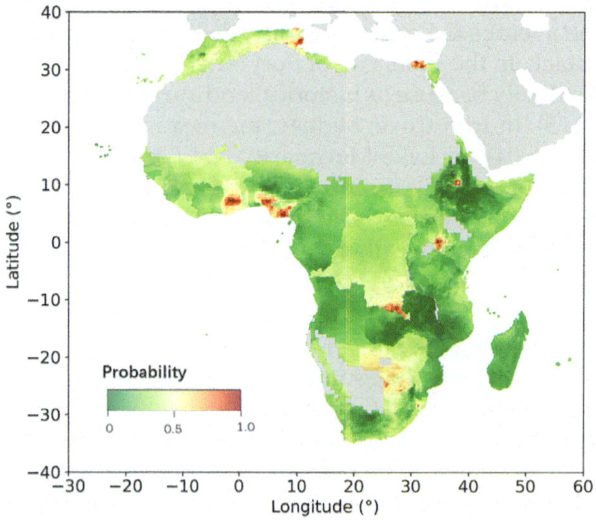

FIGURE 3.7 Soil pollution by cadmium in Africa. *Red* indicates a high probability of exceeding 1 mg/kg, whereas *green* indicates a low probability of exceedance.

of Nigeria the average Cd concentration reached 8.4 mg/kg, with hotspots existing in both densely populated and agricultural areas (Odewande & Abimbola, 2008).

3.3.3 Influencing factors of cadmium pollution

Anthropogenic processes, such as industrial, agricultural, and household activities, are major drivers of Cd pollution in soil. For instance, zinc mining can lead to serious Cd contamination because of the cooccurrence of Cd in zinc mines. In Thailand, paddy soils polluted by zinc mining operations contained Cd concentrations up to 142 mg/kg, nearly 1000 times higher than background levels (Khaokaew et al., 2011). The spatial distribution of soil Cd is influenced by both chronological anthropogenic input and leaching-led efflux, which is controlled by the mineralogy of deposits (Rausch et al., 2005a). A study near Cu–Ni mining and smelting facilities in Finland revealed that Cd concentrations in soil profiles did not match historical atmospheric deposition patterns, reflecting the dissolution of metal-bearing particles (Rausch et al., 2005b). Agricultural inputs, such as fertilizer application, represent a significant source of Cd in agricultural soils. However, the annual inputs and outputs of Cd in agricultural soils are typically three to four orders of magnitude lower than the bulk soil Cd pool (Imseng et al., 2018), unless phosphorus fertilizer containing high levels of Cd is used. The intensity of these anthropogenic inputs is highly spatially dependent, resulting in heterogeneous Cd distribution even on small scales.

The spatial distribution of Cd is also strongly influenced by soil characteristics and environmental factors. In soil matrix, Cd may be associated with various mineral phases, including carbonates, kaolinite, ferrihydrite, Mn-(oxyhydr)oxides, humic acid, and Cd-sulfide (Khaokaew et al., 2011; Van Groeningen et al., 2021; Zhai et al., 2019). Environmental factors, such as soil pH, Eh, and organic matter, affect the partitioning and mobility of Cd. Natural organic matter influences Cd enrichment by affecting its mobility (Li et al., 2015). Soil pH is the most notable factor affecting Cd mobility and ecological risks. Low pH conditions favor Cd dissolution, as protons compete for cation exchange sites and suppress the formation of metal–organic complexes (Rausch et al., 2005b). At low pH, Cd mobility is mainly driven by soil acidification, whereas at pH levels above 6, metal–organic complexes increasingly control Cd mobility (Fest et al., 2005; Floroiu et al., 2001).

In paddy soils, over 50% of Cd may be bound to humic acid under dry and initial flooding conditions. However, under prolonged flooding, Cd-carbonates become the dominant species, and Cd-sulfide may form under low Eh conditions, reducing Cd mobility (Khaokaew et al., 2011). The solubility of Cd in flooded paddy soils is reduced because of Cd sulfide formation; however, upon drainage, it readily transforms into more soluble species. The mobilization of Cd during drainage is controlled by a decrease in pH and the release of Cd from Fe and Mn (oxyhydro)oxides (Wang et al., 2019). This mechanism elevates the risk of Cd uptake by rice crops during the grain-filling stage (Wang et al., 2019), especially if the presence of other metals and low sulfate contents limit Cd sulfide formation (Fulda et al., 2013). Secondary Fe(II) minerals formed in organic-rich and reducing environments also serve as a sink for Cd in soil (Muehe et al., 2013).

Distinguishing between heavy metals originating from anthropogenic and natural sources is not straightforward (Zhao et al., 2015). Traditionally, researchers have employed the spatial variation of Cd concentrations and enrichment factors relative to natural crustal elements to distinguish natural versus anthropogenic inputs. More recently, scholars have found that anthropogenic processes can induce Cd isotopic fractionations, resulting in variation of $^{114}Cd/^{110}Cd$ at the per mil (‰) level, enabling source tracing through isotopic analysis (Cloquet et al., 2006; Wang et al., 2021).

Soil erosion plays a crucial role in the lateral transport of Cd, especially in karst regions (Liu et al., 2019). The redistribution of Cd in soil depends on its partitioning between solid particulates and aqueous solutions, which occurs in two phases: an initial rapid uptake/release within hours, followed by slower adsorption/desorption over days to weeks (Comans & Van Dijk, 1988). The mobility of Cd in the aqueous phase is influenced by complex dissolution and readsorption processes, which are governed by redox conditions and the relative abundance of reduced sulfur species, Fe-oxides, and Mn-oxides (Sun et al., 2020; Van Groeningen et al., 2021; Wang et al., 2019). Therefore the duration of soil saturation, which is affected by climatic conditions and terrain, can impact the spatial distribution of Cd pollution. Moreover, Cd mobility increases in the presence of zinc and lead, as these metals preferentially adsorb to iron oxyhydroxide, thereby reducing Cd sorption (Carroll et al., 1998).

The spatial distribution of soil Cd, as well as its ecological and health risks, is also shaped by a wide range of biogeochemical processes. Willow trees in the Colorado Rocky Mountains have been found to accumulate Cd at levels two orders of magnitude higher than soil concentrations, causing irreversible tubular damages in some bird species (Larison et al., 2000). The plant-pump effect can translocate Cd from deep soil to top soil, resulting in elevated Cd content in areas with higher primary productivity (Imseng et al., 2018). Under sulfate-reducing conditions, soil microorganisms may facilitate the formation of submicron colloidal particles, which enhance the transport of Cd, a metal traditionally considered immobile (Weber et al., 2009). Understanding these complex biogeochemical processes can greatly facilitate the assessment and mitigation of soil Cd contamination.

3.4 Conclusions

Cd is biotoxic and bioaccumulative, posing a threat to both human health and ecosystems. To optimize the effectiveness of risk management associated with soil Cd pollution, it is essential to integrate knowledge regarding Cd emissions, as well as how soil texture, mineralogy, and biogeochemistry affect its migration, distribution, and ecological risks. However, there is a general lack of data to estimate the magnitude and chorological sequence of the influx and efflux of soil Cd. Consequently, it is difficult to quantitatively allocate pollution source contribution and construct future trajectories. In this study, we present a quantitative estimate of soil quality exceedance and spatially explicit map of soil Cd pollution on a global scale. This can serve as a screening tool for policymakers and farmers, enabling more groundwork to better understand the sources, migration, and distribution of soil Cd. On the other hand, because of the lack of global data, significant uncertainties remain in the prediction. As global data on Cd concentration, along with environmental and social covariates, become available at higher spatial resolutions and with greater accuracy, future models will be more powerful in predicting spatiotemporal patterns of soil pollution.

Acknowledgments

This work was supported by the National Natural Science Foundation of China (grant no. 42225703).

References

Abbasi, A. M., Iqbal, J., Khan, M. A., & Shah, M. H. (2013). Health risk assessment and multivariate apportionment of trace metals in wild leafy vegetables from Lesser Himalayas, Pakistan. *Ecotoxicology and Environmental Safety, 92*, 237–244. doi:10.1016/j.ecoenv.2013.02.011.

Adedeji, O. H., Olayinka, O. O., & Tope-Ajayi, O. O. (2019). Spatial distribution and health risk assessment of soil pollution by heavy metals in Ijebu-Ode, Nigeria. *Journal of Distribution Science, 17*(6), 1–14.

https://www.journalhealthpollution.org/doi/pdf/10.5696/2156-9614-9.22.190601. doi:10.5696/2156-9614-9.22.190601.

Adewumi, A. J. (2022). Heavy metals in soils and road dust in Akure City, Southwest Nigeria: Pollution, sources, and ecological and health risks. *Exposure and Health, 14*(2), 375–392. doi:10.1007/s12403-021-00456-y.

Alloway, B. J. (2013). *Heavy metals in soils: Trace metals and metalloids in soils and their bioavailability* (3rd ed.). Springer.

Amini, M., Afyuni, M., Khademi, H., Abbaspour, K., & Schulin, R. (2005). Mapping risk of cadmium and lead contamination to human health in soils of Central Iran. *Science of the Total Environment, 347*(1–3), 64–77. doi:10.1016/j.scitotenv.2004.12.015.

Antoniadis, V., & Golia, E. E. (2021). Spatial and temporal assessment of cadmium and chromium contamination in soils in the Karditsa region (Central Greece). *Environmental Science and Pollution Research, 28*(4), 3820–3827. https://link.springer.com/journal/11356. doi:10.1007/s11356-020-09273-3.

Arao, T., Kawasaki, A., Baba, K., & Matsumoto, S. (2011). Effects of arsenic compound amendment on arsenic speciation in Rice Grain. *Environmental Science & Technology, 45*(4), 1291–1297. doi:10.1021/es1033316.

ATSDR. (2020). *ATSDR's substance priority list*. Agency for Toxic Substances and Disease Registry.

Barać, N., Škrivanj, S., Bukumirić, Z., Živojinović, D., Manojlović, D., Barać, M., Petrović, R., & Ćorac, A. (2016). Distribution and mobility of heavy elements in floodplain agricultural soils along the Ibar River (Southern Serbia and Northern Kosovo). Chemometric investigation of pollutant sources and ecological risk assessment. *Environmental Science and Pollution Research, 23*(9), 9000–9011. doi:10.1007/s11356-016-6142-2.

Borgese, L., Federici, S., Zacco, A., Gianoncelli, A., Rizzo, L., Smith, D. R., Donna, F., Lucchini, R., Depero, L. E., & Bontempi, E. (2013). Metal fractionation in soils and assessment of environmental contamination in Vallecamonica, Italy. *Environmental Science and Pollution Research, 20*(7), 5067–5075. doi:10.1007/s11356-013-1473-8.

Boutron, C. F., Görlach, U., Candelone, J. P., Bolshov, M. A., & Delmas, R. J. (1991). Decrease in anthropogenic lead, cadmium and zinc in Greenland snows since the late 1960s. *Nature, 353*(6340), 153–156. doi:10.1038/353153a0.

Bravo, D., Leon-Moreno, C., Martínez, C. A., Varón-Ramírez, V. M., Araujo-Carrillo, G. A., Vargas, R., Quiroga-Mateus, R., Zamora, A., & Rodríguez, E. A. G. (2021). The first national survey of cadmium in cacao farm soil in Colombia. *Agronomy, 11*(4). https://www.mdpi.com/2073-4395/11/4/761/pdf. doi:10.3390/agronomy11040761.

Carroll, S. A., O'Day, P. A., & Piechowski, M. (1998). Rock-water interactions controlling zinc, cadmium, and lead concentrations in surface waters and sediments, U.S. Tri-State Mining District. 2. Geochemical interpretation. *Environmental Science and Technology, 32*(7), 956–965. doi:10.1021/es970452k.

Chen, H., Teng, Y., Lu, S., Wang, Y., & Wang, J. (2015). Contamination features and health risk of soil heavy metals in China. *Science of the Total Environment, 512–513*, 143–153. doi:10.1016/j.scitotenv.2015.01.025.

Cheng, K. E, Wang, Y., Tian, H., Gao, X., Zhang, Y., Wu, X., Zhu, C., & Gao, J. (2015a). Atmospheric emission characteristics and control policies of five precedent-controlled toxic heavy metals from anthropogenic sources in China. *Environmental Science & Technology, 49*(2), 1206–1214. doi:10.1021/es5037332.

Cheng, Z., Paltseva, A., Li, I., Morin, T., Huot, H., Egendorf, S., Su, Z., Yolanda, R., Singh, K., Lee, L., Grinshtein, M., Liu, Y., Green, K., Wai, W., Wazed, B., & Shaw, R. (2015b). Trace metal contamination in New York City garden soils. *Soil Science, 180*(4–5), 167–174. http://journals.lww.com/soilsci. doi:10.1097/SS.0000000000000126.

Cloquet, C., Carignan, J., Libourel, G., Sterckeman, T., & Perdrix, E. (2006). Tracing source pollution in soils using cadmium and lead isotopes. *Environmental Science & Technology, 40*(8), 2525–2530. doi:10.1021/es052232+.

Comans, R. N. J., & Van Dijk, C. P. J. (1988). Role of complexation processes in cadmium mobilization during estuarine mixing. *Nature, 336*(6195), 151–154. doi:10.1038/336151a0.

Cui, W., Meng, Q., Feng, Q., Zhou, L., Cui, Y., & Li, W. (2019). Occurrence and release of cadmium, chromium, and lead from stone coal combustion. *International Journal of Coal Science & Technology, 6*(4), 586–594. doi:10.1007/s40789-019-00281-4.

da Silva, F. B. V., do Nascimento, C. W. A., Araújo, P. R. M., da Silva, F. L., & Lima, L. H. V. (2017). Soil contamination by metals with high ecological risk in urban and rural areas. *International Journal of Environmental Science and Technology, 14*(3), 553–562. doi:10.1007/s13762-016-1170-5.

da Silva, F. B. V., do Nascimento, C. W. A., Araújo, P. R. M., da Silva, L. H. V., & da Silva, R. F. (2016). Assessing heavy metal sources in sugarcane Brazilian soils: An approach using multivariate analysis. *Environmental Monitoring and Assessment, 188*(8). www.wkap.nl/journalhome.htm/0167-6369. doi:10.1007/s10661-016-5409-x.

da Silva Júnior, F. M. R., Ramires, P. F., dos Santos, M., Seus, E. R., Soares, M. C. F., Muccillo-Baisch, A. L., Mirlean, N., & Baisch, P. R. M. (2019). Distribution of potentially harmful elements in soils around a large coal-fired power plant. *Environmental Geochemistry and Health, 41*(5), 2131–2143. www.wkap.nl/journalhome.htm/0269-4042. doi:10.1007/s10653-019-00267-w.

de Lima, M. W., Hamid, S. S., de Souza, E. S., Teixeira, R. A., da Conceição Palheta, D., do Carmo Freitas Faial, K., & Fernandes, A. R. (2020). Geochemical background concentrations of potentially toxic elements in soils of the Carajás Mineral Province, southeast of the Amazonian Craton. *Environmental Monitoring and Assessment, 192*(10). https://link.springer.com/journal/10661. doi:10.1007/s10661-020-08611-9.

FAO. (2021). *Global assessment of soil pollution*. Food and Agriculture Organization.

Fest, E. P. M. J., Temminghoff, E. J. M., Griffioen, J., & Van Riemsdijk, W. H. (2005). Proton buffering and metal leaching in sandy soils. *Environmental Science and Technology, 39*(20), 7901–7908. doi:10.1021/es0505806.

Floroiu, R. M., Davis, A. P., & Torrents, A. (2001). Cadmium adsorption on aluminum oxide in the presence of polyacrylic acid. *Environmental Science and Technology, 35*(2), 348–353. doi:10.1021/es9913479.

Franklin, R. E., Duis, L., Smith, B. R., Brown, R., & Toler, J. E. (2003). Elemental concentrations in soils of South Carolina. *Soil Science, 168*(4), 280–291. doi:10.1097/01.ss.0000064891.94869.e1.

Fulda, B., Voegelin, A., & Kretzschmar, R. (2013). Redox-controlled changes in cadmium solubility and solid-phase speciation in a paddy soil as affected by reducible sulfate and copper. *Environmental Science & Technology, 47*(22), 12775–12783. doi:10.1021/es401997d.

Golia, E. E., Dimirkou, A., & Floras, S. A. (2009). Monitoring the variability of nitrogen and cadmium concentrations in soils and irrigation water in the Almyros Area of Central Greece. *Communications in Soil Science and Plant Analysis, 40*(1–6), 376–390. doi:10.1080/00103620802646902.

Golia, E. E., Dimirkou, A., & Floras, S. t. A. (2015). Spatial monitoring of arsenic and heavy metals in the Almyros area, Central Greece. Statistical approach for assessing the sources of contamination. *Environmental Monitoring and Assessment, 187*(7). doi:10.1007/s10661-015-4624-1.

Gupta, N., Yadav, K. K., Kumar, V., Krishnan, S., Kumar, S., Nejad, Z. D., Majeed Khan, M. A., & Alam, J. (2021). Evaluating heavy metals contamination in soil and vegetables in the region of North India: Levels, transfer and potential human health risk analysis. *Environmental Toxicology and Pharmacology, 82*. www.elsevier.com/locate/etap. doi:10.1016/j.etap.2020.103563.

Ha, H., Olson, J. R., Bian, L., & Rogerson, P. A. (2014). Analysis of heavy metal sources in soil using kriging interpolation on principal components. *Environmental Science and Technology, 48*(9), 4999–5007. http://pubs.acs.org/journal/esthag. doi:10.1021/es405083f.

Harris, N. S., & Taylor, G. J. (2013). Cadmium uptake and partitioning in durum wheat during grain filling. *BMC Plant Biology, 13*(1). http://www.biomedcentral.com/1471-2229/13/103. doi:10.1186/1471-2229-13-103 Canada.

Hellström, L., Persson, B., Brudin, L., Grawé, K. P., Öborn, I., & Järup, L. (2007). Cadmium exposure pathways in a population living near a battery plant. *Science of the Total Environment, 373*(2–3), 447–455. doi:10.1016/j.scitotenv.2006.11.028.

Hjortenkrans, D. S. T., Bergbäck, B. G., & Häggerud, A. V. (2007). Metal emissions from brake linings and tires: Case studies of Stockholm, Sweden 1995/1998 and 2005. *Environmental Science and Technology, 41*(15), 5224–5230. doi:10.1021/es070198o.

Hou, D. (2023). Sustainable soil management for food security. *Soil Use and Management, 39*(1), 1–7. doi:10.1111/sum.12883.

Hou, D., Jia, X., Wang, L., McGrath, S. P., Zhu, Y. G., Hu, Q., Zhao, F. J., Bank, M. S., O'Connor, D., & Nriagu, J. (2025). Global soil pollution by toxic metals threatens agriculture and human health. *Science, 388*(6744), 316–321. https://www.science.org/toc/science/388/6744. doi:10.1126/science.adr5214.

Hou, D., O'Connor, D., Igalavithana, A. D., Alessi, D. S., Luo, J., Tsang, D. C. W., Sparks, D. L., Yamauchi, Y., Rinklebe, J., & Ok, Y. S. (2020). Metal contamination and bioremediation of agricultural soils for food safety and sustainability. *Nature Reviews Earth and Environment, 1*(7), 366–381. nature.com/natrevearthenviron/. doi:10.1038/s43017-020-0061-y.

Hou, D., O'Connor, D., Nathanail, P., Tian, L. i, & Ma, Y. (2017). Integrated GIS and multivariate statistical analysis for regional scale assessment of heavy metal soil contamination: A critical review. *Environmental Pollution, 231*(Pt 1), 1188–1200. doi:10.1016/j.envpol.2017.07.021.

Hu, Y., & Cheng, H. (2013). Application of stochastic models in identification and apportionment of heavy metal pollution sources in the surface soils of a large-scale region. *Environmental Science & Technology, 47*(8), 3752–3760. doi:10.1021/es304310k.

Hung, W. C., Adams, N., Ibrahim-Watkins, Z. R., Nguyen, D., Jain, T., Wang, Y. H., & Jay, J. A. (2023). Incorporating field-based research into remote learning: An assessment of soil lead pollution in different land-use types in Los Angeles. *Environmental Research, 216*. https://www.sciencedirect.com/journal/environmental-research. doi:10.1016/j.envres.2022.114480.

Imseng, M., Wiggenhauser, M., Keller, A., Müller, M., Rehkämper, M., Murphy, K., Kreissig, K., Frossard, E., Wilcke, W., & Bigalke, M. (2018). Fate of Cd in agricultural soils: A stable isotope approach to anthropogenic impact, soil formation, and soil-plant cycling. *Environmental Science and Technology, 52*(4), 1919–1928. http://pubs.acs.org/journal/esthag. doi:10.1021/acs.est.7b05439.

Kara, E. E., Pirlak, U., & Özdilek, H. G. (2004). Evaluation of heavy metals' (Cd, Cu, Ni, Pb, and Zn) distribution in sowing regions of potato fields in the Province of Niğde, Turkey. *Water, Air, and Soil Pollution, 153*(1–4), 173–186. doi:10.1023/B:WATE.0000019942.37633.31.

Kara, M., Dumanoğlu, Y., Altıok, H., Elbir, T., Odabası, M., & Bayram, A. (2014). Spatial distribution and source identification of trace elements in topsoil from heavily industrialized region, Aliaga, Turkey. *Environmental Monitoring and Assessment, 186*(10), 6017–6038. doi:10.1007/s10661-014-3837-z.

Kashyap, R., Sharma, R., & Uniyal, S. K. (2019). Distribution of heavy metals in habitation land-use soils with high ecological risk in urban and peri-urban areas. *International Journal of Environmental Science and Technology, 16*(12), 8093–8106. doi:10.1007/s13762-018-02203-4.

Khan, K., Mohsin, A., Sharif, H. M. A., Maryam, A., Ali, J., Li, X., Ibrahim, S. M., Ayaz, M., Zhou, Y., & Younas, M. (2022). Heavy metal pollution in the soil of a riverine basin: Distribution, source, and potential hazards. *Environmental Monitoring and Assessment, 194*(9). https://link.springer.com/journal/10661. doi:10.1007/s10661-022-10287-2.

Khaokaew, S., Chaney, R. L., Landrot, G., Ginder-Vogel, M., & Sparks, D. L. (2011). Speciation and release kinetics of cadmium in an alkaline paddy soil under various flooding periods and draining conditions. *Environmental Science and Technology, 45*(10), 4249–4255. doi:10.1021/es103971y.

Kormoker, T., Proshad, R., Islam, S., Ahmed, S., Chandra, K., Uddin, M., & Rahman, M. (2021). Toxic metals in agricultural soils near the industrial areas of Bangladesh: Ecological and human health risk assessment. *Toxin Reviews, 40*(4), 1135–1154. http://www.tandfonline.com/loi/itxr20. doi:10.1080/15569543.2019.1650777.

Kubier, A., Wilkin, R. T., & Pichler, T. (2019). Cadmium in soils and groundwater: A review. *Applied Geochemistry, 108*, 104388. doi:10.1016/j.apgeochem.2019.104388.

Larison, J. R., Likens, G. E., Fitzpatrick, J. W., & Crock, J. G. (2000). Cadmium toxicity among wildlife in the Colorado Rocky Mountains. *Nature, 406*(6792), 181–183. doi:10.1038/35018068.

Lee, K., Han, C., Hong, S. B., Jun, S. J., Han, Y., Xiao, C., Du, Z., Hur, S. D., Lee, J. I., Boutron, C. F., & Hong, S. (2019). A 300-year high-resolution greenland ice record of large-scale atmospheric pollution by arsenic in the Northern Hemisphere. *Environmental Science and Technology, 53*(22), 12999–13008. https://pubs.acs.org/doi/pdf/10.1021/acs.est.9b01805. doi:10.1021/acs.est.9b01805.

Li, H., Luo, N., Li, Y. W., Cai, Q. Y., Li, H. Y., Mo, C. H., & Wong, M. H. (2017). Cadmium in rice: Transport mechanisms, influencing factors, and minimizing measures. *Environmental Pollution, 224*, 622–630. www.elsevier.com/inca/publications/store/4/0/5/8/5/6. doi:10.1016/j.envpol.2017.01.087.

Li, X., Wang, L., Huang, C., Hou, R., & Hou, D. (2025). Long-term soil remediation using layered double hydroxides: Field evidence for simultaneous immobilization of both cations and oxyanions. *Environmental Pollution, 366*, 125417. doi:10.1016/j.envpol.2024.125417.

Li, Y., Duan, Z., Liu, G., Kalla, P., Scheidt, D., & Cai, Y. (2015). Evaluation of the possible sources and controlling factors of toxic metals/metalloids in the Florida Everglades and their potential risk of exposure. *Environmental Science & Technology, 49*(16), 9714–9723. doi:10.1021/acs.est.5b01638.

Liu, M., Zhang, Q., Ge, S., Mason, R. P., Luo, Y., He, Y., Xie, H., Sa, R., Chen, L., & Wang, X. (2019). Rapid increase in the lateral transport of trace elements induced by soil erosion in major karst regions in China. *Environmental Science and Technology, 53*(8), 4206–4214. http://pubs.acs.org/journal/esthag. doi:10.1021/acs.est.8b06143.

Lund, L. J., Betty, E. E., Page, A. L., & Elliott, R. A. (1981). Occurrence of naturally high cadmium levels in soils and its accumulation by vegetation. *Journal of Environmental Quality, 10*(4), 551–556. doi:10.2134/jeq1981.00472425001000040027x.

Maliki, A. Al, Bruce, D., & Owens, G. (2015). Spatial distribution of Pb in urban soil from Port Pirie, South Australia. *Environmental Technology & Innovation, 4*, 123–136. doi:10.1016/j.eti.2015.05.002.

Meharg, A. A., & Meharg, C. (2021). The pedosphere as a sink, source, and record of anthropogenic and natural arsenic atmospheric deposition. *Environmental Science and Technology, 55*(12), 7757–7769. http://pubs.acs.org/journal/esthag. doi:10.1021/acs.est.1c00460.

Miller, J. R., Hudson-Edwards, K. A., Lechler, P. J., Preston, D., & Macklin, M. G. (2004). Heavy metal contamination of water, soil and produce within riverine communities of the Rıó Pilcomayo basin, Bolivia. *Science of the Total Environment, 320*(2–3), 189–209. doi:10.1016/j.scitotenv.2003.08.011.

Moghadam, H. S., Griffin, W. L., Li, X. H., Santos, J. F., Karsli, O., Stern, R. J., Ghorbani, G., Gain, S., Murphy, R., & O'Reilly, S. Y. (2017). Crustal evolution of NW Iran: Cadomian Arcs, Archean fragments and the Cenozoic magmatic flare-up. *Journal of Petrology, 58*(11), 2143–2190. http://petrology.oxfordjournals.org/. doi:10.1093/petrology/egy005.

Muehe, E. M., Adaktylou, I. J., Obst, M., Zeitvogel, F., Behrens, S., Planer-Friedrich, B., Kraemer, U., & Kappler, A. (2013). Organic carbon and reducing conditions lead to cadmium immobilization by secondary fe mineral formation in a pH-neutral soil. *Environmental Science and Technology, 47*(23), 13430–13439. doi:10.1021/es403438n.

Murphy, D., & Hutchinson, D. (2015). Cadmium, road dust and rheumatoid arthritis: An alternative hypothesis to general air pollution. *Journal of Inflammation, 12*(1). doi:10.1186/s12950-015-0103-2.

Ngo, H. T. T., Watchalayann, P., Nguyen, D. B., Doan, H. N., & Liang, L. (2021). Evaluation of heavy metal exposure pathways on children from an informal e-waste processing village in Viet Nam. *Human and Ecological Risk Assessment: An International Journal, 27*(9–10), 2342–2358. doi:10.1080/10807039.2021.2000362.

Nriagu, J. O. (1989). A global assessment of natural sources of atmospheric trace metals. *Nature, 338*(6210), 47–49. doi:10.1038/338047a0.

Nriagu, J. O., & Pacyna, J. M. (1988). Quantitative assessment of worldwide contamination of air, water and soils by trace metals. *Nature, 333*(6169), 134–139. doi:10.1038/333134a0.

Odewande, A. A., & Abimbola, A. F. (2008). Contamination indices and heavy metal concentrations in urban soil of Ibadan metropolis, southwestern Nigeria. *Environmental Geochemistry and Health, 30*(3), 243–254. doi:10.1007/s10653-007-9112-2.

Özkaraova Güngör, E. B., & Orkun, M. O. (2014). Fractionation, mobility, and sources of selected heavy metals in topsoils from the Middle Black Sea Region of Turkey. *Clean—Soil, Air, Water, 42*(7), 986–993. http://onlinelibrary.wiley.com/journal/10.1002/%28ISSN%291863-0669. doi:10.1002/clen.201200219.

Quezada-Hinojosa, R. P., Matera, V., Adatte, T., Rambeau, C., & Föllmi, K. B. (2009). Cadmium distribution in soils covering Jurassic oolitic limestone with high Cd contents in the Swiss Jura. *Geoderma, 150*(3–4), 287–301. doi:10.1016/j.geoderma.2009.02.013.

Ramos-Ruiz, A., Wilkening, J. V., Field, J. A., & Sierra-Alvarez, R. (2017). Leaching of cadmium and tellurium from cadmium telluride (CdTe) thin-film solar panels under simulated landfill conditions. *Journal of Hazardous Materials, 336*, 57–64. www.elsevier.com/locate/jhazmat. doi:10.1016/j.jhazmat.2017.04.052.

Rausch, N., Nieminen, T., Ukonmaanaho, L., Le Roux, G., Krachler, M., Cheburkin, A. K., Bonani, G., & Shotyk, W. (2005a). Comparison of atmospheric deposition of copper, nickel, cobalt, zinc, and cadmium recorded by Finnish peat cores with monitoring data and emission records. *Environmental Science and Technology, 39*(16), 5989–5998. doi:10.1021/es050260m.

Rausch, N., Ukonmaanaho, L., Nieminen, T. M., Krachler, M., & Shotyk, W. (2005b). Porewater evidence of metal (Cu, Ni, Co, Zn, Cd) mobilization in an acidic, ombrotrophic bog impacted by a smelter, Harjavalta, Finland and comparison with reference sites. *Environmental Science and Technology, 39*(21), 8207–8213. doi:10.1021/es0509026.

Rehman, I., Ishaq, M., Ali, L., Muhammad, S., Din, I. U., Yaseen, M., & Ullah, H. (2020). Potentially toxic elements' occurrence and risk assessment through water and soil of Chitral urban environment, Pakistan: A case study. *Environmental Geochemistry and Health, 42*(12), 4355–4368. www.wkap.nl/journalhome.htm/0269-4042. doi:10.1007/s10653-020-00531-4.

Reimann, C., Birke, M., Demetriades, A., & Filzmoser, P. O. 'C. (2014). *Chemistry of Europe's agricultural soils, part A. Methodology and Interpretation of the GEMAS data set*. Schweizerbart Science Publishers.

Reimann, C., & de Caritat, P. (2017). Establishing geochemical background variation and threshold values for 59 elements in Australian surface soil. *Science of the Total Environment, 578*, 633–648. doi:10.1016/j.scitotenv.2016.11.010.

Reimann, C., de Caritat, P., Albanese, S., Andersson, M., Arnoldussen, A., Baritz, R., Batista, M. J., Bel-lan, A., Birke, M., Cicchella, D., Demetriades, A., Dinelli, E., De Vivo, B., De Vos, W., Duris, M., Dusza-Dobek, A., Eggen, O. A., Eklund, M., Ernstsen, V., . . . Wygralak, A. (2012). New soil composition data for Europe and Australia: Demonstrating comparability, identifying continental-scale processes and learning lessons for global geochemical mapping. *Science of the Total Environment, 416*, 239–252. doi:10.1016/j.scitotenv.2011.11.019.

Saglam, M., Dengiz, O., Özyazici, M. A., & Kizilkaya, R. (2011). Application of geostatistical methods to heavy metals status in Çarşamba plain soils. *Asian Journal of Chemistry, 23*(8), 3454–3460. http://www.asianjournalofchemistry.co.in/Journal/ViewArticle.aspx?ArticleID=23_8_37.

Salmanzadeh, M., Hartland, A., Stirling, C. H., Balks, M. R., Schipper, L. A., Joshi, C., & George, E. (2017). Isotope tracing of long-term cadmium fluxes in an agricultural soil. *Environmental Science and Technology, 51*(13), 7369–7377. http://pubs.acs.org/journal/esthag. doi:10.1021/acs.est.7b00858.

Schaider, L. A., Senn, D. B., Brabander, D. J., Mccarthy, K. D., & Shine, J. P. (2007). Characterization of zinc, lead, and cadmium in mine waste: Implications for transport, exposure, and bioavailability. *Environmental Science and Technology, 41*(11), 4164–4171. doi:10.1021/es0626943.

Setia, R., Dhaliwal, S. S., Singh, R., Singh, B., Kukal, S. S., & Pateriya, B. (2023). Ecological and human health risk assessment of metals in soils and wheat along Sutlej river (India). *Chemosphere, 312*(Pt 1), 137331. doi:10.1016/j.chemosphere.2022.137331.

Shi, J., Du, P., Luo, H., Wu, H., Zhang, Y., Chen, J., Wu, M., Xu, G., & Gao, H. (2022). Soil contamination with cadmium and potential risk around various mines in China during 2000–2020. *Journal of Environmental Management, 310*, 114509. doi:10.1016/j.jenvman.2022.114509.

Six, L., & Smolders, E. (2014). Future trends in soil cadmium concentration under current cadmium fluxes to European agricultural soils. *Science of the Total Environment, 485–486*(1), 319–328. doi:10.1016/j.scitotenv.2014.03.109.

Škrbić, B., & Đurišić-Mladenović, N. (2010). Chemometric interpretation of heavy metal patterns in soils worldwide. *Chemosphere, 80*(11), 1360–1369. doi:10.1016/j.chemosphere.2010.06.010.

Slavković, L., Škrbić, B., Miljević, N., & Onjia, A. (2004). Principal component analysis of trace elements in industrial soils. *Environmental Chemistry Letters, 2*(2), 105–108. doi:10.1007/s10311-004-0073-8.

Smith, D. B., Cannon, W. F., Woodruff, L. G., Solano, F., Kilburn, J. E., & Fey, D. L. (2013). *Geochemical and mineralogical data for soils of the conterminous United States. U.S. Geological Survey Data Series 801*. US Geological Survey.

Snowdon, R., & Birch, G. F. (2004). The nature and distribution of copper, lead, and zinc in soils of a highly urbanised sub-catchment (Iron Cove) of Port Jackson, Sydney. *Australian Journal of Soil Research, 42*(3), 329–338. doi:10.1071/sr03017.

Sofianska, E., & Michailidis, K. (2015). Chemical assessment and fractionation of some heavy metals and arsenic in agricultural soils of the mining affected Drama plain, Macedonia, northern Greece. *Environmental Monitoring and Assessment, 187*(3). doi:10.1007/s10661-015-4335-7.

Sosa, D., Hilber, I., Buerge-Weirich, D., Faure, R., Escobar, A., & Bucheli, T. D. (2022). Heavy metals in soils of Mayabeque, Cuba: Multifaceted and hardly discernable contributions from pedogenic and anthropogenic sources. *Environmental Monitoring and Assessment, 194*(6). https://link.springer.com/journal/10661. doi:10.1007/s10661-022-10097-6.

Suchara, I., & Sucharová, J. (2004). Distribution of 36 element deposition rates in a historic mining and smelting area as determined through fine-scale biomonitoring techniques part II: Relative long-term accmulated atmospheric deposition levels. *Water, Air, and Soil Pollution, 153*(1–4), 229–252. doi:10.1023/B:WATE.0000019915.70739.bf.

Sun, Q., Cui, P., Wu, S., Liu, C., Fan, T., Alves, M. E., Cheng, H., Huang, M., Zhou, D., & Wang, Y. (2020). Role of Reduced Sulfur in the Transformation of Cd(II) Immobilized by δ-MnO_2. *Environmental Science and Technology, 54*(23), 14955–14963. http://pubs.acs.org/journal/esthag. doi:10.1021/acs.est.0c02936.

Szwalec, A., Mundała, P. ł, Kędzior, R., & Pawlik, J. (2020). Monitoring and assessment of cadmium, lead, zinc and copper concentrations in arable roadside soils in terms of different traffic conditions. *Environmental Monitoring and Assessment, 192*(3). doi:10.1007/s10661-020-8120-x.

Taylor, M. P., Mackay, A. K., Hudson-Edwards, K. A., & Holz, E. (2010). Soil Cd, Cu, Pb and Zn contaminants around Mount Isa city, Queensland, Australia: Potential sources and risks to human health. *Applied Geochemistry, 25*(6), 841–855. doi:10.1016/j.apgeochem.2010.03.003.

Tian, H., Liu, K., Zhou, J., Lu, L., Hao, J., Qiu, P., Gao, J., Zhu, C., Wang, K., & Hua, S. (2014). Atmospheric emission inventory of hazardous trace elements from China's coal-fired power plants—Temporal trends and spatial variation characteristics. *Environmental Science & Technology, 48*(6), 3575–3582. doi:10.1021/es404730j.

Tóth, G., Hermann, T., Szatmári, G., & Pásztor, L. (2016). Maps of heavy metals in the soils of the European Union and proposed priority areas for detailed assessment. *Science of the Total Environment, 565*, 1054–1062. doi:10.1016/j.scitotenv.2016.05.115.

Van Groeningen, N., Christl, I., & Kretzschmar, R. (2021). The effect of aeration on Mn(II) sorbed to clay minerals and its impact on Cd retention. *Environmental Science & Technology, 55*(3), 1650–1658. doi:10.1021/acs.est.0c05875.

Velleux, M. L., Julien, P. Y., Rojas-Sanchez, R., Clements, W. H., & England, J. F. (2006). Simulation of metals transport and toxicity at a mine-impacted watershed: California Gulch, Colorado. *Environmental Science and Technology, 40*(22), 6996–7004. doi:10.1021/es0608592.

Vives, I., Müller, A., Umlauf, G., Christoph, E. H., Mariani, G., Skejo, H., Cenci, R. M., Sena, F., & Beone, G. M. (2008). Levels of PCDD/Fs and trace elements in superficial soils of Pavia Province (Italy). *Environment International, 34*(7), 994–1000. www.elsevier.com/locate/envint. doi:10.1016/j.envint.2008.03.003.

Wang, J., Wang, P. M., Gu, Y., Kopittke, P. M., Zhao, F. J., & Wang, P. (2019). Iron-manganese (oxyhydro)oxides, rather than oxidation of sulfides, determine mobilization of Cd during soil drainage in paddy soil systems. *Environmental Science and Technology, 53*(5), 2500–2508. http://pubs.acs.org/journal/esthag. doi:10.1021/acs.est.8b06863.

Wang, L., Jin, Y., Weiss, D. J., Schleicher, N. J., Wilcke, W., Wu, L., Guo, Q., Chen, J., O'Connor, D., & Hou, D. (2021). Possible application of stable isotope compositions for the identification of metal sources in soil. *Journal of Hazardous Materials, 407*. www.elsevier.com/locate/jhazmat. doi:10.1016/j.jhazmat.2020.124812.

Weber, F. A., Voegelin, A., Kaegi, R., & Kretzschmar, R. (2009). Contaminant mobilization by metallic copper and metal sulphide colloids in flooded soil. *Nature Geoscience, 2*(4), 267–271. doi:10.1038/ngeo476.

WHO. (2020). *10 Chemicals of public health concern*. WHO https://www.who.int/news-room/photo-story/detail/10-chemicals-of-public-health-concern.

Yalcin, M. G., Battaloglu, R., & Ilhan, S. (2007). Heavy metal sources in Sultan Marsh and its neighborhood, Kayseri, Turkey. *Environmental Geology, 53*(2), 399–415. doi:10.1007/s00254-007-0655-4.

Zhai, H., Wang, L., Hövelmann, J., Qin, L., Zhang, W., & Putnis, C. V. (2019). Humic acids limit the precipitation of cadmium and arsenate at the brushite-fluid interface. *Environmental Science and Technology, 53*(1), 194–202. http://pubs.acs.org/journal/esthag. doi:10.1021/acs.est.8b05584.

Zhai, L., Liao, X., Chen, T., Yan, X., Xie, H., Wu, B., & Wang, L. (2008). Regional assessment of cadmium pollution in agricultural lands and the potential health risk related to intensive mining activities: A case study in Chenzhou City, China. *Journal of Environmental Sciences, 20*(6), 696–703. doi:10.1016/s1001-0742(08)62115-4.

Zhang, L., Zhang, L., Wang, L., & Hou, D. (2025). Biogeochemical interaction between thallium (Tl) and schwertmannite in acidic environment and the anti-dissolution mechanisms of Tl(I)-coprecipitated schwertmannite. *Journal of Hazardous Materials, 484*, 136764. doi:10.1016/j.jhazmat.2024.136764.

Zhao, F. J., Ma, Y., Zhu, Y. G., Tang, Z., & McGrath, S. P. (2015). Soil contamination in China: Current status and mitigation strategies. *Environmental Science and Technology, 49*(2), 750–759. http://pubs.acs.org/journal/esthag. doi:10.1021/es5047099.

CHAPTER

4

Emissions and atmospheric deposition of cadmium

Zaijin Sun[a], Wenkai Lei[b,c], Jiawei Li[d], Lu Bai[e] and Yuepeng Pan[c]

[a]Technical Centre for Soil, Agriculture and Rural Ecology and Environment, Ministry of Ecology and Environment, Beijing, China [b]The Analysis and Test Center, Capital Normal University, Beijing, China [c]State Key Laboratory of Atmospheric Environment and Extreme Meteorology, Institute of Atmospheric Physics, Chinese Academy of Sciences, Beijing, China [d]State Key Laboratory of Earth System Numerical Modeling and Application, Institute of Atmospheric Physics, Chinese Academy of Sciences, Beijing, China [e]State Environmental Protection Key Laboratory of Eco-Industry, Chinese Research Academy of Environmental Sciences, Beijing, China

4.1 Introduction

Emissions to the atmosphere and deposition processes are key pathways for the global biogeochemical cycling of cadmium (Cd), serving as critical links between anthropogenic activities and environmental contamination. Cd emissions arise from both natural and anthropogenic sources, with the latter dominating global contributions. Natural sources include soil dust and sea spray (Wu et al., 2020), whereas anthropogenic emissions predominantly originate from industrial activities such as nonferrous metal smelting, fossil fuel combustion, and waste incineration (Nriagu, 1979). These emissions are particularly significant in regions with high levels of industrialization (Shao et al., 2013), reflecting variation in energy use structures and pollution control measures globally.

Once emitted, Cd enters the atmosphere, predominantly associated with fine particulate matter, enabling its transport over long distances. Atmospheric deposition, encompassing both wet and dry processes, facilitates the transfer of Cd from the atmosphere to terrestrial and aquatic ecosystems. Dry deposition involves the gravitational settling or direct impaction of particles onto surfaces, whereas wet deposition refers to the incorporation of particles and gases into precipitation, followed by their removal during rain, snow, or other forms of precipitation

Cadmium Contamination in Soils and Plants: Mechanisms and Mitigation.
DOI: https://doi.org/10.1016/B978-0-443-34104-5.00021-9

TABLE 4.1 Estimates of global anthropogenic emissions of Cd to the atmosphere from different studies.

References (Data source)	Cd emissions (t)	Year of estimate
Nriagu (1979)	7300	1979
Nriagu and Pacyna (1988)	7570	1983
Pacyna and Pacyna (2001)	2983	1995
Zhu et al. (2020)	2246	2012

Cd, cadmium.

(Feng et al., 2019; Sakata et al., 2008; Wu et al., 2018). These processes not only contribute to soil contamination but also amplify risks to water quality and agricultural productivity.

Addressing the challenges posed by atmospheric Cd deposition requires an integrated approach combining emission reduction strategies with advanced monitoring and modeling efforts. High-emission regions must prioritize cleaner industrial processes, adopt advanced dust removal technologies, and enforce stricter regulatory frameworks (Cheng et al., 2015). At the same time, establishing global monitoring networks and refining chemical transport models (CTMs) are essential for capturing the complexities of Cd transport and deposition. International collaboration is crucial for facilitating technology transfer and fostering coordinated policy initiatives, particularly in rapidly industrializing regions. Significant progress can be achieved in mitigating the environmental and health impacts of atmospheric Cd deposition by integrating scientific advancement with evidence-based policy formulation, ultimately contributing to global environmental sustainability.

4.2 Emissions of cadmium to the atmosphere

4.2.1 Cadmium emissions by region

Emissions of Cd to the atmosphere exhibit significant regional disparities. These variations are primarily influenced by the degree of industrialization, energy use structures, and regional pollution control policies. Currently the widely used method for estimating atmospheric Cd emissions is the emission factor approach, which calculates emissions (E) using the formula $E = A \times B$, where A represents activity data (e.g., fuel consumption and industrial production) and B denotes the emission factor (emissions per unit of activity). This method is favored for its low data requirements and ease of implementation, making it widely applicable for compiling emission inventories at national and regional scales. However, its accuracy is constrained by uncertainties associated with emission factors and activity data. Table 4.1 summarizes global anthropogenic atmospheric Cd emissions from various studies. According to Nriagu (1979), global Cd emissions in 1979 were estimated at 7300 t, which decreased to 2246 t by 2012 (Zhu et al., 2020)—a nearly 70% reduction. This trend is closely linked to the strengthened environmental pollution control efforts and optimized energy structures in developed countries, particularly in Europe and North America. However, this downward trend is less pronounced in developing countries. For example, Cd emissions have continued to rise in regions such as China.

TABLE 4.2 Global spatial distribution pattern of anthropogenic emissions of Cd to the atmosphere in 2012.

Continents	Cd emissions (t)	Global contribution (%)
Asia	949	42.2
South America	563	25.1
North America	114	5.1
Europe	340	15.1
Africa	272	12.1
Oceania	9	0.4
Global	2246	100

Cd, cadmium. *Data are compiled from* Zhu et al. (2020).

TABLE 4.3 Emissions of Cd to the atmosphere from the top 10 countries in the world in 2012.

Country	Continents	Cd emission (t)	Contribution (%)
China	Asia	492	21.9
Chile	South America	456	20.3
India	Asia	179	8.0
Russia	Europe	157	7.0
Zambia	Africa	119	5.3
Congo	Africa	115	5.1
Poland	Europe	106	4.7
Mexico	North America	94	4.2
Kazakhstan	Asia	81	3.6
Peru	South America	64	2.8

Cd, cadmium. *Data are compiled from* Zhu et al. (2020).

The spatial distribution of global atmospheric Cd emissions in 2012 shows marked disparities (Table 4.2). Cd emissions were predominantly concentrated in developing regions such as Asia and South America. Asia contributed 42.2% of global Cd emissions, representing two to three times the emissions of Europe and Africa, and eight times that of North America. Table 4.3 lists the top 10 nations by atmospheric Cd emissions globally. China emerged as the predominant emitter, releasing nearly 500 t annually, which constituted 21.9% of the global total. With emissions of 456 t, Cd emissions of Chile were slightly lower than China, accounting for 20.3% of the global total. Other countries, such as India, Russia, and Zambia, also emitted substantial amounts of Cd, highlighting the critical role of emerging industrialized nations in global Cd emissions.

TABLE 4.4 Emissions of Cd to the atmosphere from different sources in the world.

Region	Sectoral	Cd emissions (t)	Contribution (%)
Global	Coal combustion	277.3	12.34
	Liquid fuel combustion	66.6	2.96
	Ferrous metal smelting	67.5	3.00
	Nonferrous metal smelting	1731.7	77.09
	Nonmetallic minerals manufacturing	93.7	4.17
	Municipal solid waste incineration	1.1	0.05
	Brake wear	8.4	0.37

Cd, cadmium. *Data are compiled from* Zhu et al. (2020).

4.2.2 Cadmium emissions by sector

The primary sources of Cd emissions are fossil fuel combustion and industrial activities, with their distribution reflecting regional economic structures and energy utilization patterns. Globally, nonferrous metal smelting is the largest source of Cd emissions, accounting for 77.1% of the total, with annual emissions of 1732 t (Table 4.4). This predominantly arises from the release of Cd during high-temperature processing in the smelting of zinc, lead, and copper (Tian et al., 2015). Coal combustion ranks second, contributing 12.3% of the global total Cd emissions, amounting to 277 t, and is primarily associated with coal-fired power generation and industrial boiler applications. Although urban waste incineration and brake wear contribute less than 1% to the total emissions, their localized impacts in urban areas cannot be overlooked.

The structural characteristics of Cd emission sources in China broadly align with the global pattern, albeit with different emphases. Notably, coal-related emissions account for a higher proportion (Table 4.5). Among these, coal-fired industrial boilers are a significant source, accounting for 22.8% of the national Cd emissions, with annual emissions reaching 104 t. In addition, coal-fired power plants and residential coal use contribute 3.4% and 0.8%, respectively. Nonferrous metal smelting remains the largest source of Cd emissions in China, accounting for 44.0%, with zinc and lead smelting being primary contributors. Biomass burning, which is prevalent in some rural areas, contributes 8.3% of Cd emissions, reflecting the reliance on biomass for energy in rural areas.

Nonferrous metal smelting is the dominant source of Cd emissions globally, whereas China exhibits a higher proportion of coal-related emissions. This disparity highlights China's reliance on coal as an energy source and the global concentration of metallurgical industries. Future strategies for Cd emission mitigation should be tailored according to regional characteristics.

4.2.3 Cadmium emissions by year

Fig. 4.1 provides the long-term trends in atmospheric Cd emissions across major regions, including China, Europe, and the United States. Since 1990, Cd emissions in China have increased rapidly, reaching approximately 550 t by 2012 (Tian et al., 2015). Furthermore, some researches indicate that Cd emissions in China reached 791 t by 2015 but subsequently declined

TABLE 4.5 Emissions of Cd to the atmosphere from different sources in China.

Region	Sectoral	Cd emissions (t)	Contribution (%)
China	Coal-fired power plant	15.5	3.40
	Coal-fired industrial boiler	104	22.82
	Coal-fired residential sector	3.7	0.81
	Coal-fired other sectors	25.7	5.64
	Liquid fuel combustion	9.6	2.11
	Nonferrous metal smelting	200.6	44.02
	Ferrous metal smelting	28.3	6.21
	Nonmetallic mineral products	28.8	6.32
	Municipal solid waste incineration	0.3	0.07
	Brake pad and tire abrasion	1.2	0.26
	Biomass burning	38	8.34

Cd, cadmium. *Data are compiled from* Tian et al. (2015).

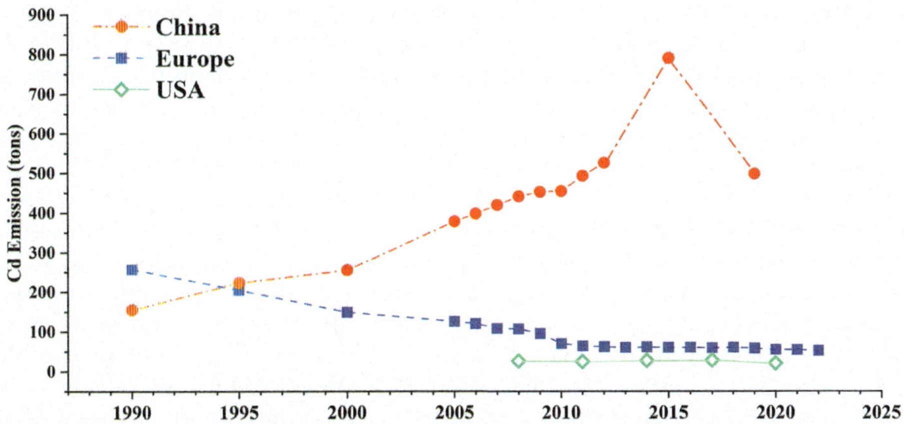

FIGURE 4.1 Temporal variation of Cd emissions to the atmosphere in major regions of the world. *Data are compiled from* EEA (2024); EPA (2022); Liu et al. (2018); Tian et al. (2015); Zhang et al. (2023)). *Cd*, Cadmium.

to 499 t by 2019 (Liu et al., 2018; Zhang et al., 2023). The growth was predominantly driven by accelerated industrialization and urbanization, with coal combustion and nonferrous metal smelting identified as the primary sources. In contrast, Europe has shown a substantial reduction in Cd emissions, declining from 250 t in 1990 to around 50 t by 2020 (EEA, 2024). Emissions in the United States have remained consistently below 30 t since the early 21st century (EPA, 2022). This progress is attributable to stringent environmental regulations implemented in the United States and Europe, the transition from coal-fired power generation to cleaner energy sources, and the technological retrofitting and relocation of high-pollution industries. Particularly critical to their success in achieving significant Cd emission reductions has been the shift in energy consumption from coal to natural gas. Although China's recent environmental policies have begun to yield positive outcomes, its total Cd emissions remain significantly higher than those of Europe and the United States, reflecting the considerable challenges associated with mitigating emissions in a rapidly industrializing economy.

To achieve a sustained decline in Cd emissions, China must expedite its transition to cleaner energy sources, intensify regulatory oversight of high-emission industries, and leverage the experiences of Europe and the United States. At the global level, enhanced multilateral collaboration is imperative to facilitate technology transfer and foster coordinated policy frameworks. Such efforts will not only support developing countries in addressing pollution challenges but also contribute to the long-term reduction of Cd emissions on a global scale, thereby promoting ecological sustainability and environmental resilience.

4.3 Atmospheric processes of cadmium

4.3.1 Global patterns of cadmium concentrations in rain

The spatial distribution of Cd concentrations in rainwater underscores stark regional disparities across the globe, driven by variations in industrialization levels, energy structures, and pollution control measures (Fig. 4.2). Asia emerges as a prominent hotspot for elevated Cd concentrations, particularly in China, where cities such as Changchun (18.4 μg L^{-1}), Chengdu (2.1 μg L^{-1}), and Xi'an (1.9 μg L^{-1}) exhibited persistently high levels in the monitoring period from 2006 to 2008 (Bai et al., 2010; Wang et al., 2010; Yang et al., 2009). These concentrations are predominantly attributed to rapid industrialization, with coal combustion and nonferrous metal smelting serving as the primary sources of Cd emissions. During the winter months, northern Chinese cities experience a notable increase in atmospheric Cd levels because of intensified coal use for residential heating. In contrast, developed Asian countries such as Japan (0.1 μg L^{-1}) and South Korea (0.07 μg L^{-1}) reported lower concentrations of Cd in rainwater (Kim et al., 2012; Sakata & Asakura, 2009). These lower values reflect the efficacy of long-term pollution control policies, stringent emission standards, and the adoption of clean energy technologies. However, developing nations within Asia, such as Turkey, show more variable trends. Cities such as Mersin (0.81 μg L^{-1}) and İzmir (3.1 μg L^{-1}) reported Cd levels comparable to or exceeding those observed in some Chinese industrial cities (Muezzinoglu & Cizmecioglu, 2006; Özsoy & Örnektekin, 2009), illustrating the ongoing challenges faced by countries attempting to balance rapid economic growth with environmental sustainability.

In Europe and North America, Cd concentrations in rainwater are markedly lower compared to Asian countries, with Europe emerging as one of the regions with the lowest recorded values globally. For instance the reported Cd concentrations in Germany (0.013 μg L^{-1}), the United Kingdom (0.005 μg L^{-1}), and France (0.023 μg L^{-1}) were all significantly lower than the average for China (0.24 μg L^{-1}) (EMEP, 2024; Ma, 2023). These achievements can largely be attributed to stringent environmental regulations, such as the Protocol on Heavy Metals—the seventh to enter into effect under the Convention on Long-range Transboundary Air Pollution of the UN Economic Commission for Europe (ECE)—and the widespread transition to cleaner energy systems. However, there are notable exceptions within Europe. Southern European nations, such

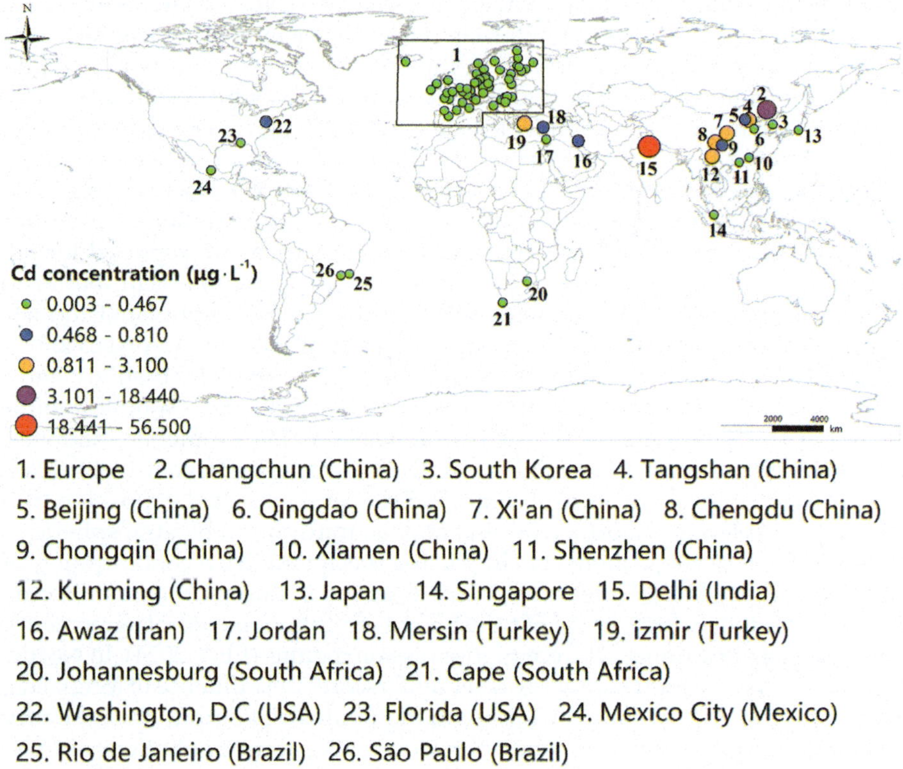

FIGURE 4.2 Global spatial pattern of Cd concentrations in precipitation from 2000 to 2022. *Cd*, Cadmium. *Data are compiled from Al-Momani (2003), Hu (2003), Báez et al. (2007), Melaku et al. (2008), Muezzinoglu and Cizmecioglu (2006), Toledo et al. (2008), Fontenele et al. (2009), Bai et al. (2010), EMEP (2024), Gichuki and Mason (2013), Jiang et al. (2018), Katoch et al. (2024), Kim et al. (2012), Kong et al. (2022), Landing et al. (2010), Li et al. (2012), Moaref et al. (2014), Özsoy and Örnektekin (2009), Peng et al. (2014), Sakata and Asakura (2009), Shen et al. (2016), Wang et al. (2010), Yang et al. (2009), Zhang et al. (2019), Zhang (2015). Map lines delineate study areas and do not necessarily depict accepted national boundaries.*

as Spain (0.174 µg L^{-1}), reported slightly elevated concentrations, likely because of localized industrial emissions. Similarly, some Central European countries, including Slovakia (0.265 µg L^{-1}) and Hungary (0.052 µg L^{-1}), demonstrate elevated levels compared to Western Europe, such as the Netherlands (0.022 µg L^{-1}) and Belgium (0.014 µg L^{-1}), reflecting disparities in industrial practices and pollution controls. In North America, Cd concentrations in rainwater vary considerably by region. Washington, DC (0.59 µg L^{-1}), and Mexico City (0.41 µg L^{-1}) reported moderate levels, which were influenced by localized industrial activities, vehicular emissions, and urbanization patterns (Báez et al., 2007; Melaku et al., 2008). In contrast, other parts of the United States, such as Florida (0.007 µg L^{-1}), reported extremely low concentrations, indicative of effective pollution mitigation efforts and a reduced reliance on heavy industry (Landing et al., 2010).

In other regions of the world, such as Africa and South America (Fontenele et al., 2009; Gichuki & Mason, 2013), Cd concentrations in rainwater remain relatively low, as evidenced by data from Johannesburg (0.04 μg L^{-1}) and Cape Point (0.02 μg L^{-1}) in South Africa and São Paulo in Brazil (0.17 μg L^{-1}). These low levels reflect lower industrial activities compared to Asia, Europe, and North America. However, the ongoing industrialization and urbanization of certain regions within Africa and South America pose risks for localized increases in Cd concentrations in the future. For example, Rio de Janeiro (0.4 μg L^{-1}) already exhibits relatively high levels, suggesting that industrial activities and urban density contribute significantly to local atmospheric Cd concentrations (Toledo et al., 2008). Among emerging economies, India represents an outlier. A study revealed that the mean Cd concentration in monsoon rainfall across Delhi was 56.5 μg L^{-1}, markedly exceeding the measurements reported for other major cities globally (Katoch et al., 2024). The high Cd concentration in rainfall occurs because the Delhi region is subject to atmospheric pollutant transport from the densely populated, industrially and agriculturally developed northern Indo-Gangetic Plain during the monsoon season.

The chemical nature of Cd in rainwater also influences its spatial variability and transport behavior. Cd exists predominantly in soluble forms such as Cd^{2+} or as soluble complexes with anions such as Cl^-, SO_4^{2-}, and NO_3^-. Cd in rainwater is highly mobile, allowing it to be transported over long distances (Deboudt et al., 2004; Nriagu & Pacyna, 1988). This property amplifies the role of atmospheric transport mechanisms in shaping Cd deposition patterns. For instance, elevated Cd levels in northern China may be transported by atmospheric circulation patterns, such as the East Asian monsoon, impacting precipitation Cd concentrations in regions downwind (Liang et al., 2018). Similar trends are observed in Europe, where uniformly low Cd concentrations reflect the effectiveness of transboundary pollution control agreements, limiting the impact of atmospheric transport from high-emission regions (EEA, 2024). In North America, regional variability in Cd concentrations further underscores the influence of localized industrial activities and atmospheric dynamics. Cities with high industrial output or dense urbanization tend to show elevated concentrations, whereas regions with effective pollution controls maintain significantly lower levels. Consequently the global variability in rainwater Cd concentrations highlights not only the distribution of pollution sources but also the effectiveness of environmental governance and the distinct mechanisms of Cd atmospheric transport, discussed in Section 4.4.4 below.

4.3.2 Global pattern of cadmium concentrations in aerosol

The global distribution of Cd concentrations in aerosols, much like its presence in precipitation, reveals pronounced regional variation. These differences are shaped by industrial activity, energy consumption patterns, pollution control measures, and atmospheric transport mechanisms (Fig. 4.3). Spatial analysis of atmospheric Cd demonstrates consistently higher burdens across monitoring sites in Asia relative to Europe and North America between 2000 and 2020. Notably, several Chinese metropolitan areas—including Beijing (8.5 ng m^{-3}), Xi'an (3.19 ng m^{-3}), and Zhengzhou (2.22 ng m^{-3})—exhibit significantly higher atmospheric Cd levels (Xu et al., 2016; Yan et al., 2019; Yuan et al., 2019). These high concentrations are largely driven by coal combustion, nonferrous metal smelting, and liquid fuel utilization, which remain key sources of Cd emissions in the region. The seasonality of these emissions is also notable, as northern Chinese cities experience higher atmospheric Cd levels during winter months because of increased coal usage for residential heating. In contrast, some countries in Asia reported substantially lower Cd concentrations, exemplified by Tokushima (0.422 ng m^{-3}) in Japan (Yamamoto et al., 2020) and Manila (0.44 ng m^{-3}) in the Philippines (Gonzalez et al., 2021).

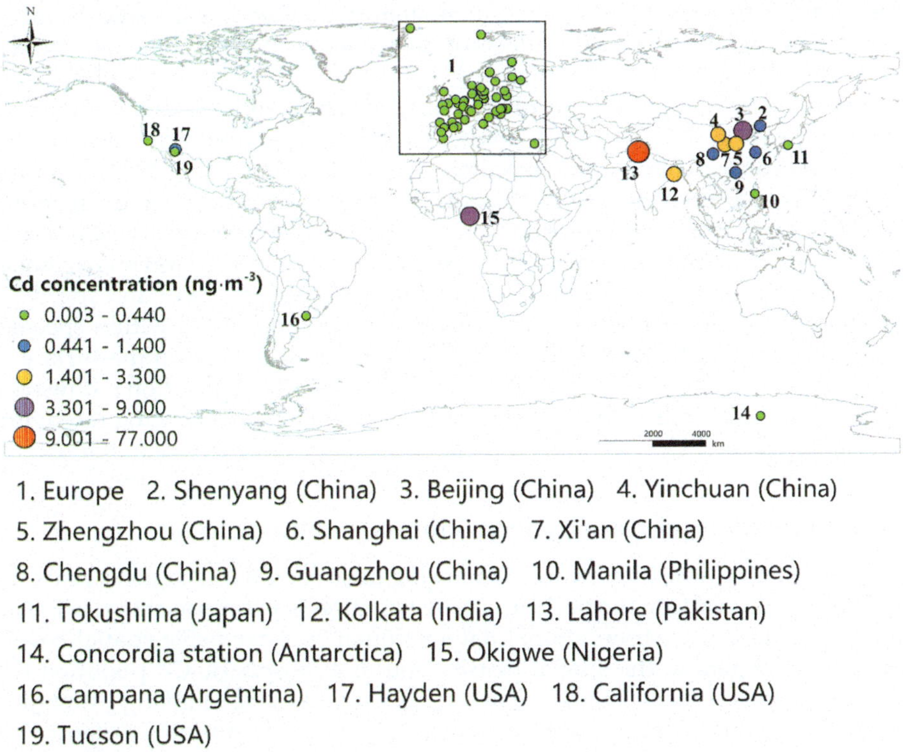

FIGURE 4.3 Global distribution of cadmium concentrations in aerosols from 2000 to 2022. Cd, Cadmium. *Data are compiled from Chiemeka (2010), Karar et al. (2006), Nie et al. (2018), Qi et al. (2017), Smichowski et al. (2005), von Schneidemesser et al. (2010), Xu et al. (2016), Zhou (2018), EMEP (2024), Gonzalez et al. (2021), Illuminati et al. (2021), Jiang et al. (2019), Wang et al. (2019), Yamamoto et al. (2020), Yan et al. (2019), Yuan et al. (2019). Map lines delineate study areas and do not necessarily depict accepted national boundaries.*

However, certain regions in South Asia, such as Lahore (77 ng m^{-3}) in Pakistan (von Schneidemesser et al., 2010), reported anomalously high Cd levels, likely because of inadequacy of emission control measures. These discrepancies further highlight the urgent need for enhanced pollution management strategies tailored to regional circumstances.

Although Cd concentrations in aerosols are generally lower in Europe and North America, elevated levels have been recorded in areas of frequent industrial activity. For example, high concentrations of aerosol Cd (1.14 ng m^{-3}) were recorded in Hayden, Arizona, United States—a town affected by metallurgical activities. Across Europe, countries such as Germany (0.05 ng m^{-3}), France (0.04 ng m^{-3}), and Finland (0.029 ng m^{-3}) reported Cd concentrations significantly below the European average (0.065 ng m^{-3}) (EMEP, 2024). Nevertheless, notable disparities exist within Europe itself. For example, Eastern and Central European countries such as Hungary (0.116 ng m^{-3}) reported higher concentrations, reflecting their continued reliance on coal-based energy production and the regional concentration of metallurgical industries. Similarly, Southern European nations such as Italy (0.054 ng m^{-3}) reported slightly elevated Cd levels, likely influenced by industrial emissions. Conversely, Nordic countries such as Sweden

(0.022 ng m^{-3}) and Denmark (0.023 ng m^{-3}) consistently reported the lowest Cd concentrations globally. These findings emphasize the effects of clean energy adoption, low industrial emission intensity, and effective air quality governance.

The transport characteristics of atmospheric Cd significantly influence its spatial distribution. Cd primarily adheres to fine particulate matter in the atmosphere, facilitating its long-range transport via atmospheric circulation (Hamilton et al., 2023; Wong et al., 2003). This association allows Cd to disperse from emission hotspots to regions hundreds or even thousands of kilometers away. Such long-range transport mechanisms underscore the transboundary nature of Cd pollution, necessitating international collaboration in monitoring and management efforts. Local meteorological conditions, including temperature inversions, low wind speeds, and high humidity, also play a critical role in contaminant accumulation of Cd within specific regions (Xu et al., 2018). For example, temperature inversions trap pollutants near the surface, resulting in localized spikes in Cd concentrations. These reactions alter the deposition pathways of Cd by producing particles with enhanced deposition efficiency, thereby influencing its environmental and ecological impacts.

4.3.3 Modeling cadmium transfer from emission sources to air

Numerous studies on atmospheric Cd monitoring have sought to characterize its spatiotemporal distribution and evaluate the particle size distribution of Cd-associated particulate matter (Pan et al., 2015). However, observational datasets often lack extensive spatial coverage, creating gaps in understanding the spatial distribution of atmospheric Cd pollution (Hutzell & Luecken, 2008). In this context, modeling serves as a crucial tool, enabling estimations of Cd concentration in regions without direct observational data. Despite this potential, relatively few numerical simulation studies have focused on the dispersion and behavior of heavy metals such as Cd in the atmosphere (Adani et al., 2015; Bai et al., 2021; Liu et al., 2019a; Schröder et al., 2010). Models also play an indispensable role in elucidating the dynamics and chemical transformations of Cd in the atmosphere, while also providing policymakers with critical insights for managing sustainable economic development and mitigating air pollution.

Simulations of atmospheric Cd concentration primarily rely on CTMs, including CMAQ, FRAME-HM, GEOS-Chem, and WRF-Chem (Dore et al., 2014; Liu et al., 2019a; Wu et al., 2022). These models integrate emission inventories, meteorological fields, and atmospheric physicochemical processes to quantitatively evaluate Cd migration, transformation, and deposition. Accurate simulations necessitate high-resolution emission inventories incorporating data from diverse sources, such as industrial activities, transportation, windblown dust, and natural emissions, alongside dynamic meteorological parameters such as wind fields, humidity, and precipitation. Although recent advancements have incorporated more complex processes, such as below-cloud scavenging and in-cloud reactions, limitations in emission inventories (e.g., the omission of dust and secondary sources) and low-resolution meteorological fields continue to hinder model accuracy (van der Swaluw et al., 2011). Consequently, considerable discrepancies between simulated and observed Cd concentrations remain.

Model validation typically involves comparing simulated results with observational data to assess the accuracy of Cd concentration and deposition flux predictions. For example, Dore et al. (2014) utilized the FRAME-HM model to simulate Cd deposition fluxes in Europe and found that the simulated dry and wet deposition fluxes differed from observational data by factors of 2–10. These discrepancies were attributed to the omission of essential processes, such as below-cloud scavenging of atmospheric Cd, and inaccuracies in both emission inventories and meteorological inputs. In addition, some models assume a uniform distribution of Cd across all particle sizes, a simplification that neglects the influence of particle size on transport and deposition behavior. This assumption can lead to significant underestimations of regional transport and spatiotemporal variability. Chen et al. (2013) demonstrated that Cd distribution among different particle sizes is strongly linked to emission sources, with fine particles dominating emissions from metallurgical activities. Cai et al. (2017) further emphasized that ignoring particle size distribution can markedly increase simulation errors, especially in regions where particle-driven deposition processes dominate. Moreover, localized meteorological phenomena, such as coastal land–sea breezes and mountain–valley wind systems, can significantly alter Cd dispersion patterns, necessitating higher model resolution to capture these effects accurately.

Future research on simulations of Cd concentrations must prioritize improvements in input data quality and representation of physical and chemical processes. Dynamic meteorological fields with higher spatial and temporal resolution, along with refined emission inventories that include underrepresented sources such as windblown dust, are essential for enhancing model precision. Furthermore, accounting for particle size distributions in simulations by integrating detailed observational datasets will improve the characterization of Cd migration and deposition dynamics, particularly in regions with significant long-range transport. Leveraging multisource observational data, such as high-resolution ground measurements, will enhance model validation and reduce uncertainties. These advancements will not only refine our understanding of Cd spatiotemporal distribution in the atmosphere but also provide critical support for health risk assessments and targeted pollution mitigation strategies, fostering effective Cd management and cross-regional collaboration.

4.4 Atmospheric deposition of cadmium

4.4.1 Deposition fluxes of cadmium

The dry and wet depositions of atmospheric Cd represent important pathways for the migration of Cd from the atmosphere to terrestrial and aquatic ecosystems. The mechanisms driving Cd deposition emphasize the interplay between atmospheric transport processes, wet deposition, and dry deposition (Pan & Wang, 2015). In regions with abundant precipitation, wet deposition acts as the primary mechanism for transferring Cd from the atmosphere to terrestrial surfaces (Gunawardena et al., 2013). Conversely, in arid and semiarid regions, dry deposition becomes the dominant pathway (Liu et al., 2019b). Among these mechanisms, wet deposition is typically more efficient, occurring at rates several orders of magnitude higher than dry deposition, especially in regions with humid climates (Wu et al., 2018). Consequently, wet deposition is the dominant process for pollutant removal in many regions, underscoring the importance of understanding its flux characteristics for evaluating pollutant transport and assessing ecological risks (Feng et al., 2019). Studies on the wet deposition of trace metals, including Cd, have focused on deposition flux patterns, compositional distributions, and influencing factors, offering valuable insights into their environmental behavior. Research has demonstrated that precipitation intensity, particle size distribution, and solubility are critical determinants of wet deposition fluxes, which exhibit significant temporal and spatial variability (Pan & Wang, 2015).

The global distribution of Cd deposition fluxes exhibits substantial regional disparities, driven by variations in emission source intensity and prevailing meteorological conditions. Observational data reveal that Cd deposition levels are significantly higher in Asia, particularly in rapid industrializing countries such as China and India, compared to other regions (Fig. 4.4A). In contrast, regions such as Europe, the Americas, and Oceania show markedly lower Cd deposition levels. For instance, deposition fluxes of Cd in Finland (0.009 mg m^{-2} $year^{-1}$) and Germany (0.007 mg m^{-2} $year^{-1}$) exemplify the effectiveness of pollution control measures and optimized industrial practices in Europe (EMEP, 2024). Similarly, regions in the Americas, such as the Caribbean region of Cuba (0.03 mg m^{-2} $year^{-1}$) and Reston in the United States (0.054 mg m^{-2} $year^{-1}$), show relatively low levels of Cd deposition, indicating the diminished impact of industrial emissions in these areas (Conko et al., 2004; Morera-Gómez et al., 2019). The elevated atmospheric Cd deposition fluxes observed in the São Paulo region (0.657 mg m^{-2} $year^{-1}$) of Brazil (Fontenele et al., 2010) and Mexico City (0.44 mg m^{-2} $year^{-1}$) of Mexico (Báez et al., 2007), relative to European and North American regions, are primarily attributable to intensified Cd emissions from transportation and fossil fuel combustion driven by high population density. Furthermore, industrial Cd discharges in Latin American nations significantly exceed those in Europe and North America, constituting a key contributing factor because of lenient emission standards and outdated end-of-pipe treatment technologies. Oceania, represented by New Zealand (0.02 mg m^{-2} $year^{-1}$), reports some of the lowest Cd deposition fluxes globally, underscoring the benefits of geographic isolation from major industrial centers and the widespread use of clean energy technologies (Gray et al., 2003).

Within Asia, Cd deposition levels show notable intraregional variability. In China (Fig. 4.4B), Hunan province, particularly Xiangtan, recorded the highest observed deposition flux (6.5 mg m^{-2} $year^{-1}$) in 2016, primarily driven by clusters of nonferrous metal smelting facilities and intensive industrial activities (Feng et al., 2019). In contrast, Beijing (0.24 mg m^{-2} $year^{-1}$) and Tianjin (0.19 mg m^{-2} $year^{-1}$) display lower levels, with Cd deposition being influenced by coal-based heating, vehicular emissions, and industrial production (Cong et al., 2008; Wang et al., 2017). Coastal economic zones, such as the Yangtze River Delta (0.26 mg m^{-2} $year^{-1}$) and Jiaozhou Bay (0.2 mg m^{-2} $year^{-1}$), exhibit elevated deposition levels because of port-related industrial activities and enhanced wet deposition rates (Ma et al., 2019; Xing et al., 2017). Similarly, deposition fluxes in Japan's Noto region (0.3 mg m^{-2} $year^{-1}$) are lower than those in China's industrial hubs but higher than in most other Asian countries, reflecting a combination of local industrial emissions and transboundary atmospheric transport (Sakata & Asakura, 2009). Singapore, with a deposition level of 0.78 mg m^{-2} $year^{-1}$ in 2000, ranks among the highest in the region, primarily because of its dense urbanization and significant maritime activities (Hu, 2003). Conversely, regions such as Hainan Island in southern China (0.09 mg m^{-2} $year^{-1}$), Kathmandu in Nepal (0.1 mg m^{-2} $year^{-1}$), and Oral in Kazakhstan (0.06 mg m^{-2} $year^{-1}$) report substantially lower levels of Cd deposition, attributable to lower industrial activity, sparse pollution sources, and favorable atmospheric dispersion conditions (Cherednichenko et al., 2021; Jiang et al., 2014; Tripathee et al., 2014).

The regional distribution of Cd deposition in Europe also illustrates significant variability (EMEP, 2024). Western European countries, such as the Netherlands (0.021 mg m^{-2} $year^{-1}$) and

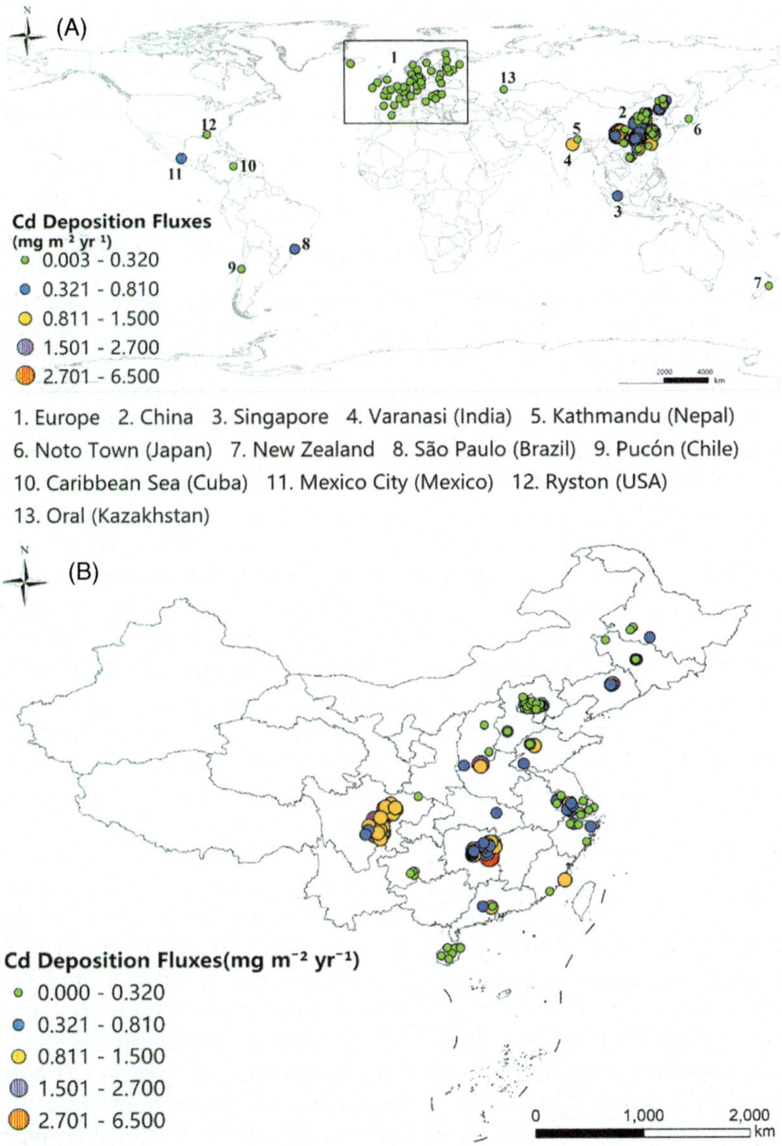

FIGURE 4.4 Monitoring of atmospheric dry and wet deposition of cadmium in (A) the world and in (B) China from 2000 to 2022. Cd, Cadmium. *Data are compiled from (A) Gray et al. (2003), Hu (2003), Báez et al. (2007), Cherednichenko et al. (2021), Cong et al. (2008), Conko et al. (2004), EMEP (2024), Feng et al. (2019), Fontenele et al. (2010), Guo et al. (2023), Jiang et al. (2014), Ma et al. (2019), Morera-Gómez et al. (2019), Rueda-Holgado et al. (2016), Sakata and Asakura (2009), Sharma et al. (2008), Tripathee et al. (2014), Wang et al. (2010), Xing et al. (2017). (B) Cong et al. (2008), Feng et al. (2019), Guo et al. (2023), Jiang et al. (2014), Wang et al. (2010), Ma et al. (2019), Xing et al. (2017). Map lines delineate study areas and do not necessarily depict accepted national boundaries.*

France (0.02 mg m^{-2} $year^{-1}$), show relatively low Cd deposition fluxes, benefiting from long-standing implementation of the protocol on heavy metals and continuous investments in clean energy technologies. Eastern and Central European countries, such as Slovakia (0.136 mg m^{-2} $year^{-1}$), report higher Cd fluxes. These elevated levels are attributed to the continued reliance on coal and the presence of metallurgical industries.

4.4.2 Advances in modeling atmospheric deposition of cadmium

To better understand and quantify deposition processes, CTMs have been developed, including LOTOS–EUROS (van der Swaluw et al., 2011) and EMEP MSC-W (Simpson et al., 2012). These models integrate emission inventories, atmospheric dynamics, and chemical processes to simulate the transport, transformation, and deposition of atmospheric contaminants. Some advanced CTMs also incorporate cloud and subcloud scavenging mechanisms to capture the complexity of wet deposition processes more accurately. In parallel, statistical models have been widely employed to estimate wet deposition fluxes based on observational data (Huang et al., 2018; Mamun et al., 2022). However, despite these advancements, significant discrepancies exist between model predictions and observed values. These discrepancies often arise from limitations in emission inventories, which may lack spatial and temporal resolution, as well as oversimplified representations of particle size distributions and deposition dynamics. For example, many models assume uniform particle compositions and neglect the heterogeneous chemical interactions that can alter particle solubility and scavenging efficiency.

Improving the accuracy and reliability of deposition models requires addressing several critical challenges. First, more accurate parameterizations in models for the physical and chemical features of Cd such as size distribution, solubility, and hygroscopicity are required. The size of particles influences not only the distance they travel but also their deposition pathways. Smaller particles are more likely to undergo long-range transport, whereas larger particles are more efficiently removed by wet deposition. Incorporating observational data on particle size distributions and chemical compositions into models can help address these limitations. For example, integrating ground-based measurements with simulation data can provide a more comprehensive understanding of the spatial and temporal variability of wet deposition fluxes (Li et al., 2023). Long-term monitoring networks are also critical for validating model predictions and enhancing the reliability of wet deposition simulations. Such networks should include a combination of ground-based stations, satellite observations, and mobile monitoring platforms to capture the full spectrum of deposition processes across different regions. These efforts should be complemented by advanced statistical techniques, such as machine learning algorithms, to analyze complex datasets and identify key drivers of variability in wet deposition fluxes. Second, accurate representation of spatial and temporal variations in Cd emissions requires high-resolution emission inventories (Jiang et al., 2024). This includes accounting for small-scale emission sources, such as regional waste incineration and localized industrial activities, which can contribute disproportionately to atmospheric Cd concentrations.

In addition, dynamic meteorological fields with higher spatial and temporal resolution are necessary to capture the influence of changing weather patterns on deposition processes. For example, the interaction between wind speed, precipitation events, and atmospheric stability plays a certain influence in determining deposition fluxes, particularly in complex terrains and coastal regions (Zhang et al., 2023).

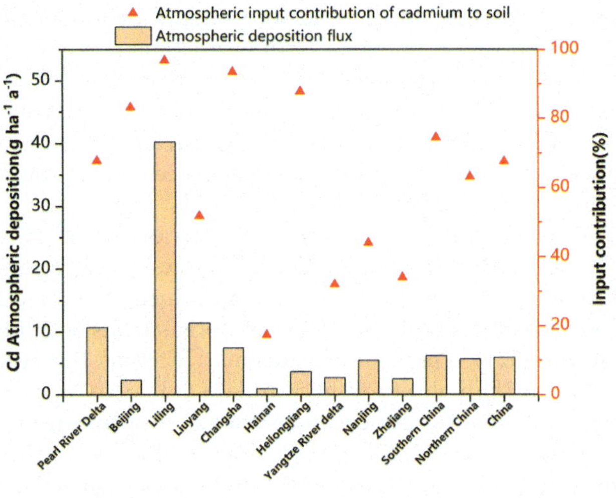

FIGURE 4.5 Input flux and contribution of atmospheric Cd deposition to soil in different regions of China from 2005 to 2018. Cd, Cadmium. *Data are compiled from Cong et al. (2008), Hou et al. (2014), Hu et al. (2013), Jiang et al. (2014), Peng et al. (2019), Shi et al. (2019), Xia et al. (2014), Yi et al. (2018), Zhou et al. (2012).*

4.4.3 Atmospheric deposition of cadmium to soil

Among the various pathways for heavy metal entry into soils, atmospheric deposition has emerged as a predominant contributor in recent years. Data on atmospheric Cd deposition fluxes and their contributions to soil inputs across different regions of China (Fig. 4.5) reveal striking spatial variability, underscoring the interplay between industrial activities, atmospheric transport, and meteorological factors. Nationally the average atmospheric Cd deposition flux in China is 0.59 mg m^{-2} $year^{-1}$, accounting for, on average, 67.6% of the total Cd inputs to soil (Peng et al., 2019). This figure highlights atmospheric deposition as a major source of Cd contamination in Chinese soils, with significant implications for soil quality and ecosystem health.

Industrialized regions such as Liling and Changsha in Hunan Province stand out with exceptionally high Cd deposition fluxes of 4.03 and 0.75 mg m^{-2} $year^{-1}$ in 2016, respectively, accounting for more than 90% of the total Cd inputs to soil (Yi et al., 2018). These levels are primarily driven by emissions from nonferrous metal smelting and the ceramic industry, which release large amounts of Cd into the atmosphere. These emissions rapidly accumulate in soil through dry and wet deposition. Similarly, in regions such as the Pearl River Delta (1.07 mg m^{-2} $year^{-1}$) and southern China (0.62 mg m^{-2} $year^{-1}$), atmospheric deposition contributes 67.8% and 74.5% of the total Cd inputs to soil, respectively, in 2007 (Zhou et al., 2012). This is attributed to a combination of industrial and vehicular emissions, compounded by high wet deposition efficiency in these humid areas. In such regions, frequent rainfall enhances the removal of atmospheric Cd through precipitation, leading to a more effective transfer of Cd into soil.

Conversely, regions such as Beijing (0.23 mg m^{-2} $year^{-1}$) and Heilongjiang (0.36 mg m^{-2} $year^{-1}$) have relatively low atmospheric deposition fluxes of Cd, yet they still contribute substantially to soil Cd inputs, at around 80% (Cong et al., 2008; Xia et al., 2014). The primary sources in these areas include coal combustion for heating and electricity generation, as well as regional dust emissions. In low-deposition areas such as Hainan, where industrial activities are low and meteorological conditions are less conducive to deposition, Cd input fluxes and atmospheric contributions are the lowest, measured at 0.091 mg m^{-2} $year^{-1}$ and 17.5%, respectively (Jiang et al., 2014). These findings illustrate the significant influence of industrial distribution and local climate on regional Cd deposition patterns. Southern China as a whole shows higher overall Cd deposition fluxes and contributions compared to northern China, largely because of its more humid climate and enhanced wet deposition efficiency. In areas with abundant precipitation, atmospheric Cd is deposited into soils with greater efficiency through wet deposition, whereas in drier northern regions, dry deposition processes are more important.

The deposition of atmospheric Cd not only exacerbates the accumulation of Cd in soils but also poses substantial risks to agricultural ecosystems and water resources (Tang et al., 2019). In high-deposition areas such as Liling and Changsha in Hunan province, prolonged accumulation of atmospheric deposition may result in soil Cd concentrations surpassing the thresholds for safe agricultural use (Yu et al., 2024). This can significantly increase the risk to food safety, particularly for crops that have a high ability to accumulate Cd, such as rice. In addition, Cd in soils can migrate to surface or even groundwater systems, amplifying the risks of contamination in aquatic ecosystems (Kubier et al., 2019). For example, the Pearl River Delta, a critical agricultural and aquacultural region in China, faces elevated Cd exposure risks through soil-to-water migration pathways, which could heighten environmental and public health concerns in the region (Wang, 2021) (Fig. 4.5).

4.4.4 Advancing integrated strategies and future perspectives for global cadmium pollution mitigation

Abatement of Cd pollution in the atmosphere demands comprehensive and coordinated mitigation strategies. Addressing this challenge necessitates a multifaceted approach centered on stringent source control, advanced monitoring and predictive capabilities, and robust international cooperation. Primary emission reduction must be prioritized, particularly in high-emission regions such as South Asia and parts of East Asia. This requires stricter enforcement of industrial emission standards targeting key sectors such as metal smelting, coal combustion, and waste incineration. Accelerating the transition to cleaner energy sources, notably solar and wind power, is fundamental to reducing reliance on Cd-emitting processes. Concurrently the widespread adoption of technological innovations, including high-efficiency dust collectors, advanced filtration systems, and optimized industrial production processes, is critical for minimizing emissions at their origin.

Beyond localized source control, establishing comprehensive regional and global atmospheric monitoring networks is imperative. These networks are essential for accurately quantifying Cd deposition fluxes, elucidating transport dynamics and deposition mechanisms (including complex interactions with meteorological conditions and topography), and assessing transboundary pollution impacts. Crucially the robust datasets generated enable the refinement of atmospheric models, particularly the development of integrated multipollutant simulation systems capable of representing complex transport and deposition pathways. Enhanced modeling facilitates more accurate predictions of the environmental behavior and fate of Cd, providing the data-driven insights necessary for evidence-based policy interventions and systematic evaluation of cross-regional transport and accumulation patterns. Understanding the relative contributions of local versus transboundary sources in high-deposition regions, such as industrialized areas of Asia, is vital for designing effective mitigation strategies.

International collaboration is pivotal across all facets of addressing atmospheric Cd pollution. Collaborative research efforts are indispensable for advancing technological innovation, including the development and dissemination of advanced composite particle tracking systems, high-resolution atmospheric modeling tools, and novel remediation technologies. Interdisciplinary research integrating atmospheric science, public health, environmental policy, and ecosystem management fosters a holistic understanding of Cd impacts, informing the design of comprehensive mitigation and soil/ecosystem management strategies. Knowledge sharing and the exchange of technological expertise between developed and developing nations accelerate capacity building and the adoption of best practices in emission reduction, regulatory enforcement, and sustainable soil management. Integrating atmospheric deposition studies with health risk assessments further strengthens the understanding of societal impacts. Aligning national actions within a framework of sustained international cooperation is paramount for tackling the inherently global nature of Cd transport and deposition.

The observed global disparities in atmospheric Cd concentrations and depositions reflect the complex interplay of industrial intensity, energy consumption patterns, regulatory rigor, and atmospheric processes. Although developed regions have made excellent progress, significant challenges remain in rapidly industrializing nations. By leveraging scientific advancements—integrating observational data, regional case studies, and improved modeling techniques—strengthening international cooperation, and implementing targeted emission control and ecosystem management measures, substantial reductions in atmospheric Cd levels and deposition can be achieved. Future research must focus on elucidating deposition interactions with anthropogenic activities and advancing multipollutant simulation capabilities. These concerted efforts at local, national, and global levels are critical for minimizing environmental and health risks, ensuring sustainable air quality improvements, safeguarding terrestrial and aquatic ecosystems, promoting global environmental sustainability, and ultimately protecting human and environmental health from the adverse effects of Cd exposure. Coordinated actions pave the way for sustainable soil management, improved agricultural productivity, quality of produce, and the preservation of biogeochemical cycles.

4.5 Summary

- Global Cd emissions exhibit significant regional and sectoral disparities—emissions in Asia (particularly China) show an increasing trend, with nonferrous metal smelting and coal combustion being the dominant emission sources, whereas Europe and North America have achieved substantial emission reductions through regulatory measures and the adoption of cleaner energy sources.
- Cd concentrations in aerosols and precipitation in certain Asian countries (e.g., China and India) are markedly higher than those in Europe and North America, as a result of regional differences in emission intensity and atmospheric transport conditions.
- Current research on atmospheric Cd transport primarily relies on atmospheric CTMs; however, simulation uncertainties remain high because of inaccuracies in emission inventories, insufficient resolution of meteorological data, and neglecting particle size effects.
- Atmospheric Cd enters soils through dry and wet deposition processes, with deposition fluxes and their contribution to soil Cd inputs being particularly prominent in specific regions of China compared to Europe and North America.

- Future global mitigation of atmospheric Cd pollution necessitates integrated strategies, including source reduction, establishing regional atmospheric monitoring networks, and fostering international collaboration to achieve sustainable cross-regional management.

Acknowledgments

This work was supported by the National Key Research and Development Plan (2022YFC3704802).

References

Adani, M., Mircea, M., D'Isidoro, M., Costa, M. P., & Silibello, C. (2015). Heavy metal modelling study over Italy: Effects of grid resolution, lateral boundary conditions and foreign emissions on air concentrations. *Water, Air, and Soil Pollution, 226*(3). http://www.kluweronline.com/issn/0049-6979/. doi:10.1007/s11270-015-2324-7.

Al-Momani, I. (2003). Trace elements in atmospheric precipitation at Northern Jordan measured by ICP-MS: Acidity and possible sources. *Atmospheric Environment, 37*(32), 4507–4515. doi:10.1016/s1352-2310(03)00562-4.

Báez, A., Belmont, R., García, R., Padilla, H., & Torres, M. C. (2007). Chemical composition of rainwater collected at a southwest site of Mexico City, Mexico. *Atmospheric Research, 86*(1), 61–75. doi:10.1016/j.atmosres.2007.03.005.

Bai, L., Wang, Z., & Huang, Y. (2010). Characteristics and source analysis of trace metals in precipitation in Xi'an district, Shaanxi Province, China (in Chinese). *Arid Land Geography, 33*(3), 385–393.

Bai, X., Luo, L., Tian, H., Liu, S., Hao, Y., Zhao, S., Lin, S., Zhu, C., Guo, Z., & Lv, Y. (2021). Atmospheric vanadium emission inventory from both anthropogenic and natural sources in China. *Environmental Science and Technology, 55*(17), 11568–11578. http://pubs.acs.org/journal/esthag. doi:10.1021/acs.est.1c04766.

Cai, J., Wang, J., Zhang, Y., Tian, H., Zhu, C., Gross, D. S., Hu, M., Hao, J., He, K., Wang, S., & Zheng, M. (2017). Source apportionment of Pb-containing particles in Beijing during January 2013. *Environmental Pollution, 226*, 30–40. https://www.journals.elsevier.com/environmental-pollution. doi:10.1016/j.envpol.2017.04.004.

Chen, B., Stein, A. F., Maldonado, P. G., Sanchez de la Campa, A. M., Gonzalez-Castanedo, Y., Castell, N., & de la Rosa, J. D. (2013). Size distribution and concentrations of heavy metals in atmospheric aerosols originating from industrial emissions as predicted by the HYSPLIT model. *Atmospheric Environment, 71*, 234–244. doi:10.1016/j.atmosenv.2013.02.013.

Cheng, K., Wang, Y., Tian, H., Gao, X., Zhang, Y., Wu, X., Zhu, C., & Gao, J. (2015). Atmospheric emission characteristics and control policies of five precedent-controlled toxic heavy metals from anthropogenic sources in China. *Environmental Science and Technology, 49*(2), 1206–1214. http://pubs.acs.org/journal/esthag. doi:10.1021/es5037332.

Cherednichenko, V. S., Cherednichenko, A. V., Cherednichenko, A. V., Zheksenbaeva, A. K., & Madibekov, A. S. (2021). Heavy metal deposition through precipitation in Kazakhstan. *Heliyon, 7*(1), e05844. doi:10.1016/j.heliyon.2020.e05844.

Chiemeka, I. U. (2010). Air aerosol metal constituent and concentration at Okigwe, Nigeria. *International Journal of Physical Sciences, 5*(4), 283–286. http://www.academicjournals.org/ijps/PDF/pdf2010/Apr/Chiemeka.pdf.

Cong, Y., Zheng, P., Chen, Y. L., & Hou, Q. Y. (2008). Ecological risk assessments of heavy metals in soils of the farmland ecosystem of Beijing, China (in Chinese). *Geological Bulletin of China, 27*(5), 681–688.

Conko, K. M., Rice, K. C., & Kennedy, M. M. (2004). Atmospheric wet deposition of trace elements to a suburban environment, Reston, Virginia, USA. *Atmospheric Environment, 38*(24), 4025–4033. doi:10.1016/j.atmosenv.2004.03.062.

Deboudt, K., Flament, P., & Bertho, M. L. (2004). Cd, Cu, Pb and Zn concentrations in atmospheric wet deposition at a coastal station in Western Europe. *Water, Air, and Soil Pollution, 151*(1–4), 335–359. doi:10.1023/B:WATE.0000009906.55895.30.

Dore, A. J., Hallsworth, S., McDonald, A. G., Werner, M., Kryza, M., Abbot, J., Nemitz, E., Dore, C. J., Malcolm, H., Vieno, M., Reis, S., & Fowler, D. (2014). Quantifying missing annual emission sources of heavy metals in the United Kingdom with an atmospheric transport model. *Science of the Total Environment, 479–480*(1), 171–180. doi:10.1016/j.scitotenv.2014.02.001.

EEA. (2024). European Union emission inventory report 1990-2022. https://www.eea.europa.eu/en/analysis/publications/european-union-emissions-inventory-report-1990-2022.

EMEP. (2024). EMEP-CCC-Report_Heavy metals and POP measurements 2022. https://emep-ccc.nilu.no/static/reports/EMEP_CCC-Report_3_2024_HMandPOP_09-09-2024_FINAL.pdf.

EPA. (2022). National Emissions Inventory (NEI) report 2008, 2011, 2014, 2017, 2020. https://www.epa.gov/air-emissions-inventories/2020-national-emissions-inventory-nei-data.

Feng, W., Guo, Z., Peng, C., Xiao, X., Shi, L., Zeng, P., Ran, H., & Xue, Q. (2019). Atmospheric bulk deposition of heavy metal(loid)s in central south China: Fluxes, influencing factors and implication for paddy soils. *Journal of Hazardous Materials, 371*, 634–642. Www.elsevier.com/locate/jhazmat. doi:10.1016/j.jhazmat.2019.02.090.

Fontenele, A. P. G., Fornaro, A., & Pedrotti, J. J. (2010). Measurements of heavy metals in dry and wet deposition in São Paulo city. *Highway and Urban Environment, 17*, 105–113. doi:10.1007/978-90-481-3043-6_12.

Fontenele, A. P. G., Pedrotti, J. J., & Fornaro, A. (2009). Evaluation of trace metals and major ions concentrations in rainwater in downtown São Paulo city. *Quimica Nova, 32*(4), 839–844.

Gichuki, S. W., & Mason, R. P. (2013). Mercury and metals in South African precipitation. *Atmospheric Environment, 79*, 286–298. doi:10.1016/j.atmosenv.2013.04.009.

Gonzalez, M. E., Stahl, C., Cruz, M. T., Bañaga, P. A., Betito, G., Braun, R. A., Azadi Aghdam, M., Cambaliza, M. O., Lorenzo, G. R., MacDonald, A. B., Simpas, J. B., Csavina, J., Sáez, A. E., Betterton, E., & Sorooshian, A. (2021). Contrasting the size-resolved nature of particulate arsenic, cadmium, and lead among diverse regions. *Atmospheric Pollution Research, 12*(3), 352–361. http://www.atmospolres.com/accesstext.html. doi:10.1016/j.apr.2021.01.002.

Gray, C. W., McLaren, R. G., & Roberts, A. H. C. (2003). Atmospheric accessions of heavy metals to some New Zealand pastoral soils. *Science of the Total Environment, 305*(1–3), 105–115. doi:10.1016/S0048-9697(02)00404-7.

Gunawardena, J., Egodawatta, P., Ayoko, G. A., & Goonetilleke, A. (2013). Atmospheric deposition as a source of heavy metals in urban stormwater. *Atmospheric Environment, 68*, 235–242. doi:10.1016/j.atmosenv.2012.11.062.

Guo, Y., Du, E., Li, B., Xia, N., Wu, X., & de Vries, W. (2023). Significant urban hotspots of atmospheric trace element deposition and potential effects on urban soil pollution in China. *Journal of Cleaner Production, 415*. https://www.journals.elsevier.com/journal-of-cleaner-production. doi:10.1016/j.jclepro.2023.137872.

Hamilton, D. S., Baker, A. R., Iwamoto, Y., Gassó, S., Bergas-Masso, E., Deutch, S., Dinasquet, J., Kondo, Y., Llort, J., Myriokefalitakis, S., Perron, M. M. G., Wegmann, A., & Yoon, J. E. (2023). An aerosol odyssey: Navigating nutrient flux changes to marine ecosystems. *Elementa, 11*(1). https://www.elementascience.org/articles/. doi:10.1525/elementa.2023.00037.

Hou, Q., Yang, Z., Ji, J., Yu, T., Chen, G., Li, J., Xia, X., Zhang, M., & Yuan, X. (2014). Annual net input fluxes of heavy metals of the agro-ecosystem in the Yangtze River delta, China. *Journal of Geochemical Exploration, 139*, 68–84. doi:10.1016/j.gexplo.2013.08.007.

Hu, G. (2003). Soluble trace metals in rainwater in Singapore. *Journal of South China University of Technology, Natural Science Edition, 31*(7), 86–90.

Hu, W., Huang, B., Shi, X., Chen, W., Zhao, Y., & Jiao, W. (2013). Accumulation and health risk of heavy metals in a plot-scale vegetable production system in a peri-urban vegetable farm near Nanjing, China. *Ecotoxicology and Environmental Safety, 98*, 303–309. doi:10.1016/j.ecoenv.2013.09.040.

Huang, M., Sun, H., Liu, H., Wang, X., Wang, B., & Zheng, D. (2018). Atmospheric arsenic deposition in the Pearl River Delta Region, South China: Influencing factors and speciation. *Environmental Science and Technology, 52*(5), 2506–2516. http://pubs.acs.org/journal/esthag. doi:10.1021/acs.est.7b04427.

Hutzell, W. T., & Luecken, D. J. (2008). Fate and transport of emissions for several trace metals over the United States. *Science of the Total Environment, 396*(2–3), 164–179. doi:10.1016/j.scitotenv.2008.02.020.

Illuminati, S., Annibaldi, A., Truzzi, C., Mantini, C., Conca, E., Malandrino, M., Giglione, G., Fanelli, M., & Scarponi, G. (2021). Determination of Cd, Pb, and Cu in the atmospheric aerosol of Central East Antarctica at Dome C (Concordia Station). *Molecules, 26*(7), 1997. doi:10.3390/molecules26071997.

Jiang, B., He, L., Chen, D., Gu, T., Liang, Z., Wang, M., ... Li, S. (2018). Wet deposition flux and sources of heavy metals in precipitation in the coastal area of Shenzhen (in Chinese). *Environmental Chemistry, 37*(7), 1460–1473.

Jiang, S., Li, W., Shi, T., Lv, J., Feng, W., Liu, P., ... Wu, Y. (2019). Pollution characteristics and health risk assessment of water-soluble heavy metals in $PM_{2.5}$ in Guangzhou, 2017 (in Chinese). *South China Journal of Preventive Medicine, 45*(2), 107–114.

Jiang, S., Zhang, Y., Yu, G., Han, Z., Zhao, J., Zhang, T., & Zheng, M. (2024). Source-resolved atmospheric metal emissions, concentrations, and deposition fluxes into the East Asian seas. *Atmospheric Chemistry and Physics, 24*(14), 8363–8381. http://www.atmos-chem-phys.net/volumes_and_issues.html. doi:10.5194/acp-24-8363-2024.

Jiang, W., Hou, Q., Yang, Z., Yu, T., Zhong, C., Yang, Y., & Fu, Y. (2014). Annual input fluxes of heavy metals in agricultural soil of Hainan Island, China. *Environmental Science and Pollution Research, 21*(13), 7876–7885. http://www.springerlink.com/content/0944-1344. doi:10.1007/s11356-014-2679-0.

Karar, K., Gupta, A. K., Kumar, A., & Biswas, A. K. (2006). Characterization and identification of the sources of chromium, zinc, lead, cadmium, nickel, manganese and Iron in PM_{10} particulates at the two sites of Kolkata, India. *Environmental Monitoring and Assessment, 120*(1–3), 347–360. doi:10.1007/s10661-005-9067-7.

Katoch, A., Yadav, S., Singh, Y., & Kulshrestha, U. (2024). Wet scavenging of trace metals and reactive nitrogen in Delhi, India. *International Journal of Environmental Science and Technology, 21*(3), 2897–2912. doi:10.1007/s13762-023-05097-z.

Kim, J. E., Han, Y. J., Kim, P. R., & Holsen, T. M. (2012). Factors influencing atmospheric wet deposition of trace elements in rural Korea. *Atmospheric Research, 116*, 185–194. doi:10.1016/j.atmosres.2012.04.013.

Kong, L. D., Qi, J. H., & Zhang, X. (2022). Concentration, solubility, and sources of metal elements in atmospheric precipitation in Qingdao. *Huanjing Kexue/Environmental Science, 43*(12), 5387–5398. https://www.hjkx.ac.cn/hjkx/ch/reader/view_abstract.aspx?file_no=20221206&flag=1. doi:10.13227/j.hjkx.202202114.

Kubier, A., Wilkin, R. T., & Pichler, T. (2019). Cadmium in soils and groundwater: A review. *Applied Geochemistry, 108*, 104388. doi:10.1016/j.apgeochem.2019.104388.

Landing, W. M., Caffrey, J. M., Nolek, S. D., Gosnell, K. J., & Parker, W. C. (2010). Atmospheric wet deposition of mercury and other trace elements in Pensacola, Florida. *Atmospheric Chemistry and Physics, 10*(10), 4867–4877. doi:10.5194/acp-10-4867-2010.

Li, D., Zheng, J., Yang, M., Meng, Y., Yu, X., Zhou, H., Tong, L., Wang, K., Li, Y. F., Wang, X., & Xiao, H. (2023). Atmospheric wet deposition of trace metal elements: Monitoring and modelling. *Science of the Total Environment, 893*. Www.elsevier.com/locate/scitotenv. doi:10.1016/j.scitotenv.2023.164880.

Li, Y. M., Pan, Y. P., Wang, Y. S., Wang, Y. F., & Li, X. R. (2012). Chemical characteristics and sources of trace metals in precipitation collected from a typical industrial city in Northern China (in Chinese). *Huanjing Kexue/Environmental Science, 33*(11), 3712–3717.

Liang, X., Tian, C., Zong, Z., Wang, X., Jiang, W., Chen, Y., Ma, J., Luo, Y., Li, J., & Zhang, G. (2018). Flux and source-sink relationship of heavy metals and arsenic in the Bohai Sea, China. *Environmental Pollution, 242*, 1353–1361. Www.elsevier.com/inca/publications/store/4/0/5/8/5/6. doi:10.1016/j.envpol.2018.08.011.

Liu, S., Zhu, C., Tian, H., Wang, Y., Zhang, K., Wu, B., Liu, X., Hao, Y., Liu, W., Bai, X., Lin, S., Wu, Y., Shao, P., & Liu, H. (2019a). Spatiotemporal variations of ambient concentrations of trace elements in a highly polluted region of China. *Journal of Geophysical Research: Atmospheres, 124*(7), 4186–4202. http://agupubs.onlinelibrary.wiley.com/hub/jgr/journal/10.1002/(ISSN)2169-8996/. doi:10.1029/2018JD029562.

Liu, Y., Fu, B., Shen, Y., Liu, C., Liu, H., Zhao, Z., & Zhang, L. (2019b). Environmental effect of heavy metals deposition in arid city. *Global Nest Journal, 21*(3), 405–409. https://journal.gnest.org/sites/default/files/Submissions/gnest_03143/gnest_03143_published.pdf. doi:10.30955/gnj.003143.

Liu, Y., Xing, J., Wang, S., Fu, X., & Zheng, H. (2018). Source-specific speciation profiles of $PM_{2.5}$ for heavy metals and their anthropogenic emissions in China. *Environmental Pollution, 239*, 544–553. Www.elsevier.com/inca/publications/store/4/0/5/8/5/6. doi:10.1016/j.envpol.2018.04.047.

Ma, X. (2023). *Characteristics and sources of nitrogen, phosphorus and heavy metals bulk deposition in China and their ecological risk assessment (in Chinese, Ph.D thesis)*. China Agricultural University.

Ma, Y., Tang, Y., Xu, H., Zhang, X., Liu, H., Wang, S., & Zhang, W. (2019). Bulk/wet deposition of trace metals to rural, industrial, and urban areas in the Yangtze River Delta, China. *Ecotoxicology and Environmental Safety, 169*, 185–191. http://www.elsevier.com/inca/publications/store/6/2/2/8/1/9/index.htt. doi:10.1016/j.ecoenv.2018.11.002.

Mamun, A. A., Cheng, I., Zhang, L., Celo, V., Dabek-Zlotorzynska, E., & Charland, J. P. (2022). Estimation of atmospheric dry and wet deposition of particulate elements at four monitoring sites in the Canadian Athabasca oil sands region. *Journal of Geophysical Research: Atmospheres, 127*(3). http://agupubs.onlinelibrary.wiley.com/hub/jgr/journal/10.1002/(ISSN)2169-8996/. doi:10.1029/2021JD035787.

Melaku, S., Morris, V., Raghavan, D., & Hosten, C. (2008). Seasonal variation of heavy metals in ambient air and precipitation at a single site in Washington, DC. *Environmental Pollution, 155*(1), 88–98. doi:10.1016/j.envpol.2007.10.038.

Moaref, S., Sekhavatjou, M. S., & Hosseini Alhashemi, H. (2014). Determination of trace elements concentration in wet and dry atmospheric deposition and surface soil in the largest industrial city, southwest of Iran. *International Journal of Environmental Research, 8*(2), 335–346. doi:10.1016/j.hrthm.2009.04.015.

Morera-Gómez, Y., Santamaría, J. M., Elustondo, D., Lasheras, E., & Alonso-Hernández, C. M. (2019). Determination and source apportionment of major and trace elements in atmospheric bulk deposition in a Caribbean rural area. *Atmospheric Environment, 202*, 93–104. Www.elsevier.com/locate/atmosenv. doi:10.1016/j.atmosenv.2019.01.019.

Muezzinoglu, A., & Cizmecioglu, S. C. (2006). Deposition of heavy metals in a Mediterranean climate area. *Atmospheric Research, 81*(1), 1–16. doi:10.1016/j.atmosres.2005.10.004.

Nie, L., Li, Y., Hua, Z., & Cui, Z. (2018). Pollution characteristics and related health risk of heavy metals in $PM_{2.5}$ in Shenyang city of Liaoning province (in Chinese). *Chinese Journal of Public Health, 34*(04), 574–576.

Nriagu, J. O. (1979). Global inventory of natural and anthropogenic emissions of trace metals to the atmosphere. *Nature, 279*(5712), 409–411. doi:10.1038/279409a0.

Nriagu, J. O., & Pacyna, J. M. (1988). Quantitative assessment of worldwide contamination of air, water and soils by trace metals. *Nature, 333*(6169), 134–139. doi:10.1038/333134a0.

Özsoy, T., & Örnektekin, S. (2009). Trace elements in urban and suburban rainfall, Mersin, Northeastern Mediterranean. *Atmospheric Research, 94*(2), 203–219. doi:10.1016/j.atmosres.2009.05.017.

Pacyna, J. M., & Pacyna, E. G. (2001). An assessment of global and regional emissions of trace metals to the atmosphere from anthropogenic sources worldwide. *Environmental Reviews, 9*(4), 269–298. doi:10.1139/a01-012.

Pan, Y., Tian, S., Li, X., Sun, Y., Li, Y., Wentworth, G. R., & Wang, Y. (2015). Trace elements in particulate matter from metropolitan regions of Northern China: Sources, concentrations and size distributions. *Science of the Total Environment, 537*, 9–22. Www.elsevier.com/locate/scitotenv. doi:10.1016/j.scitotenv.2015.07.060.

Pan, Y. P., & Wang, Y. S. (2015). Atmospheric wet and dry deposition of trace elements at 10 sites in Northern China. *Atmospheric Chemistry and Physics, 15*(2), 951–972. doi:10.5194/acp-15-951-2015.

Peng, H., Chen, Y., Weng, L., Ma, J., Ma, Y., Li, Y., & Islam, M. S. (2019). Comparisons of heavy metal input inventory in agricultural soils in North and South China: A review. *Science of the Total Environment, 660*, 776–786. Www.elsevier.com/locate/scitotenv. doi:10.1016/j.scitotenv.2019.01.066.

Peng, Y. L., Wang, Y. M., Qin, C. Q., & Wang, D. Y. (2014). Concentrations and deposition fluxes of heavy metals in precipitation in core urban areas, Chongqing. *Huanjing Kexue/Environmental Science*, (7), 2490–2496. http://www.hjkx.ac.cn/hjkx/ch/index.aspx. doi:10.13227/j.hjkx.2014.07.008.

Qi, A., Zhang, Y., Ding, Y., & Yang, H. (2017). Characteristics and source of metal elements in ambient $PM_{2.5}$ in urban area of Yinchuan (in Chinese). *Journal of Environment and Health, 34*(07), 591–594.

Rueda-Holgado, F., Calvo-Blázquez, L., Cereceda-Balic, F., & Pinilla-Gil, E. (2016). Temporal and spatial variation of trace elements in atmospheric deposition around the industrial area of Puchuncaví-Ventanas (Chile) and its influence on exceedances of lead and cadmium critical loads in soils. *Chemosphere, 144*, 1788–1796. doi:10.1016/j.chemosphere.2015.10.079.

Sakata, M., & Asakura, K. (2009). Factors contributing to seasonal variations in wet deposition fluxes of trace elements at sites along Japan Sea coast. *Atmospheric Environment, 43*(25), 3867–3875. doi:10.1016/j.atmosenv.2009.05.001.

Sakata, M., Tani, Y., & Takagi, T. (2008). Wet and dry deposition fluxes of trace elements in Tokyo Bay. *Atmospheric Environment, 42*(23), 5913–5922. doi:10.1016/j.atmosenv.2008.03.027.

Schröder, W., Holy, M., Pesch, R., Harmens, H., Ilyin, I., Steinnes, E., Alber, R., Aleksiayenak, Y., Blum, O., Coşkun, M., Dam, M., De Temmerman, L., Frolova, M., Frontasyeva, M., Miqueo, L. G., Grodzińska, K., Jeran, Z., Korzekwa, S., Krmar, M., … Zechmeister, H. G. (2010). Are cadmium, lead and mercury concentrations in mosses across Europe primarily determined by atmospheric deposition of these metals? *Journal of Soils and Sediments, 10*(8), 1572–1584. http://www.springerlink.com/content/1439-0108. doi:10.1007/s11368-010-0254-y.

Shao, X., Cheng, H., Li, Q., & Lin, C. (2013). Anthropogenic atmospheric emissions of cadmium in China. *Atmospheric Environment, 79*, 155–160. doi:10.1016/j.atmosenv.2013.05.055.

Sharma, R. K., Agrawal, M., & Marshall, F. M. (2008). Atmospheric deposition of heavy metals (Cu, Zn, Cd and Pb) in Varanasi City, India. *Environmental Monitoring and Assessment, 142*(1–3), 269–278. doi:10.1007/s10661-007-9924-7.

Shen, Y., Shi, Z., Yang, F., Liu, G., & Su, B. (2016). Research on characteristics of heavy metal of precipitation in suburbs of Kunming (in Chinese). *Journal of Water Resources and Water Engineering, 27*(3), 112–116.

Shi, T., Ma, J., Wu, F., Ju, T., Gong, Y., Zhang, Y., Wu, X., Hou, H., Zhao, L., & Shi, H. (2019). Mass balance-based inventory of heavy metals inputs to and outputs from agricultural soils in Zhejiang Province, China. *Science of the Total Environment, 649*, 1269–1280. Www.elsevier.com/locate/scitotenv. doi:10.1016/j.scitotenv.2018.08.414.

Simpson, D., Benedictow, A., Berge, H., Bergström, R., Emberson, L. D., Fagerli, H., Flechard, C. R., Hayman, G. D., Gauss, M., Jonson, J. E., Jenkin, M. E., Nyíri, A., Richter, C., Semeena, V. S., Tsyro, S., Tuovinen, J.-P., Valdebenito, Á., & Wind, P. (2012). The EMEP MSC-W chemical transport model—Technical description. *Atmospheric Chemistry and Physics, 12*(16), 7825–7865. doi:10.5194/acp-12-7825-2012.

Smichowski, P., Marrero, J., & Gómez, D. (2005). Inductively coupled plasma optical emission spectrometric determination of trace element in PM_{10} airborne particulate matter collected in an industrial area of Argentina. *Microchemical Journal, 80*(1), 9–17. doi:10.1016/j.microc.2004.07.023.

Tang, L., Deng, S., Tan, D., Long, J., & Lei, M. (2019). Heavy metal distribution, translocation, and human health risk assessment in the soil-rice system around Dongting Lake area, China. *Environmental Science and Pollution Research, 26*(17), 17655–17665. http://www.springerlink.com/content/0944-1344. doi:10.1007/s11356-019-05134-w.

Tian, H. Z., Zhu, C. Y., Gao, J. J., Cheng, K., Hao, J. M., Wang, K., Hua, S. B., Wang, Y., & Zhou, J. R. (2015). Quantitative assessment of atmospheric emissions of toxic heavy metals from anthropogenic sources in China: Historical trend, spatial distribution, uncertainties, and control policies. *Atmospheric Chemistry and Physics, 15*(17), 10127–10147. doi:10.5194/acp-15-10127-2015.

Toledo, V. E., Almeida, P. B., Quiterio, S. L., Arbilla, G., Moreira, A., Escaleira, V., & Moreira, J. C. (2008). Evaluation of levels, sources and distribution of toxic elements in PM_{10} in a suburban industrial region, Rio de Janeiro, Brazil. *Environmental Monitoring and Assessment, 139*(1–3), 49–59. doi:10.1007/s10661-007-9815-y.

Tripathee, L., Kang, S., Huang, J., Sharma, C. M., Sillanpää, M., Guo, J., & Paudyal, R. (2014). Concentrations of trace elements in wet deposition over the central Himalayas, Nepal. *Atmospheric Environment, 95*, 231–238. Www.elsevier.com/locate/atmosenv. doi:10.1016/j.atmosenv.2014.06.043.

van der Swaluw, E., Schaap, M., Sauter, F., Banzhaf, S., & Manders, A. (2011). *Measuring and modeling wet deposition fluxes in the Netherlands and Europe* (pp. 193–198). Springer Science and Business Media LLC. doi:10.1007/978-94-007-1359-8_33.

von Schneidemesser, E., Stone, E. A., Quraishi, T. A., Shafer, M. M., & Schauer, J. J. (2010). Toxic metals in the atmosphere in Lahore, Pakistan. *Science of the Total Environment, 408*(7), 1640–1648. http://www.sciencedirect.com. doi:10.1016/j.scitotenv.2009.12.022.

Wang, F. (2021). *Study on the migration and transformation of cadmium in soil under the background of sea-land interaction in the Pearl River Delta Plain (in Chinese, Ph.D thesis).* China University of Geosciences.

Wang, H., Ma, N., Yang, X., Yang, K., & Han, G. (2010). Heavy metal characteristics of rainwater in Chengdu (in Chinese). *Earth and Environment, 38*(1), 49–53.

Wang, W., Cao, S., Li, G., & Zhang, Y. (2017). Sedimentation flux and its evaluation of dry and wet atmospheric deposition of heavy metal elements in north Tianjin (in Chinese). *Environmental Science and Management, 42*(5), 46–51.

Wang, Y., Gan, X., & Lang, C. (2019). Chemical speciation of heavy metals in $PM_{2.5}$ and optimization of extraction conditions in spring and winter in Chengdu (in Chinese). *Guangdong Chemical Industry, 46*(2), 15–17.

Wong, C. S. C., Li, X. D., Zhang, G., Qi, S. H., & Peng, X. Z. (2003). Atmospheric deposition of heavy metals in the Pearl River Delta, China. *Atmospheric Environment, 37*(6), 767–776. doi:10.1016/s1352-2310(02)00929-9.

Wu, J., Kong, S., Yan, Y., Cheng, Y., Yan, Q., Liu, D., … Qi, S. (2022). The toxicity emissions and spatialized health risks of heavy metals in $PM_{2.5}$ from biomass fuels burning. *Atmospheric Environment, 284*. Www.elsevier.com/locate/atmosenv. doi:10.1016/j.atmosenv.2022.119178.

Wu, Y., Lin, S., Tian, H., Zhang, K., Wang, Y., Sun, B., Liu, X., Liu, K., Xue, Y., Hao, J., Liu, H., Liu, S., Shao, P., Luo, L., Bai, X., Liu, W., Wu, B., & Zhao, S. (2020). A quantitative assessment of atmospheric emissions and spatial distribution of trace elements from natural sources in China. *Environmental Pollution, 259*. https://www.journals.elsevier.com/environmental-pollution. doi:10.1016/j.envpol.2020.113918.

Wu, Y., Zhang, J., Ni, Z., Liu, S., Jiang, Z., & Huang, X. (2018). Atmospheric deposition of trace elements to Daya Bay, South China Sea: Fluxes and sources. *Marine Pollution Bulletin, 127*, 672–683. Www.elsevier.com/locate/marpolbul. doi:10.1016/j.marpolbul.2017.12.046.

Xia, X., Yang, Z., Cui, Y., Li, Y., Hou, Q., & Yu, T. (2014). Soil heavy metal concentrations and their typical input and output fluxes on the southern Song-nen Plain, Heilongjiang Province, China. *Journal of Geochemical Exploration., 139*, 85–96. doi:10.1016/j.gexplo.2013.06.008.

Xing, J., Song, J., Yuan, H., Wang, Q., Li, X., Li, N., Duan, L., & Qu, B. (2017). Atmospheric wet deposition of dissolved trace elements to Jiaozhou Bay, North China: Fluxes, sources and potential effects on aquatic environments. *Chemosphere, 174*, 428–436. Www.elsevier.com/locate/chemosphere. doi:10.1016/j.chemosphere.2017.02.004.

Xu, X., González, J. E., Shen, S., Miao, S., & Dou, J. (2018). Impacts of urbanization and air pollution on building energy demands—Beijing case study. *Applied Energy, 225*, 98–109. http://www.elsevier.com/inca/publications/store/4/0/5/8/9/1/index.htt. doi:10.1016/j.apenergy.2018.04.120.

Xu, X., Zeng, L., Zhang, Y., & Fan, C. (2016). The pollution status analysis of atmospheric heavy metal elements during winter in Huairou District of Beijing (in Chinese). *Environmental Chemistry, 35*(12), 2460–2468.

Yamamoto, Y., Sanagawa, Y., & Imai, S. (2020). Determination of trace elements in aerosols at a rural mountainous area and a local city of eastern Shikoku region, Japan. *Analytical Sciences, 36*(5), 637–641. doi:10.2116/analsci.20sbn06.

Yan, G., Zhang, P., Wang, C., Song, X., Gao, Y., Zhang, J., … Wang, Y. (2019). Source apportionment and health risk assessment of heavy metals of $PM_{2.5}$ in heating and non-heating period in Zhengzhou. *Huanjing Kexue Xuebao/Acta Scientiae Circumstantiae, 39*(8), 2811–2820. http://www.actasc.cn/. doi:10.13671/j.hjkxxb.2019.0095.

Yang, Z. P., Lu, W. X., Long, Y. Q., & Liu, X. R. (2009). Current situation of pH and wet deposition of heavy metals in precipitation in Changchun city, China (in Chinese). *Jilin Daxue Xuebao (Diqiu Kexue Ban)/Journal of Jilin University (Earth Science Edition), 39*(5), 887–892.

Yi, K., Fan, W., Chen, J., Jiang, S., Huang, S., Peng, L., Zeng, Q., & Luo, S. (2018). Annual input and output fluxes of heavy metals to paddy fields in four types of contaminated areas in Hunan Province, China. *Science of the Total Environment, 634*, 67–76. Www.elsevier.com/locate/scitotenv. doi:10.1016/j.scitotenv.2018.03.294.

Yu, P., Shao, X., Wang, M., Zhu, Z., Tong, Z., Peng, J., Deng, Y., & Huang, Y. (2024). Effects of atmospheric deposition on heavy metal contamination in paddy field systems under different functional areas in ChangZhuTan, Hunan Province, China. *Science of the Total Environment, 933*. https://www.sciencedirect.com/science/journal/00489697. doi:10.1016/j.scitotenv.2024.172953.

Yuan, H., Yang, M., Zhang, Y., & Wang, J. (2019). Assessment of pollution characteristics and health risks of heavy metals in winter atmospheric particulate matters in Xi'an (in Chinese). *Environmental Engineering, 37*(1), 176–181.

Zhang, G. Z., Pan, Y. P., Tian, S. L., Wang, Y. H., Xiong, Q. L., Li, G., Gu, M. N., Lü, X. M., Ni, X., He, Y. X., Huang, W., Liu, B. W., & Wang, Y. S. (2019). Concurrent measurement of wet and bulk deposition of trace metals in Urban Beijing (in Chinese). *Huanjing Kexue/Environmental Science, 40*(6), 2493–2500.

Zhang, J., Liu, X., Wang, J., He, H., Yao, X., & Gao, H. (2023). Atmospheric dry deposition fluxes of trace metals over the Eastern China Marginal Seas: Impact of emission controls. *Science of the Total Environment, 873*. Www.elsevier.com/locate/scitotenv. doi:10.1016/j.scitotenv.2023.162117.

Zhang, Y. (2015). Heavy metals in atmospheric precipitation from a typical seaport in southern China (in Chinese). *Environmental Science & Technology, 38*(3), 131–135.

Zhou, Y. (2018). *Pollution characteristic, source and health risks of heavy metals in urban atmospheric $PM_{2.5}$ (in Chinese, Ph.D thesis)*. East China Normal University.

Zhou, Y., Shen, W., Li, Y., Dou, L., Li, W., Lai, Q., Du, H., Zhong, L., & Liang, T. (2012). A study of prediction and early-warning forecast on geochemical accumulation of soil heavy metals based on flux model in pearl river delta economic zone, China (in Chinese). *Advances in Earth Science, 27*(10), 1115–1125.

Zhu, C., Tian, H., & Hao, J. (2020). Global anthropogenic atmospheric emission inventory of twelve typical hazardous trace elements, 1995–2012. *Atmospheric Environment, 220*. Www.elsevier.com/locate/atmosenv. doi:10.1016/j.atmosenv.2019.117061.

CHAPTER

5

Inputs and outputs of cadmium in agricultural systems

Xinping Yang, Fang-Jie Zhao

College of Resources and Environmental Sciences, Nanjing Agricultural University, Nanjing, China

5.1 Introduction

Cadmium (Cd) is a nonessential metal naturally present in all soils. Cd in surface soils originates from both parent materials and anthropogenic activities (Peng et al., 2019). In uncontaminated soils, Cd concentrations generally increase with the proportion of clay, reflecting its preferential association with fine soil particles (Zhao et al., 2007). Current Cd levels in most soils are higher than pristine (geogenic) background concentrations worldwide, largely because of anthropogenic inputs (Nicholson et al., 2003). The major anthropogenic sources of Cd in agricultural soils include atmospheric deposition, irrigation water, and the application of mineral and organic fertilizers, such as phosphate fertilizers, manure, sewage sludge, and other soil amendments (Six & Smolders, 2014). Cd is removed from arable soils primarily through crop uptake, surface runoff, and leaching (Sterckeman et al., 2018). The Cd mass balance in agricultural soils reflects the complex interplay between these inputs and outputs and is influenced by numerous environmental and management factors. Understanding this balance is essential for predicting long-term changes in soil Cd concentrations and for developing effective policies to regulate Cd emissions and safeguard agricultural sustainability.

5.2 Cadmium inputs in agricultural systems

5.2.1 Atmospheric deposition

Atmospheric deposition of Cd is discussed only briefly here, as it is covered in more detail in Chapter 4. Cd inputs from atmospheric deposition vary widely, from negligible levels to more than 100 g ha^{-1} year^{-1}, depending on proximity to emission sources (Fan et al., 2022). Fluxes reported for different countries are summarized in Table 5.1. For consistency across regions,

Cadmium Contamination in Soils and Plants: Mechanisms and Mitigation.
DOI: https://doi.org/10.1016/B978-0-443-34104-5.00019-0

TABLE 5.1 Cadmium fluxes from atmospheric deposition in different countries and regions.

		Cd fluxes (g ha^{-1} $year^{-1}$)		
Area	**Year**	**Mean**	**Min–Max, *n***	**References**
Australia	2013	0.2	–, –	de Vries and McLaughlin (2013)
China	2008	4.0	0.4–25, $n = 118$	Luo et al. (2009)
China	2019	6.5	0.2–380, $n = 625$	Peng et al. (2019)
Egypt	2014	0.55	0.52–0.57, $n = 2$	Salman et al. (2017)
England and Wales	2000	1.9	0.7–6.1, $n = 68$	Nicholson et al. (2003)
Europe	2014	0.35	0.1–1, $n = 33$	Six and Smolders (2014)
France	2018	0.27	–, $n = 67$	Sterckeman et al. (2018)
Italy	2003	1.53	1.2–2.2, $n = 3$	Morselli et al. (2006)
Japan	2002	2.41	–, $n = 25$	Kikuchi et al. (2007b)
Netherlands	2004	0.5	–, –	Bonten et al. (2008)
New Zealand	2003	0.2	0.09–0.36, $n = 7$	Gray et al. (2003)
Switzerland	2015	0.11	–, $n = 15$	Imseng et al. (2018)

Cd, cadmium.

Chinese data exclude sites that are influenced by industrial or mining pollution, aligning with the European and Japanese datasets shown that represent background conditions. On average the Cd flux from atmospheric deposition in China (6.5 g ha^{-1} $year^{-1}$) is higher than values reported for England and Wales, Italy, and Japan and much higher than those from Australia, France, and the Netherlands.

5.2.2 Irrigation water

Irrigation water can contribute substantial amounts of Cd to agricultural soils (Zhao et al., 2015). Cd concentrations in irrigation water show considerable spatial heterogeneity across major irrigation zones (Table 5.2). For example, irrigation water in the North China Plain typically contains <0.5 μg L^{-1} Cd (Gu et al., 2019), whereas some streams in the Xiangjiang River Basin—impacted by mining and smelting—may contain 20–40 μg L^{-1} Cd (Du et al., 2013), exceeding the Chinese Agricultural Irrigation Water Quality Standard (≤10 μg L^{-1}) by two- to fourfold (Ministry of Ecology and Environment of the People's Republic of China, 2021). Globally, regulatory thresholds for Cd in irrigation water vary considerably: the World Health Organization (2013) recommends 10 μg L^{-1}, the US Environmental Protection Agency (EPA) allows 20 μg L^{-1} (United States Environmental Protection Agency, 2024), India permits up to 50 μg L^{-1} (River Water User Association (India), 2021), whereas the European Union sets a stricter limit of 5 μg L^{-1}, derived from drinking water directives (though not uniformly applied to agriculture) (The European Parliament and of the Council, 2020). This regulatory divergence reflects regional differences in ecological risk management.

TABLE 5.2 Cadmium concentrations in irrigation water and estimated Cd inputs to agricultural soils.

Country	Crop type	Irrigation water consumption (m^3 ha^{-1} $year^{-1}$)	Cd concentration of irrigation water (μg L^{-1})		Cd input (mean or min–max) (g ha^{-1})	Proportion of total Cd input (%)	References
			Mean ± SD	Min–Max, *n*			
China	Paddy rice	–	–	0.09–0.87	0.19–1.12	1.1–4.0	Wang et al. (2021)
	Paddy rice	443–8755	–	0.21–1.77	0.19–15.49	–	Fu et al. (2021), Yang et al. (2022), Yao et al. (2024)
	Dryland	4350	0.21 ± 0.09	–	0.93	7.1	Yao et al. (2024)
	Wheat	2130	2.25	–	4.64–4.79	5.1	Yan et al. (2020)
Spain	Paddy rice	7010	–	nd–0.4	2.80	–	Moreno-Jiménez et al. (2014)
Pakistan	Vegetables	–	140–440	–	–	–	Khan et al. (2013)
Bangladesh	Vegetables	–	0.7–1.0	–	–	–	Ahmed et al. (2018)

Cd, cadmium.

Crop water demand is a key factor affecting Cd input through irrigation. Paddy rice, a water-intensive crop, requires 1200–1600 m^3 ha^{-1}, far exceeding the irrigation demands of upland crops such as wheat (250–600 m^3 ha^{-1}) and maize (300–400 m^3 ha^{-1}) (Ali, 2010). Climatic conditions further accentuate these differences (Table 5.2). For example, the Heihe River Basin in China, which primarily cultivates maize and vegetables, uses 5811 m^3 ha^{-1} $year^{-1}$ of irrigation water (Li et al., 2021). In paddy fields downstream of the Zhangshi irrigation area (Shenyang, China), irrigation contributed 15.5 g ha^{-1} $year^{-1}$ of Cd—making it the dominant input pathway—based on an annual irrigation volume of 8755 m^3 ha^{-1} and Cd concentrations of 0.91–2.72 μg L^{-1} (Fu et al., 2021).

Double-rice cropping systems in southern China require particularly high irrigation volumes, leading to considerable Cd inputs. Based on measured irrigation volumes of ~5000 m^3 ha^{-1} per year (Dai et al., 2021), the input of Cd into the paddy system would amount to 50 and 200 g ha^{-1}, respectively, with Cd concentrations in water of 10 and 40 μg L^{-1} [these concentrations correspond to the irrigation water limit and the highest concentration reported by Du et al. (2013), respectively]. These inputs could increase topsoil Cd concentrations (0–20 cm depth) by 0.02–0.08 mg kg^{-1} within a single year if no losses occurred. In regions where irrigation water is a major Cd source, stricter limits (e.g., <10 μg L^{-1}) may be necessary.

In many water-scarce regions, wastewater is used for irrigation, further elevating Cd risks. For instance, in Pakistan, untreated wastewater with Cd concentrations up to 440 μg L^{-1} was used for irrigation (Khan et al., 2013). In Shenyang, China, long-term irrigation with industrial wastewater (1962–92) raised topsoil Cd concentrations from ~0.10 to 1.75 mg kg^{-1} ($n = 21$) (range 0.42–3.93 mg kg^{-1}) (Li et al., 2009). More recently, Cd levels in irrigation water downstream of the Baoshan River (southern China) ranged from 7.5 to 26.3 μg L^{-1}, with an average concentration of 13.9 μg L^{-1} (Wei et al., 2024), exceeding national standards (<10 μg L^{-1}). Treatment systems combining plant ponds, constructed wetlands, and ecological ditches successfully reduced Cd concentrations in irrigation water to below regulatory limits, significantly lowering Cd accumulation in both soils and rice grain (Wei et al., 2024).

5.2.3 Fertilizers

Nitrogen, potassium, and sulfur fertilizers generally contain negligible amounts of Cd and therefore contribute little to Cd inputs. By contrast, phosphate (P) fertilizers are a major Cd source in agricultural systems because phosphate rock, the raw material for P fertilizer production, often contains trace elements, including Cd, that remain in the final product (Gupta et al., 2014). In China a survey of 171 P fertilizer samples revealed Cd concentrations ranging from trace levels to 27.2 mg kg^{-1} fertilizer, with 68% containing <1 mg kg^{-1} and only one sample exceeding the national limit of 10 mg kg^{-1} fertilizer (National Standardization Administration., 2019; Yu et al., 2018). A broader literature review reported a mean Cd concentration of 0.75 mg kg^{-1} fertilizer, indicating relatively low Cd levels in Chinese P fertilizers (Li et al., 2020). In contrast, European P fertilizers generally contain more Cd. A survey of 196 samples used in 12 EU countries found Cd concentrations of <0.1–120 mg kg^{-1} P_2O_5 (equivalent to <0.7–42 mg kg^{-1} fertilizer), with a mean of 36 mg kg^{-1} P_2O_5 (7.4 mg kg^{-1} fertilizer) (Six & Smolders, 2014). Variation in Cd content reflects differences in the geological origin of phosphate rock. Igneous deposits, such as those from Kola (Russia), South Africa, and Brazil, typically contain <1–4 mg Cd kg^{-1}, whereas sedimentary deposits—common in Morocco, Algeria, Jordan, and the United States—often contain much higher concentrations (up to 150 mg kg^{-1}) (Kouzbour et al., 2019;

TABLE 5.3 Regulatory limits for Cd concentrations in phosphate fertilizers in different countries and regions.

Country	Limit/(mg Cd kg^{-1} P_2O_5 or fertilizer)	References
China	10 mg Cd kg^{-1} fertilizer	National Standardization Administration (2019)
Japan	146	Ulrich (2019)
Kenya	30	
South Africa	20	
New Zealand	122	
Australia	131	
Norway	43	The European Parliament and the Council of the European Union (2019)
Sweden	44	
Austria	20	
Denmark	48	
Finland	22	
Hungary	20	
Slovakia	20	
European Union proposal	60	
United States	10 mg kg^{-1} per 1% P_2O_5	Association of American Plant Food Control Officials (n.d.)
Canada	4 kg Cd ha^{-1}	Canadian Food Inspection Agency (n.d.)

Cd, cadmium.

McLaughlin et al., 2021; Roberts, 2014). Regulatory limits for Cd in P fertilizers vary widely (Table 5.3). The European Union established a limit of 60 mg Cd kg^{-1} P_2O_5 (137 mg Cd kg^{-1} P), following more than two decades of debate (Smolders, 2016). This threshold was based on the "standstill principle" (zero net accumulation in soils under the current Cd input-output scenario) rather than health risk assessment, and several member states argued for stricter national limits tailored to local soil and crop conditions (McLaughlin et al., 2021). As global demand for P fertilizers grows and low-Cd phosphate rock becomes increasingly scarce, risks associated with Cd in fertilizers require careful assessment.

Repeated applications of high-Cd fertilizers can lead to Cd accumulation in soils (Bracher et al., 2021; Hu et al., 2024; Suciu et al., 2022). For example, Gramlich et al. (2017) reported Cd concentrations of 235 mg kg^{-1} P in an NPK compound fertilizer used in cacao systems; application rates of 2.6–5.2 kg P ha^{-1} $year^{-1}$ contributed up to 1.2 g Cd ha^{-1} $year^{-1}$.

Bracher et al. (2021) found Cd concentration in the soil increased from 79.5 ± 0.8 to 86.9 ±1.4 μg kg^{-1} when the past mineral P fertilizer application rates increased from 33% to 167% of the Swiss theoretical amount of fertilizer applied, leading to the increases of Cd concentration in wheat shoots from 47.3 ± 2.8 to 57.9 ±3.3 μg kg^{-1}. In grains, Cd concentrations increased from 30.6 ± 1.3 μg kg^{-1} in 33% treatment to 44.6 ± 1.7 μg kg^{-1} in 167% treatment. Long-term field studies showed that soil Cd concentration increases in proportion to P fertilizer application rates, although some of the added Cd becomes immobilized in nonbioavailable pools (Rocha de Carvalho et al., 2024). Hamon et al. (1998) examined changes in total and phytoavailable Cd in a soil that had received inputs of Cd because of single superphosphate fertilizer used at various rates for 47 years. Accumulation of Cd in the soil was highly correlated with the application rate of P fertilizer. Increases in soil Cd led to an increase in the Cd concentration of wheat that was grown in the soil. However, a significant proportion of the added Cd was found to exist in a nonbioavailable pool in the soil, and Cd was being fixed in this soil at a rate of 1%–1.5% of the total added Cd per year. Though only a small proportion of the total Cd was being immobilized in this soil per year, it is apparent that in the long term this might lead to a significant proportion of the total soil Cd being excluded from the equilibrium reactions between the solid and solution phases, which are important determinants of plant uptake. Table 5.4 shows the concentrations of Cd in fertilizer samples collected from different locations as well as the estimated Cd inputs from fertilizer applications.

Modeling studies suggest that when Cd concentrations in P fertilizers are limited to <20 mg kg^{-1} P_2O_5, Cd inputs to soils from typical fertilizer applications remain <2 g ha^{-1} $year^{-1}$ (Carne et al., 2021). Nevertheless, excessive fertilizer use can lead to significant Cd accumulation. Moreover, heavy P applications may induce Zn deficiency in crops, thereby promoting greater Cd uptake (Roberts, 2014). Optimizing P application rates through soil testing, combined with appropriate timing and placement strategies, can reduce Cd inputs while sustaining crop productivity.

The main raw material for P fertilizer production is phosphate rock, in which phosphorus occurs primarily as hydroxyapatite [$Ca_{10}(PO_4)_6(OH)_2$] or fluorapatite [$Ca_{10}(PO_4)_6F_2$]. These minerals are not readily available to plants and require further industrial processing to make them accessible (Cheira et al., 2019). The production of phosphate fertilizers requires an intermediate step with the production of phosphoric acid. There are two main production processes of phosphoric acid from phosphate ores: the thermal process and the wet process (Da Silva & Kulay, 2005). The thermal process is much more expensive and requires about twice as much utility (such as steam and water), labor, and overhead rates as the wet process to produce the same amount of P_2O_5. Therefore the wet process remains the most widely used method, accounting for about 90% of global phosphoric acid production (Kouzbour et al., 2019). The wet process involves the digestion of the phosphate rocks using strong mineral acid. Wet-process phosphoric acid contains impurities, including Cd, that are difficult to remove. Decadmiation, that is, removal of Cd, in wet phosphoric acid leads to a 2%–15% increase in fertilizer price (Oosterhuis et al., 2000). For example, using resin to reduce Cd concentration in 1000 m^3 day^{-1} of 28% P_2O_5 wet phosphoric acid from 15 mg L^{-1} to <2 mg L^{-1} is estimated to increase the P_2O_5 price by approximately 13% (Kouzbour et al., 2019). Decadmiation from industrial H_3PO_4 remains a technical challenge because of the strong complexation of the acid for Cd, high impurity content, and the lack of efficient scalable processes (Merroune et al., 2024). During the last decades, different strategies for the purification of wet process phosphoric acid have been developed, including methods such as precipitation, flotation, solvent extraction, separation

TABLE 5.4 Cadmium concentrations in phosphate fertilizers and estimated Cd inputs under typical application rates.

			Cd concentration (mg Cd kg^{-1} P_2O_5 or fertilizer)				
Country	**Crop**	**Fertilizer type**	**Mean ± SD**	**Min–Max, *n***	**Application rate of P fertilizer (kg P_2O_5 ha^{-1} $year^{-1}$)**	**Cd input (mean/min–max) (g ha^{-1} $year^{-1}$)**	**References**
Iran	Large regional scale	NPK	2 ± 0.07 mg Cd kg^{-1} fertilizer	0.7–3.8, $n = 18$	–	–	Latifi and Jalali (2018)
		TSP	11.5 ± 8.7 mg Cd kg^{-1} fertilizer	2.7–23.7, $n = 8$	–	–	
		DAP	3.8 ± 2.9 mg Cd kg^{-1} fertilizer	1.3–9.3, $n = 7$	–	–	
India	Large regional scale	DAP	13.29 ± 4.76	$n = 3$	–	–	Lugon-Moulin et al. (2006)
Turkey	Large regional scale	MAP	6.27 ± 10.86	$n = 3$	–	–	
Greece	Large regional scale	NPK	28.49 ± 0.67	$n = 3$	–	–	
EU-18 countries	Large regional scale	Phosphate fertilizer	36	0.1–120, $n = 196$,	22	0.80	Six and Smolders (2014)
China	Paddy	Compound fertilizer	0.16 ± 0.07 mg Cd kg^{-1} fertilizer	–	103	0.14	Fu et al. (2021)
	Large regional scale	Phosphate fertilizer (83% DAP and MAP)	0.75 mg Cd kg^{-1} fertilizer	nd–27.2, $n = 1222$	12–99	0.03/0.009–0.05	Li et al. (2020)

Cd, cadmium; *DAP*, diammonium phosphate; *MAP*, monoammonium phosphate; *nd*, not detectable; *NPK*, compound nitrogen, phosphorus, and potassium fertilizer; *TSP*, triple superphosphate.

using membrane technology, crystallization, ion exchange, and other adsorption-like processes. These techniques present various drawbacks, such as high energy input, environmental pollution because of byproducts, high cost of reactants, or the need to use expensive organic solvents and resins (Kouzbour et al., 2019; Omri & Batis, 2013). Among various methods, solvent extraction is the most common method implemented at the industrial scale for the extraction and separation of Cd from phosphoric acid. Solvent extraction processes incur high operating costs, approximately 75%–100% of those for wet-process acid production, primarily because of the expense of solvents and resins (Davister, 1996).

Calcination is the only known method for the removal of Cd from phosphates in the rock state. This technique consisted of heating at 750°C–950°C under controlled pressure to volatilize Cd (Cichy et al., 2014). Calcination can reduce the phosphorite activity and lead to the emission of Cd to the atmosphere. In addition, it is a costly method. A rough estimate of the extra cost for calcination is about US $6.7 per t of ore or US $20 per t of P_2O_5 (Kouzbour et al., 2019). Industrially, reducing Cd content in phosphate fertilizers—whether by lowering Cd concentrations in the raw material (phosphate rock) or in the phosphoric acid intermediate—remains both a technical and economic challenge.

5.2.4 Manures

Animal manure is a valuable source of plant nutrients and organic matter and has long been applied to soils as part of nutrient recycling practices. Most Cd in manure originates from animal feed (Fan et al., 2022). At the European scale, Six and Smolders (2014) estimated the Cd inputs from manures by considering only Cd derived from animal feed. Based on this data a net Cd input of about 0.006–0.014 g Cd ha^{-1} $year^{-1}$ was estimated. In China, manure Cd concentrations reported between 2000 and 2014 were considerably higher than those in Europe (Leclerc & Laurent, 2017). For example, Wang et al. (2013) estimated Cd inputs of 10 g ha^{-1} $year^{-1}$ to soils receiving pig manure at a rate equivalent to 250 kg N ha^{-1} $year^{-1}$—seven times higher than values reported for the United Kingdom at the same application rate (Nicholson et al., 2003). Luo et al. (2009) concluded that livestock manure accounted for ~55% of total Cd inputs to Chinese farmland. Elevated Cd levels were attributed to heavy use of feed supplements (particularly Zn salts containing Cd) in intensive animal production (Wang et al., 2013). In response, China introduced strict regulations on Cd in livestock feed and manure compost (≤3 mg kg^{-1} on a manure dry matter basis) (National Standardization Administration., 2019). A survey of >200 manure compost samples reported mean and median Cd concentrations of 0.67 and 0.14 mg kg^{-1}, respectively, comparable to European values (Yang et al., 2017). Assuming a typical manure compost application rate of 3000 kg ha^{-1} $year^{-1}$ the median Cd input would be 0.42 g ha^{-1} $year^{-1}$. However, Cd inputs ultimately depend on both manure Cd concentration and application rate. Because manure contains lower nutrient concentrations than mineral fertilizers, larger quantities are often applied, which can increase the Cd loading (Mu et al., 2020). Regional differences in application rates further influence Cd inputs. In Chongqing (southwest China), manure applied to paddy fields (3000 kg ha^{-1} $year^{-1}$, Cd = 0.11 mg kg^{-1}) resulted in a Cd input of 0.33 g ha^{-1} $year^{-1}$. In dryland fields, application rates averaged ~10,500 kg ha^{-1} $year^{-1}$—about three times higher, producing Cd inputs of 1.16 g ha^{-1} $year^{-1}$ (Yao et al., 2024). Table 5.5 shows the concentrations of Cd in animal manures from surveys conducted in different countries.

Whether manure application increases soil Cd depends on the total Cd added. Long-term field studies in China provide clear evidence of Cd accumulation. In Jilin, 17 years of pig manure

TABLE 5.5 Cadmium concentrations in livestock manure and corresponding Cd inputs to croplands.

		Cd concentration (mg kg^{-1}, dry weight)			Cd input (g ha^{-1} year^{-1})		
Country	**Manure type**	**Mean**	**Min–Max, *n***	**Dosage (kg ha^{-1})**	**Mean ± SD**	**Min–Max, n**	**References**
China	Manure	0.11	–	3000 10,500	0.33 ± 1.16	–	Yao et al. (2024)
	Pig manure	0.54	–	10,935–43,740	–	5.9–23.6	Huang et al. (2018)
	Chicken manure	0.41	–	5610–22,440	–	2.3–9.2	
	Manure compost	0.67	0.012–8.72, *n* = 212	3000	2.01	0.036–26.16	Yang et al. (2017)
Europe	Poultry manure	–	–	–	2.10 ± 5.65	0.01–18.80, *n* = 11	Fan et al. (2022)
Austria	Cattle manure	0.54	0.13–1.0	–	–	–	Leclerc and Laurent (2017)
Germany	Pig manure	0.38	0.28–0.48	–	–	–	
United Kingdom	Poultry manure	0.67	0.20–1.16	–	–	–	
Portugal	Manure	0.12	0.009–0.24, *n* = 3	–	–	–	Felipe-Sotelo et al. (2025)
England and Wales	Manure	–	0.2–1.3	270–8450	–	0.11–1.69	Nicholson et al. (2003)

Cd, cadmium.

applications (equivalent to 115–150 kg N ha^{-1} $year^{-1}$) increased soil Cd from 0.087 to 1.47–1.64 mg kg^{-1} (Wu et al., 2012). In the same study, manure application raised DTPA-extractable Cd by ~300 μg kg^{-1}, whereas mineral fertilizer application alone caused little change. The high increases were attributed to Cd contamination of animal feed. Similarly, in Hunan, chicken manure applied at ~5.3 t ha^{-1} $year^{-1}$ for 32 years raised soil Cd by 0.35 mg kg^{-1}, compared with only 0.05 mg kg^{-1} with chemical fertilizer in a double-rice system (Xu et al., 2018).

5.2.5 Sewage sludge

Sewage sludge, or biosolids, is the semisolid byproduct of municipal wastewater treatment. It contains substantial amounts of nutrients, organic matter, and beneficial microorganisms but may also introduce pollutants, including heavy metals (Kumar & Chopra, 2016). When properly managed, sewage sludge can improve soil fertility and crop productivity; however, its application requires careful monitoring to minimize risks of Cd accumulation (Zhao et al., 2015). Cd concentrations in sewage sludge vary widely across regions. In the Czech Republic, sludge contained about 10 times more Cd than organic fertilizers (Čásová et al., 2009). In Brazil, municipal biosolids averaged 8.2 mg kg^{-1} Cd (Sampaio et al., 2016). In South Africa, analysis of 18 wastewater treatment plants showed Cd concentrations below 8 mg kg^{-1} in most sludge samples, with only one exceeding the pollutant threshold of 40 mg kg^{-1} for agricultural use (Badza et al., 2020; Sedlak, 2019). The use of municipal liquid sludge in agriculture has been widely practiced, especially in the water-scarce countries like South Africa (Badza et al., 2020; Sedlak, 2019). Regulatory limits also vary. The EU Sewage Sludge Directive (86/278/EEC, 1986) sets permissible Cd concentrations at 20–40 mg kg^{-1}, though several member states enforce stricter thresholds (e.g., 10 mg kg^{-1} in Germany, 20 mg kg^{-1} in France and Italy, and 0.7–10 mg kg^{-1} in Austria, depending on federal state and quality class) (Hudcová et al., 2019). According to the US EPA the permissible limit for Cd in biosolids for agricultural soils is 85 mg kg^{-1} (Code of Federal Regulations, 2025). The composition of municipal wastewater and sewage sludge reflects the use and emissions of elements and contaminants by society. In Sweden, biosolid Cd concentrations have declined by ~90% since the 1970s because of environmental programs and statutory limits on metals in sludge and soil (Kirchmann et al., 2017), whereas Switzerland has banned land application of sewage sludge since 2006 (Bigalke et al., 2017).

The extent of Cd accumulation in soils depends on both sludge Cd content and application rate. A 3-year wheat field study found that sewage sludge applications of 10, 25, and 40 t ha^{-1} increased soil Cd by 6%, 22%, and 42%, respectively (Cocarta et al., 2017). In a 4-year trial in a southern Chinese paddy, repeated sludge applications (Cd = 2.46 mg kg^{-1}) (below China's limit of 5 mg kg^{-1}) raised topsoil Cd concentrations by 0.13–0.35 mg kg^{-1} (Huang et al., 2018). Even when sludge meets regulatory standards, Cd accumulation may occur if inputs exceed outputs.

5.2.6 Other agricultural additives

Lime is widely applied to neutralize soil acidity. Common liming materials include calcium carbonate ($CaCO_3$), calcium hydroxide ($Ca(OH)_2$), calcium oxide (CaO), and dolomite. Large quantities may be required; for example, raising soil pH by one unit in a medium-textured soil typically requires ~7 t $CaCO_3$ ha^{-1}. Thereafter, smaller maintenance doses (1–2 t ha^{-1}) are applied to sustain optimal pH. Liming generally reduces Cd bioavailability and crop uptake

TABLE 5.6 Cadmium concentrations in lime and pesticides/PGR and estimated Cd inputs under typical application rates.

Country (year)	Additive type	Cd concentration (mean/range), n ($mg\ kg^{-1}$)	Dosage ($kg\ ha^{-1}\ year^{-1}$)	Cd input ($g\ ha^{-1}\ year^{-1}$)	References
China (2018)	Pesticide	0.004–0.21, $n = 8$	0.03–4.5	0.0002	Yan et al. (2020)
	Plant growth regulator	0.21	0.38	0.00008	
China (2016)	Lime[a]	1.05	4.2	0.004	Chen et al. (2018)
England and Wales (2000)	Lime	0.3	–	0.28	Nicholson et al. (2003)
Europe (2014)	Lime	0.35	250	0.0875	Six and Smolders (2014)
France (2018)	Lime	0.35	54	0.0189	Sterckeman et al. (2018)
New Zealand (2021)	Lime	0.02–0.5, $n = 12$	280	0.0644	Gray and Cavanagh (2023)

[a] The dosage of lime is expressed in terms of calcium oxide (CaO) as kg CaO ha–[1] year–[1]. *Cd*, cadmium.

(Chen et al., 2018). However, lime itself can contain Cd, particularly when derived as a byproduct of the processing of Zn and Pb ores (Chaney, 2012). Reported Cd concentrations in lime are generally low. For example, Kabata-Pendias (2001) showed that the concentrations of Cd in agricultural limestone varied in the range of 0.04–0.1 mg kg^{-1}. Six and Smolders (2014) estimated that applying lime containing 0.35 mg Cd kg^{-1} CaO (average value in Swedish lime) at 250 kg CaO ha^{-1} would add only 0.09 g Cd ha^{-1} $year^{-1}$—less than inputs from atmospheric deposition or fertilizers. A recent New Zealand survey found lime Cd concentrations ranging from 0.02 to 0.50 mg kg^{-1} (median 0.15, mean 0.23 mg kg^{-1}) across 12 quarries (Gray & Cavanagh, 2023). Based on a lime application rate of 500 kg ha^{-1} $year^{-1}$ (with 100% $CaCO_3$), Cd inputs would range from 0.01 to 0.25 g ha^{-1} $year^{-1}$, again lower than inputs from P fertilizers (Gray & Cavanagh, 2023). Table 5.6 provides representative examples of Cd inputs from lime.

Other agricultural inputs, such as pesticides and plant growth regulators, are applied at much smaller doses and typically contain negligible Cd (Table 5.6). For instance, reported Cd inputs from pesticides and growth regulators are in the order of 10^{-4} g ha^{-1} $year^{-1}$ (Yan et al., 2020), and thus their contribution to overall Cd loading is minimal.

5.3 Cadmium outputs in agricultural systems

5.3.1 Crop removal

Crop uptake and harvesting represent a major pathway for Cd export from soils. The amount of Cd removed depends on multiple factors, including soil Cd concentration and bioavailability, plant species, growth stage, and harvested plant parts. Soil properties, such as pH, organic matter content, clay content, and redox status, strongly influence Cd availability (see Chapters 6 and 7). Plant species vary widely in their Cd accumulation capacity. Kuboi et al. (1986) compared Cd uptake by 34 plant species belonging to 9 families in a sandy soil amended with a range of Cd concentrations. Based on the relationship between plant and soil Cd concentrations, they grouped the plant species tested into three categories: (1) low Cd accumulation, Leguminosae (including soybean, pea, and kidney bean); (2) moderate Cd accumulation, Gramineae (rice, wheat, barley, maize, and ryegrass), Liliaceae (onion and leek), Cucurbitaceae (cucumber, pumpkin, and watermelon), and Umbelliferae (carrot and parsley); and (3) high Cd accumulation, Chenopodiaceae (sugar beet and Swiss chard), Cruciferae (turnip, radish, cabbage, and oilseed rape), Solanaceae (pepper, eggplant, and tomato), and Compositae (lettuce). In addition, within Umbelliferae, Cucurbitaceae, and Cruciferae, the steepest increases in Cd uptake with rising soil Cd were shown, whereas plant species in other six families displayed weaker responses.

The total Cd output via crop removal is determined by both biomass and Cd concentration in the harvested material. Straw removal is widely practiced in moderately contaminated soils to accelerate Cd removal. In a typical double-rice cropping system in southern China, assuming biomass yields of 12 t ha^{-1} for both grain and straw and reported Cd concentrations ranging from 0.05 to 0.5 mg kg^{-1} in grain and 0.4 to 4 mg kg^{-1} in straw, the total annual crop removal of Cd varies from 5 to 55 g ha^{-1}, depending on soil contamination levels (Zhao et al., 2015). Using a metaanalysis of field data, Zhao et al. (2021) estimated the potential of rice straw removal for reducing Cd concentration in soil. The regression equation obtained and average straw biomass were used to estimate the amounts of Cd removal in straw from soils with 0.5–5 mg Cd kg^{-1}. Straw Cd removal represents 0.23%–0.37% (mean 0.27%) of the total Cd in the plow layer. Including Cd removal in rice grain increases the removal rate to approximately 0.30%. In acid soils with a high Cd bioavailability, crop Cd removal could be up to 1%.

Table 5.7 summarizes the total output of Cd from grain harvesting and straw removal. There are differences among different countries because of factors such as crop species and soil conditions. In rice–wheat rotation systems, such as the Yangtze River delta region of China, the average Cd removal through wheat grain was 0.21 g ha^{-1} $year^{-1}$ (range 0.12–0.52 g ha^{-1} $year^{-1}$), whereas the average Cd removal through rice grain was 0.45 g ha^{-1} $year^{-1}$ (range 0.05–1.25 g ha^{-1} $year^{-1}$) (Hou et al., 2014). In wheat–maize rotations, outputs were 0.25 g ha^{-1} $year^{-1}$ (wheat grain) and 0.16 g ha^{-1} $year^{-1}$ (maize grain) (Yan et al., 2020). In rice systems, straw typically contains four to eight times more Cd than grain (Zhao et al., 2015), highlighting the importance of residue management. Integrating winter oilseed rape with summer rice has been proposed as a more effective removal strategy than rice monoculture because of the higher Cd uptake of oilseed rape (Cao et al., 2019; Lu et al., 2020).

5.3.2 Surface runoff

Rainfall is the primary driver of surface runoff in agricultural systems. When the precipitation intensity exceeds soil infiltration capacity, excess water flows overland, transporting both soluble and particle-bound contaminants, including Cd, to downstream ecosystems. Losses of Cd through runoff are influenced by rainfall intensity, Cd concentrations in soil and rainwater, and a range of site-specific factors such as soil pH, redox potential, moisture content, organic matter, slope, vegetation cover, and hydrogeochemical conditions (Elrashidi et al., 2007). Cd mobility is particularly associated with water-soluble and exchangeable fractions in soils (Fan et al., 2022; Liu et al., 2014).

TABLE 5.7 Cadmium removal via crop harvest in different cropping systems.

		Cd output flux (mean or range, n) (g ha^{-1} year^{-1})				
		Crop				
Country	**Area**	**Rice grain**	**Wheat grain**	**Maize grain**	**Straw**	**References**
China	Sichuan	0.62–3.89, $n = 276$	–	–	–	Yang et al. (2008)
	Yangtze River delta	0.05–1.25, $n = 141$	0.12–0.52, $n = 126$	–	–	Hou et al. (2014)
	Hebei	–	0.25	0.16	0.24 (wheat), 0.25 (maize)	Yan et al. (2020)
	Shenyang	0.25–5.63, $n = 50$	–	–	1.66–27.2, $n = 50$	Fu et al. (2021)
	Guangxi	0.52–3.64, $n = 6591$	–	–	–	Li et al. (2022)
Europe	Large regional scale	–	0.20	–	–	Six and Smolders (2014)
France	Versailles	–	0.58	–	–	Azimi et al. (2004)
	Large regional scale	0.87			–	Sterckeman et al. (2018)
Japan	Tokyo (2001)	0.00–1.60, $n = 6$	–	–	0.007–4.74, $n = 6$	Kikuchi et al. (2007b)
	Tokyo (2002)	0.00–0.32, $n = 15$	–	–	0.00–1.64, $n = 15$	

Cd, cadmium.

Field experiments on runoff are challenging because of their dependence on natural rainfall patterns and long monitoring periods. Rainfall simulation experiments provide an alternative by allowing controlled investigation of runoff responses under varying precipitation intensities (Liu et al., 2014). For example, in soils containing 0.10–3.88 mg Cd kg^{-1}, simulated rainfall of 20–80 mm produced runoff with Cd concentrations of 0.15 ± 0.01–0.41 ± 0.13 μg L^{-1}, demonstrating clear dilution effects under high-intensity events (Yan et al., 2020).

Cd fluxes via surface runoff vary regionally. Reported values include 0.71 g ha^{-1} $year^{-1}$ in the Songnen Plain (Heilongjiang province) (Xia et al., 2014), 2.16 g ha^{-1} $year^{-1}$ in Shenyang (Liaoning province) (Fu et al., 2021), and 0.78–3.59 g ha^{-1} $year^{-1}$ in the Changsha–Zhuzhou–Xiangtan urban agglomeration (Hunan province) (Xie et al., 2020). Differences among studies partly reflect calculation methods. For instance, Xia et al. (2014) estimated Cd fluxes by multiplying Cd concentrations in streams by the water flux given in the local hydrology yearbook, whereas Fu et al. (2021) assumed surface runoff accounted for 30% of total water input (rainfall + irrigation) in rice paddies. Xie et al. (2020) directly measured runoff flow rates and Cd concentrations in field plots, reporting highest fluxes in industrial/mining areas (3.59 g ha^{-1} $year^{-1}$), intermediate fluxes in suburban areas (1.61 g ha^{-1} $year^{-1}$), and lowest in rural areas (0.78 g ha^{-1} $year^{-1}$), consistent with corresponding soil Cd concentrations (2.20, 0.57, and 0.44 mg kg^{-1}, respectively). These findings highlight the importance of local soil, hydrology, and land-use conditions in determining Cd losses via runoff.

5.3.3 Leaching

Leaching is an important Cd output pathway in agricultural soils, particularly under conditions that enhance Cd mobility. Cd can be transported downward either in dissolved form, which readily moves through the vadose zone into groundwater, or as colloidal suspensions (organo-mineral complexes). Cd is strongly adsorbed by soil and only a small fraction resides in soil solution. Thus vertical Cd losses are usually minor unless soils are acidic and low in organic matter. These interfacial reactions between Cd and soil components are modulated by biotic factors (e.g., microbial activity and root exudates), physical parameters (e.g., texture and porosity), redox oscillations, and adsorption–desorption kinetics (see Chapters 6 and 7).

The annual Cd loss through leaching is typically <0.1% of total topsoil Cd stocks (Imseng et al., 2018). Detecting such small fluxes directly is difficult, as they often fall below the resolution of mass balance studies spanning only a few decades. Leaching does not sufficiently change the isotope ratio of the naturally present stable isotopes to be useful to infer long-term leaching fluxes (Imseng et al., 2018). The most straightforward method for determining Cd leaching relies on measuring Cd concentrations in the leachates through continuous collection of leachates to capture the complete annual leaching flux (McLaughlin et al., 2021). The accurate estimation of the minute quantities of Cd that leaves soil by vertical leaching is difficult for technical reasons. Suction cups are commonly used for point sampling of soil porewater, which, when combined with drainage estimates, can provide leaching flux estimates (Imseng et al., 2018; Six & Smolders, 2014). Alternatively, modeling approaches predict porewater Cd concentrations from soil properties and multiply them by drainage volumes to estimate leaching fluxes (Six & Smolders, 2014). Both methods carry significant uncertainties: modeled data may overestimate fluxes if desorption is slower than water movement or underestimate fluxes if colloid-bound Cd is excluded by filtration. Porewater Cd concentration derived from modeling may overestimate leaching when nonequilibrium occurs; desorption of Cd from soil may be slower than the water transport, and preferential water flow may occur. In contrast, modeled porewater Cd data may also underestimate Cd leaching when porewater colloids are filtered prior to analysis, thereby underestimating the colloidal transport of Cd. However, measuring leaching flux accurately remains challenging. Most existing studies have been conducted under controlled laboratory conditions, and there are only a few measured field data of Cd leaching. Furthermore, significant spatial and temporal variations complicate the assessment. Both measured and modeled Cd leaching fluxes are characterized by a considerable degree of uncertainty.

Paddy soils show strong seasonal variation in Cd leaching because of shifts in redox status. Under flooded, anoxic conditions, microbial sulfate reduction generates sulfide, which precipitates Cd as CdS (see Chapter 7), thereby reducing Cd solubility (Huang et al., 2023). When fields are drained and soil conditions become oxic, Cd solubility increases and the leaching risk rises. Soil pH is another key factor affecting leaching of Cd, as it affects Cd solubility in the soil. In paddy soils with low pH and low soil organic matter content, leaching of Cd is elevated. In neutral and alkaline soils, Cd leaching was found to be negligible (<0.1 g Cd ha^{-1} $year^{-1}$) (Mao et al., 2021; McLaughlin et al., 2021).

In upland soils, rainfall and irrigation are the critical drivers of Cd leaching. Cd in upland soils migrates primarily through rainfall-induced erosion and irrigation water infiltration. Key soil properties, including soil texture, structure, pH, and organic matter content, collectively govern both the rate and extent of Cd leaching. Notably, sandy soils facilitate rapid Cd transport because of their high hydraulic conductivity and weak Cd sorption, whereas clay-rich soils exhibit strong Cd adsorption capacities that substantially retard leaching processes.

Table 5.8 summarizes Cd leaching fluxes in paddy and upland fields. Reported Cd leaching fluxes vary widely across regions. Field measurements in Jiangsu and Zhejiang provinces, China, showed a wide range of Cd leaching of 0.13–8.49 and 0.09–3.68 g ha^{-1} $year^{-1}$, respectively, accounting for 9%–96% and 10%–77% of total Cd outputs (Hou et al., 2014). In contrast, fluxes in Hunan (Yi et al., 2018) and Sichuan (Yang et al., 2008), China, were much lower (≤ 0.21 and ≤ 3.31 g ha^{-1} $year^{-1}$, respectively). The small leaching fluxes reported in the study in Hunan were attributed to a slower soil percolation rate. In regions featuring mixed cultivation systems, where paddy fields and uplands coexist, leaching fluxes are notably elevated. This pattern indicates that soil type heterogeneity may amplify Cd leaching complexity through interactions among multiple drivers, including soil moisture dynamics under alternating wet–dry cycles, irrigation regime variability, and spatial heterogeneity of soil texture. Overall, Cd leaching remains one of the most uncertain components of Cd mass balance because of strong spatial and temporal variability and the technical difficulty of field measurements.

5.4 Mass balance of cadmium in agricultural soil

The mass balance of Cd in agricultural soils reflects the dynamic interplay between inputs (e.g., atmospheric deposition, irrigation water, fertilizers, manures, sewage sludge, and other amendments) and outputs (e.g., crop removal, surface runoff, and leaching). Understanding this balance is critical for assessing whether soils act as net sinks or sources of Cd and for predicting long-term changes in soil Cd concentrations. At the field scale, annual Cd inputs often exceed outputs, leading to gradual accumulation in soils. However, the magnitude and direction of Cd balance vary substantially across regions because of differences in input sources, soil properties, cropping systems, and climate. A number of countries, including France (Sterckeman et al., 2019)

TABLE 5.8 Cadmium leaching fluxes in agricultural soils across different regions.

			Cd leaching flux (g ha^{-1} $year^{-1}$)			
Country	Area	Soil conditions	Mean	Min–Max, n	Proportion of estimated total Cd output (%)	References
China	Sichuan	Predominantly paddy soil	1.66	0.00–3.31, $n = 121$	0–70	Yang et al. (2008)
	Jiangsu	Paddy soil and dry soil	4.31	0.13–8.49, $n = 31$	8.86–95.8	Hou et al. (2014)
	Shanghai	Paddy soil and dry soil	0.73	$n = 2$	64.6	
	Zhejiang	Paddy soil and dry soil	1.89	0.09–3.68, $n = 18$	10.2–77.0	
	Hunan	Paddy soil	0.11	0.00–0.21, $n = 40$	0.01–0.43	Yi et al. (2018)
Europe	Large regional scale	Dry soil	2.56	$n = 151$	–	Six and Smolders (2014)
New Zealand	Large regional scale	Dry soil	0.33	–	–	Gray et al. (2024)
Japan	Tokyo	Paddy soil	4.77	3.86–5.67	46.4–65.5	Kikuchi et al. (2007b)

Cd, cadmium.

and Switzerland (Wiggenhauser et al., 2024), have instituted comprehensive input and output inventories to evaluate the accumulation and removal of Cd in agricultural soils. Some studies have also predicted long-term changes in soil Cd concentrations based on mass balance (de Vries & McLaughlin, 2013). Table 5.9 lists representative literature from different countries or regions, including specific data on various input and output pathways.

The relative importance of Cd input sources exhibits substantial regional variation (Table 5.9). In European and African agricultural systems, fertilizers appear to be the dominant pathway for Cd input into soils. Data indicates fertilizer-derived Cd accounts for 75%, 88%, and 96% of the total Cd input in France, Switzerland, and selected African nations, respectively. By contrast, atmospheric deposition generally dominates Cd inputs in Chinese agricultural regions. Provincial-level studies reveal atmospheric Cd deposition rates exceed European benchmarks by approximately tenfold (Fu et al., 2021; Xie et al., 2020), a disparity attributable to different stages of development, industrial configurations, and environmental regulations between these regions. Notably, irrigation water constitutes a secondary but important pathway of Cd input in some areas in China. Two primary factors drive this phenomenon: the prevalent use of flood irrigation systems with suboptimal water quantity, particularly in rice cultivation systems, and contamination of water resources in localized areas with industrial effluent. Cross-regional comparisons show that aggregate Cd inputs in European countries remain markedly lower than those observed in other regions (Table 5.9), highlighting fundamental differences in pollution prevention frameworks.

Significant regional variations exist in Cd outputs. European and Oceanian agricultural systems exhibit only small contributions from irrigation water inputs and surface runoff outputs. Leaching dominates Cd export pathways in Australia (de Vries & McLaughlin, 2013), most European nations (Sterckeman et al., 2019), and Japan (Kikuchi et al., 2007a,2007b). Nevertheless, leaching fluxes constitute the most uncertain component of Cd mass balance models because of two primary limitations: heavy reliance on model extrapolations rather than in situ field measurements, and exclusion of colloidal/particulate-bound Cd in existing calculations. Empirical studies reveal that modeled leaching rates may exceed measured values by up to two orders of magnitude , particularly in neutral-pH soils (McLaughlin et al., 2021). Studies in Huanan and Liaoning provinces of China and in Egypt showed that crop harvesting accounts for >90% of the total Cd removal.

Some studies in Switzerland, Egypt, Australia, and most provinces in China show a net positive balance of Cd flux, suggesting that soil Cd concentration is likely to increase gradually, whereas a negative balance was reported in some studies in France, Switzerland, and Japan.

Mass balance studies also highlight the importance of management practices. For example, excessive irrigation with wastewater or long-term use of high-Cd phosphate fertilizers can result in substantial Cd accumulation, whereas adoption of low-Cd fertilizers, stricter standards for irrigation water, optimized residue management, and improved nutrient recycling can mitigate Cd loading. Priority mitigation efforts should target regions with positive Cd net fluxes through customized prevention strategies based on comprehensive input–output inventories. For China specifically , it is imperative to enhance regulatory oversight of industrial entities to curtail Cd emissions (see Chapter 4). Moreover, the implementation of stringent control measures for Cd in irrigation water is paramount in contaminated regions. With respect to output pathways the removal and safe disposal of crop residues can have a significant impact on soil Cd levels. In some studies, whether crop residues are removed determines whether the Cd input–output balance is positive or negative (Xie et al., 2020; Yan et al., 2020; Yi et al., 2018).

TABLE 5.9 Soil Cd mass balances in different studies.

Country (year)		Inputs ($g\ ha^{-1}\ year^{-1}$)					Outputs ($g\ ha^{-1}\ year^{-1}$)				Net balance	Change in soil Cd[a]	
		Atmospheric deposition	Fertilization	Irrigation	Manure, sludge, and compost	Total input	Crop offtake	Runoff	Leaching	Total output	($g\ ha^{-1}\ year^{-1}$)	(%)	References
Australia (2013)		0.2	1.2	–	5	6.4	0.2	–	1	1.2	5.2	3.92	de Vries and McLaughlin (2013)
China	Nationwide (2008)	4	0.9	6.4	–	11.3	1.5	–	–	1.5	9.8	2.20	Luo et al. (2009)
	Nationwide (2019)	6.5	0.1	1.3	1.7	9.6	0.34			0.34	9.26	2.01	Peng et al. (2019)
	Hunan (2016-2019)	7.12–10.4	0.01–2.46	0.96–4.58	–	8.18–17.4	9.84–25.1	0.04–0.78	0.02	10.6–25.2	−2.4–−10.7	−0.05–−4.27	Xie et al. (2020), Yang et al. (2021), Yi et al. (2018)
	Liaoning (2016)	11.6	0.14	15.5	0.02	27.2	18.7	2.16	–	20.8	6.39	0.24	Fu et al. (2021)
	Yangtze River Delta (2014)	2.46	0.09	1.30	–	3.85	0.34	–	0.59	0.93	2.92	0.55	Shao et al. (2016)

(*continued on next page*)

Egypt (2014)	0.55	14.9	0.01	–	15.5	10.1	C.01	–	10.1	5.35	0.11	Salman et al. (2017)
Europe (2014)	0.35	0.76	–	0.15	1.26	0.2	–	2.56	2.76	−1.5	−0.22	Six and Smolders (2014)
France (2018)	0.2	1.42	–	0.27	1.89	0.93	–	2.83	3.76	−1.87	−0.22	Sterckeman et al. (2019)
France (2021)	0.27	1.05		1.73	3.05	0.32		0.06	0.37	2.67	0.30	Zhong et al. (2022)
Japan (2003)	2.41	0.56	0.91		3.88	2.03		3.86	5.89	−2.01	−0.07	Kikuchi et al. (2007a, 2007b)
Switzerland (2015)	0.11	0.60		0.45	1.16	1.72		0.35	2.07	−0.91	−0.06	Imseng et al. (2018)
Switzerland (2021)	–	1.69	–	0.22	1.91	0.18	–	0.12	0.3	1.61	0.09	Wiggenhauser et al. (2024)

[a] The change in soil Cd concentration was calculated based on the net mass balance, regional soil Cd concentration, and topsoil mass (2.6×10^6 kg ha^{-1}). This topsoil mass was estimated assuming that Cd accumulation is concentrated within the top 20 cm of soil and adopting a soil bulk density of 1.30 g cm^{-3}, yielding an approximate value of 2.6×10^6 kg ha^{-1}. *Cd*, cadmium.

Despite progress, large uncertainties remain because of methodological challenges in quantifying certain fluxes—particularly leaching and runoff—and the high spatial variability of Cd concentrations in soils, inputs, and outputs. Long-term monitoring and standardized methodologies are essential for improving mass balance estimates and for developing models that can predict future soil Cd dynamics under different agricultural and environmental scenarios.

5.5 Conclusions

Comprehensive assessment of Cd inputs and outputs is essential for understanding the long-term dynamics of Cd in agricultural soils and for guiding management strategies to minimize risks to human health and the environment. Cd inputs to croplands arise primarily from atmospheric deposition, irrigation water, phosphate fertilizers, manure, sewage sludge, and, to a lesser extent, lime and other soil amendments. Outputs occur mainly through crop removal, surface runoff, and leaching.

The relative importance of these pathways varies regionally. In China, atmospheric deposition and irrigation water are dominant inputs, whereas in Europe, phosphate fertilizers are the major source. In both regions, crop removal accounts for only a small fraction of Cd losses, whereas leaching and runoff play variable but often significant roles. Mass balance studies generally indicate positive Cd budgets, suggesting ongoing accumulation in many agricultural soils.

Effective mitigation requires region-specific approaches. Key strategies include enforcing stricter limits on Cd concentrations in phosphate fertilizers, irrigation water, livestock feed, and manure; optimizing fertilizer and manure application rates; and improving irrigation and residue management. Long-term monitoring and the development of standardized methodologies are urgently needed to reduce uncertainties in Cd flux estimates. Integrating field measurements with modeling approaches will provide more robust predictions of soil Cd dynamics and help design sustainable agricultural systems that protect both crop safety and environmental quality.

5.6 Summary

- Cd enters agricultural soils through both natural processes and human activities, primarily via atmospheric deposition, wastewater irrigation, fertilizers, and manure.
- Soil Cd is lost mainly through leaching and crop removal, processes strongly influenced by soil Cd levels, pH, redox status, texture, plant species, and cropping patterns.
- Many studies report a positive Cd input–output balance, indicating that soil Cd concentrations are likely to continue increasing in the future.
- Estimates of Cd inputs and outputs remain highly uncertain. More field-based studies are needed to clarify the controlling factors and to support the development of predictive models for future soil Cd dynamics.

Acknowledgments

This work was supported by the National Key R&D Program of China (2022YFC3704803).

AI disclosure

During the preparation of this work the author(s) used ChatGPT to improve the readability of the text. After using this tool/service the author(s) reviewed and edited the content as needed and took full responsibility for the content of the publication.

References

Ahmed, M., Matsumoto, M., & Kurosawa, K. (2018). Heavy metal contamination of irrigation water, soil, and vegetables in a multi-industry district of Bangladesh. *International Journal of Environmental Research, 12*(4), 531–542. doi:10.1007/s41742-018-0113-z.

Ali, M. H. (2010). Crop water requirement and irrigation scheduling. In M. H. Ali (Ed.), *Fundamentals of irrigation and on-farm water management: Volume 1* (pp. 399–452). Springer. doi:10.1007/978-1-4419-6335-2_92010.

Association of American Plant Food Control Officials. (n.d.). *Statement of Uniform Interpretation and Policy (SUIP) #25—The "heavy metal rule"*. https://www.aapfco.org/rules.html see 2.3 Table5.3 SUIP #25 - The Heavy Metal Rule (aapfco.org)(AAPFCO)

Azimi, S., Cambier, P., Lecuyer, I., & Thevenot, D. (2004). Heavy metal determination in atmospheric deposition and other fluxes in northern France agrosystems. *Water, Air, and Soil Pollution, 157*(1–4), 295–313. doi:10.1023/B:WATE.0000038903.25700.5a.

Badza, T., Tesfamariam, E. H., & Cogger, C. G. (2020). Agricultural use suitability assessment and characterization of municipal liquid sludge: Based on South Africa survey. *Science of the Total Environment, 72*, 137658. doi:10.1016/j.scitotenv.2020.137658.

Bigalke, M., Ulrich, A., Rehmus, A., & Keller, A. (2017). Accumulation of cadmium and uranium in arable soils in Switzerland. *Environmental Pollution, 22*, 85–93. doi:10.1016/j.envpol.2016.11.035.

Bonten, L. T. C., Römkens, P. F. A. M., & Brus, D. J. (2008). Contribution of heavy metal leaching from agricultural soils to surface water loads. *Environmental Forensics, 9*(2–3), 252–257. doi:10.1080/15275920802122981.

Bracher, C., Frossard, E., Bigalke, M., Imseng, M., Mayer, J., & Wiggenhauser, M. (2021). Tracing the fate of phosphorus fertilizer derived cadmium in soil-fertilizer-wheat systems using enriched stable isotope labeling. *Environmental Pollution, 28*, 117314. doi:10.1016/j.envpol.2021.117314.

Canadian Food Inspection Agency. (n.d.). *T-4-93–Safety standards for fertilizers and supplements*. https://inspection.canada.ca/en/plant-health/fertilizers/trade-memoranda/t-4-93 see 2.3 Table5.3 T-4-93–Safety standards for fertilizers and supplements.

Cao, X., Wang, X., Tong, W., Gurajala, H. K., Lu, M., Hamid, Y., Feng, Y., He, Z., & Yang, X. (2019). Distribution, availability and translocation of heavy metals in soil-oilseed rape (*Brassica napus* L.) system related to soil properties. *Environmental Pollution, 25*, 733–741. https://www.journals.elsevier.com/environmental-pollution. doi:10.1016/j.envpol.2019.05.147.

Carne, G., Leconte, S., Sirot, V., Breysse, N., Badot, P.-M., Bispo, A., Deportes, I. Z., Dumat, C., Rivière, G., & Crépet, A. (2021). Mass balance approach to assess the impact of cadmium decrease in mineral phosphate fertilizers on health risk: The case-study of French agricultural soils. *Science of the Total Environment, 76*, 143374. doi:10.1016/j.scitotenv.2020.143374.

Čásová, K., Černý, J., Száková, J., Balík, J., & Tlustoš, P. (2009). Cadmium balance in soils under different fertilization managements including sewage sludge application. *Plant, Soil and Environment, 55*(8), 353–361. doi:10.17221/142/2009-pse.

Chaney, R. L. (2012). Chapter two—Food safety issues for mineral and organic fertilizers. In D. L. Sparks (Ed.). In *Advances in agronomy: 117* (pp. 51–116). Academic Press. doi:10.1016/B978-0-12-394278-4.00002-7.

Cheira, M. F., Rashed, M. N., Mohamed, A. E., Hussein, G. M., & Awadallah, M. A. (2019). Removal of some harmful metal ions from wet-process phosphoric acid using murexide-reinforced activated bentonite. *Materials Today Chemistry, 1*, 100176. doi:10.1016/j.mtchem.2019.06.002.

Chen, H., Zhang, W., Yang, X., Wang, P., McGrath, S. P., & Zhao, F. J. (2018). Effective methods to reduce cadmium accumulation in rice grain. *Chemosphere, 20*, 699–707. Www.elsevier.com/locate/chemosphere. doi:10.1016/j.chemosphere.2018.05.143.

Cichy, B., Jaroszek, H., Paszek, A., & Tarnowska, A. (2014). Kadm w nawozach fosforowych; aspekty ekologiczne i ekonomiczne | Cadmium in phosphate fertilizers; ecological and economical aspects. *Chemik Science-Technique-Market, 68*(10), 837–842.

Cocarta, D. M., Subtirelu, V. R., & Badea, A. (2017). Effect of sewage sludge application on wheat crop productivity and heavy metal accumulation in soil and wheat grain. *Environmental Engineering and Management Journal, 16*(5), 1093–1100. Www.eemj.eu. doi:10.30638/eemj.2017.112.

Code of Federal Regulations. (2025). 40 CFR 503.13. https://www.ecfr.gov/current/title-40/chapter-I/subchapter-O/part-503/subpart-B/section-503.13.

Dai, Y. T., Fu, K. D., Yang, Y., Wang, M. E., & Chen, W. P. (2021). Simulation cadmium (Cd) accumulation in typical paddy soils in South China. *Huanjing Kexue/Environmental Science, 42*(1), 353–358. http://www.hjkx.ac.cn/hjkx/ch/index.aspx. doi:10.13227/j.hjkx.202006121.

Da Silva, G. A., & Kulay, L. A. (2005). Environmental performance comparison of wet and thermal routes for phosphate fertilizer production using LCA—A Brazilian experience. *Journal of Cleaner Production, 13*(13–14), 1321–1325. doi:10.1016/j.jclepro.2005.05.004.

Davister, A. (1996). Studies and research on processes for the elimination of cadmium from phosphoric acid. Fertilizers as a source of cadmium, OECD Cadmium Workshop. Paris, Saltsjöbaden, Sweden: Organisation for Economic Co-operation and Development, OECD.

de Vries, W., & McLaughlin, M. J. (2013). Modeling the cadmium balance in Australian agricultural systems in view of potential impacts on food and water quality. *Science of the Total Environment, 461–46*, 240–257. doi:10.1016/j.scitotenv.2013.04.069.

Du, Y., Hu, X. F., Wu, X. H., Shu, Y., Jiang, Y., & Yan, X. J. (2013). Affects of mining activities on Cd pollution to the paddy soils and rice grain in Hunan province, Central South China. *Environmental Monitoring and Assessment, 185*(12), 9843–9856. doi:10.1007/s10661-013-3296-y.

Elrashidi, M. A., Hammer, D., Fares, A., Seybold, C. A., Ferguson, R., & Peaslee, S. D. (2007). Loss of heavy metals by runoff from agricultural watersheds. *Soil Science, 172*(11), 876–894. doi:10.1097/ss.0b013e31814cec7b.

Fan, T., Long, T., Lu, Y., Yang, L., Mi, N., Xia, F., Wang, X., Deng, S., Hu, Q., & Zhang, F. (2022). Meta-analysis of Cd input-output fluxes in agricultural soil. *Chemosphere, 303*(Pt 2), 134974. doi:10.1016/j.chemosphere.2022.134974.

Felipe-Sotelo, M., Brandtner, M., Ward, N. I., Al Sid Cheikh, M., Chambers, M. A., Rab, G., Rosel, A. C., Woegerbauer, M., Hassan, M. M., La Ragione, R., Tenson, T., Kisand, V., Jeremejeva, J., Kořínková, M., Drahošová, Z., de Menezes, A. B., Gajda, A., Gbylik-Sikorska, M., & Caniça, M. (2025). Impact of manure and artificial fertilizer application on metal and metalloid distributions in agricultural soils and crops. *Chemosphere, 37*, 144243. doi:10.1016/j.chemosphere.2025.144243.

Fu, Y., Li, F., Guo, S., & Zhao, M. (2021). Cadmium concentration and its typical input and output fluxes in agricultural soil downstream of a heavy metal sewage irrigation area. *Journal of Hazardous Materials, 41*, 125203. doi:10.1016/j.jhazmat.2021.125203.

Gramlich, A., Tandy, S., Andres, C., Paniagua, J. C., Armengot, L., Schneider, M., & Schulin, R. (2017). Cadmium uptake by cocoa trees in agroforestry and monoculture systems under conventional and organic management. *Science of the Total Environment, 58*, 677–686. doi:10.1016/j.scitotenv.2016.12.014.

Gray, C., Johnstone, P., Norris, M., & Wallace, D. (2024). Cadmium leaching from some New Zealand cropping soils. *New Zealand Journal of Crop and Horticultural Science, 52*(2), 181–191. doi:10.1080/01140671.2022.2105902.

Gray, C. W., & Cavanagh, J. A. E. (2023). The state of knowledge of cadmium in New Zealand agricultural systems: 2021. *New Zealand Journal of Agricultural Research, 66*(4), 285–335. https://www.tandfonline.com/loi/tnza20. doi:10.1080/00288233.2022.2069130.

Gray, C. W., McLaren, R. G., & Roberts, A. H. C. (2003). Atmospheric accessions of heavy metals to some New Zealand pastoral soils. *Science of the Total Environment, 305*(1–3), 105–115. doi:10.1016/S0048-9697(02)00404-7.

Gu, X., Xiao, Y., Yin, S., Liu, H., Men, B., Hao, Z., Qian, P., Yan, H., Hao, Q., Niu, Y., Huang, H., & Pei, Q. (2019). Impact of long-term reclaimed water irrigation on the distribution of potentially toxic elements in soil: An in-situ experiment study in the North China Plain. *International Journal of Environmental Research and Public Health, 16*(4), 649. doi:10.3390/ijerph16040649.

Gupta, D. K., Chatterjee, S., Datta, S., Veer, V., & Walther, C. (2014). Role of phosphate fertilizers in heavy metal uptake and detoxification of toxic metals. *Chemosphere, 10*, 134–144. doi:10.1016/j.chemosphere.2014.01.030.

Hamon, R. E., Mclaughlin, M. J., Naidu, R., & Correll, R. (1998). Long-term changes in cadmium bioavailability in soil. *Environmental Science and Technology, 32*(23), 3699–3703. http://pubs.acs.org/journal/esthag. doi:10.1021/es980198b.

Hou, Q., Yang, Z., Ji, J., Yu, T., Chen, G., Li, J., Xia, X., Zhang, M., & Yuan, X. (2014). Annual net input fluxes of heavy metals of the agro-ecosystem in the Yangtze River delta, China. *Journal of Geochemical Exploration, 13*, 68–84. doi:10.1016/j.gexplo.2013.08.007.

Hu, J., Wang, Z., Williams, G. D. Z., Dwyer, G. S., Gatiboni, L., Duckworth, O. W., & Vengosh, A. (2024). Evidence for the accumulation of toxic metal(loid)s in agricultural soils impacted from long-term application of phosphate fertilizer. *Science of the Total Environment, 90*. Www.elsevier.com/locate/scitotenv. doi:10.1016/j.scitotenv.2023.167863.

Huang, H., Lv, Y., Tian, K., Shen, Y., Zhu, Y., Lu, H., Li, R., & Han, J. (2023). Influence of sulfate reducing bacteria cultured from the paddy soil on the solubility and redox behavior of Cd in a polymetallic system. *Science of the Total Environment, 90*, 166369. doi:10.1016/j.scitotenv.2023.166369.

Huang, Q., Yu, Y., Wan, Y., Wang, Q., Luo, Z., Qiao, Y., Su, D., & Li, H. (2018). Effects of continuous fertilization on bioavailability and fractionation of cadmium in soil and its uptake by rice (*Oryza sativa* L.). *Journal of Environmental Management, 21*, 13–21. doi:10.1016/j.jenvman.2018.03.036.

Hudcová, H., Vymazal, J., & Rozkošný, M. (2019). Present restrictions of sewage sludge application in agriculture within the European Union. *Soil and Water Research, 14*(2), 104–120. doi:10.17221/36/2018-swr.

Imseng, M., Wiggenhauser, M., Keller, A., Müller, M., Rehkämper, M., Murphy, K., Kreissig, K., Frossard, E., Wilcke, W., & Bigalke, M. (2018). Fate of Cd in agricultural soils: A stable isotope approach to anthropogenic impact, soil formation, and soil-plant cycling. *Environmental Science & Technology, 52*(4), 1919–1928. doi:10.1021/acs.est.7b05439.

Kabata-Pendias, A. (2001). *Trace elements in soils and plants*. CRC Press.

Khan, M. U., Malik, R. N., & Muhammad, S. (2013). Human health risk from Heavy metal via food crops consumption with wastewater irrigation practices in Pakistan. *Chemosphere, 93*(10), 2230–2238. Www.elsevier.com/locate/chemosphere. doi:10.1016/j.chemosphere.2013.07.067.

Kikuchi, T., Okazaki, M., Toyota, K., Motobayashi, T., & Kato, M. (2007a). Corrigendum to "The input–output balance of cadmium in a paddy field of Tokyo" [Chemosphere 67(5) (2007) 920–927]. *Chemosphere, 68*(11), 2181–2182. doi:10.1016/j.chemosphere.2007.03.057.

Kikuchi, T., Okazaki, M., Toyota, K., Motobayashi, T., & Kato, M. (2007b). The input–output balance of cadmium in a paddy field of Tokyo. *Chemosphere, 67*(5), 920–927. doi:10.1016/j.chemosphere.2006.11.018.

Kirchmann, H., Börjesson, G., Kätterer, T., & Cohen, Y. (2017). From agricultural use of sewage sludge to nutrient extraction: A soil science outlook. *Ambio, 46*(2), 143–154. http://www.bioone.org/loi/ambi. doi:10.1007/s13280-016-0816-3.

Kouzbour, S., Gourich, B., Gros, F., Vial, C., Allam, F., & Stiriba, Y. (2019). Comparative analysis of industrial processes for cadmium removal from phosphoric acid: A review. *Hydrometallurgy, 18*, 222–247. doi:10.1016/j.hydromet.2019.06.014.

Kuboi, T., Noguchi, A., & Yazaki, J. (1986). Family-dependent cadmium accumulation characteristics in higher plants. *Plant and Soil, 92*(3), 405–415. doi:10.1007/bf02372488.

Kumar, V., & Chopra, A. K. (2016). Agronomical performance of high yielding cultivar of eggplant (*Solanum melongena* L.) grown in sewage sludge amended soil. *Research in Agriculture, 1*(1), 1. doi:10.22158/ra.v1n1p1.

Latifi, Z., & Jalali, M. (2018). Trace element contaminants in mineral fertilizers used in Iran. *Environmental Science and Pollution Research, 25*(32), 31917–31928. doi:10.1007/s11356-018-1810-z.

Leclerc, A., & Laurent, A. (2017). Framework for estimating toxic releases from the application of manure on agricultural soil: National release inventories for heavy metals in 2000–2014. *Science of the Total Environment, 590–59*, 452–460. doi:10.1016/j.scitotenv.2017.01.117.

Li, C., Zhang, C., Yu, T., Liu, X., Xia, X., Hou, Q., Yang, Y., Yang, Z., & Wang, L. (2022). Annual net input fluxes of cadmium in paddy soils in karst and non-karst areas of Guangxi, China. *Journal of Geochemical Exploration, 24*, 107072. doi:10.1016/j.gexplo.2022.107072.

Li, H., Yang, Z., Dai, M., Diao, X., Dai, S., Fang, T., & Dong, X. (2020). Input of Cd from agriculture phosphate fertilizer application in China during 2006–2016. *Science of the Total Environment, 69*, 134149. doi:10.1016/j.scitotenv.2019.134149.

Li, J., Shang, S., Jiang, H., Song, J., Rahman, K. U., & Adeloye, A. J. (2021). Simulation-based optimization for spatiotemporal allocation of irrigation water in arid region. *Agricultural Water Management, 25*. http://www.journals.elsevier.com/agricultural-water-management/. doi:10.1016/j.agwat.2021.106952.

Li, P., Wang, X., Allinson, G., Li, X., & Xiong, X. (2009). Risk assessment of heavy metals in soil previously irrigated with industrial wastewater in Shenyang, China. *Journal of Hazardous Materials, 161*(1), 516–521. doi:10.1016/j.jhazmat.2008.03.130.

Liu, R., Wang, J., Shi, J., Chen, Y., Sun, C., Zhang, P., & Shen, Z. (2014). Runoff characteristics and nutrient loss mechanism from plain farmland under simulated rainfall conditions. *Science of the Total Environment, 468–469*, 1069–1077. doi:10.1016/j.scitotenv.2013.09.035.

Lu, H., Dong, Y., Feng, Y., Bai, Y., Tang, X., Li, Y., Yang, L., & Liu, J. (2020). Paddy periphyton reduced cadmium accumulation in rice (*Oryza sativa*) by removing and immobilizing cadmium from the water–soil interface. *Environmental Pollution, 26*, 114103. doi:10.1016/j.envpol.2020.114103.

Lugon-Moulin, N., Ryan, L., Donini, P., & Rossi, L. (2006). Cadmium content of phosphate fertilizers used for tobacco production. *Agronomy for Sustainable Development, 26*(3), 151–155. doi:10.1051/agro:2006010.

Luo, L., Ma, Y., Zhang, S., Wei, D., & Zhu, Y. G. (2009). An inventory of trace element inputs to agricultural soils in China. *Journal of Environmental Management, 90*(8), 2524–2530. https://www.sciencedirect.com/science/journal/03014797. doi:10.1016/j.jenvman.2009.01.011.

Mao, Y., Fan, W., Yan, Y., Xiang, W., Hu, S., & Yan, S. (2021). Cd adsorption by iron–organic associations: Implications for Cd mobility and fate in natural and contaminated environments. *Bulletin of Environmental Contamination and Toxicology, 106*(1), 109–114. https://link.springer.com/journal/128. doi:10.1007/s00128-020-02962-2.

McLaughlin, M. J., Smolders, E., Zhao, F. J., Grant, C., & Montalvo, D. (2021). Managing cadmium in agricultural systems. *Advances in Agronomy, 16*, 1–129. http://www.elsevier.com/wps/find/bookdescription.cws_home/675816/description#description. doi:10.1016/bs.agron.2020.10.004.

Merroune, A., Brahim, J. A., Achiou, B., Kada, C., Mazouz, H., & Beniazza, R. (2024). Closed-loop purification process of industrial phosphoric acid: Selective recovery of heavy metals and rare earth elements via solvent extraction. *Desalination, 58*, 117515. doi:10.1016/j.desal.2024.117515.

Ministry of Ecology and Environment of the People's Republic of China. (2021). *Standard for irrigation water quality (GB 5084-2021) (in Chinese)*. 7 21 2021 7 21 2026 2 12 SEPA,(C., 2021). https://www.mee.gov.cn/ywgz/fgbz/bz/bzwb/shjbh/shjzlbz/202102/t20210209_821075.shtml

Moreno-Jiménez, E., Meharg, A. A., Smolders, E., Manzano, R., Becerra, D., Sánchez-Llerena, J., Albarrán, A., & López-Piñero, A. (2014). Sprinkler irrigation of rice fields reduces grain arsenic but enhances cadmium. *Science of the Total Environment, 485–486*(1), 468–473. Www.elsevier.com/locate/scitotenv. doi:10.1016/j.scitotenv.2014.03.106.

Morselli, L., Bernardi, E., Passarini, F., & Tesini, E. (2006). Critical loads for Cd and Pb in the province of Bologna. *Annali di Chimica, 96*(11–12), 697–705. doi:10.1002/adic.200690072.

Mu, H. Y., Zhuang, Z., Li, Y. M., Qiao, Y. H., Chen, Q., Xiong, J., Guo, L. L., Jiang, R. F., & Li, H. F. (2020). Heavy metal contents in animal manure in China and the related soil accumulation risks. *Huanjing Kexue/Environmental Science, 41*(2), 986–996. http://www.hjkx.ac.cn/hjkx/ch/index.aspx. doi:10.13227/j.hjkx.201903078.

National Standardization Administration. (2019). 12 17 2019 12 17 2026 2 12 NSA,(C., 2019). *Chinese National Standardization Administration: Limitation requirements of toxic and harmful substance in fertilizers (GB38400-2019) (in Chinese)*. https://openstd.samr.gov.cn/bzgk/gb/newGbInfo?hcno=76E30A68743B8F552A3EBE6334C7BC07

Nicholson, F. A., Smith, S. R., Alloway, B. J., Carlton-Smith, C., & Chambers, B. J. (2003). An inventory of heavy metals inputs to agricultural soils in England and Wales. *Science of the Total Environment, 311*(1–3), 205–219. doi:10.1016/s0048-9697(03)00139-6.

Omri, H., & Batis, N. H. (2013). Removal of Cd(II) from Phosphoric Acid Solution by Adsorbents: Equilibrium and Kinetic Studies. *Chemical Science Transactions, 2*(2), 357–366. doi:10.7598/cst2013.285.

Oosterhuis, F. H., Brouwer, F. M., & Wijnants, H. J. (2000). A possible EU wide charge on cadmium in phosphate fertilisers: Economic and environmental implications. *(IVM Report); (No. E-00/02)*. Institute for Environmental Studies.

Peng, H., Chen, Y., Weng, L., Ma, J., Ma, Y., Li, Y., & Islam, M. S. (2019). Comparisons of heavy metal input inventory in agricultural soils in North and South China: A review. *Science of the Total Environment, 66*, 776–786. Www.elsevier.com/locate/scitotenv. doi:10.1016/j.scitotenv.2019.01.066.

River Water User Association (India). (2021). *General Standards for Discharge of Environmental Polluters (Schedule VI)*. https://rwua.org.in/download/General%20Standards%20For%20Discharge%20Of%20Environmental%20Pollutants.pdf see 2.2 (RWUA).

Roberts, T. L. (2014). Cadmium and phosphorous fertilizers: The issues and the science. *Procedia Engineering, 8*, 52–59. doi:10.1016/j.proeng.2014.09.012.

Rocha de Carvalho, M., Adorno de Almeida, T., Zorgdrager Van Opbergen, G. A., Alves Bispo, F. H., Botelho, L., Boari de Lima, A., Ribeiro Marchiori, P. E., & Guimarães Guilherme, L. R. (2024). Arsenic, cadmium, and chromium concentrations in contrasting phosphate fertilizers and their bioaccumulation by crops: Towards a green label?

Environmental Research, 263(Part 2), 120171. https://doi.org/10.1016/j.envres.2024.120171. doi:10.1016/j.envres.2024.120171.

Salman, S. A., Elnazer, A. A., & El Nazer, H. A. (2017). Integrated mass balance of some heavy metals fluxes in Yaakob village, south Sohag, Egypt. *International Journal of Environmental Science and Technology, 14*(5), 1011–1018. doi:10.1007/s13762-016-1200-3.

Sampaio, T. F., Guerrini, I. A., Otero, X. L., Vazquez, F. M., Bogiani, J. C., Oliveira, F. C., Gava, J. L., Ciol, M. A., Littke, K. M., & Harrison, R. B. (2016). The impact of biosolid application on soil and native plants in a degraded Brazilian Atlantic Rainforest Soil. *Water, Air, and Soil Pollution, 227*(1). http://www.kluweronline.com/issn/0049-6979/. doi:10.1007/s11270-015-2689-7.

Sedlak, D. L. (2019). The food-environment nexus. *Environmental Science and Technology, 53*(12), 6597–6598. http://pubs.acs.org/journal/esthag. doi:10.1021/acs.est.9b02959.

Shao, D., Zhan, Y., Zhou, W., & Zhu, L. (2016). Current status and temporal trend of heavy metals in farmland soil of the Yangtze River Delta Region: Field survey and meta-analysis. *Environmental Pollution, 21*, 329–336. doi:10.1016/j.envpol.2016.10.023.

Six, L., & Smolders, E. (2014). Future trends in soil cadmium concentration under current cadmium fluxes to European agricultural soils. *Science of the Total Environment, 485–486*(1), 319–328. doi:10.1016/j.scitotenv.2014.03.109.

Smolders, E. (2016). *Scientific aspects underlying the regulatory framework in the area of fertilisers—State of play and future reforms*. Think Tank. European Parliament 2016, 12 15 2016 12 15 2026 2 12, 2016) https://www.europarl.europa.eu/thinktank/en/document/IPOL_ID 595354.

Sterckeman, T., Gossiaux, L., Guimont, S., Sirguey, C., & Lin, Z. (2018). Cadmium mass balance in French soils under annual crops: Scenarios for the next century. *Science of the Total Environment, 63*, 1440–1452. doi:10.1016/j.scitotenv.2018.05.225.

Sterckeman, T., Gossiaux, L., Guimont, S., Sirguey, C., & Lin, Z. (2019). Corrigendum to "Cadmium mass balance in French soils under annual crops: Scenarios for the next century" [Sci. Total Environ. 639 (2018) 1440–1452]. *Science of the Total Environment, 650*(Pt 2), 3180–3188. doi:10.1016/j.scitotenv.2018.09.179.

Suciu, N. A., De Vivo, R., Rizzati, N., & Capri, E. (2022). Cd content in phosphate fertilizer: Which potential risk for the environment and human health? *Current Opinion in Environmental Science & Health, 3*, 100392. doi:10.1016/j.coesh.2022.100392.

The European Parliament and of the Council. (2020). *On the quality of water intended for human consumption*. http://data.europa.eu/eli/dir/2020/2184/oj see 2.2.

Ulrich, A. E. (2019). Cadmium governance in Europe's phosphate fertilizers: Not so fast? *Science of the Total Environment, 65*, 541–545. Www.elsevier.com/locate/scitotenv. doi:10.1016/j.scitotenv.2018.09.014.

United States Environmental Protection Agency. (2024). *Water Quality Standards Handbook* https://www.epa.gov/wqs-tech/water-quality-standards-handbook see 2.2.

Wang, F., Peng, L., Zhou, X., Zeng, Q., & Luo, S. (2021). Typical sources of Cd to paddy fields in different contaminated areas and their impacts on Cd accumulation in topsoil and rice in Changzhutan, China. *Environmental Research, 19*, 110523. doi:10.1016/j.envres.2020.110523.

Wang, H., Dong, Y., Yang, Y., Toor, G. S., & Zhang, X. (2013). Changes in heavy metal contents in animal feeds and manures in an intensive animal production region of China. *Journal of Environmental Sciences, 25*(12), 2435–2442. doi:10.1016/s1001-0742(13)60473-8.

Wei, X., Yang, D., Yin, X., Yang, H., Fang, Y., Chen, N., Zhang, H., & Hu, Z. (2024). Comparative study of efficiencies of purification of cadmium contaminated irrigation water by different purification systems. *Science of the Total Environment, 90*, 167941. doi:10.1016/j.scitotenv.2023.167941.

Wiggenhauser, M., Illmer, D., Spiess, E., Holzkämper, A., Prasuhn, V., & Liebisch, F. (2024). Cadmium, zinc, and copper leaching rates determined in large monolith lysimeters. *Science of the Total Environment, 92*, 171482. doi:10.1016/j.scitotenv.2024.171482.

The European Parliament and the Council of the European Union. (2019). *Laying down rules on the making available on the market of EU fertilising products and amending Regulations (EC) No 1069/2009 and (EC) No 1107/2009 and repealing Regulation (EC) No 2003/2003*. http://data.europa.eu/eli/reg/2019/1009/oj see 2.3 Table5.3 Regulation (EU) 2019/1009

World Health Organization. (2013). *Guidelines for the safe use of wastewater, excreta and greywater, Volume 2*. 4 12 2013 4 12 2026 2 12, https://www.who.int/publications/i/item/9241546832

Wu, L., Tan, C., Liu, L., Zhu, P., Peng, C., Luo, Y., & Christie, P. (2012). Cadmium bioavailability in surface soils receiving long-term applications of inorganic fertilizers and pig manure. *Geoderma, 173–174*, 224–230. doi:10.1016/j.geoderma.2011.12.003.

Xia, X., Yang, Z., Cui, Y., Li, Y., Hou, Q., & Yu, T. (2014). Soil heavy metal concentrations and their typical input and output fluxes on the southern Song-nen Plain. *Journal of Geochemical Exploration, 13*, 85–96. doi:10.1016/j.gexplo.2013.06.008.

Xie, Y. H., Zhu, J., Liu, S. H., Pan, S. F., & Ji, X. H. (2020). Input and output of cadmium (Cd) for paddy soil in central south China: Fluxes, mass balance, and model predictions. *Environmental Science and Pollution Research, 27*(17), 21847–21858. https://link.springer.com/journal/11356. doi:10.1007/s11356-020-08519-4.

Xu, Y., Tang, H., Liu, T., Li, Y., Huang, X., & Pi, J. (2018). Effects of long-term fertilization practices on heavy metal cadmium accumulation in the surface soil and rice plants of double-cropping rice system in Southern China. *Environmental Science and Pollution Research, 25*(20), 19836–19844. doi:10.1007/s11356-018-2175-z.

Yan, X., Zhao, W., Yang, X., Liu, C., & Zhou, Y. (2020). Input–output balance of cadmium in typical agriculture soils with historical sewage irrigation in China. *Journal of Environmental Management, 27*, 111298. doi:10.1016/j.jenvman.2020.111298.

Yang, X., Li, Q., Tang, Z., Zhang, W., Yu, G., Shen, Q., & Zhao, F. J. (2017). Heavy metal concentrations and arsenic speciation in animal manure composts in China. *Waste Management, 6*, 333–339. Www.elsevier.com/locate/wasman. doi:10.1016/j.wasman.2017.03.015.

Yang, Y., Li, Y., Dai, Y., Wang, M., Chen, W., & Wang, T. (2021). Historical and future trends of cadmium in rice soils deduced from long-term regional investigation and probabilistic modeling. *Journal of Hazardous Materials, 41*. Www.elsevier.com/locate/jhazmat. doi:10.1016/j.jhazmat.2021.125746.

Yang, Y., Xiao, C., Wang, F., Peng, L., Zeng, Q., & Luo, S. (2022). Assessment of the potential for phytoremediation of cadmium polluted soils by various crop rotation patterns based on the annual input and output fluxes. *Journal of Hazardous Materials, 423*(Pt B), 127183. doi:10.1016/j.jhazmat.2021.127183.

Yang, Z., Hou, Q., Yu, T., Chen, J., Xia, X., Huang, Y., & Li, K. (2008). Eco-geochemical assessment of agroecosystems: Cd in the Chengdu Economical Region, Sichuan, China. *Earth Science Frontiers, 15*(5), 23–35. https://doi.org/10.1016/s1872-5791(09)60002-6.

Yao, C., Yang, Y., Li, C., Shen, Z., Li, J., Mei, N., Luo, C., Wang, Y., Zhang, C., & Wang, D. (2024). Heavy metal pollution in agricultural soils from surrounding industries with low emissions: Assessing contamination levels and sources. *Science of the Total Environment, 91*, 170610. doi:10.1016/j.scitotenv.2024.170610.

Yi, K., Fan, W., Chen, J., Jiang, S., Huang, S., Peng, L., Zeng, Q., & Luo, S. (2018). Annual input and output fluxes of heavy metals to paddy fields in four types of contaminated areas in Hunan Province, China. *Science of the Total Environment, 63*, 67–76. doi:10.1016/j.scitotenv.2018.03.294.

Yu, Y., Zhu, L. N., Guo, T. L., Huang, Q. Q., Wang, Q., Chen, Q., & Li, H. F. (2018). Risk assessment of cadmium and arsenic in phosphate fertilizer. *Journal of Agro-Environment Science, 37*(7), 1326–1331. http://www.aes.org.cn/nyhjkxxben/ch/reader/create_pdf.aspx?file_no=20180703&year_id=2018&quarter_id=7&falg=1. doi:10.11654/jaes.2018-0715.

Zhao, F. J., Ma, Y., Zhu, Y. G., Tang, Z., & McGrath, S. P. (2015). Soil contamination in China: Current status and mitigation strategies. *Environmental Science and Technology, 49*(2), 750–759. http://pubs.acs.org/journal/esthag. doi:10.1021/es5047099.

Zhao, F. J., McGrath, S. P., & Merrington, G. (2007). Estimates of ambient background concentrations of trace metals in soils for risk assessment. *Environmental Pollution, 148*(1), 221–229. doi:10.1016/j.envpol.2006.10.041.

Zhao, F. J., Zhao, X. Y., Tao, Y. M., Wang, H. Q., & Wang, P. (2021). Straw removal has a limited effect on decreasing cadmium concentration in soil. *Journal of Agro-Environment Science, 40*(3), 693–699. http://www.aes.org.cn/nyhjkxxben/ch/reader/view_abstract.aspx?file_no=20210202&flag=1. doi:10.11654/jaes.2020-1248.

Zhong, X., Joimel, S., Schwartz, C., & Sterckeman, T. (2022). Assessing the future trends of soil trace metal contents in French urban gardens. *Environmental Science and Pollution Research, 29*(3), 3900–3917. https://doi.org/10.1007/s11356-021-15679-4.

CHAPTER

6

Modeling chemical speciation of cadmium in soils

Hui Gao[a], *Yingxuan Deng*[b], *Wanli Lian*[b], *Jinbo Li*[c], *Long Wang*[d], *Yifan Huang*[a] *and Liping Weng*[a]

[a] Soil Chemistry Group, Wageningen University & Research, Wageningen, The Netherlands [b] School of Environment, Henan Normal University, Xinxiang, China [c] School of Grassland Science, Beijing Forestry University, Beijing, China [d] School of Resources and Environment, Henan Agricultural University, Zhengzhou, China

6.1 Introduction

6.1.1 Chemical speciation of cadmium in soils

The environmental impact of cadmium (Cd) depends not only on its total concentration in soil but also on its distribution among various chemical forms, that is, chemical speciation (Fig. 6.1). The partitioning of metals between the solid and solution phases is a key factor influencing their bioavailability and mobility (Weng et al., 2001a). Cd is one of the most mobile heavy metals in the environment (Jackson & Alloway, 2017). This chapter reviews the development of geochemical models for simulating Cd partitioning in soils, sediments, and water, focusing on the adsorption models, including both semiempirical and advanced mechanistic surface complexation models (SCMs). In addition, attention is also given to mine-waste-contaminated, calcareous, and anoxic soils, where Cd-bearing minerals may form. In addition to equilibrium modeling, the chapter also discusses the kinetics of Cd adsorption/desorption processes.

In the solution phase, chemical species of Cd include the free Cd^{2+} ion as well as inorganic and organic soluble complexes (Sauvé et al., 2000a). In soil solution, as well as in ground- and surface waters, a substantial fraction of soluble Cd is present as free Cd^{2+} ion, for example, 55%–90% in groundwater (Kumar et al., 2024) and 44%–91% in the solution of two sandy soils with varying pH (Weng et al., 2001b). The proportion of free Cd^{2+} relative to total soluble Cd is generally higher compared to other heavy metals such as Pb and Cu, as a result of weaker binding of Cd with dissolved organic matter (DOM). Inorganic anions that may play a significant role in forming soluble Cd complexes include Cl^-, CO_3^{2-}, PO_4^{3-}, OH^-, and S^{2-} (Table 6.1). Natural DOM is the major organic ligand forming soluble complexes with Cd in soil

Cadmium Contamination in Soils and Plants: Mechanisms and Mitigation.
DOI: https://doi.org/10.1016/B978-0-443-34104-5.00024-4

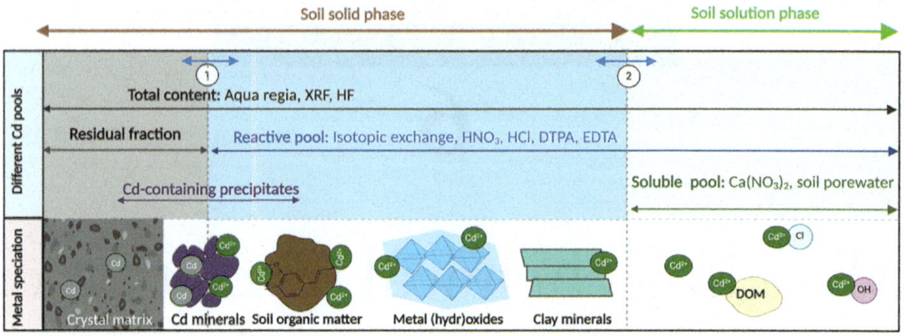

FIGURE 6.1 Conceptual illustration of different Cd pools in soil, with different species within the soil solid phase and soil solution phase, and corresponding extraction or analytical methods (Gao, 2025). ① The release of Cd from the inert fraction to the reactive pool is slow because of the slow weathering of soil minerals. The "total" Cd concentration in soil can be measured by aqua regia digestion, hydrofluoric acid (HF) dissolution, or by X-ray fluorescence (XRF) methods. ② The equilibrium between the reactive pool and the soluble pool is assumed to be relatively fast via adsorption/desorption processes. Depending on the kinetics of precipitation/dissolution reactions, Cd-containing precipitates can be regarded as part of the inert or residual fraction when reactions are slow or can be considered part of the reactive pool when reactions proceed rapidly. The size of the reactive Cd pool can be determined using the isotopic dilution method or by extracting soil with, for example, 0.43 M HNO_3, 0.005 M DTPA, 0.1 M HCl, or 0.05 M EDTA, whereas the soluble fraction can be determined by extracting soil with, for example, 0.01 M $Ca(NO_3)_2$, or by collecting soil porewater either in situ by soil moisture samplers or by centrifuging moist soil. *Cd*, cadmium; *DOM*, dissolved organic matter. *This Figure was reproduced with permission from Gao (2025).*

solution and water. Synthetic ligands in soil and water introduced through anthropogenic inputs (e.g., ethylenediaminetetraacetic acid (EDTA) from chelated fertilizers) can also form soluble complexes containing Cd (Table 6.1). The concentration of soluble Cd in soil can be measured in soil porewater or in extractions using water or electrolyte solutions (e.g., $Ca(NO_3)_2$).

In the solid phase, Cd can be adsorbed to soil reactive surfaces, precipitated or coprecipitated with soil minerals, or incorporated as an impurity within mineral structures. A significant fraction of Cd in soil is adsorbed to natural organic matter (NOM), metal (hydr)oxides, and phyllosilicate clay minerals (Fig. 6.1) (Weng et al., 2001a). Cd adsorption to these surfaces is driven by specific chemical binding and/or nonspecific electrostatic interactions. NOM is often the predominant adsorbent for Cd in soil (Weng et al., 2001a). Most of the adsorbed Cd can be released in extractions containing neutral electrolytes of bivalent metal ions (e.g., $MgCl_2$); this fraction is defined operationally as the "exchangeable" metal pool (Cui & Weng, 2015). The reactively adsorbed amount of heavy metals, including Cd, is often estimated using acid (e.g., 0.43 M HNO_3) extraction, chelate (e.g., 0.005 M DTPA) extraction, or isotope dilution (Groenenberg et al., 2017; Marzouk et al., 2013).

In most soils the precipitates of Cd with CO_3^{2-}, PO_4^{3-}, or OH^- can hardly persist because of undersaturation ($\log K_{sp}$ $Cd(OH)_2 = -14.35$, $CdCO_3 = -12.1$, $Cd_3(PO_4)_2 = -32.6$) (Table 6.1), except in highly contaminated soils with elevated concentrations of Cd and associated anions. Nevertheless, coprecipitation or occlusion of Cd in various minerals may occur in soils contaminated with mine waste and in calcareous soils (Gao et al., 2022). A special case applies to water-saturated soils such as paddy and wetland soils. Under submerged and anoxic conditions, when redox potential (E_h) is below about −100 mV, sulfate (SO_4^{2-}) can be reduced to sulfide

TABLE 6.1 Thermodynamic constants for aqueous species formation and precipitation reactions of Cd. log**K** and log**K**$_{sp}$ values are reported at ***I*** = 0°C and 25°C.

Complexation	**log*K***
$Cd^{2+} + Cl^- \rightleftharpoons CdCl^+$	1.98
$Cd^{2+} + 2Cl^- \rightleftharpoons CdCl_2$	2.6
$Cd^{2+} + 3Cl^- \rightleftharpoons CdCl_3^-$	2.4
$Cd^{2+} + OH^- \rightleftharpoons CdOH^+$	3.9
$Cd^{2+} + 2OH^- \rightleftharpoons Cd(OH)_2$	7.7
$Cd^{2+} + CO_3^{2-} \rightleftharpoons CdCO_3$	4.09
$Cd^{2+} + H^+ + CO_3^{2-} \rightleftharpoons CdHCO_3^+$	12.42
$Cd^{2+} + NO_3^- \rightleftharpoons CdNO_3^+$	0.31
$Cd^{2+} + HPO_4^{2-} \rightleftharpoons CdHPO_4$	14.55
$Cd^{2+} + DTPA^{5-} \rightleftharpoons CdDTPA^{3-}$	21.14
$Cd^{2+} + H^+ + DTPA^{5-} \rightleftharpoons CdHDTPA^{2-}$	25.96
$Cd^{2+} + EDTA^{4-} \rightleftharpoons CdEDTA^{2-}$	18.03
$Cd^{2+} + H^+ + EDTA^{4-} \rightleftharpoons CdHEDTA^-$	21.4
Precipitation	logK_{sp}
$Cd(OH)_2[s] \rightleftharpoons Cd^{2+} + 2OH^-$	−14.35
$CdCO_3[s] \rightleftharpoons Cd^{2+} + CO_3^{2-}$	−12.1
$CdS[s] \rightleftharpoons Cd^{2+} + S^{2-}$	−27.4
$Cd_3(PO_4)_2[s] \rightleftharpoons 3Cd^{2+} + 2PO_4^{3-}$	−32.6

*The log*K* represents the equilibrium constant for a complexation reaction, whereas the logK_{sp} represents the solubility product for a precipitate. *I* denotes the ionic strength of the solution, and *I* = 0 indicates an ideal solution with no interference from other ions. The log*K* values were taken from Lindsay (1979). The logK_{sp} of $CdCO_3$ was obtained from Stipp et al. (1993), whereas that of $Cd(OH)_2$ was taken from Lindsay (1979). The logK_{sp} of $Cd_3(PO_4)_2$ and CdS were retrieved from the Visual MINTEQ database. *Cd*, cadmium; *CdS*, cadmium sulfide.

(S^{2-}) and precipitation of Cd sulfide (CdS) (greenockite) may take place. With its low solubility (logK_{sp} = −27.4) the precipitation of CdS strongly decreases the soluble Cd concentration, thereby reducing Cd uptake and accumulation in plants such as rice (Yao et al., 2021) (see Chapter 7). The residual Cd fraction after sequential extraction or dilute acid extraction mainly consists of Cd incorporated into mineral structures as impurities, which is only slowly released through long-term mineral weathering. The residual fraction of Cd in soil is generally smaller than that of other heavy metals (Sungur et al., 2014).

6.1.2 Importance of adsorption for cadmium distribution in soils

Soil contains various particles of different sizes and properties, to which Cd can adsorb. True adsorption refers to the process where a chemical (adsorbate, e.g., Cd) accumulates on the surface of another substance (adsorbent, e.g., clay mineral) because of forces of attraction. In addition to adsorption a chemical may also accumulate in the solid or suspended particle phase through diffusion into the solid matrix or through surface precipitation, forming an adherent phase composed of species derived from both the aqueous solution and the dissolution of solid (Sposito, 1989). Sorption is a broader term than adsorption, which includes accumulation of a chemical in the (mostly) solid phase via adsorption, diffusion into the bulk, and surface precipitation (Sposito, 1989). Various forces can drive the processes of adsorption, depending on the properties of both the adsorbate and adsorbent, as well as on the solution chemistry. When a chemical reaction occurs between the adsorbate and reactive surface sites, a surface complexation reaction takes place, resulting in the formation of surface complexes.

Over the past decades, considerable attention has been devoted to investigating the adsorption of metal ions, including Cd, onto reactive surfaces in soils. One of the motivations behind this effort is driven by the importance of adsorption process for metal speciation in soils and other environmental systems. Adsorbed Cd is the dominant fraction of Cd in soils. For instance, in the study of Cui and Weng (2015), 40%–97% of the total Cd in 16 soils, including grassland, wetland, peat, paddy, and mine-impacted soils, was adsorbed. Adsorption and desorption are critical processes controlling the bioavailability and mobility of Cd in many soils (Loganathan et al., 2012). Only in highly contaminated soils or under strongly reducing conditions, Cd-containing precipitates may dominate (Naidu et al., 1997). Even in such cases a portion of Cd remains adsorbed, maintaining equilibrium between free Cd^{2+} (regulated by the solubility of the precipitate) and Cd adsorbed to reactive surfaces.

Another motivation for investigating Cd adsorption lies in the fact that adsorbed Cd constitutes the major reactive metal pool. This pool can immobilize or release Cd rapidly and responds to changes in solution parameters, such as pH and ionic strength, much faster than processes such as precipitation or mineral weathering. Therefore adsorbed Cd is highly relevant for assessing the environmental risk of Cd and effectiveness of remediation practices. However, predicting and understanding surface complexation reactions in soils remains challenging. By contrast the formation of soluble inorganic complexes (e.g., $CdCl^+$) and precipitation reactions (e.g., $Cd_3(PO_4)_2$) are better understood and can be more readily calculated using thermodynamic principles.

6.1.3 Factors influencing cadmium adsorption to soil

The partitioning and speciation of Cd in soils are influenced by factors such as total Cd concentration, solution chemistry (e.g., pH, ionic strength, major cations and anions, soluble ligands, especially Cl^- and DOM), and composition of the soil solid phase (e.g., the amount and reactivity of soil organic matter (SOM), metal (hydr)oxides, clay, and carbonate) (Loganathan et al., 2012). Among these factors, soil pH is the single most dominant factor influencing Cd speciation in soils (Buekers et al., 2007; Degryse et al., 2009). Sauvé et al. (2000a) showed that with an increase of one pH unit, the concentration of soluble Cd decreases by about fivefold. An increase in pH enhances the negative charge of the organic matter molecules and other solid particles with pH-dependent functional sites. The decrease in proton competition with increasing pH favors Cd binding to soil solid phase, thereby decreasing Cd concentration in soil solution.

Another factor is the loading of Cd relative to the adsorption capacity of the soil. SOM, (hydr)oxides of Al, Fe, and Mn, and clay minerals are the main Cd sorbents in soil. Among these, SOM is frequently identified as the most important component determining the strength of Cd retention (Degryse & Smolders, 2006). Geochemical modeling also predicts that the majority of Cd is bound to solid-phase organic matter in soils, especially at pH < ~6.5, whereas metal (hydr)oxides become the most important adsorptive constituents at pH > ~6.5 (Gao et al., 2022; Ren et al., 2015). According to a regression equation derived by Sauvé et al. (2000a) ($\log K_d = -1.47 + 0.82 \log SOM + 0.48\ pH$), a 16.6-fold increase in SOM content can result in a 10-fold increase in K_d, thus a 10-fold decrease in soluble Cd concentration. In addition to organic matter in soil solid phase, formation of soluble species with DOM can strongly increase the Cd concentration in soil solution, especially at relatively high pH. Thus any factors (e.g., application of manure and liming) that induce variations in concentration and composition of DOM may significantly affect the distribution of Cd.

In addition, the distribution and adsorption of Cd in soil can be influenced by coexisting cations and anions (Loganathan et al., 2012; Tiberg & Gustafsson, 2016). An increase in cation concentrations, particularly of multivalent cations, has been shown to reduce Cd adsorption (Boekhold et al., 1993; Lagerwerff & Brower, 1972). Naidu et al. (1994) reported that Cd adsorption on a sandy soil was reduced by approximately 80% in 0.01 M $Ca(NO_3)_2$ compared to 0.03 M $NaNO_3$. Similar results were reported by other studies (Lagerwerff & Brower, 1972; Naidu et al., 1994). This effect is attributed to the stronger competitive adsorption of Ca^{2+} than Na^+ against Cd^{2+} on soil adsorbents. Coexisting anions such as phosphate ($(H_x)PO_{4x-3}$) can enhance Cd immobilization in soils, which can be attributed to increased Cd sorption induced by electrostatic synergy with phosphate and possibly also by formation of ternary surface complexes, including both Cd and phosphate (Bolan et al., 2003). At high concentrations of Cd and phosphate the precipitation of $Cd_3(PO_4)_2$ ($\log K_{sp} = -32.6$) may occur, which can further decrease the concentration of soluble Cd in soil (Siebers et al., 2013).

6.2 Development of surface complexation models

6.2.1 A brief description of soil reactive surfaces for cadmium binding

Micro- and nanometer-sized organic and mineral particles in soil possess a large surface area, onto which metal ions, including Cd, may interact and be adsorbed. Among these geocolloidal particles, NOM contains a large number of acidic functional groups, that is, carboxylic and phenolic groups. Once deprotonated, these acidic functional groups become negatively charged, which facilitates electrostatic interaction between Cd and NOM. Furthermore, Cd binds chemically to reactive functional groups on NOM. This specific binding is generally more important than electrostatic interaction, especially at low Cd concentration and high pH (Milne et al., 2003). NOM in soil is heterogeneous in its molecular structure. The operationally defined humic acid (HA) and fulvic acid (FA) fractions in NOM have a higher density of acidic functional groups than the humin fraction (Stevenson, 1994) and therefore play a more important role in Cd adsorption by NOM. The content of HA and FA relative to SOM varies among soils, and the metal binding capacity of HA and FA from different soils also varies.

Metal (hydr)oxides are omnipresent in soils and sediments. The most abundant metal (hydr)oxides in soils are iron and aluminum forms. However, some studies suggest that Mn (hydr)oxides have a greater affinity for Cd binding than iron or aluminum (hydr)oxides (Basta et al., 2005). Protonation and deprotonation of hydroxy groups on metal (hydr)oxide surfaces generate variable surface charges. Within the typical range of soil pH the surface of

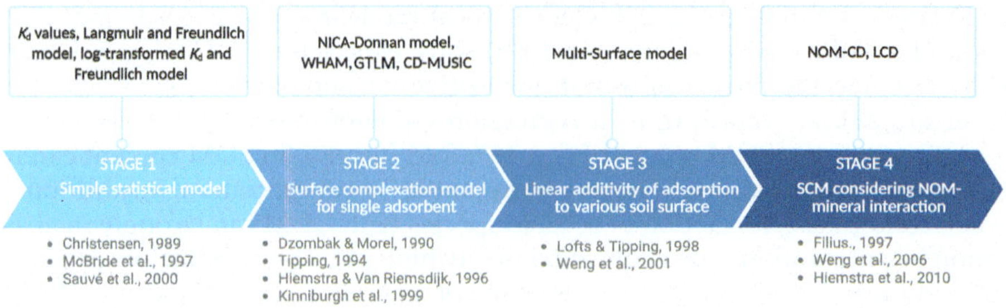

FIGURE 6.2 Developments in modeling Cd and other metal ion distribution and adsorption in soil. *Cd*, Cadmium; *CD-MUSIC*, charge distribution-multisite complexation; *LCD*, Ligand and Charge Distribution; *NICA*, nonideal competitive adsorption; *NOM-CD*, natural organic matter-charge distribution; *SCM*, surface complexation model; *WHAM*, Windermere humic aqueous model.

iron and aluminum (hydr)oxides is positively charged. Despite the unfavorable electrostatic conditions, Cd can still bind to metal (hydr)oxides via specific chemical binding (Ponthieu et al., 2006). Poorly crystalline oxides possess a higher specific surface area (SSA) and thus have a larger Cd binding capacity than crystalline oxides.

Phyllosilicate clay is another major component in soils for cation adsorption. The silanol planar surface of clay is permanently negatively charged because of isomorphic substitution, allowing Cd to be adsorbed through electrostatic attraction. The alumina face and edges of clay contain hydroxy functional groups, which can bind Cd specifically (Camino-Serrano et al., 2018; Gräfe et al., 2014; Lackovic et al., 2003). The relative importance of these micro- and nanoparticles in Cd adsorption depends on conditions, such as their relative content, pH, and Cd concentration.

6.2.2 A general description of surface complexation model development

To improve our understanding of interfacial processes, SCMs have been developed over decades. The development of models describing the adsorption of Cd and other metal ions to soils can be categorized into four stages (Fig. 6.2).

1. Simple statistical models. Initially, most of the focus was on modeling solid-solution distribution of Cd. The distribution or partitioning coefficient (K_d), which is the ratio of Cd between the solid and solution phase, is widely used (Degryse et al., 2009). Simple semiempirical functions, such as the Langmuir and Freundlich isotherms, have also been widely applied to describe metal ion binding to soil components such as NOM, oxides, and clay, and to soil as a whole (Barrow & Whelan, 1998; Elzinga et al., 1999). After logarithmic transformation, the Freundlich equation becomes linear and can be extended to incorporate additional variables beyond Cd concentration, accounting for factors such as pH and NOM content (Sauvé et al., 2000a; Temminghoff et al., 1995). These statistically derived empirical relationships, also referred to as pedo-transfer functions (McBratney et al., 2002; Pachepsky & Rawls, 2003) are simple and easy to apply. However, caution is warranted when applying these models to soils outside the range for which they were calibrated.
2. SCMs for a single type of adsorbent. Advancements in analytical techniques (e.g., spectroscopy) and theoretical calculations (e.g., quantum chemistry) have facilitated the development of more advanced mechanistic SCMs. The widely accepted nonideal competitive

adsorption (NICA)–Donnan model has been developed for ion binding to NOM, which encompasses chemical heterogeneity, variable stoichiometry, electrostatic interaction, and competition between ions (Kinniburgh et al., 1999). For metal (hydr)oxides the CD-MUSIC (charge distribution-multisite complexation) model (Hiemstra & Van Riemsdijk, 1996) is becoming the mainstream SCM, which is based on the structure of the mineral surface, the structure of adsorbed species, and the structure of the electrostatic potential profile near the mineral–water interface. These SCMs have proven effective in describing ion adsorption to purified NOM and synthesized oxide minerals. These advanced SCMs combine site binding isotherms with electrostatic submodels and reflect as much as possible the physicochemical processes in the binding reactions. The mechanistic nature of these models permits them to be more widely applicable, and their model parameters remain generic when solution chemistry changes.

3. Multisurface model (MSM). By ascribing Cd bound to soil to one dominant type of surface, SCM developed for single type of surface is applied to simulating Cd binding to the whole soil. For example, Shi et al. (2007) applied the Windermere humic aqueous model (WHAM) (Tipping, 1994) to predict Cd adsorption to 14 noncalcareous soils, considering Cd binding to reactive organic matter. But in addition to NOM, metal (hydr)oxides and silicate clays are also reactive toward Cd binding. The MSM approach combines SCM developed for different types of reactive surface and assumes that the overall Cd adsorption to soil is the sum of Cd bound to all surfaces (Weng et al., 2001a). As MSM does not consider interactions between NOM and minerals, instead it adopts a linearly additive approach. Despite this simplification the MSM has proven effective in describing the speciation and distribution of metal ions across a range of soils (Dijkstra et al., 2009; Duffner et al., 2014; Gao et al., 2022). Recently the MSM has been integrated with machine learning techniques to predict regional-scale Cd speciation patterns in European and Chinese topsoils, where Cd solid–solution distributions simulated using MSM were used to train the machine learning models (Zhang et al., 2025).
4. SCM considering NOM–mineral interaction. Recent developments in SCM of ion adsorption on geo-colloidal surfaces have focused on the effects of NOM–mineral interactions. The NOM-CD (natural organic matter-charge distribution) model (Hiemstra et al., 2010), which is an extension of the CD-MUSIC model, includes site competition and electrostatic effects of NOM present at the intimate surface of oxides on the adsorption of other ions to oxides. Alternatively the LCD (Ligand and Charge Distribution) model (Filius et al., 1997). Weng et al. (2006) integrate the NICA–Donnan and CD-MUSIC models, which allow a more detailed description of NOM–oxide interactions and their effects on ion adsorption to both NOM and oxides. A few attempts have been made to incorporate NOM-CD or LCD model into geochemical modeling of Cd speciation in soils (Deng et al., 2023; Li et al., 2022). Because of site competition, electrostatic interactions, and formation of ternary complexes, Cd binding to NOM–mineral assemblages may deviate from the linear additivity assumption (Deng et al., 2023; Yi et al., 2022).

6.3 Empirical modeling

6.3.1 Common empirical models

Empirical models for Cd distribution in soils are typically based on relationships between Cd concentrations in the solution and solid phases at equilibrium. These models, often referred to as pedo-transfer functions, are generally derived through regression analysis. Commonly used pedo-transfer functions for heavy metal distribution include the solid–solution partitioning coefficient (K_d), as well as simple isotherm equations such as the Langmuir and Freundlich isotherms and their extensions.

6.3.1.1 K_d values (linear model)

The solid-solution partitioning coefficient (K_d) is the simplest empirical model that expresses the ratio between Cd in the solid phase and in the equilibrium solution phase, as given in Eq. (6.1):

$$K_d = \frac{Q}{C} \tag{6.1}$$

where Q is the content of Cd in the solid phase (mol/kg or mg/kg), and C is the Cd concentration in the solution phase (mol/L or mg/L). Therefore the value of K_d typically has a unit of L/kg.

Sauvé et al. (2000b) evaluated the K_d values for several heavy metals (Cd, Cu, Pb, Ni, and Zn), based on data compiled from over 70 published studies. In that study the values of K_d collected ($N = 830$) for Cd varied over six orders of magnitude, ranging between 0.44 and 192,000 L/kg at a solid-to-solution ratio of 1:10, with an average of 2869 L/kg and a median of 390 L/kg. The median K_d values found for Zn, Cu, Ni, and Pb in the same study are 1731, 2120, 2333, and 102,410 L/kg, respectively, higher than that of Cd. A lower K_d value of Cd indicates low retention of the metal ion by the solid phase through, for example, adsorption reactions and, consequently, a higher bioavailability and mobility, thus a higher risk to food and ecological safety. Although K_d values exhibit substantial variability, a constant value of K_d is still frequently employed in biogeochemical modeling to facilitate computational processes. This simplification has a great impact on the accuracy of the estimation.

6.3.1.2 Langmuir model

The well-known Langmuir equation assumes a monolayer adsorption over a homogeneous adsorbent surface with a fixed number of reactive sites. The basic form of the Langmuir model is given in Eq. (6.2):

$$Q = Q_{\max}\frac{K_L C}{1 + K_L C} \tag{6.2}$$

where $Q_{\max}$ (mol/kg or mg/kg) is the maximum amount of adsorbate adsorbed, K_L (L/kg) is the reaction constant, and C (mol/L or mg/L) is the equilibrium concentration of adsorbate in solution.

The $Q_{\max}$ for Cd adsorption in 49 soils reported by Fan et al. (2022) ranges between 0.66 and 4.90 g/kg soil, showing a variation of more than sevenfold. The Langmuir reaction constant (K_L) reported by Esfandbod et al. (2011) ranges between 10 and 180 L/kg, varying almost 20-fold. Both parameters are dependent on the properties of the soil as well as on solution chemistry, such as pH and presence of competitive ions. The Langmuir model can be extended by including ion competition (multicomponent Langmuir) and more site types (combinations of multiple Langmuir isotherms).

6.3.1.3 *Freundlich model*

Another empirical model is the Freundlich model (Freundlich, 1906), which considers the nonlinear adsorption behavior, as given in Eq. (6.3):

$$Q = K_F C^n \tag{6.3}$$

where Q (mol/kg or mg/kg) is the amount of adsorbate adsorbed, K_F ($mol^{1-n}L^n/kg$) is the reaction constant, C (mol/L or mg/L) is the equilibrium concentration of adsorbate in solution, and n is an empirical constant ($0 < n < 1$). A smaller n indicates a higher surface site heterogeneity. The Freundlich equation can be converted into a linear expression through a logarithmic transformation, as given in Eq. (6.4):

$$\log Q = \log K_F + n \log C \tag{6.4}$$

The linear Freundlich model and its extensions have been used extensively in calculating Cd and other metals' distributions in soil (Degryse et al., 2009).

6.3.2 Extension of empirical models

To expand the applicability of these empirical models, model extensions have been introduced to incorporate variations in soil properties such as pH, content of SOM, Fe/Al (hydr)oxides and clay, and DOM concentration, which influence the retention of metals in soils (de Vries et al., 2011).

6.3.2.1 *Extended* K_d *model*

To consider the effect of soil properties the K_d values have been related to soil and soil solution properties (X_i), as given in Eq. (6.5):

$$\log K_d = \alpha_0 + \sum \alpha_i \log X_i \tag{6.5}$$

Based on 830 K_d values collected from literature, Sauvé et al. (2000a) included SOM and pH in the linear model as given in Eq. (6.5) for $\log K_d$ of Cd ($\log K_d = -1.47 + 0.48\,pH + 0.82\,\log SOM$), which shows an increase in $\log K_d$ with the increase of SOM content and soil pH. Based on the results of 36 Australian soils, de Vries et al. (2011) extended the linear model by taking DOC (dissolved organic carbon) into account ($\log K_d = 0.88 + 0.36\,pH + 0.67\,\log SOC - 0.94\,\log DOC$), showing a decrease of $\log K_d$ of Cd with an increase in DOC concentration.

6.3.2.2 *Extended Freundlich model*

Similar to the extended K_d model, Freundlich equations can be established as a function of particular soil characteristics, such as pH, SOM, and clay content, utilizing the extended Freundlich model, as given in Eq. (6.6):

$$\log Q = \log K_F + n \log C + \sum \alpha_i \log X_i \tag{6.6}$$

The amount of adsorbed Cd (Q) is sometimes replaced with total Cd content or reactive Cd content. Eq. (6.6) has been presented in different formats after rearrangement with either $\log Q$ or $\log K_F$ and $\log C$ as dependent variable (Groenenberg et al., 2010; Tye et al., 2003).

Different soil factors have been considered the explanatory variables (X_i) in previous studies. Selected pedo-transfer functions for Cd in soil can be found in It is clear from these equations that the most important variables to predict the solubility of Cd in soil are soil pH and SOM (or SOC) content (Table 6.2).

6.3.3 Applicability and limitations of empirical models

Compared to mechanistic models (see Section 6.4), empirical models or pedo-transfer requires fewer input parameters and significantly less computational time (Bonten et al., 2008; Groenenberg et al., 2012). However, pedo-transfer functions reported in the literature often show substantial variability (Table 6.2), which can be attributed to several factors: (1) differences in the range of soil properties and metal concentrations in the datasets; (2) differences in methodologies for extracting the soil solution, retrieving metals from the solid phase, and distinguishing between total Cd concentrations and free metal ion activities in solution; (3) the specific soil and soil solution properties included; and (4) whether the functions are based on sorption or desorption data (Groenenberg et al., 2010).

Römkens et al. (2004) and Groenenberg et al. (2010) evaluated the pedo-transfer functions for heavy metals and found that the transfer function in the format of $\log K_F$, including soil properties, is the most suitable for predicting metal concentrations in solution from solid phase concentrations. Further the transfer function is more reliable for predicting free metal ion activity in solution than total soluble metal concentration, with the predicted free metal ion activities within one order of magnitude of the observations from the 0.43 M HNO_3 extractable metal content using only pH and SOM as explanatory variables. Groenenberg et al. (2010) and Liang et al. (2013) further demonstrated that when Cd solubility is controlled by adsorption rather than precipitation, its distribution between the soil solid and solution phases can be effectively described by an extended Freundlich function.

6.4 Mechanistic modeling

6.4.1 Modeling cadmium binding to individual soil reactive constituents

Differing from empirical models, mechanistic or process-based adsorption models are based on the physicochemical processes governing surface complexation reactions. These models aim to reflect, as accurately as possible, the types of surface sites, the surface species formed, and the charge behavior and electrostatic interactions between the adsorbent and the adsorbate.

6.4.1.1 Modeling cadmium binding to natural organic matter

NOM in soil and water plays an important role in Cd adsorption. One of the challenges in modeling Cd adsorption to NOM arises from the chemical heterogeneity of NOM. HAs and FAs fractions of NOM contain a large amount of acidic functional groups, thus they are the most reactive NOM fractions in relation to metal ion adsorption. In both HA and FA, carboxylic and phenolic types of acidic groups contribute to metal ion adsorption. The NICA–Donnan model (Kinniburgh et al., 1999) and the WHAM model (Tipping, 1994) are the most widely used advanced models for ion binding to humic materials. Both models take chemical heterogeneity, competition, variable reaction stoichiometry, and electrostatic interactions into account. Both WHAM and NICA–Donnan models have adopted a Donnan approach to separate the "intrinsic" affinity from the electrostatic effects. In the WHAM model the distribution of the intrinsic

TABLE 6.2 Linear regression equations relating soluble Cd concentration, value of K_d and K_F to different soil properties.

Soil samples	Regression equations	R^2	N	References
Nanjing, China	$\log[Cd_{soluble}] = 4.49 - 0.59$ pH	0.59	160	Luo et al. (2006)
Lake Macquarie, Australia	$\log[Cd_{soluble}] = 1.675 + 1.186 \log[Cd_{total}]$	0.83	32	Kim et al. (2009)
	$\log[Cd_{soluble}] = 4.551 - 0.435$ pH $+ 1.144 \log[Cd_{total}]$	0.95	32	
	$\log[Cd_{soluble}] = 6.44 - 0.47$ pH $+ 1.23 \log[Cd_{total}] - 0.66 \log[SOM]$	0.96	32	
America and Canada	$\log[Cd_{soluble}] = 3.23 - 0.54$ pH $+ 0.77 \log[Cd_{total}]$	0.76	64	Sauvé et al. (2000a)
Netherlands, Britain, and France	$\log[Cd_{soluble}] = 3.62 - 0.50$ pH $+ 0.96 \log[Cd_{total}] - 0.45 \log[SOM]$	0.87	31	McBride et al. (1997)
Portugal	$\log[Cd_{soluble}] = 1.58 - 0.55$ pH $+ 1.1 \log[Cd_{reactive}] - 0.58 \log[SOC]$	0.86	84	Rodrigues et al. (2013)
Germany	$\log[Cd_{soluble}] = 4.14 - 0.57$ pH $+ 1.06 \log[Cd_{adsorbed}] - 0.45 \log[SOC]$	0.98	17	Filius et al. (1998)
Multiple Sites	$\log[Cd_{soluble}] = 4.66 - 0.41$ pH $+ 1.11 \log[Cd_{reactive}] - 0.62 \log[SOM] - 0.39 [AlFe] + 0.29 \log[DOC]$	0.82	241	Groenenberg et al. (2012)
Netherlands and United Kingdom	$\log K_F = -2.88 + 0.41$ pH $+ 0.84 \log[SOM]$	0.82	216	Groenenberg et al. (2010)
India and England	$\log K_F = 1.07 + 0.21$ pH $+ 0.0003 [SOM] + 0.004 CaCO_3$	0.85	12	Hooda and Alloway (1998)
Netherlands	$\log K_F = -5.43 + 0.27$ pH $+ 0.58 \log[SOM] + 0.28 \log[Clay]$	0.79	118	Römkens et al. (2004)
	$\log K_F = -5.36 + 0.26$ pH $+ 0.61 \log[SOM] + 0.29 \log[Clay] - 0.05 \log[DOC]$	0.80	118	
Denmark	$\log K_d = -0.74 + 0.53$ pH	0.72	63	Christensen (1989)
Multiple Sites	$\log K_d = -1.47 + 0.48$ pH $+ 0.82 \log[SOM]$	0.61	751	Sauvé et al. (2000b)
Denmark	$\log K_d = -1.35 + 0.59$ pH $+ 0.016 [SOM]$	0.79	63	Christensen (1989)
Nanjing, China	$\log K_d = -1.52 + 0.59$ pH $+ 0.82 \log[Cd_{total}]$	0.70	40	Luo et al. (2006)
Jiaodong Peninsula, China	$\log K_d = 1.02 + 0.2$ pH $+ 1.22 \log[Cd_{total}]$	0.48	18	Li et al. (2014)
Australian	$\log K_d = 0.88 + 0.36$ pH $+ 0.67 \log[SOC] - 0.94 \log[DOC]$	0.82	36	de Vries et al. (2011)

*$Cd_{soluble}$ (μg/L) is the concentration of soluble Cd in soil solution; Cd_{total} (mg/kg), $Cd_{reactive}$ (mg/kg), and $Cd_{adsorbed}$ (mg/kg) are, respectively, total, reactive, and adsorbed content of Cd in soil; SOM (g/kg) and SOC (g/kg) are, respectively, the content of organic matter and organic carbon in soil; clay (%) is soil clay content; DOC (mg/L) is concentration of dissolved organic carbon; AlFe is the sum of oxalate extractable Al and Fe (mmol/kg); $CaCO_3$ (%) is the $CaCO_3$ content in soil; K_F ($mol^{1-n}L^n$/kg) is Freundlich reaction constant; and K_d (L/kg) is the solid-solution partitioning coefficient. *Cd*, cadmium; *DOC*, dissolved organic carbon; *SOM*, soil organic matter.

affinity is treated as a discrete distribution, whereas NICA–Donnan model assumes a continuous distribution in ion binding affinity for each type of group. The site-binding NICA equation for either carboxylic or phenolic type of group can be found in Eq. (6.7).

$$Q_i = Q_{\max}\frac{n_i}{n_H}\frac{(\tilde{K}_iC_i)^{n_i}}{\sum(\tilde{K}_iC_i)^{n_i}}\frac{\left\{\sum(\tilde{K}_iC_i)^{n_i}\right\}^p}{1+\left\{\sum(\tilde{K}_iC_i)^{n_i}\right\}^p} \tag{6.7}$$

where Q_i is the amount of component i bound (mol/kg), $Q_{\max}$ (mol/kg) is the total amount of carboxylic or phenolic type of groups on NOM, $\tilde{K}_i$ is the median affinity constant, and C_i (mol/L)is the local concentration of component i. The parameter n_i ($0 < n_i \leq 1$) is related to the degree of correlation between the affinity distribution of the proton and the metal ion i and is also related to the stoichiometry of the binding reaction of the ion. The width of the affinity distribution, indexed as p ($0 < p \leq 1$), represents the intrinsic heterogeneity of the ligands. The total amount of a metal ion bound is the sum of the ions adsorbed to the carboxylic and phenolic types of groups and accumulated in the Donnan phase. Note that the NICA model simplifies to a competitive Langmuir model when all n_i values and p are equal to 1.

Generic NICA–Donnan model parameters for protons and a range of cations have been derived by Milne et al. (2001, 2003). In Table 6.3 the generic parameters for H^+, Cd^{2+}, Ca^{2+}, and Cu^{2+} are listed for HA and FA. Based on the NICA–Donnan model, Cd forms predominantly monodentate complexes on the carboxylic and phenolic types of groups. Milne et al. (2003) pointed out that Cd binding to HA and FA is weaker than Ni, Zn, Cu, and Pb, especially in the acidic pH range, which is also demonstrated by the model simulation we carried out for metal ion binding to HA using the NICA–Donnan model with generic parameters (Fig. 6.3A).

6.4.1.2 Modeling cadmium binding to metal (hydr)oxides

Ion adsorption to metal (hydr)oxides can be described with the CD-MUSIC model (Hiemstra & Van Riemsdijk, 1996). For goethite, two types of surface sites are considered reactive, and the Langmuir equation was used to describe ion binding to these two types of surface sites. An electrostatic model, for example, the extended triple layer model, has been applied to describe the electrostatic profile near the surface, which has two Stern layers and two outer electrostatic planes (1- and 2-plane). According to the model the primary charging behavior of goethite is mainly caused by protonation and deprotonation of the singly ($\equiv FeOH^{-0.5}$) and triply ($\equiv Fe_3O^{-0.5}$) coordinated surface oxygens (Eq. 6.8).

$$\equiv FeOH^{-0.5} + H^+ = \equiv FeOH_2^{+0.5} \tag{6.8a}$$

$$\equiv Fe_3O^{-0.5} + H^+ = \equiv Fe_3OH^{+0.5} \tag{6.8b}$$

$$2\equiv FeOH^{-0.5} + Cd^{2+} = \equiv(FeOH)_2Cd^{+1} \tag{6.8c}$$

$$2\equiv FeOH^{-0.5} + Cd^{2+} + OH^- = \equiv(FeOH)_2CdOH^0 \tag{6.8d}$$

The intrinsic proton affinities are assumed to be equal to the point of zero charge of goethite. Indifferent electrolyte ions (e.g., Na^+ and NO_3^-) can form ion pairs with both singly and triply coordinate surface sites. Two bidentate ($\equiv(FeOH)_2Cd$, $\equiv(FeOH)_2CdOH$, Eq. (6.8-c) and (6.8-d)) Cd surface species have been proposed to describe Cd adsorption to goethite. To describe strong

TABLE 6.3 Generic NICA–Donnan model parameters for fulvic and humic acids (Milne et al., 2001, 2003).

Fulvic acids (FA) Q_{max1} **5.88 mol/kg;** Q_{max2} **1.86 mol/kg** p_1 **0.59;** p_2 **0.70** b **0.57**					**Humic acids (HA)** Q_{max1} **3.15 mol/kg;** Q_{max2} **2.55 mol/kg** p_1 **0.62;** p_2 **0.41** b **0.49**				
Ion	$\log\tilde{K}_1$	n_1	$\log\tilde{K}_2$	n_2	Ion	$\log\tilde{K}_1$	n_1	$\log\tilde{K}_2$	n_2
H	2.34	0.66	8.60	0.76	H	2.93	0.81	8.00	0.63
Ca	−2.13	0.85	−3.0	0.80	Ca	−1.37	0.78	−0.43	0.75
Cd	−0.97	0.68	0.50	0.50	Cd	−0.20	0.73	2.37	0.54
Cu	0.26	0.53	8.26	0.36	Cu	2.23	0.56	6.85	0.34
Ni	−2.07	0.65	2.03	0.53	Ni	−0.26	0.64	1.0	0.55
Pb	−1.16	0.60	6.92	0.69	Pb	1.25	0.60	4.84	0.69
Zn	−3.84	0.67	−0.73	0.61	Zn	0.11	0.67	2.39	0.27

Data were taken from Milne et al. (2001, 2003).
Cd, cadmium; *NICA*, nonideal competitive adsorption.

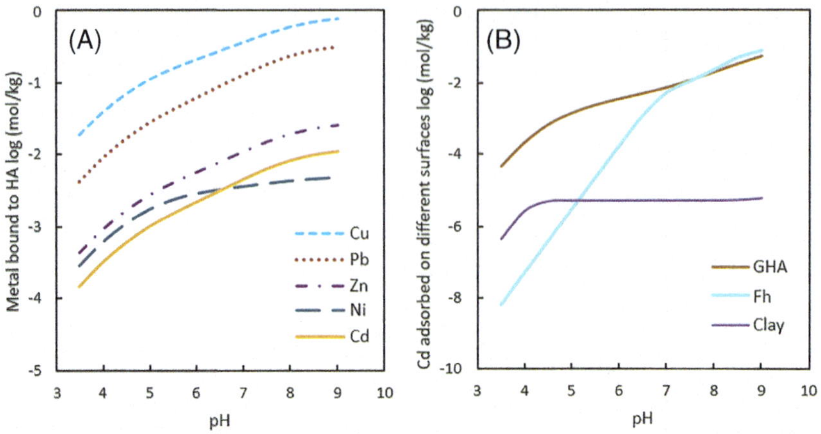

FIGURE 6.3 (A) Adsorption of Cd to generic humic acids (GHA) in comparison with other heavy metals (Cu, Pb, Zn, and Ni). (B) Comparison of pH effect on Cd adsorption to generic HA (GHA, representing SOM), to ferrihydrite (representing iron (hydr)oxides), and to clay minerals. For panels (A) and (B), the activity of free metal ion was fixed at 0.1 μM in a background of 2 mM $Ca(NO_3)_2$. Metal adsorption to GHA was calculated with NICA–Donnan model using generic HA parameters from Milne et al. (2001, 2003) (Table 6.3). Adsorption of Cd to ferrihydrite (Fh) was calculated with the generalized two-layer model (GTLM) using parameters taken from Dzombak and Morel (1990) (Table 6.4). Adsorption of Cd to clay was calculated using a Donnan model (Tipping, 1994) using a charge density of 0.25 mol/kg and a Donnan volume of 1 L/kg. *Cd*, Cadmium; *HA*, humic acid; *NICA*, nonideal competitive adsorption; *SOM*, soil organic matter.

adsorption at low Cd concentrations, part of the $\equiv FeOH^{-0.5}$ site is assumed to be the high-affinity site ($\equiv Fe_hOH^{-0.5}$) for metal ion binding.

For ferrihydrite, CD-MUSIC parameters for Cd binding based on the most recent CD-MUSIC modeling of ferrihydrite (Hiemstra & Zhao, 2016) are still lacking. Alternatively, in simulating metal ion adsorption in soils, metal ion binding to amorphous metal (hydr)oxides has been treated in a similar way to that of goethite after correcting for the difference in SSA or adopting the Generalized Two-Layer Model (GTLM) (Dzombak & Morel, 1990) (Table 6.4). On a mass basis, more Cd is adsorbed on NOM compared to metal (hydr)oxides and clay at a relatively low pH (Fig. 6.3B). However, the increase of Cd adsorption to oxides with increasing pH is steeper compared to that of NOM. At a high pH, comparable or even more Cd can be adsorbed on oxides than on NOM and clay at the same mass content.

6.4.1.3 Modeling cadmium binding to clay

Adsorption of Cd to clay silicates involves two distinct mechanisms. At low pH, nonspecific cation exchange dominates, driven by electrostatic interactions between Cd^{2+} and negatively charged clay surfaces. In contrast, specific adsorption, involving chemical binding of Cd to edge sites of clay minerals, becomes increasingly important at higher pH (Schindler et al., 1987). The cation-exchange process on clay can be described using a Donnan approach (Tipping, 1994). The advantage of such a model is that there is no need to select ion-exchange coefficients from the literature, because they are implicit in the model. The concentration of Cd in the Donnan phase is related to the concentration in the solution according to the Boltzmann accumulation factor.

TABLE 6.4 CD-MUSIC model parameters for goethite and GTLM model parameters for ferrihydrite. $Fe_hOH^{-0.5}$ represents 10% of the $FeOH^{-0.5}$ site and is assumed to act as the high-affinity site for Cd binding, accounting for strong adsorption at low Cd concentrations.

CD-MUSIC model parameters for goethite						**GTLM model parameters**	
$FeOH^{-0.5}$ ($Fe_hOH^{-0.5}$) site: 3.45 nm^{-2} **$Fe_3O^{-0.5}$ site: 2.70 nm^{-2}**			$C_1 = 0.93$ F/m^2; $C_2 = 0.75$ F/m^2			**S_1-OH site: 0.056 nm^{-2}** **S_2-OH site: 2.255 nm^{-2}**	
Surface species	**log*K***	**Site**	**0-Plane**	**1-Plane**	**2-Plane**	**Surface species**	**Log*K***
$FeOH^{-0.5}$	0	$1FeOH^{-0.5}$	0	0	0	S_1-OH	0.00
$FeOH_2^{+0.5}$	9.0	$1FeOH^{-0.5}$	1	0	0	S_1-O^-	−8.93
$Fe_3O^{-0.5}$	0	$1Fe_3O^{-0.5}$	0	0	0	S_1-OH_2^+	7.29
$Fe_3OH^{+0.5}$	9.0	$1Fe_3O^{-0.5}$	1	0	0	S_2-OH	0.00
$FeOH_2NO_3^{-0.5}$	8.32	$1FeOH^{-0.5}$	1	−1	0	S_2-O^-	−8.93
$Fe_3OHNO_3^{-0.5}$	8.32	$1Fe_3O^{-0.5}$	1	−1	0	S_2-OH_2^+	7.29
$FeOHNa^{+0.5}$	−0.6	$1FeOH^{-0.5}$	0	1	0	S_2-$OHCa^{2+}$	−5.85
$Fe_3ONa^{+0.5}$	−0.6	$1Fe_3O^{-0.5}$	0	1	0	S_1-$OHCa^{2+}$	4.97
$FeOHCa^{+1.5}$	3.23	$1FeOH^{-0.5}$	0.31	1.69	0	S_1-OCd^+	0.47
$FeOHCaOH^{+0.5}$	−6.42	$1FeOH^{-0.5}$	0.31	0.69	0	S_2-OCd^+	−2.90
$FeOHCa^{+1.5}$	1.8	$1FeOH^{-0.5}$	0	2	0	S_1-PO_4^{2-}	17.72
$Fe_3OCa^{+1.5}$	1.8	$1Fe_3O^{-0.5}$	0	2	0	S_1-HPO_4^-	25.39
$(FeO)_2PO_2^{-2}$	29.72	$2FeOH^{-0.5}$	0.46	−1.46	0	S_1-H_2PO_4	31.29
$FeOPO_2OH^{-1.5}$	27.63	$1FeOH^{-0.5}$	0.28	−1.28	0	S_2-PO_4^{2-}	17.72
$(FeOH)_2Cd^{+1}$	6.98	$2FeOH^{-0.5}$	0.71	1.29	0	S_2-HPO_4^-	25.39
$(FeOH)_2CdOH^0$	−2.94	$2FeOH^{-0.5}$	0.71	0.29	0	S_2-H_2PO_4	31.29
$(Fe_hOH)_2Cd^{+1}$	6.98	$2Fe_hOH^{-0.5}$	0.71	1.29	0		
$(Fe_hOH)_2CdOH^0$	−0.32	$2Fe_hOH^{-0.5}$	0.71	0.29	0		

Parameters for CD-MUSIC model were taken from Hiemstra et al. (2013) and Weng et al. (2001a), whereas the parameters for GTLM model were taken from Dzombak and Morel (1990). *Cd*, cadmium; *CD-MUSIC*, charge distribution-multisite complexation.

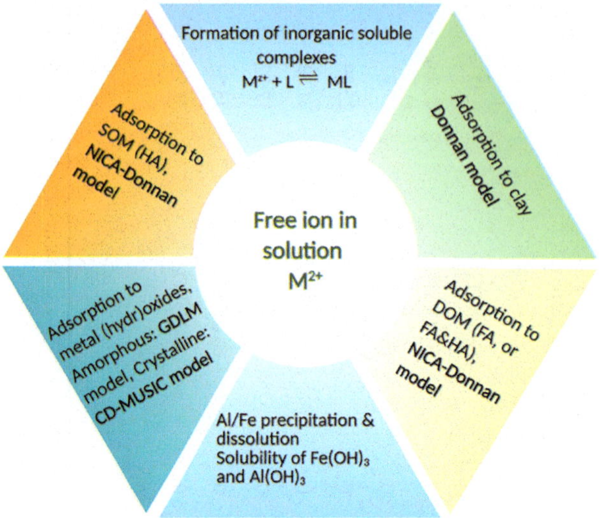

FIGURE 6.4 Conceptual structure of the multisurface model (MSM), showing the reactive surfaces considered for metal adsorption in the solid phase (SOM, metal (hydr)oxides, and clay) and in the solution phase (DOM), as well as the corresponding surface complexation model used. In addition, formation of inorganic soluble species is considered in the model. The mineral phase of Fe and Al (hydr)oxides is introduced to control the activity of free Al^{3+} and Fe^{3+}, which can compete with metal ions of concern in their adsorption to soil reactive surfaces. Precipitation of Cd^- containing minerals can also be integrated into the model calculation by using the solubility product of the corresponding mineral. *Cd*, Cadmium; *CD-MUSIC*, charge distribution-multisite complexation; *DOM*, dissolved organic matter; *NICA*, nonideal competitive adsorption; *SOM*, soil organic matter.

6.4.2 Modeling cadmium binding to soil

6.4.2.1 Multisurface modeling

Soil is a complex system in which Cd can interact with various reactive surfaces. To apply SCMs developed for individual soil reactive components the approach of MSM was proposed (Weng et al., 2001a). The MSM considers soil as a set of independent adsorbent surfaces, that is, SOM, DOM, metal (hydr)oxides, and clay silicates. The MSM assumes no interaction between these components, and the amounts of Cd adsorbed to SOM, metal (hydr)oxides, and clay silicates are summed to derive the total amount adsorbed to the solid phase of soils. For each reactive surface a representative adsorbent is selected, and a specific SCM is used to simulate ion binding (Fig. 6.4). For instance the NICA–Donnan model can be used to describe Cd adsorption to SOM using generic parameters for HA listed in Table 6.3. The reactive fraction of SOM can be quantified through fractionation procedures following International Humic Substances Society (IHSS) (Aiken, 1985) or van Zomeren and Comans (2007), or by assuming a fixed proportion, typically 30%–50%, of total SOM as reactive (Groenenberg and Lofts, 2014).

In the MSM, Cd binding to metal (hydr)oxides is typically simulated based on iron (hydr)oxides for practical reasons, with aluminum (hydr)oxides treated similarly after summing Fe and Al (hydr)oxides on a molar basis. The amorphous oxides and the sum of amorphous and crystalline oxides are commonly estimated using acid ammonium oxalate extraction (Schwertmann, 1964) and dithionite-citrate-carbonate extraction, respectively (Hiemstra et al., 2010). In some studies, metal ion adsorption to amorphous and crystalline oxides is modeled separately using the GTLM or CD-MUSIC model (Weng et al., 2001a), whereas in other studies, the surface area of amorphous and crystalline oxides is combined and simulated with either GTLM or CD-MUSIC model (Gao et al., 2022; Van Eynde et al., 2022).

For simplicity, only the electrostatic binding of metal ions to clay is considered, which can be described using the Donnan model (Tipping, 1994). The charge of clay can be estimated based on the cation exchange capacity (CEC) of the clay minerals or from the soil CEC after subtracting charge contribution from SOM (Weng et al., 2001a). A Donnan volume of 1 L/kg is commonly used (McBride, 1994).

In addition to metal ion adsorption to reactive surfaces in the solid phase, the formation of soluble complexes, including both inorganic and organic complexes, is also considered in the MSM (Fig. 6.4). Metal binding to DOM can be simulated using the NICA–Donnan model, with FA or a combination of HA and FA representing the metal-reactive fraction of DOM (Groenenberg & Lofts, 2014). The reactive fraction of DOM can be determined using fractionation methods of IHSS (Aiken, 1985; van Zomeren & Comans, 2007), similar to that of SOM, or by assuming a certain fraction (typically 50%–100%) of DOM as above. Furthermore, competition from ions such as Ca^{2+}, Al^{3+}, and Fe^{3+} with metal ions of concern in their adsorption needs also needs to be considered. In many cases the activity of free Al^{3+} and Fe^{3+} is regulated in the model by the solubility of Al and Fe (hydr)oxides in soils.

The MSM approach has been widely applied and has proven effective in predicting the distribution of metals such as Cd, Cu, Ni, and Zn in various soils (Dijkstra et al., 2009; Duffner et al., 2014; Gao et al., 2022; Ren et al., 2015; Van Eynde et al., 2022; Zhao et al., 2018).

6.4.2.2 Considering surface interaction (natural organic matter-charge distribution, Ligand and Charge Distribution model)

The MSM approach does not account for the effect of particle interactions, such as NOM mineral interactions, on ion binding to each type of soil surface. However, strong interactions between NOM and soil minerals are commonly observed. In particular the metal-reactive HA and FA fractions exhibit a high affinity for binding to soil metal (hydr)oxides. To account for this phenomenon the NOM-CD model (Hiemstra et al., 2010, 2013) and the LCD model (Filius et al., 1997; Weng et al., 2006) have been developed. The NOM-CD model is an extension of the CD-MUSIC model. The model includes a virtual component, that is, $HNOM^{-1}$, which is equivalent to two carboxylic groups, to represent NOM adsorbed in the Stern layer at the surface of oxides. This virtual organic molecule can form an inner-sphere, an outer-sphere, and a protonated inner-sphere complex with reactive sites on oxides (Hiemstra et al., 2013). The LCD model integrates the NICA model and the CD-MUSIC model to describe the surface complexation reactions on soil minerals under the influence of NOM adsorption. In this model the NOM present on the surface of oxides is represented with HA or FA, and their charge carried and metal ion adsorption to their acidic groups are calculated with the NICA model. In addition, oxides–NOM–metal and oxides–metal–NOM ternary surface complexes can be accounted for (Weng et al., 2011).

Li et al. (2022) modeled Cd binding to a mixture of HA and goethite and to soils using NOM-CD and NICA–Donnan models. In their approach, HA or SOM is divided into two fractions: one resides within the Stern layer on the oxide surface, whereas the other extends beyond the Stern layer and remains unaffected by the oxides (Fig. 6.5). The fraction within the Stern layer influences Cd binding to oxides primarily through electrostatic effects, which are accounted for by the NOM-CD model. Cd binding to the remaining of HA and SOM (free from the influence of oxides) is calculated with the NICA–Donnan model, similar to the approach in the MSM. The results show that for Cd, calculations using the combination of NOM-CD and NICA–Donnan models are comparable to those calculated with the linear additive MSM (Li et al., 2022), indicating that the effect of NOM–oxide interaction on the overall amount of Cd adsorbed to soils is small. This limited effect can be attributed to the dominance of NOM as a metal ion sorbent in soils. At relatively high NOM-to-oxide ratios, only a small fraction of NOM is in direct contact with oxides, whereas most extends farther from the mineral surface. Consequently the influence

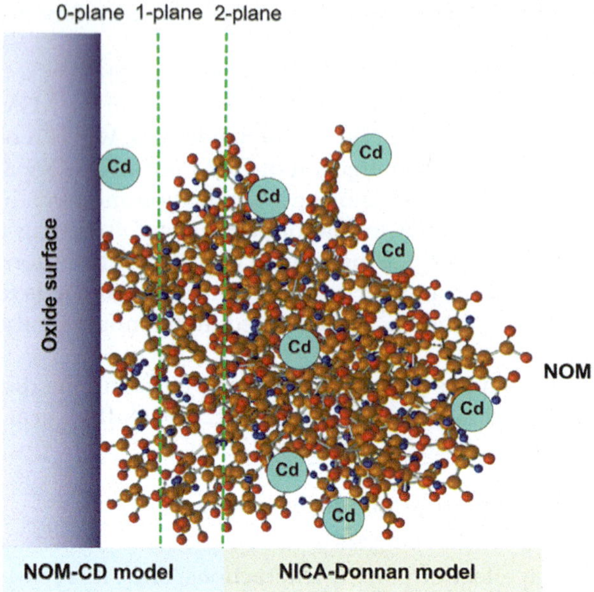

FIGURE 6.5 Schematic illustration of modeling Cd adsorption to the organo-mineral assemblage. In this approach, NOM is divided into two fractions: one fraction resides in the Stern layer close to the mineral surface (between the 0-plane and 2-plane), whereas the remaining fraction extends out of the Stern layer. It is assumed that NOM present in the Stern layer influences Cd adsorption to the mineral surface, whereas Cd adsorption to the remaining NOM is not influenced by the underlying mineral. The NOM-CD model can be applied to simulate Cd adsorption on mineral surfaces, accounting for mineral–organic matter interactions. The NICA–Donnan model can be used to simulate Cd adsorption to NOM extruding out of the Stern layer. *Cd*, Cadmium; *NICA*, nonideal competitive adsorption; *NOM-CD*, natural organic matter-charge distribution. *This illustration was modified from Li et al. (2022) under the CC-BY 4.0 license.*

of oxide interactions on metal binding to NOM is minimal. This explains the success of MSM in describing and predicting metal ion partitioning in soils, despite its simplified, linear-additive assumptions.

6.5 Modeling cadmium speciation in soils with cadmium-bearing minerals

6.5.1 Modeling cadmium speciation in mine-waste-contaminated soils

Although some soils naturally contain Cd, elevated concentrations of this metal in soils are primarily attributed to anthropogenic activities, such as metal mining and smelting operations (Hooda, 2010; Wei et al., 2025; Zhao et al., 2015). Waste from mining and smelting of zinc and lead ores typically contains Cd (Anju & Banerjee, 2012; Schaider et al., 2007). Cd is also present in other mine wastes, such as waste from copper, gold, mercury, silver, and coal production. One characteristic of Cd in soil contaminated with mine waste is its very high concentration. For example, levels as high as 64.7 mg/kg have been detected in agricultural soils impacted by Zn smelter waste (Gao et al., 2022). Another characteristic relates to the presence of Cd-bearing minerals. In sulfide-containing mine waste, CdS (greenockite) can be present. Other forms of Cd-containing minerals include CdO, $Cd(OH)_2$, and $CdCO_3$. In addition, Cd can also be present in minerals associated with other metals (Loganathan et al., 2012).

Although the Cd-bearing precipitates mentioned above may not persist under typical soil conditions (see Section 6.1.1), they can occur in mine-waste-contaminated soils because of elevated Cd and/or carbonate concentrations and the slow dissolution of certain mineral phases. In principle, as long as the Cd-containing minerals remain, their solubility, rather than adsorption/desorption, controls free Cd^{2+} concentration in the soil solution. However, if dissolution is substantially slower than Cd loss through processes such as plant uptake or leaching, desorption may still dominate the control of soluble Cd^{2+}, whereas the minerals act as kinetically inert phases (Loganathan et al., 2012). In any case, whether governed by the dissolution of Cd-bearing minerals or the desorption of Cd from reactive soil surfaces, a certain fraction of Cd remains adsorbed to soil reactive surfaces.

The MSM approach has been effectively applied to predict Cd partitioning and speciation in soils contaminated by smelting activities and mine waste (Gao et al., 2022; Ren et al., 2015). One critical consideration in assessing and modeling Cd speciation in mine-waste-contaminated soils is the choice of extractant used to estimate the reactively adsorbed Cd pool. Nitric acid (0.43 M) extraction has been employed to approximate the reactively adsorbed pool of metal ions (Fig. 6.1). Gao et al. (2022) found that MSM predictions were more accurate when 0.005 M DTPA-extractable Cd was used as the input for the reactively adsorbed fraction in high-pH mine-waste-contaminated soils. In contrast, 0.43 M HNO_3-extractable Cd tended to overestimate the adsorbed Cd fraction, likely because of the partial dissolution of Cd-bearing minerals (Degryse et al., 2009; Gao et al., 2022; Groenenberg et al., 2017). Another important consideration in modeling Cd adsorption in such soils relates to their distinct physicochemical properties compared to nearby uncontaminated soils. For example, Gao et al. (2022) reported an average metal-reactive SOM fraction of 24% in total SOM among 12 soil samples contaminated by mine waste, which is notably lower than previously reported values (31%–87%) (Groenenberg & Lofts, 2014). The lower reactive SOM fraction in the contaminated soils may be because of a relatively (1) low SOM status and (2) low humification rate of fresh organic matter. In addition, mine waste may contain large quantities of iron and manganese (hydr)oxides, which can enhance the contribution of metal (hydr)oxides to Cd adsorption in soil.

6.5.2 Modeling cadmium speciation in calcareous soils

Previous studies have shown that calcareous soils, characterized by high pH and the presence of calcium carbonate, exhibit a strong ability to immobilize Cd. In such soils, calcium carbonate may play an important role in Cd immobilization because of sorption and precipitation of Cd by calcium carbonate (Moral et al., 2005). Although the precipitation of $CdCO_3$ (otavite, $\log K_{sp} = -12.1$) (Lindsay, 1979) in bulk soil solution is unlikely under normal Cd concentrations, it may occur at high Cd levels. Because of the similar hydrated ionic radii of Cd^{2+} and Ca^{2+} (Martin-Garin et al., 2003; Zachara & Smith, 1994), heteroepitaxial growth of otavite can occur at calcite ($CaCO_3$) surfaces, facilitated by lattice consistency (Prieto et al., 2003). Cd can thus precipitate in the form of $Cd_{(1-x)}Ca_xCO_3$ with a calcite-type structure at calcite surfaces (Stipp et al., 1992; Van et al., 2019). The rapid initial removal is often attributed to chemisorption, whereas the slower subsequent removal is associated with surface precipitation or coprecipitation (Prieto et al., 2003). In addition, under alkaline conditions, Cd adsorption to reactive surfaces, such as SOM, metal (hydr)oxides, and clay, is enhanced, further promoting Cd immobilization.

Nevertheless, quantifying Cd sorption and surface precipitation on calcium carbonate remains challenging. It is unclear whether Cd–carbonate interactions or other processes predominantly govern soluble and free Cd^{2+} concentrations in calcareous soils. In current geochemical modeling of Cd partitioning in soil, sorption and surface precipitation of Cd on calcium carbonate are mostly ignored. Still the MSM has been applied to predict Cd partitioning in calcareous soils with reasonable success (Gao et al., 2022) by considering Cd adsorption to SOM, metal (hydr)oxides, and clay as well as to DOM. Similar to the mine-waste-contaminated soils (see Section 6.5.1), attention has to be paid to choosing an appropriate method to measure reactively adsorbed Cd. As the MSM does not incorporate Cd associated with calcium carbonate, it is preferable to use methods that exclude this fraction. Marzouk et al. (2013) found that the reactively adsorbed Cd determined with 0.43 M HNO_3 extraction was about twice that determined with isotopic dilution in three calcareous soils. As noted above, this may be because of dissolution of Cd-bearing carbonate minerals during acid extraction. Therefore, when applying the MSM to calcareous soils without modeling Cd associated with carbonates, the reactive Cd pool is better estimated using isotopic dilution or DTPA extraction (Gao et al., 2022; Marzouk et al., 2013).

6.5.3 Modeling cadmium speciation in paddy soil

Unlike upland soils, lowland rice paddy soils undergo alternating flooding and draining cycles, resulting in shifts between anoxic and oxic conditions that alter biological and chemical properties, particularly pH and redox potential (E_h) (Kögel-Knabner et al., 2010). These changes can markedly affect Cd speciation in paddy soils (see Chapter 7). Flooding typically increases the pH of acidic soils because of proton consumption during reduction processes, whereas it lowers the pH in alkaline soils because of the accumulation of carbon dioxide and organic acids produced during the fermentation of organic matter (Du Laing et al., 2009; Kögel-Knabner et al., 2010; Li et al., 2020). Changes in pH and reductive dissolution of iron (hydr)oxides can also result in fluctuations of DOM concentration and composition because of impacts on the dissolution and desorption of organic matter. Changes in DOM contribute as well to the dynamics of Cd partitioning in paddy soil via the formation of DOM-complexed Cd.

Adsorption of Cd to soil reactive surfaces generally increases with increasing pH. Ding et al. (2019) developed a mathematical model to simulate the dynamics of pH changes under the influence of cycles of flooding and drainage based on field monitoring data collected in 20 paddy soils. Li et al. (2020) incorporated the pH dynamic model of Ding et al. (2019) into the MSM to account for drastic pH changes in paddy soils. After soil drainage, reduced substances (e.g., NH_4^+, Fe(II), Mn(II), and S^{2-}) are rapidly oxidized by oxygen, resulting in substantial proton release and subsequent Cd desorption from soil surfaces (de Livera et al., 2011; Du Laing et al., 2009; Kögel-Knabner et al., 2010).

In addition to variations in soil pH and DOM during a flooding and draining cycle the reduction of sulfate can enhance Cd retention via the formation of CdS (greenockite) precipitates during soil flooding (Hashimoto et al., 2016), whereas the reoxidation of CdS occurs during the soil drainage and thereby releases Cd, facilitating Cd remobilization (Wang et al., 2019) (see Chapter 7). To model metal (including Cd) speciation under flooded conditions, Pan et al. (2016) incorporated CdS precipitation into the MSM framework. They estimated free sulfide activity using measured soil sulfate concentrations and E_h values and determined free Cd^{2+} activity based on the solubility of CdS. Their results highlighted the critical role of Cd-sulfide precipitation in maintaining low Cd solubility under anoxic conditions.

6.6 Adsorption and desorption kinetics

6.6.1 Kinetics in adsorption and desorption

The adsorption models described in Sections 6.3 and 6.4 assume chemical equilibrium among different species of Cd in both the solution and solid phases. However, this assumption of equilibrium may not always be appropriate as a result of fast and frequent variations in soil environmental conditions (Shi et al., 2005). The dynamic behavior of heavy metals may be affected by multiple coupled physical, chemical, and biological processes, such as mass flow, diffusion, chemical reactions such as adsorption/desorption and precipitation/dissolution, and uptake by plant roots (Buffle et al., 2009; Peng et al., 2018; Selim, 2014). The study of kinetics provides insights into the possible mechanisms of adsorption along with the reaction pathways (Gupta & Bhattacharyya, 2011). The choice of either following an equilibrium or kinetic approach regarding adsorption modeling depends on the rate of the process under consideration relative to the rate of adsorption and desorption reactions.

Adsorption of Cd can be divided into fast and slow stages, with a transition point observed between 0.5 and 1 hour. For instance, Elbana and Selim (2019) concluded that the adsorption of Cd to soils is generally fast, and pseudo-equilibrium can be obtained within about 1 hour. Lu et al. (2022) reported that Cd adsorption in six paddy soils increased significantly within the first 30 minutes and approached pseudo-equilibrium after 1 hour. Similarly, Zheng et al. (2022) observed that Cd adsorption to reclaimed soils exhibited a rapid phase within the first hour, followed by a slower phase lasting up to 12 hours.

Similarly a fast and slow stage is common for the desorption kinetic process, often with a long tail in the timescale. In the biphasic process the rapid release phase and the slower phase are attributed to exchangeable and specifically adsorbed Cd fractions, respectively (Wang et al., 2009). In addition, adsorption–desorption hysteresis exists for Cd in soils. This hysteresis suggests that the amount of Cd desorbed from the soil is less than that originally adsorbed at the same concentration in solution. This behavior reflects irreversible adsorption, potentially because of pore blockage or strong chemical binding. Elbana and Selim (2019) investigated Cd adsorption and desorption in a range of soils differing in properties. After 7 days of adsorption, only 2%–70% Cd can be desorbed in the following 7 days. A high degree of hysteresis was observed in calcareous soils but also in some of the acidic soils. In general, lower desorption rates, lower reversibility, and higher hysteresis of Cd are associated with soils exhibiting higher adsorption capacities. For instance, Oh and Shin (2010) observed that the desorbed fraction of Cd decreased with increasing soil CEC and surface area (irrespective of adsorption aging time). Zhang et al. (2017) found that Cd desorption rates were negatively correlated with SOM, CEC, free iron oxides, and manganese oxides, consistent with findings of Rezaei Rashti et al. (2014), who identified pH, CEC, SOM, and oxalate-extractable Fe as key factors controlling Cd desorption behavior in subtropical soils.

6.6.2 Modeling kinetic processes in adsorption and desorption

Empirical kinetic models, such as the pseudo-first-order, pseudo-second-order, and Elovich equations, are widely used to describe the adsorption/desorption processes of metal ions in soils (Gupta & Bhattacharyya, 2011). Lu et al. (2021) reported that Cd adsorption in six paddy soils was well described by both the pseudo-first-order and pseudo-second-order kinetic models ($R^2 > 0.85$), whereas the Elovich and intraparticle diffusion models showed poor fits to the data ($R^2 < 0.57$). Regarding desorption, Wang et al. (2009) found that Cd desorption under simulated acid rain conditions in two soils (Ultisol and Oxisol) was well described by both the first-order and Elovich equations. To address the complexity of soil, some studies assume multiple adsorption–desorption sites in soils, each with distinct kinetic rates (Elbana & Selim, 2010; Liao & Selim, 2010). For instance, Kandpal et al. (2005) investigated the desorption kinetics of Cd from acidic (Inceptisol) and neutral (Mollisol) soils spiked with either inorganic salts or acidic sludge, and they found that in general, Cd release from these contaminated soils conforms to a multisite first-order kinetic model.

Predicting the kinetics of heavy metal adsorption and desorption in soil requires consideration of multiple heterogeneous soil binding sites and variations in reaction chemistry conditions. The above-mentioned approach relies on equation fitting, and the resulting model is empirical. Shi et al. (2005, 2007, 2013) attempted to couple the mechanistic SCMs with kinetic models to generate a unified adsorption–desorption equilibrium-kinetics model. They first focused on metal binding to SOM and found reasonable predictions of Cd adsorption and desorption kinetics by integrating the equilibrium model WHAM into a simple two-site kinetics model (Shi et al., 2005, 2007, 2013). Later, this approach was extended by including CD-MUSIC model to describe Cd binding to metal (hydr)oxides (Peng et al., 2018). These attempts are promising and lay the foundation for future mechanistic modeling of Cd adsorption and desorption kinetics in soils.

6.7 Understanding immobilization effects of soil additives using chemical speciation modeling

6.7.1 Immobilization mechanisms of soil additives

Various technologies have been developed to remediate Cd-contaminated soils. Among these, in situ immobilization technology has emerged as a particularly promising strategy, especially for agricultural soils, because of its efficiency and feasibility (Hamid et al., 2020; Palansooriya et al., 2022; Yuan et al., 2021; Zhang et al., 2021; Zhao et al., 2022) (see Chapter 18). Immobilization in this case refers to the reduction of soluble Cd concentration in soil solution via decreasing its solubility (Cd in aqueous phase/solid phase). This technique involves the application of remediation materials, such as lime compounds, clay minerals, zero-valent iron, phosphates, biochar, and organic fertilizers, to Cd-contaminated soil. The primary objective is to modify the chemical form of Cd within the soil, thereby decreasing its mobility and bioavailability.

The mechanisms of various remediation materials for Cd in contaminated soils can be systematically categorized as follows (Hamid et al., 2020; Yan et al., 2022; Zhang et al., 2021), in line with the factors controlling Cd speciation and distribution discussed above (Section 6.1.3).

1. **Elevation of soil pH**: The addition of materials, such as lime and biochar, can raise soil pH substantially, thereby decreasing Cd activity. The increase in pH promotes deprotonation of soil particles and increases the negative surface charge, enhancing Cd adsorption. Furthermore, in soils containing Cd-bearing minerals, these minerals are generally less soluble at higher pH.
2. **Expanding adsorption capacity**: Materials, such as clay minerals, zero-valent iron, biochar, and organic fertilizers, can effectively adsorb Cd. Their introduction directly increases Cd adsorption sites, enhances the soil's CEC, and reduces soluble Cd concentrations.
3. **Coadsorption with anions**: Anionic remediation materials, such as phosphate, can form ternary coadsorption structures with Cd at soil mineral interfaces (e.g., metal (hydr)oxides) or promote Cd adsorption through electrostatic synergy by lowering the electrostatic potential of soil colloids.

4. **Precipitation formation**: Carbonate- and phosphate-containing materials can promote the formation of Cd precipitates. For instance, Cd can be immobilized via surface precipitation on calcite. At high concentrations of Cd, carbonate, or phosphate, $CdCO_3$ (otavite) and $Cd_3(PO_4)_2$ may also precipitate. In addition, certain immobilizing agents, such as biochar and HAs, can promote electron transfer under flooded conditions, facilitating the formation of CdS precipitates.

6.7.2 Examples of mechanistic model simulation of cadmium immobilization by soil additives

The mechanistic model can be employed to quantify the contributions of different mechanisms involved in Cd immobilization by passivating agents. This approach is instrumental in elucidating the mode of action of passivating agents and predicting their effects for efficient remediation. Currently the use of mechanistic models to investigate the impact of amendments on Cd dynamics in soil is in its early stages. However, some insightful findings have emerged. The MSM discussed in Section 6.4.2.1 has shown considerable reliability in predicting the impacts of amendments such as compost, Zn-containing materials, and lime on soil Cd immobilization. Because of the mechanistic nature of the SCMs within MSM the effects of pH changes on Cd adsorption, as well as competitive and synergistic ion interactions, can be simulated using existing model parameters without altering the model structure. For instance, Li et al. (2022) utilized MSM to predict changes of dissolved Cd with pH in soil amended with lime and its relationship with Cd uptake by wheat. The study demonstrated that MSM could predict the accumulation of Cd by wheat over a wide concentration range, regardless of whether amendments were applied. Similarly, Qin et al. (2024) employed the MSM to calculate Cd activity in soils amended with lime and compost. The results indicated that the model could adequately capture changes in dissolved Cd because of the chemical alterations induced by the amendments.

The synergistic effects of anionic agents such as phosphate on Cd adsorption occur mainly on metal (hydr)oxides. Deng et al. (2023) and Li et al. (2022) successfully predicted soluble Cd in soil using the NOM-CD model when phosphate was added. Their modeling considered the formation of ternary complexes between phosphate and Cd on oxides. The calculations revealed that phosphate addition promoted Cd immobilization primarily through electrostatic enhancement, rather than the formation of ternary complexes, at concentrations relevant to mildly contaminated soils. In addition to the effects on adsorption, anionic agents such as phosphate can immobilize Cd via precipitation at high concentrations. We assessed Cd speciation in soil under varying pH and E_h with the presence of phosphate and sulfate via MSM (Fig. 6.6) for two levels of total Cd (0.2 and 20 mg/kg) and phosphate (90 and 300 mg/kg) contents. The MSM simulation showed that adsorption of Cd onto NOM (Cd–NOM) predominates under moderately acidic to neutral pH conditions combined with high E_h (well aerated soil). As pH increases, adsorption onto metal (hydr)oxides (Cd–oxide) becomes increasingly significant. The addition of phosphorus enhances Cd adsorption onto oxide surfaces under low Cd conditions. Cd phosphate precipitation (Cd–P) is absent at low Cd concentrations but occurs prominently at elevated Cd and phosphate levels under high pH. Notably, CdS precipitation formation occurs under reducing conditions, especially in combination with acidic pH.

In the above model simulations on the effects of soil amendments on Cd partitioning resulting from changes in pH, E_h, and phosphate concentration, existing geochemical models can be

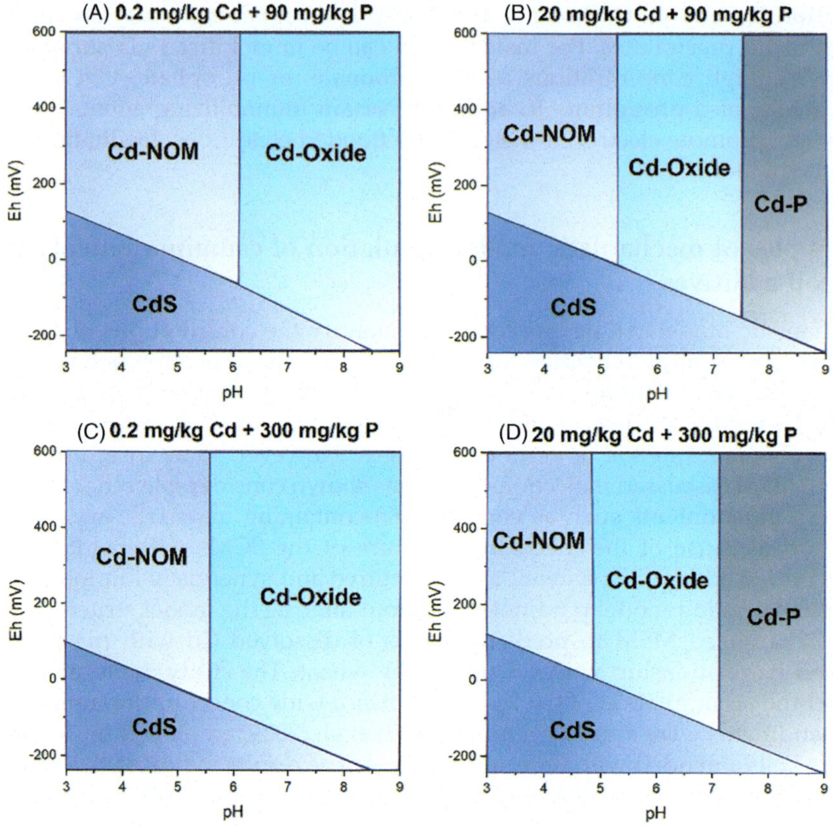

FIGURE 6.6 pH-E_h predominance diagram of Cd speciation in soil under varying total Cd and phosphate (P) levels: (A) Cd 0.2 mg/kg, P 90 mg/kg; (B) Cd 20 mg/kg, P 90 mg/kg; (C) Cd 0.2 mg/kg, P 300 mg/kg; and (D) Cd 20 mg/kg, P 300 mg/kg. Colored fields indicate the dominant Cd species under specific pH-E_h conditions: Cd-NOM (Cd associated with natural organic matter); Cd-Oxide (Cd adsorbed onto metal (hydr)oxides); Cd-P (Cd phosphate precipitates, $\log K_{sp} = -32.6$); and CdS (Cd sulfide precipitates, $\log K_{sp} = -27.4$). Calculations were performed using the multisurface model. Adsorptive surfaces included soil organic matter (represented by humic acid (HA), 20 g organic matter/kg soil, and 50% is regarded as HA), metal (hydr)oxides (represented by goethite, 10 g/kg soil), and clay minerals (50 g/kg soil, represented by kaolinite) (negligible contribution, not depicted). Adsorption to soil organic matter was simulated with NICA–Donnan model using parameters (Milne et al., 2001, 2003) for generic HA (Table 6.3). Adsorption to metal (hydr)oxides was simulated with CD-MUSIC model using parameters of Table 6.4 (Weng et al., 2001a). The influence of coexisting cations (Ca^{2+}, Fe^{3+}, and Al^{3+}) was incorporated, with Ca^{2+} fixed at 2 mmol/L and Fe^{3+}/Al^{3+} activities governed by the solubility of their respective hydroxide minerals. Total sulfur content is 300 mg/kg. *Cd*, Cadmium; *CD-MUSIC*, charge distribution-multisite complexation; *NICA*, nonideal competitive adsorption.

directly applied. However, for amendments that themselves bind Cd, speciation modeling is only feasible if models with relevant parameters for Cd binding to those materials are available. For simplicity, organic amendments resembling natural SOM (e.g., humic substances and manure) can be modeled using the NICA–Donnan approach with generic parameters, provided that the reactive fractions of the materials are determined. For oxide-based materials such as nanozero-valent iron the CD-MUSIC model can be used after characterizing the SSA of the oxides. More accurate assessment can be carried out once the model and model parameters are validated and, if necessary, optimized using batch adsorption experiments in which Cd adsorption to the material under concern is measured.

Immobilization agents often operate through multiple mechanisms, for which mechanistic modeling offers considerable application potential. A typical example is biochar and its modified derivatives. Beyond its Cd-binding capacity, biochar contains alkaline and mineral components (e.g., carbonate and phosphate) that can raise soil pH and promote Cd immobilization via enhanced adsorption and precipitation. In addition, biochar may facilitate electron transfer, accelerating sulfate reduction and thereby promoting CdS precipitation under flooded conditions (Wang et al., 2022). Recently the generic three-site NICA–Donnan model parameters for proton and Cd adsorption by biochar were derived (Huang et al., 2024, 2025; Xiong et al., 2021), providing a basis for modeling Cd adsorption to biochar. Modified biochar can also be included in mechanistic calculations. Once available, such models allow differentiation between liming, adsorption, and precipitation effects of biochar on Cd immobilization. The modeling results can be used to optimize the amount of biochar required for remediation.

6.8 Summary and future perspectives

- The environmental impact of Cd is influenced not only by its total concentration but also by its chemical speciation across solid (adsorbed, precipitated, and mineral-occluded) and solution (free ion, inorganic, and organic complexes) phases, with adsorption often being the dominant controlling process on its distribution.
- Both empirical (e.g., K_d, Langmuir, and Freundlich) and mechanistic models (e.g., MSM, NOM-CD, and LCD) have been developed to quantify Cd adsorption and distribution in soils, with mechanistic models offering greater predictive power by considering physicochemical nature of surface reactions.
- Bulk precipitation, surface, or coprecipitation of, for example, $CdCO_3$, Cd_2PO_4, and CdS may occur under elevated Cd concentration and appropriate soil conditions (e.g., pH and E_h). Accurate modeling of adsorption requires the prevention of the dissolution of Cd-bearing minerals during the extraction of reactively adsorbed Cd, particularly in mine-waste-contaminated and calcareous soils.
- Future research directions may include (1) advancing the understanding and modeling of ternary surface complexation of Cd with other ligands (e.g., phosphate) on mineral surfaces, (2) developing approaches to identify key molecular properties for better predicting the reactivity of soil surfaces toward Cd adsorption, and (3) integrating models of organic carbon–mineral interactions and anion/cation adsorption into a unified modeling framework.

AI disclosure

During the preparation of this work the author(s) used ChatGPT to improve readability. After using this tool/service the author(s) reviewed and edited the content as needed and take(s) full responsibility for the content of the publication.

References

Aiken, G. R. (1985). Isolation and concentration techniques for aquatic humic substances. In G. R. Aiken, D. M. McKnight, R. L. Wershaw, & P. MacCarthy (Eds.), *Humic substances in soil, sediment, and water: Geochemistry and isolation*. Wiley-Interscience.

Anju, M., & Banerjee, D. K. (2012). Multivariate statistical analysis of heavy metals in soils of a Pb–Zn mining area, India. *Environmental Monitoring and Assessment, 184*(7), 4191–4206. doi:10.1007/s10661-011-2255-8.

Barrow, N. J., & Whelan, B. R. (1998). Comparing the effects of pH on the sorption of metals by soil and by goethite, and on uptake by plants. *European Journal of Soil Science, 49*(4), 683–692. doi:10.1046/j.1365-2389.1998.4940683.x.

Basta, N. T., Ryan, J. A., & Chaney, R. L. (2005). Trace element chemistry in residual-treated soil: Key concepts and metal bioavailability. *Journal of Environmental Quality, 34*(1), 49–63. doi:10.2134/jeq2005.0049dup.

Boekhold, A. E., Temminghoff, T. J. M., & Van der Zee, S. E. A. T. M. (1993). Influence of electrolyte composition and pH on cadmium sorption by an acid sandy soil. *Journal of Soil Science, 44*(1), 85–96. doi:10.1111/j.1365-2389.1993.tb00436.x.

Bolan, N. S., Adriano, D. C., Duraisamy, P., Mani, A., & Arulmozhiselvan, K. (2003). Immobilization and phytoavailability of cadmium in variable charge soils. I. Effect of phosphate addition. *Plant and Soil, 250*(1), 83–94. doi:10.1023/a:1022826014841.

Bonten, L. T. C., Groenenberg, J. E., Weng, L., & van Riemsdijk, W. H. (2008). Use of speciation and complexation models to estimate heavy metal sorption in soils. *Geoderma, 146*(1–2), 303–310. doi:10.1016/j.geoderma.2008.06.005.

Buekers, J., Van Laer, L., Amery, F., Van Buggenhout, S., Maes, A., & Smolders, E. (2007). Role of soil constituents in fixation of soluble Zn, Cu, Ni and Cd added to soils. *European Journal of Soil Science, 58*(6), 1514–1524. doi:10.1111/j.1365-2389.2007.00958.x.

Buffle, J., Wilkinson, K. J., & Van Leeuwen, H. P. (2009). Chemodynamics and bioavailability in natural waters. *Environmental Science and Technology, 43*(19), 7170–7174. http://pubs.acs.org/doi/pdfplus/10.1021/es9013695. doi:10.1021/es9013695. Switzerland.

Camino-Serrano, M., Guenet, B., Luyssaert, S., Ciais, P., Bastrikov, V., De Vos, B., Gielen, B., Gleixner, G., Jornet-Puig, A., Kaiser, K., Kothawala, D., Lauerwald, R., Peñuelas, J., Schrumpf, M., Vicca, S., Vuichard, N., Walmsley, D., & Janssens, I. A. (2018). ORCHIDEE-SOM: Modeling soil organic carbon (SOC) and dissolved organic carbon (DOC) dynamics along vertical soil profiles in Europe. *Geoscientific Model Development, 11*(3), 937–957. http://www.geosci-model-dev.net/volumes_and_issues.html. doi:10.5194/gmd-11-937-2018.

Christensen, T. H. (1989). Cadmium soil sorption at low concentrations: VIII. Correlation with soil parameters. *Water, Air, and Soil Pollution, 44*(1–2), 71–82. doi:10.1007/BF00228779.

Cui, Y., & Weng, L. (2015). Interpretation of heavy metal speciation in sequential extraction using geochemical modelling. *Environmental Chemistry, 12*(2), 163–173. doi:10.1071/en13216.

Degryse, F., & Smolders, E. (2006). Mobility of Cd and Zn in polluted and unpolluted Spodosols. *European Journal of Soil Science, 57*(2), 122–133. doi:10.1111/j.1365-2389.2005.00721.x.

Degryse, F., Smolders, E., & Parker, D. R. (2009). Partitioning of metals (Cd, Co, Cu, Ni, Pb, Zn) in soils: Concepts, methodologies, prediction and applications—A review. *European Journal of Soil Science, 60*(4), 590–612. doi:10.1111/j.1365-2389.2009.01142.x.

de Livera, J., McLaughlin, M. J., Hettiarachchi, G. M., Kirby, J. K., & Beak, D. G. (2011). Cadmium solubility in paddy soils: Effects of soil oxidation, metal sulfides and competitive ions. *Science of the Total Environment, 409*(8), 1489–1497. doi:10.1016/j.scitotenv.2010.12.028.

Deng, Y., Ren, C., Chen, N., Huang, Y., Zhu, G., Zhang, X., Weng, L., & Li, Y. (2023). Effects of pH and phosphate on cadmium adsorption onto goethite and a paddy soil: Experiments and NOM-CD model. *Journal of Soils and Sediments, 23*(5), 2072–2082. doi:10.1007/s11368-023-03481-3.

de Vries, W., McLaughlin, M. J., & Groenenberg, J. E. (2011). Transfer functions for solid–solution partitioning of cadmium for Australian soils. *Environmental Pollution, 159*(12), 3583–3594. doi:10.1016/j.envpol.2011.08.006.

Dijkstra, J. J., Meeussen, J. C. L., & Comans, R. N. J. (2009). Evaluation of a generic multisurface sorption model for inorganic soil contaminants. *Environmental Science and Technology, 43*(16), 6196–6201. http://pubs.acs.org/doi/pdfplus/10.1021/es900555g. doi:10.1021/es900555g.

Ding, C., Du, S., Ma, Y., Li, X., Zhang, T., & Wang, X. (2019). Changes in the pH of paddy soils after flooding and drainage: Modeling and validation. *Geoderma, 337*, 511–513. doi:10.1016/j.geoderma.2018.10.012.

Duffner, A., Weng, L., Hoffland, E., & Van Der Zee, S. E. A. T. M. (2014). Multi-surface modeling to predict free zinc ion concentrations in low-zinc soils. *Environmental Science and Technology, 48*(10), 5700–5708. http://pubs.acs.org/journal/esthag. doi:10.1021/es500257e.

Du Laing, G., Rinklebe, J., Vandecasteele, B., Meers, E., & Tack, F. M. G. (2009). Trace metal behaviour in estuarine and riverine floodplain soils and sediments: A review. *Science of the Total Environment, 407*(13), 3972–3985. doi:10.1016/j.scitotenv.2008.07.025.

Dzombak, D., & Morel, F. (1990). *Surface complexation modelling: Hydrous ferric oxide.*

Elbana, T. A., & Selim, H. M. (2010). Cadmium transport in alkaline and acidic soils: Miscible displacement experiments. *Soil Science Society of America Journal, 74*(6), 1956–1966. doi:10.2136/sssaj2010.0146.

Elbana, T. A., & Selim, H. M. (2019). Modeling of cadmium and nickel release from different soils. *Geoderma, 338*, 78–87. www.elsevier.com/inca/publications/store/5/0/3/3/3/2. doi:10.1016/j.geoderma.2018.11.041.

Elzinga, E. J., Van Grinsven, J. J. M., & Swartjes, F. A. (1999). General purpose Freundlich isotherms for cadmium, copper and zinc in soils. *European Journal of Soil Science, 50*(1), 139–149. doi:10.1046/j.1365-2389.1999.00220.x.

Esfandbod, M., Forghani, A., Adhami, E., & Rashti, M. R. (2011). The role of CEC and pH in Cd retention from soils of North of Iran. *Soil and Sediment Contamination: An International Journal, 20*(8), 908–920. doi:10.1080/15320383.2011.620044.

Fan, T. T., Sun, Q., Cui, P. X., Xuan, L., & Wang, Y. J. (2022). Sorption mechanism of cadmium on soils: A combination of batch experiment, path analysis, and EXAFS techniques. *Geoderma, 422*. www.elsevier.com/inca/publications/store/5/0/3/3/3/2. doi:10.1016/j.geoderma.2022.115950.

Filius, A., Streck, T., & Richter, J. (1998). Cadmium sorption and desorption in limed topsoils as influenced by pH: Isotherms and simulated leaching. *Journal of Environmental Quality, 27*(1), 12–18.

Filius, J. D., Hiemstra, T., & Van Riemsdijk, W. H. (1997). Adsorption of small weak organic acids on goethite: Modeling of mechanisms. *Journal of Colloid and Interface Science, 195*(2), 368–380. http://www.elsevier.com/inca/publications/store/6/2/2/8/6/1/index.htt. doi:10.1006/jcis.1997.5152.

Freundlich, H. (1906). Over the adsorption in solution. *Journal of Physical Chemistry, 57*, 385–470.

Gao, H. (2025). *Heterogeneity in composition and reactivity of dissolved organic matter and metal partitioning in soil.* PhD thesis, Wageningen University & Research, Wageningen, the Netherlands.

Gao, H., Koopmans, G. F., Song, J., Groenenberg, J. E., Liu, X., Comans, R. N. J., & Weng, L. (2022). Evaluation of heavy metal availability in soils near former zinc smelters by chemical extractions and geochemical modelling. *Geoderma, 423*. doi:10.1016/j.geoderma.2022.115970.

Gräfe, M., Donner, E., Collins, R. N., & Lombi, E. (2014). Speciation of metal(loid)s in environmental samples by X-ray absorption spectroscopy: A critical review. *Analytica Chimica Acta, 822*, 1–22. www.elsevier.com/locate/aca. doi:10.1016/j.aca.2014.02.044.

Groenenberg, J. E., Dijkstra, J. J., Bonten, L. T. C., De Vries, W., & Comans, R. N. J. (2012). Evaluation of the performance and limitations of empirical partition-relations and process based multisurface models to predict trace element solubility in soils. *Environmental Pollution, 166*, 98–107. doi:10.1016/j.envpol.2012.03.011.

Groenenberg, J. E., & Lofts, S. (2014). The use of assemblage models to describe trace element partitioning, speciation, and fate: A review. *Environmental Toxicology and Chemistry, 33*(10), 2181–2196. http://www.interscience.wiley.com/jpages/0730-7268. doi:10.1002/etc.2642.

Groenenberg, J. E., Römkens, P. F. A. M., Comans, R. N. J., Luster, J., Pampura, T., Shotbolt, L., Tipping, E., & De Vries, W. (2010). Transfer functions for solid-solution partitioning of cadmium, copper, nickel, lead and zinc in soils: Derivation of relationships for free metal ion activities and validation with independent data. *European Journal of Soil Science, 61*(1), 58–73. doi:10.1111/j.1365-2389.2009.01201.x.

Groenenberg, J. E., Römkens, P. F. A. M., Zomeren, A. V., Rodrigues, S. M., & Comans, R. N. J. (2017). Evaluation of the single dilute (0.43 M) nitric acid extraction to determine geochemically reactive elements in soil. *Environmental Science and Technology, 51*(4), 2246–2253. http://pubs.acs.org/journal/esthag. doi:10.1021/acs.est.6b05151.

Gupta, S. S., & Bhattacharyya, K. G. (2011). Kinetics of adsorption of metal ions on inorganic materials: A review. *Advances in Colloid and Interface Science, 162*(1–2), 39–58. doi:10.1016/j.cis.2010.12.004.

Hamid, Y., Tang, L., Hussain, B., Usman, M., Lin, Q., Saqib Rashid, M., He, Z., & Yang, X. (2020). Organic soil additives for the remediation of cadmium contaminated soils and their impact on the soil-plant system: A review. *Science of the Total Environment, 707*, 136121. doi:10.1016/j.scitotenv.2019.136121.

Hashimoto, Y., Furuya, M., Yamaguchi, N., & Makino, T. (2016). Zerovalent iron with high sulfur content enhances the formation of cadmium sulfide in reduced paddy soils. *Soil Science Society of America Journal, 80*(1), 55–63. doi:10.2136/sssaj2015.06.0217.

Hiemstra, T., Antelo, J., van Rotterdam, A. M. D., & van Riemsdijk, W. H. (2010). Nanoparticles in natural systems II: The natural oxide fraction at interaction with natural organic matter and phosphate. *Geochimica et Cosmochimica Acta, 74*(1), 59–69. doi:10.1016/j.gca.2009.10.019.

Hiemstra, T., Mia, S., Duhaut, P. B., & Molleman, B. (2013). Natural and pyrogenic humic acids at goethite and natural oxide surfaces interacting with phosphate. *Environmental Science and Technology, 47*(16), 9182–9189. doi:10.1021/es400997n.

Hiemstra, T., & Van Riemsdijk, W. H. (1996). A surface structural approach to ion adsorption: The charge distribution (CD) model. *Journal of Colloid and Interface Science, 179*(2), 488–508. doi:10.1006/jcis.1996.0242.

Hiemstra, T., & Zhao, W. (2016). Reactivity of ferrihydrite and ferritin in relation to surface structure, size, and nanoparticle formation studied for phosphate and arsenate. *Environmental Science: Nano, 3*(6), 1265–1279. doi:10.1039/C6EN00061D.

Hooda, P. S. (2010). *Trace elements in soils.* Wiley. doi:10.1002/9781444319477.

Hooda, P. S., & Alloway, B. J. (1998). Cadmium and lead sorption behaviour of selected English and Indian soils. *Geoderma, 84*(1–3), 121–134. doi:10.1016/S0016-7061(97)00124-9.

Huang, Y., Koopmans, G. F., Groenenberg, J. E., Song, Z., Weng, L., & Comans, R. N. J. (2024). Generic proton binding parameters of a three-site NICA-Donnan model for plant-based biochars. *ACS Sustainable Resource Management, 1*(4), 707–720. https://pubs.acs.org/journal/asrmcd. doi:10.1021/acssusresmgt.3c00104.

Huang, Y., Koopmans, G. F., Song, Z., Comans, R. N. J., & Weng, L. (2025). Modelling cadmium binding to plant-based biochars using the three-site NICA-Donnan model. *Applied Geochemistry, 186*, 106404. doi:10.1016/j.apgeochem.2025.106404.

Jackson, A. P., & Alloway, B. J. (2017). The transfer of cadmium from agricultural soils to the human food chain. In *Biogeochemistry of trace metals: Advances in trace substances research* (pp. 109–158). CRC Press. https://www.routledge.com/Revival-Biogeochemistry-of-Trace-Metals-1992-Advances-In-Trace-Substances/Adriano/p/book/9781138557727.

Kandpal, G., Srivastava, P. C., & Ram, B. (2005). Kinetics of desorption of heavy metals from polluted soils: Influence of soil type and metal source. *Water, Air, and Soil Pollution, 161*(1–4), 353–363. doi:10.1007/s11270-005-5548-0.

Kim, K. R., Owens, G., & Naidu, R. (2009). Heavy metal distribution, bioaccessibility, and phytoavailability in long-term contaminated soils from Lake Macquarie, Australia. *Australian Journal of Soil Research., 47*(2), 166–176. doi:10.1071/SR08054.

Kinniburgh, D. G., Van Riemsdijk, W. H., Koopal, L. K., Borkovec, M., Benedetti, M. F., & Avena, M. J. (1999). Ion binding to natural organic matter: Competition, heterogeneity, stoichiometry and thermodynamic consistency. *Colloids and Surfaces A: Physicochemical and Engineering Aspects, 151*(1–2), 147–166. www.elsevier.com/locate/colsurfa. doi:10.1016/S0927-7757(98)00637-2.

Kögel-Knabner, I., Amelung, W., Cao, Z., Fiedler, S., Frenzel, P., Jahn, R., Kalbitz, K., Kölbl, A., & Schloter, M. (2010). Biogeochemistry of paddy soils. *Geoderma, 157*(1–2), 1–14. doi:10.1016/j.geoderma.2010.03.009.

Kumar, U., Jha, A. K., & Kumar, N. (2024). Cadmium toxicity in the environment: Sources, issues, remediation, and challenges. In *Cadmium toxicity: Challenges and solutions* (pp. 1–28). Springer Nature. https://link.springer.com/book/9783031656118. doi:10.1007/978-3-031-65611-8_1.

Lackovic, K., Johnson, B. B., Angove, M. J., & Wells, J. D. (2003). Modeling the adsorption of citric acid onto Muloorina illite and related clay minerals. *Journal of Colloid and Interface Science, 267*(1), 49–59. http://www.elsevier.com/inca/publications/store/6/2/2/8/6/1/index.htt. doi:10.1016/S0021-9797(03)00693-3.

Lagerwerff, J. V., & Brower, D. L. (1972). Exchange adsorption of trace quantities of cadmium in soils treated with chlorides of aluminum, calcium and sodium. *Soil Science Society of America Journal, 36*(5), 734–737. doi:10.2136/sssaj1972.03615995003600050017x.

Li, J., Weng, L., Deng, Y., Ma, J., Chen, Y., & Li, Y. (2022). NOM-mineral interaction: Significance for speciation of cations and anions. *Science of the Total Environment, 820*, 153259. doi:10.1016/j.scitotenv.2022.153259.

Li, L., Wu, H., Van Gestel, C. A. M., Peijnenburg, W. J. G. M., & Allen, H. E. (2014). Soil acidification increases metal extractability and bioavailability in old orchard soils of Northeast Jiaodong Peninsula in China. *Environmental Pollution, 188*, 144–152. doi:10.1016/j.envpol.2014.02.003.

Li, S., Chen, S., Wang, M., Lei, X., Zheng, H., Sun, X., Wang, L., & Han, Y. (2020). Iron fractions responsible for the variation of Cd bioavailability in paddy soil under variable pe+pH conditions. *Chemosphere, 251*, 126355. doi:10.1016/j.chemosphere.2020.126355.

Liang, Z., Ding, Q., Wei, D., Li, J., Chen, S., & Ma, Y. (2013). Major controlling factors and predictions for cadmium transfer from the soil into spinach plants. *Ecotoxicology and Environmental Safety, 93*, 180–185. doi:10.1016/j.ecoenv.2013.04.003.

Liao, L., & Selim, H. M. (2010). Transport of nickel in different soils: Column experiments and kinetic modeling. *Soil Science Society of America Journal, 74*(6), 1946–1955. doi:10.2136/sssaj2010.0006.

Lindsay, W. L. (1979). *Chemical equilibria in soils*. John Wiley and Sons Ltd.

Loganathan, P., Vigneswaran, S., Kandasamy, J., & Naidu, R. (2012). Cadmium sorption and desorption in soils: A review. *Critical Reviews in Environmental Science and Technology, 42*(5), 489–533. doi:10.1080/10643389.2010.520234.

Lu, H. l., Li, K. w., Nkoh, J. N., He, X., Xu, R. k., Qian, W., Shi, R. y., & Hong, Z. n. (2022). Effects of pH variations caused by redox reactions and pH buffering capacity on Cd(II) speciation in paddy soils during submerging/draining alternation. *Ecotoxicology and Environmental Safety, 234*. http://www.elsevier.com/inca/publications/store/6/2/2/8/1/9/index.htt. doi:10.1016/j.ecoenv.2022.113409.

Lu, W., Liu, Y., Ye, H., Lin, D., Li, G., Zhao, Y., Deng, T., Li, H., & Wang, R. (2021). Adsorption and desorption characteristics of cadmium on different contaminated paddy soil types: Kinetics, isotherms, and the effects of soil properties. *Sustainability, 13*(13), 7052. doi:10.3390/su13137052.

Luo, X. S., Zhou, D. M., Liu, X. H., & Wang, Y. J. (2006). Solid/solution partitioning and speciation of heavy metals in the contaminated agricultural soils around a copper mine in eastern Nanjing city, China. *Journal of Hazardous Materials, 131*(1–3), 19–27. doi:10.1016/j.jhazmat.2005.09.033.

Martin-Garin, A., Van Cappellen, P., & Charlet, L. (2003). Aqueous cadmium uptake by calcite: A stirred flow-through reactor study. *Geochimica et Cosmochimica Acta, 67*(15), 2763–2774. doi:10.1016/s0016-7037(03)00091-7.

Marzouk, E. R., Chenery, S. R., & Young, S. D. (2013). Measuring reactive metal in soil: A comparison of multi-element isotopic dilution and chemical extraction. *European Journal of Soil Science, 64*(4), 526–536. doi:10.1111/ejss.12043.

McBratney, A. B., Minasny, B., Cattle, S. R., & Vervoort, R. W. (2002). From pedotransfer functions to soil inference systems. *Geoderma, 109*(1–2), 41–73. doi:10.1016/S0016-7061(02)00139-8.

McBride, M., Sauvé, S., & Hendershot, W. (1997). Solubility control of Cu, Zn, Cd and Pb in contaminated soils. *European Journal of Soil Science, 48*(2), 337–346. www.blacksci.co.uk/~cgilib/jnlpage.bin?Journal=ejss&File=ejss&Page=aims. doi:10.1111/j.1365-2389.1997.tb00554.x.

McBride, M. B. (1994). *Environmental chemistry of soils*. Oxford University Press.

Milne, C. J., Kinniburgh, D. G., & Tipping, E. (2001). Generic NICA-Donnan model parameters for proton binding by humic substances. *Environmental Science and Technology, 35*(10), 2049–2059. doi:10.1021/es000123j.

Milne, C. J., Kinniburgh, D. G., Van Riemsdijk, W. H., & Tipping, E. (2003). Generic NICA—Donnan model parameters for metal-ion binding by humic substances. *Environmental Science and Technology, 37*(5), 958–971. doi:10.1021/es0258879.

Moral, R., Gilkes, R. J., & JordÁn, M. M. (2005). Distribution of heavy metals in calcareous and non-calcareous soils in Spain. *Water, Air, & Soil Pollution, 162*(1–4), 127–142. doi:10.1007/s11270-005-5997-5.

Naidu, R., Bolan, N. S., Kookana, R. S., & Tiller, K. G. (1994). Ionic-strength and pH effects on the sorption of cadmium and the surface charge of soils. *European Journal of Soil Science, 45*(4), 419–429. doi:10.1111/j.1365-2389.1994.tb00527.x.

Naidu, R., Kookana, R. S., Sumner, M. E., Harter, R. D., & Tiller, K. G. (1997). Cadmium sorption and transport in variable charge soils: A review. *Journal of Environmental Quality, 26*(3), 602–617. doi:10.2134/jeq1997.00472425002600030004x.

Oh, S., & Shin, W. S. (2010). Effect of ageing on desorption of lead and cadmium from sediments: Kinetics and desorption-resistance. *Journal of Environmental Science and Health, Part A, 45*(9), 1150–1168. doi:10.1080/10934529.2010.486351.

Pachepsky, Y. A., & Rawls, W. J. (2003). Soil structure and pedotransfer functions. *European Journal of Soil Science, 54*(3), 443–452. doi:10.1046/j.1365-2389.2003.00485.x.

Palansooriya, K. N., Li, J., Dissanayake, P. D., Suvarna, M., Li, L., Yuan, X., Sarkar, B., Tsang, D. C. W., Rinklebe, J., Wang, X., & Ok, Y. S. (2022). Prediction of soil heavy metal immobilization by biochar using machine learning. *Environmental Science and Technology, 56*(7), 4187–4198. http://pubs.acs.org/journal/esthag. doi:10.1021/acs.est.1c08302.

Pan, Y., Bonten, L. T. C., Koopmans, G. F., Song, J., Luo, Y., Temminghoff, E. J. M., & Comans, R. N. J. (2016). Solubility of trace metals in two contaminated paddy soils exposed to alternating flooding and drainage. *Geoderma, 261*, 59–69. www.elsevier.com/inca/publications/store/5/0/3/3/3/2. doi:10.1016/j.geoderma.2015.07.011.

Peng, L., Liu, P., Feng, X., Wang, Z., Cheng, T., Liang, Y., Lin, Z., & Shi, Z. (2018). Kinetics of heavy metal adsorption and desorption in soil: Developing a unified model based on chemical speciation. *Geochimica et Cosmochimica Acta, 224*, 282–300. doi:10.1016/j.gca.2018.01.014.

Ponthieu, M., Juillot, F., Hiemstra, T., van Riemsdijk, W. H., & Benedetti, M. F. (2006). Metal ion binding to iron oxides. *Geochimica et Cosmochimica Acta, 70*(11), 2679–2698. doi:10.1016/j.gca.2006.02.021.

Prieto, M., Cubillas, P., & Fernández-Gonzalez, Á (2003). Uptake of dissolved Cd by biogenic and abiogenic aragonite: A comparison with sorption onto calcite. *Geochimica et Cosmochimica Acta, 67*(20), 3859–3869. doi:10.1016/s0016-7037(03)00309-0.

Qin, Y., Groenenberg, J. E., Viala, Y., Alves, S., & Comans, R. N. J. (2024). Optimizing multi-surface modelling of available cadmium as measured in soil pore water and salt extracts of soils amended with compost and lime: The role of organic matter and reactive metal. *Science of the Total Environment, 957*. https://www.sciencedirect.com/science/journal/00489697. doi:10.1016/j.scitotenv.2024.177769.

Ren, Z. L., Sivry, Y., Dai, J., Tharaud, M., Cordier, L., & Benedetti, M. F. (2015). Multi-element stable isotopic dilution and multi-surface modelling to assess the speciation and reactivity of cadmium and copper in soil. *European Journal of Soil Science, 66*(6), 973–982. doi:10.1111/ejss.12298.

Rezaei Rashti, M., Esfandbod, M., Adhami, E., & Srivastava, P. (2014). Cadmium desorption behaviour in selected sub-tropical soils: Effects of soil properties. *Journal of Geochemical Exploration, 144*, 230–236. doi:10.1016/j.gexplo.2014.01.023.

Rodrigues, S. M., Cruz, N., Coelho, C., Henriques, B., Carvalho, L., Duarte, A. C., Pereira, E., & Römkens, P. F. A. M. (2013). Risk assessment for Cd, Cu, Pb and Zn in urban soils: Chemical availability as the central concept. *Environmental Pollution, 183*, 234–242. doi:10.1016/j.envpol.2012.10.006.

Römkens, P. F. A. M., Groenenberg, J. E., Bonten, L. T. C., de Vries, W., & Bril, J. (2004). *Derivation of partition relationships to calculate Cd, Cu, Ni, Pb, Zn solubility and activity in soil solutions (1566-7197, Alterra report)*. Alterra.

Sauvé, S., Norvell, W. A., McBride, M., & Hendershot, W. (2000a). Speciation and complexation of cadmium in extracted soil solutions. *Environmental Science and Technology, 34*(2), 291–296. doi:10.1021/es990202z.

Sauvé, S., Hendershot, W., & Allen, H. E. (2000b). Solid-solution partitioning of metals in contaminated soils: Dependence on pH, total metal burden, and organic matter. *Environmental Science and Technology, 34*(7), 1125–1131. doi:10.1021/es9907764.

Schaider, L. A., Senn, D. B., Brabander, D. J., Mccarthy, K. D., & Shine, J. P. (2007). Characterization of zinc, lead, and cadmium in mine waste: Implications for transport, exposure, and bioavailability. *Environmental Science and Technology, 41*(11), 4164–4171. doi:10.1021/es0626943.

Schindler, P. W., Liechti, P., & Westall, J. C. (1987). Adsorption of copper, cadmium and lead from aqueous solution to the kaolinite/water interface. *Netherlands Journal of Agricultural Science, 35*(3), 219–230. doi:10.18174/njas.v35i3.16720.

Schwertmann, U. (1964). Differenzierung der Eisenoxide des Bodens durch Extraktion mit Ammoniumoxalat-Lösung. *Zeitschrift für Pflanzenernährung, Düngung, Bodenkunde, 105*(3), 194–202. doi:10.1002/jpln.3591050303.

Selim, H. M. (2014). *Transport & fate of chemicals in soils: Principles & applications* (pp. 1–326). CRC Press. https://www.taylorfrancis.com/books/9781466557956. doi:10.1201/b17405.

Shi, Z., Allen, H. E., Di Toro, D. M., Lee, S. Z., Flores Meza, D. M., & Lofts, S. (2007). Predicting cadmium adsorption on soils using WHAM VI. *Chemosphere, 69*(4), 605–612. www.elsevier.com/locate/chemosphere. doi:10.1016/j.chemosphere.2007.03.001.

Shi, Z., Di Toro, D. M., Allen, H. E., & Ponizovsky, A. A. (2005). Modeling kinetics of Cu and Zn release from soils. *Environmental Science and Technology, 39*(12), 4562–4568. doi:10.1021/es048554f.

Shi, Z., Di Toro, D. M., Allen, H. E., & Sparks, D. L. (2013). A general model for kinetics of heavy metal adsorption and desorption on soils. *Environmental Science and Technology, 47*(8), 3761–3767. doi:10.1021/es304524p.

Siebers, N., Kruse, J., & Leinweber, P. (2013). Speciation of phosphorus and cadmium in a contaminated soil amended with bone char: Sequential fractionations and XANES spectroscopy. *Water, Air, & Soil Pollution, 224*(5). doi:10.1007/s11270-013-1564-7.

Sposito, G. (1989). Surface reactions in natural aqueous colloidal systems. *CHIMIA, 43*(6), 169. doi:10.2533/chimia.1989.169.

Stevenson, F. J. (1994). *Humus chemistry: Genesis, composition, reactions*. John Wiley & Sons.

Stipp, S. L., Hochella, M. F., Parks, G. A., & Leckie, J. O. (1992). Cd2+ uptake by calcite, solid-state diffusion, and the formation of solid-solution: Interface processes observed with near-surface sensitive techniques (XPS, LEED, and AES). *Geochimica et Cosmochimica Acta, 56*(5), 1941–1954. doi:10.1016/0016-7037(92)90321-9.

Stipp, S. L. S., Parks, G. A., Nordstrom, D. K., & Leckie, J. O. (1993). Solubility-product constant and thermodynamic properties for synthetic otavite, $CdCO_3(s)$, and aqueous association constants for the Cd(II)-CO_2-H_2O system. *Geochimica et Cosmochimica Acta, 57*(12), 2699–2713. doi:10.1016/0016-7037(93)90384-9.

Sungur, A., Soylak, M., Yilmaz, E., Yilmaz, S., & Ozcan, H. (2014). Characterization of heavy metal fractions in agricultural soils by sequential extraction procedure: The relationship between soil properties and heavy metal fractions. *Soil and Sediment Contamination: An International Journal, 24*(1), 1–15. doi:10.1080/15320383.2014.907238.

Temminghoff, E. J. M., Van Der Zee, S. E. A. T. M., & De Haan, F. A. M. (1995). Speciation and calcium competition effects on cadmium sorption by sandy soil at various pHs. *European Journal of Soil Science, 46*(4), 649–655. doi:10.1111/j.1365-2389.1995.tb01361.x.

Tiberg, C., & Gustafsson, J. P. (2016). Phosphate effects on cadmium(II) sorption to ferrihydrite. *Journal of Colloid and Interface Science, 471*, 103–111. http://www.elsevier.com/inca/publications/store/6/2/2/8/6/1/index.htt. doi:10.1016/j.jcis.2016.03.016.

Tipping, E. (1994). WHAMC-A chemical equilibrium model and computer code for waters, sediments, and soils incorporating a discrete site/electrostatic model of ion-binding by humic substances. *Computers and Geosciences, 20*(6), 973–1023. doi:10.1016/0098-3004(94)90038-8.

Tye, A. M., Young, S. D., Crout, N. M. J., Zhang, H., Preston, S., Barbosa-Jefferson, V. L., Davison, W., McGrath, S. P., Paton, G. I., Kilham, K., & Resende, L. (2003). Predicting the activity of Cd2+ and Zn2+ in soil pore water from the radio-labile metal fraction. *Geochimica et Cosmochimica Acta, 67*(3), 375–385. doi:10.1016/s0016-7037(02)01138-9.

Van, H. T., Nguyen, L. H., Nguyen, V. D., Nguyen, X. H., Nguyen, T. H., Nguyen, T. V., Vigneswaran, S., Rinklebe, J., & Tran, H. N. (2019). Characteristics and mechanisms of cadmium adsorption onto biogenic aragonite shells-derived biosorbent: Batch and column studies. *Journal of Environmental Management, 241*, 535–548. https://www.sciencedirect.com/science/journal/03014797. doi:10.1016/j.jenvman.2018.09.079.

Van Eynde, E., Groenenberg, J. E., Hoffland, E., & Comans, R. N. J. (2022). Solid-solution partitioning of micronutrients Zn, Cu and B in tropical soils: Mechanistic and empirical models. *Geoderma, 414*. www.elsevier.com/inca/publications/store/5/0/3/3/3/2. doi:10.1016/j.geoderma.2022.115773.

van Zomeren, A., & Comans, R. N. J. (2007). Measurement of humic and fulvic acid concentrations and dissolution properties by a rapid batch procedure. *Environmental Science & Technology, 41*(19), 6755–6761. doi:10.1021/es0709223.

Wang, J., Wang, P. M., Gu, Y., Kopittke, P. M., Zhao, F. J., & Wang, P. (2019). Iron-manganese (oxyhydro)oxides, rather than oxidation of sulfides, determine mobilization of Cd during soil drainage in paddy soil systems. *Environmental Science and Technology, 53*(5), 2500–2508. http://pubs.acs.org/journal/esthag. doi:10.1021/acs.est.8b06863.

Wang, J., Yuan, R., Zhang, Y., Si, T., Li, H., Duan, H., Li, L., & Pan, G. (2022). Biochar decreases Cd mobility and rice (*Oryza sativa* L.) uptake by affecting soil iron and sulfur cycling. *Science of the Total Environment, 836*, 155547. doi:10.1016/j.scitotenv.2022.155547.

Wang, Y., Tang, X., Chen, Y., Zhan, L., Li, Z., & Tang, Q. (2009). Adsorption behavior and mechanism of Cd(II) on loess soil from China. *Journal of Hazardous Materials, 172*(1), 30–37. doi:10.1016/j.jhazmat.2009.06.121.

Wei, N., Gu, X., Wen, Y., Guo, C., & Ji, J. (2025). Geochemical speciation and activation risks of Cd, Ni, and Zn in soils with naturally high background in karst regions of southwestern China. *Journal of Hazardous Materials, 486*, 137100. doi:10.1016/j.jhazmat.2025.137100.

Weng, L., Temminghoff, E. J. M., & Van Riemsdijk, W. H. (2001a). Contribution of individual sorbents to the control of heavy metal activity in sandy soil. *Environmental Science and Technology, 35*(22), 4436–4443. doi:10.1021/es010085j.

Weng, L., Temminghoff, E. J. M., & Van Riemsdijk, W. H. (2001b). Determination of the free ion concentration of trace metals in soil solution using a soil column Donnan membrane technique. *European Journal of Soil Science, 52*(4), 629–637. doi:10.1046/j.1365-2389.2001.00416.x.

Weng, L., Van Riemsdijk, W. H., Koopal, L. K., & Hiemstra, T. (2006). Ligand and charge distribution (LCD) model for the description of fulvic acid adsorption to goethite. *Journal of Colloid and Interface Science, 302*(2), 442–457. doi:10.1016/j.jcis.2006.07.005.

Weng, L., Vega, F. A., & Van Riemsdijk, W. H. (2011). Competitive and synergistic effects in pH dependent phosphate adsorption in soils: LCD modeling. *Environmental Science and Technology, 45*(19), 8420–8428. doi:10.1021/es201844d.

Xiong, J., Xu, J., Zhou, M., Zhao, W., Chen, C., Wang, M., Tan, W., & Koopal, L. (2021). Quantitative characterization of the site density and the charged state of functional groups on biochar. *ACS Sustainable Chemistry & Engineering., 9*(6), 2600–2608. doi:10.1021/acssuschemeng.0c09051.

Yan, Y., Wan, B., Mansor, M., Wang, X., Zhang, Q., Kappler, A., & Feng, X. (2022). Co-sorption of metal ions and inorganic anions/organic ligands on environmental minerals: A review. *Science of the Total Environment, 803,* 149918. doi:10.1016/j.scitotenv.2021.149918.

Yao, A., Liu, Y., Luo, X., Liu, C., Tang, Y., Wang, S., Huang, X., & Qiu, R. (2021). Mediation effects of different sulfur forms on solubility, uptake and accumulation of Cd in soil-paddy rice system induced by organic carbon and liming. *Environmental Pollution, 279,* 116862. doi:10.1016/j.envpol.2021.116862.

Yi, S., Chen, X., Cao, X., Yi, B., & He, W. (2022). Influencing factors and environmental effects of interactions between goethite and organic matter: A critical review. *Frontiers in Environmental Science, 10.* doi:10.3389/fenvs.2022.1023277.

Yuan, X., Suvarna, M., Low, S., Dissanayake, P. D., Lee, K. B., Li, J., Wang, X., & Ok, Y. S. (2021). Applied machine learning for prediction of CO_2 adsorption on biomass waste-derived porous carbons. *Environmental Science and Technology, 55*(17), 11925–11936. http://pubs.acs.org/journal/esthag. doi:10.1021/acs.est.1c01849.

Zachara, J. M., & Smith, S. C. (1994). Edge complexation reactions of cadmium on specimen and soil-derived smectite. *Soil Science Society of America Journal, 58*(3), 762–769. doi:10.2136/sssaj1994.03615995005800030018x.

Zhang, N., Lv, C., Li, Y., Panagos, P., Ballabio, C., Man, J., Gu, X., Zhao, F.-J., Wang, P., Liu, X., Qian, Y., Cui, P., Wu, T., Huang, M., Liu, C., & Wang, Y. (2025). Geochemical-integrated machine learning approach predicts the distribution of cadmium speciation in European and Chinese topsoils. *Communications Earth & Environment, 6*(1). doi:10.1038/s43247-025-02516-6.

Zhang, Q., Li, Z., Huang, B., Luo, N., Long, L., Huang, M., Zhai, X., & Zeng, G. (2017). Effect of land use pattern change from paddy soil to vegetable soil on the adsorption-desorption of cadmium by soil aggregates. *Environmental Science and Pollution Research, 24*(3), 2734–2743. doi:10.1007/s11356-016-7853-0.

Zhang, Y., Zhang, Y., Udochukwu Akakuru, O., Xu, X., & Wu, A. (2021). Research progress and mechanism of nanomaterials-mediated in-situ remediation of cadmium-contaminated soil: A critical review. *Journal of Environmental Sciences, 104,* 351–364. doi:10.1016/j.jes.2020.12.021.

Zhao, F. J., Ma, Y., Zhu, Y. G., Tang, Z., & McGrath, S. P. (2015). Soil contamination in China: Current status and mitigation strategies. *Environmental Science and Technology, 49*(2), 750–759. http://pubs.acs.org/journal/esthag. doi:10.1021/es5047099.

Zhao, F. J., Tang, Z., Song, J. J., Huang, X. Y., & Wang, P. (2022). Toxic metals and metalloids: Uptake, transport, detoxification, phytoremediation, and crop improvement for safer food. *Molecular Plant, 15*(1), 27–44. https://www.cell.com/molecular-plant/home. doi:10.1016/j.molp.2021.09.016.

Zhao, X., Jiang, Y., Gu, X., Gu, C., Taylor, J. A., & Evans, L. J. (2018). Multisurface modeling of Ni bioavailability to wheat (*Triticum aestivum* L.) in various soils. *Environmental Pollution, 238,* 590–598. www.elsevier.com/inca/publications/store/4/0/5/8/5/6. doi:10.1016/j.envpol.2018.03.064.

Zheng, Y., Zhang, Z., Chen, Y., An, S., Zhang, L., Chen, F., Ma, C., & CAi, W. (2022). Adsorption and desorption of Cd in reclaimed soil under the influence of humic acid: Characteristics and mechanisms. *International Journal of Coal Science & Technology, 9*(1). doi:10.1007/s40789-022-00480-6.

CHAPTER

7

Transformation of cadmium in paddy soils

Peng Wang[a,b], *Hui Huang*[c] *and Fang-Jie Zhao*[a]

[a]College of Resources and Environmental Sciences, Nanjing Agricultural University, Nanjing, China [b]College of Sciences, Nanjing Agricultural University, Nanjing, China [c]College of Ecology and Environment, Nanjing Forestry University, Nanjing, China

7.1 Introduction

Cadmium (Cd) contamination in paddy soils poses severe risks to food safety, as rice readily accumulates Cd, threatening the health of rice-dependent populations. Paddy soils' unique flooding-drainage cycles drive dynamic shifts in redox potential (E_h), pH, and microbial activity, which govern Cd speciation, solubility, and bioavailability, with these processes distinct from upland systems.

This chapter focuses on Cd transformation in paddy soils, building on Chapter 6's coverage of fundamental Cd adsorption–desorption mechanisms. It first contextualizes global and regional Cd contamination patterns, emphasizing that bioavailability, rather than total Cd concentration, determines rice Cd accumulation. Core sections explore paddy-specific biogeochemical drivers: pH-dependent Cd mobility, redox-controlled CdS precipitation and dissolution, and novel mechanisms such as the voltaic effect and free radical-mediated Cd release. Finally, it links mechanistic insights to practical mitigation strategies, providing a targeted framework for mitigating Cd risks in paddy agroecosystems.

7.2 Cadmium contamination in soil–rice systems

Cd in paddy soil derives from both natural sources (e.g., parent materials and volcanic emissions) that contribute to background concentrations and anthropogenic inputs (e.g., mining, smelting, phosphate fertilizer application, and wastewater irrigation) that lead to Cd accumulation (Smolders & Mertens, 2013) (see Chapter 5). Natural background Cd concentrations in soils typically range from 0.01 to 1.0 mg kg^{-1}, with regional variability intimately associated with

Cadmium Contamination in Soils and Plants: Mechanisms and Mitigation.
DOI: https://doi.org/10.1016/B978-0-443-34104-5.00006-2

geological and pedological characteristics. For instance, the average background Cd concentration in China was documented as 0.097 mg kg^{-1} in 1990 (Chen et al., 2015), whereas in Japan background levels ranged from 0.05 to 0.4 mg kg^{-1} (Ministry of the Environment of Japan, 1984).

In China a nationwide soil survey ($n = 38{,}393$) revealed paddy soil Cd concentrations ranging from 0.058 to 0.505 mg kg^{-1} (10th–90th percentiles), with a geometric mean of 0.225 mg kg^{-1} and a median of 0.140 mg kg^{-1}. The proportion of Cd-contaminated paddy soils is increasing, particularly in southern China. Notably, 19.4% of the soil samples from China's cropland are contaminated with various pollutants, with Cd exhibiting the highest exceedance rate (7%) among all pollutants (MEP and MLR, 2014). Anthropogenic activities remain the primary driver of Cd accumulation in paddy soils, and the expansion of contaminated croplands poses substantial threats to rice safety and ecosystem health.

The accumulation of Cd in rice grains is primarily governed by Cd bioavailability in paddy soil rather than total soil Cd concentrations. Unlike certain regions where food crops seldom exceed safety limits despite elevated soil Cd, rice produced in some areas in southern China accumulates Cd to unsafe levels as a result of high Cd bioavailability in acidic paddy soils (Wang et al., 2019b; Zhao & Wang, 2020). Extensive large-scale surveys show that 2%–10% of rice grains exceed the Chinese food safety limit (0.2 mg Cd kg^{-1}) across most regions of China, whereas the exceedance rate reaches 60%–70% in some severely polluted southern areas (Chen et al., 2018b, 2021; Qian et al., 2010; Zhu et al., 2016). The mean Cd concentration of rice grains in these contaminated areas (0.43 mg kg^{-1}, n = 39,642) even surpasses that in Japan's Jinzu River Basin (0.38 mg kg^{-1}), the site of the itai-itai disease outbreak (Nogawa, 1981; Zhu et al., 2016).

For populations dependent on rice as a staple food, dietary intake constitutes the principal route of Cd exposure. Rice accounts for 56% of the total dietary Cd intake for the Chinese population, ranging from 38% in northern regions to 65% in southern areas (Song et al., 2017). Chronic Cd ingestion poses significant health hazards, including renal dysfunction, osteoporosis, cardiovascular diseases, and several types of cancers (Elinder et al., 1985; Jin et al., 2004; Revis & Osborne, 1984; Staessen et al., 1991) (see Chapter 15). Despite substantial research efforts, the mechanistic basis underlying regional disparities in Cd bioavailability in paddy soils remains insufficiently elucidated. Key factors, including soil pH, redox oscillations, Fe/Mn (oxy)hydroxides, and sulfur cycling, are increasingly acknowledged as dominant regulators of Cd solubility and mobility in paddy soils (Huang et al., 2021c; Wang et al., 2019b). A deeper understanding of these paddy-specific biogeochemical processes is pivotal for devising targeted mitigation strategies to reduce Cd accumulation in rice and minimize dietary Cd exposure.

7.3 Transformation processes of cadmium during soil flooding

Rice paddies undergo episodic flooding-draining cycles, and these hydrological fluctuations strongly regulate Cd speciation, solubility, and bioavailability. Flooding-induced shifts in E_h, pH, and microbial activity jointly control Cd transformation, with paddy-specific biogeochemical processes governing Cd solubility dynamics (Chen et al., 2021; Ford, 2007; Huang et al., 2023; Wang et al., 2017; Yang et al., 2024).

7.3.1 Cadmium speciation change and solubility dynamics

Paddy fields feature a long-term flooding phase, during which E_h gradient variation exerts a significant directional driving effect on Cd mobilization. Initially, as paddy soils transition from an oxic state (redox potential $E_h \sim$ 300–500 mV) to suboxic conditions ($E_h \sim$ 200–300 mV), oxygen is rapidly depleted. This shift is accompanied by a modest rise in pH because of the consumption of protons (H^+) during microbial respiration and abiotic reactions (Borch et al., 2010), which enhances Cd adsorption to Fe/Mn oxides and organic matter, lowering initial Cd solubility (de Livera et al., 2011). Under these conditions, exchangeable Cd^{2+}, the more bioavailable fraction in oxidized soils, partly transforms into carbonate-bound (in highly contaminated soils) or colloid-bound forms associated with clays and soil organic matter (SOM), leading to a modest decline in Cd bioavailability. Speciation modeling shows that SOM binds Cd primarily at pH < 6.5 during this phase, whereas Fe/Mn oxides become dominant adsorbents at higher pH (Argüello et al., 2019; He et al., 2022; Khaokaew et al., 2011; Yang et al., 2021).

As E_h drops below 200 mV, manganese-reducing bacteria and iron-reducing bacteria drive the reductive dissolution of Fe(III) and Mn(III) (oxyhydr)oxides. This process releases Fe^{2+}/Mn^{2+} into the soil solution, along with previously adsorbed or coprecipitated Cd^{2+}, temporarily elevating dissolved Cd concentrations (Bao et al., 2018; Borch et al., 2010). Dissolved Cd generally represents only a small fraction of total soil Cd ($\lesssim$1.5% even in acidic contaminated paddy soils), but this pool is the most mobile and bioavailable for rice root uptake, as rice roots efficiently absorb Cd^{2+} from soil solution (Christensen, 1984; Wang et al., 2019a; Zhao & Wang, 2020).

When E_h drops to $\sim$−100 to −200 mV, sulfate-reducing bacteria convert SO_4^{2-} to sulfide (S^{2-}/HS^-), which readily reacts with free Cd^{2+} to form insoluble CdS ($K_{sp} = 8 \times 10^{-27}$), which is the dominant precipitation pathway for Cd in sulfate-sufficient paddy soils (Fig. 7.1) (Fulda et al., 2013a; Furuya et al., 2016; Hashimoto et al., 2016). This process drastically immobilizes Cd, offsetting the temporary Cd release from Fe/Mn oxide dissolution and lowering dissolved Cd to levels even below the ICP-MS detection limit in many Cd-contaminated paddy soils (Wang et al., 2019a). Using high-resolution transmission electron microscopy (HRTEM–EDX–SAED), Huang et al., 2021a, Huang et al., 2024b) confirmed the formation of nanoscale metal-sulfide particles (including CdS, ZnS, and FeS) during flooding, which further reduce Cd solubility but may enhance colloidal transport of Cd in soil pores (Fig. 7.2) (Rozan et al., 2000; Weber et al., 2009). Empirical models show that doubling SOM content can decrease solution Cd concentration by $\sim$35%, with Cd solubility proportional to total soil Cd content (Argüello et al., 2019).

7.3.2 Competitive sulfide precipitation and pH-dependent cadmium solubility

Sulfide can form precipitates with chalcophilic metals (Cu^{2+}, Zn^{2+}, Pb^{2+}, and Hg^{2+}). Because of their lower solubility products (Table 7.1), CuS, PbS, and HgS form preferentially, suppressing CdS precipitation when metal concentrations exceed that of sulfide (Fulda, 2013; Fulda et al., 2013b). In addition, the high concentration of Fe(II) in paddy soils promotes FeS formation ($K_{sp} = 6.3 \times 10^{-18}$) via mass effect, often dominating over CdS precipitation despite its higher K_{sp} (Table 7.1). Furthermore, in the Fe- and Mn-rich paddy soils, elevated Fe^{2+}/Mn^{2+} may undergo

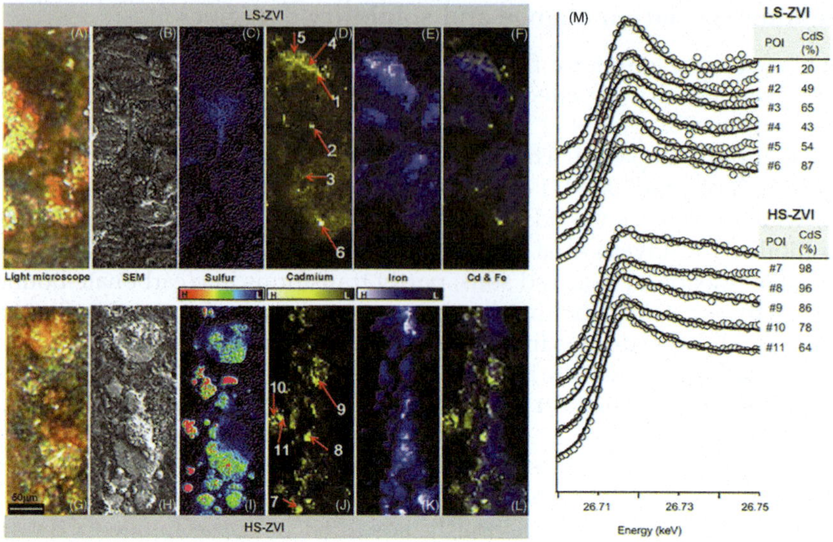

FIGURE 7.1 Formation of CdS in paddy soil. Low-S zerovalent iron (LS-ZVI) (top) and high-S zerovalent iron (HS-ZVI) (bottom) amended soil thin sections analyzed by (A and G) a light microscope and (B and H) a scanning electron microscope and element maps of (C and I) S, (D and J) Cd, and (E and K) Fe and (F and L) an overlaid element map of Cd–Fe. The numbers on (D and J) indicate the spots where Cd K-edge micro-X-ray absorption near edge structure spectra were collected, with the analysis results shown in (M). *From Hashimoto et al. (2016), with permission.*

TABLE 7.1 Solubility products of various trace metal sulfides corresponding to the equilibrium MeS(s) + H^+ ↔ Me^{2+}(aq) + HS^-(aq). Data are obtained from the Minteq.v4 database (Allison et al., 1991).

Metal sulfide	Mineral	K_{sp}
CdS	Greenockite	8×10^{-27}
ZnS	Sphalerite	2.9×10^{-25}
CuS	Covellite	6.3×10^{-36}
PbS	Galena	8×10^{-28}
HgS	Cinnabar	4×10^{-53}
FeS	Mackinawite	6.3×10^{-18}

Data from Allison, J. D., Brown, D. S., & Novo-Gradac, K. J. (1991). MINTEQA2/PRODEFA2, A geochemical assessment model for environmental systems: Version 3.0 user's manual. Environmental Research Laboratory, Office of Research and Development, U.S. Environmental Protection Agency (U.S. EPA).

local reoxidation to form secondary Fe/Mn (oxyhydr)oxides, creating new adsorption sites that capture Cd (Wang et al., 2019a). In alkaline and severely Cd-contaminated paddy soils, $CdCO_3$ precipitation (K_{sp} = 10^{-12}) can complement CdS formation to further reduce Cd solubility, though this requires a total soil Cd > 100 mg kg^{-1} which is a rare scenario in agricultural soils (Khaokaew et al., 2011; Smolders & Mertens, 2013). During flooding, Cd solubility in paddy soils follows the relationship: Cd solubility = 44.5 × exp(−1.77 × porewater pH) (R^2 = 0.76,

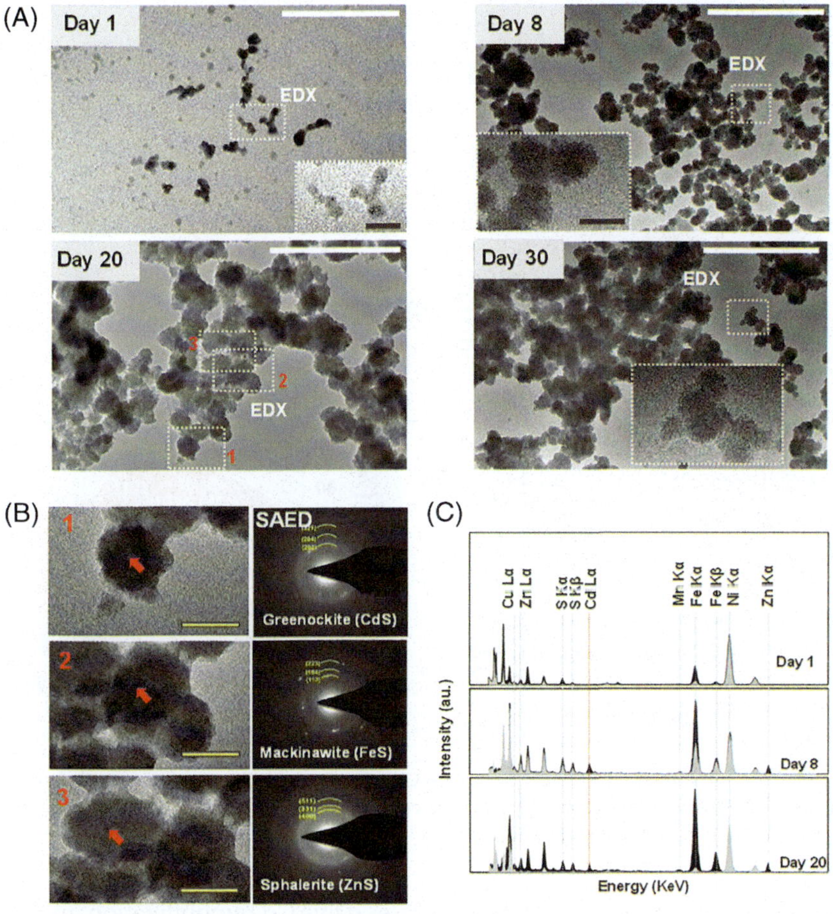

FIGURE 7.2 Formation of metal-sulfide colloidal particles in flooded paddy soil. (A) TEM images of colloids isolated from a paddy soil at different time points after flooding. (B) Identification of colloidal particles via high-resolution TEM (HRTEM) combined with selected area electron diffraction (SAED) analyses. (C) Energy-dispersive X-ray (EDX) spectra acquired from the regions marked by rectangles in (A). *White scale* bars stand at 500 nm; *yellow scale* bars indicate 50 nm. *From Huang et al. (2021a), with permission.*

$P < .001$), with a 1-unit increase in pH reducing solution Cd concentration by 3–4 times (Fig. 7.3A) (Argüello et al., 2019; Wang et al., 2019a).

7.4 Transformation processes of cadmium during soil drainage

Rice paddies are typically drained 2–3 weeks before harvest to promote root respiration and facilitate harvesting, a period that coincides with the grain-filling stage, when over 90% of Cd in rice grains is absorbed by roots (Huang et al., 2022a). Drainage-induced shifts in soil redox

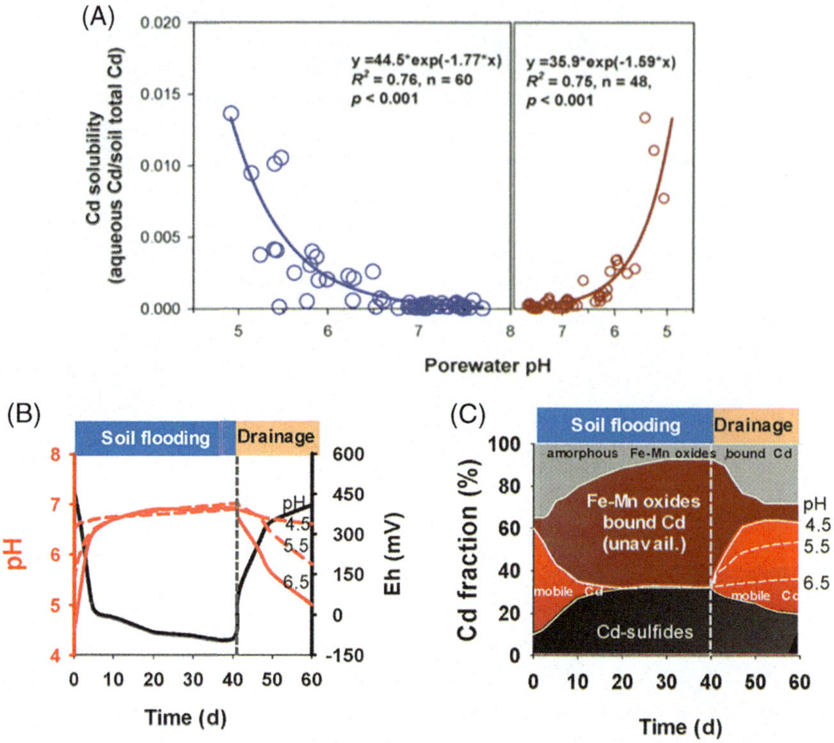

FIGURE 7.3 The Cd solubility (solution Cd/soil total Cd) as influenced by pore water pH during the reduction phase (left) and the reoxidation phase (right) (A) and the effects of pH dynamics coupling with redox potential (E_h) from flooding to drainage on Cd fractions (B and C). *From Wang et al. (2019a), with permission.*

status, pH, and associated biogeochemical processes drive Cd remobilization, making this stage critical for Cd accumulation in rice.

7.4.1 Redox and pH dynamics during drainage

Drainage allows oxygen intrusion into soil pores, increasing E_h and triggering sequential oxidative reactions. This reoxidation releases H^+ from Fe(II) and S(–II) oxidation, lowering soil pH, particularly in acidic paddy soils, from the near-neutral range achieved during flooding back to initial acidic levels (Nieder et al., 2018; Wang et al., 2019a). Soil pH exerts a dominant control on Cd solubility during drainage: Cd solubility follows the relationship Cd solubility $= 35.9 \times \exp(-1.59 \times$ porewater pH) ($R^2 = 0.75$, $P < .001$), with a 1-unit decrease in porewater pH increasing Cd solubility by 4.3-fold (Fig. 7.3B and C) (Wang et al., 2019a). Lower pH weakens Cd adsorption to Fe/Mn oxides and organic matter, promoting Cd desorption into the soil solution, where it exists as the highly bioavailable Cd^{2+} pool (Argüello et al., 2019; Zhu et al., 2016).

In field experiments, spatial variations in E_h and pH during drainage (e.g., in sloped paddy fields) led to six- to sevenfold differences in rice grain Cd concentrations, with

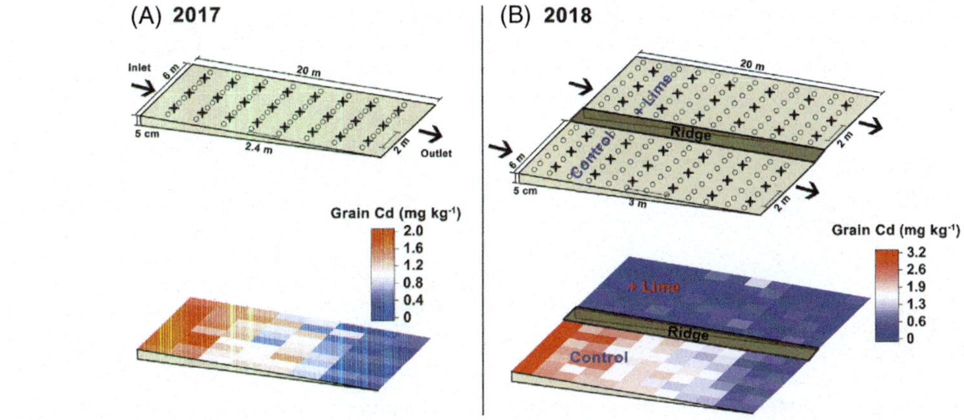

FIGURE 7.4 Variation in grain Cd concentration in sloped paddy plots (6.0 × 20 m^2, 0.15 slope). (A) Experiment in 2017; (B) Experiment in 2018 with a control and a liming treatment. The slope was created from the irrigation inlet to the outlet according to local farming practice. *From Chen et al. (2020), with permission.*

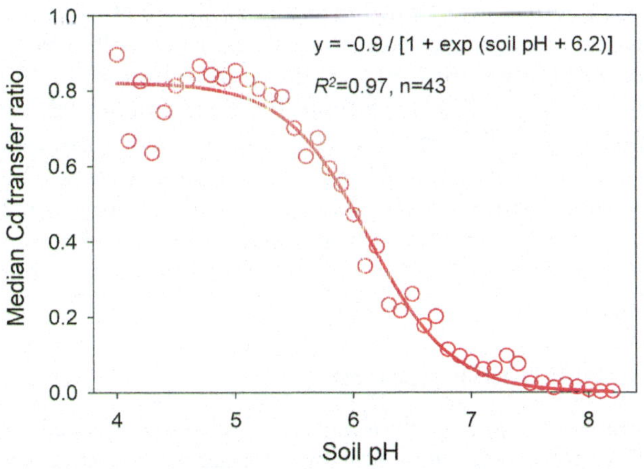

FIGURE 7.5 Median Cd transfer ratio (rice grain Cd concentration/soil Cd concentration) as a function of soil pH. *Data from Zhu et al. (2016), with permission.*

higher levels near irrigation inlets because of more pronounced oxidative conditions (Fig. 7.4) (Chen et al., 2020). In addition, anthropogenic drivers have accelerated soil acidification globally. Long-term nitrogen fertilizer use, industrial emissions, and acid rain have caused substantial pH declines (Cai et al., 2015; Tian & Niu, 2015; Sainju et al., 2019; Zeng et al., 2017). In China, mean pH in agricultural soils has fallen by 0.1–0.8 unit over the past three decades, between the 1980s and 2010 (Guo et al., 2010), and some double-cropping regions in Hunan province experienced nearly a full unit decrease because of cumulative nitrogen fertilizer inputs (Zhu et al., 2016). Such acidification markedly enhances Cd solubility and its accumulation in rice grain. For example, median Cd transfer factor (i.e., rice grain Cd concentration/soil Cd concentration) increases from ~0.1 at soil pH 7.0 to ~1.0 when pH drops below 5.5, representing a 10-fold increase in Cd bioavailability (Fig. 7.5) (Zhu et al., 2016).

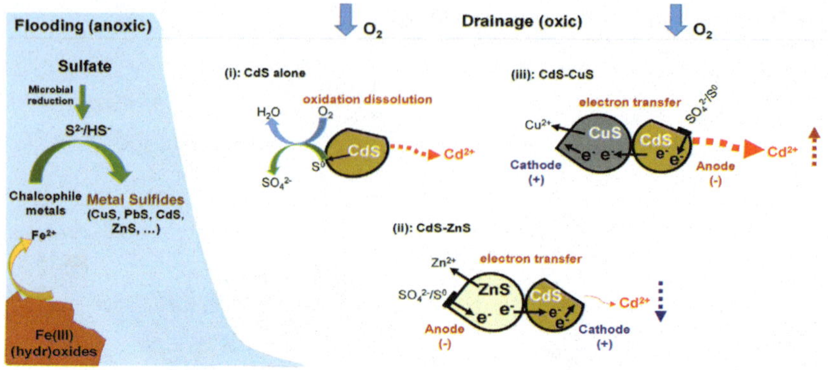

FIGURE 7.6 Conceptual model depicting the voltaic effect controlling the remobilization of Cd in paddy soils during flooding and subsequent drainage. *From Huang et al. (2021a), with permission.*

7.4.2 The voltaic effect and cadmium transformation

During flooding, Cd is immobilized as CdS (accounting for over 60% of available Cd in some soils) alongside other metal sulfides (ZnS, CuS, PbS, and MnS) (Furuya et al., 2016; Hashimoto et al., 2016; Huang et al., 2021a). Upon drainage, these metal sulfide minerals form voltaic cells based on differences in electrochemical potential, regulating the oxidative dissolution of CdS. Metal sulfides with lower potential act as anodes and preferentially oxidize, whereas those with higher potential (cathodes) are protected (Fig. 7.6) (Huang et al., 2021a). For example, ZnS (electrochemical potential = 243 mV) and MnS act as sacrificial anodes, suppressing CdS (electrochemical potential 361 = mV) oxidation; in contrast, CuS (electrochemical potential = 427 mV) acts as a cathode, forcing CdS to oxidize and accelerate Cd remobilization. This mechanism explains the large variability in Cd release rates among paddy soils during drainage, as the composition of metal sulfides formed during flooding varies by soil type.

7.4.3 The free radical effect and cadmium transformation

Flooding-induced reductive dissolution of amorphous Fe(III) (oxyhydr)oxides generates abundant Fe^{2+}, which reacts with S^{2-}/HS^{-} to form FeS/FeS_2. During drainage, FeS oxidation modulates CdS stability through three sequential mechanisms (Huang et al., 2021b): (1) early-stage voltaic cell formation (FeS, electrochemical potential = 49 mV, as anode) protects CdS from oxidation; (2) progressive oxidation of FeS generates hydroxyl radicals (•OH), altering FeS electrochemical potential and reversing the protective effect; (3) •OH directly oxidizes CdS, becoming the dominant Cd remobilization pathway as drainage proceeds (Fig. 7.7). Soils with high FeS production during flooding exhibit accelerated CdS oxidation and enhanced Cd release because of this free radical effect, further increasing dissolved Cd concentrations in the soil solution (Huang et al., 2021c, 2024c).

7.4.4 Cadmium partitioning and solubility during drainage

The Cd released from CdS oxidation partitions between soil solid and liquid phases, governed primarily by pH-dependent adsorption to Fe/Mn (oxyhydr)oxides (Wang et al., 2019a). A

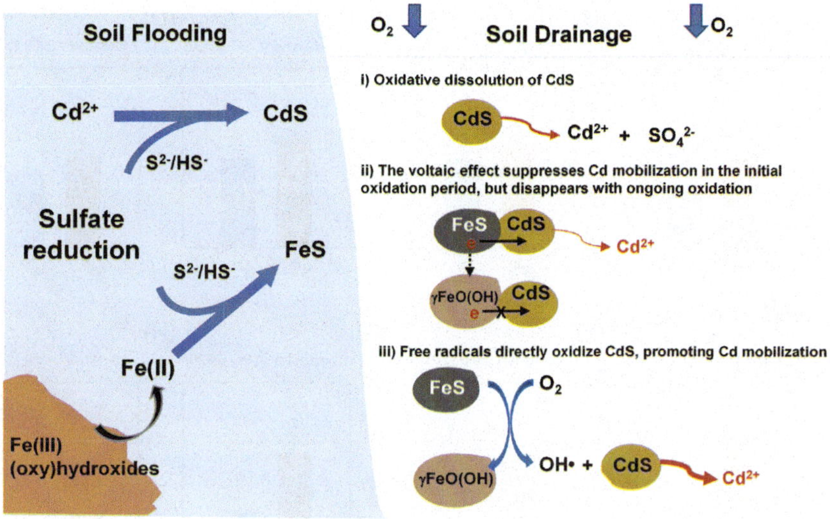

FIGURE 7.7 Conceptual model depicting the effects of FeS formed during soil flooding on the oxidative dissolution of CdS during the subsequent soil drainage. *From Huang et al. (2021b), with permission.*

1-unit decrease in pH typically increases solution Cd by 3–4 folds, as lower pH reduces the adsorption capacity of Fe/Mn oxides (Fig. 7.3A). Dissolved Cd concentrations, though still a small fraction of total soil Cd, rise significantly during drainage; they often exceed preflooding levels because of the combined effects of oxidative dissolution of CdS and pH-dependent desorption (Christensen, 1984; Wang et al., 2019a). This bioavailable Cd pool is efficiently taken up by rice roots during the grain-filling stage, directly contributing to Cd accumulation in rice grains.

Collectively, the interplay of the voltaic effect, free radical oxidation, and pH-dependent desorption drives Cd remobilization during drainage. These paddy-specific processes reactivate previously immobilized Cd, increasing its bioavailability for rice uptake and highlighting the importance of drainage management for controlling grain Cd accumulation.

7.5 Strategies for regulating cadmium transformation in paddy soils

7.5.1 Dynamic regulation through biogeochemical processes

Efforts to regulate Cd transformation in paddy soils focus on reducing Cd bioavailability and suppressing its mobilization, with soil physicochemical improvement as the core strategy (Huang et al., 2021c; Wang et al., 2019b; Zhao & Wang, 2020).

- Soil pH adjustment: The partitioning of Cd^{2+} between the soil solid phase and solution is strongly pH–dependent. Liming is highly effective in acidic soils, where low pH promotes Cd desorption from Fe/Mn oxides and SOM. Application of limestone powder ($CaCO_3$) to raise pH to ~6.5 markedly reduces Cd solubility, with effects lasting more than 3 years, and has been widely adopted in southern China (Chen et al., 2018a; Zhu et al., 2016). However,

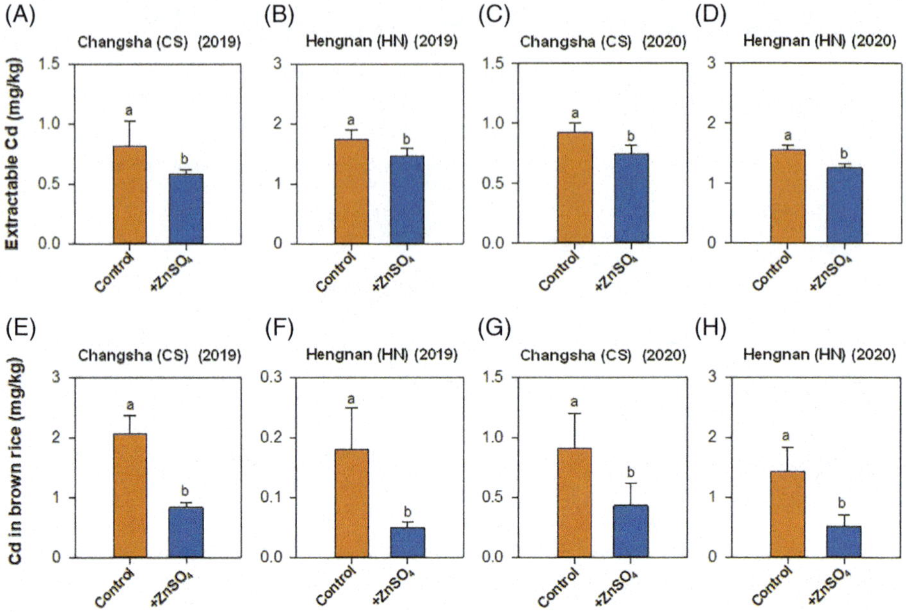

FIGURE 7.8 Effects of one-time addition of $ZnSO_4$ on the concentrations of extractable Cd in soils during the grain filling period (A–D) and Cd in brown rice (E–H) in the continuous 2-year field trials at two Cd-contaminated paddy fields (Changsha and Hengnan, Southern China). *From Huang et al. (2022b),* with permission.

in alkaline soils, liming has little effect on Cd bioavailability (Khaokaew et al., 2011; Scherger et al., 2021).

- Water management: Paddy water status during the grain-filling stage is critical, as this is the period when rice grain accumulates most Cd. Maintaining flooded conditions or delaying drainage lowers soil E_h, thereby suppressing the oxidative dissolution of CdS and the •OH-mediated oxidation triggered by FeS reoxidation (Honma et al., 2016; Huang et al., 2024a). Both strategies can substantially reduce Cd accumulation in grains. Intermittent flooding and drainage cycles also decrease grain Cd; for example, alternating 3 days of flooding with 5 days of drainage reduced grain Cd concentration by 56% (Honma et al., 2016; Huang et al., 2022b). However, paddy water management is constrained by farming practices and rainfall patterns. Furthermore, it is worth noting that continuous flooding often increases arsenic (As) accumulation in grains, so water management should be applied cautiously in paddy soils contaminated with both Cd and As (Arao et al., 2009).
- Leveraging the voltaic effect: Amendments with Zn or Mn fertilizers (e.g., $ZnSO_4$ or $MnSO_4$) exploit the voltaic cell effect between metal sulfides. Under flooding, ZnS and MnS form and, upon drainage, act as sacrificial anodes and preferentially oxidize to protect CdS from dissolution. Field trials demonstrate that a single application of 75 kg Zn ha^{-1} (as $ZnSO_4$) can reduce Cd bioavailability for over 2 years while also enhancing Zn accumulation in rice grains, benefiting Zn nutrition for humans (Fig. 7.8) (Huang et al., 2021a, 2022b). Mn-containing amendments are also effective in reducing Cd accumulation in rice grains (Fang et al., 2021). This may be attributed to two potential mechanisms: first, the voltaic cell effect

by MnS can reduce Cd mobilization in soil; second, the addition of Mn oxides increases Mn^{2+} concentrations in the soil solution, and Mn^{2+} can inhibit Cd^{2+} uptake by rice roots through competitive effects because the two cations share the transporters for uptake into rice roots (see Chapter 10). Because free radical oxidation may enhance Cd release, strategies that combine •OH suppression with voltaic protection represent a particularly promising avenue.

7.5.2 Reduction- and removal-oriented strategies

Dynamic regulation strategies described above aim to reduce Cd mobility without removal of Cd from the soil. However, long-term remediation requires strategies that actively remove Cd from soils.

- Straw removal: Removing rice straw is a widely recommended safe-utilization measure in China. However, its capacity to reduce soil Cd is limited: each harvest removes only 0.2%–0.3% of soil Cd, meaning ~20 croppings are required to lower Cd from 1.0 to 0.9 mg kg^{-1} (Zhao et al., 2021) (see also Chapter 5).
- High-Cd-accumulating rice varieties: A more radical remediation strategy involves cultivating rice varieties with high Cd accumulation capacity, particularly those carrying natural variation in loss of function alleles of *OsHMA3*. Normally, *OsHMA3* encodes a heavy-metal ATPase that sequesters Cd^{2+} into root vacuoles, thereby restricting its translocation to shoots. Loss-of-function variants of *OsHMA3* disrupt this sequestration, leading to enhanced Cd transport to aboveground tissues, especially straw (Sui et al., 2019; Ueno et al., 2010; Yan et al., 2016) (see Chapter 10). As a result, Cd accumulates primarily in straw biomass, which can be removed during harvest, enabling net extraction of Cd from soil. In paddy soils managed with dry farming or intermittent water management (to increase available Cd that rice can absorb adequately), the Cd bioconcentration factor (BCF) of these cultivars can reach 20–40 (Murakami et al., 2009). Assuming an average BCF of 30 and biomass yields of 8 t ha^{-1} each for straw and grain, the Cd removal efficiency per season can be as high as ~18%. Under this scenario, six cropping seasons would reduce soil Cd concentrations from 1.0 to 0.3 mg kg^{-1}. Moreover, rice does not necessarily need to reach maturity, and with two crop cycles per year, remediation could be accelerated (Fig. 7.9) (Tang et al., 2025).

7.6 Concluding remarks

Cd transformation in paddy soils is governed by interconnected biogeochemical processes, including pH-dependent adsorption, redox-controlled sulfide dynamics (e.g., CdS precipitation during flooding and its oxidative dissolution during drainage), voltaic interactions among coexisting metal sulfides, and free radical oxidation mediated by FeS. Collectively, these mechanisms regulate Cd solubility and bioavailability, thereby determining Cd accumulation in rice grains—a critical food safety issue in rice-based diets.

A range of mitigation strategies has demonstrated effectiveness. Liming is a reliable long-term measure for reducing Cd solubility in acidic soils. Water management practices, such as sustained flooding or delayed drainage, can suppress Cd mobilization, though they must be carefully balanced against the risk of enhanced As accumulation. Soil amendments with

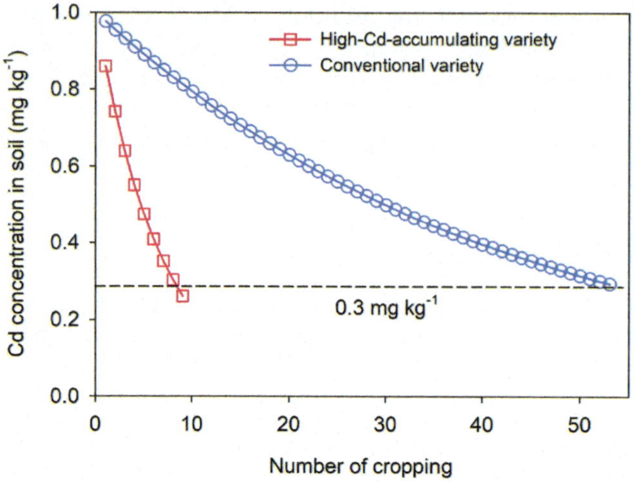

FIGURE 7.9 Predicted duration required for remediation of Cd-contaminated soils using different rice cultivars. The dashed line indicates the soil pollution risk screening value for agricultural land specified in GB15618-2018 (China). *Data from Tang et al. (2025).*

Zn or Mn fertilizers exploit the voltaic effect to stabilize Cd sulfides while simultaneously supplying essential micronutrients. In parallel, genetic approaches, including the use of low-Cd-accumulating cultivars and high-Cd-accumulating lines for phytoremediation, provide complementary tools for reducing Cd transfer into rice grain or gradually depleting soil Cd stocks.

Despite this progress, significant knowledge gaps remain. The mechanistic basis for regional differences in Cd solubility and bioavailability is still poorly understood, and the field-scale feasibility, sustainability, and ecological trade-offs of some emerging technologies require further validation. Moving forward, research should prioritize integrated, site-specific management frameworks that couple soil amendments, redox regulation, and crop genetic strategies. Such holistic approaches will be essential to reduce Cd risk while maintaining rice yield and quality, thereby securing both food safety and environmental sustainability in Cd-contaminated paddy agroecosystems.

7.7 Summary

- Flooding of paddy soil reduces soil E_h, stimulating microbial sulfate reduction to generate sulfide that forms insoluble CdS, resulting in immobilization of Cd; concurrent reductive dissolution of Fe/Mn (oxyhydr)oxides temporarily releases adsorbed Cd but is offset by CdS precipitation.
- Draining of paddy soil controls Cd mobilization via the voltaic effect (competitive oxidation between coexisting metal sulfides like ZnS/CuS and CdS) or reactivates Cd via free radicals from FeS oxidation, whereas pH decline weakens Cd adsorption to Fe/Mn oxides, enhancing its solubility.
- Mitigation strategies include liming to raise soil pH, optimized water management (e.g., delayed drainage), Zn/Mn fertilizer amendments (exploiting the voltaic effect), and phytoremediation using high-Cd-accumulating rice cultivars.

Acknowledgments

This work was supported by the National Natural Science Foundation of China (grants no. 42325701, W2411035, 42307025).

References

Allison, J., Brown, D. S., & Novo-Gradac, K. J. (1991). *MINTEQA2/PRODEFA2, A geochemical assessment model for environmental systems: Version 3.0 user's manual*. U.S. Environmental Protection Agency (U.S. EPA)Environmental Research Laboratory, Office of Research and Development.

Arao, T., Kawasaki, A., Baba, K., Mori, S., & Matsumoto, S. (2009). Effects of water management on cadmium and arsenic accumulation and dimethylarsinic acid concentrations in Japanese rice. *Environmental Science & Technology, 43*(24), 9361–9367. doi:10.1021/es9022738.

Argüello, D., Chavez, E., Lauryssen, F., Vanderschueren, R., Smolders, E., & Montalvo, D. (2019). Soil properties and agronomic factors affecting cadmium concentrations in cacao beans: A nationwide survey in Ecuador. *Science of the Total Environment, 649*, 120–127. doi:10.1016/j.scitotenv.2018.08.292.

Bao, P., Li, G. X., Sun, G. X., Xu, Y. Y., Meharg, A. A., & Zhu, Y. G. (2018). The role of sulfate-reducing prokaryotes in the coupling of element biogeochemical cycling. *Science of the Total Environment, 613–614*, 398–408. http://www.elsevier.com/locate/scitotenv. doi:10.1016/j.scitotenv.2017.09.062.

Borch, T., Kretzschmar, R., Kappler, A., Van Cappellen, P., Ginder-Vogel, M., Voegelin, A., & Campbell, K. (2010). Biogeochemical redox processes and their impact on contaminant dynamics. *Environmental Science & Technology, 44*(1), 15–23. doi:10.1021/es9026248.

Cai, Z. J., Wang, B. R., Xu, M. G., Zhang, H. M., He, X. H., Zhang, L., & Gao, S. D. (2015). Intensified soil acidification from chemical N fertilization and prevention by manure in an 18-year field experiment in the red soil of southern China. *Journal of Soils and Sediments, 15*(2), 260–270. http://www.springerlink.com/content/1439-0108. doi:10.1007/s11368-014-0989-y.

Chen, H. P., Tang, Z., Wang, P., & Zhao, F. J. (2018b). Geographical variations of cadmium and arsenic concentrations and arsenic speciation in Chinese rice. *Environmental Pollution, 238*, 482–490. http://www.elsevier.com/inca/publications/store/4/0/5/8/5/6. doi:10.1016/j.envpol.2018.03.048.

Chen, H. P., Wang, P., Chang, J. D., Kopittke, P. M., & Zhao, F. J. (2021). Producing Cd-safe rice grains in moderately and seriously Cd-contaminated paddy soils. *Chemosphere, 267*. http://www.elsevier.com/locate/chemosphere. doi:10.1016/j.chemosphere.2020.128893.

Chen, H. P., Wang, P., Gu, Y., Kretzschmar, R., Kopittke, P. M., & Zhao, F. J. (2020). The within-field spatial variation in rice grain Cd concentration is determined by soil redox status and pH during grain filling. *Environmental Pollution, 261*. https://www.journals.elsevier.com/environmental-pollution. doi:10.1016/j.envpol.2020.114151.

Chen, H. P., Zhang, W. W., Yang, X. P., Wang, P., McGrath, S. P., & Zhao, F. J. (2018a). Effective methods to reduce cadmium accumulation in rice grain. *Chemosphere, 207*, 699–707. http://www.elsevier.com/locate/chemosphere. doi:10.1016/j.chemosphere.2018.05.143.

Chen, H. Y., Teng, Y. G., Lu, S. J., Wang, Y. Y., & Wang, J. S. (2015). Contamination features and health risk of soil heavy metals in China. *Science of the Total Environment, 512–513*, 143–153. http://www.elsevier.com/locate/scitotenv. doi:10.1016/j.scitotenv.2015.01.025.

Christensen, T. H. (1984). Cadmium soil sorption at low concentrations: I. Effect of time, cadmium load, pH, and calcium. *Water, Air, and Soil Pollution, 21*(1–4), 105–114. doi:10.1007/BF00163616.

de Livera, J., McLaughlin, M. J., Hettiarachchi, G. M., Kirby, J. K., & Beak, D. G. (2011). Cadmium solubility in paddy soils: Effects of soil oxidation, metal sulfides and competitive ions. *Science of the Total Environment, 409*(8), 1489–1497. doi:10.1016/j.scitotenv.2010.12.028.

Elinder, C. G., Kjellstrom, T., Hogstedt, C., Andersson, K., & Spang, G. (1985). Cancer mortality of cadmium workers. *Occupational and Environmental Medicine, 42*(10), 651–655. doi:10.1136/oem.42.10.651.

Fang, X., Wang, J., Chen, H. P., Christl, I., Wang, P., Kretzschmar, R., & Zhao, F. J. (2021). Two-year and multi-site field trials to evaluate soil amendments for controlling cadmium accumulation in rice grain. *Environmental Pollution, 289*. https://www.journals.elsevier.com/environmental-pollution. doi:10.1016/j.envpol.2021.117918.

Ford, R. G. (2007). *Monitored natural attenuation of inorganic contaminants in ground water*. National Risk Management Research Laboratory, Office of Research and Development, U.S. Environmental Protection Agency.

Fulda, B. (2013). *Changes in copper and cadmium solubility and speciation induced by soil redox dynamics: Competitive metal sulfide formation and interactions with natural organic matter*. ETH Zurich.

Fulda, B., Voegelin, A., Ehlert, K., & Kretzschmar, R. (2013a). Redox transformation, solid phase speciation and solution dynamics of copper during soil reduction and reoxidation as affected by sulfate availability. *Geochimica et Cosmochimica Acta, 123*, 385–402. doi:10.1016/j.gca.2013.07.017.

Fulda, B., Voegelin, A., & Kretzschmar, R. (2013b). Redox–controlled changes in cadmium solubility and solid-phase speciation in a paddy soil as affected by reducible sulfate and copper. *Environmental Science & Technology, 47*(22), 12775–12783. doi:10.1021/es401997d.

Furuya, M., Hashimoto, Y., & Yamaguchi, N. (2016). Time-course changes in speciation and solubility of cadmium in reduced and oxidized paddy soils. *Soil Science Society of America Journal, 80*(4), 870–877. doi:10.2136/sssaj2016.03.0062.

Guo, J. H., Liu, X. J., Zhang, Y., Shen, J. L., Han, W. X., Zhang, W. F., Christie, P., Goulding, K. W. T., Vitousek, P. M., & Zhang, F. S. (2010). Significant acidification in major Chinese croplands. *Science, 327*(5968), 1008–1010. doi:10.1126/science.1182570.

Hashimoto, Y., Furuya, M., Yamaguchi, N., & Makino, T. (2016). Zerovalent iron with high sulfur content enhances the formation of cadmium sulfide in reduced paddy soils. *Soil Science Society of America Journal, 80*(1), 55–63. doi:10.2136/sssaj2015.06.0217.

He, L., Wang, B., Cui, H., Yang, S., Wang, Y., Feng, Y., Sun, X., & Feng, Y. (2022). Clay-hydrochar composites return to cadmium contaminated paddy soil: Reduced Cd accumulation in rice seed and affected soil microbiome. *Science of the Total Environment, 835*. http://www.elsevier.com/locate/scitotenv. doi:10.1016/j.scitotenv.2022.155542.

Honma, T., Ohba, H., Kaneko-Kadokura, A., Makino, T., Nakamura, K., & Katou, H. (2016). Optimal soil Eh, pH, and water management for simultaneously minimizing arsenic and cadmium concentrations in rice grains. *Environmental Science & Technology, 50*(8), 4178–4185. doi:10.1021/acs.est.5b05424.

Huang, B. Y., Zhao, F. J., & Wang, P. (2022a). The relative contributions of root uptake and remobilization to the loading of Cd and As into rice grains: Implications in simultaneously controlling grain Cd and As accumulation using a segmented water management strategy. *Environmental Pollution, 293*. https://www.journals.elsevier.com/environmental-pollution. doi:10.1016/j.envpol.2021.118497.

Huang, G. X., Wu, Y., Cheng, L. X., Zhou, D. M., Wang, X. X., Ding, M. J., Wang, P., & Wang, Y. R. (2024a). Spatial heterogeneity of soil moisture caused by drainage and its effects on cadmium variation in rice grain within individual fields. *Science of the Total Environment, 946*. https://www.sciencedirect.com/science/journal/00489697. doi:10.1016/j.scitotenv.2024.174500.

Huang, H., Chen, H. P., Kopittke, P. M., Kretzschmar, R., Zhao, F. J., & Wang, P. (2021a). The voltaic effect as a novel mechanism controlling the remobilization of cadmium in paddy soils during drainage. *Environmental Science & Technology, 55*(3), 1750–1758. http://pubs.acs.org/journal/esthag. doi:10.1021/acs.est.0c06561.

Huang, H., Ge, L., Zhang, X., Chen, H., Shen, Y., Xiao, J., Lu, H., Zhu, Y., Han, J., & Li, R. (2024b). Rice straw biochar and lime regulate the availability of heavy metals by managing colloid-associated- but dissolved-heavy metals. *Chemosphere, 349*. https://www.sciencedirect.com/science/journal/00456535. doi:10.1016/j.chemosphere.2023.140813.

Huang, H., Ji, X. B., Cheng, L. Y., Zhao, F. J., & Wang, P. (2021b). Free radicals produced from the oxidation of ferrous sulfides promote the remobilization of cadmium in paddy soils during drainage. *Environmental Science & Technology, 55*(14), 9845–9853. http://pubs.acs.org/journal/esthag. doi:10.1021/acs.est.1c00576.

Huang, H., Lv, Y. W., Tian, K. K., Shen, Y., Zhu, Y. L., Lu, H. Y., Li, R. H., & Han, J. G. (2023). Influence of sulfate reducing bacteria cultured from the paddy soil on the solubility and redox behavior of Cd in a polymetallic system. *Science of the Total Environment, 901*. http://www.elsevier.com/locate/scitotenv. doi:10.1016/j.scitotenv.2023.166369.

Huang, H., Tang, Z. X., Qi, H. Y., Ren, X. T., Zhao, F. J., & Wang, P. (2022b). Soil amendments with $ZnSO_4$ or $MnSO_4$ are effective at reducing Cd accumulation in rice grain: An application of the voltaic cell principle. *Environmental Pollution, 294*. https://www.journals.elsevier.com/environmental-pollution. doi:10.1016/j.envpol.2021.118650.

Huang, H., Tian, Z. Q., Guo, D., Tang, Z. X., Li, R. H., Ali, A., Cao, Z. X., Lu, H. Y., Shen, Y., Zhu, Y. L., & Han, J. G. (2024c). Rice straw returning enhances cadmium activation by accelerating iron cycling thus hydroxyl radical production in paddy soils during drainage. *Science of the Total Environment, 923*. https://www.sciencedirect.com/science/journal/00489697. doi:10.1016/j.scitotenv.2024.171543.

Huang, H., Zhao, D., & Wang, P. (2021c). Biogeochemical control on the mobilization of Cd in soil. *Current Pollution Reports, 7*(2), 194–200. doi:10.1007/s40726-021-00180-w.

Jin, T. Y., Nordberg, G., Ye, T. T., Bo, M. H., Wang, H. F., Zhu, G. Y., Kong, Q. H., & Bernard, A. (2004). Osteoporosis and renal dysfunction in a general population exposed to cadmium in China. *Environmental Research, 96*(3), 353–359. doi:10.1016/j.envres.2004.02.012.

Khaokaew, S., Chaney, R. L., Landrot, G., Ginder-Vogel, M., & Sparks, D. L. (2011). Speciation and release kinetics of cadmium in an alkaline paddy soil under various flooding periods and draining conditions. *Environmental Science & Technology, 45*(10), 4249–4255. doi:10.1021/es103971y.

MEP and MLR. (2014). *The report on national general survey on soil contamination*. Ministry of Environmental Protection and Ministry of Land and Resources of China.

Ministry of the Environment of Japan. (1984). *National soil survey report*. Ministry of the Environment of Japan.

Murakami, M., Nakagawa, F., Ae, N., Ito, M., & Arao, T. (2009). Phytoextraction by rice capable of accumulating Cd at high levels: Reduction of Cd content of rice grain. *Environmental Science & Technology, 43*(15), 5878–5883. doi:10.1021/es8036687.

Nieder R., Benbi D.K., & Reichl F.X. (Eds.). Soil components and human health role of potentially toxic elements in soils. Springer.

Nogawa, K. (2026). Itai-itai disease and follow-up studies. In J. O. Nriagu (Ed.), *Cadmium in the environment Part 2: Health effects*. A Wiley-Interscience Publication.

Revis, N. W., & Osborne, T. R. (1984). Dietary protein effects on cadmium and metallothionein accumulation in the liver and kidney of rats. *Environmental Health Perspectives, 54*, 83–91. doi:10.1289/ehp.845483.

Rozan, T. F., Lassman, M. E., Ridge, D. P., & Luther, G. W. (2000). Evidence for iron, copper and zinc complexation as multinuclear sulphide cluster in oxic rivers. *Nature, 406*(6798), 879–882. doi:10.1038/35022561.

Sainju, U. M., Ghimire, R., & Pradhan, G. P. (2026). Nitrogen fixation nitrogen fertilization I: Impact on crop, soil, and environment. In E. C. Rigobelo, & A. P. Serra (Eds.), *Nitrogen in agricultural systems [working title]*. IntechOpen. doi:10.5772/intechopen.86028.

Scherger, L. E., Luengo, C. V., & Lexow, C. (2021). Comparative study of cadmium adsorption on three contrasting soils from a highly industrialized region. *Environmental Nanotechnology, Monitoring & Management, 16*. http://www.journals.elsevier.com/environmental-nanotechnology-monitoring-and-management/. doi:10.1016/j.enmm.2021.100605.

Smolders, E., & Mertens, J. (2026). Cadmium. In B. J. Alloway (Ed.), *Heavy metals in soils: Trace metals and metalloids in soils and their bioavailability*. Springer.

Song, Y., Wang, Y. B. N., Mao, W. F., Sui, H. X., Yong, L., Yang, D. J., Jiang, D. G., Zhang, L., & Gong, Y. Y. (2017). Dietary cadmium exposure assessment among the Chinese population. *PLoS One, 12*(5). http://journals.plos.org/plosone/article/file?id=10.1371/journal.pone.0177978&type=printable. doi:10.1371/journal.pone.0177978.

Staessen, J., Amery, A., Bernard, A., Bruaux, P., Buchet, J. P., Bulpitt, C. J., Claeys, F., De Plaen, P., Ducoffre, G., Fagard, R., Lauwerys, R. R., Lijnen, P., Nick, L., Remy, A. S., Roels, H., Rondia, D., Sartor, F., & Thijs, L. (1991). Blood pressure, the prevalence of cardiovascular diseases, and exposure to cadmium: A population study. *American Journal of Epidemiology, 134*(3), 257–267. doi:10.1093/oxfordjournals.aje.a116079.

Sui, F. Q., Zhao, D. K., Zhu, H. T., Gong, Y. F., Tang, Z., Huang, X. Y., Zhang, G. Q., & Zhao, F. J. (2019). Map-based cloning of a new total loss-of-function allele of OsHMA3 causes high cadmium accumulation in rice grain. *Journal of Experimental Botany, 70*(10), 2857–2871. http://jxb.oxfordjournals.org/. doi:10.1093/jxb/erz093.

Tang, Z. X., Dong, G., Tang, Z., Zhao, F. J., & Wang, P. (2025). Cadmium reduction and remediation technology in soil based on a high-Cd-accumulating rice cultivar. *Acta Pedologica Sinica, 62*(6), 1689–1698. http://pedologica.issas.ac.cn/trxben/article/issue/2025_62_6. doi:10.11766/trxb202410080387.

Tian, D. S., & Niu, S. L. (2015). A global analysis of soil acidification caused by nitrogen addition. *Environmental Research Letters, 10*. http://iopscience.iop.org/1748-9326/10/2/024019/pdf/1748-9326_10_2_024019.pdf. doi:10.1088/1748-9326/10/2/024019.

Ueno, D., Yamaji, N., Kono, I., Huang, C. F., Ando, T., Yano, M., & Ma, J. F. (2010). Gene limiting cadmium accumulation in rice. *Proceedings of the National Academy of Sciences, 107*(38), 16500–16505. doi:10.1073/pnas.1005396107.

Wang, J., Wang, P. M., Gu, Y., Kopittke, P. M., Zhao, F. J., & Wang, P. (2019a). Iron-manganese (oxyhydro)oxides, rather than oxidation of sulfides, determine mobilization of Cd during soil drainage in paddy soil systems. *Environmental Science & Technology, 53*(5), 2500–2508. http://pubs.acs.org/journal/esthag. doi:10.1021/acs.est.8b06863.

Wang, P., Chen, H. P., Kopittke, P. M., & Zhao, F. J. (2019b). Cadmium contamination in agricultural soils of China and the impact on food safety. *Environmental Pollution, 249,* 1038–1048. https://www.journals.elsevier.com/environmental-pollution. doi:10.1016/j.envpol.2019.03.063.

Wang, P., Kopittke, P. M., McGrath, S. P., & Zhao, F. J. (2026). Cadmium transfer from soil to plants and its potential risk to human health. In B. R. Singh, M. J. McLaughlin, & E. Brevik (Eds.), *The nexus of soils, plants, animals and human health.* Catena-Schweizerbart.

Weber, F. A., Voegelin, A., Kaegi, R., & Kretzschmar, R. (2009). Contaminant mobilization by metallic copper and metal sulphide colloids in flooded soil. *Nature Geoscience, 2*(4), 267–271. doi:10.1038/ngeo476.

Yan, J. L., Wang, P. T., Wang, P., Yang, M., Lian, X. M., Tang, Z., Huang, C. F., Salt, D. E., & Zhao, F. J. (2016). A loss-of-function allele of OsHMA3 associated with high cadmium accumulation in shoots and grain of Japonica rice cultivars. *Plant Cell & Environment, 39*(9), 1941–1954. http://www.blackwellpublishing.com/journals/PCE. doi:10.1111/pce.12747.

Yang, X., Pan, H., Shaheen, S. M., Wang, H. L., & Rinklebe, J. (2021). Immobilization of cadmium and lead using phosphorus-rich animal-derived and iron-modified plant-derived biochars under dynamic redox conditions in a paddy soil. *Environment International, 156,* 106628. doi:10.1016/j.envint.2021.106628.

Yang, Y., Liu, T. X., Borch, T., Fang, L. P., Hu, S. W., Chi, W. T., Chen, G. J., Cheng, K., Wang, Q., Li, X. M., Yuan, X., & Li, F. B. (2024). Iron biogeochemical redox cycling dominantly controls cadmium availability in acidic paddy soils. *Geochimica et Cosmochimica Acta, 378,* 186–202. https://www.sciencedirect.com/science/journal/00167037. doi:10.1016/j.gca.2024.06.029.

Zeng, M. F., Vries, W. D., Bonten, L. T. C., Zhu, Q. C., Hao, T. X., Liu, X. J., Xu, M. G., Shi, X. J., Zhang, F. S., & Shen, J. B. (2017). Model-based analysis of the long-term effects of fertilization management on cropland soil acidification. *Environmental Science & Technology, 51*(7), 3843–3851. http://pubs.acs.org/journal/esthag. doi:10.1021/acs.est.6b05491.

Zhao, F. J., & Wang, P. (2020). Arsenic and cadmium accumulation in rice and mitigation strategies. *Plant and Soil, 446,* 1–21. http://www.wkap.nl/journalhome.htm/0032-079X. doi:10.1007/s11104-019-04374-6.

Zhao, F. J., Zhao, X. Y., Tao, Y. M., Wang, H. Q., & Wang, P. (2021). Straw removal has a limited effect on decreasing cadmium concentration in soil. *Journal of Agro-Environment Science, 40*(3), 693–699. http://www.aes.org.cn/nyhjkxxben/ch/reader/view_abstract.aspx?file_no=20210202&flag=1. doi:10.11654/jaes.2020-1248.

Zhong Qian, Y., Chen, C., Zhang, Q., Li, Y., Jun Chen, Z., & Li, M. (2010). Concentrations of cadmium, lead, mercury and arsenic in Chinese market milled rice and associated population health risk. *Food Control, 21*(12 SUPPL.), 1757–1763. http://www.sciencedirect.com/. doi:10.1016/j.foodcont.2010.08.005.

Zhu, H. H., Chen, C., Xu, C., Zhu, Q. H., & Huang, D. Y. (2016). Effects of soil acidification and liming on the phytoavailability of cadmium in paddy soils of central subtropical China. *Environmental Pollution, 219,* 99–106. http://www.elsevier.com/inca/publications/store/4/0/5/8/5/6. doi:10.1016/j.envpol.2016.10.043.

CHAPTER

8

Impacts of cadmium on soil microorganisms and fauna

Steve McGrath[a], *Fang-Jie Zhao*[b]

[a] Sustainable Soils and Crops, Rothamsted Research, Harpenden, United Kingdom
[b] College of Resources and Environmental Sciences, Nanjing Agricultural University, Nanjing, China

8.1 Introduction

Soil microbes and other small organisms that live for the most part unseen in the soil are vitally important for the continued functions and health of soils and our planet. Large numbers of archaea, bacteria, fungi, and protozoa are present in soils, along with micro/mesofauna, including mites, springtails, nematodes, and earthworms. Together they form a very biodiverse ecosystem, providing important soil functions. These include decomposition and transformation of organic matter into stable humus and cycling nutrients by breaking down dead plants and animals, returning vital nutrients for plants to use, as well as engineering good soil structure, which is key for water infiltration and filtration. Their roles in carbon cycling and gas and water fluxes are very important for regulating climate. Specialized bacteria have evolved symbiotic relationships with some plant species such as legumes, enabling nitrogen fixation from the atmosphere, or in the case of symbiotic mycorrhizal fungi (see Chapter 13), more efficiently supplying phosphorus and nitrogen from the soil to plants. For healthy soils and a healthy planet, it is therefore important to protect soil organisms from the toxic effects of cadmium (Cd).

As Cd is not an essential element, it is taken up by organisms via pathways that normally take up essential elements (see Chapter 10), and it is then transported within the organism and often incorporated into metabolites and enzymes that usually contain chemically similar elements such as zinc (Zn). This renders such metabolites/enzymes nonfunctional in many cases, which is linked to the toxic impacts of Cd. However, some organisms have evolved mechanisms to tolerate relatively high Cd exposure as a result of evolutionary processes in environments naturally rich in Cd or, in shorter periods, by recent adaptations to anthropogenic pollution.

There are at least four important issues to take into consideration related to any assessment of Cd toxicity in soils. First, Cd almost never occurs alone in nature or in polluted areas, and so areas with elevated Cd concentrations in soil are always multimetal. This makes it difficult

Cadmium Contamination in Soils and Plants: Mechanisms and Mitigation.
DOI: https://doi.org/10.1016/B978-0-443-34104-5.00001-3

to distinguish the impacts that are because of Cd from other commonly cooccurring metals such as Zn, lead (Pb), and copper (Cu). For example, a review of Cd toxicity to microorganisms by Vig et al. (2003) included four field studies that had multimetal contamination where concentrations of Zn, Pb, Cu, and Ni were up to 51, 18, 12, and 3 times those of Cd, respectively. It is therefore likely that the other metals alone or acting together may have been more toxic than the Cd present. A further example is a study of the impact of Cd on the bacterial community structure in an arable soil polluted by mining in Hunan, China. In this case no other cocontaminants were analyzed, despite the likely impacts of long-term Pb and Zn and smelting activities in the province (Yu et al., 2021). This chapter attempts to overcome this issue by concentrating mainly on studies in which only Cd has been applied or, in the case of multicontaminant studies, focusing on those treatments where Cd has been applied alone as part of a multifactor experimental design. Although this restricts the number of sources or information, it is necessary to separate out the actual impacts of Cd on soil microorganisms and fauna. So-called multiple metal mixture models have been used to attribute element-specific toxicity to some extent in aquatic systems but are difficult to utilize in soils. For example, a study added five metals singly and in mixtures (but not Cd) to five different soils and found that it was difficult to assign toxicity to each specific metal reproduction by *Oppia nitens* mites in the metal mixtures (Jegede et al., 2020).

Second, the time for which a soil has been contaminated has important implications for assessment of Cd toxicity in at least two aspects. Soil organisms can be selected that are tolerant to Cd (Section 8.4), and although this can be rapid if the selection pressure (i.e., toxicity) is high, it is less likely to occur in the usual very short-term studies. In the latter type of study the original soil microbial or faunal population may have been suddenly reduced in size by the sudden imposition of Cd stress, and the study may not be addressing the size or activity of that initial population, but rather only those that resist the sudden challenge. Another factor is changes in the chemical form in which Cd was added. Whether the original source of contamination was soluble salts, relatively insoluble minerals, or organically bound, the forms change with time according to the chemical environment of the soil. Such changes are more likely to have taken place in longer field studies than short-term laboratory incubations of soil. This has resulted in the introduction of so-called lab-to-field or aging factors in toxicity risk assessments (e.g., Smolders et al., 2009) (see Chapter 16).

A third issue is the form and amount of metal present in the system studied. Fully soluble Cd salts are likely to have a greater impact on soil microorganisms and fauna than Cd present in less soluble forms in minerals or Cd bound to organic matter or precipitated as Cd sulfide (see Chapters 6 and 7). For example, Ren et al. (2021) assume that the soil studied from the Qinling Mountains, Shaanxi Province, China, was not contaminated with other metals, but they evaluated the effect of Cd on the composition and function of soil microbes by adding 10, 30, and 60 mg kg^{-1} of Cd using soluble Cd sulfate for 45 days. These concentrations are very high compared with the normal concentration range of Cd in soils and therefore are relevant only to highly polluted soils. In Chapter 3, this is illustrated by the fact that the Chenzhou region in Hunan province of China, well known for its mining activities, has a geometric mean of Cd concentrations in soil of 1.4 mg kg^{-1}.

The fourth issue that must be borne in mind is that soil properties influence bioavailability and therefore the toxicity of Cd. Important factors are pH and redox status of soils, because high pH decreases the solubility of Cd and low redox potential results in Cd sulfide formation, which also reduces the solubility of Cd (see Chapters 6 and 7).

The toxicity of Cd to soil microorganisms and their activity has been reviewed by McGrath (1999) and Vig et al. (2003). Therefore this chapter focuses largely on more recent publications since 2003.

8.2 Impacts of cadmium on soil microbial biomass, enzyme activities, soil respiration, and symbiotic activities

8.2.1 Laboratory incubation studies

Cd is toxic and can decrease soil enzyme activities and soil respiration and can also have impacts on symbiotic activities, microbial diversity, and abundance in soils. The EU Risk Assessment, as part of Registration, Evaluation, Authorization, and Restriction of Chemicals in the European Community, collected microbial toxicity data mainly based on parameters relevant at a population level, for example, survival, growth, and activities such as soil respiration and nitrification, and this is reviewed in Chapter 16. However, molecular endpoints (e.g., enzyme activity and biomarkers) are not generally included in risk assessments.

Wang et al. (2019) studied the impact of Cd alone on enzymatic activities and microbial community structure in soil from the Shandong Agricultural University farmland. The nominal amount of Cd added was 45 mg kg^{-1}, but recalculation shows that the actual amount was 36 mg kg^{-1} soil. Inhibition of the enzymes urease, phosphatase, sucrose, and catalase averaged over measurements at 7, 14, 21, and 28 days was, respectively, 14.9%, 23.8%, 15.0%, and 9.4%. Biolog EcoPlates were used at 14 and 21 days to assess use of substrates and showed that the use of amino acids, saccharides, and metabolites by the microbial community was lower than in the control at 14 days. But the utilization after 21 days of amino acids, aliphatics, and metabolites was higher than control, which may indicate that some recovery and adaptation to Cd had occurred after 14 days. The study also shows that the substrate utilization in Biolog EcoPlates was decreased by Cd contamination alone after all incubation periods.

Respiration of the microbial community (as measured by cumulative oxygen consumption) after 28 days of exposure was not affected by a Cd dose (as sulfate salt) as high as 100 mg kg^{-1}, but it decreased significantly in 1000 and 5000 mg kg^{-1} treatments. All of these are extremely large Cd additions compared to likely environmental exposure levels, and there was no significant effect on respiration when 0.1, 1, and 10 mg kg^{-1} were added (Cáliz et al., 2013). This was a sandy forest soil from Spain with no history of heavy metal pollution and so unlikely to contain microorganisms with enhanced tolerance to Cd. This example suggests that the microbial community is diverse and contains much functional redundancy in some general activities like respiration (Chen et al., 2022).

Using sophisticated analysis of bacterial community abundances in a KEGG approach (Kyoto encyclopedia of genes and genomes), homologous gene clusters (KO) derived from a cluster analysis showed that the abundances of all functional bacterial groups were lower than a control when 1 mg Cd kg^{-1} was added to soil, except for the iron complex outer membrane receptor protein and glutathione S-transferase (Duan et al., 2020). The abundance of microbial functional genes was also lower with 5 and 100 mg Cd kg^{-1}, but higher at 25, 50, and 75 mg kg^{-1}. This lack of a uniform response was explained by the authors as being because of Cd decreasing functions at 1 and 5 mg kg^{-1}, but that selection for tolerance increased function at 25, 50, and 75, but at the highest treatment, 100 mg kg^{-1}, metabolic functioning was severely inhibited (little tolerance). It is worth emphasizing that Cd concentrations above 5 mg kg^{-1} are rare, except for highly polluted sites and in studies examining the development of tolerance in the community (Section 8.4) and for studies on restoration of highly contaminated sites.

Impacts of Cd on microbial community composition, diversity, and function were assessed in a soil from a farm near Chengdu City, Sichuan, China, with a Cd salt added to give a 2.5 mg Cd kg^{-1} treatment that was equilibrated for 4 weeks (You et al., 2024). Using 16S rRNA sequencing of extracted DNA and the amplicon sequence variant method for assessing bacterial diversity, it was found that the diversity indices Chao1, observed species, and Shannon did not show significant differences between the control and Cd-treated soil. However, Simpson's and Faith's diversity indexes for the Cd treatment were significantly greater than the control. The authors then used PICRUSt2 analysis to predict the presence of genes involved in a comprehensive range of microbial functional pathways, but only genes related to lipid metabolism showed a significant decrease because of Cd.

In a laboratory incubation study with soil from an uncultivated field in central Iran the addition of a Cd salt at 30 mg Cd kg^{-1} decreased soil microbial biomass, whereas respiration was little affected (Raiesi & Sadeghi, 2019). Microbial qCO2 (soil respiration expressed as a percentage of the soil microbial C concentration) in the Cd-treated soil was higher than that in the control after 30 days of incubation, likely indicating metal stress. The activities of the catalase (CAT), fluorescein diacetate hydrolysis (FDA), alkaline phosphomonoesterase (ALP), and arylsulfatase (ARY) enzymes all decreased by more than control in the Cd treatment. CAT and FDH activities show the total metabolic function of the soil microbial community and therefore mirror the soil microbial biomass. ALP and ARY activity influence the availability of phosphorus and sulfur, respectively, which are essential for the cycling and supply of these important plant nutrients, which could be compromised in Cd-polluted soil.

Sewage sludge is often applied to land to recycle organic matter and the plant nutrients it contains. However, sludges can contain metals such as Cd, but evaluating the toxicity of Cd is challenging. A number of lab studies have addressed the issue of whether toxicity is higher or lower with the added organic matter by adding sewage sludge or Cd salts to soils. Again the presence of other heavy metals at higher concentrations in the sewage sludge can be a confounding factor. One study used a high pH (8.5) soil from southern Spain and additions of $CdSO_4$ (0, 25, 50, 100, 250, 500, 1000, 2000, 4000, 6000, and 8000 mg Cd kg^{-1}) to either control soil or the same soil with sewage sludge added at 5% w/w with low content of other heavy metals (Moreno et al., 2002). Their estimates of ED_{10} (effective dose inhibiting microbial indicators by 10%) after 40 days of incubation show that the organic matter in sludge appears to have increased the toxicity of Cd to a dehydrogenase assay and soil respiration, both of which reflect the oxidative and metabolic potential of the soil microbial community. On the other hand, sludge appears to have decreased the toxicity (i.e., higher fitted ED_{10} values) for ATP and soil microbial biomass, which both relate to the size of the soil microbial community (Table 8.1). However, the concentrations of Cd added to obtain the dose response curves were very high, and the fitting of curves and resulting ED_{10} estimates were highly uncertain in some cases. Even the lowest concentration of Cd added in this study was above the 90th percentile (0.43 mg kg^{-1}) and maximum observed concentration (7.5 mg kg^{-1}) in 2209 ploughed agricultural soils sampled in Europe (Birke et al., 2017).

Renella et al. (2002) used sewage sludge that had been artificially enriched with Cd or Zn added to an arable or a grassland soil in the lab to reach 12 or 300 mg kg^{-1} of the metals,

TABLE 8.1 Estimated ED_{10} values for soil microbial measurements after 40 days of incubation with Cd of either a pH 8.5 control soil or the same soil with sewage sludge added.

Measurements and units	Treatment	R^2	ED_{10} (mg Cd kg^{-1})
Dehydrogenase (g *p*-iodonitrotetrazolium formazan/G soil)	Soil	0.92	92.6
	Soil + sewage sludge	0.94	58.5
Respiration (mg CO_2-C/kg soil/d)	Soil	0.91	78.2
	Soil + sewage sludge	0.51	5.1
ATP content (ng ATP/g soil)	Soil	0.97	157.2
	Soil + sewage sludge	0.93	283.4
Soil microbial biomass (mg C/g soil)	Soil	0.51	7.0
	Soil + sewage sludge	0.38	307.8

Data from Moreno et al. (2002).
Cd, cadmium; *ED*, effective dose.

respectively. They concluded that even though the concentration of Cd exceeded the EU limit for Cd in soil derived from sludge by four times, there was no evidence of an impact on soil microbial biomass or its activity. They contrasted this with earlier evidence from field experiments that other metals (Cu and Zn) could affect soil microbes at concentrations above and below the corresponding EU limits.

8.2.2 Field experiments

A large program of field experiments was started in 1994–97 in the United Kingdom, part of which specifically examined the impact of the form of Cd (and Cu and Zn) applied at rates above and below the EU concentration limits for soils receiving sewage sludge. The forms tested were Cd carbonate salt only, or low-metal liquid sludge that had been enriched with sulfate salts at 3 sites, or, at 9 sites, digested sewage cakes from different sewage treatment works that contained naturally high concentrations of metals because of the activity of specific local industries. Impacts on a number of soil microbial parameters or crop yields were measured on 4 occasions for up to 8 years after the last applications of metals, making the findings from these long-term field experiments unique and therefore valuable for understanding the ecotoxicology of Cd (Chaudri et al., 2007, 2008). Concentrating mainly on the results for Cd, Table 8.2 shows the concentrations of Cd in the enriched sludges and those of Zn and Cu, which were also present. The concentrations of Zn and Cu in the sludge were generally within the background range (550–650 mg kg^{-1}) for sewage sludges, apart from the Cd-rich sludge cake, which was also high in Zn. The two "uncontaminated" sludges were low in Cd. The target concentrations for total soil Cd concentrations built up after four applications of salt or sludges added yearly (1994–97) in the field were 1, 2, 3, and 4 mg kg^{-1}, which were considered to be relevant potential upper limits for Cd contamination in agricultural soils receiving sewage sludge.

Five different microbial endpoints (soil microbial biomass, respiration, qCO_2, soil biomass C expressed as a percentage of total soil organic C, and rhizobial numbers) were measured in

TABLE 8.2 Mean concentrations of metals in sludges applied to field experiments in the United Kingdom during 1994–97.

		Mean metal concentration (mg kg^{-1} dry solids)		
	Sludge type	**Cadmium**	**Zinc**	**Copper**
1.	Cd-amended liquid sludge	**285**	610	645
2.	"Uncontaminated" liquid sludge	1.7	590	590
3.	Cadmium-rich (digested cake)	**44**	**1100**	540
4.	"Uncontaminated" digested cake	1.8	560	590

Data from Gibbs et al. (2026a, b).
Values that would generally be regarded as high for sewage sludges are highlighted in bold. *Cd*, cadmium.

1999, 2001, 2003, and 2005. Note that a decrease in these endpoints is regarded as an indicator of toxicity, apart from qCO_2, where metal toxicity generally causes the respiration per unit soil biomass to increase, indicating that the microbial community is under stress. This response occurs despite the total soil respiration rate often decreasing in highly contaminated soils. The main results regarding the effects of Cd are summarized as follows and in Table 8.3.

1. Only one site showed significant impacts (Watlington) in the Cd-only inorganic Cd salt treatments. Soil respiration decreased in 1999, but the effect was reversed in 2003. The qCO_2 at this site decreased in 2003, which is the opposite of the assumed response of this parameter to metal toxicity. So, metal toxicity was unlikely to have occurred at this site.
2. The Cd-amended liquid sludge showed more impacts on soil microbial endpoints, but these were sporadic and, in one case, reversed on different measurement times in Rosemaund. At this site the biomass increased with increasing Cd concentrations in soil in 2005, which cannot be explained by the toxicity of Cd.
3. In the sludge cake plots, impacts on microbial endpoints were more frequent and often included increased qCO_2 and decreased numbers of rhizobia in the soil. The latter were observed more frequently some 4–8 years after the last addition of sewage cake but not after Cd salts or Cd-amended liquid sludge was applied at the field sites.

It seems highly likely that these differences reflect the low impact of Cd on soil microbes at concentrations less than 4 mg kg^{-1} in soil and that the observed impacts in the sludge cake experiments were mainly because of the high Zn concentration present, because the controls for each of sludge experiments all had similar Cu concentrations (Table 8.2). Most probable numbers of rhizobia (MPN; of *Rhizobium leguminosarum* bv. *trifolii*) in many of the sewage cake experiments across different sites decreased, and Chaudri et al. (2008) attributed this to the Zn present in this sludge cake treatment. The symbiosis of rhizobia with legumes is a particularly important function, as it supplies N fixed from the atmosphere to legumes (McGrath et al., 1988) and saves on inputs of inorganic fertilizers and their environmental impacts.

Table 8.4 shows that impacts on the yields of wheat and grass in these long-term field experiments were few, especially in the Cd salt and liquid sludge amended with Cd. But in strong contrast, Cd concentrations in wheat grain or grass herbage increased in most years, in all sites and experiments (with only two exceptions—grain in 2005 a Shirburn, which is a calcareous

TABLE 8.3 Responses of five microbial endpoints in the field between 1999 and 2005 to increasing Cd concentrations in soil because of additions of: (a) inorganic cadmium salt, (b) liquid sewage sludge amended with Cd salt, or (c) sewage sludge cake naturally high in Cd.

Site	Respiration	Biomass C	qCO_2	BQ	Rhizobial MPN
a. Cadmium—inorganic salt					
Woburn	NS	NS	NS	NS	NS
Watlington	↓ 1999 ↑ 2003	NS	↓ 1999	NS	NS
Rosemaund	NS	NS	NS	NS	NS
b. Sewage sludge—liquid, Cd-salt-amended					
Woburn	NS	NS	↓ 2001	NS	NS
Watlington	↑ 1999, 2003	NS	↑ 2003	NS	NS
Rosemaund	NS	↑ 2005	↑ 2001 ↓ 2005	NS	NS
c. Sewage sludge cake, high in Cd					
Woburn	↓ 2003	↑ 1999	↓ 2003	NS	↓ 2005
Watlington	NS	NS	NS	NS	NS
Rosemaund	NS	NS	NS	NS	↓ 2003
Pwllpeiran	NS	NS	↑ 2003	↓ 2001	↓ 2005
Gleadthorpe	NS	↓ 2001, 2003	↓ 2001 ↑ 2005	↓ 1999	↓ 2001, 2005
Shirburn	↑ 1999	NS	↑ 2005	NS	↓ 2001, 2003
Hartwood	↓ 2003	NS	NS	NS	↓ 2005
Bridgets	NS	NS	NS	NS	NS
Auchincruive	NS	NS	NS	NS	NS

Data from Chaudri et al. (2008); Department for Environment, Food and Rural Affairs (DEFRA) (2002, 2006); UK Water Industry Research (UKWIR) (2004, 2007).

qCO_2 = soil respiration expressed as a percentage of the soil microbial C concentration; BQ = soil biomass C expressed as a percentage of total soil organic C, rhizobial MPN is the most probable number of the N2-fixing symbiont *Rhizobium leguminosarum* biovar *trifolii* in soil; ↑ (year[s]) = significant increase ($P < .05$) with Cd; ↓ (year[s]) = significant decrease ($P < .05$) with Cd; NS = no significant Cd effect detected in any year. *Cd*, cadmium.

site, and grass in 2003 at Auchincruive). This consistent effect demonstrates clearly that although Cd may not be toxic to soil microbial populations and their activity at concentrations up to 4 mg kg^{-1} in many different soils, Cd is mobile and taken up by plants at levels of concern (Chaudri et al., 2007) for animals and humans consuming those plants (Chapters 15 and 16).

8.3 Impacts of cadmium on soil fauna

8.3.1 Bioaccumulation of cadmium

Invertebrates living in soil are in close contact with soil contaminants such as Cd. Accumulation of Cd by invertebrates has important implications not only for the potential toxicity to the

TABLE 8.4 Effects on yields and Cd concentrations in plant tissues in field experiments between 1999 and 2005 with (a) inorganic cadmium salt, (b) liquid sewage sludge amended with Cd salt, or (c) sewage sludge cake naturally high in Cd.

Site	Yield	Cd in plant tissues
a. Cadmium—inorganic salt		
Woburn	↓ Grain 2001	↑ Grain and grass 1999–2005
Watlington	NS	↑ Grain and grass 1999–2005
Rosemaund	NS	↑ Grain and grass 1999–2005
b. Sewage sludge—liquid, Cd-salt-amended		
Woburn	NS	↑ Grain and grass 1999–2005
Watlington	NS	↑ Grain and grass 1999–2005
Rosemaund	NS	↑ Grain and grass 1999–2005
c. Sewage sludge cake, high in Cd		
Woburn	NS	↑ Grain and grass 1999–2005
Watlington	↑ Grass 2000	↑ Grain and grass 1999–2005
Rosemaund	NS	↑ Grain and grass 1999–2005
Pwllpeiran*	NS	↑ Grass 1999–2005
Gleadthorpe	↑ Grain 1999	↑ Grain and grass 1999–2005
Shirburn	NS	↑ Grain and grass 1999–2005
Hartwood*	↑ Grass 1999	↑ Grass 1999–2005
Bridgets	↑ Grass 2000	↑ Grain and grass 1999–2005
Auchincruive*	NS	↑ Grass 1999–2005

Data from Chaudri et al. (2007); Department for Environment, Food and Rural Affairs(DEFRA) (2002, 2006); UK Water Industry Research (UKWIR) (2004, 2007).
↑ (year[s]) = significant increase ($P < .05$) with Cd; ↓ (year[s]) = significant decrease ($P < .05$) with Cd; NS = no significant Cd effect detected in any year; * indicates only grass was grown at three sites. *Cd*, cadmium.

animals themselves but also for its trophic transfer to predators (e.g., birds and mammals) and detritivores in terrestrial food webs (Richardson et al., 2020).

Earthworms (Lumbricidae) are one of the most important soil fauna because of their ability to alter soil structure and the chemical and biological environment of the soil. Earthworms can take up Cd from soil and porewater through their skin (dermal) and by ingestion (oral). A study with *Lumbricus rubellus* showed that the dermal uptake is the main route, accounting for 70%–80% of the total Cd uptake, with ingestion contributing the rest (Vijver et al., 2003). In a modeling study on metal uptake by *Eisenia andrei*,Saxe et al. (2001) showed that >96% of Cd uptake by the earthworm was through the dermal exposure route.

The ability of earthworms to accumulate metals is reflected by the bioaccumulation factor (BAF), which is the ratio of metal concentration in the earthworm body to that in the soil. In

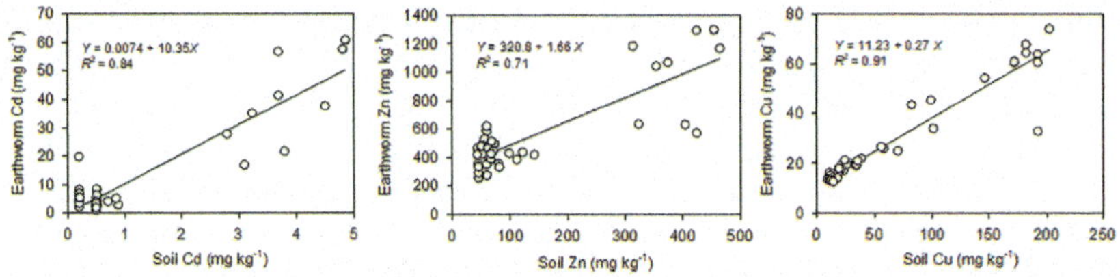

FIGURE 8.1 Relationship between earthworm and soil concentrations of Cd, Zn, and Cu in a long-term field experiment with sewage sludge and metal salt application at Woburn, United Kingdom (unpublished data). *Cd*, Cadmium.

a meta-analysis of 56 studies containing 951 paired data sets of soil–earthworm trace metal concentrations, the median BAF (dry weight basis) for Cd is the highest among eleven trace metals or metalloids (9.49, compared with 0.14–3.44 for As, Cr, Cu, Hg, Mn, Ni, Pb, Sb, U, and Zn) (Richardson et al., 2020). Moreover, the median earthworm tissue concentration of Cd in these studies was above the concentration potentially hazardous for consumption of rodents and fowl, suggesting a significant risk to terrestrial food webs (Richardson et al., 2020). The high bioaccumulation of Cd by earthworms is an important reason why secondary poisoning to mammals is a key factor in the risk assessment of Cd in terrestrial environments (see Chapter 16).

The ability of earthworms to bioaccumulate Cd can also be seen in Fig. 8.1 from a long-term field experiment testing the effects of metals (Cd, Zn, or Cu) applied in sewage cake, liquid sludge, or metal salts (Chaudri et al., 2008). This experiment established concentration gradients of Cd (0.2–5 mg kg^{-1}), Zn (50–450 mg kg^{-1}), or Cu (15–200 mg kg^{-1}) in the topsoil (0–20 cm) of field plots. Earthworms were collected from some of the treatments 10 years after the establishment of the experiment. The concentration of Cd in the earthworm bodies increased linearly with soil Cd concentration with a slope of 10.35. This slope is much greater than that for Zn (1.66) and Cu (0.27). The intercept in the regression equation between earthworms versus soil metal concentrations was negligible for Cd (0.0074 mg kg^{-1}) but considerable for Zn (320.8 mg kg^{-1}) and Cu (11.23 mg kg^{-1}). The intercepts for Zn and Cu can be seen as representing the basal requirement of earthworms for these two essential trace metals. In contrast, Cd is nonessential, and earthworms appear to lack specific mechanisms to regulate its accumulation (Spurgeon & Hopkin, 1999).

Bioaccumulation of Cd by earthworms is influenced by factors such as earthworm genus, ecophysiological group, soil pH, and organic matter content (Nahmani et al., 2007; Richardson et al., 2020). In the meta-analysis by Richardson et al. (2020), Lumbricidae genera (*Allolobophora, Aporrectodea, Dendrobaena*, and *Dendrodrilus*) were found to bioaccumulate Cd at greater rates than Megascolecidae genera (*Metaphire* and *Pheretima*). Among the different ecophysiological groups of earthworms the median BAF of Cd follows the order of endogeic > epigeic > anecic, likely reflecting their habitat and food consumption pattern (Richardson et al., 2020). In contrast, Huang et al. (2021) and (Audusseau et al., 2020) reported that epigeic and anecic earthworms bioaccumulated more soil Cd than endogeic earthworms. In a study comparing Cd accumulation by *Eisenia fetida* (epigeic), *Pheretima guillelmi* (endogeic), and *Metaphire guillelmi* (anecic), *P. guillelmi* showed the highest Cd accumulation, likely because of their highly efficient geophagous feeding strategy (Zhang et al., 2018). Because earthworms take up Cd mainly through dermal exposure, soil properties that affect Cd solubility and speciation in the porewater, such as pH and organic matter content, also influence Cd accumulation by earthworms (Oste et al., 2001; Saxe et al., 2001; Spurgeon et al., 2006).

Cd bioaccumulation varies between different soil invertebrates. Based on a literature study, Heikens et al. (2001) found that Cd bioaccumulation generally increases with soil Cd concentration and follows the order of Lumbricidae > Arachnida > Diplura > Diplopoda > Collembola > Coleoptera. In contrast, Cd accumulation by Isopoda, Formicidae, and Chilopoda appears to be independent of soil Cd concentration, although Isopoda also accumulates high concentrations of Cd in their bodies.

8.3.2 Toxicity of cadmium

Many studies have investigated the toxicity of Cd to soil invertebrates, especially earthworms and collembola (see Chapter 16). The toxicity threshold varies widely among the test organisms used, the endpoints measured (e.g., growth, mortality, or reproduction), soil properties, and whether Cd was freshly added or allowed to age in soil. Short-term exposure (e.g., 14 days) tests acute toxicity such as mortality, whereas longer-term assays evaluate the effects on growth and reproduction, which are more relevant to ecological functions. Acute toxicity thresholds for Cd are usually much higher than those of chronic toxicity. In a study on the acute toxicity (mortality after 14 days of Cd exposure) to *E. fetida* with 18 soils collected from different regions in China, the median lethal concentration (LC_{50}) of Cd varied from 441 to 1520 mg kg^{-1} in freshly spiked soils (Liu et al., 2018). This range is similar to that reported by Lock and Janssen (2001). Soil pH and organic matter content were the two major factors affecting Cd toxicity. Aging of Cd in soil by 180 or 360 days reduced Cd bioavailability and increased LC_{50} (i.e., lower toxicity) to some extent (Liu et al., 2018). In a test using three OECD artificial soils spiked with Cd, acute LC_{50} based on 14-day mortality of *E. fetida* varied from 354 mg Cd kg^{-1} in two soils to no mortality at the highest dose of Cd (625 mg kg^{-1}) in one soil (Irizar et al., 2015). In contrast, earthworm reproduction after 28 days of Cd exposure was five to seven times more sensitive to Cd exposure than mortality, with the effect concentration that causes 50% inhibition (EC_{50}) in the range of 10.5–51 mg Cd kg^{-1} (Irizar et al., 2015). For *Folsomia candida* (collembola), the 10% lethal concentrations (LC_{10}) of Cd after 7 days of exposure varied from 68.6 to >1000 mg Cd kg^{-1} among 18 soils collected from China, whereas reproduction of *F. candida* after 28 days of exposure was more sensitive to Cd, with EC_{50} ranging from 41.4 to 146.8 mg Cd kg^{-1} (Liu et al., 2019). In toxicity tests in three Dutch soils the EC_{50} for added Cd on growth of *F. candida* ranged from 203 to 401 mg Cd kg^{-1}, and on reproduction 43.3–103 mg Cd kg^{-1} after 6 weeks (Van Gestel & Mol, 2003). In three forest soils, Bur et al. (2010) showed that the median inhibition effect concentration of Cd (EC_{50}) on reproduction of *F. candida* ranged from 107 to 182 mg Cd kg^{-1}. Compared with other soil invertebrates, *F. candida* appears to be less sensitive to Cd (Bur et al., 2010).

The toxicity thresholds of tens to hundreds mg Cd kg^{-1} for soil invertebrates suggest that these organisms are not particularly sensitive to Cd, as soil Cd concentrations rarely reach the levels of these thresholds. Although earthworms bioaccumulate Cd to high levels, they are able to detoxify Cd internally through complexation with metallothioneins, which are cysteine-rich proteins strongly induced by Cd exposure (Stürzenbaum et al., 2001). Using immunohistochemistry with a polyclonal antibody raised against the metallothionein wMT-2, Stürzenbaum et al. (2004) showed that wMT-2 is strongly expressed in the gut epithelium and the typhlosole of the earthworm *L. rubellus*, as well as in the apexes of the chloragocytes, which likely serve as Cd storage cells.

8.3.3 Impact on earthworm communities and diversity

Many studies have investigated the impact of metal pollution on the communities of soil fauna, especially earthworms, using transects or field plots that vary in metal concentrations because of varying distances from the pollution source. The results from these studies are highly relevant ecologically, although the soils are usually contaminated with multiple metals, making it difficult to attribute the observed effects to a single metal.

At 22 sites around a metal smelter at Avonmouth, southwest England, the metal-sensitive earthworm species *Aporrectodea rosea*, *Aporrectodea caliginosa*, and *Allolobophora chlorotica* were absent near the smelter, whereas metal-tolerant *L. rubellus*, *Lumbricus castaneus*, and *Lumbricus terrestris* were present. The soils near the smelter were contaminated by Cd, Cu, Pb, and Zn, although Zn was likely the limiting factor that impacted the distribution of earthworms (Spurgeon & Hopkin, 1996). In field plots near a Zn smelter at Scarpe-Escaut, northern France, contamination of Zn, Cd, and Pb drastically decreased the population densities of *Eiseniella tetraedra*, *A. rosea*, and *A. chlorotica*, whereas *L. castaneus*, *Dendrobaena rubida*, and *L.* rubellus were able to maintain their populations (Nahmani et al., 2003). In an urban wasteland in France contaminated with Pb, Zn, Cu, and Cd, species richness of earthworms decreased along a pollution gradient (Audusseau et al., 2020). Two field studies at metal-contaminated sites in Hunan, China, suggested that Cd pollution, with soil Cd concentration varying in the range of 1.4–6.6 and 0.8–17.8 mg kg^{-1}, was the main factor contributing to the decrease in total earthworm density and biomass, or the species diversity Simpson index (Huang et al., 2021; Wang et al., 2018). With increasing pollution, epigeic earthworms became more dominant than anecic and endogeic earthworms, likely because the former live on the surface and feed on the litter layer with less contact with or ingestion of soil (Huang et al., 2021). Furthermore, Cd contamination altered the age structure of the earthworm community by decreasing the proportion of juveniles to adults (Wang et al., 2018). These studies indicate that pollution of metals, including Cd, can have a detrimental effect on earthworm density and diversity.

8.4 Sensitivity and tolerance to cadmium

Using plate count culturing methods to enumerate colony forming units (CFU) of *Pseudomonas aeruginosa*, actinomycetes, and fungi in soil from near an e-waste site in Changzhou, China, there was little impact of 1 and 2 mg Cd kg^{-1} additions on the inhibition of *P. aeruginosa* over time. However, there was a progressive decrease in CFUs of all three groups in the 4, 6, 8, and 10 mg Cd kg^{-1} treatments over 35 days of incubation (Han et al., 2016). The authors concluded that the order of sensitivity to added Cd was *P. aeruginosa*> actinomycetes> fungi. An unexplained result was that the CFUs of actinomycetes showed decreasing inhibition with time at the highest (8 and 10 mg kg^{-1}) Cd additions. Whether this was because of the development of tolerance to Cd was not investigated. One methodological issue with plate counts is that they assess only a very small proportion of the microbes present—those that can grow on a culture plate. It has been estimated that around 99% of microbial taxa are not assessed using this method (Dickinson, 2021).

Metagenomics techniques have great promise for characterizing impacts on nonculturable microbes present in the soil community, thus avoiding what has been called the "great plate count anomaly." These molecular approaches are not only useful for assessing the presence of microbial taxa and cells in soil but also their activity using metatranscriptomics. However, to date few studies have used these techniques for specifically assessing the impacts of Cd (Duan et al., 2020; Li et al., 2022; You et al., 2024).

However, other biochemical assays have been used to detect impacts and more subtle changes because of Cd exposure. Some examples of these are enzyme activity assays, demonstrated for microbes in Section 8.2.1. For earthworms the antioxidant enzymes superoxide dismutase (SOD), catalase (CAT), and malondialdehyde (MDA), which are produced when free radicals attack the polyunsaturated fatty acids in cell membranes, all increased in the worm *E. fetida* in soil with 5 mg Cd kg^{-1} added but not with 1 mg kg^{-1} (Xu et al., 2023). ATP concentrations in *E. fetida* decreased similarly, only with 5 mg Cd kg^{-1} added to soil. The authors concluded from these results that the toxic mechanism of Cd was associated with oxidative stress and mitochondrial dysfunction, which would activate antioxidant enzymes and induce energy metabolism disturbance. Furthermore, they found that, with only 1 mg Cd kg^{-1} added, reactive oxygen species in coelomocytes (specialized immune cells found in the body cavity of earthworms, acting as a defense system like white blood cells in vertebrates) and the proportion of those undergoing stages of cell death (apoptosis) were higher than control. Using advanced genomic and transcriptomic analyses, You et al. (2024) showed that the addition of 2.5 mg Cd kg^{-1} to a soil from agricultural land near Chengdu City, China, resulted in decreased expression of genes in *E. fetida* associated with both the structure and function of cells and increases in the expression of genes associated with the activity of some enzymes, particularly those involved in antioxidant function. There was no mortality, but the growth in the Cd treatment was about 25% less than control after 28 days.

Importantly, these results indicate that oxidative stress and consequent impacts on metabolism occur at Cd concentrations that are one or two orders of magnitude below the lethal or reproductive endpoints discussed in Section 8.3.2. However, it is not known how these metabolic changes affect the long-term growth and survival of the organisms. Similar considerations surround the use of metagenomic and metatranscriptomic data for soil microorganisms and genomic/transcriptomic data for soil fauna, as the impacts on functions are usually not known. This is a common knowledge gap in the use of such data for the purpose of ecological risk assessment.

One phenomenon that may change the long-term growth and survival of the soil organisms is the evolution of tolerance to Cd (Duan et al., 2020; You et al., 2024; Yu et al., 2021). However, this may not be considered an acceptable outcome of pollution. As shown above, pollution decreases the number and diversity of some soil communities. It is therefore unlikely that a robust and diverse soil ecosystem would survive if all of its constituent organisms needed to be tolerant to high levels of Cd. So, tolerance may mainly be of more interest in the rehabilitation of very polluted sites.

However, tolerance to Cd (or other metals) may not be a linear function of the exposure concentration. For example, using DNA-based analysis, it has been shown that diversity can exhibit an inverted "U"-shaped curve when graphed against exposure concentrations (Giller et al., 1998). This is called the intermediate stress hypothesis, where low environmental stress allows fewer genotypes to survive under ideal "control" (unstressed) conditions because of competitive dominance. Release of his competition for niches allows more genotypes to survive under intermediate stress. This is followed by a decrease in genotypes as concentrations become much higher, because of lack of tolerance in much of the population and selection of fewer strongly tolerant genotypes. Therefore interpretation of microbial diversity data with increasing Cd exposure may not be straightforward.

8.5 Concluding remarks and future perspectives

Most of the published evidence shows relatively few impacts of Cd additions on soil microbial and faunal endpoints at levels that are realistic in soils (e.g., <5–10 mg kg^{-1}). In field-contaminated sites, there are reports of Cd impacting soil microbial and/or faunal diversity and community structure. However, the presence of multimetals, especially Zn, Cu, and Pb, which are usually present at much higher concentrations than Cd, makes it difficult to attribute the effects to Cd solely. At a typical Cd:Zn ratio of 1:200 in soil (Smolders & Mertens, 2013), Zn contamination is likely to produce greater toxicity to soil organisms than Cd.

Under scenarios of metal contamination in field conditions, metals (including Cd) are usually added to soils gradually, which allows metals to age and become less bioavailable and soil organisms to adapt and evolve tolerance mechanisms. Short-term laboratory incubations using soluble metal salts may overestimate the toxicity of Cd and other metals.

More data are becoming available using the latest metagenomic and metatranscriptomic techniques for soil microorganisms and genomic/transcriptomic techniques for soil fauna. Such data could prove useful for revealing the impacts of Cd or other metals on the community structure, diversity, and expression of functional genes. However, the links to potential impacts on function and survival at the population level are usually not known. This is a common knowledge gap in the use of such data for purposes of ecological risk assessment.

Despite the presence of much literature on soil microbial and faunal endpoints, relatively little information exists on impacts of Cd on symbiotic activities important to ecosystems such as rhizobial nitrogen fixation or colonization of plants by mycorrhizal fungi and their functions in soil–plant systems. More studies are needed to investigate the impact of Cd on keystone taxa of soil organisms that are pivotal to soil ecological functions.

Despite few impacts of Cd on the ecotoxicology endpoints, some soil fauna, especially earthworms, are able to bioaccumulate Cd to levels that are potentially harmful to predators in the food web. The transfer of Cd from soil to invertebrates and subsequently the higher trophic levels is of great concern when assessing its ecological risk. Evidence from field experiments clearly shows that Cd is mobile into plants and food at these levels in soil, which are sufficient to cause concern and possible threats in the food chain.

Cd is now used in engineered nanomaterials (such as Cd quantum dots), which are emitted into the environment, and there are few data on their impacts on soil microbes and fauna. It is important to establish their uptake by organisms and the magnitude of food chain transfer, as well as their long-term toxic effects, to provide a sound basis for ecological risk assessment of engineered nanomaterials (Wang et al., 2024).

8.6 Summary

- Additions of soluble Cd in laboratory incubation experiments can decrease soil microbial and faunal populations and activity, although few impacts are observed at levels that are typical of field-contaminated soils (<5–10 mg Cd kg^{-1}).

- Studies on field-contaminated sites reveal adverse effects of multimetal contamination, including Cd, on the community structure and diversity of soil microorganisms and fauna.
- Spiking of soluble Cd into soil in laboratory incubation experiments tends to overestimate the toxicity of Cd, as gradual additions of Cd at field-contaminated sites allow the metal to become less bioavailable and soil organisms to adapt and evolve tolerance mechanisms over time.
- Some soil fauna, especially earthworms, are able to bioaccumulate Cd to levels that are potentially harmful to animals at the higher trophic levels in the food web.

References

Audusseau, H., Vandenbulcke, F., Dume, C., Deschins, V., Pauwels, M., Gigon, A., Bagard, M., & Dupont, L. (2020). Impacts of metallic trace elements on an earthworm community in an urban wasteland: Emphasis on the bioaccumulation and genetic characteristics in *Lumbricus castaneus*. *Science of the Total Environment, 718*, 137259. doi:10.1016/j.scitotenv.2020.137259.

Birke, M., Reimann, C., Rauch, U., Ladenberger, A., Demetriades, A., Jähne-Klingberg, F., Oorts, K., Gosar, M., Dinelli, E., & Halamić, J. (2017). GEMAS: Cadmium distribution and its sources in agricultural and grazing land soil of Europe—Original data versus clr-transformed data. *Journal of Geochemical Exploration, 173*, 13–30. doi:10.1016/j.gexplo.2016.11.007.

Bur, T., Probst, A., Bianco, A., Gandois, L., & Crouau, Y. (2010). Determining cadmium critical concentrations in natural soils by assessing Collembola mortality, reproduction and growth. *Ecotoxicology and Environmental Safety, 73*(3), 415–422. doi:10.1016/j.ecoenv.2009.10.010.

Cáliz, J., Montserrat, G., Martí, E., Sierra, J., Chung, A. P., Morais, P. V., & Vila, X. (2013). Emerging resistant microbiota from an acidic soil exposed to toxicity of Cr, Cd and Pb is mainly influenced by the bioavailability of these metals. *Journal of Soils and Sediments, 13*(2), 413–428. http://www.springerlink.com/content/1439-0108. doi:10.1007/s11368-012-0609-7.

Chaudri, A., McGrath, S., Gibbs, P., Chambers, B., Carlton-Smith, C., Godley, A., Bacon, J., Campbell, C., & Aitken, M. (2007). Cadmium availability to wheat grain in soils treated with sewage sludge or metal salts. *Chemosphere, 66*(8), 1415–1423. www.elsevier.com/locate/chemosphere. doi:10.1016/j.chemosphere.2006.09.068.

Chaudri, A., McGrath, S., Gibbs, P., Chambers, B., Carlton-Smith, C., Bacon, J., Campbell, C., & Aitken, M. (2008). Population size of indigenous *Rhizobium leguminosarum* biovar *trifolii* in long-term field experiments with sewage sludge cake, metal-amended liquid sludge or metal salts: Effects of zinc, copper and cadmium. *Soil Biology and Biochemistry, 40*(7), 1670–1680. doi:10.1016/j.soilbio.2008.01.026.

Chen, H., Ma, K., Lu, C., Fu, Q., Qiu, Y., Zhao, J., Huang, Y., Yang, Y., Schadt, C. W., & Chen, H. (2022). Functional redundancy in soil microbial community based on metagenomics across the globe. *Frontiers in Microbiology, 13*. https://www.frontiersin.org/journals/microbiology#. doi:10.3389/fmicb.2022.878978.

Department for Environment, Food and Rural Affairs (DEFRA). (2002). *Effects of inorganic metal salt additions to agricultural soils on soil microbial activity and long-term soil fertility—Phase II (Report SP0126)*. Department for Environment, Food and Rural Affairs (DEFRA).

Department for Environment, Food and Rural Affairs (DEFRA). (2006). *Effects of inorganic metal salt additions to agricultural soils on soil microbial activity and long-term soil fertility—Phase III (Report SP0133)*. Department for Environment, Food and Rural Affairs (DEFRA).

Dickinson, I. (2021). Sequencing the unculturable majority. *Nature Portfolio, Milestones*. https://www.nature.com/articles/d42859-020-00102-8.

Duan, C., Liu, Y., Zhang, H., Chen, G., & Song, J. (2020). Cadmium pollution impact on the bacterial community of Haplic Cambisols in Northeast China and inference of resistant genera. *Journal of Soil Science and Plant Nutrition, 20*(3), 1156–1170. https://www.springer.com/life+sciences/plant+sciences/journal/42729. doi:10.1007/s42729-020-00201-5.

Gibbs, P. A., Chambers, B. J., Chaudri, A. M., McGrath, S. P., Carlton-Smith, C. H., Bacon, J. R., Campbell, C. D., & Aitken, M. N. (2006a). Initial results from a long-term, multi-site field study of the effects on soil fertility and microbial activity of sludge cakes containing heavy metals. *Soil Use and Management, 22*(1), 11–21. doi:10.1111/j.1475-2743.2006.00003.x.

Gibbs, P. A., Chambers, B. J., Chaudri, A. M., McGrath, S. P., & Carlton-Smith, C. H. (2006b). Initial results from long-term field studies at three sites on the effects of heavy metal-amended liquid sludges on soil microbial activity. *Soil Use and Management, 22*(2), 180–187. doi:10.1111/j.1475-2743.2006.00009.x.

Giller, K. E., Witter, E., & Mcgrath, S. P. (1998). Toxicity of heavy metals to microorganisms and microbial processes in agricultural soils: A review. *Soil Biology and Biochemistry, 30*(10–11), 1389–1414. www.elsevier.com/inca/publications/store/3/3/2. doi:10.1016/S0038-0717(97)00270-8.

Han, Z., Wang, N., Yang, X., & Zhou, R. (2016). Toxic effects of the joint exposure of CD and decabromodiphenyl ether on soil microbial population. *Fresenius Environmental Bulletin, 25*(12), 5280–5286. http://www.psp-parlar.de/feb_auswahl.asp.

Heikens, A., Peijnenburg, W. J. G. M., & Hendriks, A. J. (2001). Bioaccumulation of heavy metals in terrestrial invertebrates. *Environmental Pollution, 113*(3), 385–393. doi:10.1016/s0269-7491(00)00179-2.

Huang, C., Ge, Y., Yue, S., Qiao, Y., & Liu, L. (2021). Impact of soil metals on earthworm communities from the perspectives of earthworm ecotypes and metal bioaccumulation. *Journal of Hazardous Materials, 406*. www.elsevier.com/locate/jhazmat. doi:10.1016/j.jhazmat.2020.124738.

Irizar, A., Rodríguez, M. P., Izquierdo, A., Cancio, I., Marigómez, I., & Soto, M. (2015). Effects of soil organic matter content on cadmium toxicity in *Eisenia fetida*: Implications for the use of biomarkers and standard toxicity tests. *Archives of Environmental Contamination and Toxicology, 68*(1), 181–192. doi:10.1007/s00244-014-0060-4.

Jegede, O. O., Awuah, K. F., Renaud, M. J., Cousins, M., Hale, B. A., & Siciliano, S. D. (2020). Single metal and metal mixture toxicity of five metals to *Oppia nitens* in five different Canadian soils. *Journal of Hazardous Materials, 392*. www.elsevier.com/locate/jhazmat. doi:10.1016/j.jhazmat.2020.122341.

Li, D., Chen, J., Zhang, H., Zhang, X., & Li, J. (2022). Toxic factors of lead and cadmium fit in the ecological risk assessment for microorganisms. *Frontiers in Microbiology, 13*. https://www.frontiersin.org/journals/microbiology#. doi:10.3389/fmicb.2022.927947.

Liu, H., Li, M., Zhou, J., Zhou, D., & Wang, Y. (2018). Effects of soil properties and aging process on the acute toxicity of cadmium to earthworm *Eisenia fetida*. *Environmental Science and Pollution Research, 25*(4), 3708–3717. http://www.springerlink.com/content/0944-1344. doi:10.1007/s11356-017-0739-y.

Liu, H., Xuan, L., Zhou, J., Zhou, D., & Wang, Y. (2019). Effects of soil properties on cadmium toxicity to *Folsomia candida* (Collembola). *Bulletin of Environmental Contamination and Toxicology, 103*(1), 90–97. link.springer.de/link/service/journals/00128/index.htm. doi:10.1007/s00128-018-2514-2.

Lock, K., & Janssen, C. R. (2001). Cadmium toxicity for terrestrial invertebrates: Taking soil parameters affecting bioavailability into account. *Ecotoxicology, 10*(5), 315–322. doi:10.1023/A:1016767519556.

McGrath, S. P., Brookes, P. C., & Giller, K. E. (1988). Effects of potentially toxic metals in soil derived from past applications of sewage sludge on nitrogen fixation by *Trifolium repens* L. *Soil Biology and Biochemistry, 20*(4), 415–424. doi:10.1016/0038-0717(88)90052-1.

McGrath, S. P. (1999). Cadmium in soils, plants and the food chain. In M. J. McLaughlin, & B. R. Singh (Eds.), *Adverse effects of cadmium on soil microflora and fauna* (pp. 119–218). Kluwer Academic Publishers.

Moreno, J. L., Hernández, T., Pérez, A., & García, C. (2002). Toxicity of cadmium to soil microbial activity: Effect of sewage sludge addition to soil on the ecological dose. *Applied Soil Ecology, 21*(2), 149–158. doi:10.1016/S0929-1393(02)00064-1.

Nahmani, J., Lavelle, P., Lapied, E., & Van Oort, F. (2003). Effects of heavy metal soil pollution on earthworm communities in the north of France: The 7th international symposium on earthworm ecology Cardiff Wales 2002. *Pedobiologia, 47*(5–6), 663–669. www.urbanfischer.de/journals/pedo/pedobiol.htm. doi:10.1078/0031-4056-00243.

Nahmani, J., Hodson, M. E., & Black, S. (2007). A review of studies performed to assess metal uptake by earthworms. *Environmental Pollution, 145*(2), 402–424. doi:10.1016/j.envpol.2006.04.009.

Oste, L. A., Dolfing, J., Ma, W. C., & Lexmond, T. M. (2001). Cadmium uptake by earthworms as related to the availability in the soil and the intestine. *Environmental Toxicology and Chemistry, 20*(8), 1785–1791. http://www.interscience.wiley.com/jpages/0730-7268. doi:10.1002/etc.5620200823.

Raiesi, F., & Sadeghi, E. (2019). Interactive effect of salinity and cadmium toxicity on soil microbial properties and enzyme activities. *Ecotoxicology and Environmental Safety, 168*, 221–229. doi:10.1016/j.ecoenv.2018.10.079.

Ren, C., Teng, Y., Chen, X., Shen, Y., Xiao, H., & Wang, H. (2021). Impacts of earthworm introduction and cadmium on microbial communities composition and function in soil. *Environmental Toxicology and Pharmacology, 83*. www.elsevier.com/locate/etap. doi:10.1016/j.etap.2021.103606.

Renella, G., Brookes, P. C., & Nannipieri, P. (2002). Cadmium and zinc toxicity to soil microbial biomass and activity. *Developments in Soil Science, 28*(2), 267–273. doi:10.1016/S0166-2481(02)80024-1.

Richardson, J. B., Görres, J. H., & Sizmur, T. (2020). Synthesis of earthworm trace metal uptake and bioaccumulation data: Role of soil concentration, earthworm ecophysiology, and experimental design. *Environmental Pollution, 262*, 114126. doi:10.1016/j.envpol.2020.114126.

Saxe, J. K., Impellitteri, C. A., Peijnenburg, W. J. G. M., & Allen, H. E. (2001). Novel model describing trace metal concentrations in the earthworm, *Eisenia andrei. Environmental Science and Technology, 35*(22), 4522–4529. doi:10.1021/es0109038.

Smolders, E., Oorts, K., Van Sprang, P., Schoeters, I., Janssen, C. R., McGrath, S. P., & McLaughlin, M. J. (2009). Toxicity of trace metals in soil as affected by soil type and aging after contamination: Using calibrated bioavailability models to set ecological soil standards. *Environmental Toxicology and Chemistry, 28*(8), 1633–1642. doi:10.1897/08-592.1.

Smolders, E., & Mertens, J. (2013). Chapter 10: Cadmium. In B. J. Alloway (Ed.), *Heavy metals in soils: Trace metals and metalloids in soils and their bioavailability*. Springer Science+Business Media.

Spurgeon, D. J., Lofts, S., Hankard, P. K., Toal, M., McLellan, D., Fishwick, S., & Svendsen, C. (2006). Effect of pH on metal speciation and resulting metal uptake and toxicity for earthworms. *Environmental Toxicology and Chemistry, 25*(3), 788–796. doi:10.1897/05-045R1.1 07307268.

Spurgeon, D. J., & Hopkin, S. P. (1999). Comparisons of metal accumulation and excretion kinetics in earthworms (*Eisenia fetida*) exposed to contaminated field and laboratory soils. *Applied Soil Ecology, 11*(2–3), 227–243. doi:10.1016/s0929-1393(98)00150-4.

Spurgeon, D. J., & Hopkin, S. P. (1996). The effects of metal contamination on earthworm populations around a smelting works: Quantifying species effects. *Applied Soil Ecology, 4*(2), 147–160. www.elsevier.com/inca/publications/store/5/2/4/5/1/8/index.htt. doi:10.1016/0929-1393(96)00109-6.

Stürzenbaum, S. R., Georgiev, O., Morgan, A. J., & Kille, P. (2004). Cadmium detoxification in earthworms: From genes to cells. *Environmental Science and Technology, 38*(23), 6283–6289. doi:10.1021/es049822c.

Stürzenbaum, S. R., Winters, C., Galay, M., Morgan, A. J., & Kille, P. (2001). Metal ion trafficking in earthworms. Identification of a cadmium-specific metallothionein. *Journal of Biological Chemistry, 276*(36), 34013–34018. doi:10.1074/jbc.M103605200.

UK Water Industry Research (UKWIR). (2004). *Effects of sewage sludge applications to agricultural soils on soil microbial activity and the implications for agricultural productivity and long-term soil fertility, Phase II. Report ref 03/SL/01/4*. https://ukwir.org/reports/03-SL-01-4/66928/Effects-of-Sewage-Sludge-Applications-to-Agricultural-Soils-on-Soil-Microbial-Activity-and-the-Implications-for-Agricultural-Productivity-and-Long-Term-Soil-Fertility.

UK Water Industry Research (UKWIR). (2007). *Effects of sewage sludge applications to agricultural soils on soil microbial activity and the implications for agricultural productivity and long-term soil fertility: Phase III. Report Ref: SP0130; CSA 6222*. https://randd.defra.gov.uk/ProjectDetails?ProjectId=10677

Van Gestel, C. A. M., & Mol, S. (2003). The influence of soil characteristics on cadmium toxicity for *Folsomia candida* (Collembola: Isotomidae). *Pedobiologia, 47*(4), 387–395. www.urbanfischer.de/journals/pedo/pedobiol.htm. doi:10.1078/0031-4056-00202.

Vig, K., Megharaj, M., Sethunathan, N., & Naidu, R. (2003). Bioavailability and toxicity of cadmium to microorganisms and their activities in soil: A review. *Advances in Environmental Research, 8*(1), 121–135. http://elsevier.com/locate/aer. doi:10.1016/S1093-0191(02)00135-1.

Vijver, M. G., Vink, J. P. M., Miermans, C. J. H., & Van Gestel, C. A. M. (2003). Oral sealing using glue: A new method to distinguish between intestinal and dermal uptake of metals in earthworms. *Soil Biology and Biochemistry, 35*(1), 125–132. doi:10.1016/S0038-0717(02)00245-6.

Wang, C., Chen, L., Xu, J., Zhang, L., Yang, X., Zhang, X., Zhang, C., Gao, P., & Zhu, L. (2024). Environmental behaviors and toxic mechanisms of engineered nanomaterials in soil. *Environmental Research, 242*. https://www.sciencedirect.com/science/journal/00139351. doi:10.1016/j.envres.2023.117820.

Wang, K., Qiao, Y., Zhang, H., Yue, S., Li, H., Ji, X., & Crowley, D. (2018). Influence of cadmium-contaminated soil on earthworm communities in a subtropical area of China. *Applied Soil Ecology, 127*, 64–73. www.elsevier.com/inca/publications/store/5/2/4/5/1/8/index.htt. doi:10.1016/j.apsoil.2018.02.026.

Wang, L., Zhang, W., Wang, J., Zhu, L., Wang, J., Yan, S., & Ahmad, Z. (2019). Toxicity of enrofloxacin and cadmium alone and in combination to enzymatic activities and microbial community structure in soil. *Environmental Geochemistry and Health, 41*(6), 2593–2606. www.wkap.nl/journalhome.htm/0269-4042. doi:10.1007/s10653-019-00307-5.

Xu, Q., Shi, Y., Ke, L., Qian, L. i, Zhou, X., & Shao, X. (2023). Ciprofloxacin enhances cadmium toxicity to earthworm *Eisenia fetida* by altering the gut microorganism composition. *Environmental Pollution, 333*, 122106. doi:10.1016/j.envpol.2023.122106.

You, R., Li, H., Li, X., Luo, L., Wang, P., Xia, H., & Zhou, Y. (2024). Ecotoxicological impacts of cadmium on soil microorganisms and earthworms *Eisenia foetida*: From gene regulation to physiological processes. *Frontiers in Environmental Science, 12*. journal.frontiersin.org/journal/environmental-science. doi:10.3389/fenvs.2024.1479500.

Yu, X., Zhao, J. T., Liu, X., Sun, L. X., Tian, J., & Wu, N. (2021). Cadmium pollution impact on the bacterial community structure of arable soil and the isolation of the cadmium resistant bacteria. *Frontiers in Microbiology, 12*. https://www.frontiersin.org/journals/microbiology#. doi:10.3389/fmicb.2021.698834.

Zhang, L., He, N., Chang, D., Liu, X., Zhang, X., Xu, Y., Zhao, C., Sun, J., Li, W., Li, H., Hu, F., & Xu, L. (2018). Does ecotype matter? The influence of ecophysiology on benzo[a]pyrene and cadmium accumulation and distribution in earthworms. *Soil Biology and Biochemistry, 121*, 24–34. www.elsevier.com/inca/publications/store/3/3/2. doi:10.1016/j.soilbio.2018.02.020.

CHAPTER

9

Assessment of cadmium bioavailability in soils

Taige Li[a], *Hao Zhang*[b], *Xing Liu*[a], *Jun Luo*[a]

[a] **State Key Laboratory of Water Pollution Control and Green Resource Recycling, School of Environment, Nanjing University, Nanjing, China** [b] **Lancaster Environment Centre, Lancaster University, Lancaster, United Kingdom**

9.1 Introduction

9.1.1 The importance of assessing the bioavailability of cadmium in soils

As a well-known toxic metal, cadmium (Cd) has garnered considerable attention in environmental research. Total concentrations of Cd in soils are usually determined. However, it is well established that the total concentration does not necessarily reflect its ecological risk (Wu et al., 2020), as only the fraction that is bioavailable to organisms exerts toxic effects (Riaz et al., 2021). Therefore assessing the bioavailability of Cd in soil is particularly important. Bioavailability refers to the extent to which a pollutant can be absorbed by organisms and exert toxic effects (Li et al., 2021; Sun et al., 2023). The bioavailability of Cd from soil depends on various factors, including the chemical forms of Cd present in the soil, the physicochemical properties of the soil, and the physiological characteristics of the organisms (Zhao et al., 2023). The imperative for assessing bioavailability is predicated on several key aspects:

1. Risk assessment: By assessing the bioavailability of Cd the risk of soil Cd pollution to human and ecological health can be more accurately assessed (Gujre et al., 2021).
2. Selection of remediation technology: Different remediation techniques have different mitigation effects on different forms of Cd. Understanding the bioavailability of Cd helps to select appropriate remediation techniques (Cao et al., 2024).
3. Determination of remediation endpoints: Different evaluation systems are commonly applied based on land use types. Accurate assessment of bioavailability plays a critical role in guiding the determination of remediation endpoints for contaminated sites.
4. Safety of agricultural products: Assessing the bioavailability of Cd in farmland soils helps to assess the risk of Cd accumulation in agricultural products and to ensure food safety (Mubeen et al., 2023).

Cadmium Contamination in Soils and Plants: Mechanisms and Mitigation.
DOI: https://doi.org/10.1016/B978-0-443-34104-5.00003-7

Despite its significance, accurately assessing Cd bioavailability remains challenging because of the confounding influence of multiple variables, including soil pH (Yan et al., 2023), organic matter content (Jin et al., 2021), and moisture conditions (Li et al., 2023a), all of which affect Cd speciation, mobility, and biological accessibility.

9.2 Key factors and mechanisms influencing cadmium bioavailability in soil

9.2.1 Soil pH

Soil pH is a key factor influencing the bioavailability of Cd. It directly affects Cd solubility, mobility, and plant uptake. The influence of pH on Cd bioavailability in soil mainly involves three aspects (Bandara et al., 2021b) (see also Chapter 6).

- Soil pH can affect the solubility of Cd. In acidic soils, Cd has a higher solubility and is more easily absorbed by plants. Cd compounds readily dissociate into ionic forms, thereby enhancing their mobility. At higher pH, Cd may precipitate in the form of insoluble compounds such as Cd hydroxide or Cd carbonate, reducing its bioavailability (Biswash et al., 2024).
- Soil pH significantly influences the surface charge characteristics of soil colloids, thereby altering the cation exchange capacity (CEC). As soil pH increases, negative charge of soil organic matter and other solid phases increases, resulting in increased capacity for Cd adsorption. Under acidic conditions the concentration of H^+ ions in the soil is relatively high, and they compete with Cd for the adsorption sites on soil colloids. This usually leads to a decrease in the adsorption of Cd, thereby causing more Cd to dissolve in the soil solution, increasing its bioavailability (Jia et al., 2022).
- A low pH may also promote the dissolution of soil minerals (such as iron and aluminum oxides), releasing bound Cd (Beri et al., 2020; Gong et al., 2021).

9.2.2 Soil organic matter

Organic matter is an important part of soil and has significant influence on the physical, chemical, and biological properties of soil. It has a complex effect on the bioavailability of heavy metals such as Cd (Xu et al., 2022). The mechanisms by which soil organic matter affects the bioavailability of Cd mainly include sorption, complexation, microbial action, and soil structure change. Complexation involves the coordination of a central atom with one or more ligands (molecules or ions containing lone pairs of electrons) through coordinate bonds, forming a complex with a complicated structure and making Cd less likely to be absorbed by plant roots (Arabi et al., 2017). On the other hand, organic acids produced during the microbial decomposition of organic matter can lower the pH of the soil solution, thereby enhancing the solubility of Cd. In addition, organic matter can improve soil structure, increase soil porosity, promote oxygen transmission, and facilitate microbial activity, thus indirectly affecting the bioavailability of Cd. Organic matter has strong adsorption capacity, which could adsorb Cd^{2+} in soil and reduce its concentration in soil solution, therefore reducing its uptake by plants. Under certain conditions, organic matter and Cd^{2+} can form insoluble organo–metal complexes and reduce the bioavailability of Cd (Liu et al., 2021). Organic matter can bind to mineral particles in the soil to form organic-inorganic complexes that combine with Cd in the soil and reduce its mobility (Tong et al., 2023). Different types of organic matter have different adsorption and complexation capacities for Cd. Humic acid has a greater adsorption capacity for Cd than fulvic acid. The higher the organic matter content, the stronger the adsorption of Cd. The organic matter bound to the minerals will affect the interaction between the minerals and Cd, thereby promoting the formation of secondary minerals and enhancing the binding between organic matter and Cd (Fariña et al., 2018).

9.2.3 Soil texture

Soil texture refers to the relative content of particles of different sizes (sand, loam, and clay) in the soil. The difference in soil texture directly affects the physical and chemical properties of the soil, thus affecting the bioavailability of heavy metals (such as Cd) (Khaledian et al., 2017). Soil texture affects the CEC of the soil. Clay particles have higher CEC than coarse particles and can adsorb more cations, thus reducing the bioavailability of Cd. The CEC of sand grains is low, and the adsorption capacity of Cd^{2+} is weak, so Cd is more easily taken up by plants in sandy soils (Shaghaleh et al., 2024).

In addition, soil texture determines soil structure. The compact structure and small porosity of clay soils restrict the movement of water and air. Clay soil is characterized by small pore sizes, low permeability, and poor drainage, which can lead to prolonged water saturation and the development of anaerobic conditions. Sandy soil has a loose structure and high porosity, facilitating water and air exchange and promoting Cd migration.

9.2.4 Redox conditions

Soil redox condition is one of the important factors affecting the bioavailability of Cd. The oxidation-reduction potential (E_h) of soil directly affects the form, solubility, and migration of Cd in soil, thus affecting its bioavailability (see also Chapter 7). Under anoxic conditions, sulfate-reducing bacteria proliferate in the soil and reduce sulfate to sulfide, which can react with Cd to form CdS precipitate. During transition from anoxic to oxic conditions, dissolved oxygen reacts with Fe(II) to produce reactive oxygen species (ROS), such as hydroxyl radicals ($^{\bullet}$OH), via Fenton-type reactions, which can promote the oxidation of CdS and increase Cd bioavailability in soil (Huang et al., 2021; Li et al., 2023b). Specifically, ROS can oxidize the sulfide component (S^{2-}) of sparingly soluble CdS to sulfate ($SO_4{}^{2-}$), destabilizing the CdS precipitate and leading to the release of mobile Cd^{2+} into the soil solution.

Reductive processes generally consume protons, resulting in increase in soil pH to near neutrality under anoxic conditions, whereas oxidative processes produce protons and decrease soil pH to the initial value (Kirk, 2004). The pH changes associated with redox fluctuations can have a large influence on Cd bioavailability.

In addition, organic matter decomposes more quickly under oxic conditions than under anoxic conditions. The complex formed by organic matter and Cd is more stable under anoxic conditions, thus reducing the bioavailability of Cd (Zhang et al., 2024).

Under oxic conditions the Cd concentration in soil solution and thus bioavailability is usually much higher than under anoxic conditions (Wang et al., 2019). This explains why rice accumulates much more Cd when grown under aerobic than under flooded conditions (Huang et al., 2022). In plants such as rice that are adapted to flooded environments, the root system releases oxygen into the rhizosphere, creating an aerobic microenvironment. This causes the high concentration of Fe^{2+} diffusing from the reducing soil to be oxidized on the root surface, forming iron oxyhydroxide precipitates, which are encapsulated on the root surface and form a layer of iron plaque. By controlling the redox condition of soil through water management or additions of organic matter the speciation and bioavailability of Cd can be manipulated (Huang et al., 2021) (see Chapter 18).

9.2.5 Competitive ions

Competitive ions are those that have similar chemical properties to Cd^{2+} in soil solution and can compete with Cd^{2+} for adsorption and complexation, thus affecting the bioavailability of Cd. Competing ions compete with Cd for adsorption sites on soil particles such as clay minerals, iron oxyhydroxides, and organic matter. When the concentration of competitive ions is high, they will preferentially occupy adsorption sites, thereby increasing the concentration of Cd in the soil solution and, consequently, enhancing its bioavailability. Competitive ions and Cd can form complexes with organic acids and humic acids in the soil. The competition depends on both the affinity with the ligand and the concentrations of the ions. Ca^{2+} and Mg^{2+}, which are common cations in soil, have similar chemical properties to many heavy metal ions and compete for the adsorption of heavy metal ions. In general, when the concentrations of Ca^{2+} and Mg^{2+} are high in soil solution, the bioavailability of Cd will increase. The binding capacity of Cu^{2+} and Zn^{2+} to soil organic matter and clay mineral surfaces is stronger than that of Cd^{2+}. Thus when the concentration of Cu^{2+} in soil is high, it preferentially occupies the adsorption sites. The binding ability of Mn^{2+} to soil particles is relatively weak. At high concentrations the quantity advantage of Mn^{2+} enables it to compete with Cd^{2+}.

9.3 Methods for assessing the bioavailability of cadmium in soils

The methods for evaluating the bioavailability of Cd in soil are mainly divided into chemical extraction methods, in situ techniques, and biological evaluation methods (Bandara et al., 2021a; Ma et al., 2022).

9.3.1 Chemical extraction methods

9.3.1.1 Single-step extraction methods

1. Chelating agents, for example, EDTA (ethylenediaminetetraacetic acid) and DTPA (diethylene triamine pentaacetic acid). Chelating agents are chemical compounds that can form stable ring structure with metal ions. Different forms of Cd can be selectively extracted from soil by chelating agents to evaluate the extractability of Cd. The structure of EDTA is relatively simple, and the complex formed with Cd^{2+} is stable, but the selectivity is poor. The structure of DTPA is more complex than that of EDTA, and the complex formed with Cd^{2+} has higher stability and better selectivity (Feng et al., 2005).
2. Multielement extractants, for example, Mehlich-3, AB–DTPA (NH_4HCO_3 + DTPA), and $CaCl_2$-DTPA (DTPA + $CaCl_2$ + TEA). The Mehlich-3 extract consists of 0.2 M NH_4NO_3, 0.013 M HNO_3, 0.015 M NH_4F, 0.25 M CH_3COOH, and 0.001 M EDTA, commonly used to assess

plant-available nutrients (e.g., phosphorus, potassium, and micronutrients) in soil samples. It can extract metals in exchangeable, carbonate-bound, and partial oxide-bound forms from soil. Although combination extractants are often employed to characterize various forms of Cd in soil, caution is required when interpreting these results as representing "bioavailability." Chemical extractability does not inherently equate to biological availability, and these extraction techniques often lack consistent validation against actual biological uptake. Therefore, without concurrent biological assays, such extraction data should be viewed as a measure of chemical solubility rather than confirmed bioavailability. Furthermore, the extraction results are affected by temperature and pH, and the optimization of extraction conditions is complicated (Novak et al., 2018).

Ammonium bicarbonate–DTPA (AB–DTPA) is a widely used extractant for soil analysis, with its components being NH_4HCO_3 and DTPA. The final concentrations of the solution are 1 M NH_4HCO_3 and 0.005 M DTPA. The AB–DTPA extraction solution is a buffer solution. It can form stable chelates with Cd^{2+} in the soil in ionic, carbonate-bound, humus-bound, and some oxide-bound states, thereby extracting these forms of Cd from the solid phase of the soil. NH_4^+ in the extract can exchange with the adsorbed Cd^{2+} on soil colloids, thereby helping to release some of the adsorbed Cd. The combined action of AB–DTPA keeps the pH of the extraction solution at a weakly alkaline or near-neutral level (typically around pH 7.3).

The $CaCl_2$-DTPA method (typically consisting of 0.005 M DTPA, 0.1 M TEA, and 0.01 M $CaCl_2$ at pH 7.3) is another method for extracting the potentially available Cd in soil. Specifically the 0.01 M $CaCl_2$ component is chosen because its ionic strength approximates that of typical soil solutions (which generally ranges from 0.001 to 0.05 M). $CaCl_2$ provides a stable ionic strength, simulating the real environment of soil solutions, and helps to desorb Cd^{2+} from the exchange sites on the soil.

Single-step extraction methods have the advantages of simple operation, fast, and low cost and are the most commonly used method for chemical analysis at present. However, there are also significant disadvantages, as the extraction results are greatly affected by the extraction agent type, soil properties, extraction conditions, and other factors and do not fully reflect the bioavailability of Cd (Ma et al., 2022).

9.3.1.2 Sequential extraction methods

1. Tessier method

Tessier sequential extraction is a classic chemical extraction method for conditional classification of different fractions of metals (including Cd) in soil (Tessier et al., 1979). By using a series of extractants (Table 9.1) the Cd in the soil is divided into six operationally defined fractions: (1) exchangeable, which mainly refers to the weak adsorption on the soil particle surface or easy exchange with other cations; (2) carbonate binding, which mainly refers to Cd combined with the carbonate in the soil; (3) Mn-oxide-binding, which mainly refers to the Cd combined with Mn oxides in the soil; (4) Fe-oxide-binding, which mainly refers to the Cd combined with Fe oxides in the soil; (5) organic binding, which mainly refers to the Cd combined with soil organic matter; and (6) residual, which refers to Cd that remains unextracted, usually bound to the soil mineral lattice. By measuring the amount of Cd in the extract at each step the behaviors of Cd in the soil may be understood, and its potential bioavailability and environmental risks could be

TABLE 9.1 The Tessier sequential extraction procedure (Tessier et al., 1979; Khanmirzaei et al., 2013; Leleyter & Probst, 1999; Sulkowski & Hirner, 2006).

Step	Fraction	Extractant	pH	Temperature (°C)
1	Soluble + Exchangeable	20 mL 1 M NH_4OAC	7	25
2	Associated with carbonates	20 mL 1 M NaOAc	5	25
3	Associated with Mn oxides	20 mL 0.1 M $NH_2OH\bullet HCl$ in 0.1 M HNO_3		25
4	Associated with Fe oxides	20 mL 0.04 M $NH_2OH\bullet HCl$ in 25% w/v HOAc	2	96
5	Associated with organic matter	5 mL 0.1 M HNO_3		85
6	Residual	Aqua regia		25

Reproduced with permission from Sulkowski and Hirner (2006).

assessed (Orecchio et al., 2016). However, it should be noted that the so-called different fractions of Cd are operationally defined and do not represent distinct chemical forms, and there is often overlap between different steps of extractions.

The Tessier sequential extraction method extracts soil samples in five steps in air-dried soils that have been sieved through a 2-mm sieve. The extraction solution is centrifuged and filtered to separate the solid and liquid phases. The filtrate is appropriately diluted and determined using inductively coupled plasma optical emission spectrometer, inductively coupled plasma mass spectrometry (ICP–MS), or atomic absorption spectroscopy for measuring concentrations of Cd (He et al., 2013) (see also Chapter 2).

2. BCR sequential extraction method

The BCR sequential extraction procedure is a soil extraction method for heavy metals proposed by the European Community Metrology Agency (BCR). Compared to the Tessier method, the BCR protocol is a more streamlined and standardized procedure. The BCR method divides Cd in soil into four fractions (Table 9.2): (1) acid-extractable fraction mainly refers to Cd weakly adsorbed on the surface of soil particles or can be easily exchangeable with other cations and carbonate-bound heavy metals; (2) reducible fraction mainly refers to heavy metals that combine with iron and manganese oxides in soil; (3) oxidizable fraction refers primarily to heavy metals that combine with soil organic matter and sulfides; and (4) residual fraction refers to the heavy metal that remains in the soil, usually bound to the mineral lattice of the soil (Qureshi et al., 2020; Rosado et al., 2016).

9.3.1.3 Isotope labeling method

Isotope labeling is a method that uses isotopes as tracers to study the behavior, transport, and transformation of substances in chemical reactions and biological processes in environmental systems. By introducing isotopes into a substance to be studied and then tracking the distribution and changes of the isotopes, much important information about that substance could be obtained. The radioactive isotope ^{109}Cd or the stable isotope ^{111}Cd is used to track Cd in the soil

TABLE 9.2 The BCR-optimized sequential extraction procedure for high-carbonate soils (Sulkowski & Hirner, 2006).

Fraction	Extractant	Temperature
Soluble + Exchangeable + Carbonate	40 mL 0.11 M CH_3COOH	25°C
Iron and manganese oxyhydroxides	40 mL 0.5 M $NH_2OH\bullet HCl$ pH 1.5 with HNO_3	85°C
organic matter + sulfides	10 mL 8.8 M H_2O_2 50 mL CH_3COONH_4 1 M	25°C
Residual	Aqua regia	

Reproduced with permission from Sulkowski and Hirner (2006).

and plants (Degryse et al., 2004; Hamon et al., 2008; Huang et al., 2020). The *E*-value (isotopically exchangeable Cd) represents the total amount of soil Cd that can readily exchange with a labeled Cd isotope (^{109}Cd or ^{111}Cd) added to the soil. This value is determined under nonsteady-state conditions, typically by adding the isotope to a soil suspension and allowing it to equilibrate for a relatively short 24 hours (Midwood, 2007). The *E*-value essentially measures the pool of Cd that is in a dynamic equilibrium with the soil solution. This pool includes not only the Cd in the soil water but also the solid-phase Cd that can quickly desorb and become available. The *E*-value is considered a more accurate predictor of a metal's immediate bioavailability than traditional chemical extractants, as it directly quantifies the fraction of Cd that is mobile and can interact with the soil solution. Some researchers have used the isotopes of Cd to track the movement of Cd in the soil-wheat system, and an *L*-value was introduced, which represents the total amount of soil Cd that is bioavailable for plant uptake during their growth period and is considered the standard for assessing bioavailability because it directly measures what the plant absorbs. The *L*-value is determined by growing plants in soil spiked with a known amount of a Cd isotope (Ng et al., 2015). After the growing period the total amount of Cd and the amount of the added isotope in the plant tissue are measured. The *L*-value is then calculated using the isotopic dilution principle. This method allows for the precise tracing of the fate of Cd in soil–plant systems (Bracher et al., 2021).

9.3.1.4 Limitations of chemical methods

There are a number of limitations of chemical methods for assessing Cd bioavailability in soil, including:

- Temporal limitation: Measurement at a single time point cannot reflect bioavailability over the whole growth period of the organisms.
- Spatial discrepancy: Bulk soil measurements are not the environment that organisms experience. Plant roots actively modify their immediate vicinity (the rhizosphere) through the secretion of organic acids, protons, and enzymes, creating a chemical environment significantly different from the bulk soil used in standard chemical extractions. Therefore extractants applied to bulk soil may underestimate or overestimate the metal fractions actually accessible to roots.

- Operational definition limitation: The fractions defined by methods like BCR (e.g., "reducible" or "oxidizable") are operationally defined by the reagents used, not by strict mineralogical phases. Chemical reagents often lack perfect selectivity; for instance, a reagent targeting manganese oxides might also attack iron oxides or specific organic complexes, leading to an inaccurate categorization of metal forms compared to biological reality.
- Readsorption artifacts: During sequential extraction procedures, metals released from one solid phase during a specific extraction step can readsorb onto remaining solid surfaces before the liquid and solid phases are separated. This redistribution can lead to an underestimation of the mobile fractions and an overestimation of the more recalcitrant fractions.
- Lack of biological feedback: Unlike the *L*-value method, which uses the plant itself as a "sensor," chemical extractants cannot mimic the active uptake mechanisms or exclusion strategies employed by different plant species.

9.3.2 In situ techniques

9.3.2.1 High-resolution diffusive gradients in thin films

Diffusive gradients in thin films (DGT) are an in situ dynamic sampling technology. It quantitatively determines the labile and available speciation of target pollutants in environmental media by simulating the absorption, diffusion, and uptake process of pollutants by organisms. DGT devices are usually composed of three layers: filter membrane, diffusive gel, and specific binding layer. Cd is diffused through the diffusion layer and accumulated in the binding layer, following Fick's first law of diffusion. For detecting Cd in soil, Chelex is commonly used as the binding layer, as it efficiently adsorbs labile Cd from soil pore water. By measuring the amount of Cd accumulated on the binding layer the labile concentration of Cd in the soil can be quantified (Luo et al., 2018).

Because of its measurement principle based on the diffusive flux of target substances and the size-selective capability provided by its diffusion layer featuring nanoscale pores, the DGT technique offers unique advantages when applied to soil and sediment environments. Its measurement results reflect not only the concentration of pollutants in pore water and their replenishment capability from the solid phase (Zhang et al., 2001) but also exclude the contribution of colloids in soil and sediment pore water that are difficult for organisms to absorb directly (Van Der Veeken et al., 2008). Consequently, DGT has gained wide attention as an analytical tool for determining bioavailable concentrations of heavy metals in soils and sediments.

In recent years the DGT technique has achieved substantial progress in the field of in situ high-resolution (HR) analysis. Particularly, its coupling with laser ablation–ICP–MS (LA–ICP–MS) has been widely applied to study the migration and transformation of heavy metals in sediment or plant rhizosphere environments (Fang et al., 2021; Kreuzeder et al., 2018). A two-dimensional (2D) rhizobox system was employed to characterize the spatial heterogeneity of labile solutes in the rhizosphere. The setup procedure involves preculturing plants in a wedge-shaped soil block to establish a root system, which is then transferred into a specialized transparent rhizobox. During deployment the DGT device is assembled such that a protective glass-etched membrane is placed directly against the exposed soil profile. This membrane serves as the interface to separate the underlying binding gel from soil particles while allowing ion diffusion. A transparent front plate is clamped to seal the assembly, ensuring intimate physical

FIGURE 9.1 Detailed procedures for in situ measurement of bioavailable metals using two-dimensional (2D) rhizobox-DGT system (Fang et al., 2021). *DGT*, Diffusive gradients in thin films. *From Fang et al. (2021), with permission.*

contact between the glass-etched membrane, the soil surface, and the developing root system for in situ 2D mapping. This combined technique enables HR multielemental imaging of samples (Fig. 9.1).

The rhizosphere is an important zone for plant uptake of both nutrients and contaminants. The physicochemical conditions in the rhizosphere drastically differ from those in the bulk soil. Particularly in the rice rhizosphere, root activities, including radial oxygen loss and root exudation, can induce significant changes of redox, pH, and microbial activities, thus altering the Cd bioavailability. However, a complete understanding about how root activities influence the bioavailability of Cd is limited by traditional methods, which have constraints such as (1) inability to make measurements at suitable spatial sensitivities and scales and (2) disturbing the local environment in which they are sampling. HR-DGT has been shown to accurately resolve features in soil or sediment solute chemistries in the 2D scale at resolutions of ~100 μm (Fig. 9.2) (Fang et al., 2020). This has led to the technique being more widely applied for 2D mapping of the potential availability of Cd within plant rhizospheres and microscale interfaces (Hinsinger et al., 2005; Hoefer et al., 2015).

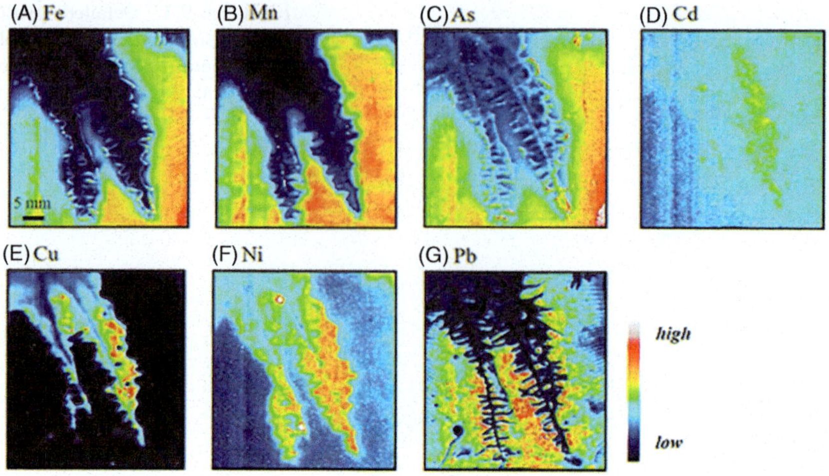

FIGURE 9.2 High-resolution mapping of (A) Fe, (B) Mn, (C) As, (D) Cd, (E) Cu, (F) Ni, and (G) Pb in the rhizosphere of rice by DGT (Fang et al., 2021). *DGT*, Diffusive gradients in thin films. *From Fang et al. (2021), with permission.*

9.3.2.2 Donnan membrane technique

Donnan membrane technique (DMT) is a method that can be combined with electrochemical methods to determine the concentration of free ions of Cd^{2+}. Based on the Donnan equilibrium principle the solution to be measured is separated from the reference solution by an ion exchange membrane, and the activity or concentration of metal ions in the solution to be measured is calculated by the balance of charges on both sides of the membrane. In the DMT system, Cd^{2+} ions diffuse toward a reference solution through an ion exchange membrane. Because of the fixed charge on the ion exchange membrane an electric potential difference is formed between the two sides of the membrane. By measuring this potential difference, combined with the Nernst equation, the activity of Cd ions in the solution under test can be calculated (Temminghoff et al., 2000). The DMT has been used successfully to measure in situ free trace metal concentrations in surface waters. Researchers applied this technique in a column experiment using flooded paddy soil contaminated with trace metals and found that DMT measurements were generally consistent with predictions from a geochemical speciation model under aerobic conditions (Pan et al., 2015). The DMT also has some disadvantages, including:

- Long equilibration time: Unlike rapid extraction methods, reaching Donnan equilibrium is a slow kinetic process that typically requires 2–7 days. This makes DMT unsuitable for monitoring rapid dynamic changes in bioavailability or for high-throughput routine analysis.
- Ionic strength limitations: The accuracy of DMT calculations relies heavily on the ionic strength of the donor and acceptor solutions. Very high ionic strengths (e.g., in saline soils) can compress the electrical double layer on the membrane or interfere with the Donnan equilibrium principle, leading to calculation errors.
- Underestimation of the bioavailable pool: DMT specifically measures free metal ions (e.g., Cd^{2+}). However, if the organism can also take up labile complexes (e.g., weak organic ligands) or if the uptake is kinetically limited (as described by the dynamic flux), DMT

may underestimate the total pool of metal available to the organism compared to dynamic techniques like DGT.

9.3.3 Bioassay methods

9.3.3.1 Plant indication method

The method of using plants as indicators of Cd bioavailability refers to a technique that assesses the degree of Cd contamination in soil or water bodies based on the characteristics of plant absorption, accumulation, and morphological changes related to Cd. The plant indication method is a direct and reliable approach for assessing the bioavailability of Cd in soil. It is based on the core principle that the amount of Cd absorbed by plants can most directly reflect the degree to which Cd is available to living organisms. Different plant species vary widely in the ability to take up and tolerate Cd. Based on the relationship between plant and soil metal (e.g., Cd) concentrations, plant species are categorized into three types: excluders that restrict metal uptake; indicators that take up metals proportionally to their available concentrations in the soil; and (hyper)accumulators that exhibit extraordinary abilities to accumulate metals even at low concentrations of available metals in the soil (Baker, 1981). Indicators are therefore the most appropriate plants to be used to assess metal bioavailability in soil and constitute the majority of plant species of interest. Typical Cd indicator plant species include cruciferous plants such as rapeseed and cabbage and *Solanaceae* plants such as tomato, pepper, and eggplant. In contrast, legumes and gramineous plants are relatively low Cd accumulators (Kuboi et al., 1986), but these are important if they are the target species in agriculture and food systems.

The degree of phytotoxicity may also be used as a qualitative indicator of Cd bioavailability in soil, taking into account the wide variation in Cd tolerance among plant species. However, this applies only to soils with Cd concentrations that can cause toxicity to plants. For the majority of agricultural soils, bioavailable Cd concentrations are usually not high enough to cause obvious phytotoxicity but can be problematic for food safety and human health.

9.3.3.2 Microbiological indicator method

The microbial indicator method monitors and evaluates the degree of Cd bioavailability by using the sensitivity and tolerance of microorganisms to Cd and the change of community structure. Different species of microorganisms have different degrees of tolerance to Cd, and some microorganisms are highly tolerant to Cd. Therefore, by analyzing changes in microbial communities, the extent of Cd bioavailability in the environment can be inferred (Melloni et al., 2023). The microbial indicator method can also measure the growth rate, enzyme activity, and respiration rate of microorganisms to evaluate the effect of Cd bioavailability. However, there are some limitations of the microbial indicator methods, as the results are strongly influenced by the characteristics of the existing soil microbial community or by Cd tolerance of the particular microbial species used in the assays.

9.3.3.3 Animal indication method

The animal indicator method of Cd bioavailability refers to a method to monitor and evaluate the degree of Cd bioavailability by using the characteristics of animal sensitivity, accumulation, and physiological and biochemical reaction to Cd (Schludermann et al., 2003). It assesses whether the environment is contaminated with Cd by observing changes in the animal's growth, development, reproduction, behavior, and Cd accumulation in its body (Moafi et al., 2011). Different species of animals have different sensitivity to Cd. Some animals are sensitive to Cd, and when the concentration of Cd in the soil increases, the growth, development, and reproduction of these animals are inhibited. Other animals have a strong tolerance to Cd and can survive in high concentrations, but they accumulate large amounts of Cd in their bodies. By analyzing the Cd content in these animals the level of Cd bioavailability can be assessed.

Soil animal indicators include earthworms, snails, insects, among others. These animals live in the soil and ingest Cd through the food chain or through dermal contact with soil solution (Dallinger et al., 1992). Earthworms are commonly used as a biological indicator to reflect the bioavailability of Cd in soil. Biological markers at the individual and population levels mainly refer to the biological response indicators that occur in individual or population of earthworms under certain doses of pollutant exposure. The individual and population-level biological markers include behavioral and morphological indicators, reproductive indicators, and growth indicators. Behavior and morphological indicators mainly include changes in the body size, weight, and color of earthworms, the presence or absence of loosening, ulceration, or detachment of segments, the presence or absence of behavioral twisting, curling or paralysis, and mortality (Calisi et al., 2011).

The accumulation of Cd in earthworms is also an important biological marker. Earthworms accumulate Cd from the soil mainly through passive diffusion and feeding (see Chapter 8). The former is when Cd passes through the body surface from the soil solution into the earthworms' bodies; the latter is when Cd is ingested by the earthworms and then absorbed by the internal organs (Booth & O'Halloran, 2001).

Earthworm enzymes and proteins have also been used as indicators of Cd bioavailability. Studies have shown that the biosynthesis of cytochrome P450 in earthworms' bodies is enhanced by exposure to long-term low Cd concentration or short-term high Cd concentration but inhibited by exposure to long-term high concentration of Cd, with a dose–response to the increase of Cd stress (Feng et al., 2025). Cd exposure can induce the production of $^{\bullet}OH$ free radicals in earthworms. A significant induction occurred at a low Cd concentration of 0.5 mg kg^{-1}, and the earthworms exhibited an oxidative stress response, which can serve as an indicator to reflect the bioavailability of Cd (Kumar et al., 2022).

Disadvantages of bioassays include the following:

- Time-consuming and labor-intensive: Unlike chemical extractions, which can be completed in hours, bioassays require maintaining living organisms over days, weeks, or even months (e.g., full plant growth cycles). This makes them unsuitable for rapid screening of large numbers of samples or for emergency response scenarios.
- Species-specificity limits extrapolation: A bioassay reflects the bioavailability only for the specific test species used (e.g., an earthworm or a specific plant species). Because of differences in uptake mechanisms and physiological barriers, results from one species (e.g., *Eisenia fetida*) may not accurately predict the risk to other organisms (e.g., rice or humans), limiting the universal applicability of the data.
- Toxicity thresholds: In highly contaminated soils the test organisms may suffer from acute toxicity and die before bioaccumulation data can be obtained, rendering the accumulation-based indicator method ineffective.

9.4 The practical value of assessing the bioavailability of cadmium

The results of Cd pollution assessment are usually expressed in terms of Cd concentration, bioavailability, ecological risk (see Chapter 16), and other indicators (Wieczorek et al., 2018). Assessing the bioavailability of Cd in soil holds important practical value, especially in environmental management, agricultural production, and protection of human health (Li et al., 2016). Measuring the total amount of Cd in the soil alone cannot accurately reflect its potential harm to living organisms. From an ecotoxicological perspective, only the bioavailable fraction of Cd that can be absorbed by organisms is considered biologically relevant and poses a direct risk (Wieczorek et al., 2018). Measurement of the total Cd concentration in the soil alone may overestimate the risk, leading to unnecessary and costly soil remediation projects; conversely, it may also underestimate the risk, missing the best opportunity to take intervention measures (López-Rojo et al., 2020).

Humans are the top consumers in the food chain. The Cd ingested through food is the main route for human exposure to Cd (see Chapter 15). Cd has the characteristic of a long half-life in human bodies. Long-term low-dose exposure can lead to health problems. The ultimate and most important goal of assessing the bioavailability of Cd is to reduce the amount of Cd ingested by humans through food at the source (Abadi et al., 2024). When conducting environmental health risk assessments, bioavailability data can more accurately evaluate the actual exposure levels of the population and potential health risks compared to total concentration data. This allows for a more targeted and rational approach to managing Cd-contaminated sites, prioritizing those that pose the greatest actual risk to human health.

Assessments of Cd bioavailability provide scientific basis for environmental protection. By assessing the Cd bioavailability, we can conduct more precise risk classification for agricultural soil. Areas with high bioavailability should be classified as high-risk zones, and large-scale restoration may be necessary, while areas with low biological availability, even if the total concentration is slightly higher, may still be allowed for safe utilization under specific agricultural practices. For remediation of contaminated soil, bioavailability should also be considered when selecting appropriate remediation technologies.

9.5 Future outlook

9.5.1 Development of new methods and techniques

In recent years, advancements in science and technology have continuously refined methods for assessing Cd bioavailability. In particular the development of advanced characterization techniques has facilitated the emergence of in situ approaches, such as HR-DGT, which represent a new direction for Cd bioavailability determination. Advanced speciation techniques, such as X-ray absorption near-edge structure (XANES) and extended X-ray absorption fine structure (EXAFS), are powerful tools for evaluating the molecular-level forms of Cd. It provides detailed information on the oxidation state, coordination environment, bond lengths, and the types and quantities of neighboring atoms of Cd. This allows for the identification of whether Cd is associated with sulfur (forming CdS), oxygen (forming oxides or $CdCO_3$), or other atoms, as well as the relative strength of these bonds. This enables the factors that affect the bioavailability of Cd at the atomic level to be explained.

New methods for testing the bioavailability of Cd in soil will no longer be limited to macroscopic extraction and analysis but will delve into the atomic and molecular levels. Through in situ characterization and powerful theoretical calculations, it will reveal the microscopic behavior and chemical transformation mechanisms of Cd bioavailability at the soil–plant interface.

9.5.2 Improvement of cadmium bioavailability assessment

Accurate assessment of the bioavailability of Cd is of great significance for environmental risk assessment and ecological restoration. By improving the accuracy of Cd bioavailability assessments the impacts of pollutants on ecosystems and human health can be more effectively predicted, leading to more scientific and effective environmental management measures. Recognizing the limitations of extractable Cd, future assessments are moving toward an integrated, multiindicator approach, integrating chemical extraction, DGT, plant indicators, and atomic-level morphological information to form a more comprehensive and reliable bioavailability evaluation system.

Improving the accuracy of Cd bioavailability assessments is a complex task that requires constant exploration and innovation. Through interdisciplinary technological innovation and standardization construction, the environmental behavior and ecological effects of Cd can be better understood, and scientific basis for environmental protection and ecological restoration can be provided.

9.6 Conclusions

Assessing Cd bioavailability is crucial because the total concentration of Cd in soil does not accurately reflect its ecological risk or potential harm to organisms. This assessment is essential for precise risk evaluation, the selection of appropriate remediation techniques, and the development of effective environmental standards. The bioavailability of Cd is governed by a complex interplay of factors, primarily soil pH, organic matter content, redox conditions, and the presence of competitive ions, which collectively determine the chemical speciation and mobility of Cd at the soil–organism interface.

Current assessment methodologies have evolved from traditional bulk chemical extractions (e.g., Tessier and BCR sequential extractions) to more advanced in situ techniques such as HR-DGT and DMT, alongside direct biological indicators. Although chemical methods offer rapid screening capabilities, they are often limited by readsorption artifacts and spatial discrepancies compared to the actual rhizosphere environment. Conversely, in situ and biological methods provide higher ecological relevance but have limitations regarding equilibration times and species specificity.

Looking forward the future of Cd bioavailability assessment lies in the integration of multi-indicator approaches. Combining macroscopic extraction data with in situ HR sampling (DGT) and molecular-level spectroscopic characterization (e.g., XANES) will be key to establishing a comprehensive evaluation system that accurately predicts the environmental behavior and biological effects of Cd.

9.7 Summary

- Cd bioavailability denotes the fractions of soil Cd existing in forms that are sufficiently mobile and chemically reactive to be taken up by organisms (e.g., plants, microbes, and soil fauna) and to exert biological or toxic impacts.
- Assessing Cd bioavailability is crucial because the total concentration of Cd in soil does not accurately reflect its ecological risk or potential harm to organisms.
- Single-step and sequential extraction methods have been used to assess Cd bioavailability but are limited by spatial discrepancies, readsorption artifacts, and insufficient validation with biological responses.
- In situ techniques, such as HR-DGT and DMT, provide in situ, time-integrated measurements of Cd, offering a more realistic assessment of bioavailability than bulk concentrations, although DMT is constrained by long equilibration times.
- Biological evaluation methods provide relevant assessment of Cd bioavailability but can be time-consuming, labor-intensive, and restricted by species specificity and toxicity thresholds.

Acknowledgment

This work was funded by the National Natural Science Foundation of China (no. 42277375).

References

Abadi, H. T., Alemayehu, T., & Berhe, B. A. (2024). Heavy metal's pollution health risk assessment and source appraisal of groundwater and surface water in Irob catchment, Tigray, Northern Ethiopia. *Applied Water Science*, *14*(9). https://www.springer.com/journal/13201. doi:10.1007/s13201-024-02237-9.

Arabi, Z., Homaee, M., Asadi, M. E., & Kapourchal, S. A. (2017). Cadmium removal from Cd-contaminated soils using some natural and synthetic chelates for enhancing phytoextraction. *Chemistry and Ecology*, *33*(5), 389–402. www.tandf.co.uk/journals/titles/02757540.asp. doi:10.1080/02757540.2017.1308501.

Baker, A. J. M. (1981). Accumulators and excluders—Strategies in the response of plants to heavy metals. *Journal of Plant Nutrition*, *3*(1–4), 643–654. doi:10.1080/01904168109362867.

Bandara, T., Franks, A., Xu, J., Chathurika, J. B. A. J., & Tang, C. (2021a). Biochar aging alters the bioavailability of cadmium and microbial activity in acid contaminated soils. *Journal of Hazardous Materials*, *420*, 126666. doi:10.1016/j.jhazmat.2021.126666.

Bandara, T., Chathurika, J. B. A. J., Franks, A., Xu, J., & Tang, C. (2021b). Interactive effects of biochar type and pH on the bioavailability of As and Cd and microbial activities in co-contaminated soils. *Environmental Technology & Innovation*, *23*, 101767. doi:10.1016/j.eti.2021.101767.

Beri, W. T., Gesessew, W. S., & Tian, S. (2020). Maize cultivars relieve health risks of Cd-polluted soils: In vitro Cd bioaccessibility and bioavailability. *Science of the Total Environment*, *703*. www.elsevier.com/locate/scitotenv. doi:10.1016/j.scitotenv.2019.134852.

Biswash, M. R., Li, K. w., Xu, R. k., Uwiringiyimana, E., Guan, P., Lu, H. l., Li, J. y., Jiang, J., Hong, Z. n., & Shi, R (2024). Alteration of soil pH induced by submerging/drainage and application of peanut straw biochar and its impact on Cd(II) availability in an acidic soil to indica-japonica rice varieties. *Environmental Pollution*, *356*. https://www.sciencedirect.com/science/journal/02697491. doi:10.1016/j.envpol.2024.124361.

Booth, L. H., & O'Halloran, K. (2001). A comparison of biomarker responses in the earthworm *Aporrectodea caliginosa* to the organophosphorus insecticides diazinon and chlorpyrifos. *Environmental Toxicology and Chemistry*, *20*(11), 2494–2502. http://onlinelibrary.wiley.com/journal/10.1002/(ISSN)1552-8618. doi:10.1002/etc.5620201115.

Bracher, C., Frossard, E., Bigalke, M., Imseng, M., Mayer, J., & Wiggenhauser, M. (2021). Tracing the fate of phosphorus fertilizer derived cadmium in soil-fertilizer-wheat systems using enriched stable isotope labeling. *Environmental Pollution, 287*, 117314. doi:10.1016/j.envpol.2021.117314.

Calisi, A., Lionetto, M. G., & Schettino, T. (2011). Biomarker response in the earthworm *Lumbricus terrestris* exposed to chemical pollutants. *Science of the Total Environment, 409*(20), 4456–4464. doi:10.1016/j.scitotenv.2011.06.058.

Cao, J., Lv, C., Zhang, C., Yin, F., Gao, Z., Wei, L., & Wang, L. (2024). Asynchronous synergetic remediation strategy for Cd-contaminated soil via passivation and phytoremediation technology. *Agronomy, 14*(9), 1913. doi:10.3390/agronomy14091913.

Dallinger, R., Berger, B., & Birkel, S. (1992). Terrestrial isopods: Useful biological indicators of urban metal pollution. *Oecologia, 89*(1), 32–41. doi:10.1007/bf00319012.

Degryse, F, Buekers, J, & Smolders, E (2004). Radio-labile cadmium and zinc in soils as affected by pH and source of contamination. *European Journal of Soil Science, 55*(1), 113–122. doi:10.1046/j.1351-0754.2003.0554.x.

Fang, W., Shi, X., Yang, D., Hu, X., Williams, P. N., Shi, B., Liu, Z., & Luo, J. (2020). In situ selective measurement based on diffusive gradients in thin films technique with mercapto-functionalized mesoporous silica for high-resolution imaging of SbIII in soil. *Analytical Chemistry, 92*(5), 3581–3588. http://pubs.acs.org/journal/ancham. doi:10.1021/acs.analchem.9b04225.

Fang, W., Yang, Y., Wang, H., Yang, D., Luo, J., & Williams, P. N. (2021). Rice rhizospheric effects on the bioavailability of toxic trace elements during land application of biochar. *Environmental Science and Technology, 55*(11), 7344–7354. http://pubs.acs.org/journal/esthag. doi:10.1021/acs.est.0c07206.

Fariña, A. O., Peacock, C. L., Fiol, S., Antelo, J., & Carvin, B. (2018). A universal adsorption behaviour for Cu uptake by iron (hydr)oxide organo-mineral composites. *Chemical Geology, 479*, 22–35. http://www.sciencedirect.com/science/journal/00092541. doi:10.1016/j.chemgeo.2017.12.022.

Feng, M. H., Shan, X. Q., Zhang, S., & Wen, B. (2005). A comparison of the rhizosphere-based method with DTPA, EDTA, $CaCl_2$, and $NaNO_3$ extraction methods for prediction of bioavailability of metals in soil to barley. *Environmental Pollution, 137*(2), 231–240. doi:10.1016/j.envpol.2005.02.003.

Feng, Q., Wang, X., Wang, X., & Xu, M. (2025). A bibliometric analysis and literature review of earthworms used as bio-indicators for assessing soil heavy metal pollution. *Journal of Environmental Sciences*. doi:10.1016/j.jes.2025.06.052.

Gong, L., Wang, J., Abbas, T., Zhang, Q., Cai, M., Tahir, M., Wu, D., & Di, H. (2021). Immobilization of exchangeable Cd in soil using mixed amendment and its effect on soil microbial communities under paddy upland rotation system. *Chemosphere, 262*, 127828. doi:10.1016/j.chemosphere.2020.127828.

Gujre, N., Rangan, L., & Mitra, S. (2021). Occurrence, geochemical fraction, ecological and health risk assessment of cadmium, copper and nickel in soils contaminated with municipal solid wastes. *Chemosphere, 271*, 129573. doi:10.1016/j.chemosphere.2021.129573.

Hamon, R. E., Parker, D. R., & Lombi, E. (2008). Chapter 6: Advances in isotopic dilution techniques in trace element research. A review of methodologies, benefits, and limitations. *Advances in Agronomy, 99*, 289–343. doi:10.1016/S0065-2113(08)00406-9.

He, Q., Ren, Y., Mohamed, I., Ali, M., Hassan, W., & Zeng, F. (2013). Assessment of trace and heavy metal distribution by four sequential extraction procedures in a contaminated soil. *Soil and Water Research, 8*(2), 71–76. doi:10.17221/20/2012-swr.

Hinsinger, P., Gobran, G. R., Gregory, P. J., & Wenzel, W. W. (2005). Rhizosphere geometry and heterogeneity arising from root-mediated physical and chemical processes. *New Phytologist, 168*(2), 293–303. doi:10.1111/j.1469-8137.2005.01512.x.

Hoefer, C., Santner, J., Puschenreiter, M., & Wenzel, W. W. (2015). Localized metal solubilization in the rhizosphere of *Salix smithiana* upon sulfur application. *Environmental Science and Technology, 49*(7), 4522–4529. http://pubs.acs.org/journal/esthag. doi:10.1021/es505758j.

Huang, B. Y., Zhao, F. J., & Wang, P. (2022). The relative contributions of root uptake and remobilization to the loading of Cd and As into rice grains: Implications in simultaneously controlling grain Cd and As accumulation using a segmented water management strategy. *Environmental Pollution, 293*. https://www.journals.elsevier.com/environmental-pollution. doi:10.1016/j.envpol.2021.118497.

Huang, H., Ji, X. B., Cheng, L. Y., Zhao, F. J., & Wang, P. (2021). Free radicals produced from the oxidation of ferrous sulfides promote the remobilization of cadmium in paddy soils during drainage. *Environmental Science and Technology, 55*(14), 9845–9853. http://pubs.acs.org/journal/esthag. doi:10.1021/acs.est.1c00576.

Huang, Y. i, Zhang, S., Chen, Y., Wang, L. i, Long, Z., Hughes, S. S., Ni, S., Cheng, X., Wang, J., Li, T., Wang, R., & Liu, C. (2020). Tracing Pb and possible correlated Cd contamination in soils by using lead isotopic compositions. *Journal of Hazardous Materials, 385*, 121528. doi:10.1016/j.jhazmat.2019.121528.

Jia, Y., Li, J., Zeng, X., Zhang, N., Wen, J., Liu, J., Jiku, M. A. S, Wu, C., & Su, S. (2022). The performance and mechanism of cadmium availability mitigation by biochars differ among soils with different pH: Hints for the reasonable choice of passivators. *Journal of Environmental Management, 312*. doi:10.1016/j.jenvman.2022.114903.

Jin, Y. u, Zhang, B., Chen, J., Mao, W., Lou, L., Shen, C., & Lin, Q. I (2021). Biofertilizer-induced response to cadmium accumulation in *Oryza sativa* L. grains involving exogenous organic matter and soil bacterial community structure. *Ecotoxicology and Environmental Safety, 211*, 111952. doi:10.1016/j.ecoenv.2021.111952.

Khaledian, Y., Brevik, E. C., Pereira, P., Cerdà, A., Fattah, M. A., & Tazikeh, H. (2017). Modeling soil cation exchange capacity in multiple countries. *Catena, 158*, 194–200. www.elsevier.com/inca/publications/store/5/2/4/6/0/9. doi:10.1016/j.catena.2017.07.002.

Khanmirzaei, A., Bazargan, K., Amir Moezzi, A., Richards, B. K., & Shahbazi, K. (2013). Single and sequential extraction of cadmium in some highly calcareous soils of Southwestern Iran. *Journal of Soil Science and Plant Nutrition, 13*(1). http://www.scielo.cl/pdf/jsspn/v13n1/aop1413.pdf. doi:10.4067/s0718-95162013005000014.

Kirk, G. (2004). *The biogeochemistry of submerged soils* (pp. 1–282). John Wiley & Sons. doi:10.1002/047086303X.

Kreuzeder, A., Santner, J., Scharsching, V., Oburger, E., Hoefer, C., Hann, S., & Wenzel, W. W. (2018). In situ observation of localized, sub-mm scale changes of phosphorus biogeochemistry in the rhizosphere. *Plant and Soil, 424*(1-2), 573–589. http://www.wkap.nl/journalhome.htm/0032-079X. doi:10.1007/s11104-017-3542-0.

Kuboi, T., Noguchi, A., & Yazaki, J. (1986). Family-dependent cadmium accumulation characteristics in higher plants. *Plant and Soil, 92*(3), 405–415. doi:10.1007/bf02372488.

Kumar P., Didawat R.K, Roy A., Tigga P., Bag K., Kumar S., Ritambhara, Advances in agricultural biotechnologyearthworms bio markers soil pollution assessment. In: S. Mishra, S. Kumar, Advances in agricultural biotechnology. AkiNik Publications.

Leleyter, L., & Probst, J. L. (1999). A new sequential extraction procedure for the speciation of particulate trace elements in river sediments. *International Journal of Environmental Analytical Chemistry, 73*(2), 109–128. www.tandf.co.uk/journals/titles/03067319.asp. doi:10.1080/03067319908032656.

Li, C., Yang, Z., Yu, T., Hou, Q., Liu, X. u, Wang, J., Zhang, Q., & Wu, T. (2021). Study on safe usage of agricultural land in karst and non-karst areas based on soil Cd and prediction of Cd in rice: A case study of Heng County, Guangxi. *Ecotoxicology and Environmental Safety, 208*, 111505. doi:10.1016/j.ecoenv.2020.111505.

Li, H., Li, Z., Huang, L., Mao, X., Dong, Y., Fu, S., Su, R., Chang, Y., & Zhang, C. (2023a). Environmental factors influence the effects of biochar on the bioavailability of Cd and Pb in soil under flooding condition. *Water, Air, & Soil Pollution, 234*(2). doi:10.1007/s11270-023-06130-0.

Li, S.-W., Sun, H.-J., Li, H.-B. o, Luo, J., & Ma, L. Q. (2016). Assessment of cadmium bioaccessibility to predict its bioavailability in contaminated soils. *Environment International, 94*, 600–606. doi:10.1016/j.envint.2016.06.022.

Li, T., Li, W., Yu, S., Zhao, C., Liu, L., Li, Y. C., Gao, B., Chen, H., Wang, B., Wang, X., Wang, S., & Rinklebe, J. (2023b). Biochar inhibited hydrogen radical-induced Cd bioavailability in a paddy soil. *Science of the Total Environment, 892*, 164521. doi:10.1016/j.scitotenv.2023.164521.

Liu, N. A, Lou, X., Li, X., Shuai, Z., Liu, H., Jiang, Z., & Wei, S. (2021). Rhizosphere dissolved organic matter and iron plaque modified by organic amendments and its relations to cadmium bioavailability and accumulation in rice. *Science of the Total Environment, 792*, 148216. doi:10.1016/j.scitotenv.2021.148216.

López-Rojo, N., Pérez, J., Alonso, A., Correa-Araneda, F., & Boyero, L. (2020). Microplastics have lethal and sublethal effects on stream invertebrates and affect stream ecosystem functioning. *Environmental Pollution, 259*. https://www.journals.elsevier.com/environmental-pollution. doi:10.1016/j.envpol.2019.113898.

Luo, J., Yin, D., Cheng, H., Davison, W., & Zhang, H. (2018). Plant induced changes to rhizosphere characteristics affecting supply of Cd to *Noccaea caerulescens* and Ni to *Thlaspi goesingense*. *Environmental Science & Technology, 52*(9), 5085–5093. doi:10.1021/acs.est.7b04844.

Ma, P., Tian, T., Dai, Z., Shao, T., Zhang, W., & Liu, M. (2022). Assessment of Cd bioavailability using chemical extraction methods, DGT, and biological indicators in soils with different aging times. *Chemosphere, 296*. www.elsevier.com/locate/chemosphere. doi:10.1016/j.chemosphere.2022.133931.

Melloni, E. G. P., Melloni, R, Pastor-Jáuregui, R, Aguilar-Garrido, A., Martín-Peinado, F. J, & Basiliko, N. (2023). Microbiological indicators as sensitive indicators in the assessment of areas contaminated by heavy metals. *Soil Research, 61*(7), 663–673. doi:10.1071/sr23012.

Midwood A.J., Stable isotope methods for estimating the labile metal content of soils. In: N. Willey, Phytoremediation: Methods and reviews. Humana Press.

Moafi, H. F., Shojaie, A. F., & Zanjanchi, M. A. (2011). Titania and titania nanocomposites on cellulosic fibers: Synthesis, characterization and comparative study of photocatalytic activity. *Chemical Engineering Journal, 166*(1), 413–419. doi:10.1016/j.cej.2010.10.074.

Mubeen, S., Ni, W., He, C., & Yang, Z. (2023). Agricultural strategies to reduce cadmium accumulation in crops for food safety. *Agriculture, 13*(2), 471. doi:10.3390/agriculture13020471.

Ng, J. C., Juhasz, A., Smith, E., & Naidu, R. (2015). Assessing the bioavailability and bioaccessibility of metals and metalloids. *Environmental Science and Pollution Research, 22*(12), 8802–8825. http://www.springerlink.com/content/0944-1344. doi:10.1007/s11356-013-1820-9.

Novak, J. M., Ippolito, J. A., Ducey, T. F., Watts, D. W., Spokas, K. A., Trippe, K. M., Sigua, G. C., & Johnson, M. G. (2018). Remediation of an acidic mine spoil: Miscanthus biochar and lime amendment affects metal availability, plant growth, and soil enzyme activity. *Chemosphere, 205*, 709–718. doi:10.1016/j.chemosphere.2018.04.107.

Orecchio, S., Amorello, D., Barreca, S., & Pettignano, A. (2016). Speciation of vanadium in urban, industrial and volcanic soils by a modified Tessier method. *Environmental Science: Processes & Impacts, 18*(3), 323–329. doi:10.1039/C5EM00596E.

Pan, Y., Koopmans, G. F., Bonten, L. T. C., Song, J., Luo, Y., Temminghoff, E. J. M., & Comans, R. N. J. (2015). In-situ measurement of free trace metal concentrations in a flooded paddy soil using the Donnan membrane technique. *Geoderma, 241–242*, 59–67. www.elsevier.com/inca/publications/store/5/0/3/3/3/2. doi:10.1016/j.geoderma.2014.11.003.

Qureshi, A. A., Kazi, T. G., Baig, J. A., Arain, M. B., & Afridi, H. I. (2020). Exposure of heavy metals in coal gangue soil, in and outside the mining area using BCR conventional and vortex assisted and single step extraction methods. Impact on orchard grass. *Chemosphere, 255*. www.elsevier.com/locate/chemosphere. doi:10.1016/j.chemosphere.2020.126960.

Riaz, M., Kamran, M., Rizwan, M., Ali, S., Parveen, A., Malik, Z., & Wang, X. (2021). Cadmium uptake and translocation: Selenium and silicon roles in Cd detoxification for the production of low Cd crops: A critical review. *Chemosphere, 273*, 129690. doi:10.1016/j.chemosphere.2021.129690.

Rosado, D., Usero, J. é, & Morillo, J (2016). Ability of 3 extraction methods (BCR, Tessier and protease K) to estimate bioavailable metals in sediments from Huelva estuary (southwestern Spain). *Marine Pollution Bulletin, 102*(1), 65–71. doi:10.1016/j.marpolbul.2015.11.057.

Schludermann, C., Konecny, R., Laimgruber, S., Lewis, J. W., Schiemer, F., Chovanec, A., & Sures, B. (2003). Fish macroparasites as indicators of heavy metal pollution in river sites in Austria. *Parasitology., 126*(7), S61. doi:10.1017/s0031182003003743.

Shaghaleh, H., Rana, S., Zia-ur-Rehman, M., Usman, M., Ali, M., Alharby, H. F., Majrashi, A., Alamri, A. M., Abu Zeid, I. M., & Hamoud, Y. A. (2024). Bioassessment of Cd and Pb at multiple growth stages of wheat grown in texturally different soils using diffusive gradients in thin films and traditional extractants: A comparative study. *Plants, 13*(17). www.mdpi.com/journal/plants. doi:10.3390/plants13172445.

Sulkowski, M., & Hirner, A. V. (2006). Element fractionation by sequential extraction in a soil with high carbonate content. *Applied Geochemistry, 21*(1), 16–28. doi:10.1016/j.apgeochem.2005.09.016.

Sun, N., Shi, H., Li, X., Gao, C., & Liu, R. (2023). Combined toxicity of micro/nanoplastics loaded with environmental pollutants to organisms and cells: Role, effects, and mechanism. *Environment International, 171*, 107711. doi:10.1016/j.envint.2022.107711.

Temminghoff, E. J. M., Plette, A. C. C., Van Eck, R., & Van Riemsdijk, W. H. (2000). Determination of the chemical speciation of trace metals in aqueous systems by the Wageningen Donnan Membrane Technique. *Analytica Chimica Acta, 417*(2), 149–157. doi:10.1016/S0003-2670(00)00935-1.

Tessier, A., Campbell, P. G. C., & Bisson, M. (1979). Sequential extraction procedure for the speciation of particulate trace metals. *Analytical Chemistry, 51*(7), 844–851. doi:10.1021/ac50043a017.

Tong, F., Huang, Q., Liu, L., Fan, G., Shi, G., Lu, X., & Gao, Y. (2023). Interactive effects of inorganic–organic compounds on passivation of cadmium in weakly alkaline soil. *Agronomy, 13*(10), 2647. doi:10.3390/agronomy13102647.

Van Der Veeken, P. L. R., Pinheiro, J. P., & Van Leeuwen, H. P. (2008). Metal speciation by DGT/DET in colloidal complex systems. *Environmental Science and Technology, 42*(23), 8835–8840. doi:10.1021/es801654s.

Wang, J., Wang, P. M., Gu, Y., Kopittke, P. M., Zhao, F. J., & Wang, P. (2019). Iron-manganese (oxyhydro)oxides, rather than oxidation of sulfides, determine mobilization of Cd during soil drainage in paddy soil systems. *Environmental Science and Technology, 53*(5), 2500–2508. http://pubs.acs.org/journal/esthag. doi:10.1021/acs.est.8b06863.

Wieczorek, J., Baran, A., Urbański, K., Mazurek, R., & Klimowicz-Pawlas, A. (2018). Assessment of the pollution and ecological risk of lead and cadmium in soils. *Environmental Geochemistry and Health, 40*(6), 2325–2342. doi:10.1007/s10653-018-0100-5.

Wu, Z., Jiang, Q. i, Yan, T., Zhang, X., Xu, S., Shi, H., Deng, T., Li, F., Du, Y., Du, R., Hu, C., Wang, X., & Wang, F. (2020). Ammonium nutrition mitigates cadmium toxicity in rice (*Oryza sativa* L.) through improving antioxidase system and the glutathione-ascorbate cycle efficiency. *Ecotoxicology and Environmental Safety, 189*, 110010. doi:10.1016/j.ecoenv.2019.110010.

Xu, M., Dai, W., Zhao, Z., Zheng, J., Huang, F., Mei, C., Huang, S., Liu, C., Wang, P., & Xiao, R. (2022). Effect of rice straw biochar on three different levels of Cd-contaminated soils: Cd availability, soil properties, and microbial communities. *Chemosphere, 301*, 134551. doi:10.1016/j.chemosphere.2022.134551.

Yan, Z., Ding, W., Xie, G., Yan, M., Li, J., Han, Y. A, Xiong, X., & Wang, C. (2023). Identification of cadmium phytoavailability in response to cadmium transformation and changes in soil pH and electrical conductivity. *Chemosphere, 342*, 140042. doi:10.1016/j.chemosphere.2023.140042.

Zhang, H., Zhao, F. J., Sun, B., Davison, W., & Mcgrath, S. P. (2001). A new method to measure effective soil solution concentration predicts copper availability to plants. *Environmental Science and Technology, 35*(12), 2602–2607. doi:10.1021/es000268q.

Zhang, T., Sun, Y., Parikh, S. J., Colinet, G., Garland, G., Huo, L., Zhang, N., Shan, H., Zeng, X., & Su, S. (2024). Water-fertilizer regulation drives microorganisms to promote iron, nitrogen and manganese cycling: A solution for arsenic and cadmium pollution in paddy soils. *Journal of Hazardous Materials, 477*, 135244. doi:10.1016/j.jhazmat.2024.135244.

Zhao, P., Huang, P., Yan, X., Chukwuma, A., Yang, S., Yang, Z., Li, H., & Yang, W. (2023). Inhibitory effect of exogenous mineral elements (Si, P, Zn, Ca, Mn, Se, Fe, S) on rice Cd accumulation and soil Cd bioavailability in Cd-contaminated farmlands: A meta-analysis. *Chemosphere, 343*, 140282. doi:10.1016/j.chemosphere.2023.140282.

CHAPTER

10

Molecular mechanisms of cadmium uptake and translocation in plants

Fang-Jie Zhao, Xin-Yuan Huang

College of Resources and Environmental Sciences, Nanjing Agricultural University, Nanjing, China

10.1 Introduction

Cadmium (Cd) is absorbed by plant roots and translocated to aerial parts through a series of coordinated processes. These include uptake into root cells, radial transport across root tissues, xylem loading, long-distance transport via the xylem and phloem, and eventual transfer into seeds or other storage organs (Clemens & Ma, 2016; Zhao & Wang, 2020). Throughout these processes, Cd repeatedly crosses cellular membranes, entering (influx) and exiting (efflux) various cell types. In addition, Cd is transported across membranes of subcellular organelles such as vacuoles, mitochondria, and chloroplasts. These movements are facilitated by membrane-bound transporter proteins. Understanding the mechanisms underlying Cd uptake and internal transport in plants is essential for developing strategies to either limit Cd accumulation in food crops—to protect human health—or to enhance Cd uptake for phytoremediation purposes (see Chapters 19 and 20).

Cd is primarily absorbed by roots in the form of divalent cation (Cd^{2+}). Although Cd is nonessential and toxic to plants, it can be taken up via transporters that normally mediate the uptake of essential metal ions with similar physicochemical properties, particularly those with comparable ionic radius and charge (Table 10.1). This similarity explains why transporters for micronutrients such as Fe^{2+}, Zn^{2+}, and Mn^{2+} also facilitate Cd^{2+} transport (Clemens & Ma, 2016; Zhao et al., 2022). Cd^{2+} shares similar ionic characteristics with Ca^{2+} as well (Table 10.1). However, because of its classification as a soft metal, Cd preferentially binds sulfur- and nitrogen-containing ligands, whereas Ca, a hard metal, primarily coordinates with oxygen-containing ligands (Williams, 2001). Competition between metal ions for the same transporter can occur, and the outcome depends on the transporter's affinity for each ion and their respective

Cadmium Contamination in Soils and Plants: Mechanisms and Mitigation.
DOI: https://doi.org/10.1016/B978-0-443-34104-5.00009-8

TABLE 10.1 Ionic radii of Cd and some other metal ions and their typical concentrations in soil solution.

Ion		Cd^{2+}	Zn^{2+}	Fe^{2+}	Mn^{2+}	Ca^{2+}
Radius (pm)		95	74	78	83	100
Typical concentrations in soil solution (μM)[a]	Oxic soil	0.001–1.0	0.2–50	0.5–5.0	0.4–500	100–5000
	Anoxic soil	0.0001–0.05	0.2–3.0	50–3000	10–1500	100–5000

[a] Including all soluble metal species in the soil solution. Fe concentration includes both ferric and ferrous Fe soluble in the soil solution. Based on the soil solution data of 12 paddy soils under flooded and nonflooded conditions from Wang et al., (2019).

concentrations. Since Cd concentrations in soil solutions are generally much lower than those of competing ions (Table 10.1), a transporter must exhibit relatively high affinity for Cd^{2+} to meaningfully contribute to its uptake.

This chapter focuses on the molecular mechanisms governing Cd uptake and translocation in plants. It discusses spatial expression patterns of genes involved in Cd transport, regulatory responses to micronutrient deficiencies and Cd exposure, subcellular localization of transporter proteins, and phenotypic effects resulting from gene knockout or overexpression. It also briefly covers natural genetic variation among cultivars or accessions that influences Cd accumulation, providing insight into breeding strategies for reduced Cd uptake or enhanced phytoremediation.

10.2 Approaches to dissect genetic and molecular mechanisms of cadmium transport

10.2.1 Genetic dissection of cadmium accumulation traits

Two primary approaches are commonly employed to identify genes underlying specific traits such as Cd accumulation: forward genetics and reverse genetics (Fig. 10.1).

- **Forward genetics** begins with phenotypic variation—such as differences in Cd accumulation among different germplasm of a plant species or between a mutant and its wild type—and proceeds to identify the causative gene responsible for that variation.
- **Reverse genetics**, in contrast, starts with a candidate gene and examines the phenotypic effects resulting from its disruption or altered expression.

Cd accumulation can vary substantially among different cultivars or accessions of the same species [e.g., (Duan et al., 2017; Ueno et al., 2009)]. For example, Duan et al. (2017) reported 10- to 32-fold differences in grain Cd concentrations among 471 high-yielding rice cultivars grown in the same paddy field. Genetic dissection of Cd accumulation trait can be conducted using mapping populations derived from **bi-parental crosses** between lines showing contrasting phenotypes (e.g., high vs. low Cd accumulation; see Fig. 10.1A). For instance, Ueno et al. (2010) mapped shoot Cd accumulation to a locus on chromosome 7 in a rice population derived from the high-Cd-accumulating cultivar Anjana Dhan and the low-Cd cultivar Nipponbare, identifying *OsHMA3* as the causative gene (see below for details).

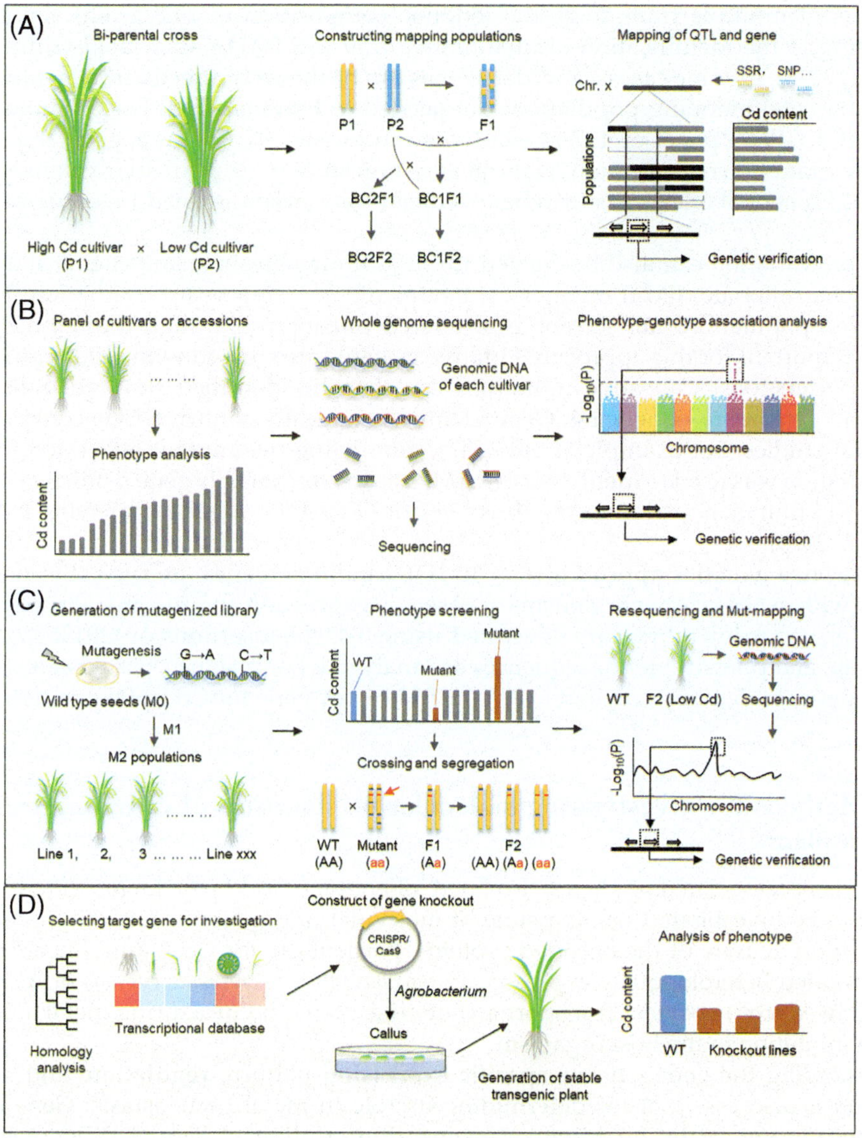

FIGURE 10.1 Approaches commonly used for genetic dissection of Cd-related phenotypes. (A–C) Forward genetics using bi-parental mapping population (A), genome-wide association mapping study (GWAS) (B), and mutagenesis and identification of causative gene for the mutant phenotype (C); (D) reverse genetics. *Diagram drawn by Dr. Han-Qing Wang.*

Another strategy for dissecting the genetic basis of Cd accumulation involves genome-wide association studies (GWAS) (Fig. 10.1B), which correlate phenotype with genotype across a large and genetically diverse panel of accessions. Genotypic data typically consist of single nucleotide polymorphisms or insertions/deletions obtained through genome resequencing. GWAS enables the analysis of multiple traits (e.g., the ionomic composition of rice grain) simultaneously, often leading to the identification of quantitative trait loci (QTL) associated with those traits (Yang et al., 2018). In some cases, candidate genes can be inferred directly from genomic data; in others, bi-parental mapping populations are needed to fine-map loci. For example, Yan et al. (2019) used a GWAS panel of 127 rice cultivars to identify *OsCd1* as a gene responsible for variation in grain Cd concentration. Natural variation studies are particularly valuable because they reveal functional alleles that can be leveraged for marker-assisted breeding of improved cultivars.

Genotypic variation can also be created through **mutagenesis**, using chemical agents (e.g., ethyl methanesulfonate, EMS) or physical treatments (e.g., ion beam irradiation; Fig. 10.1C). Mutagenized populations are screened for altered phenotypes; a high-throughput screening method is important for this approach. Putative mutants are crossed with wild type to generate mapping populations. Causative mutations can then be identified through bulk segregant analysis and genome resequencing. Gene identity is typically confirmed via complementation or knockout studies. For example, a low-Cd-accumulating rice mutant generated through ion beam irradiation was used to identify *OsNRAMP5* as the responsible gene (Ishikawa et al., 2012), whereas EMS mutagenesis led to the discovery of *OsCADT1*, a gene conferring Cd tolerance (Chen et al., 2020).

In the **reverse genetics approach** (Fig. 10.1D), candidate genes are typically inferred from homology with genes in other organisms or based on expression and functional annotation data. Knockout or knockdown lines are generated using T-DNA insertions or CRISPR/Cas9-based gene editing, and phenotypic consequences are analyzed. For instance, Sasaki et al. (2012) used reverse genetics to demonstrate that *OsNRAMP5* is a key gene for Cd uptake in rice roots.

10.2.2 Methods for investigating molecular mechanisms of cadmium transport in plants

Once a gene associated with a Cd-related phenotype has been identified, its biological function can be investigated using a suite of molecular and genetic tools. To characterize the metal transport activity of the encoded protein—particularly for Cd^{2+} and other metal ions—the gene is often heterologously expressed in *Saccharomyces cerevisiae* (yeast) or *Xenopus laevis* (African clawed frog) oocytes. Transport activity is assessed by measuring metal accumulation or growth inhibition in the host organism.

Understanding the gene's **tissue-specific expression pattern, regulation**, and **subcellular localization** is also essential for determining its role in metal homeostasis. Gene expression is commonly examined by quantitative reverse transcription PCR (qRT-PCR) or by creating transgenic plants expressing reporter genes such as β-glucuronidase or green fluorescent protein (GFP) under the control of the gene's promoter. Expression may be influenced by Cd exposure or by the supply of essential nutrients, which can provide insight into the gene's physiological relevance under different conditions.

Subcellular localization of the encoded transporter is determined by fusing the gene with a fluorescent reporter such as GFP and expressing the fusion construct in stably transformed plants, isolated protoplasts, or epidermal cells. Confocal laser scanning microscopy is used to observe the localization. Alternatively, localization can be assessed using antibodies specific to the transporter or the reporter protein.

The most definitive test of a gene's involvement in Cd transport is functional analysis through **gene knockout or knockdown**, followed by quantitative measurements of Cd uptake, distribution, and accumulation in comparison with wild-type plants. Knockout typically results in complete loss of gene function, either through frame-shift mutations, premature stop codons, or a total loss of promoter activity. Knockdown refers to reduced expression, often achieved via RNA interference (RNAi) or insertion of T-DNA in the promoter region of the target gene. CRISPR/Cas9 genome editing now enables efficient, targeted knockout of specific genes, provided that the plant species is amenable to genetic transformation.

In some studies, **overexpression** strategies are used to assess gene function. This involves expressing the gene under a strong constitutive or tissue-specific promoter, which can result in ectopic expression—that is, expression in tissues where the gene is not normally active. However, such expression may not fully reflect the gene's native function, and care must be taken when interpreting results. Importantly, multiple independent transgenic lines should be generated—whether knockout, knockdown, or overexpression lines—to account for potential off-target effects of transformation.

For physiologically and environmentally relevant conclusions, Cd uptake and distribution should be tested under realistic Cd concentrations—typically submicromolar to low micromolar—in hydroponic systems. Field or greenhouse studies using naturally Cd-contaminated soils offer further validation of gene function under realistic agronomic and environmental conditions.

In some cases, **functional redundancy** between related genes may obscure phenotypic effects. When two or more genes perform overlapping functions, disruption of a single gene may not produce a detectable phenotype. In such cases, multiple gene knockouts may be required. Gene duplication and redundancy are especially common in polyploid species such as wheat (*Triticum aestivum*), which contain multiple homologous copies of many genes.

Using these tools, researchers have identified numerous membrane transporters and Cd-binding peptides involved directly or indirectly in Cd transport in plants—particularly in the model dicot *Arabidopsis thaliana* and the monocot crop rice (*Oryza sativa*) (Table 10.2). The functions of these transporters and peptides are discussed in more detail in the following sections.

10.3 Cadmium uptake by roots

10.3.1 Cadmium uptake by dicotyledonous plants

In *A. thaliana* the iron (Fe^{2+}) transporter IRT1 also mediates the uptake of several other divalent cations, including Zn^{2+}, Mn^{2+}, and Cd^{2+} (Eide et al., 1996; Korshunova et al., 1999; Vert et al., 2002) (Fig. 10.2A). Knockout of the *IRT1* gene in *A. thaliana* results in markedly reduced Cd uptake under Fe-deficient conditions and confers increased tolerance to Cd stress (Vert et al., 2002). IRT1 belongs to the ZIP (ZRT–IRT-like protein) family and is localized to the plasma membrane. Its expression is restricted to the outer root cell layers and is strongly induced under Fe deficiency. This explains the observed increase in Cd uptake under Fe-deficient conditions in *A. thaliana* (Rodecap et al., 1994; Xu et al., 2024). In wild-type *A. thaliana*, short-term Cd uptake by roots under normal Fe supply is significantly higher than that in the *irt1* mutant, which shows only about 60% of the uptake rate. After a 10-day Fe starvation treatment, short-term Cd uptake increased by 11.5-fold in wild-type plants but only by 75% in the *irt1* mutant. Under these

TABLE 10.2 Membrane transporters or Cd-binding peptides involved in Cd uptake, sequestration, and translocation in plants (McLaughlin et al., 2021).

Transporter or peptide	Plant species	Functions	References
AtIRT1	*Arabidopsis thaliana*	Uptake of Fe(II), Zn, Mn, and Cd into root cells under Fe-deficient conditions	Connolly et al. (2002), Eide et al. (1996), Korshunova et al. (1999), Vert et al. (2002)
AtNRAMP1	*A. thaliana*	Uptake of Fe(II), Mn, and possibly Cd into root cells under Fe-replete conditions	Cailliatte et al. (2010), Castaings et al. (2016)
OsNRAMP5	Rice	Uptake of Mn and Cd into root cells	Ishikawa et al. (2012), Ishimaru et al. (2012), Sasaki et al. (2012), Yang et al. (2014)
OsNRAMP1	Rice	Uptake of Mn and Cd into root cells	Chang et al. (2020)
OsCd1	Rice	Uptake of Cd, and possibly also Mn, into root cells	Yan et al. (2019)
OsZIP5, OsZIP9	Rice	Uptake of Zn and Cd into root cells	Tan et al. (2020)
OsCAX2	Rice	Uptake of Cd and Ca into root cells	Liu et al. (2025)
HvNRAMP5	Barley	Uptake of Mn and Cd into root cells	Wu et al. (2016)
AtPDR8 (AtABCG36)	*A. thaliana*	Efflux of Cd out of root cells for tolerance	Kim et al. (2007)
OsABCG36	Rice	Efflux of Cd out of root cells for tolerance	Fu et al. (2019)
AtHMA3	*A. thaliana*	Tonoplast transporter for Cd, Zn, Co, and Pb	Chao et al. (2012), Morel et al. (2009)
OsHMA3	Rice	Tonoplast transporter for Cd, Zn	Liu et al. (2020), Miyadate et al. (2011), Sui et al. (2019), Ueno et al. (2010), Yan et al. (2016)
GmHMA3	Soybean	Tonoplast transporter for Cd, Zn	Wang et al. (2012)
TdHMA3	Durum wheat	Tonoplast transporter for Cd, Zn	Maccaferri et al. (2019)
BrHMA3	*Brassica rapa*	Tonoplast transporter for Cd	Zhang et al. (2019)
AtABCC1, AtABCC2	*A. thaliana*	Tonoplast transporters for phytochelatin–Cd conjugates	Park et al. (2012)
AtHMA2	*A. thaliana*	Efflux transporter for xylem loading of Zn and Cd	Hussain et al. (2004), Wong and Cobbett (2009)

(*continued on next page*)

TABLE 10.2 Membrane transporters or Cd-binding peptides involved in Cd uptake, sequestration, and translocation in plants (McLaughlin et al., 2021)—cont'd

Transporter or peptide	Plant species	Functions	References
AtHMA4	*A. thaliana*	Efflux transporter for xylem loading of Zn and Cd	Chen et al. (2018), Hussain et al. (2004), Mills et al. (2005), Mills et al. (2003), Verret et al. (2004), Wong and Cobbett (2009)
OsHMA2	Rice	Efflux (?) transporter mediating root-to-shoot translocation and intervascular transfer of Zn and Cd in the nodes	Satoh-Nagasawa et al. (2012), Takahashi et al. (2012), Yamaji et al. (2013)
HvHMA2	Barley	Possible involvement in xylem loading of Zn and Cd	Mills et al. (2012)
OsZIP2	Rice	Influx transporter for Cd mediating root-to-shoot translocation and intervascular transfer of Cd in the nodes	Li et al. (2024)
OsZIP7	Rice	Influx (?) transporter mediating root-to-shoot translocation and intervascular transfer of Zn and Cd in the nodes	Tan et al. (2019)
OsLCT1	Rice	Efflux transporter mediating intervascular transfer of Cd	Uraguchi et al. (2011)
OsCCX2	Rice	Possible efflux transporter mediating intervascular transfer of Cd	Hao et al. (2018)
OsCAL1	Rice	Defensin-like protein binds Cd and facilitates its secretion to the apoplast in the root stele.	Luo et al. (2018)
Defensin 8 (OsDEF8)	Rice	Similar to OsCAL1 but is involved in the phloem transport of Cd to rice grain	Gu et al. (2023)
AtHIPP20, AtHIPP21, AtHIPP22	*A. thaliana*	Binding of Cd, affecting Cd tolerance and accumulation in roots	Tehseen et al. (2010)
OsHIPP16	Rice	Possible binding of Cd, affecting Cd tolerance and accumulation	Cao et al. (2022)
OsHIPP9	Rice	Binding of Cd and Cu, affecting Cd distribution between leaves and panicles	Xiong et al. (2023)

Cd, cadmium. Modified from McLaughlin et al. (2021) with permission.

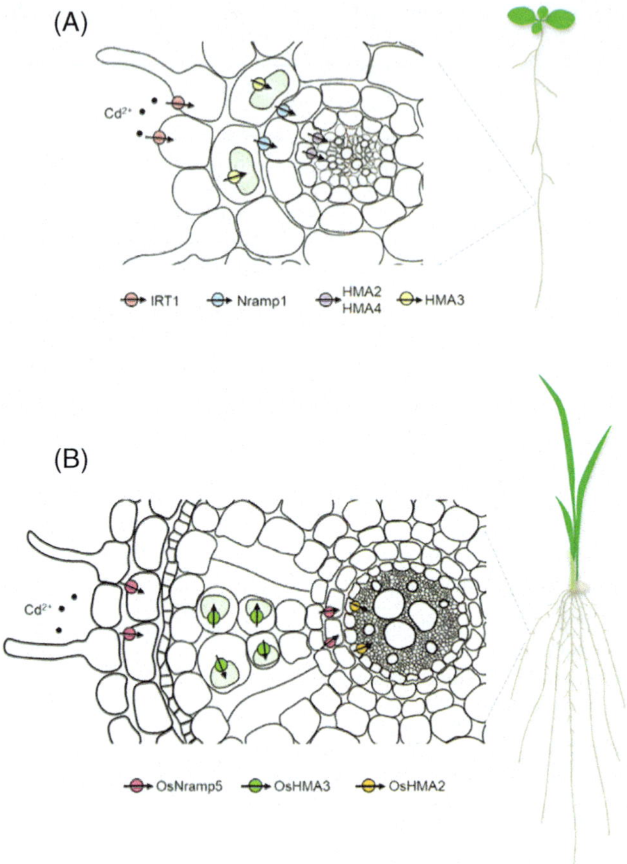

FIGURE 10.2 Key transporters for uptake, xylem loading, and vacuole sequestration of Cd in *Arabidopsis thaliana* (A) and rice (B). The organelles in *light green* are vacuoles. *Diagram drawn by Dr. Han-Qing Wang.*

conditions, Cd uptake by *irt1* was less than 10% of that in wild-type plants, clearly indicating that IRT1 is the major transporter responsible for Cd uptake under Fe-deficient conditions (Xu et al., 2024).

Iron deficiency has similarly been shown to enhance Cd uptake in other dicotyledonous species, including pea, peanut, lettuce, pak choi, and Chinese cabbage. This enhancement is likely because of upregulation of *IRT1* homologs in the roots (Cohen et al., 1998; Su et al., 2014; Xu et al., 2024). For instance, after 10 days of Fe deprivation, *IRT1* expression levels in the roots of lettuce, pak choi, and Chinese cabbage increased significantly, and short-term Cd uptake was elevated by 1.8-, 4.9-, and 9.2-fold, respectively, compared to Fe-sufficient controls (Xu et al., 2024). These results highlight the importance of adequate Fe supply in limiting Cd accumulation in dicot crops. In calcareous soils with neutral or alkaline pH, where Fe availability is often low, the application of Fe-chelate fertilizers has been shown to significantly reduce shoot Cd accumulation in these vegetables (Xu et al., 2024).

Besides IRT1, other transporters may contribute to Cd uptake, particularly under Fe-sufficient conditions when *IRT1* expression is low. For example, *A. thaliana* NRAMP1 (natural resistance–associated macrophage protein 1) is a plasma membrane-localized transporter essential for

Mn^{2+} uptake under low Mn conditions (Cailliatte et al., 2010). Mutations in *NRAMP1* enhance Cd tolerance, suggesting reduced Cd uptake via this transporter (Fig. 10.2A). Zinc transporters may also play a role in Cd uptake; however, conclusive evidence is lacking, possibly because of functional redundancy among multiple Zn transporter genes.

10.3.2 Cadmium uptake by monocotyledonous plants

Although rice possesses *IRT1* and *IRT2* genes, knockout of either gene has minimal impact on Cd uptake under normal growth conditions, indicating that these transporters do not play a major role in Cd uptake in rice roots (Chang et al., 2023; Xu et al., 2024). Instead, Cd uptake in rice is primarily mediated by OsNRAMP5, a member of the NRAMP family (Ishikawa et al., 2012; Sasaki et al., 2012; Yang et al., 2014) (Fig. 10.2B). OsNRAMP5 is capable of transporting Mn^{2+} and Cd^{2+} and also exhibits transport activity for Pb^{2+} and Co^{2+} (Chang et al., 2022; Huang et al., 2025). *OsNRAMP5* is highly expressed in rice roots, and its expression is largely unaffected by external Mn or Cd concentrations, though it is upregulated under Fe deficiency to some extent (Sasaki et al., 2012; Yang et al., 2014). The protein is polarly localized to the distal side of the plasma membrane in the exodermal and endodermal cells, facilitating efficient uptake (influx) of Mn and Cd from the rhizosphere into the root symplast. In short-term uptake assays, knockout of *OsNRAMP5* almost completely abolished Mn and Cd absorption by the roots (Sasaki et al., 2012). When grown in Cd-contaminated soil, *osnramp5* mutants accumulated <0.05 mg Cd kg^{-1} in the grain—compared to 1.73 mg Cd kg^{-1} in wild-type plants—representing a 97% reduction in grain Cd concentration (Ishikawa et al., 2012). Short-term uptake kinetic studies revealed that the apparent K_m of OsNRAMP5 is 1.08 μM for Mn and 0.38 μM for Cd, indicating that the transporter has a higher affinity for Cd (Sasaki et al., 2012). However, in flooded paddy soils, Mn concentrations are typically much higher than Cd concentrations (Table 10.1). Under Mn-deficient conditions (<2 μM), *osnramp5* mutants show severe Mn deficiency and growth inhibition. This phenotype can be rescued under higher Mn supply (>8 μM), likely through compensation by other low-affinity Mn transporters (Yang et al., 2014).

Kuramata et al. (2022) identified a mutant harboring a glutamine to lysine substitution at the 337th position (Q337K) in OsNRAMP5, which resulted in reduced Mn and Cd transport activity. In field trials, this mutant accumulated approximately 50% less Cd and 30% less Mn in the grain compared to the wild type, without compromising grain yield. Remarkably, this mutant tolerated Mn concentrations as low as 0.1 μM in nutrient solution without visible deficiency symptoms.

HvNRAMP5, the barley ortholog of OsNRAMP5, also plays a major role in Mn and Cd uptake (Wu et al., 2016). It is predominantly expressed in the outer root layers and encodes a plasma membrane-localized transporter. RNAi knockdown of *HvNRAMP5* significantly reduced both Mn and Cd uptake. The key transporters for Cd uptake in wheat remain largely unknown.

In addition to OsNRAMP5, several other transporters—including OsNRAMP1, OsCd1, OsZIP5, OsZIP9, and OsCAX2—also contribute to Cd uptake in rice, albeit to a lesser extent under normal conditions (Chang et al., 2020; Tan et al., 2020; Yan et al., 2019). OsNRAMP1, like OsNRAMP5, transports Mn^{2+} and Cd^{2+} but is expressed at lower levels and, unlike OsNRAMP5, lacks polar localization (Chang et al., 2020). Its expression is induced by both Cd exposure and Fe deficiency, suggesting that its contribution to Cd uptake may become larger under high levels of Cd exposure or Fe deficiency. Knockout of *OsNRAMP1* reduces Cd uptake and accumulation in the grain.

OsZIP5 and OsZIP9 are members of the ZIP family and transport both Zn^{2+} and Cd^{2+}. Under Zn-deficient conditions, their expression is strongly upregulated, and their contribution to Cd uptake becomes more pronounced (Chang et al., 2023). Knockout of *OsZIP9* alone or of both *OsZIP9* and *OsZIP5*, but not of *OsZIP5* alone, significantly reduced grain Cd concentration, suggesting that OsZIP9 plays a greater role in Cd uptake by rice roots (Tan et al., 2020).

OsCd1 belongs to the major facilitator superfamily and also mediates Cd and possibly Mn uptake. Knockout of *OsCd1* led to 20%–50% reductions in root and shoot Cd levels and a 15%–30% decrease in grain Cd concentration. A single amino acid polymorphism at position 449 (aspartate vs. valine) influences Cd transport efficiency, with the Asp variant, which is more common in indica cultivars, showing higher Cd uptake (Yan et al., 2019).

OsCAX2 is a cation/H^+ exchanger of the Ca^{2+}/cation antiporter family. The gene is predominantly expressed in lateral roots, and the protein is localized to the plasma membrane. Knockout of *OsCAX2* significantly reduced Cd uptake and accumulation in shoots and grain (Liu et al., 2025).

Although the transporters described above are all **influx** transporters responsible for Cd entry into root cells, plants also possess **efflux** transporters that pump Cd out of root cells to reduce toxicity. These include AtPDR8 (also known as AtABCG36) in *A. thaliana* and OsABCG36 in rice (Fu et al., 2019; Kim et al., 2007). These adenosine triphosphate (ATP)-binding cassette (ABC) transporters are upregulated in response to Cd exposure. Knockdown of *AtPDR8* increased sensitivity to Cd and Pb, while knockout of *OsABCG36* specifically reduced Cd tolerance in rice. However, *OsABCG36* knockout did not affect Cd accumulation in shoots, indicating that its role is primarily in conferring tolerance to high external Cd levels.

10.4 Transport of cadmium into vacuoles

After Cd enters root cells, a portion is transported into vacuoles—a key detoxification mechanism that reduces Cd toxicity and limits its translocation to aerial parts. Vacuolar sequestration is primarily mediated by HMA3 (heavy metal ATPase 3), a member of the P1B-type ATPase family. HMA3 is localized to the tonoplast (vacuolar membrane) and uses the energy from ATP hydrolysis to actively pump Cd^{2+} and other divalent metal ions from the cytosol into vacuoles. This mechanism is conserved across diverse plant species, including *A. thaliana*, rice (*O. sativa*), durum wheat (*Triticum turgidum* ssp. *durum*), barley (*Hordeum vulgare*), maize (*Zea mays*), soybean (*Glycine max*), and Chinese cabbage (*Brassica rapa*). HMA3 is a key target for breeding low-Cd-accumulating crop varieties through marker-assisted selection or genetic engineering (see Chapter 19).

In *A. thaliana*, AtHMA3 mediates transport of Cd, Zn, Co, and Pb into vacuoles (Gravot et al., 2004; Morel et al., 2009) (Fig. 10.2A). Loss-of-function mutations of *AtHMA3* increase plant sensitivity to Cd and Zn, whereas overexpression enhances tolerance to multiple heavy metals. Overexpression of *AtHMA3* results in a two- to three-fold increase in root Cd accumulation. Natural variation in Cd accumulation among *A. thaliana* accessions is largely explained by polymorphisms in the *AtHMA3* coding sequence. Haplotypes of AtHMA3 with weak or no Cd transport activity are less effective at sequestering Cd in roots, resulting in higher Cd accumulation in leaves (Chao et al., 2012).

In rice, OsHMA3 performs a similar function—sequestering Cd and Zn into vacuoles (Cai et al., 2019; Miyadate et al., 2011; Sasaki et al., 2014; Ueno et al., 2010; Yan et al., 2016) (Fig. 10.2B). It is predominantly expressed in roots. Knockout or knockdown of *OsHMA3* leads to heightened Cd sensitivity and increased root-to-shoot translocation of Cd. Conversely, overexpression enhances Cd tolerance and dramatically reduces shoot Cd accumulation (Lu et al., 2019; Sasaki et al., 2014; Ueno et al., 2010). Allelic variation in *OsHMA3* among rice cultivars strongly correlates with differences in shoot and grain Cd accumulation. Weak or nonfunctional alleles of *OsHMA3* encode proteins with low Cd transport activity, leading to inefficient vacuolar sequestration and higher Cd translocation. In some cases a single amino acid substitution significantly alters transporter activity (Miyadate et al., 2011; Ueno et al., 2010; Yan et al., 2016), while in others, deletions—such as a 14-amino-acid loss in the ATP-binding domain—completely abolish function (Sui et al., 2019). Strong alleles confer effective Cd sequestration and lower shoot accumulation and can be used for breeding low Cd cultivars. While OsHMA3 transports both Cd and Zn into vacuoles, overexpression markedly decreases Cd accumulation in shoots and grain but has little impact on Zn accumulation (Lu et al., 2019; Sasaki et al., 2014). The explanation for the contrasting effects is that cellular Zn homeostasis is tightly controlled; increased vacuolar sequestration of Zn leads to upregulation of several genes involved in Zn uptake and translocation.

In durum wheat a major QTL (*Cdu-B1*) on chromosome 5B accounts for over 80% of the phenotypic variation in grain Cd concentration (Wiebe et al., 2010). This variation reflects differences in root-to-shoot translocation, and *TdHMA3-B1*, the ortholog of *OsHMA3*, was identified as the causal gene (Maccaferri et al., 2019). Two alleles exist: The functional *TdHMA3-B1a*, which can transport Cd^{2+} and Zn^{2+}, and the nonfunctional *TdHMA3-B1b*, which cannot. The frequency of the nonfunctional allele increased during domestication, partly explaining the high Cd accumulation in modern durum cultivars.

In maize, *ZmHMA3* was identified through GWAS and bi-parental mapping as a key gene controlling Cd accumulation in seeds (Tang et al., 2021). Among five haplotypes of ZmHMA3 in 454 inbred lines, those with transposon insertions or deletions within *ZmHMA3* showed loss of Cd transport function and higher seed Cd concentrations.

In soybean a major QTL explains 57% of the variation in seed Cd content between high- and low-Cd cultivars (Jegadeesan et al., 2010). The causal gene, *GmHMA3*, contains a single amino acid polymorphism at position 608: the low-Cd allele has glycine, whereas the high-Cd allele has glutamic acid. The glycine variant shows Cd^{2+} and Zn^{2+} transport activities when expressed in yeast, whereas the glutamic acid variant does not (Wang et al., 2012).

In *B. rapa*, which includes widely consumed vegetables such as Chinese cabbage and pak choi, variation in the coding sequence in *BrHMA3* explains over 80% of the variation in Cd translocation from roots to shoots (Zhang et al., 2019). Five full-length and 4 truncated protein haplotypes exist among 64 accessions. Only full-length variants conferred Cd transport activity in yeast. Accessions with truncated variants exhibit higher root-to-shoot Cd translocation and shoot Cd concentrations (Zhang et al., 2019).

Most of the variation described above is because of differences in the **coding sequence** of HMA3, which affects transport activity for Cd^{2+} and other metal ions. However, differences in **promoter activity** can also influence HMA3 function. Liu et al. (2020) found that a promoter variation in *OsHMA3* between the low-Cd cultivar PA64s and the high-Cd cultivar 93-11 results in differential expression. Introducing the PA64s allele into 93-11 reduced grain Cd concentration by 37% without affecting agronomic performance.

In barley, differences in grain Cd content between high- and low-Cd cultivars were linked to *HvHMA3* expression rather than protein function (Lei et al., 2020). The low-Cd cultivar had approximately twice the root expression level of *HvHMA3*. This was attributed to a 3.3-kb Sukkula-like transposable element inserted upstream of the gene, functioning as an additional promoter. Introgression of this element into an elite cultivar decreased grain Cd accumulation under field conditions without yield penalty.

In *A. thaliana*, two tonoplast-localized ABC transporters—AtABCC1 and AtABCC2—also contribute to vacuolar Cd sequestration, likely transporting Cd–phytochelatin (PC) complexes (Park et al., 2012). These proteins also transport arsenite–PC complexes (Song et al., 2010). Double mutants (*atabcc1* and *atabcc2*) showed hypersensitivity to Cd, and Cd was largely retained in the cytosol of protoplasts, rather than being sequestered in vacuoles. Notably, these studies used the Columbia-0 ecotype, which carries a hypofunctional *AtHMA3* allele (Morel et al., 2009), suggesting that ABCC-mediated detoxification is more critical when HMA3 function is weak. In rice, which typically has a functional *OsHMA3*, knockout of *OsABCC1* impaired arsenic tolerance but had little effect on Cd accumulation, consistent with HMA3 playing the dominant role in Cd detoxification (Song et al., 2014).

10.5 Translocation of cadmium from roots to shoots

The root-to-shoot translocation of Cd is a critical determinant of Cd accumulation in aerial plant parts, including edible organs such as grains. The initial step in this process is the loading of Cd into the xylem vessels within the roots, from where it is transported upward by the transpiration stream.

Xylem loading involves the efflux of Cd from the pericycle and xylem parenchyma cells in the root stele into the xylem vessels. This process requires active transport mediated by efflux transporters. In *A. thaliana*, two P1B-type ATPases—AtHMA2 and AtHMA4—are the principal transporters responsible for this function (Hussain et al., 2004; Mills et al., 2003; Verret et al., 2004; Wong & Cobbett, 2009) (Fig. 10.2A). Both genes are highly expressed in the vascular tissues of roots, stems, and leaves. Using radioactive ^{109}Cd tracing, Wong and Cobbett (2009) showed that knockout of *AtHMA2* alone had minimal impact on Cd translocation, whereas knockout of *AtHMA4* reduced Cd movement to the shoots by 40%. The double knockout of *AtHMA2* and *AtHMA4* nearly abolished Cd translocation (98% reduction), indicating a degree of functional redundancy, with *AtHMA4* playing the more dominant role. These findings imply that, under environmentally relevant Cd concentrations (e.g., 0.5 μM), translocation is almost entirely dependent on these two HMA transporters. The findings also imply that the contribution of the apoplastic pathway—where Cd might move through the apoplast to the xylem in the root regions where the Casparian strip has not yet formed or is broken by lateral root branching—is negligible under these conditions. The apoplastic pathway may become more significant at higher external Cd concentrations or where the Casparian strip is disrupted (Lux et al., 2010). The double *hma2 hma4* knockout mutants also display pronounced Zn deficiency symptoms because of impaired Zn translocation to shoots (Hussain et al., 2004), whereas overexpression of *AtHMA4* enhances shoot accumulation of both Zn and Cd (Verret et al., 2004).

In tobacco, which possesses two *HMA4* homologs, knockout of both genes led to a >90% reduction in Cd translocation but also severely inhibited plant growth because of Zn deficiency, indicating the crucial role of the HMA4 transporter in loading Cd and Zn into the xylem (Liedschulte et al., 2017).

In rice, OsHMA2 is the functional ortholog involved in the translocation of both Zn and Cd (Chang et al., 2023; Satoh-Nagasawa et al., 2012; Takahashi et al., 2012; Yamaji et al., 2013) (Fig. 10.2B). It is expressed in roots and nodes, with the encoded protein localized to the plasma membrane of pericycle cells in the roots and phloem cells in the nodes (Yamaji et al., 2013). Mutations of *OsHMA2* significantly reduce root-to-shoot translocation of both Zn and Cd, with a more pronounced effect on Cd (Chang et al., 2023; Satoh-Nagasawa et al., 2012). There is some debate regarding the direction of transport mediated by OsHMA2. Some studies propose that OsHMA2 acts as an efflux transporter, loading Zn and Cd into the xylem (Satoh-Nagasawa et al., 2012; Takahashi et al., 2012), whereas others suggest it functions as an influx transporter for metal uptake into pericycle and phloem cells for redistribution (Yamaji et al., 2013). Based on its classification as a P-type ATPase, which hydrolyzes ATP on the cytosolic side to pump metal ions out of the cytoplasm, OsHMA2 is more likely to function as an efflux transporter (Sterckeman & Thomine, 2020). This role is consistent with the observed decrease in root-to-shoot Cd and Zn translocation in knockout mutants.

Interestingly, increasing Zn supply often enhances Cd translocation in rice, seemingly contradicting the notion of competition between Zn and Cd for OsHMA2. One explanation is that high levels of Zn supply upregulate *OsHMA2* expression in roots, increasing the abundance of the transporter and allowing more Cd to be loaded into the xylem, particularly if the transporter shows a higher affinity for Cd than for Zn (Chang et al., 2023). Unlike AtHMA4, which transports Zn and Cd, OsHMA4 in rice is a Cu transporter localized to the tonoplast and is not involved in Cd transport (Huang et al., 2016).

In barley and wheat the HMA2 orthologs—HvHMA2 and TaHMA2, respectively—are also implicated in Cd and Zn translocation (Mills et al., 2012; Tan et al., 2013). However, functional evidence is based primarily on heterologous expression studies in yeast, *Arabidopsis*, or rice, and additional data on expression patterns, subcellular localization, and knockout phenotypes are still lacking.

Several ZIP family members also participate in Cd translocation. For example, OsZIP2 is a Cd transporter in rice, strongly expressed in the parenchyma cells of the stele and the enlarged vascular bundles (EVBs) in nodes (Li et al., 2024). Localized to the plasma membrane, OsZIP2 functions to retrieve Cd from the apoplast into parenchyma cells, thereby limiting its loading into the xylem. Knockout of *OsZIP2* enhances root-to-shoot Cd translocation, whereas overexpression reduces it.

OsZIP7 is another ZIP transporter with activity for both Zn and Cd. The protein localizes to the plasma membrane of pericycle and vascular parenchyma cells (Tan et al., 2019). Its expression is strongly induced by Zn deficiency. Knockout of *OsZIP7* reduces root-to-shoot translocation of both Cd and Zn. Although OsZIP7 is suggested to act as an influx transporter, this is inconsistent with the observed phenotype of reduced shoot Cd accumulation upon gene knockout.

The role of OsNRAMP5 in Cd uptake has been described in the previous section. Unexpectedly, overexpression of *OsNRAMP5* driven by strong and constitutive promoters of *OsActin1* or maize *Ubiquitin* increases Cd and Mn uptake by roots but decreases Cd accumulation in shoots, while having minimal effect on Mn distribution (Chang et al., 2020). This is likely because of disruption of the protein's normal polar localization pattern by strong and ectopic expression driven by constitutive promoters, impairing radial Cd transport into the stele. Using forward genetics, Yu et al. (2022) discovered a tandem duplication of *OsNRAMP5* in the low-Cd-accumulating rice cultivar Pokkali. This duplication enhances *OsNRAMP5* expression and uptake of Cd and Mn by roots but reduces Cd loading into the xylem, possibly because of competitive inhibition by Mn. The duplicated allele holds promise for breeding low-Cd rice cultivars (see Chapter 19).

In addition to transporter proteins, some nontransporter proteins also influence Cd translocation. Two defensin-like proteins, OsCAL1 and OsDEF8, are expressed in the xylem parenchyma and pericycle cells of rice roots. These proteins chelate Cd in vitro and promote its efflux to the apoplast for translocation (Gu et al., 2023; Luo et al., 2018). Although OsCAL1 influences shoot Cd accumulation without affecting grain Cd levels, OsDEF8 is involved in both root-to-shoot and phloem-mediated translocation of Cd to the grain.

Chelation of Cd by peptides such as PCs may also affect Cd mobility. *A. thaliana* mutants deficient in PC synthesis exhibit increased root-to-shoot Cd translocation, indicating that PC–Cd complexes are less mobile (Wong & Cobbett, 2009). In contrast, expression of wheat *TaPCS1* in PC-deficient *A. thaliana* increased Cd translocation (Gong et al., 2003). In rice, *OsPCS1* mutants accumulated less Cd in leaves and grain, suggesting that PCs may enhance Cd mobility in this species (Uraguchi et al., 2017).

Heavy metal–associated isoprenylated plant proteins (HIPPs) also contribute to Cd detoxification and transport. HIPPs are metallochaperones capable of binding metals such as Cu, Cd, and Pb. In *A. thaliana*, the *hipp20/21/22* triple mutant shows heightened Cd sensitivity and reduced Cd accumulation (Tehseen et al., 2010). In rice, *OsHIPP16* is expressed in roots and leaf vasculature and is induced by Cd exposure. Knockout of *OsHIPP16* increases Cd sensitivity and leaf Cd levels, whereas overexpression reduces grain Cd (Cao et al., 2022). *OsHIPP9*, which binds both Cu and Cd, is expressed in the xylem region of EVBs. Its knockout results in greater Cd accumulation in upper nodes and panicles but less in leaves (Xiong et al., 2023).

10.6 Translocation of cadmium to seeds

Seeds accumulate both photosynthetic products and mineral nutrients that are delivered from vegetative tissues via the phloem. Using the radioactive ^{109}Cd tracer, it has been estimated that 91%–100% of the Cd found in rice grains is delivered through the phloem (Tanaka et al., 2007). Positron-emitting tracer imaging further demonstrated that ^{107}Cd applied to rice roots reaches the panicles more rapidly than it appears in the leaf blades (Fujimaki et al., 2010). The ^{107}Cd tracer accumulates intensively in rice nodes, which serve as translocation hubs regulating the distribution of both nutrients and Cd to aerial tissues such as leaves and panicles. Nodes in cereal crops are structurally adapted to facilitate this process, containing various types of EVBs and diffuse vascular bundles (DVBs). EVBs connect to the vascular systems of lower nodes and leaves, whereas DVBs surround the EVBs and connect to upper nodes or panicles (Yamaji & Ma, 2017) (Fig. 10.3). Cd and nutrients are transported upward via xylem in the EVBs to a node, where part of the flow is diverted into DVBs and then redistributed to the upper node or panicle—a process termed **intervascular transfer**. During this transfer, solutes are released from the xylem, absorbed by surrounding parenchyma cells, and loaded into companion cells and phloem sieve elements for delivery to the panicle, a process termed xylem-to-phloem transfer (Yamaji & Ma, 2017). This mechanism explains the predominance of phloem-mediated Cd delivery to rice grains. A similar phenomenon is observed in potatoes, where Cd accumulates

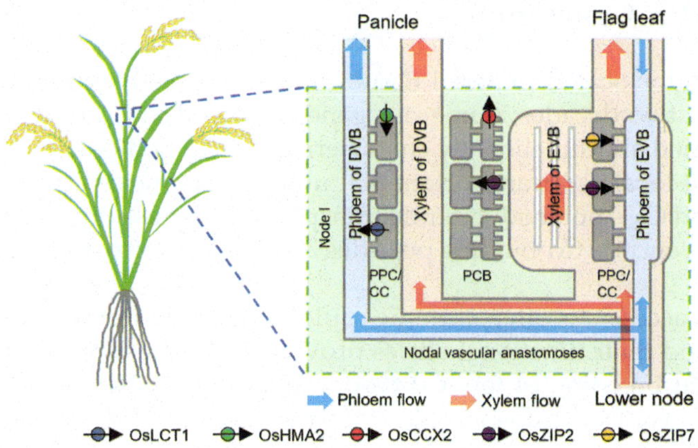

FIGURE 10.3 Key transporters for intervascular transfer of Cd in the nodes of rice. *CC*, Companion cell; *DVB*, diffuse vascular bundle; *EVB*, enlarged vascular bundle; *PCB*, parenchyma cell bridge; *PPC*, phloem parenchyma cell. *Modified from Li et al. (2024) with permission.*

in tubers primarily through phloem translocation from aboveground tissues, rather than direct uptake by stolons or tubers (Reid et al., 2003).

Several membrane transporters are essential for intervascular and xylem-to-phloem transfer of Cd in rice nodes (Fig. 10.3):

- **OsLCT1** (low-affinity cation transporter 1): The first identified transporter involved in Cd intervascular transfer, *OsLCT1* is expressed in both EVBs and DVBs (Uraguchi et al., 2011). It likely mediates Cd efflux from parenchyma cells. RNAi knockdown of *OsLCT1* significantly reduced Cd concentration in phloem sap and halved grain Cd content compared to wild-type plants, without affecting levels of essential nutrients (Ca, Mg, Mn, and K) or grain yield—making it a promising target for breeding low-Cd rice cultivars (see Chapter 19).
- **OsHMA2, OsZIP2, OsZIP7, and OsCCX2**: These transporters (except *OsCCX2*) are expressed in both the root stele (see Section 10.5) and nodal vascular bundles.
- OsHMA2 is localized to the phloem regions of both EVBs and DVBs (Yamaji et al., 2013). Its knockout leads to substantial reductions in Zn and Cd accumulation in panicles and grains but also impairs seed set and grain yield, likely because of Zn deficiency.
- *OsZIP7* is expressed in nodal parenchyma cells and similarly contributes to Zn and Cd transport. Knockout of *OsZIP7* reduces both metal accumulation in grains and overall yield (Tan et al., 2019).
- *OsZIP2* is highly expressed in the vascular bundles of nodes and internodes, especially EVBs (Li et al., 2024). Knockout of *OsZIP2* significantly reduces grain Cd while increasing Cd in flag leaves, without compromising yield.
- *OsCCX2*, a cation/Ca exchanger, is expressed in the xylem region of EVBs. The protein is localized to the plasma membrane, and its knockout greatly reduces grain Cd levels (Hao et al., 2018), suggesting a role in intervascular Cd transfer.
- OsHIPP9, a metallochaperone capable of binding Cu and Cd, *OsHIPP9* is expressed in the xylem regions of EVBs within the nodes (Xiong et al., 2023). Knockout mutants show increased Cd accumulation in upper nodes and panicles but reduced Cd levels in expanded leaves, indicating that OsHIPP9 affects Cd partitioning within the shoot.

10.7 Conclusion

Cd is absorbed by plant roots and translocated to aerial tissues, including seeds, through multiple sequential steps. Over the past two decades, numerous genes involved in Cd uptake and internal transport have been identified and functionally characterized. Most of these genes encode plasma membrane- or tonoplast-localized transporters that mediate the movement of essential nutrients such as Fe, Mn, Zn, and Ca—but inadvertently transport Cd because of shared physicochemical properties. Other genes encode Cd-binding peptides involved in detoxification and regulation of Cd distribution.

Allelic variation among cultivars and accessions often underlies differences in Cd uptake and accumulation. Identifying such genes and their functional alleles provides valuable targets for genetic engineering and marker-assisted breeding of low-Cd crop varieties (see Chapter 19). Understanding the molecular mechanisms of Cd transport also informs agronomic strategies, such as applying Mn-rich soil amendments to suppress Cd uptake in rice (Fang et al., 2021) or using Fe-chelates to prevent Fe deficiency and reduce Cd accumulation in dicotyledonous vegetables (Xu et al., 2024) (see Chapter 18).

To date, mechanistic insights into Cd uptake and translocation have focused primarily on *A. thaliana* and rice, owing to their tractable genomes and genetic tools. Future research should expand to other major crops, including wheat, maize, and barley. In addition, deeper investigation is needed into the regulatory networks that coordinate Cd uptake and transport at the transcriptional, translational, and posttranslational levels.

10.8 Summary

- Cd accumulation in plants involves sequential processes: uptake into roots, vacuolar sequestration, xylem loading, and xylem-to-phloem transfer.
- Cd is inadvertently transported via nutrient metal transporters because of similarities in ion properties and imperfect substrate specificity.
- Knowledge of Cd transport mechanisms supports agronomic strategies to limit Cd accumulation in edible crop parts.
- Natural variation in gene sequence can cause variation in Cd uptake and translocation among cultivars and accessions.
- Identified transporters and regulatory genes are valuable targets for genetic modification and marker-assisted breeding to develop low-Cd-accumulating crop varieties.

Acknowledgments

This work was funded by the National Natural Science Foundation of China (no. 42430704).

AI disclosure

During the preparation of this work the author(s) used ChatGPT to polish the English language. After using this tool/service the author(s) reviewed and edited the content as needed and take(s) full responsibility for the content of the publication.

References

Cai, H., Huang, S., Che, J., Yamaji, N., & Ma, J. F. (2019). The tonoplast-localized transporter OsHMA3 plays an important role in maintaining Zn homeostasis in rice. *Journal of Experimental Botany, 70*(10), 2717–2725. http://jxb.oxfordjournals.org/. doi:10.1093/jxb/erz091.

Cailliatte, R., Schikora, A., Briat, J. F., Mari, S., & Curie, C. (2010). High-affinity manganese uptake by the metal transporter nramp1 is essential for *Arabidopsis* growth in low manganese conditions. *Plant Cell, 22*(3), 904–917. http://www.plantcell.org/cgi/reprint/22/3/904. doi:10.1105/tpc.109.073023.

Cao, H. W., Zhao, Y. N., Liu, X. S., Rono, J. K., & Yang, Z. M. (2022). A metal chaperone OsHIPP16 detoxifies cadmium by repressing its accumulation in rice crops. *Environmental Pollution, 311*. https://www.journals.elsevier.com/environmental-pollution. doi:10.1016/j.envpol.2022.120058.

Castaings, L., Caquot, A., Loubet, S. té, & Curie, C. (2016). The high-affinity metal transporters NRAMP1 and IRT1 team up to take up iron under sufficient metal provision. *Scientific Reports, 6*(1). doi:10.1038/srep37222.

Chang, J. D., Huang, S., Wiseno, I., Sui, F. Q., Feng, F., Zheng, L., Ma, J. F., & Zhao, F. J. (2023). Dissecting the promotional effect of zinc on cadmium translocation from roots to shoots in rice. *Journal of Experimental Botany, 74*(21), 6790–6803. http://jxb.oxfordjournals.org/. doi:10.1093/jxb/erad330.

Chang, J. D., Huang, S., Yamaji, N., Zhang, W., Ma, J. F., & Zhao, F. J. (2020). OsNRAMP1 transporter contributes to cadmium and manganese uptake in rice. *Plant Cell and Environment, 43*(10), 2476–2491. http://onlinelibrary.wiley.com/journal/10.1111/(ISSN)1365-3040. doi:10.1111/pce.13843.

Chang, J. D., Gao, W., Wang, P., & Zhao, F. J. (2022). OsNRAMP5 is a major transporter for lead uptake in rice. *Environmental Science and Technology, 56*(23), 17481–17490. http://pubs.acs.org/journal/esthag. doi:10.1021/acs.est.2c06384.

Chang, J. D., Huang, S., Konishi, N., Wang, P., Chen, J., Huang, X. Y., Ma, J. F., & Zhao, F. J. (2020). Overexpression of the manganese/cadmium transporter OsNRAMP5 reduces cadmium accumulation in rice grain. *Journal of Experimental Botany, 71*(18), 5705–5715. http://jxb.oxfordjournals.org/. doi:10.1093/jxb/eraa287.

Chao, D. Y., Silva, A., Baxter, I., Huang, Y. S., Nordborg, M., Danku, J., Lahner, B., Yakubova, E., & Salt, D. E. (2012). Genome-wide association studies identify heavy metal ATPase3 as the primary determinant of natural variation in leaf cadmium in *Arabidopsis thaliana*. *PLoS Genetics, 8*(9). http://www.plosgenetics.org/article/fetchObjectAttachment.action?uri=info%3Adoi%2F10.1371%2Fjournal.pgen.1002923&representation=PDF. doi:10.1371/journal.pgen.1002923.

Chen, J., Huang, X. Y., Salt, D. E., & Zhao, F. J. (2020). Mutation in OsCADT1 enhances cadmium tolerance and enriches selenium in rice grain. *New Phytologist, 226*(3), 838–850. http://onlinelibrary.wiley.com/journal/10.1111/(ISSN)1469-8137. doi:10.1111/nph.16404.

Chen, Z. R., Kuang, L., Gao, Y. Q., Wang, Y. L., Salt, D. E., & Chao, D. Y. (2018). AtHMA4 drives natural variation in leaf Zn concentration of *Arabidopsis thaliana*. *Frontiers in Plant Science, 9*. http://journal.frontiersin.org/article/10.3389/fpls.2018.00270/full. doi:10.3389/fpls.2018.00270.

Clemens, S., & Ma, J. F. (2016). Toxic heavy metal and metalloid accumulation in crop plants and foods. *Annual Review of Plant Biology, 67*, 489–512. http://arjournals.annualreviews.org/loi/arplant. doi:10.1146/annurev-arplant-043015-112301.

Cohen, C. K., Fox, T. C., Garvin, D. F., & Kochian, L. V. (1998). The role of iron-deficiency stress responses in stimulating heavy-metal transport in plants. *Plant Physiology, 116*(3), 1063–1072. http://www.plantphysiol.org/. doi:10.1104/pp.116.3.1063.

Connolly, E. L., Fett, J. P., & Guerinot, M. L. (2002). Expression of the IRT1 metal transporter is controlled by metals at the levels of transcript and protein accumulation. *Plant Cell, 14*(6), 1347–1357. http://www.plantcell.org/. doi:10.1105/tpc.001263.

Duan, G., Shao, G., Tang, Z., Chen, H., Wang, B., Tang, Z., Yang, Y., Liu, Y., & Zhao, F. J. (2017). Genotypic and environmental variations in grain cadmium and arsenic concentrations among a panel of high yielding rice cultivars. *Rice, 10*(1). http://www.springerlink.com/content/1939-8425. doi:10.1186/s12284-017-0149-2.

Eide, D., Broderius, M., Fett, J., & Guerinot, M. L. (1996). A novel iron-regulated metal transporter from plants identified by functional expression in yeast. *Proceedings of the National Academy of Sciences of the United States of America, 93*(11), 5624–5628. doi:10.1073/pnas.93.11.5624.

Fang, X., Wang, J., Chen, H., Christl, I., Wang, P., Kretzschmar, R., & Zhao, F. J. (2021). Two-year and multi-site field trials to evaluate soil amendments for controlling cadmium accumulation in rice grain. *Environmental Pollution, 289*, 117918. doi:10.1016/j.envpol.2021.117918.

Fu, S., Lu, Y., Zhang, X., Yang, G., Chao, D., Wang, Z., Shi, M., Chen, J., Chao, D. Y., Li, R., Ma, J. F., & Xia, J. (2019). The ABC transporter ABCG36 is required for cadmium tolerance in rice. *Journal of Experimental Botany, 70*(20), 5909–5918. http://jxb.oxfordjournals.org/. doi:10.1093/jxb/erz335.

Fujimaki, S., Suzui, N., Ishioka, N. S., Kawachi, N., Ito, S., Chino, M., & Nakamura, S. i. (2010). Tracing cadmium from culture to spikelet: Noninvasive imaging and quantitative characterization of absorption, transport, and accumulation of cadmium in an intact rice plant. *Plant Physiology, 152*(4), 1796–1806. http://www.plantphysiol.org/cgi/reprint/152/4/1796. doi:10.1104/pp.109.151035.

Gong, J. M., Lee, D. A., & Schroeder, J. I. (2003). Long-distance root-to-shoot transport of phytochelatins and cadmium in *Arabidopsis*. *Proceedings of the National Academy of Sciences of the United States of America, 100*(17), 10118–10123. doi:10.1073/pnas.1734072100.

Gravot, A., Lieutaud, A., Verret, F., Auroy, P., Vavasseur, A., & Richaud, P. (2004). AtHMA3, a plant P 1B -ATPase, functions as a Cd/Pb transporter in yeast. *FEBS Letters, 561*(1-3), 22–28. doi:10.1016/s0014-5793(04)00072-9.

Gu, T. Y., Qi, Z. A., Chen, S. Y., Yan, J., Fang, Z. J., Wang, J. M., & Gong, J. M. (2023). Dual-function DEFENSIN 8 mediates phloem cadmium unloading and accumulation in rice grains. *Plant Physiology, 191*(1), 515–527. https://academic.oup.com/plphys/article/191/1/515/6695224. doi:10.1093/plphys/kiac423.

Hao, X., Zeng, M., Wang, J., Zeng, Z., Dai, J., Xie, Z., Yang, Y., Tian, L., Chen, L., & Li, D. (2018). A node-expressed transporter OsCCX2 is involved in grain cadmium accumulation of rice. *Frontiers in Plant Science, 9*. doi:10.3389/fpls.2018.00476.

Huang, H., Yamaji, N., Huang, S., & Ma, J. F. (2025). Uptake and accumulation of cobalt is mediated by OsNramp5 in rice. *Plant, Cell and Environment, 48*(1), 3–14. http://onlinelibrary.wiley.com/journal/10.1111/(ISSN)1365-3040. doi:10.1111/pce.15130.

Huang, X. Y., Deng, F., Yamaji, N., Pinson, S. R. M., Fujii-Kashino, M., Danku, J., Douglas, A., Guerinot, M. L., Salt, D. E., & Ma, J. F. (2016). A heavy metal P-type ATPase OsHMA4 prevents copper accumulation in rice grain. *Nature Communications, 7*. http://www.nature.com/ncomms/index.html. doi:10.1038/ncomms12138.

Hussain, D., Haydon, M. J., Wang, Y., Wong, E., Sherson, S. M., Young, J., Camakaris, J., Harper, J. F., & Cobbett, C. S. (2004). P-type ATPase heavy metal transporters with roles in essential zinc homeostasis in *Arabidopsis*. *Plant Cell, 16*(5), 1327–1339. http://www.plantcell.org/. doi:10.1105/tpc.020487.

Ishikawa, S., Ishimaru, Y., Igura, M., Kuramata, M., Abe, T., Senoura, T., Hase, Y., Arao, T., Nishizawa, N. K., & Nakanishi, H. (2012). Ion-beam irradiation, gene identification, and marker-assisted breeding in the development of low-cadmium rice. *Proceedings of the National Academy of Sciences of the United States of America, 109*(47), 19166–19171. http://www.pnas.org/content/109/47/19166.full.pdf+html. doi:10.1073/pnas.1211132109.

Ishimaru, Y., Takahashi, R., Bashir, K., Shimo, H., Senoura, T., Sugimoto, K., Ono, K., Yano, M., Ishikawa, S., Arao, T., Nakanishi, H., & Nishizawa, N. K. (2012). Characterizing the role of rice NRAMP5 in Manganese, Iron and Cadmium Transport. *Scientific Reports, 2*. doi:10.1038/srep00286.

Jegadeesan, S., Yu, K., Poysa, V., Gawalko, E., Morrison, M. J., Shi, C., & Cober, E. (2010). Mapping and validation of simple sequence repeat markers linked to a major gene controlling seed cadmium accumulation in soybean [*Glycine max* (L.) Merr]. *Theoretical and Applied Genetics, 121*(2), 283–294. doi:10.1007/s00122-010-1309-6.

Kim, D. Y., Bovet, L., Maeshima, M., Martinoia, E., & Lee, Y. (2007). The ABC transporter AtPDR8 is a cadmium extrusion pump conferring heavy metal resistance. *Plant Journal, 50*(2), 207–218. doi:10.1111/j.1365-313X.2007.03044.x.

Korshunova, Y. O., Eide, D., Clark, W. G., Guerinot, M. L., & Pakrasi, H. B. (1999). The IRT1 protein from *Arabidopsis thaliana* is a metal transporter with a broad substrate range. *Plant Molecular Biology, 40*(1), 37–44. Www.wkap.nl/journalhome.htm/0167-4412. doi:10.1023/A:1026438615520.

Kuramata, M., Abe, T., Tanikawa, H., Sugimoto, K., & Ishikawa, S. (2022). A weak allele of OsNRAMP5 confers moderate cadmium and manganese transport activity in rice. *Journal of Experimental Botany, 73*, 6475–6489. https://doi.org/10.1093/jxb/erac302.

Lei, G. J., Fujii-Kashino, M., Wu, D. Z., Hisano, H., Saisho, D., Deng, F., Yamaji, N., Sato, K., Zhao, F. J., & Ma, J. F. (2020). Breeding for low cadmium barley by introgression of a Sukkula-like transposable element. *Nature Food, 1*(8), 489–499. https://www.nature.com/natfood/. doi:10.1038/s43016-020-0130-x.

Li, M. Z., Hu, D. W., Liu, X. Q., Zhang, R., Liu, H., Tang, Z., Zhao, F. J., & Huang, X. Y. (2024). The OsZIP2 transporter is involved in root-to-shoot translocation and intervascular transfer of cadmium in rice. *Plant Cell and Environment, 47*(10), 3865–3881. http://onlinelibrary.wiley.com/journal/10.1111/(ISSN)1365-3040. doi:10.1111/pce.14993.

Liedschulte, V., Laparra, H., Battey, J. N. D., Schwaar, J. D., Broye, H., Mark, R., Klein, M., Goepfert, S., & Bovet, L. (2017). Impairing both HMA4 homeologs is required for cadmium reduction in tobacco. *Plant Cell and Environment, 40*(3), 364–377. http://onlinelibrary.wiley.com/journal/10.1111/(ISSN)1365-3040. doi:10.1111/pce.12870.

Liu, C. L., Gao, Z. Y., Shang, L. G., Yang, C. H., Ruan, B. P., Zeng, D. L., Guo, L. B., Zhao, F. J., Huang, C. F., & Qian, Q. (2020). Natural variation in the promoter of OsHMA3 contributes to differential grain cadmium accumulation between indica and japonica rice. *Journal of Integrative Plant Biology, 62*(3), 314–329. http://onlinelibrary.wiley.com/journal/10.1111/(ISSN)1744-7909. doi:10.1111/jipb.12794.

Liu, X. Q., Liu, H., Fu, M. J., Zhang, L. W., Yin, S. F., Tang, Z., Zhao, F. J., & Huang, X. Y. (2025). The cation/H+ exchanger OsCAX2 is involved in cadmium uptake and contributes to differential grain cadmium accumulation between indica and japonica rice. *Journal of Hazardous Materials, 487*. https://www.sciencedirect.com/science/journal/03043894. doi:10.1016/j.jhazmat.2025.137252.

Lu, C., Zhang, L., Tang, Z., Huang, X. Y., Ma, J. F., & Zhao, F. J. (2019). Producing cadmium-free indica rice by overexpressing OsHMA3. *Environment International, 126*, 619–626. Www.elsevier.com/locate/envint. doi:10.1016/j.envint.2019.03.004.

Luo, J. S., Huang, J., Zeng, D. L., Peng, J. S., Zhang, G. B., Ma, H. L., Guan, Y., Yi, H. Y., Fu, Y. L., Han, B., Lin, H. X., Qian, Q., & Gong, J. M. (2018). A defensin-like protein drives cadmium efflux and allocation in rice. *Nature Communications, 9*(1). http://www.nature.com/ncomms/index.html. doi:10.1038/s41467-018-03088-0.

Lux, A., Martinka, M., Vaculik, M., & White, P. J. (2010). Root responses to cadmium in the rhizosphere: A review. *Journal of Experimental Botany, 62*(1), 21–37. doi:10.1093/jxb/erq281.

Maccaferri, M., Harris, N. S., Twardziok, S. O., Pasam, R. K., Gundlach, H., Spannagl, M., Ormanbekova, D., Lux, T., Prade, V. M., Milner, S. G., Himmelbach, A., Mascher, M., Bagnaresi, P., Faccioli, P., Cozzi, P., Lauria, M., Lazzari, B., Stella, A., Manconi, A., … Cattivelli, L. (2019). Durum wheat genome highlights past domestication signatures and future improvement targets. *Nature Genetics, 51*(5), 885–895. http://www.nature.com/ng/index.html. doi:10.1038/s41588-019-0381-3.

McLaughlin, M. J., Smolders, E., Zhao, F. J., Grant, C., & Montalvo, D. (2021). Managing cadmium in agricultural systems. *Advances in Agronomy, 166*, 1–129. http://www.elsevier.com/wps/find/bookdescription.cws_home/675816/description#description. doi:10.1016/bs.agron.2020.10.004.

Mills, R. F., Krijgert, G. C., Baccarini, P. J., Hall, J. L., & Williams, L. E. (2003). Functional expression of AtHMA4, a P1B-type ATPase of the Zn/Co/Cd/Pb subclass. *Plant Journal, 35*(2), 164–176. doi:10.1046/j.1365-313X.2003.01790.x.

Mills, R. F., Francini, A., Ferreira Da Rocha, P. S. C., Baccarini, P. J., Aylett, M., Krijger, G. C., & Williams, L. E. (2005). The plant P1B-type ATPase AtHMA4 transports Zn and Cd and plays a role in detoxification of transition metals supplied at elevated levels. *FEBS Letters, 579*(3), 783–791. doi:10.1016/j.febslet.2004.12.040.

Mills, R. F., Peaston, K. A., Runions, J., & Williams, L. E. (2012). HvHMA2, a P(1B)-ATPase from barley, is highly conserved among cereals and functions in Zn and Cd transport. *PloS One, 7*, e42640.

Miyadate, H., Adachi, S., Hiraizumi, A., Tezuka, K., Nakazawa, N., Kawamoto, T., Katou, K., Kodama, I., Sakurai, K., Takahashi, H., Satoh-Nagasawa, N., Watanabe, A., Fujimura, T., & Akagi, H. (2011). OsHMA3, a P 1B-type of ATPase affects root-to-shoot cadmium translocation in rice by mediating efflux into vacuoles. *New Phytologist, 189*(1), 190–199. doi:10.1111/j.1469-8137.2010.03459.x.

Morel, M., Crouzet, J., Gravot, A., Auroy, P., Leonhardt, N., Vavasseur, A., & Richaud, P. (2009). AtHMA3, a P1B-ATPase allowing Cd/Zn/Co/Pb vacuolar storage in *Arabidopsis*. *Plant Physiology, 149*(2), 894–904. http://www.plantphysiol.org/cgi/reprint/149/2/894. doi:10.1104/pp.108.130294.

Park, J., Song, W. Y., Ko, D., Eom, Y., Hansen, T. H., Schiller, M., Lee, T. G., Martinoia, E., & Lee, Y. (2012). The phytochelatin transporters AtABCC1 and AtABCC2 mediate tolerance to cadmium and mercury. *Plant Journal, 69*(2), 278–288. doi:10.1111/j.1365-313X.2011.04789.x.

Reid, R. J., Dunbar, K. R., & McLaughlin, M. J. (2003). Cadmium loading into potato tubers: The roles of the periderm, xylem and phloem. *Plant, Cell and Environment, 26*(2), 201–206. doi:10.1046/j.1365-3040.2003.00945.x.

Rodecap, K. D., Tingey, D. T., & Lee, E. H. (1994). Iron nutrition influence on cadmium accumulation by *Arabidopsis thaliana* (L.) Heynh. *Journal of Environmental Quality, 23*(2), 239–246. https://acsess.onlinelibrary.wiley.com/loi/15372537. doi:10.2134/jeq1994.00472425002300020004x.

Sasaki, A., Yamaji, N., Yokosho, K., & Ma, J. F. (2012). Nramp5 is a major transporter responsible for manganese and cadmium uptake in rice. *Plant Cell, 24*(5), 2155–2167. http://www.plantcell.org/content/24/5/2155.full.pdf+html. doi:10.1105/tpc.112.096925.

Sasaki, A., Yamaji, N., & Ma, J. F. (2014). Overexpression of OsHMA3 enhances Cd tolerance and expression of Zn transporter genes in rice. *Journal of Experimental Botany, 65*(20), 6013–6021. http://jxb.oxfordjournals.org/. doi:10.1093/jxb/eru340.

Satoh-Nagasawa, N., Mori, M., Nakazawa, N., Kawamoto, T., Nagato, Y., Sakurai, K., Takahashi, H., Watanabe, A., & Akagi, H. (2012). Mutations in rice (*Oryza sativa*) heavy metal ATPase 2 (OsHMA2) restrict the translocation of zinc and cadmium. *Plant and Cell Physiology, 53*(1), 213–224. doi:10.1093/pcp/pcr166.

Song, W. Y., Yamaki, T., Yamaji, N., Ko, D., Jung, K. H., Fujii-Kashino, M., An, G., Martinoia, E., Lee, Y., & Feng, J. (2014). A rice ABC transporter, OsABCC1, reduces arsenic accumulation in the grain. *Proceedings of the National Academy of Sciences of the United States of America, 111*(44), 15699–15704. http://www.pnas.org/content/111/44/15699.full.pdf+html. doi:10.1073/pnas.1414968111.

Song, W. Y., Park, J., Mendoza-Cózatl, D. G., Suter-Grotemeyer, M., Shima, D., Hörtensteiner, S., Geisler, M., Weder, B., Rea, P. A., Rentsch, D., Schroeder, J. I., Lee, Y., & Martinoia, E. (2010). Arsenic tolerance in *Arabidopsis* is mediated by two ABCC-type phytochelatin transporters. *Proceedings of the National Academy of Sciences of the United States of America, 107*(49), 21187–21192. http://www.pnas.org/content/107/49/21187.full.pdf+html. doi:10.1073/pnas.1013964107.

Sterckeman, T., & Thomine, S. é (2020). Mechanisms of cadmium accumulation in plants. *Critical Reviews in Plant Sciences, 39*(4), 322–359. doi:10.1080/07352689.2020.1792179.

Su, Y., Liu, J., Lu, Z., Wang, X., Zhang, Z., & Shi, G. (2014). Effects of iron deficiency on subcellular distribution and chemical forms of cadmium in peanut roots in relation to its translocation. *Environmental and Experimental Botany, 97*, 40–48. doi:10.1016/j.envexpbot.2013.10.001.

Sui, F., Zhao, D., Zhu, H., Gong, Y., Tang, Z., Huang, X. Y., Zhang, G., & Zhao, F. J. (2019). Map-based cloning of a new total loss-of-function allele of OsHMA3 causes high cadmium accumulation in rice grain. *Journal of Experimental Botany, 70*(10), 2857–2871. http://jxb.oxfordjournals.org/. doi:10.1093/jxb/erz093.

Takahashi, R., Ishimaru, Y., Shimo, H., Ogo, Y., Senoura, T., Nishizawa, N. K., & Nakanishi, H. (2012). The OsHMA2 transporter is involved in root-to-shoot translocation of Zn and Cd in rice. *Plant, Cell and Environment, 35*(11), 1948–1957. doi:10.1111/j.1365-3040.2012.02527.x.

Tan, J., Wang, J., Chai, T., Zhang, Y., Feng, S., Li, Y., Zhao, H., Liu, H., & Chai, X. (2013). Functional analyses of TaHMA2, a P(1B)-type ATPase in wheat. *Plant Biotechnology Journal, 11*(4), 420–431. doi:10.1111/pbi.12027.

Tan, L., Zhu, Y., Fan, T., Peng, C., Wang, J., Sun, L., & Chen, C. (2019). OsZIP7 functions in xylem loading in roots and inter-vascular transfer in nodes to deliver Zn/Cd to grain in rice. *Biochemical and Biophysical Research Communications, 512*(1), 112–118. doi:10.1016/j.bbrc.2019.03.024.

Tan, L., Qu, M., Zhu, Y., Peng, C., Wang, J., Gao, D., & Chen, C. (2020). ZINC TRANSPORTER5 and ZINC TRANSPORTER9 function synergistically in zinc/cadmium uptake. *Plant Physiology., 183*(3), 1235–1249. doi:10.1104/pp.19.01569.

Tanaka, K., Fujimaki, S., Fujiwara, T., Yoneyama, T., & Hayashi, H. (2007). Quantitative estimation of the contribution of the phloem in cadmium transport to grains in rice plants (*Oryza sativa* L.). *Soil Science and Plant Nutrition, 53*(1), 72–77. doi:10.1111/j.1747-0765.2007.00116.x.

Tang, B., Luo, M., Zhang, Y., Guo, H., Li, J., Song, W., Zhang, R., Feng, Z., Kong, M., Li, H., Cao, Z., Lu, X., Li, D., Zhang, J., Wang, R., Wang, Y., Chen, Z., Zhao, Y., Zhao, J., & Cuypers, A. (2021). Natural variations in the P-type ATPase heavy metal transporter gene ZmHMA3 control cadmium accumulation in maize grains. *Journal of Experimental Botany, 72*(18), 6230–6246. doi:10.1093/jxb/erab254.

Tehseen, M., Cairns, N., Sherson, S., & Cobbett, C. S. (2010). Metallochaperone-like genes in *Arabidopsis thaliana*. *Metallomics, 2*(8), 556–564. doi:10.1039/c003484c.

Ueno, D., Kono, I., Yokosho, K., Ando, T., Yano, M., & Ma, J. F. (2009). A major quantitative trait locus controlling cadmium translocation in rice (*Oryza sativa*). *New Phytologist, 182*(3), 644–653. doi:10.1111/j.1469-8137.2009.02784.x.

Ueno, D., Yamaji, N., Kono, I., Huang, C. F., Ando, T., Yano, M., & Ma, J. F. (2010). Gene limiting cadmium accumulation in rice. *Proceedings of the National Academy of Sciences of the United States of America, 107*(38), 16500–16505. http://www.pnas.org/content/107/38/16500.full.pdf+html. doi:10.1073/pnas.1005396107.

Uraguchi, S., Kamiya, T., Sakamoto, T., Kasai, K., Sato, Y., Nagamura, Y., Yoshida, A., Kyozuka, J., Ishikawa, S., & Fujiwara, T. (2011). Low-affinity cation transporter (OsLCT1) regulates cadmium transport into rice grains. *Proceedings of the National Academy of Sciences of the United States of America, 108*(52), 20959–20964. http://www.pnas.org/content/108/52/20959.full.pdf+html. doi:10.1073/pnas.1116531109Japan.

Uraguchi, S., Tanaka, N., Hofmann, C., Abiko, K., Ohkama-Ohtsu, N., Weber, M., Kamiya, T., Sone, Y., Nakamura, R., Takanezawa, Y., Kiyono, M., Fujiwara, T., & Clemens, S. (2017). Phytochelatin synthase has contrasting effects on cadmium and arsenic accumulation in rice grains. *Plant and Cell Physiology, 58*(10), 1730–1742. doi:10.1093/pcp/pcx114.

Verret, F. rédé, Gravot, A., Auroy, P., Leonhardt, N., David, P., Nussaume, L., Vavasseur, A., & Richaud, P. (2004). Overexpression of AtHMA4 enhances root-to-shoot translocation of zinc and cadmium and plant metal tolerance. *FEBS Letters, 576*(3), 306–312. doi:10.1016/j.febslet.2004.09.023.

Vert, G., Grotz, N., Dédaldéchamp, F., Gaymard, F., Guerinot, M. L., Briat, J. F., & Curie, C. (2002). IRT1, an *Arabidopsis* transporter essential for iron uptake from the soil and for plant growth. *Plant Cell, 14*(6), 1223–1233. https://academic.oup.com/plcell/issue. doi:10.1105/tpc.001388.

Wang, J., Wang, P. M., Gu, Y., Kopittke, P. M., Zhao, F. J., & Wang, P. (2019). Iron-manganese (oxyhydro)oxides, rather than oxidation of sulfides, determine mobilization of Cd during soil drainage in paddy soil systems. *Environmental Science and Technology, 53*(5), 2500–2508. https://doi.org/10.1021/acs.est.8b06863

Wang, Y., Yu, K. F., Poysa, V., Shi, C., & Zhou, Y. H. (2012). A single point mutation in GmHMA3 affects cadimum (Cd) translocation and accumulation in soybean seeds. *Molecular Plant, 5*(5), 1154–1156. doi:10.1093/mp/sss069.

Wiebe, K., Harris, N. S., Faris, J. D., Clarke, J. M., Knox, R. E., Taylor, G. J., & Pozniak, C. J. (2010). Targeted mapping of Cdu1, a major locus regulating grain cadmium concentration in durum wheat (*Triticum turgidum* L. var durum). *Theoretical and Applied Genetics, 121*(6), 1047–1058. doi:10.1007/s00122-010-1370-1.

Williams, R. J. P. (2001). Chemical selection of elements by cells. *Coordination Chemistry Reviews, 216–217*, 583–595. doi:10.1016/s0010-8545(00)00398-2.

Wong, C. K. E., & Cobbett, C. S. (2009). HMA P-type ATPases are the major mechanism for root-to-shoot Cd translocation in *Arabidopsis thaliana*. *New Phytologist, 181*(1), 71–78. doi:10.1111/j.1469-8137.2008.02638.x.

Wu, D., Yamaji, N., Yamane, M., Kashino-Fujii, M., Sato, K., & Ma, J. F. (2016). The HvNramp5 transporter mediates uptake of cadmium and manganese, but not iron. *Plant Physiology, 172*(3), 1899–1910. http://www.plantphysiol.org/content/172/3/1899.full.pdf. doi:10.1104/pp.16.01189.

Xiong, S., Kong, X., Chen, G., Tian, L., Qian, D., Zhu, Z., & Qu, L. Q. (2023). Metallochaperone OsHIPP9 is involved in the retention of cadmium and copper in rice. *Plant Cell and Environment, 46*(6), 1946–1961. http://onlinelibrary.wiley.com/journal/10.1111/(ISSN)1365-3040. doi:10.1111/pce.14576.

Xu, Z. R., You, T. T., Liu, W. Y., Ye, K., Zhao, F. J., & Wang, P. (2024). Mitigating cadmium accumulation in dicotyledonous vegetables by iron fertilizer through inhibiting Fe transporter IRT1-mediated Cd uptake. *Chemosphere, 346*. Www.elsevier.com/locate/chemosphere. doi:10.1016/j.chemosphere.2023.140559.

Yamaji, N., Xia, J., Mitani-Ueno, N., Yokosho, K., & Ma, J. F. (2013). Preferential delivery of zinc to developing tissues in rice is mediated by P-type heavy metal ATPase OsHMA2. *Plant Physiology, 162*(2), 927–939. http://www.plantphysiol.org/content/162/2/927.full.pdf+html. doi:10.1104/pp.113.216564. Japan.

Yamaji, N., & Ma, J. F. (2017). Node-controlled allocation of mineral elements in *Poaceae*. *Current Opinion in Plant Biology, 39*, 18–24. Www.elsevier.com/inca/publications/store/6/0/1/3/1/4/index.htt. doi:10.1016/j.pbi.2017.05.002.

Yan, J., Wang, P., Wang, P., Yang, M., Lian, X., Tang, Z., Huang, C. F., Salt, D. E., & Zhao, F. J. (2016). A loss-of-function allele of OsHMA3 associated with high cadmium accumulation in shoots and grain of Japonica rice cultivars. *Plant Cell and Environment, 39*(9), 1941–1954. http://www.blackwellpublishing.com/journals/PCE. doi:10.1111/pce.12747.

Yan, H. L., Xu, W. X., Xie, J. Y., Gao, Y. W., Wu, L. L., Sun, L., Feng, L., Chen, X., Zhang, T., Dai, C. H., Li, T., Lin, X., Zhang, Z. Y., Wang, X. Q., Li, F. M., Zhu, X. Y., Li, J. J., Li, Z., Chen, C. Y., … He, Z. Y. (2019). Variation of a major facilitator superfamily gene contributes to differential cadmium accumulation between rice subspecies. *Nature Communications, 10*(1). doi:10.1038/s41467-019-10544-y.

Yang, M., Lu, K., Zhao, F. J., Xie, W., Ramakrishna, P., Wang, G., Du, Q., Liang, L., Sun, C., Zhao, H., Zhang, Z., Liu, Z., Tian, J., Huang, X. Y., Wang, W., Dong, H., Hu, J., Ming, L., Xing, Y., … Lian, X. (2018). Genome-wide association studies reveal the genetic basis of ionomic variation in rice. *Plant Cell, 30*(11), 2720–2740. http://www.plantcell.org/content/30/11/2720.full.pdf. doi:10.1105/tpc.18.00375.

Yang, M., Zhang, Y., Zhang, L., Hu, J., Zhang, X., Lu, K., Dong, H., Wang, D., Zhao, F. J., Huang, C. F., & Lian, X. (2014). OsNRAMP5 contributes to manganese translocation and distribution in rice shoots. *Journal of Experimental Botany, 65*(17), 4849–4861. http://jxb.oxfordjournals.org/. doi:10.1093/jxb/eru259.

Yu, E., Wang, W., Yamaji, N., Fukuoka, S., Che, J., Ueno, D., Ando, T., Deng, F., Hori, K., Yano, M., Shen, R. F., & Ma, J. F. (2022). Duplication of a manganese/cadmium transporter gene reduces cadmium accumulation in rice grain. *Nature Food, 3*(8), 597–607. https://www.nature.com/natfood/. doi:10.1038/s43016-022-00569-w.

Zhang, L., Wu, J., Tang, Z., Huang, X. Y., Wang, X., Salt, D. E., & Zhao, F. J. (2019). Variation in the BrHMA3 coding region controls natural variation in cadmium accumulation in *Brassica rapa* vegetables. *Journal of Experimental Botany, 70*(20), 5865–5878. http://jxb.oxfordjournals.org/. doi:10.1093/jxb/erz310.

Zhao, F. J., Tang, Z., Song, J. J., Huang, X. Y., & Wang, P. (2022). Toxic metals and metalloids: Uptake, transport, detoxification, phytoremediation, and crop improvement for safer food. *Molecular Plant, 15*(1), 27–44. https://www.cell.com/molecular-plant/home. doi:10.1016/j.molp.2021.09.016.

Zhao, F. J., & Wang, P. (2020). Arsenic and cadmium accumulation in rice and mitigation strategies. *Plant and Soil, 446*(1-2), 1–21. doi:10.1007/s11104-019-04374-6.

CHAPTER

11

Toxicity and detoxification of cadmium in plants

Zhong Tang, Fang-Jie Zhao

College of Resources and Environmental Sciences, Nanjing Agricultural University, Nanjing, China

11.1 Introduction

Cadmium (Cd) is a nonessential element for nearly all living organisms, with the exception of certain marine diatoms. These diatoms can utilize Cd as a catalytic metal in Cd carbonic anhydrase, an enzyme that facilitates the reversible hydration of CO_2, thereby contributing to inorganic carbon acquisition for photosynthesis (Lane et al., 2005). In contrast, excessive Cd accumulation in plants poses a major agricultural challenge worldwide. Plants absorb Cd from soils through their roots, leading to its accumulation in various tissues (see Chapter 10), which can impair plant growth and reduce productivity. Cd toxicity disrupts multiple metabolic processes in plants and activates stress response pathways. High levels of Cd in plants also pose risks to animals and humans through the food chain. Plants have also evolved sophisticated detoxification mechanisms, and the threshold for Cd phytotoxicity varies widely among species, ecotypes, and cultivars. This chapter reviews the effects of Cd toxicity on plant growth, physiology, and metabolism and the mechanisms of Cd detoxification in plants.

11.2 Physiological and biochemical effects of cadmium toxicity in plants

11.2.1 Impact of cadmium on plant growth and development

Cd is highly toxic to most plants, with toxicity symptoms typically emerging when Cd concentrations in plant tissues reach 3–30 mg kg^{-1} dry weight depending on plant species (Guo et al., 2016; Lux et al., 2011). In some nonhyperaccumulator species, leaf Cd levels of just 5–10 mg Cd g^{-1} dry weight can be lethal (Haider et al., 2021). However, the threshold for Cd-induced phytotoxicity varies widely among plant species and depends heavily on experimental conditions (Kopittke et al., 2010). Cd hyperaccumulators can tolerate Cd up to hundreds to thousands of milligrams per kilogram without toxicity (see Chapter 12).

Cadmium Contamination in Soils and Plants: Mechanisms and Mitigation.
DOI: https://doi.org/10.1016/B978-0-443-34104-5.00011-6

In nutrient culture systems the median toxic Cd concentration in the nutrient solution is around 5.0 μM (Kopittke et al., 2010; Tang et al., 2023). Sensitivity to Cd varies significantly among plant species. For example, the Cd concentration required to reduce growth by 50% (EC_{50}) ranges from as low as 27 μM in *Lactuca sativa* to as high as 940 μM in *Acacia holosericea* (Lamb et al., 2010). Soil-based experiments also reveal substantial variation. A Cd concentration of 5 mg kg^{-1} in soil was reported as toxic to many plant species (He et al., 2017; Hernández-Allica et al., 2008; Van Slycken et al., 2013), with a general phytotoxic range between 1 and 10 mg kg^{-1}. However, it is the bioavailable Cd fraction—rather than total soil Cd—that primarily drives toxicity. Cd concentrations in soil solutions typically range from less than 0.1 to 5 μg L^{-1} but can exceed 10 μg L^{-1} in contaminated soil (Alloway, 2012; Huang et al., 2022; Norini et al., 2019). This variability is influenced by soil pH, organic matter content, and contamination levels.

Plant roots are the first tissues affected by Cd exposure, as roots generally accumulate more Cd than shoots (Xu et al., 2017; Zare et al., 2018). Cd toxicity severely inhibits root elongation and growth and alters root morphology in numerous species (Bruno et al., 2017; Chen et al., 2020; Daud et al., 2009; Wang et al., 2012). In *Arabidopsis thaliana*, Cd damages the quiescent center, disrupts postembryonic root development, and reduces primary root elongation by diminishing root meristem size—all linked to disrupted auxin homeostasis (Fattorini et al., 2017; Yuan & Huang, 2016). Prolonged Cd exposure can cause visible root damage, including mucilaginous, brown, rigid, and twisted main roots (Fig. 11.1) (Javed et al., 2017; Rascio et al., 2008). These changes are accompanied by abnormal cell division and the enlargement of epidermal and cortical cell layers in the root tip (Javed et al., 2017). Cd also inhibits lateral root formation by impairing meristem activity and disrupting the dynamics of the lateral root cap, leading to reduced periodicity of programmed cell death (PCD)-driven oscillations (Wang et al., 2021; Xie et al., 2019). Furthermore, Cd interferes with auxin distribution by downregulating the expression of *PIN-FORMED* (*PIN*) genes, which encode auxin efflux carriers. This suppression hampers the formation of lateral root primordia and the emergence of lateral roots in rice (Wang et al., 2021; Fig. 11.1).

Cd stress in plant leaves can cause a range of visible symptoms, including stunted growth, chlorosis, desiccation, and necrosis. These effects are often more pronounced in young leaves compared to older ones. Excessive Cd accumulation has been associated with significant reductions in total leaf area and the dry weight of various plant organs, including leaves, stems, roots, and seeds (Jinadasa et al., 2016; Rizwan et al., 2017).

Reproductive tissues are generally more sensitive to Cd toxicity than vegetative tissues, with male organs, such as pollen, showing greater vulnerability than female organs (Shirkhani et al., 2020). Because of its high mobility within the plant, Cd can be translocated to reproductive organs such as flowers and pollen, impairing their development and negatively affecting seed formation. Cd exposure significantly reduces pollen germination rates and inhibits pollen tube growth (Sabrine et al., 2010; Sawidis, 2008; Wang et al., 2014). For instance, in *Picea wilsonii*, Cd stress disrupted endomembrane organelles, inhibited endo/exocytosis, and induced the formation of acidic vacuoles, resulting in swollen pollen tube tips and irregular tube morphology (Wang et al., 2014). Similarly, in *Datura stramonium* L., Cd exposure led to decreased pollen viability and degeneration of embryo sac cells (Shirkhani et al., 2020). Despite these findings the effects of Cd toxicity on reproductive tissues, particularly on pollen and ovule development, remain insufficiently studied and warrant further investigation.

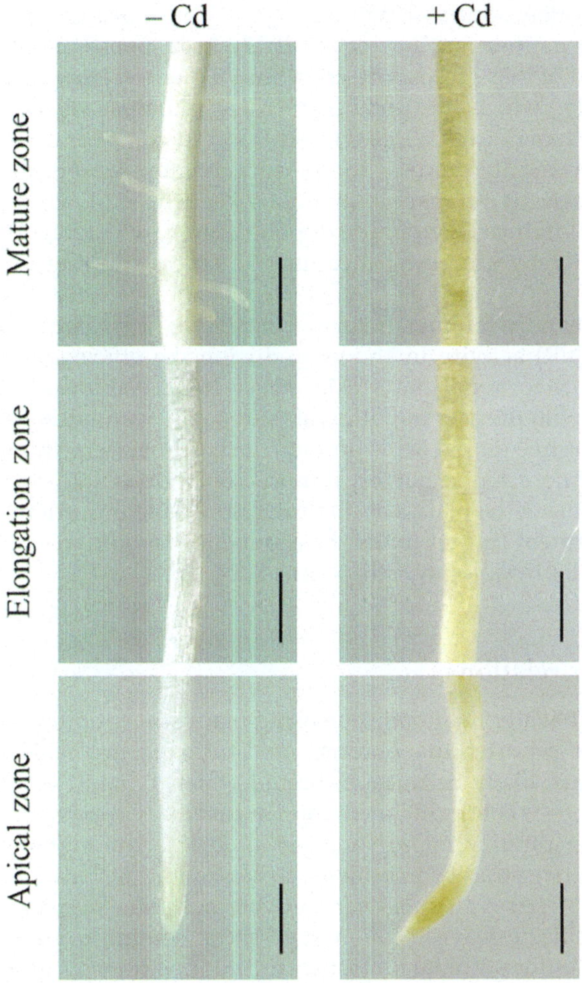

FIGURE 11.1 Cadmium exposure inhibits lateral root emergence and causes brown colorization in rice root. Rice roots were exposed to 1 μM Cd in half-strength Kimura B nutrient solution for 3 days, bar = 1 mm. *Cd*, Cadmium.

11.2.2 Disruption of nutrient uptake

One of the primary mechanisms by which Cd exerts its toxicity in plants is through the disruption of nutrient uptake and homeostasis. Although no Cd-specific transporter has been identified, Cd can enter plant roots by hitchhiking on transporters for the uptake of essential nutrients such as calcium (Ca), iron (Fe), manganese (Mn), and zinc (Zn) (Clemens et al., 2013; Zhao et al., 2022) (see Chapter 10). Key transporters implicated in Cd uptake include members of the ZIP/IRT (zinc-regulated transporter/iron-regulated transporter-like protein) family (Clemens & Ma, 2016) and NRAMP5 (natural resistance-associated macrophage protein 5); the latter is a plasma membrane transporter primarily responsible for Mn uptake (Sasaki et al., 2012; Sui et al., 2018; Tang et al., 2017). Cd can also be taken up in complexed forms such as Fe–Cd-siderophores or Cd-nicotianamine through Yellow Stripe1-Like (YSL) proteins

(Feng et al., 2017). As a result, Cd uptake often occurs at the expense of essential nutrients because of competition for shared transport pathways (Zhao et al., 2022). In response, plants may downregulate the expression or activity of these transporters to limit Cd entry. However, this defensive mechanism can inadvertently reduce the uptake of essential metals like Fe, further exacerbating nutrient deficiencies (Zhang et al., 2023b). In addition to competition at the transporter level, acute Cd stress can physically damage the plasma membrane, increasing its permeability and compromising its selective transport functions (Genchi et al., 2020). LI et al. (2012) demonstrated that Cd (50 μM) induces membrane depolarization, which blocks the inward conductance of Ca^{2+} and K^{+}, thereby impairing their uptake and disturbing ion homeostasis in rice.

Beyond metal ions, Cd toxicity also affects nitrogen metabolism. Cd inhibits nitrate uptake in roots and suppresses nitrate reductase (NR) activity in shoots, reducing the efficiency of nitrate assimilation and transport from roots to aerial parts (Ma et al., 2017; Wahid et al., 2007). Cd exposure has also been associated with the accumulation of endogenous ammonium, possibly because of increased deamination of free amino acids and enhanced protease activity (Chiraz et al., 2003; Khan et al., 2016). Furthermore, Cd negatively impacts root morphology and architecture, reducing root biomass and impairing the plant's ability to explore soil for nutrients (Haider et al., 2021). This root system impairment further limits the acquisition of both macro- and micronutrients, compounding the overall physiological stress caused by Cd.

11.2.3 Disturbance in the plant–water relations

Cd exposure can significantly impair plant–water relations, increasing plant susceptibility to water stress. For instance, Bekkai et al. (2024) reported that chronic, mild Cd exposure heightened dehydration sensitivity in tomato plants, likely because of altered water dynamics and impaired stomatal function. Similarly a study on wheat cultivars demonstrated that Cd exposure led to reduced stomatal conductance and transpiration rates (Naeem et al., 2018). Although the reduction in transpiration may act as a protective strategy to limit Cd uptake and translocation, it also compromises water use efficiency. Cd stress in roots can activate systemic signaling pathways that influence stomatal behavior in the leaves. This root-to-shoot communication, often mediated by abscisic acid (ABA), can induce stomatal closure and reduce transpiration as a coordinated defense response to Cd exposure (Deng et al., 2022; Pérez Chaca et al., 2014). In addition, Guo et al. (2023) observed that Cd exposure often leads to anatomical changes in stomata, including decreased stomatal area and increased stomatal density, both of which can impact gas exchange and water loss. The extent of stomatal response to Cd varies across plant species and cultivars. For example, Gálusová et al. (2020) found that different soybean cultivars exhibited varying degrees of stomatal sensitivity to Cd, indicating inherent genetic variability in tolerance mechanisms. At the cellular level, Cd may interfere with aquaporins—membrane proteins essential for water transport—further disrupting water uptake and distribution within the plant (Rucińska-Sobkowiak, 2016). In addition, Cd-induced alterations in root system architecture—such as reduced root length, surface area, and biomass—can significantly limit the plant's capacity to absorb water, exacerbating water stress (Zulfiqar et al., 2022). Overall, Cd exposure disrupts both anatomical and physiological aspects of water regulation in plants, affecting multiple processes from root water uptake to stomatal behavior and cellular water transport.

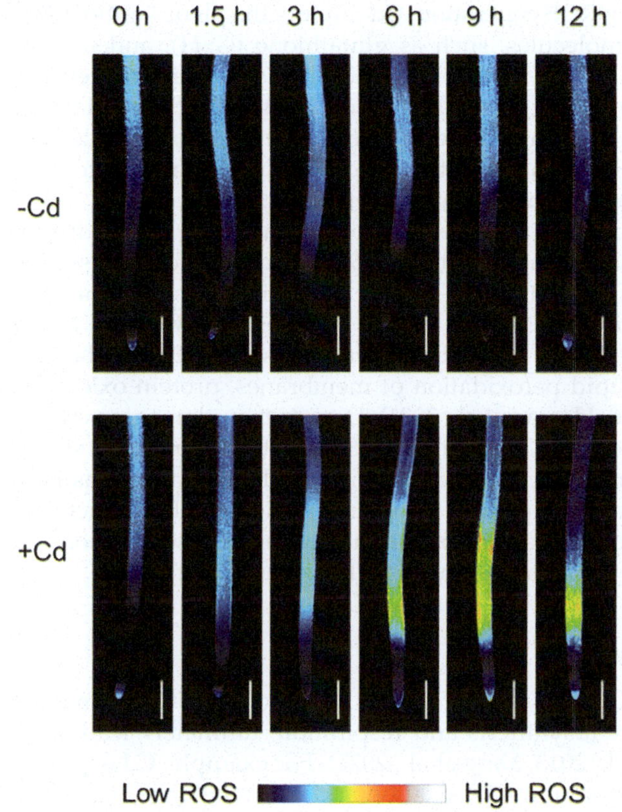

FIGURE 11.2 Time course of Cd-induced ROS production in rice root. Rice roots were exposed to 20 μM Cd in half-strength Kimura B nutrient solution for 12 hours (bottom panel). Top panel, control without Cd exposure. ROS was visualized by staining the roots with 10 μM 2′,7′-dichlorodihydrofluorescein diacetate (H_2DCFDA) in the dark for 15 minutes prior to fluorescence imaging.*Cd*, Cadmium; *ROS*, reactive oxygen species. *Modified from Wang et al. (2023). With copyright permission.*

11.2.4 Oxidative stress and generation of reactive oxygen species

Cd exposure induces oxidative stress in plants by triggering the excessive production of reactive oxygen species (ROS) through both direct and indirect mechanisms. Although Cd itself does not generate ROS via Fenton-like reactions—as it is not a redox-active metal—it disrupts cellular processes that lead to ROS accumulation. One primary pathway involves the disruption of the electron transport chains in chloroplasts and mitochondria (Srivastava et al., 2018). Chloroplasts, particularly during photosynthesis, are the main source of ROS in green tissues (Foyer & Hanke, 2022). Similarly, mitochondria contribute substantially to ROS generation during cellular respiration (Zhao et al., 2019). Cd-induced interference in these organelles results in electron leakage, leading to the formation of superoxide radicals and other ROS. Another major mechanism is the activation of NADPH oxidases, also known as respiratory burst oxidase homologs (RBOHs), which are plasma membrane-bound enzymes that play a central role in stress signaling (Gupta et al., 2017; Sachdev et al., 2021; Wang et al., 2023). These enzymes catalyze the production of superoxide ($O_2^{\bullet -}$) at the cell surface, which is then converted into other ROS such as hydrogen peroxide (H_2O_2) by superoxide dismutases (SODs). Recent studies have shown that Cd stress can differentially regulate various RBOH isoforms, enhancing ROS production in a tissue- and context-specific manner (Fig. 11.2) (Gupta et al., 2017;

Wang et al., 2023). In addition to promoting ROS generation, Cd impairs the plant's antioxidant defense system. It depletes antioxidant molecules, such as glutathione (GSH), and inhibits the activity of key antioxidant enzymes, thereby weakening the plant's capacity to detoxify ROS (Anjum et al., 2015; Kerchev & Van Breusegem, 2022). The extent of oxidative damage depends on the balance between ROS production and the efficiency of the antioxidant defense system (Unsal et al., 2020). When this balance is disrupted, oxidative stress can lead to lipid peroxidation, protein oxidation, DNA damage, and ultimately cell death.

The primary ROS generated under Cd stress include superoxide anion ($O_2^{\bullet -}$), hydrogen peroxide (H_2O_2), and hydroxyl radical ($^{\bullet}OH$). The extent of ROS production and the resulting oxidative stress can vary among plant tissues, with roots—being the initial site of Cd uptake—typically exhibiting more intense oxidative stress than shoots (Andresen & Küpper, 2013). ROS accumulation in response to Cd is generally dose-dependent. Excessive ROS can damage various cellular components, leading to lipid peroxidation of membranes, protein oxidation, DNA damage, and PCD (Genchi et al., 2020; Hassan et al., 2021). Despite their damaging effects, ROS also serve as important signaling molecules. Cd-induced ROS can activate a range of stress-responsive signaling pathways and defense mechanisms, illustrating their dual role as both toxic agents and essential regulators of stress adaptation (Cuypers et al., 2010). This functional complexity underscores the delicate balance plants must maintain between ROS production and scavenging to mitigate Cd-induced stress while preserving signaling functions.

11.2.5 Impairment of photosynthesis

Cd toxicity not only disrupts the uptake of essential nutrients but also significantly impairs plant carbon metabolism, particularly photosynthesis and respiration, ultimately leading to reduced plant growth (Sánchez-Pardo et al., 2013; Yang et al., 2022). For example, Cd exposure in lettuce (*L. sativa* L.) has been shown to inhibit CO_2 fixation and assimilation (Chen et al., 2022). Cd impairs photosynthesis through multiple, interconnected mechanisms that affect both the structure and function of the photosynthetic machinery. One of the earliest impacts is on chloroplast ultrastructure. Cd exposure alters chloroplast shape and size, causes thylakoid swelling, and leads to disintegration of grana, as observed in rice seedlings (Ge et al., 2013; Wu et al., 2014). These structural disruptions severely affect light harvesting and energy transfer efficiency within the chloroplasts.

Cd also reduces the content of photosynthetic pigments. At relatively low concentrations (1–5 μM) in the nutrient solution, Cd significantly decreases chlorophyll levels, whereas higher concentrations (e.g., 50 μM) can severely impair photosynthesis to the point of being undetectable (Dias et al., 2013). The loss of chlorophyll is attributed to both inhibition of biosynthesis and increased degradation (Parmar et al., 2013). In addition, Cd reduces carotenoid levels, which not only diminishes light absorption capacity but also compromises the plant's ability to dissipate excess excitation energy, increasing vulnerability to photooxidative damage (Haider et al., 2021).

Cd stress also interferes directly with the light reactions of photosynthesis, impairing both photosystem I (PSI) and photosystem II (PSII). In *Elsholtzia argyi*, a Cd-accumulator plant, Cd exposure decreased the maximum quantum yield of PSII and the effective quantum yield (ΦPSII), indicating reduced PSII efficiency (Li et al., 2015). Furthermore, Cd can displace Mg^{2+} in chlorophyll molecules, reducing light absorption (Küpper et al., 1998; Parmar et al., 2013). The Calvin cycle, which is responsible for CO_2 fixation, is also disrupted under Cd stress. Cd inhibits key enzymes in the Calvin cycle, particularly Rubisco (ribulose-1,5-bisphosphate carboxylase/oxygenase), reducing carbon fixation capacity (Song et al., 2019; Zhang et al., 2024). This inhibition results from both the direct inactivation of enzymes and the reduced availability of ATP and NADPH because of impaired light reactions. Cd also alters the intracellular CO_2/O_2 ratio, favoring the oxygenation reaction over carboxylation, which compromises the efficiency of Rubisco (Zhang et al., 2024). In addition, Cd-induced stomatal closure reduces CO_2 availability, further limiting photosynthesis (Li et al., 2015; Zhang et al., 2024). Although stomatal closure helps conserve water under Cd stress, it inadvertently restricts CO_2 uptake, compounding the inhibition of photosynthesis.

The severity of photosynthetic impairment under Cd stress is generally dose-dependent. In wheat seedlings, Cd exposure caused a progressive decline in photosynthetic rate that strongly correlated with leaf Cd accumulation (Ci et al., 2010). However, the threshold for Cd-induced photosynthetic toxicity varies widely across plant species and cultivars. Numerous studies have documented photosynthetic disruption in a variety of crops, including rapeseed, pea, maize, barley, rice, mung bean, and wheat (Abbas et al., 2018; Ali et al., 2015; Jabeen et al., 2022). Interestingly, not all studies report negative effects. In some cases, low Cd concentrations have been associated with increased chlorophyll content (Liu et al., 2010; Zhao et al., 2021) and enlarged chloroplasts (Vitória et al., 2006), indicating that mild stress may sometimes trigger compensatory or adaptive responses.

One of the principal ways Cd impairs photosynthesis is through the induction of oxidative stress. Elevated ROS production in chloroplasts under Cd stress damages chlorophyll molecules, proteins, lipids, and other components of the photosynthetic machinery (Genchi et al., 2020). Cd also displaces essential metal cofactors in photosynthesis-related enzymes, particularly those involved in chlorophyll biosynthesis and carbon fixation (Andresen & Küpper, 2013). Moreover, Cd interferes with the uptake and transport of key nutrients such as iron, magnesium, and calcium, which are critical for maintaining chlorophyll structure and overall photosynthetic function (Zulfiqar et al., 2022).

11.2.6 Alteration in enzyme activities

Cd toxicity can significantly disrupt plant metabolic processes by directly altering enzyme activities. This occurs primarily through Cd binding to sulfhydryl (–SH) groups in proteins or by replacing essential metal cofactors in metalloenzymes, thereby impairing their structure and function (Genchi et al., 2020). In addition, Cd-induced oxidative stress plays a major role in modulating enzyme activity. The overproduction of ROS under Cd stress activates a range of enzymatic and nonenzymatic antioxidant responses aimed at mitigating oxidative damage. Several studies have shown that the activities of key antioxidant enzymes such as SOD, catalase (CAT), peroxidase (POD), and ascorbate peroxidase (APX) are generally upregulated under Cd stress (Farooq et al., 2013; Hashem, 2014; Nahakpam & Shah, 2011). These enzymes are part of the plant's first line of defense against ROS, and their enhanced activity reflects an adaptive mechanism to detoxify superoxide radicals, hydrogen peroxide, and other ROS.

Cd has a pronounced effect on nitrogen assimilation pathways. Enzymes, such as NR, glutamine synthetase (GS), and glutamate synthase (GOGAT)—all essential for nitrogen uptake and assimilation—are particularly sensitive to Cd stress and often show decreased activity (Gouia et al., 2000). In contrast, the activity of glutamate dehydrogenase (GDH) tends to increase in response to elevated ammonium levels under Cd stress (Erdal & Turk, 2016; Sharma et al., 2010). This upregulation of GDH may result from de novo enzyme synthesis or activation of specific isozymes that aid in detoxifying excess ammonium (Gouia et al., 2000). The effects of Cd on nitrogen metabolism can vary significantly between plant species and cultivars. For instance, Yang et al. (2019) reported that Cd exposure altered total nitrogen content and the activities of NR, GS, and GOGAT across 20 alfalfa cultivars. Tolerant cultivars showed increased nitrogen content and enhanced enzymatic activity in shoots, whereas sensitive cultivars exhibited the opposite trend. GS, a key enzyme for ammonia assimilation, is especially vulnerable to Cd, and its inhibition can further exacerbate disruptions in nitrogen metabolism (Yin et al., 2022). In addition, Cd also affects enzymes involved in carbohydrate and cell wall metabolism. Disruption of these pathways can alter sugar allocation, modify cell wall composition, and hinder overall plant growth and development (Grobelak et al., 2019; Haider et al., 2021). Such alterations have downstream effects on energy distribution and structural integrity, further compounding Cd-induced stress.

11.2.7 Cadmium-induced genotoxicity and programmed cell death

Cd is a well-documented genotoxic agent in plants, capable of causing a broad range of adverse effects that compromise genomic integrity. Genotoxicity encompasses direct damage to DNA and interference with key cellular components, including the mitotic spindle, DNA polymerases, and DNA repair enzymes involved in cell cycle regulation (Tabassum et al., 2021). These harmful effects may result from the direct interaction of Cd with genetic material and associated proteins or indirectly through Cd-induced oxidative stress and the overproduction of ROS. Cd-induced genotoxicity has been observed in numerous plant species. For instance, short-term Cd exposure in wheat seedlings led to growth inhibition, oxidative stress, and measurable DNA damage (Çatav et al., 2020). Similarly, intersimple sequence repeats marker analysis revealed Cd-induced genomic alterations in three cultivars of cardoon (*Cynara cardunculus*) (Sorrentino et al., 2022). In broad bean (*Vicia faba*), Cd accumulation along with ROS-induced membrane lipid peroxidation was associated with a significant increase in DNA double-strand breaks and genomic instability (Chakrabarti & Mukherjee, 2022; Nouairi et al., 2019). Chromosomal abnormalities, such as bridges, chromosome loss, stickiness, polyploidy, and C-metaphase, were observed in tomato (*Solanum lycopersicum*) under Cd stress, and the frequency of these aberrations increased with Cd concentration (Aslam et al., 2014; Pizzaia et al., 2019). Although the precise mechanisms of oxidative stress-induced DNA damage by Cd remain unclear, it is hypothesized that Cd promotes ROS generation by upregulating respiratory burst oxidase genes, resulting in highly reactive radicals capable of damaging DNA and associated proteins (Tabassum et al., 2021; Wang et al., 2023).

In response to Cd-induced genotoxicity, plants activate the DNA damage response (DDR) pathway, a conserved eukaryotic mechanism for maintaining genome integrity. In plants the transcription factor (TF) SUPPRESSOR OF GAMMA RADIATION 1 (SOG1) acts as the central regulator of the DDR (Herbst et al., 2024; Pedroza-Garcia et al., 2022; Szurman-Zubrzycka et al., 2023). SOG1 promotes transcription of genes encoding cyclin-dependent kinase inhibitors and MYB3R proteins, which delay cell cycle progression—particularly the G2/M transition—to allow DNA repair (Bourbousse et al., 2018). Cd exposure has been shown to alter the expression of numerous cell cycle and DNA repair genes, many of which are regulated by SOG1, resulting in suppressed cell division and reduced endoreplication (Cui et al., 2017; Hendrix et al., 2020). Interestingly, *sog1* mutants of *A. thaliana* exhibit lower sensitivity to Cd, likely because of diminished activation of both DDR and oxidative stress-related genes, suggesting that SOG1 plays a crucial role in regulating plant responses to Cd stress (Hendrix et al., 2020). In general, DDR signaling activates transcriptional and epigenetic regulators that coordinate cell cycle arrest and DNA repair mechanisms following DNA damage (Pedroza-Garcia et al., 2022; Szurman-Zubrzycka et al., 2023). When DNA damage is repairable, DDR signaling facilitates cell cycle arrest and activation of DNA repair mechanisms. However, if the damage is too severe, the cell may undergo terminal differentiation or PCD to safeguard overall plant integrity (Hu et al., 2016).

Environmental PCD is frequently triggered in plants under excess heavy metal exposure, including Cd (Sychta et al., 2021). Cd-induced PCD is closely associated with ROS accumulation and can be suppressed by inhibitors of vacuolar processing enzymes (VPEs) with caspase-1-like activity and papain-like cysteine proteases. These findings underscore the role of plant proteases in initiating a proteolytic cascade that leads to PCD (Hara-Nishimura et al., 2005; Salguero-Linares & Coll, 2019; Sychta et al., 2020, 2021). The endoplasmic reticulum (ER)-resident chaperone BiP (binding immunoglobulin protein) is another key player in Cd-mediated PCD. In plants, BiP plays a central role in regulating the unfolded protein response and modulating the cell death response mediated by stress-induced development and cell death domain-containing N-rich proteins (DCD/NRPs) (Carvalho et al., 2013; Costa et al., 2008; Reis et al., 2011). In tobacco, BiP is implicated in modulating Cd-induced PCD, likely by controlling the accumulation of misfolded proteins and reducing ER stress (Srivastava et al., 2013; Xu et al., 2013). In soybean, Cd stress activates hallmark features of PCD, including nuclear loss, DNA fragmentation, elevated caspase-1-like activity, and widespread cell death in leaves (Quadros et al., 2022). Cd stress also induces the expression of components of the DCD/NRP-mediated PCD signaling pathway, such as NRP proteins, GmNAC81, GmNAC30, and VPE, which collectively orchestrate PCD via caspase-like activity (Pimenta et al., 2016; Quadros et al., 2022). Overexpression of *GmNAC81* heightens plant sensitivity to Cd and accelerates PCD, whereas BiP acts as a negative regulator, buffering against Cd-induced cell death by suppressing the activation of the NRP–GmNAC81/30–VPE module (Quadros et al., 2022).

11.3 Regulation of plant responses to cadmium toxicity

11.3.1 Signal transduction pathways involved in cadmium stress response

Plants have evolved intricate signal transduction networks to perceive and respond to abiotic stress, such as Cd toxicity. These pathways translate external stress signals into coordinated molecular and physiological responses that help maintain cellular homeostasis and minimize toxicity. Key signaling components include ROS, nitric oxide (NO), calcium ions (Ca^{2+}), and plant hormones.

11.3.1.1 Reactive oxygen species

Cd exposure leads to excessive production of ROS, which function not only as damaging agents but also as crucial signaling molecules in the stress response. The ROS signaling cascade typically involves activation of NADPH oxidases (RBOHs), which regulate ROS generation and mediate downstream responses such as nutrient uptake and antioxidant defense (Gupta et al., 2017; Wang et al., 2023). One of the principal downstream targets of ROS signaling is the mitogen-activated protein kinase (MAPK) pathway. Cd-induced ROS activate MAPKs, which in turn regulate genes involved in growth and stress adaptation. In rice, MAPKs influence auxin signaling and cell cycle regulation, thereby modulating root development under Cd stress (Zhao et al., 2013). Similarly, Cd stress induces the activation of ZmMPK3-1 and ZmMPK6-1 via ROS signaling in maize (Liu et al., 2019). In *A. thaliana* the activities of AtMPK3 and AtMPK6 significantly increase under Cd exposure, and this activation is suppressed by GSH, indicating ROS-dependence (Liu et al., 2010). These studies showed that the accumulation of Cd-induced ROS is necessary for the activation of MPK3 and MPK6 in both *Zea mays* and *A. thaliana*. However, neither study elaborated on the role of the downstream effects of these MAPKs in response to Cd toxicity.

11.3.1.2 Nitric oxide

Nitric oxide (NO) also plays a pivotal role in Cd stress signaling. In *A. thaliana*, Cd-induced NO accumulation activates MAPK cascades, promoting caspase-3-like protease activity and initiating PCD (Ye et al., 2013). Nitric oxide contributes to antioxidant defense by stimulating the ascorbate–GSH (AsA–GSH) cycle (Li et al., 2015), while also modulating the activity of enzymes like NR, which enhances Cd tolerance in maize (Kaya et al., 2021). Accumulation of NO in plants has been widely reported to strengthen the antioxidant defense system against excess Cd (Wei et al., 2020). The interplay between NO and ROS is bidirectional. Mutants of *A. thaliana* deficient in ROS-producing RBOHs (*rbohC, rbohD, rbohF*) exhibit altered NO levels under Cd stress (Gupta et al., 2017). In addition, nonsymbiotic hemoglobin 1 (Hb1) regulates both NO and ROS levels, influencing Cd transporter expression and tolerance in tobacco and *A. thaliana* (Bahmani et al., 2019; Lee & Hwang, 2015). Furthermore, Cd-induced NO production can activate iron-deficiency signaling pathways, exacerbating Cd toxicity (Besson-Bard et al., 2009).

11.3.1.3 Calcium signaling

Calcium acts as a universal second messenger in Cd stress signaling. Cd exposure causes rapid fluctuations in cytosolic Ca^{2+} levels, which are sensed by a suite of calcium-binding proteins such as calcium-dependent protein kinases (CDPKs), calmodulins (CaMs), CaM-like proteins (CMLs), and calcineurin B-like proteins (CBLs) (Kumar & Trivedi, 2016; Yeh et al., 2007). Recent studies have highlighted calcium-mediated regulation of Cd transporter activity. In *A. thaliana*, AtCPK21 and AtCPK23 phosphorylate AtNRAMP6 at Ser489 and Thr505, thereby inhibiting its Cd transport activity and enhancing Cd tolerance (Zhang et al., 2023c).

11.3.1.4 Plant hormones

Phytohormones are central regulators of plant development and stress responses. Cd exposure perturbs the biosynthesis and signaling of several hormones, including ABA, ethylene, auxin, salicylic acid (SA), jasmonates (JA), cytokinins, gibberellins (GA), and brassinosteroids (BR) (Waadt et al., 2022). Cd stress disrupts the biosynthesis and signaling pathways of these phytohormones, leading to profound physiological alterations (Fattorini et al., 2017; Lu et al., 2020; Luo et al., 2019; Schellingen et al., 2014; Tajti et al., 2019; Villiers et al., 2012; Zhao et al., 2016; Zhu et al., 2012). Modulating hormonal signaling and/or exogenously applying phytohormones has been shown to mitigate the adverse effects of Cd stress (Saini et al., 2021). Among the various phytohormones mentioned above, ABA and ethylene—commonly referred to as "stress hormones"—play central roles in mediating plant adaptive responses to both biotic and abiotic stresses, including Cd exposure (Li et al., 2023).

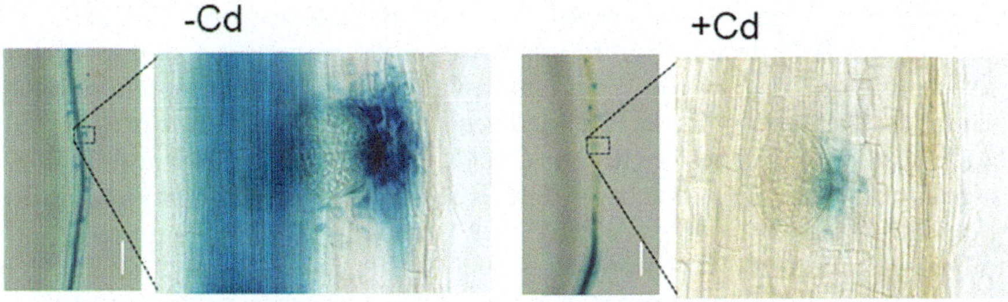

FIGURE 11.3 Effect of Cd (5 μM) on auxin accumulation in the stele and lateral root primordia in rice roots. Auxin level was visualized by the use of *DR5::GUS* reporter and β-glucuronidase (GUS) staining. *Modified from Wang et al. (2021). With copyright permission.*

Cd stress upregulates expression of genes involved in ABA production, such as *OsNCED4* and *OsNCED5* in rice (Tan et al., 2017), and increases endogenous ABA concentration in rice, *Solanum tuberosum* L., *Brassica napus* L., maize, and wheat (Hsu & Kao, 2005; Meng et al., 2024; Stroiński et al., 2013). ABA-deficient (*aba-3, aba-4, nced3*) and ABA-insensitive (*abi2-1, abi3-1*) mutants of *A. thaliana* exhibit heightened sensitivity and Cd accumulation (Fujii & Zhu, 2009; Sharma & Kumar, 2002). The TF ABI5 enhances ABA-mediated Cd tolerance by reducing Cd uptake (Zhang et al., 2019).

Cd exposure also elevates ethylene production by upregulating expression of *AtACS2* and *AtACS6*, which encode key enzymes in ethylene biosynthesis (Schellingen et al., 2014). MAPK-mediated phosphorylation stabilizes AtACS6, whereas AtPP2A dephosphorylates it to limit ethylene accumulation. Mutations in *AtPP2A* increase Cd sensitivity because of uncontrolled ethylene production (Chen et al., 2020). In ethylene signaling, AtEIN3 and AtEIL1 are crucial TFs. Loss-of-function mutants (*ein3-1, eil1-1*) display increased Cd tolerance (Ju & Chang, 2015; Kong et al., 2018). AtEIN3 directly activates *AtXTH33* and *AtLSU1*, genes involved in cell wall remodeling and sulfur metabolism, exacerbating root growth inhibition under Cd stress (Kong et al., 2018). Moreover, *ctr1-1*, a constitutive ethylene response mutant, shows heightened root sensitivity to Cd (Kong et al., 2018). These findings underscore the role of increased ethylene production in the inhibitory effects of Cd on plant growth. In addition, both ABA and ethylene also regulate apoplastic barrier formation, influencing Cd accumulation and movement within the plant (Liu et al., 2021, 2022; Tao et al., 2019).

Cd disrupts auxin homeostasis by negatively regulating key auxin-related genes, which impairs cell differentiation and inhibits root growth (Ronzan et al., 2018). This disruption affects multiple aspects of auxin signaling, including its biosynthesis, transport, and distribution. Using DR5::GUS or DII::VENUS reporter lines, Yuan and Huang (2016) demonstrated that exposure to a high Cd concentration (150 μM for 12 hours) significantly reduced auxin accumulation and transport in *A. thaliana* roots. This reduction was associated with a decrease in root meristem size and a complete arrest of primary root elongation. Further evidence of Cd-induced inhibition of root development comes from studies on the quiescent center and lateral roots. Fattorini et al. (2017) reported that Cd stress alters auxin activity and distribution at these sites, disrupting normal developmental processes. In rice, exposure to a lower concentration of Cd (5 μM for 48 hours) significantly reduced auxin levels in the stele and lateral root primordia (Fig. 11.3) (Wang et al., 2021). This was caused by a rapid downregulation of auxin efflux transporter genes, including *OsPIN1b*, *OsPIN1c*, and *OsPIN9*. A similar trend was observed in *A. thaliana*, where Cd exposure markedly reduced the expression of *AtPIN1*, *AtPIN3*, and *AtPIN7* (Bruno et al., 2017). In addition to impairing auxin transport, Cd also modulates auxin biosynthesis via the indole-3-pyruvic acid pathway, which in turn activates NO and ROS signaling cascades (Demecsová et al., 2020). These signaling molecules further influence the expression of auxin-related genes, including *OsYUCCA*, *OsPIN*, *OsARF*, and *OsIAA*, as well as cell cycle regulators, through interactions with ROS and MAPK pathways (Zhao et al., 2012, 2013).

Exogenous application of plant hormones such as JA, GA, SA, and BR has been shown to enhance plant tolerance to Cd stress. This protective effect is primarily attributed to the stimulation of antioxidant defense pathways and the downregulation of genes involved in Cd uptake and long-distance translocation (Lei et al., 2020b; Lu et al., 2018; Zhu et al., 2012). For example, in *A. thaliana*, the expression of key Cd transporter genes—*AtIRT1*, *AtHMA2*, and *AtHMA4*—was significantly upregulated in the JA biosynthesis mutant *ataos* upon exposure to 50 μM Cd compared to the wild type. Exogenous JA application suppressed the expression of these genes in both the mutant and wild-type plants, suggesting that JA negatively regulates Cd uptake and translocation pathways (Lei et al., 2020a). Similarly, exogenous GA treatment alleviated Cd toxicity in *A. thaliana* by downregulating *AtIRT1*, thereby limiting Cd entry into the plant system (Zhu et al., 2012). In rice, Cd exposure rapidly elevates endogenous BR levels in roots. Application of exogenous BR enhances Cd sequestration within the hemicellulose fraction of root cell walls, reducing Cd accumulation in aerial tissues (Sun et al., 2024).

11.3.2 Transcriptional response to cadmium stress

RNA-Seq analyses revealed extensive transcriptional reprogramming in response to Cd stress in plants. For instance, Oono et al. (2016) identified 15,850 differentially expressed genes (DEGs) in the roots of rice seedlings exposed to Cd stress (1 μM and 4 days of exposure), with over 5830 upregulated and 10,020 downregulated genes. Similarly, Yang et al. (2015) reported 2592 DEGs in the roots and 1980 DEGs in the leaves of Cd-resistant *Salix matsudana* Koidz. after exposure to 50 μM Cd for 10 hours. The number and identity of DEGs vary depending on Cd concentration, exposure duration, plant species, and tissue type. Cd stress induces complex transcriptional responses involved in adaptation, including the activation of pathways related to metal detoxification, antioxidant defense, and hormonal signaling. Key TF families, including MYB, WRKY, bZIP, NAC, and ERF, orchestrate these responses by modulating genes involved in Cd uptake, transport, chelation, and sequestration.

In *A. thaliana*, Cd exposure induces the expression of the TF *AtMYB49*, which plays a key role in regulating Cd uptake. Loss-of-function mutants of *AtMYB49* exhibit reduced Cd accumulation, whereas its overexpression leads to increased Cd uptake (Zhang et al., 2019). AtMYB49 directly binds to the promoters of *AtbHLH38* and *AtbHLH101*, two TFs that, in turn, activate the Cd uptake transporter gene *AtIRT1*. However, this pathway is modulated by the ABA-insensitive bZIP TF AtABI5, which interacts with AtMYB49 and inhibits its DNA-binding activity. By preventing AtMYB49 from activating its downstream targets, AtABI5 suppresses *AtIRT1* expression, thereby reducing Cd uptake (Zhang et al., 2019). Another regulatory mechanism involves the Cd-induced TF AtWRKY33, which enhances Cd tolerance by directly activating the transcription of *AtATL31*. *AtATL31* encodes an E3 ubiquitin ligase that targets AtIRT1 for degradation via the ubiquitin-proteasome pathway, effectively limiting Cd uptake (Zhang et al., 2023a). In addition, overexpression of *AtWRKY13* under Cd stress has been shown to decrease Cd accumulation and enhance tolerance by directly upregulating *AtPDR8*, a gene encoding a Cd efflux transporter that mediates Cd extrusion from cells (Sheng et al., 2019). In rice the key Cd-related genes, such as *OsNRAMP5* (involved in Cd uptake) and *OsHMA3* (involved in vacuolar sequestration), are not strongly affected by Cd stress. As a result the transcriptional regulation of Cd homeostasis in rice remains less understood and warrants further investigation.

In *A. thaliana*, multiple TFs contribute to Cd tolerance through diverse regulatory mechanisms involving antioxidant defense, metal chelation, cell wall modification, and hormonal signaling. AtMYB4, for example, enhances Cd tolerance by activating the antioxidant defense system and upregulating *AtPCS1* and *AtMT1C*, which are involved in phytochelatin (PC) synthesis and metal binding, respectively (Agarwal et al., 2020). In contrast, AtWRKY12 negatively affects Cd tolerance by repressing genes associated with PC biosynthesis, leading to reduced Cd detoxification capacity and higher metal sensitivity (Han et al., 2019). Cd stress downregulates *AtWRKY12* while upregulating the C2H2 zinc-finger TF *AtZAT6*, creating a regulatory balance over GSH-dependent PC production. Specifically, AtWRKY12 suppresses, whereas AtZAT6 activates, the expression of *AtGSH1*, a gene critical for GSH synthesis and downstream PC production (Chen et al., 2016; Han et al., 2019). AtWRKY45 acts as a positive regulator of Cd tolerance by directly activating *AtPCS1* and *AtPCS2*. Its overexpression leads to increased PC levels, whereas complementation of *wrky45* mutants with either *AtPCS1* or *AtPCS2* restores Cd resistance (Li et al., 2023). Members of the NAC TF family also play important roles. AtNAC102 directly targets the promoter of *WALL-ASSOCIATED KINASE-LIKE PROTEIN11* (*AtWAKL11*), a gene involved in regulating pectin metabolism in the cell wall. The *AtNAC102–AtWAKL11* module enhances Cd tolerance by increasing Cd binding in the cell wall through modified pectin composition (Han et al., 2023). Moreover, ethylene-responsive TFs AtERF34 and AtERF35 increase Cd tolerance by upregulating *AtNRT1.8*, a nitrate transporter gene. This helps counteract Cd-induced suppression of nitrate assimilation, thereby maintaining nitrogen homeostasis under Cd stress (Xie et al., 2021).

Several TFs have also been identified as key regulators of Cd tolerance in crop species beyond *A. thaliana*. In common bean (*Phaseolus vulgaris*), the Cd-responsive TF PvMTF-1 enhances Cd tolerance by activating the tryptophan biosynthesis pathway, which is strongly upregulated under Cd stress (Sun et al., 2015). The ethylene-responsive TF PvERF15 further amplifies this response by binding to a 19-bp AC-rich element in the *PvMTF-1* promoter and inducing its expression during Cd exposure, thereby contributing to improved Cd detoxification mechanisms (Lin et al., 2017). In soybean (*Glycine max*), overexpression of *GmWRKY142* significantly reduces Cd accumulation and enhances tolerance in both transgenic soybean and *A. thaliana*. GmWRKY142 binds to W-box elements in the promoters of several cysteine-rich detoxification genes—*GmCDT1-1*, *GmCDT1-2*, and *AtCDT1*—thereby upregulating their expression and facilitating Cd chelation (Cai et al., 2020). In maize (*Z. mays*), the basic helix–loop–helix (bHLH) TF ZmbHLH105 is strongly induced by Cd exposure. It activates *ZmNCED1* and *ZmNCED2*, two genes involved in ABA biosynthesis, leading to increased ABA accumulation and enhanced Cd tolerance. In addition, the overexpression of *ZmbHLH105* promotes lignin biosynthesis and cell wall thickening, which together contribute to reduced Cd uptake by roots and overall lower Cd accumulation in plant tissues (Meng et al., 2024).

MicroRNAs (miRNAs) are small noncoding RNAs that play crucial regulatory roles in plant responses to both abiotic and biotic stresses. Despite their importance, only a limited number of miRNAs have been characterized and functionally validated in the context of Cd stress responses in plants (Ding et al., 2020). In rice, *miR166* is significantly downregulated under Cd stress. Overexpression of *miR166* enhances Cd tolerance by reducing oxidative stress, limiting Cd translocation from roots to shoots, and decreasing Cd accumulation in grains (Ding et al., 2018). The homeodomain-containing TF OsHB4 has been identified as a direct target of *miR166*. *OsHB4* knockout mutants exhibit increased Cd tolerance, whereas overexpression of *OsHB4* leads to greater Cd sensitivity. Moreover, *miR166* overexpression downregulates the Zn/Cd efflux transporter gene *OsHMA2* and upregulates the tonoplast-localized Cd transporter *OsHMA3*, although it remains unclear whether OsHB4 directly controls the transcription of these transporter genes (Ding et al., 2018). *MiR535*, a highly conserved miRNA across plant species, is strongly induced by low levels of Cd. Overexpression of *miR535* in rice promotes root elongation and increases Cd uptake, whereas knockout mutants exhibit shorter roots and reduced Cd concentration under Cd stress (Yue et al., 2023). The SQUAMOSA promoter binding protein-like 7 (SPL7) TF is a target of *miR535*. SPL7 negatively regulates the expression of *OsNramp5*, a key gene involved in Cd uptake, thereby positioning the *miR535–SPL7–OsNramp5* module as a regulatory node in rice Cd tolerance (Yue et al., 2023).

Certain primary miRNAs (pri-miRNAs) have been found to encode functional regulatory short peptides, known as miRNA-encoded peptides (miPEPs). One such peptide, miPEP156e, encoded by pri-miR156e, has been identified as a modulator of *miR156* expression and its downstream targets, the *SPL* (SQUAMOSA promoter-binding protein-like) gene family (Lu et al., 2024). This regulatory module plays a significant role in *miR156*-mediated Cd tolerance in rice. Overexpression of *miPEP156e* reduced Cd uptake and accumulation, decreased ROS levels under Cd stress, and ultimately enhanced Cd tolerance—mirroring the phenotype observed in *miR156*-overexpressing lines. Conversely, *miPEP156e* loss-of-function mutants displayed heightened sensitivity to Cd stress, evidenced by increased Cd accumulation and elevated ROS levels (Lu et al., 2024). Transcriptome analysis further revealed that *miPEP156e* contributes to Cd tolerance by modulating the expression of Cd transporter genes and genes involved in ROS detoxification (Lu et al., 2024).

11.3.3 Posttranslational modifications

Posttranslational modifications, such as phosphorylation, ubiquitination, SUMOylation, and methylation, play crucial roles in regulating plant responses to Cd stress by modulating the activity, stability, and localization of Cd tolerance-related proteins. Zhang et al. (2023c) reported that the CDPKs AtCPK21 and AtCPK23 in *Arabidopsis* phosphorylate AtNRAMP6 at Ser489 and Thr505, suppressing its Cd transport activity and thereby enhancing Cd tolerance. Similarly, in wheat, Du et al. (2023) identified the receptor-like kinase TaWAK20, which interacts with and phosphorylates the TF TaSPL5, enhancing its DNA-binding activity. Overexpression of *TaWAK20* improves Cd tolerance and reduces Cd accumulation by modulating ROS production and scavenging.

In *A. thaliana* the SUMO E3 ligase AtSIZ1 positively regulates Cd tolerance through the GSH-dependent PC synthesis pathway (Hong et al., 2022; Zheng et al., 2022). AtSIZ1 activates the ubiquitin E3 ligase AtCOP1, which, in turn, promotes the degradation of the bZIP TF AtHY5. As AtHY5 is a positive regulator of sulfur assimilation—critical for providing thiol groups needed for GSH biosynthesis—its degradation relieves this regulatory constraint (Li et al., 2011; Lin et al., 2016). In response to Cd exposure, AtSIZ1 modulates the expression of key genes involved in

GSH and PC biosynthesis, including *AtGSH1*, *AtGSH2*, *AtPCS1*, and *AtPCS2*, thereby enhancing detoxification capacity and improving Cd tolerance (Hong et al., 2022; Zheng et al., 2022). In addition, protein lysine methylation has emerged as another important posttranslational modification involved in Cd stress responses. Serre et al. (2020) investigated the impact of Cd on lysine-methylated proteins and protein lysine methyltransferases in *A. thaliana*, *Arabidopsis lyrata*, and the hyperaccumulator *Arabidopsis halleri*, which display varying degrees of Cd tolerance. A knockout mutant of the calmodulin lysine methyltransferase gene showed enhanced Cd tolerance (Serre et al., 2020). These findings suggest that lysine methylation of nonhistone proteins plays a regulatory role in plant adaptation to Cd stress.

11.3.4 Epigenetic regulation and changes in chromatic structure

Cd exposure induces widespread alterations in DNA methylation patterns. In *A. thaliana*, Cd stress caused a global increase in DNA methylation levels in roots. This was attributed to the suppression of DNA demethylases ROS1, DML2, and DML3 (collectively known as RDD). Further investigation demonstrated that Cd-induced inhibition of RDD-mediated demethylation contributes to enhanced Cd tolerance, potentially by improving iron homeostasis through a feedback regulatory mechanism (Fan et al., 2020).

In rice, Cd exposure influenced the expression of DNA methylation machinery. For instance, genes encoding DNA methyltransferases, such as *OsDMT702* and *OsDMT707*, were differentially expressed under Cd stress (Feng et al., 2016). Moreover, Cd exposure led to hypermethylation or hypomethylation of several genes involved in stress response, metal transport, and transcriptional regulation—including *OsGSH2*, *OsZIP1*, and *OsIRO2*. These methylation changes were closely associated with altered transcriptional activity under Cd stress conditions (Feng et al., 2016; Liu et al., 2019). In addition to DNA methylation, Cd stress also affects RNA methylation. In barley, Cd exposure resulted in substantial changes in N6-methyladenosine (m^6A) RNA methylation in more than 400 genes, including key Cd-responsive TFs from the MAPK, WRKY, and MYB families (Su et al., 2022).

Cd exposure has also been associated with profound alterations in chromatin structure and epigenetic regulation. In *V. faba* root meristem cells, immunofluorescence analyses revealed Cd-induced modifications in histone marks, including changes in methylation and acetylation (Żabka et al., 2021). Such histone modifications can influence chromatin accessibility, thereby modulating gene expression in response to stress. Although these findings underscore the importance of chromatin-level regulation in Cd stress responses, much of the current evidence is derived from cytological observations and high-throughput sequencing. Further mechanistic studies are needed to elucidate the specific roles of epigenetic modifications—particularly DNA methylation and histone modifications—in regulating genes involved in Cd uptake, detoxification, and tolerance.

11.4 Detoxification mechanisms in plants

11.4.1 Roles of cell wall in cadmium detoxification

The plant cell wall serves as the first line of defense against Cd stress, acting as both a physical and chemical barrier that protects the protoplast from Cd-induced damage. Cell wall is primarily composed of matrix polysaccharides, structural proteins, and, in some cases, lignin and phenolic compounds (Loix et al., 2017). Suberization of root cell walls has been shown to impede Cd translocation to the xylem by altering hydrostatic pressure gradients (Tao et al., 2017). Beyond acting as a physical shield the cell wall plays a key role in adsorbing Cd via the negatively charged functional groups such as hydroxyl (–OH), carboxyl (–COOH), amino, and aldehyde groups present in polysaccharides (Peng et al., 2017). In *A. thaliana*, exposure to a high level of Cd (50 μM) resulted in the majority of Cd being bound in the hemicellulose fraction of the root cell wall (Zhu et al., 2013). Similarly, in tomato, Jia et al. (2019) found that root Cd concentrations were strongly correlated with both pectin content and the expression of pectin biosynthesis-related genes, including *SlQUA1* and *SlPME*. However, it is important to note that these findings often rely on chemical fractionation of homogenized root tissue, which may cause redistribution of Cd from intracellular compartments—particularly vacuoles—into the cell wall fraction, potentially skewing the results. To achieve more accurate and spatially resolved insights into Cd localization at the cellular and subcellular levels, in situ analytical techniques such as synchrotron-based X-ray absorption spectroscopy, secondary ion mass spectrometry, and energy-dispersive X-ray spectroscopy are recommended (Zhao et al., 2014).

11.4.2 Enzymatic antioxidants

Plants exposed to Cd stress activate the antioxidative defense system to counteract the oxidative damage caused by ROS. This defense system comprises both enzymatic and nonenzymatic antioxidants that work in concert to maintain cellular redox homeostasis. Key enzymatic antioxidants involved in the response to Cd-induced oxidative stress include SOD, CAT, APX, guaiacol peroxidase, PODs, thioredoxin (TRX), monodehydroascorbate reductase (MDAR), dehydroascorbate reductase (DHAR), GSH reductase (GR), and GSH *S*-transferases (GST) (Sachdev et al., 2021).

Among antioxidant enzymes, SOD, a metalloenzyme, serves as the first line of defense against oxidative stress by catalyzing the dismutation of superoxide radicals ($O_2^{\bullet -}$) into hydrogen peroxide (H_2O_2), thereby reducing the potential formation of highly reactive hydroxyl radicals ($^{\bullet}OH$) (Hasanuzzaman et al., 2020). SOD exists in three main isoforms, classified according to their metal cofactors: copper/zinc-SOD (Cu/Zn-SOD), found in the cytosol, chloroplasts, mitochondria, and peroxisomes; iron-SOD (Fe-SOD), primarily located in chloroplasts and peroxisomes; and manganese-SOD (Mn-SOD), which is confined to mitochondria. In soybean seedlings, exposure to Cd significantly upregulated Cu/Zn-SOD expression at both transcriptional and enzymatic activity levels in roots (Ferreira et al., 2002). Moreover, recent studies have demonstrated that overexpression of SOD isoforms enhances Cd tolerance in transgenic *Arabidopsis* and tobacco plants, further highlighting the detoxifying function of SOD in response to Cd stress (Pan et al., 2019; Pang et al., 2024).

CAT enzymes, which contain a tetrameric heme group, play a key role in detoxifying hydrogen peroxide (H_2O_2) by catalyzing its breakdown into water and molecular oxygen. In rice, maize, and *Arabidopsis*, CAT typically functions in a monomeric form, although dimeric or tetrameric forms are more common in other organisms (Anjum et al., 2016). Under Cd stress, plants also depend on low-molecular-weight (LMW) antioxidants such as ascorbate (AsA) and GSH to neutralize ROS (Jung et al., 2020). Ascorbate serves as a primary electron donor, reducing H_2O_2 via the action of APX. Oxidized ascorbate (dehydroascorbate, DHA) is regenerated through two key enzymes: MDAR, which uses NADPH or reduced ferredoxin, and

DHAR, which completes the cycle by reducing DHA back to AsA, thus maintaining cellular redox homeostasis (Foyer, 2018).

PODs play a vital role in scavenging ROS by oxidizing phenolic compounds to generate phenoxyl radicals (PhO or QA), which then react with hydrogen peroxide (H_2O_2) to form water—primarily within the apoplast and vacuole (Smirnoff & Arnaud, 2019). In addition to PODs, thiol-based PODs known as peroxiredoxins (Prxs) act as important antioxidants across various plant tissues. These enzymes utilize a cysteinyl thiolate-based catalytic cycle to reduce H_2O_2 and function in key organelles such as mitochondria, chloroplasts, and the cytosol (Liebthal et al., 2018). TRX, small redox proteins (~12 kDa) containing a conserved pair of cysteine residues in their active site, also contribute to oxidative stress tolerance. They maintain redox homeostasis by reducing oxidized 2-CysPrx proteins, particularly in chloroplasts and mitochondria (Hasanuzzaman et al., 2020). Collectively, these interlinked antioxidant systems—PODs, Prxs, TRXs, and others—ensure efficient detoxification of ROS and help maintain cellular redox balance during Cd-induced oxidative stress.

11.4.3 Nonenzymatic antioxidants and chelating agents

Nonenzymatic antioxidants are essential components of the plant defense system against Cd-induced toxicity. These include GSH, ascorbic acid (AsA), phenolic compounds, flavonoids, carotenoids, and tocopherols. They serve multiple roles, such as mitigating oxidative damage, chelating Cd ions, and maintaining cellular redox homeostasis. Working in synergy with enzymatic antioxidants, these nonenzymatic molecules are crucial for alleviating Cd-induced oxidative stress and contributing to the overall detoxification process. Within plant cells, most Cd ions are not present in free form but are complexed with LMW ligands or bound to peptides and proteins (Haas & Franz, 2009). These LMW ligands involved in Cd chelating include sulfur-containing compounds such as GSH, PCs, and oxygen-rich ligands such as organic acids (Seregin & Kozhevnikova, 2021).

PCs are among the most extensively studied heavy metal chelators in plants, playing a pivotal role in Cd tolerance. These small, cysteine-rich peptides have the general structure (γ-Glu–Cys)$_n$-Gly, where n typically ranges from 2 to 11. PCs are synthesized from GSH by the enzyme PC synthase (PCS). They bind metal ions to form stable complexes that are subsequently sequestered in vacuoles via the tonoplast transporters ABCC1/2 (Park et al., 2012). Functional studies in *Arabidopsis* demonstrated that loss of the PCS gene *AtPCS1* results in heightened Cd sensitivity, whereas overexpression of *AtPCS1* enhances Cd tolerance by promoting PC synthesis under high Cd stress (Brunetti et al., 2011; Howden et al., 1995b). In rice, two PCS genes, *OsPCS1* and *OsPCS2*, are located on chromosomes 5 and 6, respectively (Hayashi et al., 2017). Knockdown of *OsPCS1* has minimal effects on Cd sensitivity in rice seedlings (Yamazaki et al., 2018), but disruption of *OsPCS2,* also referred to as *OsPCS1* (Uraguchi et al., 2017), significantly increases Cd sensitivity (Uraguchi et al., 2017). Furthermore, enhancing PC biosynthesis through metabolic modifications in the sulfur assimilation pathway increases Cd tolerance in rice (Chen et al., 2020).

GSH, an LMW tripeptide thiol, functions as a primary antioxidant and serves as the precursor for PCs. The importance of GSH in Cd detoxification has been clearly demonstrated using mutants with impaired GSH biosynthesis. For example, the *cad2-1* mutant of *A. thaliana,* which exhibits only 15%–30% of wild-type GSH levels because of a mutation in the γ-glutamylcysteine

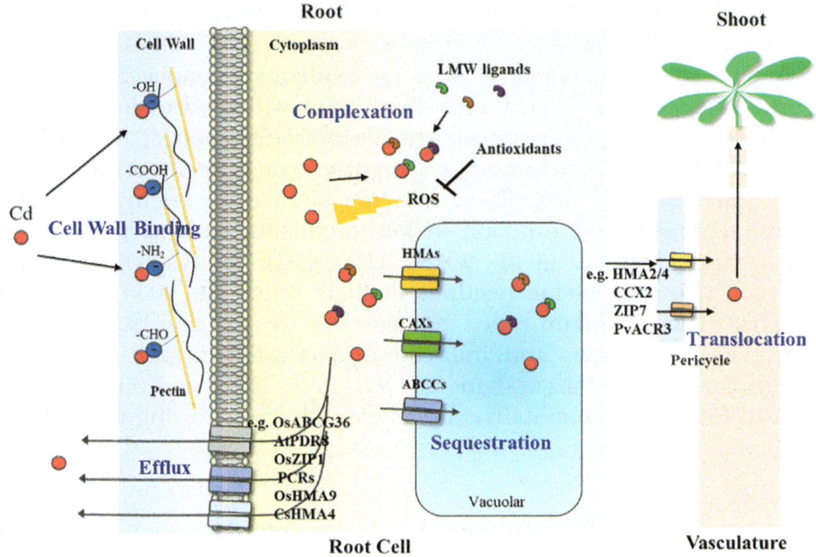

FIGURE 11.4 A schematic diagram of the mechanisms of Cd resistance in plants through exclusion or detoxification. *Modified from Tang et al. (2023). With copyright permission.*

synthetase (GSH1) gene, is both PC-deficient and hypersensitive to Cd (Howden et al., 1995a). Similarly the *pad2-1* mutant, which also exhibits reduced GSH accumulation, shows increased Cd sensitivity, underscoring the critical role of GSH as a substrate for PC biosynthesis and its broader role in stress response signaling (Parisy et al., 2007).

Compared to sulfur- and nitrogen-containing ligands, oxygen-containing ligands such as organic acids have lower affinities for Cd. However, their generally higher concentrations in plants compensate for the lower binding strength and can facilitate Cd sequestration, particularly within vacuoles. In tomatoes the secretion of oxalate from root tips alleviates Cd toxicity, whereas blocking oxalate release using anion-channel inhibitors exacerbates Cd stress (Zhu et al., 2011). The low water solubility of Cd oxalate (0.06 g L^{-1}) further reduces Cd bioavailability, contributing to its detoxifying role.

11.4.4 Sequestration in vacuoles

The vacuole plays a crucial role in Cd detoxification by serving as a primary storage site and thereby mitigating its cytotoxic effects. This sequestration process is mediated by tonoplast-localized transporters, notably members of the heavy metal ATPase (HMA), metal tolerance protein (MTP), and ATP-binding cassette C (ABCC) families, which actively transport Cd into the vacuole (Fig. 11.4) (Liu et al., 2024a; Zhao et al., 2022). Among these, HMA3, a tonoplast-localized P1B-type HMA, has been extensively characterized in various plant species (Liu et al., 2017; Ueno et al., 2010, 2011). Genome-wide association studies (GWAS) and quantitative trait loci (QTL) mapping have consistently identified *HMA3* as a key genetic determinant of shoot Cd accumulation in species, including *Arabidopsis*, rice, maize, barley, and sorghum (Chao et al., 2012; Lei et al., 2020a; Liu et al., 2020; Miyadate et al., 2011; Tang et al., 2021;

Ueno et al., 2010; Wahinya et al., 2022). In nonhyperaccumulator species such as *Arabidopsis* and rice, *HMA3* is predominantly expressed in roots, where it functions to sequester Cd into root vacuoles, thereby limiting its translocation to aerial tissues (Chao et al., 2012; Miyadate et al., 2011; Ueno et al., 2010). Overexpression of *OsHMA3* in rice has been shown to significantly enhance Cd tolerance by reducing shoot Cd accumulation (Wang et al., 2021). In addition, a recent study reported that OsMTP11, a rice MTP expressed in leaf vascular parenchyma cells, binds to the vacuolar sorting receptor OsVSR2 and is trafficked intracellularly from the *trans*-Golgi network to prevacuolar compartments and eventually to the vacuole. Therefore OsMTP11 mediates Cd sequestration into the endomembrane system, contributing further to vacuolar detoxification of Cd (Liu et al., 2024a).

Tonoplast-localized ABCC transporters play a pivotal role in vacuolar sequestration of Cd by facilitating the transport of PCs, either in their free form (apo-PCs) or as Cd–PC complexes, into the vacuole. In *A. thaliana*, AtABCC1 and AtABCC2 mediate Cd detoxification through this mechanism. Loss-of-function mutants such as *abcc1* or the double mutant *abcc1/abcc2* display heightened sensitivity to Cd, whereas overexpression of *AtABCC1* significantly enhances Cd tolerance and accumulation by promoting vacuolar sequestration of Cd–PC complexes (Park et al., 2012). AtABCC3 also contributes to vacuolar Cd detoxification. Mutants lacking *AtABCC3* exhibit increased Cd sensitivity, whereas overexpression improves tolerance. Moreover, expression of *AtABCC3* in the *abcc1/abcc2* background restores Cd tolerance, suggesting functional redundancy among these ABCC family members in mediating Cd sequestration (Brunetti et al., 2015). In rice, *OsABCC9* is a key player in Cd sequestration, particularly in the parenchyma cells within the root stele. *OsABCC9* expression is induced by Cd exposure, and the protein localizes to the tonoplast, consistent with its role in vacuolar detoxification. Knockout of *OsABCC9* results in increased Cd sensitivity and higher Cd accumulation in rice grains (Yang et al., 2021).

11.4.5 Efflux

Efflux of Cd from the cytoplasm can reduce its cytosolic concentration, thus contributing to Cd detoxification. In *A. thaliana*, AtPDR8, an ATP-binding cassette (ABC) transporter localized to the plasma membrane, is upregulated in response to Cd exposure. Loss-of-function mutants of *AtPDR8* exhibit hypersensitivity to Cd, whereas overexpression enhances Cd resistance and reduces Cd accumulation, suggesting that AtPDR8 functions as a Cd or Cd-conjugate efflux pump (Kim et al., 2007). Similarly, OsABCG36, a rice ABC transporter also localized at the plasma membrane, has been shown to participate in Cd efflux. Knockout of *OsABCG36* leads to elevated Cd accumulation in the root cell sap and increased sensitivity to Cd stress. Although the exact substrate specificity remains unclear, OsABCG36 likely facilitates the efflux of either free Cd ions or Cd–ligand complexes (Fu et al., 2019). The plant Cd resistance (PCR) protein family also contributes to Cd efflux. In *A. thaliana*, PCR1 and PCR2, both plasma membrane-localized, mediate the export of Cd out of cells and enhance Cd tolerance (Song et al., 2004, 2010). Homologs of PCR proteins have also been identified in other plant species such as rice (Wang et al., 2019), poplar (Lv et al., 2023), *B. napus* (Li et al., 2023; Liu et al., 2024b), and *Salix linearistipularis* (Hu et al., 2022), where they similarly promote Cd efflux from root cells, thereby reducing cytosolic Cd levels and mitigating Cd toxicity.

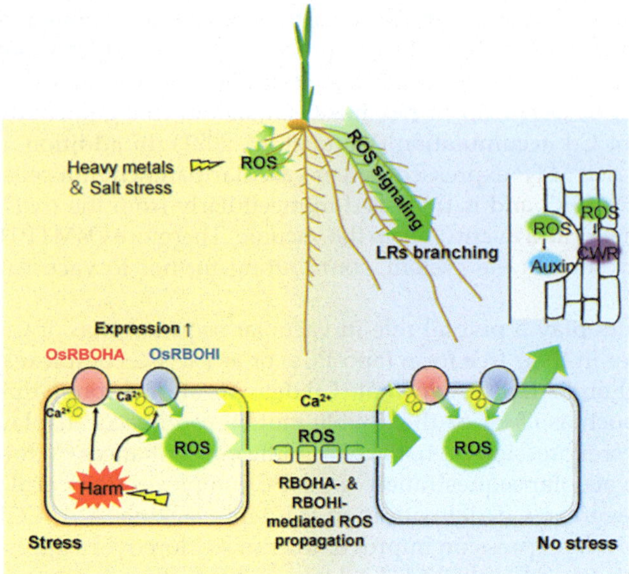

FIGURE 11.5 A proposed model for the RBOH-ROS-auxin signaling cascade involved in root avoidance of localized cadmium stress in rice (Wang et al., 2023). *From Wang et al. (2023). With copyright permission.*

11.4.6 Translocation to the above-ground tissues

The translocation of Cd from roots to shoots can serve as a detoxification strategy, particularly when Cd is more efficiently detoxified in the shoots. Transporters involved in this process play a critical regulatory role in facilitating the movement of Cd within plants (Zhao et al., 2022). This mechanism is especially prominent in Cd hyperaccumulators and hypertolerant plants (see Chapter 12), which efficiently translocate Cd to aerial parts for vacuolar sequestration. In contrast, nonhyperaccumulators typically retain most of the absorbed Cd within root tissues. Several plasma membrane-localized transporters facilitate Cd translocation by mediating its loading into the xylem and subsequent movement to shoots. Notably, HMA2 and HMA4 are major transporters for loading of Cd into xylem and intervascular translocation to aboveground tissues (Verret et al., 2004; Wong & Cobbett, 2009). Other transporters, such as OsCCX2 and OsZIP7, also contribute to Cd loading into the xylem and its transport to shoots in rice (Hao et al., 2018; Tan et al., 2019).

11.4.7 Avoidance of cadmium stress

Plants can remodel their root system architecture to avoid localized hot spots of Cd. In *Arabidopsis* a localized Cd avoidance response has been observed, where root growth is inhibited on the Cd-exposed side, whereas root biomass increases on the unexposed side. This asymmetric growth pattern is mediated by the TF GeBP-LIKE 4 (GPL4) (Khare et al., 2017). Suppression of GPL4 abolishes the root growth inhibition by localized Cd and reduces root growth on the Cd-free side. Wang et al. (2023) revealed that rice roots avoid localized Cd stress via the OsRBOH-ROS-auxin module (Fig. 11.5). When only part of the root system is exposed to Cd, rice preferentially initiates lateral root development in Cd-free area. This spatial response begins with an ROS burst in Cd-exposed root tips, triggered by the upregulation of OsRBOHA and OsRBOHI, two plasma membrane-localized NADPH oxidases. The ROS signal then propagates to the roots in the Cd-free zone, activating auxin signaling and cell wall remodeling, which together promote lateral root formation and elongation. This adaptive root architecture adjustment enables rice to maintain root function and nutrient uptake while limiting Cd absorption, serving as an important mechanism to avoid localized Cd stress.

11.5 Genetic and biotechnological approaches to enhance cadmium tolerance

11.5.1 Breeding for cadmium-resistant varieties

Breeding of Cd-resistant crop varieties aims to reduce Cd accumulation in edible plant tissues, thereby safeguarding food safety, and to enhance plant tolerance to Cd-induced stress. Strategies to improve Cd resistance can be broadly categorized into conventional breeding and modern molecular breeding approaches (see Chapter 19). Notably, substantial genetic variation in Cd resistance exists among cultivars or genotypes within the same species, providing a valuable resource for crop improvement.

Several key techniques are instrumental in identifying genes and regulatory pathways associated with Cd stress responses, including quantitative trait locus (QTL) mapping, GWAS, targeting induced local lesions in genomes (TILLING), DNA mismatch repair (MMR) pathway studies, and CRISPR/Cas9-mediated genome editing (Yu et al., 2024). GWAS has been widely applied in crops such as *Arabidopsis*, rice, wheat, barley, and maize, uncovering genetic loci linked to Cd accumulation in different plant organs (Chao et al., 2012; Wu et al., 2015; Zhao et al., 2018). In wheat, variance heterogeneity GWAS (vGWAS) has been employed to identify variance QTLs (vQTLs) associated with grain Cd accumulation (Hussain et al., 2020). TILLING has also proven effective in dissecting gene function under Cd stress. For example, in *Brassica rapa*, TILLING mutants were used to investigate the role of *HMA4a* and *CAX1a* in Cd detoxification (Navarro-León et al., 2019, 2020). In recent years, CRISPR/Cas9 genome editing has emerged as a more powerful and precise tool for functional gene analysis. For instance Zheng et al. (2022) used CRISPR/Cas9 to generate mutants in mismatch repair (MSH) genes and demonstrated that E2F TFs enhance rice tolerance to Cd by regulating MSH-mediated DNA repair. Similarly, Zhao et al. (2020) reported that MMR genes *MSH2* and *MSH6* contribute to genomic integrity under Cd stress by mediating G2/M cell cycle arrest independently of canonical MMR pathways in soybean. Furthermore, CRISPR/Cas9 has been employed to validate the roles of several genes in rice Cd tolerance. These include metal chaperones *OsHIPP16*, *OsHIPP17*, and *OsHIPP56* (Cao et al., 2022; Shi et al., 2023; Zhao et al., 2022), and the tau class GST *OsGSTU5* and *OsGSTU37* (Yang et al., 2023), all of which were found to play a role in Cd detoxification.

11.5.2 Genetic engineering strategies

Overexpression of detoxification-related genes represents a promising strategy for developing crop varieties with enhanced Cd resistance and reduced Cd accumulation. In rice, for instance, overexpression of *OsHMA3* has been shown to promote Cd sequestration into root vacuoles, thereby mitigating Cd toxicity and alleviating its inhibitory effects on lateral root development. This results in improved Cd tolerance and overall root architecture

(Lu et al., 2019; Sasaki et al., 2014; Wang et al., 2021). Similarly, in *A. thaliana*, overexpression of *AtHMA4* enhances root growth under Cd stress, contributing to better overall plant performance in Cd-contaminated environments (Verret et al., 2004). Another example is *OsHIPP29*, a gene encoding a heavy metal-associated isoprenylated plant protein, whose overexpression in rice significantly reduces Cd accumulation in both roots and shoots while improving growth under Cd exposure (Zhang et al., 2020). In addition to metal transporters, genes involved in PC biosynthesis are also considered valuable targets for enhancing Cd tolerance. Overexpression of the *PCS1* gene from *Vicia sativa* and maize has been shown to significantly improve Cd tolerance in transgenic plants by increasing PC production, which facilitates the chelation and vacuolar sequestration of Cd (Jin et al., 2022; Zhang et al., 2018).

11.6 Conclusion

Cd is a highly toxic metal to plants. Excessive accumulation of Cd affects plant growth by causing oxidative stress, inhibiting enzyme activities and photosynthesis, suppressing uptake of water and essential nutrients, and altering genetic material, ultimately leading to reduced productivity and, in severe cases, plant death. Plants have evolved various mechanisms to mitigate Cd toxicity, including modulation of Cd transporter activities, synthesis of Cd-binding peptides, regulation of antioxidant machinery components, sequestration of Cd in vacuoles, and avoidance of localized Cd stress. Many genes involved in Cd toxicity responses and detoxification have been identified, providing targets and gene resources for enhancing Cd tolerance through breeding and genetic engineering.

Cd toxicity is dose-dependent. Many studies have employed high concentrations of Cd, which leads to acute toxicity. Future studies should use environmentally relevant concentrations of Cd that likely cause low-level and chronic toxicity. Natural variation in Cd tolerance exists within the germplasm of many plant species. Understanding the genetic basis and molecular mechanisms underpinning the natural variation is another important avenue for future research, which could lead to the discovery of superior alleles for Cd tolerance that can be used in breeding of Cd-resistant crop varieties.

11.7 Summary

- Cadmium is nonessential and highly toxic to plants by disrupting root growth, nutrient uptake, photosynthesis, and development of reproductive tissues.
- Cadmium stress induces ROS generation, causing lipid peroxidation and protein/DNA damage, which results in root damage and leaf chlorosis, necrosis, reduced stomatal conductance, and impaired CO_2 assimilation.
- Plants detoxify Cd through several strategies, including limiting Cd uptake, binding Cd to cell walls, chelating it with organic molecules, sequestering it in vacuoles, and exporting it out of cells, as well as avoiding localized Cd by remodeling root system architecture.
- Cadmium stress triggers complex plant responses involving signal transduction (via ROS, NO, Ca^{2+}, and hormones) and gene regulation by key TFs and epigenetic modifications that influence gene expression.

AI disclosure

During the preparation of this work the author(s) used ChatGPT to polish English. After using this tool/service the author(s) reviewed and edited the content as needed and take(s) full responsibility for the content of the publication.

References

Abbas, T., Rizwan, M., Ali, S., Adrees, M., Zia-ur-Rehman, M., Qayyum, M. F., Ok, Y. S., & Murtaza, G. (2018). Effect of biochar on alleviation of cadmium toxicity in wheat (*Triticum aestivum* L.) grown on Cd-contaminated saline soil. *Environmental Science and Pollution Research, 25*(26), 25668–25680. doi:10.1007/s11356-017-8987-4.

Agarwal, P., Mitra, M., Banerjee, S., & Roy, S. (2020). MYB4 transcription factor, a member of R2R3-subfamily of MYB domain protein, regulates cadmium tolerance via enhanced protection against oxidative damage and increases expression of PCS1 and MT1C in *Arabidopsis. Plant Science, 29*, 110501. doi:10.1016/j.plantsci.2020.110501.

Ali, B., Gill, R. A., Yang, S., Gill, M. B., Farooq, M. A., Liu, D., Daud, M. K., Ali, S., & Zhou, W. (2015). Regulation of cadmium-induced proteomic and metabolic changes by 5-aminolevulinic acid in leaves of *Brassica napus* L. *PLoS One, 10*(4). doi:10.1371/journal.pone.0123328.

Alloway, B. J. (2012). *Heavy metals in soils: Trace metals and metalloids in soils and their bioavailability*: 2. Springer Netherlands. doi:10.1007/978-94-007-4470-7.

Andresen, E., & Küpper, H. (2013). Cadmium toxicity in plants. *Metal Ions in Life Sciences, 1*, 395–413. doi:10.1007/978-94-007-5179-8_13.

Anjum, N. A., Sharma, P., Gill, S. S., Hasanuzzaman, M., Khan, E. A., Kachhap, K., Mohamed, A. A., Thangavel, P., Devi, G. D., Vasudhevan, P., Sofo, A., Khan, N. A., Misra, A. N., Lukatkin, A. S., Singh, H. P., Pereira, E., & Tuteja, N. (2016). Catalase and ascorbate peroxidase—Representative H_2O_2-detoxifying heme enzymes in plants. *Environmental Science and Pollution Research, 23*(19), 19002–19029. doi:10.1007/s11356-016-7309-6.

Anjum, S. A., Tanveer, M., Hussain, S., Bao, M., Wang, L., Khan, I., Ullah, E., Tung, S. A., Samad, R. A., & Shahzad, B. (2015). Cadmium toxicity in maize (*Zea mays* L.): Consequences on antioxidative systems, reactive oxygen species and cadmium accumulation. *Environmental Science and Pollution Research, 22*(21), 17022–17030. doi:10.1007/s11356-015-4882-z.

Aslam, R., Ansari, M. Y. K., Choudhary, S., Bhat, T. M., & Jahan, N. (2014). Genotoxic effects of heavy metal cadmium on growth, biochemical, cyto-physiological parameters and detection of DNA polymorphism by RAPD in *Capsicum annuum* L.—An important spice crop of India. *Saudi Journal of Biological Sciences, 21*(5), 465–472. doi:10.1016/j.sjbs.2014.07.005.

Bahmani, R., Kim, D. G., Na, J. D., & Hwang, S. (2019). Expression of the tobacco non-symbiotic class 1 Hemoglobin Gene Hb1 reduces cadmium levels by modulating cd transporter expression through decreasing nitric oxide and ROS level in *Arabidopsis. Frontiers in Plant Science, 1*. doi:10.3389/fpls.2019.00201.

Bekkai, D., Chiofalo, M. T., Torre, D., Mileto, S., Genovese, G., Cimino, F., Toscano, G., Iannazzo, D., & Trifilò, P. (2024). Chronic mild cadmium exposure increases the vulnerability of tomato plants to dehydration. *Plant Physiology and Biochemistry, 21*, 109200. doi:10.1016/j.plaphy.2024.109200.

Besson-Bard, A., Gravot, A., Richaud, P., Auroy, P., Duc, C., Gaymard, F., Taconnat, L., Renou, J. P., Pugin, A., & Wendehenne, D. (2009). Nitric oxide contributes to cadmium toxicity in *Arabidopsis* by promoting cadmium accumulation in roots and by up-regulating genes related to iron uptake. *Plant Physiology, 149*(3), 1302–1315. doi:10.1104/pp.108.133348.

Bourbousse, C., Vegesna, N., & Law, J. A. (2018). SOG1 activator and MYB3R repressors regulate a complex DNA damage network in *Arabidopsis. Proceedings of the National Academy of Sciences of the United States of America, 115*(52), E12453. doi:10.1073/pnas.1810582115.

Brunetti, P., Zanella, L., De Paolis, A., Di Litta, D., Cecchetti, V., Falasca, G., Barbieri, M., Altamura, M. M., Costantino, P., & Cardarelli, M. (2015). Cadmium-inducible expression of the ABC-type transporter AtABCC3 increases phytochelatin-mediated cadmium tolerance in *Arabidopsis. Journal of Experimental Botany, 66*(13), 3815–3829. doi:10.1093/jxb/erv185.

Brunetti, P., Zanella, L., Proia, A., De Paolis, A., Falasca, G., Altamura, M. M., Sanità di Toppi, L., Costantino, P., & Cardarelli, M. (2011). Cadmium tolerance and phytochelatin content of *Arabidopsis* seedlings over-expressing the phytochelatin synthase gene AtPCS1. *Journal of Experimental Botany, 62*(15), 5509–5519. doi:10.1093/jxb/err228.

Bruno, L., Pacenza, M., Forgione, I., Lamerton, L. R., Greco, M., Chiappetta, A., & Bitonti, M. B. (2017). In *Arabidopsis thaliana* cadmium impact on the growth of primary root by altering SCR expression and auxin-cytokinin cross-talk. *Frontiers in Plant Science*. doi:10.3389/fpls.2017.01323.

Cai, Z., Xian, P., Wang, H., Lin, R., Lian, T., Cheng, Y., Ma, Q., & Nian, H. (2020). Transcription factor GmWRKY142 confers cadmium resistance by up-regulating the cadmium tolerance 1-Like Genes. *Frontiers in Plant Science, 1*. doi:10.3389/fpls.2020.00724.

Cao, H. W., Zhao, Y. N., Liu, X. S., Rono, J. K., & Yang, Z. M. (2022). A metal chaperone OsHIPP16 detoxifies cadmium by repressing its accumulation in rice crops. *Environmental Pollution, 31*. doi:10.1016/j.envpol.2022.120058.

Carvalho, H. H., Silva, P. A., Mendes, G. C., Brustolini, O. táJ. B., Pimenta, M. R., Gouveia, B. C., Valente, M. A. S., Ramos, H. J. O., Soares-Ramos, J. R. L., & Fontes, E. P. B. (2013). The endoplasmic reticulum binding protein BiP displays dual function in modulating cell death events. *Plant Physiology, 164*(2), 654–670. doi:10.1104/pp.113.231928.

Çatav, Ş. S., Genç, T. O., Oktay, M. K., & Küçükakyüz, K. (2020). Cadmium toxicity in wheat: Impacts on element contents, antioxidant enzyme activities, oxidative stress, and genotoxicity. *Bulletin of Environmental Contamination and Toxicology, 104*(1), 71–77. doi:10.1007/s00128-019-02745-4.

Chakrabarti, M., & Mukherjee, A. (2022). Metallo-adaptive response: A unique survival strategy of plants under genotoxic stress. *The Nucleus, 65*(1), 99–106. doi:10.1007/s13237-021-00360-5.

Chao, D. Y., Silva, A., Baxter, I., Huang, Y. S., Nordborg, M., Danku, J., Lahner, B., Yakubova, E., & Salt, D. E. (2012). Genome-Wide Association Studies identify heavy metal ATPase3 as the primary determinant of natural variation in leaf cadmium in *Arabidopsis thaliana*. *PLoS Genetics, 8*(9). doi:10.1371/journal.pgen.1002923.

Chen, J., Huang, X. Y., Salt, D. E., & Zhao, F. J. (2020). Mutation in OsCADT1 enhances cadmium tolerance and enriches selenium in rice grain. *New Phytologist, 226*(3), 838–850. doi:10.1111/nph.16404.

Chen, J., Yang, L., Yan, X., Liu, Y., Wang, R., Fan, T., Ren, Y., Tang, X., Xiao, F., Liu, Y., & Cao, S. (2016). Zinc-finger transcription factor ZAT6 positively regulates cadmium tolerance through the glutathione-dependent pathway in *Arabidopsis*. *Plant Physiology, 171*(1), 707–719. doi:10.1104/pp.15.01882.

Chen, X., Tao, H., Wu, Y., & Xu, X. (2022). Effects of cadmium on metabolism of photosynthetic pigment and photosynthetic system in *Lactuca sativa* L. revealed by physiological and proteomics analysis. *Scientia Horticulturae, 30*, 111371. doi:10.1016/j.scienta.2022.111371.

Chiraz, C., Houda, G., & Mohamed Habib, G. (2003). Nitrogen metabolism in tomato plants under cadmium stress. *Journal of Plant Nutrition, 26*(8), 1617–1634. doi:10.1081/PLN-120022372.

Ci, D., Jiang, D., Wollenweber, B., Dai, T., Jing, Q., & Cao, W. (2010). Cadmium stress in wheat seedlings: Growth, cadmium accumulation and photosynthesis. *Acta Physiologiae Plantarum, 32*(2), 365–373. doi:10.1007/s11738-009-0414-0.

Clemens, S., & Ma, J. F. (2016). Toxic heavy metal and metalloid accumulation in crop plants and foods. *Annual Review of Plant Biology, 6*, 489–512. doi:10.1146/annurev-arplant-043015-112301.

Clemens, S., Aarts, M. G. M., Thomine, S., & Verbruggen, N. (2013). Plant science: The key to preventing slow cadmium poisoning. *Trends in Plant Science, 18*(2), 92–99. doi:10.1016/j.tplants.2012.08.003.

Costa, M. D. L., Reis, P. A. B., Valente, M. A. S., Irsigler, A. S. T., Carvalho, C. M., Loureiro, M. E., Aragão, F. J. L., Boston, R. S., Fietto, L. G., & Fontes, E. P. B. (2008). A new branch of endoplasmic reticulum stress signaling and the osmotic signal converge on plant-specific asparagine-rich proteins to promote cell death. *Journal of Biological Chemistry, 283*(29), 20209–20219. doi:10.1074/jbc.M802654200.

Cui, W., Wang, H., Song, J., Cao, X., Rogers, H. J., Francis, D., Jia, C., Sun, L., Hou, M., Yang, Y., Tai, P., & Liu, W. (2017). Cell cycle arrest mediated by Cd-induced DNA damage in *Arabidopsis* root tips. *Ecotoxicology and Environmental Safety, 14*, 569–574. doi:10.1016/j.ecoenv.2017.07.074.

Cuypers, A., Plusquin, M., Remans, T., Jozefczak, M., Keunen, E., Gielen, H., Opdenakker, K., Nair, A. R., Munters, E., Artois, T. J., Nawrot, T., Vangronsveld, J., & Smeets, K. (2010). Cadmium stress: An oxidative challenge. *BioMetals, 23*(5), 927–940. doi:10.1007/s10534-010-9329-x.

Daud, M. K., Sun, Y., Dawood, M., Hayat, Y., Variath, M. T., Wu, Y. u-X., Raziuddin, U. M., Salahuddin, U. N., & Zhu, S. (2009). Cadmium-induced functional and ultrastructural alterations in roots of two transgenic cotton cultivars. *Journal of Hazardous Materials, 161*(1), 463–473. doi:10.1016/j.jhazmat.2008.03.128.

Demecsová, L., Zelinová, V., Liptáková, Ľ., Valentovičová, K., & Tamás, L. (2020). Indole-3-butyric acid priming reduced cadmium toxicity in barley root tip via NO generation and enhanced glutathione peroxidase activity. *Planta, 252*(3). doi:10.1007/s00425-020-03451-w.

Deng, B., Zhang, W., & Yang, H. (2022). Abscisic Acid decreases cell death in *Malus hupehensis* Rehd. Under Cd stress by reducing root Cd^{2+} influx and leaf transpiration. *Journal of Plant Growth Regulation, 41*(2), 639–646. doi:10.1007/s00344-021-10327-0.

Dias, M. C., Monteiro, C., Moutinho-Pereira, J., Correia, C., Gonçalves, B., & Santos, C. (2013). Cadmium toxicity affects photosynthesis and plant growth at different levels. *Acta Physiologiae Plantarum, 35*(4), 1281–1289. doi:10.1007/s11738-012-1167-8.

Ding, Y., Ding, L., Xia, Y., Wang, F., & Zhu, C. (2020). Emerging roles of microRNAs in plant heavy metal tolerance and homeostasis. *Journal of Agricultural and Food Chemistry, 68*(7), 1958–1965. doi:10.1021/acs.jafc.9b07468.

Ding, Y., Gong, S., Wang, Y., Wang, F., Bao, H., Sun, J., Cai, C., Yi, K., Chen, Z., & Zhu, C. (2018). MicroRNA166 modulates cadmium tolerance and accumulation in rice. *Plant Physiology, 177*(4), 1691–1703. doi:10.1104/pp.18.00485.

Du, X., Zhou, L., Zhu, B., Gu, L., Yin, H., & Wang, H. (2023). The TabHLH35–TaWAK20–TaSPL5 pathway positively regulates Cd stress in wheat. *Theoretical and Applied Genetics, 136*(7). doi:10.1007/s00122-023-04400-3.

Erdal, S., & Turk, H. (2016). Cysteine-induced upregulation of nitrogen metabolism-related genes and enzyme activities enhance tolerance of maize seedlings to cadmium stress. *Environmental and Experimental Botany, 13*, 92–99. doi:10.1016/j.envexpbot.2016.08.014.

Fan, S. K., Ye, J. Y., Zhang, L. L., Chen, H. S., Zhang, H. H., Zhu, Y. X., Liu, X. X., & Jin, C. W. (2020). Inhibition of DNA demethylation enhances plant tolerance to cadmium toxicity by improving iron nutrition. *Plant Cell and Environment, 43*(1), 275–291. doi:10.1111/pce.13670.

Farooq, M. A., Ali, S., Hameed, A., Ishaque, W., Mahmood, K., & Iqbal, Z. (2013). Alleviation of cadmium toxicity by silicon is related to elevated photosynthesis, antioxidant enzymes; suppressed cadmium uptake and oxidative stress in cotton. *Ecotoxicology and Environmental Safety, 9*, 242–249. doi:10.1016/j.ecoenv.2013.07.006.

Fattorini, L., Ronzan, M., Piacentini, D., Della Rovere, F., De Virgilio, C., Sofo, A., Altamura, M. M., & Falasca, G. (2017). Cadmium and arsenic affect quiescent centre formation and maintenance in *Arabidopsis thaliana* post-embryonic roots disrupting auxin biosynthesis and transport. *Environmental and Experimental Botany, 14*, 37–48. doi:10.1016/j.envexpbot.2017.10.005.

Feng, S., Tan, J., Zhang, Y., Liang, S., Xiang, S., Wang, H., & Chai, T. (2017). Isolation and characterization of a novel cadmium-regulated yellow stripe-like transporter (SnYSL3) in *Solanum nigrum. Plant Cell Reports, 36*(2), 281–296. doi:10.1007/s00299-016-2079-7.

Feng, S. J., Liu, X. S., Tao, H., Tan, S. K., Chu, S. S., Oono, Y., Zhang, X. D., Chen, J., & Yang, Z. M. (2016). Variation of DNA methylation patterns associated with gene expression in rice (*Oryza sativa*) exposed to cadmium. *Plant Cell and Environment, 39*(12), 2629–2649. doi:10.1111/pce.12793.

Ferreira, R. R., Fornazier, R. F., Vitória, A. P., Lea, P. J., & Azevedo, R. A. (2002). Changes in antioxidant enzyme activities in soybean under cadmium stress. *Journal of Plant Nutrition, 25*(2), 327–342.

Foyer, C. H. (2018). Reactive oxygen species, oxidative signaling and the regulation of photosynthesis. *Environmental and Experimental Botany, 15*, 134–142. doi:10.1016/j.envexpbot.2018.05.003.

Foyer, C. H., & Hanke, G. (2022). ROS production and signalling in chloroplasts: Cornerstones and evolving concepts. *The Plant Journal, 111*(3), 642–661. doi:10.1111/tpj.15856.

Fu, S., Lu, Y., Zhang, X., Yang, G., Chao, D., Wang, Z., Shi, M., Chen, J., Chao, D. Y., Li, R., Ma, J. F., & Xia, J. (2019). The ABC transporter ABCG36 is required for cadmium tolerance in rice. *Journal of Experimental Botany, 70*(20), 5909–5918. doi:10.1093/jxb/erz335.

Fujii, H., & Zhu, J. K. (2009). *Arabidopsis* mutant deficient in 3 abscisic acid-activated protein kinases reveals critical roles in growth, reproduction, and stress. *Proceedings of the National Academy of Sciences of the United States of America, 106*(20), 8380–8385. doi:10.1073/pnas.0903144106.

Gálusová, T., Piršelová, B., Rybanský, Ľ., Krasylenko, Y., Mészáros, P., Blehová, A., Bardáčová, M., Moravčíková, J., & Matušíková, I. (2020). Plasticity of soybean stomatal responses to arsenic and cadmium at the whole plant level. *Polish Journal of Environmental Studies, 29*(5), 3569–3580. doi:10.15244/pjoes/116444.

Ge, W., Jiao, Y., Yue, J., Qin, R., Jiang, W., & Liu, D. (2013). Cadmium effects on photosynthetic parameters and chloroplast ultrastructures of Populus 118. *Fresenius Environmental Bulletin, 22*(12 A), 3647–3653. http://www.psp-parlar.de/feb_auswahl.asp.

Genchi, G., Sinicropi, M. S., Lauria, G., Carocci, A., & Catalano, A. (2020). The effects of cadmium toxicity. *International Journal of Environmental Research and Public Health, 17*(11), 3782. doi:10.3390/ijerph17113782.

Gouia, H., Habib Ghorbal, M., & Meyer, C. (2000). Effects of cadmium on activity of nitrate reductase and on other enzymes of the nitrate assimilation pathway in bean. *Plant Physiology and Biochemistry, 38*(7–8), 629–638. doi:10.1016/S0981-9428(00)00775-0.

Grobelak, A., Świątek, J., Murtaś, A., & Jaskulak, M. (2019). Cadmium-induced oxidative stress in plants, cadmium toxicity, and tolerance in plants: From physiology to remediation. In *Cadmium toxicity and tolerance in plants* (pp. 213–231). Academic Press. doi:10.1016/b978-0-12-814864-8.00009-7.

Guo, B., Liu, C., Ding, N., Fu, Q., Lin, Y., Li, H., & Li, N. (2016). Silicon alleviates cadmium toxicity in two cypress varieties by strengthening the exodermis tissues and stimulating phenolic exudation of roots. *Journal of Plant Growth Regulation, 35*(2), 420–429. doi:10.1007/s00344-015-9549-y.

Guo, Z., Gao, Y., Yuan, X., Yuan, M., Huang, L., Wang, S., Liu, C. ′, & Duan, C. (2023). Effects of heavy metals on stomata in plants: A review. *International Journal of Molecular Sciences, 24*(11), 9302. doi:10.3390/ijms24119302.

Gupta, D. K., Pena, L. B., Romero-Puertas, M. C., Hernández, A., Inouhe, M., & Sandalio, L. M. (2017). NADPH oxidases differentially regulate ROS metabolism and nutrient uptake under cadmium toxicity. *Plant Cell and Environment, 40*(4), 509–526. doi:10.1111/pce.12711.

Haas, K. L., & Franz, K. J. (2009). Application of metal coordination chemistry to explore and manipulate cell biology. *Chemical Reviews, 109*(10), 4921–4960. doi:10.1021/cr900134a.

Haider, F. U., Liqun, C., Coulter, J. A., Cheema, S. A., Wu, J., Zhang, R., Wenjun, M., & Farooq, M. (2021). Cadmium toxicity in plants: Impacts and remediation strategies. *Ecotoxicology and Environmental Safety, 21*. doi:10.1016/j.ecoenv.2020.111887.

Han, G. H., Huang, R. N., Hong, L. H., Xu, J. X., Hong, Y. G., Wu, Y. H., & Chen, W. W. (2023). The transcription factor NAC102 confers cadmium tolerance by regulating WAKL11 expression and cell wall pectin metabolism in *Arabidopsis*. *Journal of Integrative Plant Biology, 65*(10), 2262–2278. doi:10.1111/jipb.13557.

Han, Y., Fan, T., Zhu, X., Wu, X., Ouyang, J., Jiang, L., & Cao, S. (2019). WRKY12 represses GSH1 expression to negatively regulate cadmium tolerance in *Arabidopsis*. *Plant Molecular Biology, 99*(1–2), 149–159. doi:10.1007/s11103-018-0809-7.

Hao, X., Zeng, M., Wang, J., Zeng, Z., Dai, J., Xie, Z., Yang, Y., Tian, L., Chen, L., & Li, D. (2018). A node-expressed transporter OsCCX2 is involved in grain cadmium accumulation of rice. *Frontiers in Plant Science*. doi:10.3389/fpls.2018.00476.

Hara-Nishimura, I., Hatsugai, N., Nakaune, S., Kuroyanagi, M., & Nishimura, M. (2005). Vacuolar processing enzyme: An executor of plant cell death. *Current Opinion in Plant Biology, 8*(4), 404–408. doi:10.1016/j.pbi.2005.05.016.

Hasanuzzaman, M., Bhuyan, M. H. M., Zulfiqar, F., Raza, A., Mohsin, S., Mahmud, J., Fujita, M., & Fotopoulos, V. (2020). Reactive oxygen species and antioxidant defense in plants under abiotic stress: Revisiting the crucial role of a universal defense regulator. *Antioxidants, 9*(8), 681. doi:10.3390/antiox9080681.

Hashem, H. A. (2014). Cadmium toxicity induces lipid peroxidation and alters cytokinin content and antioxidant enzyme activities in soybean. *Botany, 92*(1), 1–7. doi:10.1139/cjb-2013-0164.

Hassan, M., Israr, M., Mansoor, S., Hussain, S. A., Basheer, F., Azizullah, A., & Ur Rehman, S. (2021). Acclimation of cadmium-induced genotoxicity and oxidative stress in mung bean seedlings by priming effect of phytohormones and proline. *PLoS One, 16*(9). doi:10.1371/journal.pone.0257924.

Hayashi, S., Kuramata, M., Abe, T., Takagi, H., Ozawa, K., & Ishikawa, S. (2017). Phytochelatin synthase Os PCS 1 plays a crucial role in reducing arsenic levels in rice grains. *The Plant Journal, 91*(5), 840–848. doi:10.1111/tpj.13612.

He, S., Yang, X., He, Z., & Baligar, V. C. (2017). Morphological and physiological responses of plants to cadmium toxicity: A review. *Pedosphere, 27*(3), 421–438. doi:10.1016/S1002-0160(17)60339-4.

Hendrix, S., Iven, V., Eekhout, T., Huybrechts, M., Pecqueur, I., Horemans, N., Keunen, E., De Veylder, L., Vangronsveld, J., & Cuypers, A. (2020). Suppressor of gamma response 1 modulates the DNA damage response and oxidative stress response in leaves of cadmium-exposed *Arabidopsis thaliana*. *Frontiers in Plant Science, 1*. doi:10.3389/fpls.2020.00366.

Herbst, J., Li, Q. Q., & De Veylder, L. (2024). Mechanistic insights into DNA damage recognition and checkpoint control in plants. *Nature Plants, 10*(4), 539–550. doi:10.1038/s41477-024-01652-9.

Hernández-Allica, J., Becerril, J. M., & Garbisu, C. (2008). Assessment of the phytoextraction potential of high biomass crop plants. *Environmental Pollution, 152*(1), 32–40. doi:10.1016/j.envpol.2007.06.002.

Hong, Y., Chen, Y., Shi, H., Kong, X., Yao, J., Lei, M., Zhu, J. K., & Wang, Z. (2022). SUMO E3 ligase SIZ1 negatively regulates arsenite resistance via depressing GSH biosynthesis in *Arabidopsis*. *Stress Biology, 2*(1). doi:10.1007/s44154-021-00029-8.

Howden, R., Andersen, C. R., Goldsbrough, P. B., & Cobbett, C. S. (1995a). A cadmium-sensitive, glutathione-deficient mutant of *Arabidopsis thaliana*. *Plant Physiology, 107*(4), 1067–1073. doi:10.1104/pp.107.4.1067.

Howden, R., Goldsbrough, P. B., Andersen, C. R., & Cobbett, C. S. (1995b). Cadmium-sensitive, cad1 mutants of *Arabidopsis thaliana* are phytochelatin deficient. *Plant Physiology, 107*(4), 1059–1066. doi:10.1104/pp.107.4.1059.

Hsu, Y. T., & Kao, C. H. (2005). Abscisic acid accumulation and cadmium tolerance in rice seedlings. *Physiologia Plantarum, 124*(1), 71–80. doi:10.1111/j.1399-3054.2005.00490.x.

Hu, X., Wang, S., Zhang, H., Zhang, H., Feng, S., Qiao, K., Lv, F., Gong, S., & Zhou, A. (2022). Plant cadmium resistance 6 from *Salix linearistipularis* (SlPCR6) affects cadmium and copper uptake in roots of transgenic Populus. *Ecotoxicology and Environmental Safety, 24*, 114116. doi:10.1016/j.ecoenv.2022.114116.

Hu, Z., Cools, T., & De Veylder, L. (2016). Mechanisms used by plants to cope with DNA damage. *Annual Review of Plant Biology, 67*(1), 439–462. doi:10.1146/annurev-arplant-043015-111902.

Huang, B. Y., Zhao, F. J., & Wang, P. (2022). The relative contributions of root uptake and remobilization to the loading of Cd and As into rice grains: Implications in simultaneously controlling grain Cd and As accumulation using a segmented water management strategy. *Environmental Pollution, 29*. doi:10.1016/j.envpol.2021.118497.

Hussain, W., Campbell, M. T., Jarquin, D., Walia, H., & Morota, G. (2020). Variance heterogeneity genome-wide mapping for cadmium in bread wheat reveals novel genomic loci and epistatic interactions. *Plant Genome, 13*(1). doi:10.1002/tpg2.20011.

Jabeen, Z., Irshad, F., Habib, A., Hussain, N., Sajjad, M., Mumtaz, S., Rehman, S., Haider, W., & Hassan, M. N. (2022). Alleviation of cadmium stress in rice by inoculation of *Bacillus cereus*. *PeerJ, 1*, e13131. doi:10.7717/peerj.13131.

Javed, M. T., Akram, M. S., Tanwir, K., Chaudhary, H. J., Ali, Q., Stoltz, E., & Lindberg, S. (2017). Cadmium spiked soil modulates root organic acids exudation and ionic contents of two differentially Cd tolerant maize (*Zea mays* L.) cultivars. *Ecotoxicology and Environmental Safety, 14*, 216–225. doi:10.1016/j.ecoenv.2017.03.027.

Jia, H., Wang, X., Wei, T., Zhou, R., Muhammad, H., Hua, L., Ren, X., Guo, J., & Ding, Y. (2019). Accumulation and fixation of Cd by tomato cell wall pectin under Cd stress. *Environmental and Experimental Botany, 16*. doi:10.1016/j.envexpbot.2019.103829.

Jin, D., Zhang, Q., Liu, Y., Liang, M., Li, A., & Wu, J. (2022). Overexpression of the maize phytochelatin synthase gene (ZmPCS1) enhances Cd tolerance in plants. *Acta Physiologiae Plantarum, 44*(11). doi:10.1007/s11738-022-03451-1.

Jinadasa, N., Collins, D., Holford, P., Milham, P. J., & Conroy, J. P. (2016). Reactions to cadmium stress in a cadmium-tolerant variety of cabbage (*Brassica oleracea* L.): Is cadmium tolerance necessarily desirable in food crops? *Environmental Science and Pollution Research, 23*(6), 5296–5306. doi:10.1007/s11356-015-5779-6.

Ju, C., & Chang, C. (2015). Mechanistic insights in ethylene perception and signal transduction. *Plant Physiology, 169*(1), 85–95. doi:10.1104/pp.15.00845.

Jung, H. I., Lee, B. R., Chae, M. J., Lee, E. J., Lee, T. G., Jung, G. B., Kim, M. S., & Lee, J. (2020). Ascorbate-mediated modulation of cadmium stress responses: Reactive oxygen species and redox status in *Brassica napus*. *Frontiers in Plant Science, 1*. doi:10.3389/fpls.2020.586547.

Kaya, C., Polat, T., Ashraf, M., Kaushik, P., Alyemeni, M. N., & Ahmad, P. (2021). Endogenous nitric oxide and its potential sources regulate glutathione-induced cadmium stress tolerance in maize plants. *Plant Physiology and Biochemistry, 16*, 723–737. doi:10.1016/j.plaphy.2021.08.030.

Kerchev, P. I., & Van Breusegem, F. (2022). Improving oxidative stress resilience in plants. *The Plant Journal, 109*(2), 359–372. doi:10.1111/tpj.15493.

Khan, M. I. R., Iqbal, N., Masood, A., Mobin, M., Anjum, N. A., & Khan, N. A. (2016). Modulation and significance of nitrogen and sulfur metabolism in cadmium challenged plants. *Plant Growth Regulation, 78*(1), 1–11. doi:10.1007/s10725-015-0071-9.

Khare, D., Mitsuda, N., Lee, S., Song, W. Y., Hwang, D., Ohme-Takagi, M., Martinoia, E., Lee, Y., & Hwang, J. U. (2017). Root avoidance of toxic metals requires the GeBP-LIKE 4 transcription factor in *Arabidopsis thaliana*. *New Phytologist, 213*(3), 1257–1273. doi:10.1111/nph.14242.

Kim, D. Y., Bovet, L., Maeshima, M., Martinoia, E., & Lee, Y. (2007). The ABC transporter AtPDR8 is a cadmium extrusion pump conferring heavy metal resistance. *Plant Journal, 50*(2), 207–218. doi:10.1111/j.1365-313X.2007.03044.x.

Kong, X., Li, C., Zhang, F., Yu, Q., Gao, S., Zhang, M., Tian, H., Zhang, J., Yuan, X., & Ding, Z. (2018). Ethylene promotes cadmium-induced root growth inhibition through EIN3 controlled XTH33 and LSU1 expression in *Arabidopsis*. *Plant, Cell & Environment, 41*(10), 2449–2462. doi:10.1111/pce.13361.

Kopittke, P. M., Blamey, F. P. C., Asher, C. J., & Menzies, N. W. (2010). Trace metal phytotoxicity in solution culture: A review. *Journal of Experimental Botany, 61*(4), 945–954. doi:10.1093/jxb/erp385.

Kumar, S., Trivedi, P. K., & Metal, P. (2016). Heavy metal stress signaling in plants. In P. Ahmad (Ed.), *Plant metal interaction: Emerging remediation techniques*. Elsevier. doi:10.1016/B978-0-12-803158-2.00025-4.

Küpper, H., Küpper, F., & Spiller, M. (1998). In situ detection of heavy metal substituted chlorophylls in water plants. *Photosynthesis Research, 58*(2), 123–133. doi:10.1023/a:1006132608181.

Lamb, D. T., Ming, H., Megharaj, M., & Naidu, R. (2010). Relative tolerance of a range of Australian native plant species and lettuce to copper, zinc, cadmium, and lead. *Archives of Environmental Contamination and Toxicology, 59*(3), 424–432. doi:10.1007/s00244-010-9481-x.

Lane, T. W., Saito, M. A., George, G. N., Pickering, I. J., Prince, R. C., & Morel, F. M. M. (2005). A cadmium enzyme from a marine diatom. *Nature, 435*(7038), 42. doi:10.1038/435042a.

Lee, B. R., & Hwang, S. (2015). Over-expression of NtHb1 encoding a non-symbiotic class 1 hemoglobin of tobacco enhances a tolerance to cadmium by decreasing NO (nitric oxide) and Cd levels in *Nicotiana tabacum*. *Environmental and Experimental Botany, 11*, 18–27. doi:10.1016/j.envexpbot.2015.01.003.

Lei, G. J., Fujii-Kashino, M., Wu, D. Z., Hisano, H., Saisho, D., Deng, F., Yamaji, N., Sato, K., Zhao, F. J., & Ma, J. F. (2020a). Breeding for low cadmium barley by introgression of a Sukkula-like transposable element. *Nature Food, 1*(8), 489–499. doi:10.1038/s43016-020-0130-x.

Lei, G. J., Sun, L., Sun, Y., Zhu, X. F., Li, G. X., & Zheng, S. J. (2020b). Jasmonic acid alleviates cadmium toxicity in *Arabidopsis* via suppression of cadmium uptake and translocation. *Journal of Integrative Plant Biology, 62*(2), 218–227. doi:10.1111/jipb.12801.

Li, F., Deng, Y., Liu, Y., Mai, C., Xu, Y., Wu, J., Zheng, X., Liang, C., & Wang, J. (2023). *Arabidopsis* transcription factor WRKY45 confers cadmium tolerance via activating PCS1 and PCS2 expression. *Journal of Hazardous Materials, 46*, 132496. doi:10.1016/j.jhazmat.2023.132496.

Li, S., Yang, W., Yang, T., Chen, Y., & Ni, W. (2015). Effects of cadmium stress on leaf chlorophyll fluorescence and photosynthesis of *Elsholtzia argyi*—A cadmium accumulating plant. *International Journal of Phytoremediation, 17*(1), 85–92. doi:10.1080/15226514.2013.828020.

Li, S. U. N., Yu, J., Zhu, M., Zhao, F., & Luan, S. (2012). Cadmium impairs ion homeostasis by altering K^+ and Ca^{2+} channel activities in rice root hair cells. *Plant, Cell & Environment, 35*(11), 1998–2013. doi:10.1111/j.1365-3040.2012.02532.x.

Li, Z., Zhang, L., Yu, Y., Quan, R., Zhang, Z., Zhang, H., & Huang, R. (2011). The ethylene response factor AtERF11 that is transcriptionally modulated by the bZIP transcription factor HY5 is a crucial repressor for ethylene biosynthesis in *Arabidopsis*. *The Plant Journal, 68*(1), 88–99. doi:10.1111/j.1365-313x.2011.04670.x.

Liebthal, M., Maynard, D., & Dietz, K. J. (2018). Proxiredoxins and redox signaling in plants. *Antioxidants and Redox Signaling, 28*(7), 609–624. doi:10.1089/ars.2017.7164.

Lin, T., Yang, W., Lu, W., Wang, Y., & Qi, X. (2017). Transcription factors PvERF15 and PvMTF-1 form a cadmium stress transcriptional pathway. *Plant Physiology, 173*(3), 1565–1573. doi:10.1104/pp.16.01729.

Lin, X. L., Niu, D., Hu, Z. L., Kim, D. H., Jin, Y. H., Cai, B., Liu, P., Miura, K., Yun, D. J., Kim, W. Y., Lin, R., & Jin, J. B. (2016). An *Arabidopsis* SUMO E3 ligase, SIZ1, negatively regulates photomorphogenesis by promoting COP1 activity. *PLoS Genetics, 12*(4). doi:10.1371/journal.pgen.1006016.

Liu, C. L., Gao, Z. Y., Shang, L. G., Yang, C. H., Ruan, B. P., Zeng, D. L., Guo, L. B., Zhao, F. J., Huang, C. F., & Qian, Q. (2020). Natural variation in the promoter of OsHMA3 contributes to differential grain cadmium accumulation between indica and japonica rice. *Journal of Integrative Plant Biology, 62*(3), 314–329. doi:10.1111/jipb.12794.

Liu, H., Zhao, H., Wu, L., Liu, A., Zhao, F. J., & Xu, W. (2017). Heavy metal ATPase 3 (HMA3) confers cadmium hypertolerance on the cadmium/zinc hyperaccumulator *Sedum plumbizincicola*. *New Phytologist, 215*(2), 687–698. doi:10.1111/nph.14622.

Liu, P., Sun, L., Zhang, Y. u, Tan, Y., Zhu, Y., Peng, C., Wang, J., Yan, H., Mao, D., Liang, G., Liang, G., Li, X., Liang, Y., Wang, F., He, Z., Tang, W., Huang, D., & Chen, C. (2024a). The metal tolerance protein OsMTP11 facilitates cadmium sequestration in the vacuoles of leaf vascular cells for restricting its translocation into rice grains. *Molecular Plant, 17*(11), 1733–1752. doi:10.1016/j.molp.2024.09.012.

Liu, X. M., Kim, K. E., Kim, K. C., Nguyen, X. C., Han, H. J., Jung, M. S., Kim, H. S., Kim, S. H., Park, H. C., Yun, D. J., & Chung, W. S. (2010). Cadmium activates *Arabidopsis* MPK3 and MPK6 via accumulation of reactive oxygen species. *Phytochemistry, 71*(5-6), 614–618. doi:10.1016/j.phytochem.2010.01.005.

Liu, X. S., Feng, S. J., Zhang, B. Q., Wang, M. Q., Cao, H. W., Rono, J. K., Chen, X., & Yang, Z. M. (2019). OsZIP1 functions as a metal efflux transporter limiting excess zinc, copper and cadmium accumulation in rice. *BMC Plant Biology, 19*(1). doi:10.1186/s12870-019-1899-3.

Liu, Y., Tao, Q., Li, J., Guo, X., Luo, J., Jupa, R., Liang, Y., & Li, T. (2021). Ethylene-mediated apoplastic barriers development involved in cadmium accumulation in root of hyperaccumulator *Sedum alfredii*. *Journal of Hazardous Materials, 40*, 123729. doi:10.1016/j.jhazmat.2020.123729.

Liu, Y., Lu, M., Pergament Persson, D., Luo, J., Liang, Y., & Li, T. (2022). The involvement of nitric oxide and ethylene on the formation of endodermal barriers in response to Cd in hyperaccumulator *Sedum alfredii*. *Environmental Pollution, 30*, 119530. doi:10.1016/j.envpol.2022.119530.

Liu, Y., Shi, R., Xie, M., Huang, Y., Cheng, L., Song, Y., Zhang, W., & Chen, X. (2024b). Plant cadmium resistance 2.7 from *Brassica napus* (BnPCR2.7) improves copper and cadmium tolerance. *Environmental and Experimental Botany, 22*, 105919. doi:10.1016/j.envexpbot.2024.105919.

Loix, C., Huybrechts, M., Vangronsveld, J., Gielen, M., Keunen, E., & Cuypers, A. (2017). Reciprocal Interactions between cadmium-induced cell wall responses and oxidative stress in plants. *Frontiers in Plant Science*. doi:10.3389/fpls.2017.01867.

Lu, C., Zhang, L., Tang, Z., Huang, X. Y., Ma, J. F., & Zhao, F. J. (2019). Producing cadmium-free indica rice by overexpressing OsHMA3. *Environment International, 12*, 619–626. doi:10.1016/j.envint.2019.03.004.

Lu, L., Chen, X., Chen, J., Zhang, Z., Zhang, Z., Sun, Y., Wang, Y., Xie, S., Ma, Y., Song, Y., & Zeng, R. (2024). MicroRNA-encoded regulatory peptides modulate cadmium tolerance and accumulation in rice. *Plant, Cell & Environment, 47*(5), 1452–1470. doi:10.1111/pce.14819.

Lu, Q., Zhang, T., Zhang, W., Su, C., Yang, Y., Hu, D., & Xu, Q. (2018). Alleviation of cadmium toxicity in *Lemna minor* by exogenous salicylic acid. *Ecotoxicology and Environmental Safety, 14*, 500–508. doi:10.1016/j.ecoenv.2017.09.015.

Lu, Q., Chen, S., Li, Y., Zheng, F., He, B., & Gu, M. (2020). Exogenous abscisic acid (ABA) promotes cadmium (Cd) accumulation in *Sedum alfredii* Hance by regulating the expression of Cd stress response genes. *Environmental Science and Pollution Research, 27*(8), 8719–8731. doi:10.1007/s11356-019-07512-w.

Luo, Y., Wei, Y., Sun, S., Wang, J., Wang, W., Han, D., Shao, H., Jia, H., & Fu, Y. (2019). Selenium modulates the level of auxin to alleviate the toxicity of cadmium in tobacco. *International Journal of Molecular Sciences, 20*(15), 3772. doi:10.3390/ijms20153772.

Lux, A., Martinka, M., Vaculík, M., & White, P. J. (2011). Root responses to cadmium in the rhizosphere: A review. *Journal of Experimental Botany, 62*(1), 21–37. doi:10.1093/jxb/erq281.

Lv, F., Shan, Q., Qiao, K., Zhang, H., & Zhou, A. (2023). *Populus euphratica* plant cadmium resistance 2 mediates Cd tolerance by root efflux of Cd ions in poplar. *Plant Cell Reports, 42*(11), 1777–1789. doi:10.1007/s00299-023-03065-y.

Ma, Q., Cao, X., Tan, X., Si, L., & Wu, L. (2017). Effects of cadmium stress on pakchoi (*Brassica chinensis* L.) growth and uptake of inorganic and organic nitrogenous compounds. *Environmental and Experimental Botany, 13*, 49–57. doi:10.1016/j.envexpbot.2017.02.001.

Meng, Y., Li, M., Guo, Z., Chen, J., Wu, J., & Xia, Z. (2024). The transcription factor ZmbHLH105 confers cadmium tolerance by promoting abscisic acid biosynthesis in maize. *Journal of Hazardous Materials, 48*, 135826. doi:10.1016/j.jhazmat.2024.135826.

Miyadate, H., Adachi, S., Hiraizumi, A., Tezuka, K., Nakazawa, N., Kawamoto, T., Katou, K., Kodama, I., Sakurai, K., Takahashi, H., Satoh-Nagasawa, N., Watanabe, A., Fujimura, T., & Akagi, H. (2011). OsHMA3, a P 1B-type of ATPase affects root-to-shoot cadmium translocation in rice by mediating efflux into vacuoles. *New Phytologist, 189*(1), 190–199. doi:10.1111/j.1469-8137.2010.03459.x.

Naeem, A., Saifullah, M. Z.-R., Akhtar, T., Zia, M. H., & Aslam, M. (2018). Silicon nutrition lowers cadmium content of wheat cultivars by regulating transpiration rate and activity of antioxidant enzymes. *Environmental Pollution, 24*, 126–135. doi:10.1016/j.envpol.2018.06.069.

Nahakpam, S., & Shah, K. (2011). Expression of key antioxidant enzymes under combined effect of heat and cadmium toxicity in growing rice seedlings. *Plant Growth Regulation, 63*(1), 23–35. doi:10.1007/s10725-010-9508-3.

Navarro-León, E., Oviedo-Silva, J., Ruiz, J. M., & Blasco, B. (2019). Possible role of HMA4a TILLING mutants of *Brassica rapa* in cadmium phytoremediation programs. *Ecotoxicology and Environmental Safety, 18*, 88–94. doi:10.1016/j.ecoenv.2019.04.081.

Navarro-León, E., Manuel Ruiz, J., Albacete, A., & Blasco, B. (2020). Tolerance to cadmium toxicity and phytoremediation potential of three *Brassica rapa* CAX1a TILLING mutants. *Ecotoxicology and Environmental Safety, 18*, 109961. doi:10.1016/j.ecoenv.2019.109961.

Norini, M. P., Thouin, H., Miard, F., Battaglia-Brunet, F., Gautret, P., Guégan, R., Le Forestier, L., Morabito, D., Bourgerie, S., & Motelica-Heino, M. (2019). Mobility of Pb, Zn, Ba, As and Cd toward soil pore water and plants (willow and ryegrass) from a mine soil amended with biochar. *Journal of Environmental Management, 23*, 117–130. doi:10.1016/j.jenvman.2018.11.021.

Nouairi, I., Jalali, K., Essid, S., Zribi, K., & Mhadhbi, H. (2019). Alleviation of cadmium-induced genotoxicity and cytotoxicity by calcium chloride in faba bean (*Vicia faba* L. var. minor) roots. *Physiology and Molecular Biology of Plants, 25*(4), 921–931. doi:10.1007/s12298-019-00681-5.

Oono, Y., Yazawa, T., Kanamori, H., Sasaki, H., Mori, S., Handa, H., & Matsumoto, T. (2016). Genome-wide transcriptome analysis of cadmium stress in rice. *BioMed Research International, 201*, 1–9. doi:10.1155/2016/9739505.

Pan, C., Lu, H., Yu, J., Liu, J., Liu, Y., & Yan, C. (2019). Identification of cadmium-responsive *Kandelia obovata* SOD family genes and response to Cd toxicity. *Environmental and Experimental Botany, 16*, 230–238. doi:10.1016/j.envexpbot.2019.02.018.

Pang, B., Zuo, D., Yang, T., Yu, J., Zhou, L., Hou, Y., Yu, J., Ye, L., Gu, L., Wang, H., Du, X., Liu, Y., & Zhu, B. (2024). BcaSOD1 enhances cadmium tolerance in transgenic *Arabidopsis* by regulating the expression of genes related to heavy metal detoxification and arginine synthesis. *Plant Physiology and Biochemistry, 20*, 108299. doi:10.1016/j.plaphy.2023.108299.

Parisy, V., Poinssot, B., Owsianowski, L., Buchala, A., Glazebrook, J., & Mauch, F. (2007). Identification of PAD2 as a γ-glutamylcysteine synthetase highlights the importance of glutathione in disease resistance of *Arabidopsis*. *The Plant Journal, 49*(1), 159–172. doi:10.1111/j.1365-313x.2006.02938.x.

Park, J., Song, W. Y., Ko, D., Eom, Y., Hansen, T. H., Schiller, M., Lee, T. G., Martinoia, E., & Lee, Y. (2012). The phytochelatin transporters AtABCC1 and AtABCC2 mediate tolerance to cadmium and mercury. *Plant Journal, 69*(2), 278–288. doi:10.1111/j.1365-313X.2011.04789.x.

Parmar, P., Kumari, N., & Sharma, V. (2013). Structural and functional alterations in photosynthetic apparatus of plants under cadmium stress. *Botanical Studies, 54*(1). doi:10.1186/1999-3110-54-45.

Pedroza-Garcia, J. A., Xiang, Y., & De Veylder, L. (2022). Cell cycle checkpoint control in response to DNA damage by environmental stresses. *The Plant Journal, 109*(3), 490–507. doi:10.1111/tpj.15567.

Peng, J. S., Wang, Y. J., Ding, G., Ma, H. L., Zhang, Y. J., & Gong, J. M. (2017). A pivotal role of cell wall in cadmium accumulation in the *Crassulaceae* hyperaccumulator *Sedum plumbizincicola*. *Molecular Plant, 10*(5), 771–774. doi:10.1016/j.molp.2016.12.007.

Pérez Chaca, M. V., Vigliocco, A., Reinoso, H., Molina, A., Abdala, G., Zirulnik, F., & Pedranzani, H. (2014). Effects of cadmium stress on growth, anatomy and hormone contents in *Glycine max* (L.) Merr. *Acta Physiologiae Plantarum, 36*(10), 2815–2826. doi:10.1007/s11738-014-1656-z.

Pimenta, M. R., Silva, P. A., Mendes, G. C., Alves, J. R., Caetano, H. D. N., Machado, J. P. B., Brustolini, O. J. B., Carpinetti, P. A., Melo, B. P., Silva, J. C. F., Rosado, G. L., Ferreira, M. F. S., Dal-Bianco, M., Picoli, E. A. D. T., Aragao, F. J. L., Ramos, H. J. O., & Fontes, E. P. B. (2016). The stress-induced soybean NAC transcription factor GmNAC81 plays a positive role in developmentally programmed leaf senescence. *Plant and Cell Physiology, 57*(5), 1098–1114. doi:10.1093/pcp/pcw059.

Pizzaia, D., Nogueira, M. L., Mondin, M., Carvalho, M. E. A., Piotto, F. A., Rosario, M. F., & Azevedo, R. A. (2019). Cadmium toxicity and its relationship with disturbances in the cytoskeleton, cell cycle and chromosome stability. *Ecotoxicology, 28*(9), 1046–1055. doi:10.1007/s10646-019-02096-0.

Quadros, I. P. S., Madeira, N. N., Loriato, V. A. P., Saia, T. F. F., Silva, J. C., Soares, F. A. F., Carvalho, J. R., Reis, P. A. B., Fontes, E. P. B., Clarindo, W. R., & Fontes, R. L. F. (2022). Cadmium-mediated toxicity in plant cells is associated with the DCD/NRP-mediated cell death response. *Plant Cell and Environment, 45*(2), 556–571. doi:10.1111/pce.14218.

Rascio, N., Vecchia, F. D., Rocca, N. L., Barbato, R., Pagliano, C., Raviolo, M., Gonnelli, C., & Gabbrielli, R. (2008). Metal accumulation and damage in rice (cv. Vialone nano) seedlings exposed to cadmium. *Environmental and Experimental Botany, 62*(3), 267–278. doi:10.1016/j.envexpbot.2007.09.002.

Reis, P. A. A., Rosado, G. L., Silva, L. A. C., Oliveira, L. C., Oliveira, L. B., Costa, M. D. L., Alvim, F. C., & Fontes, E. P. B. (2011). The binding protein BiP attenuates stress-induced cell death in soybean via modulation of the N-RICH protein-mediated signaling pathway. *Plant Physiology, 157*(4), 1853–1865. doi:10.1104/pp.111.179697.

Rizwan, M., Ali, S., Adrees, M., Ibrahim, M., Tsang, D. C. W., Zia-ur-Rehman, M., Zahir, Z. A., Rinklebe, J., Tack, F. M. G., & Ok, Y. S. (2017). A critical review on effects, tolerance mechanisms and management of cadmium in vegetables. *Chemosphere, 18*, 90–105. doi:10.1016/j.chemosphere.2017.05.013.

Ronzan, M., Piacentini, D., Fattorini, L., Della Rovere, F., Eiche, E., Riemann, M., Altamura, M. M., & Falasca, G. (2018). Cadmium and arsenic affect root development in *Oryza sativa* L. negatively interacting with auxin. *Environmental and Experimental Botany, 15*, 64–75. doi:10.1016/j.envexpbot.2018.04.008.

Rucińska-Sobkowiak, R. (2016). Water relations in plants subjected to heavy metal stresses. *Acta Physiologiae Plantarum, 38*(11). doi:10.1007/s11738-016-2277-5.

Sabrine, H., Afif, H., Mohamed, B., Hamadi, B., & Maria, H. (2010). Effects of cadmium and copper on pollen germination and fruit set in pea (*Pisum sativum* L.). *Scientia Horticulturae, 125*(4), 551–555. doi:10.1016/j.scienta.2010.05.031.

Sachdev, S., Ansari, S. A., Ansari, M. I., Fujita, M., & Hasanuzzaman, M. (2021). Abiotic stress and reactive oxygen species: Generation, signaling, and defense mechanisms. *Antioxidants, 10*(2), 1–37. doi:10.3390/antiox10020277.

Saini, S., Kaur, N., & Kumar Pati, P. (2021). Phytohormones: Key players in the modulation of heavy metal stress tolerance in plants. *Ecotoxicology and Environmental Safety, 22*, 112578. doi:10.1016/j.ecoenv.2021.112578.

Salguero-Linares, J., & Coll, N. S. (2019). Plant proteases in the control of the hypersensitive response. *Journal of Experimental Botany, 70*(7), 2087–2095. doi:10.1093/jxb/erz030.

Sánchez-Pardo, B., Carpena, R. O., & Zornoza, P. (2013). Cadmium in white lupin nodules: Impact on nitrogen and carbon metabolism. *Journal of Plant Physiology, 170*(3), 265–271. doi:10.1016/j.jplph.2012.10.001.

Sasaki, A., Yamaji, N., Yokosho, K., & Ma, J. F. (2012). Nramp5 is a major transporter responsible for manganese and cadmium uptake in rice. *Plant Cell, 24*(5), 2155–2167. doi:10.1105/tpc.112.096925.

Sasaki, A., Yamaji, N., & Ma, J. F. (2014). Overexpression of OsHMA3 enhances Cd tolerance and expression of Zn transporter genes in rice. *Journal of Experimental Botany, 65*(20), 6013–6021. doi:10.1093/jxb/eru340.

Sawidis, T. (2008). Effect of cadmium on pollen germination and tube growth in *Lilium longiflorum* and *Nicotiana tabacum*. *Protoplasma, 233*(1-2), 95–106. doi:10.1007/s00709-008-0306-y.

Schellingen, K., Van Der Straeten, D., Vandenbussche, F., Prinsen, E., Remans, T., Vangronsveld, J., & Cuypers, A. (2014). Cadmium-induced ethylene production and responses in *Arabidopsis thaliana* rely on ACS2 and ACS6 gene expression. *BMC Plant Biology, 14*(1). doi:10.1186/s12870-014-0214-6.

Seregin, I. V., & Kozhevnikova, A. D. (2021). Low-molecular-weight ligands in plants: Role in metal homeostasis and hyperaccumulation. *Photosynthesis Research, 150*(1–3), 51–96. doi:10.1007/s11120-020-00768-1.

Serre, N. B. C., Sarthou, M., Gigarel, O., Figuet, S., Corso, M., Choulet, J., Rofidal, V., Alban, C., Santoni, V., Bourguignon, J., Verbruggen, N., & Ravanel, S. (2020). Protein lysine methylation contributes to modulating the response of sensitive and tolerant *Arabidopsis* species to cadmium stress. *Plant, Cell & Environment, 43*(3), 760–774. doi:10.1111/pce.13692.

Sharma, A., Sainger, M., Dwivedi, S., Srivastava, S., Tripathi, R. D., & Singh, R. P. (2010). Genotypic variation in *Brassica juncea* (L.) Czern. cultivars in growth, nitrate assimilation, antioxidant responses and phytoremediation potential during cadmium stress. *Journal of Environmental Biology, 31*(5), 773–780. http://www.jeb.co.in/index.php?page=journal_archives.

Sharma, S. S., & Kumar, V. (2002). Responses of wild type and abscisic acid mutants of *Arabidopsis thaliana* to cadmium. *Journal of Plant Physiology, 159*(12), 1323–1327. doi:10.1078/0176-1617-00601.

Sheng, Y., Yan, X., Huang, Y., Han, Y., Zhang, C., Ren, Y., Fan, T., Xiao, F., Liu, Y., & Cao, S. (2019). The WRKY transcription factor, WRKY13, activates PDR8 expression to positively regulate cadmium tolerance in *Arabidopsis*. *Plant, Cell & Environment, 42*(3), 891–903. doi:10.1111/pce.13457.

Shi, Y., Jiang, W., Li, M., Jiang, N., Huang, Y., Wang, M., Du, Z., Chen, J. i, Li, J., Wu, L., Zhong, M., Yang, J. u, & Huang, J. (2023). Metallochaperone protein OsHIPP17 regulates the absorption and translocation of cadmium in rice (*Oryza sativa* L.). *International Journal of Biological Macromolecules, 24*, 125607. doi:10.1016/j.ijbiomac.2023.125607.

Shirkhani, Z., Chehregani Rad, A., & Gholami, M. (2020). Effects of cadmium on perianth and anther formation in *Datura stramonium* L. *Brazilian Journal of Botany, 43*(2), 239–246. doi:10.1007/s40415-020-00595-7.

Smirnoff, N., & Arnaud, D. (2019). Hydrogen peroxide metabolism and functions in plants. *New Phytologist, 221*(3), 1197–1214. doi:10.1111/nph.15488.

Song, W. Y., Martinoia, E., Lee, J., Kim, D., Kim, D. Y., Vogt, E., Shim, D., Choi, K. S., Hwang, I., & Lee, Y. (2004). A novel family of Cys-rich membrane proteins mediates cadmium resistance in *Arabidopsis*. *Plant Physiology, 135*(2), 1027–1039. doi:10.1104/pp.103.037739.

Song, W. Y., Choi, K. S., Kim, D. Y., Geisler, M., Park, J., Vincenzetti, V., Schellenberg, M., Kim, S. H., Lim, Y. P., Noh, E. W., Lee, Y., & Martinoiaa, E. (2010). *Arabidopsis* PCR2 is a zinc exporter involved in both zinc extrusion and long-distance zinc transport. *Plant Cell, 22*(7), 2237–2252. doi:10.1105/tpc.109.070185.

Song, X., Yue, X., Chen, W., Jiang, H., Han, Y., & Li, X. u (2019). Detection of cadmium risk to the photosynthetic performance of hybrid pennisetum. *Frontiers in Plant Science, 1*. doi:10.3389/fpls.2019.00798.

Sorrentino, M. C., Giordano, S., Capozzi, F., & Spagnuolo, V. (2022). Metals induce genotoxicity in three cardoon cultivars: Relation to metal uptake and distribution in extra- and intracellular fractions. *Plants, 11*(4), 475. doi:10.3390/plants11040475.

Srivastava, R., Deng, Y., Shah, S., Rao, A. G., & Howell, S. H. (2013). Binding protein is a master regulator of the endoplasmic reticulum stress sensor/transducer bZIP28 in *Arabidopsis*. *Plant Cell, 25*(4), 1416–1429. doi:10.1105/tpc.113.110684.

Srivastava, R. K., Rajpoot, R., Pandey, P., Rani, A., & Dubey, R. S. (2018). Cadmium alters mitochondrial membrane potential, inhibits electron transport chain activity and induces callose deposition in rice seedlings. *Journal of Plant Growth Regulation, 37*(1), 335–344. doi:10.1007/s00344-017-9726-2.

Stroiński, A., Giżewska, K., & Zielezińska, M. (2013). Abscisic acid is required in transduction of cadmium signal to potato roots. *Biologia Plantarum, 57*(1), 121–127. doi:10.1007/s10535-012-0135-x.

Su, T., Fu, L., Kuang, L., Chen, D., Zhang, G., Shen, Q., & Wu, D. (2022). Transcriptome-wide m6A methylation profile reveals regulatory networks in roots of barley under cadmium stress. *Journal of Hazardous Materials, 423*(Pt A), 127140. doi:10.1016/j.jhazmat.2021.127140.

Sui, F. Q., Chang, J. D., Tang, Z., Liu, W. J., Huang, X. Y., & Zhao, F. J. (2018). Nramp5 expression and functionality likely explain higher cadmium uptake in rice than in wheat and maize. *Plant and Soil, 433*(1-2), 377–389. doi:10.1007/s11104-018-3849-5.

Sun, J. Y., Guo, R., Jiang, Q., Chen, C. Z., Gao, Y. Q., Jiang, M. M., Shen, R. F., Zhu, X. F., & Huang, J. (2024). Brassinosteroid decreases cadmium accumulation via regulating gibberellic acid accumulation and Cd fixation capacity of root cell wall in rice (*Oryza sativa*). *Journal of Hazardous Materials, 46*. doi:10.1016/j.jhazmat.2024.133862.

Sun, N., Liu, M., Zhang, W., Yang, W., Bei, X., Ma, H., Qiao, F., & Qi, X. (2015). Bean metal-responsive element-binding transcription factor confers cadmium resistance in tobacco. *Plant Physiology, 167*(3), 1136–1148. doi:10.1104/pp.114.253096.

Sychta, K., Dubas, E., Yamada, K., Słomka, A., Krzewska, M., & Kuta, E. (2020). Papain-like cysteine proteases are involved in programmed cell death in plants under heavy metal stress. *Environmental and Experimental Botany, 17*, 104041. doi:10.1016/j.envexpbot.2020.104041.

Sychta, K., Słomka, A., & Kuta, E. (2021). Insights into plant programmed cell death induced by heavy metals—Discovering a terra incognita. *Cells, 10*(1), 65. doi:10.3390/cells10010065.

Szurman-Zubrzycka, M., Jędrzejek, P., & Szarejko, I. (2023). How do plants cope with DNA damage? A concise review on the DDR pathway in plants. *International Journal of Molecular Sciences, 24*(3), 2404. doi:10.3390/ijms24032404.

Tabassum, A. S. J., & Pandey, D. (2021). Metal induced genotoxicity and oxidative stress in plants, assessment methods, and role of various factors in genotoxicity regulation. In *Induced genotoxicity and oxidative stress in plants* (pp. 133–149). Springer Nature. doi:10.1007/978-981-16-2074-4.

Tajti, J., Németh, E., Glatz, G., Janda, T., Pál, M., & Luo, Z.-B. (2019). Pattern of changes in salicylic acid-induced protein kinase (SIPK) gene expression and salicylic acid accumulation in wheat under cadmium exposure. *Plant Biology, 21*(6), 1176–1180. doi:10.1111/plb.13032.

Tan, L., Zhu, Y., Fan, T., Peng, C., Wang, J., Sun, L., & Chen, C. (2019). OsZIP7 functions in xylem loading in roots and inter-vascular transfer in nodes to deliver Zn/Cd to grain in rice. *Biochemical and Biophysical Research Communications, 512*(1), 112–118. doi:10.1016/j.bbrc.2019.03.024.

Tan, M., Cheng, D., Yang, Y., Zhang, G., Qin, M., Chen, J., Chen, Y., & Jiang, M. (2017). Co-expression network analysis of the transcriptomes of rice roots exposed to various cadmium stresses reveals universal cadmium-responsive genes. *BMC Plant Biology, 17*(1). doi:10.1186/s12870-017-1143-y.

Tang, B., Luo, M., Zhang, Y., Guo, H., Li, J., Song, W., Zhang, R., Feng, Z., Kong, M., Li, H., Cao, Z., Lu, X., Li, D., Zhang, J., Wang, R., Wang, Y., Chen, Z., Zhao, Y., Zhao, J., & Cuypers, A. (2021). Natural variations in the P-type ATPase heavy metal transporter gene ZmHMA3 control cadmium accumulation in maize grains. *Journal of Experimental Botany, 72*(18), 6230–6246. doi:10.1093/jxb/erab254.

Tang, Z., Cai, H., Li, J., Lv, Y., Zhang, W., & Zhao, F. J. (2017). Allelic variation of NtNramp5 associated with cultivar variation in cadmium accumulation in tobacco. *Plant and Cell Physiology, 58*(9), 1583–1593. doi:10.1093/pcp/pcx087.

Tang, Z., Wang, H. Q., Chen, J., Chang, J. D., & Zhao, F. J. (2023). Molecular mechanisms underlying the toxicity and detoxification of trace metals and metalloids in plants. *Journal of Integrative Plant Biology, 65*(2), 570–593. doi:10.1111/jipb.13440.

Tao, Q., Jupa, R., Luo, J., Lux, A., Ková, J., Wen, Y., Zhou, Y., Jan, J., Liang, Y., & Li, T. (2017). The apoplasmic pathway via the root apex and lateral roots contributes to Cd hyperaccumulation in the hyperaccumulator *Sedum alfredii*. *Journal of Experimental Botany, 68*(3), 739–751. doi:10.1093/jxb/erw453.

Tao, Q., Jupa, R., Liu, Y., Luo, J., Li, J., Kováč, J., Li, B., Li, Q., Wu, K., Liang, Y., Lux, A., Wang, C., & Li, T. (2019). Abscisic acid-mediated modifications of radial apoplastic transport pathway play a key role in cadmium uptake in hyperaccumulator *Sedum alfredii*. *Plant, Cell & Environment, 42*(5), 1425–1440. doi:10.1111/pce.13506.

Ueno, D., Milner, M. J., Yamaji, N., Yokosho, K., Koyama, E., Zambrano, M. C., Kaskie, M., Ebbs, S., Kochian, L. V., & Ma, J. F. (2011). Elevated expression of TcHMA3 plays a key role in the extreme Cd tolerance in a Cd-hyperaccumulating ecotype of *Thlaspi caerulescens*. *Plant Journal, 66*(5), 852–862. doi:10.1111/j.1365-313X.2011.04548.x.

Ueno, D., Yamaji, N., Kono, I., Huang, C. F., Ando, T., Yano, M., & Ma, J. F. (2010). Gene limiting cadmium accumulation in rice. *Proceedings of the National Academy of Sciences of the United States of America, 107*(38), 16500–16505. doi:10.1073/pnas.1005396107.

Unsal, V., Dalkiran, T., Çiçek, M., & Kölükçü, E. (2020). The role of natural antioxidants against reactive oxygen species produced by cadmium toxicity: A review. *Advanced Pharmaceutical Bulletin, 10*(2), 184–202. doi:10.34172/apb.2020.023.

Uraguchi, S., Tanaka, N., Hofmann, C., Abiko, K., Ohkama-Ohtsu, N., Weber, M., Kamiya, T., Sone, Y., Nakamura, R., Takanezawa, Y., Kiyono, M., Fujiwara, T., & Clemens, S. (2017). Phytochelatin synthase has contrasting effects on cadmium and arsenic accumulation in rice grains. *Plant and Cell Physiology, 58*(10), 1730–1742. doi:10.1093/pcp/pcx114.

Van Slycken, S., Witters, N., Meers, E., Peene, A., Michels, E., Adriaensen, K., Ruttens, A., Vangronsveld, J., Laing, G. D. u, Wierinck, I., Van Dael, M., Van Passel, S., & Tack, F. M. G. (2013). Safe use of metal-contaminated agricultural land by cultivation of energy maize (*Zea mays*). *Environmental Pollution, 17*, 375–380. doi:10.1016/j.envpol.2013.03.032.

Verret, F., Gravot, A., Auroy, P., Leonhardt, N., David, P., Nussaume, L., Vavasseur, A., & Richaud, P. (2004). Overexpression of AtHMA4 enhances root-to-shoot translocation of zinc and cadmium and plant metal tolerance. *FEBS Letters, 576*(3), 306–312. doi:10.1016/j.febslet.2004.09.023.

Villiers, F., Jourdain, A., Bastien, O., Leonhardt, N., Fujioka, S., Tichtincky, G., Parcy, F., Bourguignon, J., & Hugouvieux, V. (2012). Evidence for functional interaction between brassinosteroids and cadmium response in *Arabidopsis thaliana*. *Journal of Experimental Botany, 63*(3), 1185–1200. doi:10.1093/jxb/err335.

Vitória, A. P., Cunha, M. D. a, & Azevedo, R. A. (2006). Ultrastructural changes of radish leaf exposed to cadmium. *Environmental and Experimental Botany, 58*(1-3), 47–52. doi:10.1016/j.envexpbot.2005.06.014.

Waadt, R., Seller, C. A., Hsu, P. K., Takahashi, Y., Munemasa, S., & Schroeder, J. I. (2022). Plant hormone regulation of abiotic stress responses. *Nature Reviews Molecular Cell Biology, 23*(10), 680–694. doi:10.1038/s41580-022-00479-6.

Wahid, A., Ghani, A., Ali, I., & Ashraf, M. Y. (2007). Effects of cadmium on carbon and nitrogen assimilation in shoots of mungbean [Vigna radiata (L.) Wilczek] seedlings. *Journal of Agronomy and Crop Science, 193*(5), 357–365. doi:10.1111/j.1439-037X.2007.00270.x.

Wahinya, F. W., Yamazaki, K., Jing, Z., Takami, T., Kamiya, T., Kajiya-Kanegae, H., Takanashi, H., Iwata, H., Tsutsumi, N., Fujiwara, T., & Sakamoto, W. (2022). Sorghum ionomics reveals the functional SbHMA3a allele that limits excess cadmium accumulation in grains. *Plant and Cell Physiology, 63*(5), 713–728. doi:10.1093/pcp/pcac035.

Wang, H. Q., Xuan, W., Huang, X. Y., Mao, C., & Zhao, F. J. (2021). Cadmium inhibits lateral root emergence in rice by disrupting OsPIN-mediated auxin distribution and the protective effect of OsHMA3. *Plant and Cell Physiology, 62*(1), 166–177. doi:10.1093/pcp/pcaa150.

Wang, H. Q., Zhao, X. Y., Xuan, W., Wang, P., & Zhao, F. J. (2023). Rice roots avoid asymmetric heavy metal and salinity stress via an RBOH-ROS-auxin signaling cascade. *Molecular Plant, 16*(10), 1678–1694. doi:10.1016/j.molp.2023.09.007.

Wang, M., Chen, L., Chen, S., & Ma, Y. (2012). Alleviation of cadmium-induced root growth inhibition in crop seedlings by nanoparticles. *Ecotoxicology and Environmental Safety, 7*, 48–54. doi:10.1016/j.ecoenv.2011.11.044.

Wang, X., Gao, Y., Feng, Y., Li, X., Wei, Q., Sheng, X., & Bassham, D. (2014). Cadmium stress disrupts the endomembrane organelles and endocytosis during *Picea wilsonii* pollen germination and tube growth. *PLoS One, 9*(4), e94721. doi:10.1371/journal.pone.0094721.

Wang, F., Tan, H., Han, J., Zhang, Y., He, X., Ding, Y., Cheng, Z. X., & Zhu, C. (2019). A novel family of PLAC8 motif-containing/PCR genes mediates Cd tolerance and Cd accumulation in rice. *Environmental Sciences Europe, 31*(82), 1–13. doi:10.1186/s12302-019-0259-0.

Wei, L., Zhang, M., Wei, S., Zhang, J., Wang, C., & Liao, W. (2020). Roles of nitric oxide in heavy metal stress in plants: Cross-talk with phytohormones and protein S-nitrosylation. *Environmental Pollution, 25,* 113943. doi:10.1016/j.envpol.2020.113943.

Wong, C. K. E., & Cobbett, C. S. (2009). HMA P-type ATPases are the major mechanism for root-to-shoot Cd translocation in *Arabidopsis thaliana. New Phytologist, 181*(1), 71–78. doi:10.1111/j.1469-8137.2008.02638.x.

Wu, D., Sato, K., & Ma, J. F. (2015). Genome-wide association mapping of cadmium accumulation in different organs of barley. *New Phytologist, 208*(3), 817–829. doi:10.1111/nph.13512.

Wu, M., Wang, P. Y., Sun, L. G., Zhang, J. J., Yu, J., Wang, Y. W., & Chen, G. X. (2014). Alleviation of cadmium toxicity by cerium in rice seedlings is related to improved photosynthesis, elevated antioxidant enzymes and decreased oxidative stress. *Plant Growth Regulation, 74*(3), 251–260. doi:10.1007/s10725-014-9916-x.

Xie, Q., Yu, Q., Jobe, T. O., Pham, A., Ge, C., Guo, Q., Liu, J., Liu, H., Zhang, H., Zhao, Y., Xue, S., Hauser, F., & Schroeder, J. I. (2021). An amiRNA screen uncovers redundant CBF and ERF34/35 transcription factors that differentially regulate arsenite and cadmium responses. *Plant, Cell & Environment, 44*(5), 1692–1706. doi:10.1111/pce.14023.

Xie, Y., Wang, J., Zheng, L., Wang, Y., Luo, L., Ma, M., Zhang, C., Han, Y. i, Beeckman, T., Xu, G., Cai, Q., & Xuan, W. (2019). Cadmium stress suppresses lateral root formation by interfering with the root clock. *Plant, Cell & Environment, 42*(12), 3182–3196. doi:10.1111/pce.13635.

Xu, H., Xu, W., Xi, H., Ma, W., He, Z., & Ma, M. (2013). The ER luminal binding protein (BiP) alleviates Cd^{2+}-induced programmed cell death through endoplasmic reticulum stress–cell death signaling pathway in tobacco cells. *Journal of Plant Physiology, 170*(16), 1434–1441. doi:10.1016/j.jplph.2013.05.017.

Xu, Q., Wang, C., Li, S., Li, B., Li, Q., Chen, G., Chen, W., & Wang, F. (2017). Cadmium adsorption, chelation and compartmentalization limit root-to-shoot translocation of cadmium in rice (*Oryza sativa* L.). *Environmental Science and Pollution Research, 24*(12), 11319–11330. doi:10.1007/s11356-017-8775-1.

Yamazaki, S., Ueda, Y., Mukai, A., Ochiai, K., & Matoh, T. (2018). Rice phytochelatin synthases Os PCS 1 and Os PCS 2 make different contributions to cadmium and arsenic tolerance. *Plant Direct, 2*(1). doi:10.1002/pld3.34.

Yang, G., Fu, S., Huang, J., Li, L., Long, Y., Wei, Q., Wang, Z., Chen, Z., & Xia, J. (2021). The tonoplast-localized transporter OsABCC9 is involved in cadmium tolerance and accumulation in rice. *Plant Science, 30*. doi:10.1016/j.plantsci.2021.110894.

Yang, J., Li, K., Zheng, W., Zhang, H., Cao, X., Lan, Y., Yang, C., & Li, C. (2015). Characterization of early transcriptional responses to cadmium in the root and leaf of Cd-resistant *Salix matsudana* Koidz. *BMC Genomics, 16*(1). doi:10.1186/s12864-015-1923-4.

Yang, S., Zu, Y., Li, B., Bi, Y., Jia, L., He, Y., & Li, Y. (2019). Response and intraspecific differences in nitrogen metabolism of alfalfa (*Medicago sativa* L.) under cadmium stress. *Chemosphere, 22,* 69–76. doi:10.1016/j.chemosphere.2018.12.101.

Yang, Y., Li, J., Li, H., Ding, Y., Wu, W., Qin, R., Ni, J., Xu, R., Wei, P., & Yang, J. (2023). OsGSTU5 and OsGSTU37 encoding glutathione reductases are required for cadmium tolerance in rice. *International Journal of Environmental Science and Technology, 20*(9), 10253–10260. doi:10.1007/s13762-022-04550-9.

Yang, Z., Xue, B., Song, G., & Shi, S. (2022). Effects of citric acid on antioxidant system and carbon-nitrogen metabolism of Elymus dahuricus under Cd stress. *Ecotoxicology and Environmental Safety, 23,* 113321. doi:10.1016/j.ecoenv.2022.113321.

Ye, Y., Li, Z., & Xing, D. (2013). Nitric oxide promotes MPK6-mediated caspase-3-like activation in cadmium-induced *Arabidopsis thaliana* programmed cell death. *Plant, Cell & Environment, 36*(1), 1–15. doi:10.1111/j.1365-3040.2012.02543.x.

Yeh, C. M., Chien, P. S., & Huang, H. J. (2007). Distinct signalling pathways for induction of MAP kinase activities by cadmium and copper in rice roots. *Journal of Experimental Botany, 58*(3), 659–671. doi:10.1093/jxb/erl240.

Yin, H., Yang, F., He, X., Du, X., Mu, P., & Ma, W. (2022). Advances in the functional study of glutamine synthetase in plant abiotic stress tolerance response. *The Crop Journal, 10*(4), 917–923. doi:10.1016/j.cj.2022.01.003.

Yu, Y., Alseekh, S., Zhu, Z., Zhou, K., & Fernie, A. R. (2024). Multiomics and biotechnologies for understanding and influencing cadmium accumulation and stress response in plants. *Plant Biotechnology Journal, 22*(10), 2641–2659. doi:10.1111/pbi.14379.

Yuan, H.-M., & Huang, X. (2016). Inhibition of root meristem growth by cadmium involves nitric oxide-mediated repression of auxin accumulation and signalling in *Arabidopsis. Plant, Cell & Environment, 39*(1), 120–135. doi:10.1111/pce.12597.

Yue, E., Rong, F., Liu, Z., Ruan, S., Lu, T., & Qian, H. (2023). Cadmium induced a non-coding RNA microRNA535 mediates Cd accumulation in rice. *Journal of Environmental Sciences, 13*, 149–162. doi:10.1016/j.jes.2022.10.005.

Żabka, A., Gocek, N., Winnicki, K., Szczeblewski, P., Laskowski, T., & Polit, J. T. (2021). Changes in epigenetic patterns related to DNA replication in *Vicia faba* root meristem cells under cadmium-induced stress conditions. *Cells, 10*(12). doi:10.3390/cells10123409.

Zare, A. A., Khoshgoftarmanesh, A. H., Malakouti, M. J., Bahrami, H. A., & Chaney, R. L. (2018). Root uptake and shoot accumulation of cadmium by lettuce at various Cd:ZN ratios in nutrient solution. *Ecotoxicology and Environmental Safety, 14*, 441–446. doi:10.1016/j.ecoenv.2017.10.045.

Zhang, B. Q., Liu, X. S., Feng, S. J., Zhao, Y. N., Wang, L. L., Rono, J. K., Li, H., & Yang, Z. M. (2020). Developing a cadmium resistant rice genotype with OsHIPP29 locus for limiting cadmium accumulation in the paddy crop. *Chemosphere, 24*. doi:10.1016/j.chemosphere.2020.125958.

Zhang, C., Tong, C., Cao, L., Zheng, P., Tang, X., Wang, L., Miao, M., Liu, Y., & Cao, S. (2023a). Regulatory module WRKY33-ATL31-IRT1 mediates cadmium tolerance in *Arabidopsis*. *Plant, Cell & Environment, 46*(5), 1653–1670. doi:10.1111/pce.14558.

Zhang, Q., Wen, Q., Ma, T., Zhu, Q., Huang, D., Zhu, H., Xu, C., & Chen, H. (2023b). Cadmium-induced iron deficiency is a compromise strategy to reduce Cd uptake in rice. *Environmental and Experimental Botany, 20*, 105155. doi:10.1016/j.envexpbot.2022.105155.

Zhang, X., Rui, H., Zhang, F., Hu, Z., Xia, Y., & Shen, Z. (2018). Overexpression of a functional *Vicia sativa* PCS1 homolog increases cadmium tolerance and phytochelatins synthesis in *Arabidopsis*. *Frontiers in Plant Science*. doi:10.3389/fpls.2018.00107.

Zhang, X., Chen, J., Wang, W., & Zhu, L. (2024). Photosynthetic mechanisms of carbon fixation reduction in rice by cadmium and polycyclic aromatic hydrocarbons. *Environmental Pollution, 34*, 123436. doi:10.1016/j.envpol.2024.123436.

Zhang, Y., Wang, Z., Liu, Y., Zhang, T., Liu, J., You, Z., Huang, P., Zhang, Z., & Wang, C. (2023c). Plasma membrane-associated calcium signaling modulates cadmium transport. *New Phytologist, 238*(1), 313–331. doi:10.1111/nph.18698.

Zhang, P., Wang, R., Ju, Q., Li, W., Tran, L.-S. P., & Xu, J. (2019). The R2R3-MYB transcription factor MYB49 regulates cadmium accumulation. *Plant Physiology, 180*(1), 529–542. doi:10.1104/pp.18.01380.

Zhao, F. J., Moore, K. L., Lombi, E., & Zhu, Y. G. (2014). Imaging element distribution and speciation in plant cells. *Trends in Plant Science, 19*(3), 183–192. doi:10.1016/j.tplants.2013.12.001.

Zhao, F. J., Tang, Z., Song, J. J., Huang, X. Y., & Wang, P. (2022). Toxic metals and metalloids: Uptake, transport, detoxification, phytoremediation, and crop improvement for safer food. *Molecular Plant, 15*(1), 27–44. doi:10.1016/j.molp.2021.09.016.

Zhao, F. Y., Han, M. M., Zhang, S. Y., Wang, K., Zhang, C. R., Liu, T., & Liu, W. (2012). Hydrogen peroxide-mediated growth of the root system occurs via auxin signaling modification and variations in the expression of cell-cycle genes in rice seedlings exposed to cadmium stress. *Journal of Integrative Plant Biology, 54*(12), 991–1006. doi:10.1111/j.1744-7909.2012.01170.x.

Zhao, F. Y., Hu, F., Zhang, S. Y., Wang, K., Zhang, C. R., & Liu, T. (2013). MAPKs regulate root growth by influencing auxin signaling and cell cycle-related gene expression in cadmium-stressed rice. *Environmental Science and Pollution Research, 20*(8), 5449–5460. doi:10.1007/s11356-013-1559-3.

Zhao, H., Guan, J., Liang, Q., Zhang, X., Hu, H., & Zhang, J. (2021). Effects of cadmium stress on growth and physiological characteristics of sassafras seedlings. *Scientific Reports, 11*(1). doi:10.1038/s41598-021-89322-0.

Zhao, J., Yang, W., Zhang, S., Yang, T., Liu, Q., Dong, J., . . . Liu, B. (2018). Genome-wide association study and candidate gene analysis of rice cadmium accumulation in grain in a diverse rice collection. *Rice, 11*(61), 1–15. doi:10.1186/s12284-018-0254-x.

Zhao, Q., Wang, H., Du, Y., Rogers, H. J., Wu, Z., Jia, S., Yao, X., Xie, F., & Liu, W. (2020). MSH2 and MSH6 in mismatch repair system account for soybean (*Glycine max* (L.) Merr.) tolerance to cadmium toxicity by determining DNA damage response. *Journal of Agricultural and Food Chemistry, 68*(7), 1974–1985. doi:10.1021/acs.jafc.9b06599.

Zhao, R. Z., Jiang, S., Zhang, L., & Yu, Z. B. (2019). Mitochondrial electron transport chain, ROS generation and uncoupling (review). *International Journal of Molecular Medicine, 44*(1), 3–15. doi:10.3892/ijmm.2019.4188.

Zhao, S., Ma, Q., Xu, X., Li, G., & Hao, L. (2016). Tomato jasmonic acid-deficient mutant spr2 seedling response to cadmium stress. *Journal of Plant Growth Regulation, 35*(3), 603–610. doi:10.1007/s00344-015-9563-0.

Zheng, W. J., Li, W. Q., Peng, Y., Shao, Y., Tang, L., Liu, C. T., Zhang, D., Zhang, L. J., Li, J. H., Luo, W. Z., Yuan, Z. C., Zhao, B. R., & Mao, B. G. (2022). E2Fs co-participate in cadmium stress response through activation of MSHs during the cell cycle. *Frontiers in Plant Science, 1*. doi:10.3389/fpls.2022.1068769.

Zhu, X. F., Zheng, C., Hu, Y. T., Jiang, T., Liu, Y., Dong, N. Y., Yang, J. L., & Zheng, S. J. (2011). Cadmium-induced oxalate secretion from root apex is associated with cadmium exclusion and resistance in *Lycopersicon esulentum*. *Plant, Cell and Environment, 34*(7), 1055–1064. doi:10.1111/j.1365-3040.2011.02304.x.

Zhu, X. F., Wang, Z. W., Dong, F., Lei, G. J., Shi, Y. Z., Li, G. X., & Zheng, S. J. (2013). Exogenous auxin alleviates cadmium toxicity in *Arabidopsis thaliana* by stimulating synthesis of hemicellulose 1 and increasing the cadmium fixation capacity of root cell walls. *Journal of Hazardous Materials, 26*, 398–403. doi:10.1016/j.jhazmat.2013.09.018.

Zhu, X. F., Jiang, T., Wang, Z. W., Lei, G. J., Shi, Y. Z., Li, G. X., & Zheng, S. J. (2012). Gibberellic acid alleviates cadmium toxicity by reducing nitric oxide accumulation and expression of IRT1 in *Arabidopsis thaliana*. *Journal of Hazardous Materials, 239–240*, 302–307. doi:10.1016/j.jhazmat.2012.08.077.

Zulfiqar, U., Ayub, A., Hussain, S., Waraich, E. A., El-Esawi, M. A., Ishfaq, M., Ahmad, M., Ali, N., & Maqsood, M. F. (2022). Cadmium toxicity in plants: Recent progress on morpho-physiological effects and remediation strategies. *Journal of Soil Science and Plant Nutrition, 22*(1), 212–269. doi:10.1007/s42729-021-00645-3.

CHAPTER

12

Cadmium hyperaccumulation

Hendrik Küpper[a,b], *Qi Tao*[c,d], *Filis Morina*[a]

[a]Department of Plant Biophysics and Biochemistry, Biology Centre, Institute of Plant Molecular Biology, Czech Academy of Sciences, České Budějovice, Czech Republic [b]Department of Experimental Plant Biology, Faculty of Science, University of South Bohemia, České Budějovice, Czech Republic [c]College of Resources, Sichuan Agricultural University, Chengdu, China [d]New Cornerstone Science Laboratory, State Key Laboratory of Crop Gene Exploration and Utilisation in Southwest China, Sichuan Agricultural University at Wenjiang, Chengdu, Sichuan, China

12.1 Introduction: Importance and ecological role of cadmium hyperaccumulation

Since the discovery of the metal hyperaccumulator *Noccaea caerulescens* (J. Presl & C. Presl; F.K. Mey) in 1865 [at that time, hyperaccumulation was only characterized for Zn, and it was called *Thlaspi alpestre* (Sachs, 1865)], a few hundred plant species have been identified as metal hyperaccumulators. These represent far less than 1% of all known plant species. Among the known metal hyperaccumulators, only a handful are true cadmium (Cd) hyperaccumulators, the exact number depending on which treatment conditions and metal accumulation levels are accepted for the "hyperaccumulator" definition (Reeves et al., 2018). The best known are *N. caerulescens* (=*Thlaspi caerulescens* =*T. alpestre*, Vázquez et al., 1992), *Arabidopsis halleri* (L.) O'Kane & Al-Shehbaz (=*Cardaminopsis halleri*) (Küpper et al., 2000), and *Sedum alfredii* Hance (Yang et al., 2004). Later, what has originally been described and still often is referred to as the "hyperaccumulating ecotype of *S. alfredii*" has actually been suggested to be a separate species, *Sedum plumbizincicola X.H. Guo & S.B. Zhou ex L.H. Wu* (Wu et al., 2013). Generally the term "hyperaccumulators" was introduced by Brooks et al. (1977). They have been the subject of intense research, and there are several important reasons for the large interest in these rare plants.

First, metal hyperaccumulators gained interest because of their potential for cleaning up metal-polluted soils, so-called phytoremediation. The concept of removing toxic metals from soils by plants was described by Chaney (1983). It was further elaborated, and the term "phytoremediation of metals" was introduced in the early 1990s (Cunningham & Berti, 1993). The first detailed field study of this concept, in the sense of phytoextraction as the most attractive variant

Cadmium Contamination in Soils and Plants: Mechanisms and Mitigation.
DOI: https://doi.org/10.1016/B978-0-443-34104-5.00020-7

of phytoremediation, was published by Baker et al. (1994). Since then, substantial progress has been made by many studies in finding the most suitable species and ecotypes, as well as improving agricultural practices for increasing yield of biomass and metal. Case studies of using *S. plumbizincicola* to decontaminate soil Cd are described in Chapter 20. Closely linked to phytoremediation is phytomining (concept introduced by Brooks et al., 1998), where the focus is mostly on gaining metal instead of its removal from the soil, but in practice both uses can overlap. The southern French (Ganges) population of *N. caerulescens* accumulates the highest known concentrations of Cd (Lombi et al., 2000), and Cd bioaccumulation factors (i.e., the ratio of Cd concentration in shoots to that in the soil) of over 70 have been observed in the wild (Escarré et al., 2011). Phytoremediation of Cd with this species was proposed as early as 1994 (Brown et al., 1994), and its applicability in practice was demonstrated in 2003 (Zhao et al., 2003). As a result of the high bioaccumulation, recovery of the Cd from the harvested plants for phytomining has been shown to be feasible for this species (Hazotte et al., 2017). However, *N. caerulescens* has an important drawback for its investigation: it is hard to transform genetically, as classical methods such as floral dip are very inefficient in this species. Therefore its relative (also in the genus Brassicaceae), *A. halleri*, which is closest to the model plant *Arabidopsis thaliana*, has become the most popular hyperaccumulator among molecular geneticists (Corso et al., 2021; Stein et al., 2017).

Second, metal hyperaccumulators have been found to be powerful models for investigating physiological processes of metal uptake, translocation, and detoxification by sequestration. The enormous metal concentrations in their tissues make these processes much more accessible to measurements by techniques such as energy-dispersive X-ray analysis (EDX), particle-induced X-ray emission, X-ray fluorescence, or ion-selective electrodes than the normally 100 times lower concentrations in nonaccumulator plants. The underlying enormous expression level of specific genes makes them an attractive target for investigating the roles of such genes in planta. This is the basis for the detailed discussion of all those processes in the first half of this chapter.

Although the aforementioned features of hyperaccumulators are undoubtedly important for research and technology development, for many years there has been less consensus regarding the ecological role of hyperaccumulation. Several hypotheses have been proposed for the evolutionary advantage of hyperaccumulation, mostly in the work of Boyd et al. (from 1992 to 2012). Early hypotheses included osmotic regulation; however, this could never be convincingly shown. In contrast, over the decades, one hypothesis became verified for many species of hyperaccumulators of various metals, including Cd hyperaccumulators, and is therefore now the most popular. This is the use of the hyperaccumulated metals for defense against pathogens and herbivores (Boyd & Martens, 1998), in the "elemental defence" hypothesis and its extension, the "joint effects" hypothesis. Based on these hypotheses, Cd alone or in combination with Zn and/or organic defense (mainly glucosinolates have been described so far) provides efficient defense against viruses, pathogens, and some herbivores. Most of the studies were done using *A. halleri* and *N. caerulescens*, usually with a combination of very large Zn and Cd concentrations in the substrate, as found in the localities of these hyperaccumulators. We discuss the evidence toward defense-related hypotheses and the inadvertent uptake hypothesis, stating that Zn and Cd hyperaccumulation evolved because of enhanced uptake processes in nutrient-poor soils in which these plants grow. Both laboratory and field studies are needed to elucidate the complex interactions between Cd, plants, and biotic stress. In addition, we compare the cross talk between Cd and biotic stress in hyperaccumulating and nonhyperaccumulating plants, which include crops and economically important species. The potential of Cd for hormetic effects (stimulating effects of low Cd doses) and possible beneficial effects in plant defense were analyzed. The importance of applying low, environmentally relevant Cd concentrations to reveal the mechanisms of hormesis and toxicity is emphasized based on recent research (e.g., Andresen et al., 2023), as well as knowledge gaps regarding the specific Cd-binding targets (low molecular weight (MW) molecules as well as macromolecules) that can benefit the plant.

12.2 Mechanisms of cadmium hyperaccumulation

12.2.1 Symplastic and transmembrane transport

Metal hyperaccumulation in the true sense, that is, the strongly higher accumulation of metal concentrations in the shoots above the levels in the soil and the roots, requires a transport against a concentration gradient (Andresen et al., 2018; Williams et al., 2000). Biophysically, this means that such transport requires energy, either directly by coupling the splitting of high-energy bonds in adenosine triphosphate (ATP) or indirectly by using a preformed concentration gradient of another ion (typically protons). To mediate such energetic coupling, specific transport proteins are required, as explained in more detail in reviews (Andresen et al., 2018; Williams et al., 2000). Therefore, at least some steps of the transport of a metal from the soil to the final storage site in the shoots have to involve transmembrane transport. The importance of active transmembrane transport for Cd hyperaccumulation has been clearly shown in a comparison of the highly Cd-accumulating Ganges population with a low-Cd population of *N. caerulescens* (Zhao et al., 2002). This study revealed that no difference existed in the apoplastic binding between these two ecotypes, but the Ganges ecotype had a 4.5× higher active Cd uptake and showed a clear Michaelis–Menten type of concentration response typical for enzyme-mediated transport. Active transport is generally also needed for detoxification of metals whenever a sensitive cell or subcellular compartment needs to lower its toxic metal load below the level of its environment. As this is transport against the concentration gradient, it requires active transport in some way. This can be primary active (direct ATP-consuming transport of metal ions via ATPases) or secondary active (transport of metal ions made energetically "downhill" by previous active transport of other ions), as explained in detail by Andresen et al. (2018).

In the past three decades, many genes encoding such transport proteins have been identified in multiple hyperaccumulators, including those that accumulate Cd. Important early transcriptomic work revealed extreme upregulation of CPx (P_{1B}) ATPases, in particular HMA4, cation diffusion facilitators (CDFs), and ZIP transporters as key factors in hyperaccumulation (Assunção et al., 2001; Becher et al., 2004; Bernard et al., 2004; Papoyan & Kochian, 2004; Pence et al., 2000; Weber et al., 2004). Later, it was found that overexpression of the ZIP transporter ZNT1, which was first suggested to be a direct driver of hyperaccumulation (Pence et al., 2000), is actually a downstream consequence, covering the physiological needs of the mesophyll in competition with sequestration (Küpper et al., 2007b). By comparing genomes of ecotypes that have different capabilities in terms of Cd accumulation and/or resistance, it has been shown that the phenotypes of Cd-tolerance and Cd-hyperaccumulation rely on different genes and can segregate differently (Bert et al., 2003; Macnair et al., 1999). This is somewhat surprising because a hyperaccumulator living on high-metal soil cannot survive without efficient detoxification machinery. Later studies showed that there is actually an overlap of quantitative trait loci for accumulation and tolerance of both Zn and Cd (Frérot et al., 2010; Willems et al., 2010). Further,

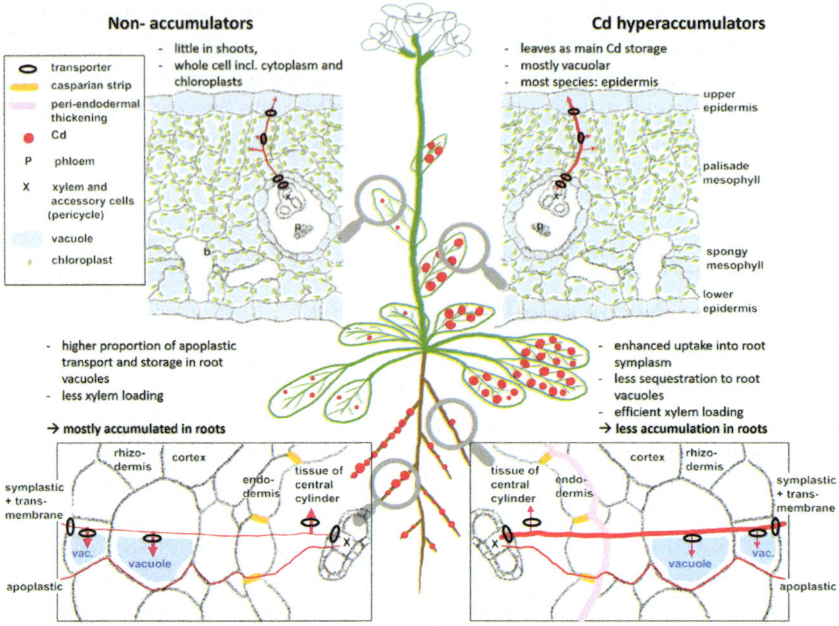

FIGURE 12.1 Illustration of the mechanism of hyperaccumulation, leading to different Cd accumulation patterns on the organ, tissue, and cellular levels in hyperaccumulators compared to nonaccumulators. The drawing is based on *Noccaea caerulescens*, but the fundamental principles are the same in other species. *Cd*, Cadmium.

as Cd is not a physiologically required metal in higher plants, it is transported by proteins that originally evolved for the transport of other metals, in particular the chemically similar zinc but also more general transporters of divalent cations. The most prominent transporters that have been found to be involved in Cd hyperaccumulation are discussed later.

In the first decisive step of hyperaccumulation, that is, the loading of the xylem in the root for transport into the shoot as opposed to retention in the root (Fig. 12.1), the CPx (=P_{1B}) ATPase HMA4 has been identified as a key player. This started in 2004 with its identification both in *A. halleri* (Becher et al., 2004; Weber et al., 2004) and *Noccaea* (formerly *Thlaspi*) *caerulescens* (Bernard et al., 2004; Papoyan & Kochian, 2004). In *A. halleri* and *N. caerulescens* the decisive change turning on hyperaccumulation with respect to this transport step is a gene multiplication (Hanikenne et al., 2008; Ó Lochlainn et al., 2011; Talke et al., 2006). Because of its high importance, HMA4 protein has also been studied in terms of its biochemistry and biophysics of metal transport. These studies showed not only its particularly high affinity for Cd but also that its activation energy of pumping decreased with increasing Cd concentration (Leitenmaier et al., 2011) and that its maximal activity was reached when all internal binding sites were saturated with Cd (Mishra et al., 2017). By using X-ray absorption spectroscopy it was found that the highest affinity binding sites inside HMA4 were two of the cysteines in its sequence (Leitenmaier et al., 2011). These were later also found to be the highest affinity sites for Zn binding in the related enzyme ZntA from *Shigella sonnei*, where this was revealed by protein crystallography (Wang et al., 2014). Although multiple studies showed the key importance of HMA4 for Cd hyperaccumulation, studies comparing ecotypes of the same species revealed that other transport proteins likely play a role as well. In *A. halleri*, comparing two metallicolous populations, of which one (PL22) was Cd-accumulating, whereas the other (i16) was Cd-excluding, PL22 had a higher expression of *IRT1*, *CATION EXCHANGER 4* (CAX4), *YELLOW STRIPE-like 3* (*YSL3*), *CNGC12*, *OLIGOPEPTIDE TRANSPORTER 3* (*OPT3*), *ZINC-INDUCED FACILITATOR 1* (*ZIF1*), and *CNGC3* (Corso et al., 2018). It remained unclear, however, which of these are directly involved in Cd hyperaccumulation and which are upregulated as a downstream consequence of it (e.g., to balance micronutrient nutrition). Knockout mutants of *CNGC12* only showed reduced root growth (Corso et al., 2018), which could be caused by both options. For IRT1 and YSL3 a lot of research has been done that shows their general importance for micronutrient balance, so their upregulation in hyperaccumulators is likely related to that (e.g., Connolly et al., 2002; Sinclair & Krämer, 2012; Waters et al., 2006). CAX4 had previously been shown to be involved in Cd detoxification by vacuolar sequestration (Korenkov et al., 2007). ZIF1 has been shown to have a similar role for the chemically related Zn but indirectly by pumping the ligand nicotianamine into the vacuole (Haydon et al., 2012). To date, the only gene shown to enhance Cd accumulation through changes in expression—although not in hyperaccumulators—is *OPT3* (Mendoza-Cózatl et al., 2014). This is likely related to its role in phloem redistribution of Cd (Zhai et al., 2014). For the other aforementioned genes, clearly further research is required to reveal their direct or indirect role in Cd (hyper)accumulation.

Once in the shoot, Cd has to be unloaded from the xylem and transported toward the final storage sites in the shoot (Fig. 12.1). For Cd unloading in the veins, again, HMA4 seems to play a decisive role, explaining why the gene is expressed also in the shoots (Bernard et al., 2004; Mishra et al., 2017). After xylem unloading, further transport toward the final storage sites occurs. These are mostly the vacuoles of epidermal storage cells in *N. caerulescens* (Leitenmaier & Küpper, 2011) and mesophyll vacuoles in *A. halleri* (Küpper et al., 2000) and *S. alfredii* (Tian et al., 2011); see more in Section 12.2.4. The accumulation in different final storage locations in *A. halleri* versus *N. caerulescens* has been shown to involve differential expression patterns of *HMA4 and HMA3* (Mishra et al., 2017). Interestingly, although microscopically *N. caerulescens* performs epidermal sequestration, on the macroscopic scale it was found that the accumulation of Cd is strongest around the veins (Morina & Küpper, 2020; Yan et al., 2022), which matches the expression patterns of *HMA3 and HMA4* (Mishra et al., 2017). On a cellular level, enhanced uptake into the large epidermal storage cells has been shown to involve ZNT5 (a member of the ZIP family) for transport over the cytoplasmic membrane (Küpper & Kochian, 2010) before vacuolar sequestration can finally occur (Fig. 12.1, see Section 12.2.4). For epidermal metal accumulation in general, there had been early suggestions that the sequestration is achieved by passive transport via the transpiration stream (Wójcik et al., 2005). However, later it could be shown by work with isolated protoplasts that it is active transport as well, leading to much higher uptake rates per volume in the epidermal storage cells compared to mesophyll cells (Leitenmaier & Küpper, 2011). As the protoplasts were without cell walls and in suspension the differences in uptake rate would be caused by differences in cell walls, or the transpiration stream could be excluded. The postulation of the latter in the earlier work (Wójcik et al., 2005) was likely because of artifacts of fixing tissues chemically before EDX analysis, which is known to cause element redistribution (van der Ent et al., 2018).

Besides delivery to the storage sites, detoxification of Cd in compartments where Cd accumulation would be particularly harmful also requires active transport, which has been studied far less. However, one CPx (P_{1B}) ATPase (SpHMA1) has been identified that exports Cd from chloroplasts of *S. plumbizincicola*, and it could be shown that this is indeed decisive for protection of photosynthesis (Zhao et al., 2019).

Despite the requirement of active transmembrane transport for hyperaccumulation as explained so far, apoplastic transport is also important, as well as apoplastic barriers, in particular in the roots. This has been revealed by many studies, of which, again, some important ones will be highlighted in the following section.

12.2.2 Apoplastic transport

The apoplast, representing the first contact between the cell and the surrounding environment, encompasses the extracellular space that includes cell walls, intercellular spaces, and xylem vessels (Sattelmacher, 2001). In the apoplast, cations bind to negatively charged cell wall components (e.g., pectin, lignin, and carboxyl groups), which can sequester metals but also act as a reservoir for gradual release. The important role of the cell wall in Cd storage in *S. plumbizincicola* was attributed to cell wall modification in the process of evolution, hence contributing to hypertolerance (Peng et al., 2017). On the contrary the isotope labeling results showed that more cell-wall-bound ^{113}Cd was available for xylem loading in Cd/Zn hyperaccumulator *S. alfredii*, indicating a higher capacity of Cd release in the apoplast compared to a nonhyperaccumulator (Li et al., 2015). Similarly a major role of the cell wall in Cd accumulation has also been suggested in *A. halleri* (Corso et al., 2018; Meyer et al., 2015). Several pectinmethylesterase genes showed lower expression in *A. halleri* with higher Cd accumulation, suggesting that the higher level of low-methylesterified pectins in the cell wall could participate in releasing Cd from the cell wall and enhancing Cd accumulation (An et al., 2023; Corso et al., 2021). The released Cd in the apoplast of hyperaccumulators will subsequently be taken up into the cell and then be translocated to the xylem and the upper shoot (Li et al., 2015).

Movement of ions from the external solution to the xylem through the apoplast is a nonmetabolic, passive process, which is defined as apoplastic transport (Lux et al., 2011). Generally, apoplastic transport occurs where endodermal barriers, including Casparian strips (CSs) and suberin lamellae (SL), have not formed or are broken by lateral root branching (Lux et al., 2011). Although symplastic transport has long been emphasized in heavy metal hyperaccumulation studies, emerging evidence highlights the underappreciated role of apoplastic transport in driving heavy metal radial transport toward the xylem and subsequent translocation to shoots (White et al., 2002; Tao et al., 2017, 2019). It was previously speculated that nearly all Zn entering the xylem had traversed the root through a symplastic pathway in *T. caerulescens*. However, the arguments presented by White et al. (2002) challenge this speculation and indicate that some Zn may reach the xylem through an apoplastic pathway, especially at higher external Zn exposure. Concentration-dependent Cd^{2+} influx in hyperaccumulators (i.e., *N. caerulescens, A. halleri*, and *S. alfredii*) normally yielded nonsaturating kinetic curves, which could be resolved into saturable and linear components (Lombi et al., 2001; Lu et al., 2008; Zhao et al., 2006). The linear component observed in short-term Cd exposure can be interpreted as the result of a nonselective apoplastic Cd flux to the xylem or low-affinity uptake into the symplast (Lu et al., 2008; Lux et al., 2011). Experimental evidence using apoplastic tracers, like trisodium-8-hydroxy-1,3,6-pyrenetrisulfonic acid, confirmed that the contribution of apoplastic transport to Cd accumulation was up to 37% at the highest external Cd exposure in *S. alfredii* (Tao et al., 2017). The relative contribution of the apoplastic transport to the total accumulation of metal ions in shoots will increase with their external concentration in a nutrient solution, as has been

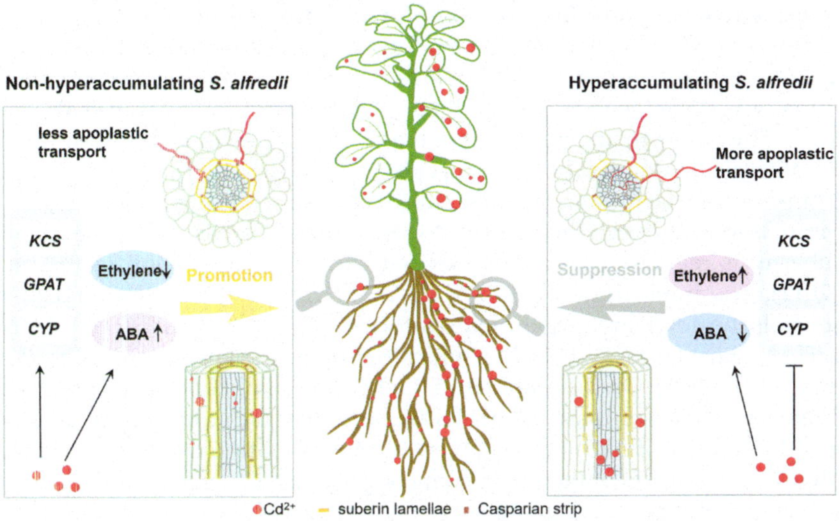

FIGURE 12.2 Regulation in development of apoplastic barriers and the contribution of apoplastic transport in nonhyperaccumulating and hyperaccumulating Sedum *alfredii*. *ABA*, Abscisic acid; *CYP*, cytochrome P450 oxygenases; *GPAT*, glycerol-3-phosphate acyltransferase; *KCS*, beta-ketoacyl-CoA synthases.

proposed for Zn in *T. caerulescens* (White et al., 2002), Na in *Oryza sativa* (Faiyue et al., 2010), and Cd in *Salix caprea* (Andresen et al., 2023; Martens & Boyd, 1994).

Unlike symplastic transport, which requires crossing plasma membranes via transporters, apoplastic transport is largely governed by CSs and SL (Chao & Chao, 2025; Lux et al., 2011). The CSs are mainly composed of lignin and surround the radial and transverse cell walls of the endodermal cells, forming a belt-like structure that hinders bypass flow (Chao & Chao, 2025; Naseer et al., 2012; Uddin et al., 2024). The SL, forming on the whole inner cell walls of endodermal cells, represents a more complete apoplastic barrier to the radial flow of ions to the xylem (Serra & Geldner, 2022; Shukla & Barberon, 2021). It is possible that Cd^{2+} or Cd-chelates could reach the xylem solely via apoplastic pathway in regions of the root lacking CSs or SL, such as root tips and initial lateral roots (Lux et al., 2004; Tao et al., 2017). Supporting the above hypothesis are observations that root tip is the most active part for Cd^{2+} influx (Piñeros et al., 1998; Tao et al., 2017). Furthermore the differences in the development of CSs and SL translate into the different capacity of Cd accumulation among species (Lux et al., 2004). Compared to a nonhyperaccumulating ecotype the Cd hyperaccumulating ecotype of *S. alfredii* exhibits delayed and less extensive CSs and SL development in roots, allowing more apoplastic Cd influx (Tao et al., 2019) (Fig. 12.2).

CSs and SL are plastic to environmental stimuli, and nonaccumulators exhibit different responses to Cd from hyperaccumulators (Barberon et al., 2016; Lux et al., 2011). When nonaccumulators are exposed to high concentrations of Cd, the formation of CSs and SL is initiated earlier in the root apex in several species, including maize (Vaculík et al., 2009), rice (Qi et al., 2020), and wheat (Liu et al., 2020). All these changes can be regarded as acclimatory responses that decrease radial movement of Cd to the xylem and upward translocation to the shoot, which are typically related to the capacity of Cd tolerance. However, in hyperaccumulator *S. alfredii*, a delayed and less extensive deposition of the CSs and SL was observed upon Cd exposure, indicating an opposite response (Tao et al., 2017, 2019). The delayed suberization upon short-term Cd treatment is controlled by the lower endogenous ABA and the higher endogenous ethylene in hyperaccumulator *S. alfredii* than in the nonhyperaccumulator ecotype (Liu et al., 2021; Tao et al., 2019) (Fig. 12.2). The downstream regulators of these hormones, which transcriptionally regulate suberin biosynthesis and degradation, have yet to be identified.

Several transcription factors and genes are involved in the regulation of CSs and SL. In the nonaccumulator *A. thaliana*, the precise deposition of CSs/SL, as well as the biosynthesis of lignin/suberin, is regulated by a complex network of transcription factors and enzymes (Barberon, 2017). The transcription factor MYB DOMAIN PROTEIN36 (MYB36) directly and positively controls the downstream gene expression of *CASPARIAN STRIP DOMAIN PROTEINs* (*CASPs*), *ENHANCED SUBERIN 1* (*ESB1*), which regulates the formation of CSs (Kamiya et al., 2015; Liberman et al., 2015). On the other hand, CYP86 subfamily of cytochrome P450 monooxygenases, 3-ketoacyl CoA synthase 20 (KCS20), and the GLYCEROL-3-PHOSPHATE SN2-ACYLTRANSFERASE 5 (GPAT5) are involved in the regulation of SL. Comparative analysis confirmed that the expression of the above homologous genes is significantly lower in hyperaccumulators than in nonhyperaccumulators, especially under Cd exposure (Tao et al., 2017, 2019) (Fig. 12.2).

12.2.3 Role of metal ligands in cadmium hyperaccumulation

In nonhyperaccumulator plants, complexation of toxic metals by strong ligands is the main strategy of detoxification. For Cd the main ligand is mostly phytochelatin, a small enzymatically synthesized oligopeptide (Grill et al., 1985, 1989). Therefore, in nonhyperaccumulators, the inhibition or knockout of the key enzyme phytochelatin synthase causes a dramatic increase of sensitivity to Cd toxicity. Hyperaccumulators, in contrast, remain Cd-resistant and Cd-accumulating even without phytochelatin synthesis (Ebbs et al., 2002; Schat et al., 2002). Correspondingly, XAS of shock-frozen hydrated tissue showed a predominant ligation of Cd by oxygen/nitrogen ligands in all three Cd hyperaccumulation model species: *N. caerulescens* (Küpper et al., 2004), *A. halleri* (Isaure et al., 2006), and *S. alfredii* (Tian et al., 2011). This was found to be independent of S supply to the plants; the preferential Cd binding to O/N ligands remained regardless of whether Cd was fed as $CdCl_2$ or $CdSO_4$ (Koren et al., 2013). In nonhyperaccumulating ecotypes of *N. caerulescens*, however, the situation was found to be the same as the pattern of nonaccumulators; that is, blockage of PC synthesis led to Cd hypersensitivity (Schat et al., 2002). This makes sense in the light of the transport-based strategy of hyperaccumulation (Fig. 12.1) and also detoxification as described in more detail in Section 12.2.4. Nevertheless, in hyperaccumulators, metal ligands also play important roles in two ways. First, even in hyperaccumulators, metals are never "free"—there are no free ions in biology contrary to technology and space (plasma). At the very least, water itself acts as a ligand, forming an aquo complex. Usually, there are stronger ligands around, and whatever the strongest ligand is in the same compartment will bind Cd if it has not formed a complex with even higher affinity with another metal (e.g., Cu). In this way, in vacuoles as main storage sites, Cd will bind to the high concentrations of organic acids and histidine that are usually present in this compartment (Küpper et al., 2004). In the cytoplasm, which Cd has to pass to reach the vacuole, it might still be complexed by phytochelatins even in hyperaccumulators. There, Cd may also be bound by the nonproteogenic amino acid nicotianamine, which has been shown to play an important role in long-distance Cd and Zn transport in the xylem and phloem (Becher et al., 2004; Deinlein et al., 2012; Van De Mortel et al., 2006, 2008; Weber et al., 2004). Knocking down nicotianamine synthase led to impaired Cd/Zn hyperaccumulation in *A. halleri* (Deinlein et al., 2012). The overexpression in a Cd-hyperaccumulating *A. halleri* ecotype of ZIF1, which shuffles nicotianamine into the vacuole, further supports the importance of nicotianamine (Corso et al., 2018).

When discussing the role of ligands in Cd speciation, problems of its analysis have to be critically considered because of the severe risk of artifacts that lead to wrong conclusions. When a plant sample dries, either when the tissue is dying or during sample preparation, all aqua complexation of metals will be replaced by other, normally stronger, ligands. Second, during drying, membranes rupture and metal in the vacuole that was separated by the tonoplast membrane from strong thiol ligands in the cytoplasm will come into contact with the latter, leading to thiol complexes. These effects were clearly seen in the direct comparison of senescent versus dead leaves (Küpper et al., 2004). It can also explain the higher percentage of thiol ligands in a later study that worked on freeze-dried instead of frozen-hydrated tissues (Yan et al., 2022)—even freeze-drying never completely avoids such artifacts of drying. No change of ligation during freeze-drying is only found if the metal was already strongly bound, for example, to the cell wall in trichomes, before drying (Isaure et al., 2006).

12.2.4 Role of vacuolar sequestration for cadmium hypertolerance

In roots, compared to nonaccumulators, hyperaccumulators have reduced vacuolar sequestration (Fig. 12.1), so that more of the Cd goes to the shoot via the xylem. This would by itself reduce tolerance, as metal is being sent into potentially more vulnerable tissues, in particular endangering photosynthesis. However, hyperaccumulators prevent such toxicity and become even hypertolerant to Cd by improved vacuolar sequestration in the leaves. This is observed in all hyperaccumulators studied so far, and several genes have been identified to play a key role in this process. *MTP1* (a Cd- and Zn-transporting member of the CDF family) was found to be highly expressed in the epidermis of *Thlaspi goesingense* (Gustin et al., 2009) and of *N. caerulescens* (Küpper & Kochian, 2010). The vacuolar CPx (P_{1B}) ATPase *HMA3* was first found to be highly expressed in *A. halleri* (Becher et al., 2004), contributing to Zn detoxification. Later it was found to be decisive for Cd tolerance in *N. caerulescens* (Ueno et al., 2011) and *S. plumbizincicola* (Liu et al., 2017). Knockdown of *HMA3* in *S. plumbizincicola* led to Cd toxicity symptoms in the young leaves and growth inhibition (Liu et al., 2017). Higher copy number and several-fold higher *HMA3* expression in the whole leaf and leaf tissues (mesophyll, epidermis) were reported for the Ganges population compared to Prayon, and even more compared to the nonaccumulator *A. thaliana* (Ueno et al., 2011), contributing to Cd hypertolerance. In *Nicotiana tabacum*, cation exchanger 4 (CAX4) has been shown to be involved in Cd detoxification by vacuolar sequestration (Korenkov et al., 2007), and in *A. thaliana*, ZIF1 has been shown to have a similar role for the chemically related Zn, but indirectly by pumping the ligand nicotianamine into the vacuole (Haydon et al., 2012). Although these earlier reports focused on nonaccumulators, both *CAX4* and *ZIF* were found to be overexpressed in the Cd-hyperaccumulator *A. halleri* (Corso et al., 2018).

Still, fundamental differences exist in which tissue Cd is finally stored in the shoot. This difference may explain the different Cd tolerance of Cd hyperaccumulator species. In the most efficient Cd hyperaccumulator known so far, the Southern French (Ganges) ecotype of *N. caerulescens*, Cd is stored mostly in large epidermal cells (Leitenmaier & Küpper, 2011), just like Zn in the same species (Küpper et al., 1999). This was confirmed with different techniques in the closely related *Noccaea praecox* (=*Thlaspi praecox*) at several Cd treatments (Koren et al., 2013). The epidermal vacuole is probably the best possible storage location, because Cd leaking out of the vacuole (no biological membrane is completely leak-proof) will not cause inhibition of photosynthesis as the most Cd-sensitive process in plants. In *A. halleri*, in contrast, Cd and Zn are mostly stored in vacuoles of the mesophyll (Küpper et al., 2000), and the same was later found for *S. alfredii* (Tian et al., 2011). The unavoidable Cd transport through the cytoplasm of those cells and/or leakage from the vacuole is then a danger for the chloroplasts and the photosynthetic machinery in them. This can explain why Cd causes a stronger inhibition of photosynthesis and ultimately growth in *A. halleri* than in *N. caerulescens* (Küpper et al., 2007a; Morina & Küpper, 2020). This is further in line with the finding of Koren et al. (2013); in *N. praecox* the preferential Cd storage is in the epidermal vacuoles (measured as "symplast," but little Cd is elsewhere in the symplast), while in the mesophyll the distribution of Cd between symplast and cell walls depended on the Cd salts and concentrations used.

12.3 Cadmium hyperaccumulation against biotic stress

What is the evolutionary advantage of metal hyperaccumulation? Several hypotheses have been postulated by Boyd and Martens (1992) and Martens and Boyd (1994) to answer this question, such as increased tolerance to excess metal in the environment (disposal by efficient sequestration and shredding of the metal-loaded tissues), defense against biotic stress, inadvertent uptake, drought tolerance, and allelopathy. Most of these could not be verified. Therefore here we focus on the well-proven defense-related hypotheses that were further developed by Boyd (2004, 2007, 2012) and review the reports on Cd hyperaccumulation and tolerance in plants challenged by viruses, pathogens, and herbivores.

12.3.1 Elemental defense and joint effects hypotheses—Benefits of cadmium hyperaccumulation against pathogens and pests

To date, metal hyperaccumulation as an efficient defense against predators and pathogens has received considerable attention. The pioneering studies of Boyd et al. showed that Ni (and Zn) hyperaccumulation protects the plants from herbivores (e.g., Boyd, 2007; Boyd & Martens, 1998; Martens & Boyd, 1994; Pollard & Baker, 1997; reviewed by Boyd, 2012). These findings were the basis for establishing the elemental defense hypothesis, also termed the inorganic defense hypothesis, stating that metal hyperaccumulation evolved because of its role in plant defense (Boyd, 2007, 2012). The hypothesis postulates that the hyperaccumulated metal is directly toxic to the herbivores and pathogens. This type of chemical defense, as described by Martens and Boyd (1994), differs from organic defense because elements are taken up from the soil and not synthesized by the plant, and they are chemically persistent, therefore not degraded by the herbivore or pathogen. In this way, investing in elemental defense may be metabolically more efficient than investing the energy for the biosynthesis of organic defense (secondary metabolites—phenolics, glucosinolates, and alkaloids) at the expense of growth. A variation of elemental defense hypothesis is the trade-off hypothesis (Boyd, 2007), postulating that investing in elemental defense results in decreased organic defense (Fig. 12.3). Accumulation

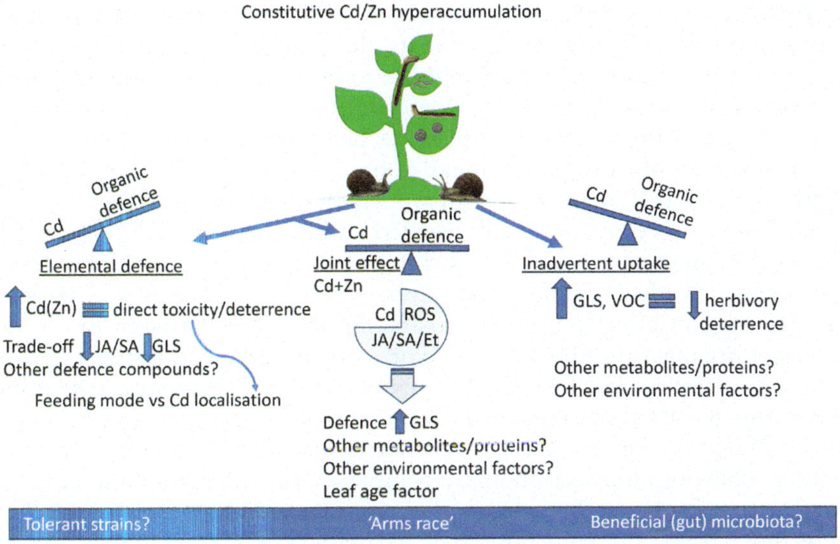

FIGURE 12.3 Beneficial effects of Cd hyperaccumulation against biotic stress and possible scenarios of interaction with organic defense and other environmental factors. *Et*, Ethylene; *GLS*, glucosinolates; *JA*, jasmonic acid; *ROS*, reactive oxygen species; *SA*, salicylic acid; *VOC*, volatile organic compounds. *This conceptual model is based on the hypotheses described by Boyd and Martens (1992). Intercept Limited.*

and sequestration of metals, here Cd, is achieved by active uptake and sequestration to either epidermal storage cells (e.g., *N. caerulescens*) (Küpper et al., 1999, 2004) or mesophyll (*A. halleri, S. alfredii*) and trichomes (Küpper et al., 2000; Tian et al., 2011). As a nonessential element, Cd is not required in proteins but only sequestered through binding to low molecular weight ligands. The whole process has energy requirements that the plant has to meet, and at the same time, balancing the (hyper)accumulation of essential metals such as Zn, as they usually share the uptake and transport routes. However, this may be more energetically affordable than biosynthesis of carbon- and nitrogen-rich secondary metabolites.

An extension of the elemental defense hypothesis is the joint effect hypothesis (Boyd, 2007, 2012). It considers the involvement of both metal accumulation and organic defense as a strategy against biotic stress. The extension means that lower concentrations of the accumulated metal can provide enhanced defense when accompanied by the accumulation of either another metal or an organic chemical defense. The joint effects can be additive, synergistic, or antagonistic: if the two chemicals have a joint effect that is equal to the sum of their individual effects, then the joint effect is additive; if the effect of two combined chemicals is greater than each chemical, the effect is synergistic. Although the first two concepts are positive, the antagonistic joint effect means that the two chemicals together have a negative interaction, and the effect is smaller than the additive effect.

Pioneering ecophysiological studies showed a negative correlation between metal hyperaccumulation and herbivory or pathogen attack, although the conclusions were not always in favor of the defense hypothesis (Davis et al., 2001; reviewed by Pollard & Baker, 1997;

Boyd, 2012; Boyd & Moar, 1999). With the advances in analyses of organic compounds in hyperaccumulators, especially metabolomics (e.g., glucosinolates) and phytohormone signaling pathways, more evidence was found toward the joint effects and trade-off hypotheses. The majority of the studies include Zn and Cd hyperaccumulation, and making a distinction between the effects of each metal individually is not always taken into account. The reason may be that Cd and Zn are usually both present in metal-contaminated soil. They usually share similar chemistry and uptake and sequestration pathways, although for the Ganges population of *N. caerulescens* the absence of competition between Cd and Zn was shown (Zhao et al., 2002).

The available data on the mechanisms of Cd hyperaccumulation and biotic stress are mostly in favor of the beneficial effects against biotic stress (Table 12.1). However, whether it is a result of direct toxicity or joint effect is not conclusive, because of variations in the experimental conditions, model organism (pathogen, herbivore, or virus), and questions addressed in the study, such as Cd accumulation, hormone pathways, and/or secondary metabolites. The distribution and chemistry of Cd binding in the plant (e.g., strong S ligands vs. weak O ligands) in relation to the pathogen (apoplast, leaf tissue, photosynthetic cells) and herbivore feeding type (phloem sucking, chewing) may additionally complicate the interpretation and comparison of data. Among the first studies to investigate the joint effects of two metals, or metal + organic defense, were those based on artificial diet (Cheruiyot et al., 2015; Jhee et al., 2006). Before that, Coleman et al. (2005) pointed to an extension of the elemental defense hypothesis by testing the toxicity of several metals in artificial diet of the herbivore *Plutella xylostella* larvae (Brassicaceae specialist). Cheruiyot et al. (2015) tested the joint effects hypothesis, exposing *Spodoptera exigua* larvae to an artificial diet with a combination of metal + metal (Co, Cu, Ni, and Zn) and metal + organic compounds [nicotine (alkaloid), tannic acid (phenolic), and mustard seed powder (glucosinolates)]. The authors reported both additive and synergistic effects in different combinations for both sublethal and lethal concentrations of the chemicals.

Some of the limitations of studies using artificial diets are not accounting for the plant metabolites and their interactions with Cd, as well as availability of Cd, which may differ between mixing Cd salts with food and Cd taken up and sequestered in the leaves from the soil, as well as effects of other environmental factors under realistic scenarios. Nevertheless, they are important to know the toxicity thresholds of Cd. That constitutively high Cd concentration in plants provides efficient defense against herbivory has been shown by several groups. However, if and to what extent organic defense was involved has not always been considered. Jiang et al. (2005) compared two *N. caerulescens* populations with high and low Cd accumulation traits, as well as the F_2 and F_3 populations from the crosses, to show a direct connection between Cd hyperaccumulation and leaf feeding damage index; however, the possibility of a joint effect was not excluded. Moreover, the Zn/Cd hyperaccumulator *S. alfredii* has been studied in relation to elemental defense against herbivory (Xu et al., 2022). The authors have shown that initially, during the first hour of infestation, aphids did not differentiate between Cd-treated and control *S. alfredii* plants; however, after 7 days, the number of aphids on high Cd plants decreased. High Cd concentration in the phloem in longer treatment and, subsequently, higher Cd concentration in the aphid stylets contributed to elemental defense through direct Cd toxicity. Under field conditions, Shimizu-Inatsugi et al. (2021) showed a direct correlation between Cd levels and herbivory defense in *A. halleri*. High Cd accumulation in *A. halleri* was more efficient against leaf damage than low Cd levels in *A. lyrata*. In contrast, in allotetraploid species *A. kamchatica*, derived from *A. halleri* and *A. lyrata*, with intermediate Cd accumulation, the Cd-based defense against herbivory was reduced. Mohiley (2021) investigated how foraging and clonal sharing

TABLE 12.1 Studies investigating the effects of Cd (and Zn) in hyperaccumulating plants and artificial diets against biotic factors, viruses, bacteria, fungal pathogens, insects, and snails. Media Cd (and Zn) concentrations, leaf Cd and Zn concentrations, parameters studied, and hypotheses tested are shown.

Plant species	Cd/Zn in the media	Cd/Zn in the shoots/leaf compartments	Biotic stress	Parameters studied	Supports hypothesis	References
Arabidopsis halleri	3 mg kg^{-1} Cd	200 µg $g^{-1}{}_{DW}$ Cd	No-choice herbivory caterpillars *Pieris brassicae* and field experiment	Herbivory damage, Cd concentration in nonmetallicolous and metallicolous populations	Elemental defense and joint effect are not addressed, e.g., constitutive differences in metabolites in the two populations	Mohiley et al. (2020)
A. halleri	1.7 mg kg^{-1} Cd, 127 mg kg^{-1} Zn	av. 300 µg $g^{-1}{}_{DW}$ Cd, 4000 µg $g^{-1}{}_{DW}$ Zn	Phloem sucking generalist aphid *Myzus persicae*, chewing polyphagous *Mamestra brassicae*	Performance of two herbivore species with different feeding modes (chewing vs. sucking) andindividual GLS and element concentrations of young and old leaves in response to different soil and herbivore treatments	Elemental defense and joint effects	Stolpe et al. (2017b)
A. halleri	1.7 mg kg^{-1} Cd, 127 mg kg^{-1} Zn	phloem sap 2.3–4.2 RU/µg $leaf_{DW}$ Cd, 75–126 RU/µg leaf $_{DW}$ Zn	*M. persicae*	Phloem metal content, GLS, aphid abundance, and survival	Joint effect, especially in younger leaves	Stolpe et al. (2017a)

(*continued on next page*)

TABLE 12.1 Studies investigating the effects of Cd (and Zn) in hyperaccumulating plants and artificial diets against biotic factors, viruses, bacteria, fungal pathogens, insects, and snails. Media Cd (and Zn) concentrations, leaf Cd and Zn concentrations, parameters studied, and hypotheses tested are shown—cont'd

Plant species	Cd/Zn in the media	Cd/Zn in the shoots/leaf compartments	Biotic stress	Parameters studied	Supports hypothesis	References
A. halleri	1.7 mg kg^{-1} Cd, 127 mg kg^{-1} Zn soil	160 µg g^{-1}_{DW} Cd, 1000 µg g^{-1}_{DW} Zn,	Necrotrophic pathogen *Alternaria brassicae*	Camalexin, Cd/Zn content	Potential trade-off: organic defense not excluded	Tewes et al. (2018)
A. halleri	5 mg kg^{-1} Cd, Zn 300 mg kg^{-1}	400 µg g^{-1}_{DW} Cd and 100 µg g^{-1}_{DW} Cd	Field experiment	Number of insects, GLS, metal content, biomass	Elemental defense, transient GLS increase, joint effect not excluded	Kazemi-Dinan et al. (2015a)
A. halleri	18.4 mg kg^{-1} Cd, 7067 mg kg^{-1} Zn	> 18 µg g^{-1}_{DW}Cd, >1000 µg g^{-1}_{DW} Zn	Brassicaceae specialists *Phaedon cochleariae* and *Athalia rosae*, and generalist *Pieris napi* L. (from metal-contaminated site)	Paired-choice feeding preference, GLS	Elemental defense, additive joint effect	Kazemi-Dinan et al. (2014)
A. halleri	0.5 µM Cd, NS and soil	100 µg g^{-1}_{DW} Cd (NS) after herbivory, >100 µg g^{-1}_{DW}Cd (soil) after wounding	Wounding, larvae of the Brassicaceae specialist *Pieris rapae*	Metal content, root transporters GE, feeding rate	Elemental, Cd is inducible by wounding; GE of vacuolar transporters in the roots downregulated	Plaza et al. (2015)

(*continued on next page*)

A. halleri	29 mg kg^{-1} Cd, 100 mg kg^{-1} Zn	100–400 µg g^{-1}_{DW} Cd	Simulated herbivory by leaf piercing and JA application	Root and shoot biomass, Cd content in 4 non-metallicolous and 4 metallicolous populations	Elemental defense, joint not considered	Mohiley et al. (2020)
A. halleri	5 mg kg^{-1} Cd, 300 mg kg^{-1} Zn	av. 100 µg g^{-1}_{DW} Cd	Field experiment	Metal accumulation, leaf damage, growth	Elemental defense, joint not considered	Shimizu-Inatsugi et al. (2021)
Noccaea (Thlaspi) caerulescens	300 mg kg^{-1} Cd	>100 µg g^{-1}_{DW} Cd	Flower thrips generalist *Frankliniella occidentalis*	Prayon and Ganges populations, number of thrips, leaf-feeding damage index, Cd content	Elemental defense, joints not excluded	Jiang et al. (2005)
N. caerulescens	10 and 30 µM, NS	200–700 µg g^{-1}_{DW} Cd, 1 and 3 µM Cd in apoplast	*Pseudomonas syringae* pv. maculicola	Bacterial growth in planta and in apoplast	Elemental defense for Zn, possible joint effect for Cd	Fones et al. (2010)

(*continued on next page*)

TABLE 12.1 Studies investigating the effects of Cd (and Zn) in hyperaccumulating plants and artificial diets against biotic factors, viruses, bacteria, fungal pathogens, insects, and snails. Media Cd (and Zn) concentrations, leaf Cd and Zn concentrations, parameters studied, and hypotheses tested are shown—cont'd

Plant species	Cd/Zn in the media	Cd/Zn in the shoots/leaf compartments	Biotic stress	Parameters studied	Supports hypothesis	References
N. caerulescens	highest 60 mg kg^{-1} Cd, 4700 Zn extractable	highest 1700 Cd, 13,500 Zn	Generalist snail *Helix aspersa*	Within-population, between-population feeding choices, GLS, Zn/Cd content	GLS levels deter snails more than Zn, inadvertent uptake hypothesis	Noret et al. (2005)
Noccaea praecox	50 μM Cd, NS	>300 μg $g^{-1}{}_{DW}$Cd	Biotrophic powdery mildew *Erysiphe cruciferarum*	Phytohormones, pathogen abundance, metal content	Direct toxicity and/or enhanced fungal-induced JA signaling	Llugany et al. (2013a)
N. praecox	50 μM Cd NS	>500 μg $g^{-1}{}_{DW}$Cd depending on plant size	Generalist snail *Cantareus aspersus*	Leaf area, snail biomass, metal content, metabolites, SA, JA	Joint effects	Llugany et al. (2019)
N. caerulescens	1 μM Cd, 10 μM Zn	>600 μg $g^{-1}{}_{DW}$Cd, >2000 Zn μg $g^{-1}{}_{DW}$	Turnip yellow mosaic virus, Brassicaceae host	Photosynthesis, metal localization, GE, virus abundance	Constitutive elemental defense, joint effect not considered	Morina et al. (2023)

(*continued on next page*)

Sedum alfredii	100 µM Cd, NS	338 µg $g^{-1}{}_{DW}$Cd (2w), 3200 µg $g^{-1}{}_{DW}$Cd (6w)	Phloem-sucking polyphagous *Aphis fabae*	Cd content in the phloem and in the aphids, number of aphids, and Cd localization in aphids by microXRF	Elemental defense, joint not considered	Xu et al. (2022)
S. alfredii	0.75-49 mg kg^{-1} Cd, 68-5990 mg kg^{-1} Zn	700 µg $g^{-1}{}_{DW}$Cd, 15,000 µg $g^{-1}{}_{DW}$DW Zn	Specialist snail *Bradybaena ravida*	VOC, metal content, snail performance	Possible trade-off between Cd accumulation and organic defense	Zhang et al. (2023)
Artificial diet	7.5–56.5 µg g^{-1} Cd	n.a.	Brassicaceae specialist *Plutella xylostella* larvae	Percent survival, toxicity	Elemental defense, possible joint effect not addressed	Coleman et al. (2005)
Artificial diet	2 µg g^{-1} Cd 100 µg g^{-1} Zn	n.a.	Generalist snail *Heliothis virescens*	Performance test, survival	Additive joint effect	Kazemi-Dinan et al. (2014)

GE, gene expression; *GLS*, glucosinolates; *JA*, jasmonic acid; *NS*, nutrient solution; *RU*, relative units; *SA*, salicylic acid; *VOC*, volatile organic compounds.

in *A. halleri* (metalliforius and nonmetalliferous populations) grown in Cd-enriched and control soils affect the response to simulated herbivory by mechanical damage and jasmonic acid (JA) addition. Cd foraging was observed in nonmetalliferous populations in response to simulated herbivory, pointing to induced defense, whereas there was no effect on constitutively higher Cd content in metalliferous populations. In addition, increased Cd sharing was observed between ramets of metallicolous *A. halleri* populations under simulated herbivory, indicating its crucial part in defense response.

Cd hyperaccumulation as a constitutive defense against viruses was shown in *N. caerulescens* (Ganges) population (Morina et al., 2023). Furthermore inducible accumulation of Cd in response to wounding in *A. halleri* has been observed by Plaza et al. (2015) by yet unknown mechanisms. Both wounding and herbivory upregulated Cd accumulation in the *A. halleri* leaves regardless of Cd background. Wounding enhanced root-to-shoot Cd transport, which was supported by downregulation of *MTP3* and *HMA3*.

Extensive work on elemental defense against a bacterial pathogen, *Pseudomonas syringae*, in *N. caerulescens* was done by Fones et al. (2010, 2013). The authors could demonstrate that Zn, Cd, and Ni hyperaccumulation in *N. caerulescens* provided efficient defense against *P. syringae*. Elemental defense by Zn accumulation was further proven using bacterial mutants with contrasting Zn tolerance. Although Cd decreased the bacterial growth in planta, it only transiently inhibited the bacterial growth in the apoplastic fluid (contrary to Zn and Ni). This result indicated possible involvement of other organic defense compounds [reactive oxygen species (ROS), volatiles] in addition to Cd (Fones et al., 2010).

The joint effects of Cd and organic defense were further tested in *N. praecox* challenged with the biotrophic fungal pathogen, *Erysiphe cruciferarum* (Llugany et al., 2013b, 2019). High Cd visually inhibited the disease, and the efficient defense was based on high Cd content in the leaves, accompanied by JA accumulation. Interestingly, solely Cd supply promoted salicylic acid (SA) accumulation, and the mechanisms behind switching toward JA under pathogen attack remained inconclusive (Llugany et al., 2013a). On the other hand a negative interaction between elemental defense (Cd and Zn) and organic defense against the necrotrophic fungal pathogen *Alternaria brassicae* on *A. halleri* pointed to a potential trade-off between two types of defenses (Tewes et al., 2018). In this model system, camalexin induction, which was inducible in response to pathogens in nonhyperaccumulators, was lower in plants grown in metal-enriched soils, supporting the trade-off hypothesis. The pathogen lifestyle (necrotrophic vs. biotrophic) results in differential induction of organic defense (jasmonic vs. salicylic) and thus may overlap with Cd defense.

Our understanding of the complex interactions between hyperaccumulators and herbivores was expanded with analyses of secondary metabolites known for their role in defense. As both *N. caerulescens* and *A. halleri* are members of the Brassicaceae, the focus was on sulfur-containing compounds, glucosinolates, as a key metabolic feature of this family. Nevertheless, the results are controversial. Noret et al. (2005) showed that glucosinolates but not Zn and Cd negatively affected feeding preferences of the snail *Helix aspersa (=Cornu aspersum)*. Regardless of Zn concentration, nonmetallicolous populations of *N. caerulescens* were better protected against herbivory than the metallicolous ones. These results led to the conclusion that the inadvertent uptake hypothesis (see Section 12.1) should be investigated further.

On the other hand, Kazemi-Dinan et al. (2014) showed that three Brassicaceae specialist herbivores preferred leaves of *A. halleri* grown in uncontaminated compared to contaminated soil. Furthermore the use of the *A. halleri* LAN 3.1 genotype from metal-contaminated sites and an *HMA4* RNAi knockdown line with the same genetic background (impaired in both

Zn and Cd accumulation) confirmed that Zn and Cd accumulation alone affected the feeding preferences of three herbivore species. The profiles of glucosinolates in wild-type *A. halleri* populations from low- and high-metal-amended sites and of transgenic plants were similar, thus excluding their effects on the feeding choice. Moreover, the additive joint effects of Zn and Cd when supplemented with the artificial diet were reported for a generalist species, *Heliothis virescens* (Kazemi-Dinan et al., 2014). The results of this extensive study showed the beneficial effects of Zn and Cd accumulation against herbivory, unrelated to the glucosinolate accumulation. Further studies from the same group analyzed Zn and Cd accumulation and glucosinolate profiles in relation to leaf age and two feeding modes (chewing vs. sucking) in *A. halleri* (Stolpe et al., 2017b). The results supported the elemental defense hypothesis in terms of the feeding preference of both herbivores. However, a possible trade-off between glucosinolates and metal (hyper)accumulation was not observed. Young leaves accumulated more Zn and Cd and had a different glucosinolate profile than older leaves in the metal-enriched plants, whereas herbivory alone induced changes in several glucosinolates. Going further to investigate the mechanisms of Cd/Zn accumulation with phloem feeding mode, higher Zn, Cd, Pb, Se, and glucosinolate content in the phloem sap of young *A. halleri* leaves grown in metal-amended soils provided enhanced defense against aphids (Stolpe et al., 2017a), supporting the joint effects hypothesis. Similarly, Cd together with glucosinolates and sugars deterred the feeding of generalist snails in *N. praecox* (Llugany et al., 2019).

Kazemi-Dinan et al. (2015a) tested whether Zn and Cd amendments in common garden conditions contributed to the defense of *A. halleri*. The authors pointed to the importance of joint-effect defenses, particularly in younger plants. Although the Cd effect alone was not considered in these studies (the contaminated soil contained both Zn and Cd), the joint effect of the two metals could be concluded, whereas the contribution of glucosinolate defense was not so clear.

The susceptibility to herbivory in the hyperaccumulating ecotype of *S. alfredii* in low metal-amended soils was elaborated by comparing the constitutive profile of volatile compounds with that of the nonhyperaccumulating ecotype. Generalist snails preferred hyperaccumulators with low metal content over nonhyperaccumulators, although the latter were rich in sugars. Constitutive downregulation of volatile organic compounds (VOCs) in the hyperaccumulating plants growing in the field could contribute to compromised herbivory defense in low-metal-amended soils (Zhang et al., 2023). Although these results imply a trade-off between organic and elemental defense, it should be considered that the VOC analyses were done on field plants, which may have already been exposed to differential environmental factors, including biotic stress.

Generally the trade-off hypothesis between Zn/Cd defense and glucosinolate accumulation has been difficult to support considering the large variability between the *A. halleri* populations (Kazemi-Dinan et al., 2015b; Putra et al., 2024). These studies have been focused on glucosinolates as the main defense organic compounds in Brassicaceae. However, the role of other metabolites, but also mechanical defenses (trichome density), has not been considered, therefore leaving space for the joint effect hypothesis of both organic and inorganic (metal + metal) defense.

From an evolutionary perspective, herbivores and pathogens thriving on hyperaccumulating plants can evolve to tolerate and even accumulate metals in the so-called arms race with the plants (Boyd, 2004; Hörger et al., 2013). Indeed, *N. caerulescens* metallicolous populations were associated with metal-tolerant bacteria (Fones et al., 2010), whereas *Melanotrichus boydi*, a specialist insect feeding on the Ni hyperaccumulator *Streptanthus polygaloides*, was found to accumulate Ni to a significant extent (Boyd & Wall, 2001). Furthermore Zhang et al. (2022) investigated

Cd/Zn hyperaccumulator *S. alfredii* grown under lower metal concentrations in the soil and in its natural habitat (mining area). Snail abundance and their gut microbiota were analyzed in two localities, agricultural remediation soil and greenhouse, whereas no snails were found in the mining area. The gut microbiota in both populations of snails, although with high biodiversity, were similar in terms of metal resistance and their function in facilitating the high metal diet in the snails, indicating their role in metal tolerance of these organisms.

The interaction between rhizosphere microorganisms and hyperaccumulating plants is a topic with expanding interest, especially for phytoremediation and phytoextraction trials and for limiting Cd uptake in the crop species. A few studies have dealt with the role of soil microarthropods, collembolas (*Folsomia candida*), in Cd uptake of *S. plumbizincicola* (Li et al., 2021; Pu et al., 2022). Collembola stimulated plant growth and Cd and Zn uptake in the roots and, at the same time, upregulated plant defense responses by increasing the biosynthesis of SA. Collembola were furthermore able to survive and promote Cd uptake in different metal-contaminated soils (Li et al., 2021). The beneficial role of bacterial strains isolated from the rhizosphere of metallicolous *A. halleri* was shown by Farinati et al. (2009). Besides enhancing Cd and Zn accumulation, microorganisms generally downregulated defense-related proteins.

In most of the studies discussed so far the main analyses involved total Cd concentration in the tissues, parameters of herbivory/pathogen success, phytohormone signaling pathways, and analyses of glucosinolates as the major class of metabolites in Brassicaceae. Analyses of other metabolites (e.g., phenolics) and metalloproteomics to reveal the role of Cd hyperaccumulation in biotic stress interactions are very limited to date.

The hypotheses discussed in this section are not mutually exclusive. The complex model systems, plant history, coaccumulation of Zn and Cd, and choice of biotic stressors discussed here show that the answer to our starting question is not an easy one. Further work integrating genetic, omics, biochemistry, and soil (microbiome) analyses may advance our knowledge on the origin of hyperaccumulation and its impact on plant defense.

12.3.2 Comparison of cadmium hyperaccumulators versus nonaccumulators in defense responses

Based on the concept of hormesis, low concentrations of toxic elements can activate defense mechanisms in plants (Poschenrieder et al., 2013; Wang et al., 2023). Acting as a potent elicitor, low doses of Cd can activate ROS and antioxidants and the accumulation of defense-related metabolites; however, Cd can also cause direct toxicity to biotic stress agents. Essential metals may be involved in plant defense responses as cofactors of enzymes directly involved in immunity and phytohormone synthesis or as triggers of redox imbalance (ROS accumulation) when present in excess, followed by activation of the immune response (reviewed by Morina et al., 2022). The former cannot be the case for Cd as a nonessential metal. Its possible physiological role in activating carbonic anhydrase has been suggested only in *N. caerulescens* (Liu et al., 2008), and it remains to be investigated further.

Signal transduction pathways in response to both metals and biotic stress are initiated with stress perception and activation of a defense response involving Ca^{2+} ions, ROS, reactive nitrogen species (RNS), phytohormones, mitogen-activated protein kinases (MAPKs), and transcription factors. The cascade of events leads to protein posttranslational modifications and changes in gene expression and/or epigenetic effects and accumulation of defense metabolites. ROS, RNS, and the activation of their "processing" (antioxidative) systems have been recognized

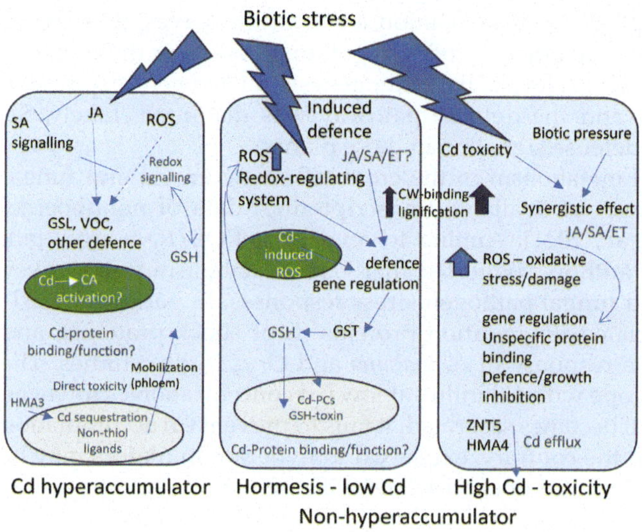

FIGURE 12.4 Cadmium and biotic stress interactions in hyper- and nonhyperaccumulating plants. Depending on the level/duration of Cd excess, hormetic (induced defense) or synergistic effects of Cd toxicity and biotic stress can be expected in nonhyperaccumulating plants. In both cases, depending on the level of Cd excess, glutathione metabolism has a strong role, mainly redox regulative in the former and as a substrate for Cd detoxification by phytochelatins in the latter. Under high Cd treatment, cell wall binding becomes the primary detoxification pathway. *CA*, Carbonic anhydrase; *ET*, ethylene; *GLS*, glucosinolates; *GSH*, glutathione; *GST*, glutathione S-transferase; *JA*, jasmonic acid; *ROS*, reactive oxygen species; *SA*, salicylic acid; *VOC*, volatile organic compounds; HMA3, HMA4, ZNT5, ABC transporters (for more details see part 2.1).

as key players in the cross talk between abiotic and biotic stress (Fujita et al., 2006; Noctor et al., 2018). Controlled ROS and RNS production and redox-derived signals can induce specific cell responses to stress. MAPK signaling mechanisms can be specific in terms of exact kinase and the kinetics of activation, as observed for Cd and Cu (Jonak et al., 2004). Nevertheless, the induction of phytohormone signaling pathways, such as JA, ethylene, and SA, has been documented in different species following Cd treatment (Romero-Puertas et al., 2021; Schellingen et al., 2015). Induced defense was mostly triggered by Cd stress in crops (Dang et al., 2019; Mittra et al., 2004); however, Cd failed to induce JA and VOC accumulation in *Zea mays* in response to herbivory (Winter et al., 2012). Metal excess can lead to increasing ROS accumulation; however, hyperaccumulating plants have an efficient constitutive redox-regulation system (Fones et al., 2013; Freeman et al., 2005; Sun et al., 2007). Under Cd treatment, glutathione (GSH) levels were depleted in nonhyperaccumulating *Thlaspi arvense*, with no effect or even enhancement in hyperaccumulating plants (Llugany et al., 2013b). This observation is in-line with Cd detoxification strategy in nonhyperaccumulators, based on using GSH and GSH-derived peptides as a substrate for phytochelatin synthesis by phytochelatin synthase (Fig. 12.4). The interactions between Cd and organic-induced defense in response to biotic stress are described in detail in 3.1. A more mechanistic interpretation of the role of Cd was shown in relation to virus infection in both hyper- and nonhyperaccumulating plants. Cd excess acted synergistically with turnip yellow mosaic virus (TYMV), aggravating photosynthetic activity in Zn-tolerant and Cd-sensitive *N. ochroleucum* (Morina et al., 2020). Tissue-specific gene expression showed upregulation of *HMA4* in mesophyll and *ZNT5* in epidermal cells, indicating active exclusion of excess Cd under systemic infection. This contrasted with *N. caerulescens*, in which Cd levels efficiently inhibited the TYMV accumulation and maintained photosynthetic activity without changing the total leaf Cd content (Morina et al., 2023).

In tobacco, a glycine-rich protein inducible by low Cd treatment inhibited systemic turnip vein clearing virus infection (Citovsky et al., 1998; Ueki & Citovsky, 2002) by an SA-independent mechanism. Uncoupling between ROS and defense signaling pathways in response to a bacterial pathogen was reported in *N. caerulescens* following sufficient Zn supply (Fones et al., 2013).

Although *N. caerulescens* plants had higher superoxide anion levels in the leaves, no signs of oxidative stress were observed, indicating an efficient redox-regulating system in these plants. Although pathogen infection increased the SA levels, the pathogenesis-related gene expression of *PR1* and *PR2* was not upregulated, and the defense pathway was disturbed. Likely, SA accumulation had a role different from defense signaling in these plants.

Further evidence of the role of GSH metabolism in the cross talk between Cd and fungal pathogens was revealed in meta-analyses of available transcriptomics data of nonhyperaccumulating plants (Romero-Puertas et al., 2021). Aiming to reveal the ROS/RNS cross talk between the two types of stresses, the authors suggested that GSH metabolism was the key player in the cross talk between Cd and fungal pathogen stress responses. In particular, GST-encoding genes, followed by ER-associated degradation proteins (heat shock proteins), and chitin-responsive genes were a common response in *A. thaliana* and *Oryza sativa* studies. The GSH/PC pathway may be sufficient to cope with Cd influx at low Cd concentrations. However, at high Cd concentrations, when cell wall becomes saturated, it fails to prevent Cd accumulation in the cytoplasm, leading to toxicity. On the contrary, excess Cd in hyperaccumulators may be sufficient to defend against biotic stress because of efficient detoxification and sequestration as discussed in Sections 12.2.3 and 12.2.4, not relying on S metabolism and phytochelatins.

12.4 Conclusion and outlook

Research over the many decades since the first discovery of metal hyperaccumulation has revealed fundamental understanding of mechanisms involved in the phenotype, which includes all steps from uptake into the root via translocation into the shoots until final storage. It turned out to involve much more transport than binding to specific ligands, in particular for the final storage. Although the genes involved in transport are mostly known by now, future research should focus on investigating their mechanism of function. Besides the symplastic and transmembrane pathways, a deeper understanding of apoplastic barrier plasticity and its function in a hyperaccumulator's capacity for Cd accumulation will also be fundamental for phytoremediation. Another topic for the future is the involvement of additional genes in Cd tolerance, which is required for maintaining a good biomass despite the high Cd accumulation, a prerequisite for efficient phytoremediation.

The interplay between Cd concentration and the types of biotic stress (and their feeding mechanisms) reveals a complex relationship. It is influenced by other environmental factors and is difficult to reproduce under controlled conditions. The specific Cd-binding protein targets involved in biotic stress response and hormesis are still not identified. Further work is needed to reveal if Cd distribution under biotic stress changes, whether it provides defense by direct toxicity, systematically increases ROS/phytohormone pathways, or regulates plant metabolism by protein binding. Understanding this relationship and underlying mechanisms of Cd hormesis and tolerance is important for developing more effective defense strategies in plants, including crops, for novel remediation methods and sustainable agricultural practices.

12.5 Summary

- Cadmium hyperaccumulation is rare; only <10 plant species have been confirmed conclusively to be Cd hyperaccumulators.

- Cadmium hyperaccumulation is mostly driven by active transport, although the apoplast plays an important role as well, in particular in the root uptake.
- Highly efficient root-to-shoot translocation and detoxification are the hallmarks of hyperaccumulators.
- Hyperaccumulated Cd is mostly stored in weakly bound form in shoot vacuoles, epidermis, or mesophyll depending on the species.
- Cadmium hyperaccumulation provides efficient defense against pathogens and some herbivores, supporting elemental and joint effect hypotheses.

Abbreviations

ABA	abscisic acid
CSs	casparian strips
GLS	glucosinolates
EDX	energy-dispersive X-ray analysis
EXAFS	extended X-ray absorption fine structure
JA	jasmonic acid
ROS	reactive oxygen species
PIXE	particle-induced X-ray emission
SA	salicylic acid
SL	suberin lamellae
XANES	X-ray absorption near edge structure
XRF	X-ray fluorescence (analysis)
VOC	volatile organic compounds

Acknowledgments

The authors are grateful for funding by the Ministry of Education of the Czech Republic with cofinancing from the EU (grant KOROLID, CZ.02.1.01/0.0/0.0/15_003/0000336), the National Natural Science Foundation of China (42011530397), Science Fund for Distinguished Young Scholars of Sichuan Province (24NSFJQ0025, Qi Tao), the Czech Academy of Sciences (grant numbers: NSFC-21-05_PZ, RVO 60077344), and FWF-GACR grant 25-19190L.

References

An, X., Totozafy, J. C., Peaucelle, A., Jones, C. Y., Willats, W. G. T., Höfte, H., Corso, M., & Verbruggen, N. (2023). Contrasting Cd accumulation of *Arabidopsis halleri* populations: A role for (1→4)-β-galactan in pectin. *Journal of Hazardous Materials, 445*. doi:10.1016/j.jhazmat.2022.130581.

Andresen, E., Flores-Sanchez, I. J, Brückner, D., Bokhari, S. NH., Falkenberg, G., & Küpper, H. (2023). Sublethal and lethal Cd toxicity in soybean roots specifically affects the metabolome, Cd binding to proteins and cellular distribution of Cd. *Journal of Hazardous Materials, 442*, 130062. doi:10.1016/j.jhazmat.2022.130062.

Andresen, E., Peiter, E., & Küpper, H. (2018). Trace metal metabolism in plants. *Journal of Experimental Botany, 69*(5), 909–954. doi:10.1093/jxb/erx465.

Assunção, A. G. L., Martins, P. D. aC., De Folter, S., Vooijs, R., Schat, H., & Aarts, M. G. M. (2001). Elevated expression of metal transporter genes in three accessions of the metal hyperaccumulator *Thlaspi caerulescens*. *Plant, Cell & Environment, 24*(2), 217–226. doi:10.1111/j.1365-3040.2001.00666.x.

Baker, A. J. M., McGrath, S. P., Sidoli, C. M. D., & Reeves, R. D. (1994). The possibility of in situ heavy metal decontamination of polluted soils using crops of metal-accumulating plants. *Resources, Conservation and Recycling, 11*(1–4), 41–49. doi:10.1016/0921-3449(94)90077-9.

Barberon, M. (2017). The endodermis as a checkpoint for nutrients. *New Phytologist, 213*(4), 1604–1610. doi:10.1111/nph.14140.

Barberon, M., Vermeer, J. E. M., De Bellis, D., Wang, P., Naseer, S., Andersen, T. G., Humbel, B. M., Nawrath, C., Takano, J., Salt, D. E., & Geldner, N. (2016). Adaptation of root function by nutrient-induced plasticity of endodermal differentiation. *Cell, 164*(3), 447–459. doi:10.1016/j.cell.2015.12.021.

Becher, M., Talke, I. N., Krall, L., & Krämer, U. (2004). Cross-species microarray transcript profiling reveals high constitutive expression of metal homeostasis genes in shoots of the zinc hyperaccumulator *Arabidopsis halleri*. *The Plant Journal, 37*(2), 251–268. doi:10.1046/j.1365-313x.2003.01959.x.

Bernard, C., Roosens, N., Czernic, P., Lebrun, M., & Verbruggen, N. (2004). A novel CPx-ATPase from the cadmium hyperaccumulator *Thlaspi caerulescens*. *FEBS Letters, 569*(1–3), 140–148. doi:10.1016/j.febslet.2004.05.036.

Bert, V., Meerts, P., Saumitou-Laprade, P., Salis, P., Gruber, W., & Verbruggen, N. (2003). Genetic basis of Cd tolerance and hyperaccumulation in *Arabidopsis halleri*. *Plant and Soil, 249*(1), 9–18. doi:10.1023/a:1022580325301.

Boyd, R. S. (2004). Ecology of metal hyperaccumulation. *New Phytologist, 162*(3), 563–567. doi:10.1111/j.1469-8137.2004.01079.x.

Boyd, R. S. (2007). The defense hypothesis of elemental hyperaccumulation: Status, challenges and new directions. *Plant and Soil, 293*(1–2), 153–176. doi:10.1007/s11104-007-9240-6.

Boyd, R. S. (2012). Plant defense using toxic inorganic ions: Conceptual models of the defensive enhancement and joint effects hypotheses. *Plant Science, 195*, 88–95. doi:10.1016/j.plantsci.2012.06.012.

Boyd, R. S., & Martens, S. N. (1992). The raison d'être for metal hyperaccumulation by plants. In A. J. M. Baker, J. Proctor, & R. D. Reeves (Eds.), *The vegetation of ultramafic (serpentine) soils*. Intercept Limited.

Boyd, R. S., & Martens, S. N. (1998). The significance of metal hyperaccumulation for biotic interactions. *Chemoecology, 8*(1), 1–7. doi:10.1007/s000490050002.

Boyd, R. S., & Moar, W. J. (1999). The defensive function of Ni in plants: Response of the polyphagous herbivore *Spodoptera exigua* (Lepidoptera: Noctuidae) to hyperaccumulator and accumulator species of *Streptanthus* (Brassicaceae). *Oecologia, 118*(2), 218–224. doi:10.1007/s004420050721.

Boyd, R. S., & Wall, M. A. (2001). Responses of generalist predators fed high-Ni *Melanotrichus boydi* (Heteroptera: Miridae): Elemental defense against the third trophic level. *American Midland Naturalist, 146*(1), 186–198. doi:10.1674/0003-0031(2001)146[0186:ROGPFH]2.0.CO;2.

Brooks, R. R., Chambers, M. F., Nicks, L. J., & Robinson, B. H. (1998). Phytomining. *Trends in Plant Science, 3*(9), 359–362. doi:10.1016/S1360-1385(98)01283-7.

Brooks, R. R., Lee, J., Reeves, R. D., & Jaffre, T. (1977). Detection of nickeliferous rocks by analysis of herbarium specimens of indicator plants. *Journal of Geochemical Exploration, 7*(C), 49–57. doi:10.1016/0375-6742(77)90074-7.

Brown, S. L., Chaney, R. L., Angle, J. S., & Baker, A. J. M. (1994). Phytoremediation potential of *Thlaspi caerulescens* and bladder campion for zinc- and cadmium-contaminated soil. *Journal of Environmental Quality, 23*(6), 1151–1157. doi:10.2134/jeq1994.00472425002300060004x.

Chaney, R. L. (1983). Plant uptake of inorganic waste constituents. In J. F. Parr, P. B. Marsh, & J. M. Kla (Eds.), *Land treatment of hazardous wastes* (pp. 50–76). Park Ridge: Noyes Data Corp.

Chao, Z.-F., & Chao, D.-Y. (2025). Barriers and carriers for transition metal homeostasis in plants. *Plant Communications, 6*(2), 101235. doi:10.1016/j.xplc.2024.101235.

Cheruiyot, D. J., Boyd, R. S., & Moar, W. (2015). Testing the joint effects hypothesis of elemental defense using spodoptera exigua. *Journal of Chemical Ecology, 41*(2), 168–177. doi:10.1007/s10886-015-0553-2.

Citovsky, V., Ghoshroy, S., Tsui, F., & Klessig, D. (1998). Non-toxic concentrations of cadmium inhibit systemic movement of turnip vein clearing virus by a salicylic acid-independent mechanism. *The Plant Journal, 16*(1), 13–20. doi:10.1046/j.1365-313x.1998.00263.x.

Coleman, C. M., Boyd, R. S., & Eubanks, M. D. (2005). Extending the elemental defense hypothesis: Dietary metal concentrations below hyperaccumulator levels could harm herbivores. *Journal of Chemical Ecology, 31*(8), 1669–1681. doi:10.1007/s10886-005-5919-4.

Connolly, E. L., Fett, J. P., & Guerinot, M. L. (2002). Expression of the IRT1 metal transporter is controlled by metals at the levels of transcript and protein accumulation. *Plant Cell, 14*(6), 1347–1357. doi:10.1105/tpc.001263.

Corso, M., An, X., Jones, C. Y., Gonzalez-Doblas, V., Schvartzman, M. S., Malkowski, E., Willats, W. G. T., Hanikenne, M., & Verbruggen, N. (2021). Adaptation of *Arabidopsis halleri* to extreme metal pollution through limited metal accumulation involves changes in cell wall composition and metal homeostasis. *New Phytologist, 230*(2), 669–682. doi:10.1111/nph.17173.

Corso, M., Schvartzman, M. S., Guzzo, F., Souard, F., Malkowski, E., Hanikenne, M., & Verbruggen, N. (2018). Contrasting cadmium resistance strategies in two metallicolous populations of *Arabidopsis halleri*. *New Phytologist, 218*(1), 283–297. doi:10.1111/nph.14948.

Cunningham, S. D., & Berti, W. R. (1993). Remediation of contaminated soils with green plants: An overview. *In Vitro Cellular & Developmental Biology—Plant, 29*(4), 207–212. doi:10.1007/BF02632036.

Dang, F., Lin, J., Chen, Y., Li, G. X, Guan, D., Zheng, S. J, & He, S. (2019). A feedback loop between CaWRKY41 and H2O2 coordinates the response to *Ralstonia solanacearum* and excess cadmium in pepper. *Journal of Experimental Botany, 70*(5), 1581–1595. doi:10.1093/jxb/erz006.

Davis, M. A., Murphy, J. F., & Boyd, R. S. (2001). Nickel increases susceptibility of a nickel hyperaccumulator to *Turnip mosaic virus*. *Journal of Environmental Quality, 30*(1), 85–90. doi:10.2134/jeq2001.30185x.

Deinlein, U., Weber, M., Schmidt, H., Rensch, S., Trampczynska, A., Hansen, T. H., Husted, S., Schjoerring, J. K., Talke, I. N., Krämer, U., & Clemens, S. (2012). Elevated nicotianamine levels in *Arabidopsis halleri* roots play a key role in zinc hyperaccumulation. *Plant Cell, 24*(2), 708–723. doi:10.1105/tpc.111.095000.

Ebbs, S., Lau, I., Ahner, B., & Kochian, L. (2002). Phytochelatin synthesis is not responsible for Cd tolerance in the Zn/Cd hyperaccumulator *Thlaspi caerulescens* (J. & C. Presl). *Planta, 214*(4), 635–640. doi:10.1007/s004250100650.

Escarré, J., Lefèbvre, C., Raboyeau, S., Dossantos, A., Gruber, W., Cleyet Marel, J. C., Frérot, H., Noret, N., Mahieu, S., Collin, C., & Van Oort, F. (2011). Heavy metal concentration survey in soils and plants of the les Malines mining district (southern France): Implications for soil restoration. *Water, Air, and Soil Pollution, 216*(1–4), 485–504. doi:10.1007/s11270-010-0547-1.

Faiyue, B., Al-Azzawi, M. J., & Flowers, T. J. (2010). The role of lateral roots in bypass flow in rice (*Oryza sativa* L.). *Plant, Cell and Environment, 33*(5), 702–716. doi:10.1111/j.1365-3040.2009.02078.x.

Farinati, S., DalCorso, G., Bona, E., Corbella, M., Lampis, S., Cecconi, D., Polati, R., Berta, G., Vallini, G., & Furini, A. (2009). Proteomic analysis of *Arabidopsis halleri* shoots in response to the heavy metals cadmium and zinc and rhizosphere microorganisms. *Proteomics, 9*(21), 4837–4850. doi:10.1002/pmic.200900036.

Fones, H., Davis, C. A. R., Rico, A., Fang, F., Smith, J. A. C., & Preston, G. M. (2010). Metal hyperaccumulation armors plants against disease. *PLoS Pathogens, 6*(9). doi:10.1371/journal.ppat.1001093.

Fones, H. N., Eyles, C. J., Bennett, M. H., Smith, J. A. C., & Preston, G. M. (2013). Uncoupling of reactive oxygen species accumulation and defence signalling in the metal hyperaccumulator plant *Noccaea caerulescens*. *New Phytologist, 199*(4), 916–924. doi:10.1111/nph.12354.

Freeman, J. L., Garcia, D., Kim, D., Hopf, A., & Salt, D. E. (2005). Constitutively elevated salicylic acid signals glutathione-mediated nickel tolerance in *Thlaspi* nickel hyperaccumulators. *Plant Physiology, 137*(3), 1082–1091. doi:10.1104/pp.104.055293.

Frérot, H., Faucon, M. P., Willems, G., Godé, C., Courseaux, A., Darracq, A., Verbruggen, N., & Saumitou-Laprade, P. (2010). Genetic architecture of zinc hyperaccumulation in *Arabidopsis halleri*: The essential role of QTL × environment interactions. *New Phytologist, 187*(2), 355–367. doi:10.1111/j.1469-8137.2010.03295.x.

Fujita, M., Fujita, Y., Noutoshi, Y., Takahashi, F., Narusaka, Y., Yamaguchi-Shinozaki, K., & Shinozaki, K. (2006). Crosstalk between abiotic and biotic stress responses: A current view from the points of convergence in the stress signaling networks. *Current Opinion in Plant Biology, 9*(4), 436–442. doi:10.1016/j.pbi.2006.05.014.

Grill, E., Löffler, S., Winnacker, E.-L., & Zenk, M. H. (1989). Phytochelatins, the heavy-metal-binding peptides of plants, are synthesized from glutathione by a specific γ-glutamylcysteine dipeptidyl transpeptidase (phytochelatin synthase). *Proceedings of the National Academy of Sciences, 86*(18), 6838–6842. doi:10.1073/pnas.86.18.6838.

Grill, E., Winnacker, E. L., & Zenk, M. H. (1985). Phytochelatins: The principal heavy-metal complexing peptides of higher plants. *Science, 230*(4726), 674–676. doi:10.1126/science.230.4726.674.

Gustin, J. L., Loureiro, M. E., Kim, D., Na, G., Tikhonova, M., & Salt, D. E. (2009). MTP1-dependent Zn sequestration into shoot vacuoles suggests dual roles in Zn tolerance and accumulation in Zn-hyperaccumulating plants. *Plant Journal, 57*(6), 1116–1127. doi:10.1111/j.1365-313X.2008.03754.x.

Hanikenne, M., Talke, I. N., Haydon, M. J., Lanz, C., Nolte, A., Motte, P., Kroymann, J., Weigel, D., & Krämer, U. (2008). Evolution of metal hyperaccumulation required cis-regulatory changes and triplication of HMA4. *Nature, 453*(7193), 391–395. doi:10.1038/nature06877.

Haydon, M. J., Kawachi, M., Wirtz, M., Hillmer, S., Hell, R. ü, & Krämer, U. (2012). Vacuolar nicotianamine has critical and distinct roles under iron deficiency and for zinc sequestration in *Arabidopsis*. *The Plant Cell, 24*(2), 724–737. doi:10.1105/tpc.111.095042.

Hazotte, C., Laubie, B., Rees, F., Morel, J. L., & Simonnot, M. O. (2017). A novel process to recover cadmium and zinc from the hyperaccumulator plant *Noccaea caerulescens*. *Hydrometallurgy, 174*, 56–65. doi:10.1016/j.hydromet.2017.09.012.

Hörger, A. C., Fones, H. N., & Preston, G. M. (2013). The current status of the elemental defense hypothesis in relation to pathogens. *Frontiers in Plant Science, 4*. doi:10.3389/fpls.2013.00395.

Isaure, M.-P., Fayard, B., Sarret, G., Pairis, S., & Bourguignon, J. (2006). Localization and chemical forms of cadmium in plant samples by combining analytical electron microscopy and X-ray spectromicroscopy. *Spectrochimica Acta Part B: Atomic Spectroscopy, 61*(12), 1242–1252. doi:10.1016/j.sab.2006.10.009.

Jhee, E. M., Boyd, R. S., & Eubanks, M. D. (2006). Effectiveness of metal-metal and metal-organic compound combinations against *Plutella xylostella*: Implications for plant elemental defense. *Journal of Chemical Ecology, 32*(2), 239–259. doi:10.1007/s10886-005-9000-0.

Jiang, R. F., Ma, D. Y., Zhao, F. J., & McGrath, S. P. (2005). Cadmium hyperaccumulation protects *Thlaspi caerulescens* from leaf feeding damage by thrips (*Frankliniella occidentalis*). *New Phytologist, 167*(3), 805–814. doi:10.1111/j.1469-8137.2005.01452.x.

Jonak, C., Nakagami, H., & Hirt, H. (2004). Heavy metal stress. Activation of distinct mitogen-activated protein kinase pathways by copper and cadmium. *Plant Physiology, 136*(2), 3276–3283. doi:10.1104/pp.104.045724.

Kamiya, T., Borghi, M., Wang, P., Danku, J. M. C., Kalmbach, L., Hosmani, P. S., Naseer, S., Fujiwara, T., Geldner, N., & Salt, D. E. (2015). The MYB36 transcription factor orchestrates Casparian strip formation. *Proceedings of the National Academy of Sciences of the United States of America, 112*(33), 10533–10538. doi:10.1073/pnas.1507691112.

Kazemi-Dinan, A., Barwinski, A., Stein, R. J., Krämer, U., & Müller, C. (2015a). Metal hyperaccumulation in Brassicaceae mediates defense against herbivores in the field and improves growth. *Entomologia Experimentalis et Applicata, 157*(1), 3–10. doi:10.1111/eea.12333.

Kazemi-Dinan, A., Sauer, J., Stein, R. J., Krämer, U., & Müller, C. (2015b). Is there a trade-off between glucosinolate-based organic and inorganic defences in a metal hyperaccumulator in the field? *Oecologia, 178*(2), 369–378. doi:10.1007/s00442-014-3218-x.

Kazemi-Dinan, A., Thomaschky, S., Stein, R. J., Krämer, U., & Müller, C. (2014). Zinc and cadmium hyperaccumulation act as deterrents towards specialist herbivores and impede the performance of a generalist herbivore. *New Phytologist, 202*(2), 628–639. doi:10.1111/nph.12663.

Koren, Š, Arčon, I., Kump, P., Nečemer, M., & Vogel-Mikuš, K. (2013). Influence of $CdCl_2$ and $CdSO_4$ supplementation on Cd distribution and ligand environment in leaves of the Cd hyperaccumulator *Noccaea (Thlaspi) praecox*. *Plant and Soil, 370*(1–2), 125–148. doi:10.1007/s11104-013-1617-0.

Korenkov, V., Hirschi, K., Crutchfield, J. D., & Wagner, G. J. (2007). Enhancing tonoplast Cd/H antiport activity increases Cd, Zn, and Mn tolerance, and impacts root/shoot Cd partitioning in *Nicotiana tabacum* L. *Planta, 226*(6), 1379–1387. doi:10.1007/s00425-007-0577-0.

Küpper, H., & Kochian, L. V. (2010). Transcriptional regulation of metal transport genes and mineral nutrition during acclimatization to cadmium and zinc in the Cd/Zn hyperaccumulator, *Thlaspi caerulescens* (Ganges population). *New Phytologist, 185*(1), 114–129. doi:10.1111/j.1469-8137.2009.03051.x.

Küpper, H., Lombi, E., Zhao, F. J., & McGrath, S. P. (2000). Cellular compartmentation of cadmium and zinc in relation to other elements in the hyperaccumulator *Arabidopsis halleri*. *Planta, 212*(1), 75–84. doi:10.1007/s004250000366.

Küpper, H., Mijovilovich, A., Meyer-Klaucke, W., & Kroneck, P. M. H. (2004). Tissue- and age-dependent differences in the complexation of cadmium and zinc in the cadmium/zinc hyperaccumulator *Thlaspi caerulescens* (Ganges ecotype) revealed by X-ray absorption spectroscopy. *Plant Physiology, 134*(2), 748–757. doi:10.1104/pp.103.032953.

Küpper, H., Parameswaran, A., Leitenmaier, B., Trtílek, M., & Šetlík, I. (2007a). Cadmium-induced inhibition of photosynthesis and long-term acclimation to cadmium stress in the hyperaccumulator *Thlaspi caerulescens*. *New Phytologist, 175*(4), 655–674. doi:10.1111/j.1469-8137.2007.02139.x.

Küpper, H., Seib, L. O., Sivaguru, M., Hoekenga, O. A., & Kochian, L. V. (2007b). A method for cellular localization of gene expression via quantitative in situ hybridization in plants. *Plant Journal, 50*(1), 159–187. doi:10.1111/j.1365-313X.2007.03031.x.

Küpper, H., Zhao, F. J., & McGrath, S. P. (1999). Cellular compartmentation of zinc in leaves of the hyperaccumulator *Thlaspi caerulescens*. *Plant Physiology, 119*(1), 305–311. doi:10.1104/pp.119.1.305.

Leitenmaier, B., & Küpper, H. (2011). Cadmium uptake and sequestration kinetics in individual leaf cell protoplasts of the Cd/Zn hyperaccumulator *Thlaspi caerulescens*. *Plant, Cell & Environment, 34*(2), 208–219. doi:10.1111/j.1365-3040.2010.02236.x.

Leitenmaier, B., Witt, A., Witzke, A., Stemke, A., Meyer-Klaucke, W., Kroneck, P. M. H., & Küpper, H. (2011). Biochemical and biophysical characterisation yields insights into the mechanism of a Cd/Zn transporting ATPase purified from the hyperaccumulator plant *Thlaspi caerulescens*. *Biochimica et Biophysica Acta—Biomembranes, 1808*(10), 2591–2599. doi:10.1016/j.bbamem.2011.05.010.

Li, T., Tao, Q., Shohag, M. J. I., Yang, X., Sparks, D. L., & Liang, Y. (2015). Root cell wall polysaccharides are involved in cadmium hyperaccumulation in *Sedum alfredii*. *Plant and Soil, 389*(1–2), 387–399. doi:10.1007/s11104-014-2367-3.

Li, Z., Ding, Y., Ke, X., Wu, L., & Christie, P. (2021). Enhancement by soil micro-arthropods of phytoextraction of metal-contaminated soils using a hyperaccumulator plant species. *Plant and Soil, 464*(1–2), 335–346. doi:10.1007/s11104-021-04958-1.

Liberman, L. M., Sparks, E. E., Moreno-Risueno, M. A., Petricka, J. J., & Benfey, P. N. (2015). MYB36 regulates the transition from proliferation to differentiation in the *Arabidopsis* root. *Proceedings of the National Academy of Sciences, 112*(39), 12099–12104. doi:10.1073/pnas.1515576112.

Liu, H., Zhao, H., Wu, L., Liu, A., Zhao, F. J., & Xu, W. (2017). Heavy metal ATPase 3 (HMA3) confers cadmium hypertolerance on the cadmium/zinc hyperaccumulator *Sedum plumbizincicola*. *New Phytologist, 215*(2), 687–698. doi:10.1111/nph.14622.

Liu, M. Q., Yanai, J., Jiang, R. F., Zhang, F., McGrath, S. P., & Zhao, F. J. (2008). Does cadmium play a physiological role in the hyperaccumulator *Thlaspi caerulescens*? *Chemosphere, 71*(7), 1276–1283. doi:10.1016/j.chemosphere.2007.11.063.

Liu, Y., Lu, M., Tao, Q. i, Luo, J., Li, J., Guo, X., Liang, Y., Yang, X., & Li, T. (2020). A comparative study of root cadmium radial transport in seedlings of two wheat (*Triticum aestivum* L.) genotypes differing in grain cadmium accumulation. *Environmental Pollution, 266*(Pt 3), 115235. doi:10.1016/j.envpol.2020.115235.

Liu, Y., Tao, Q., Li, J., Guo, X., Luo, J., Jupa, R., Liang, Y., & Li, T. (2021). Ethylene-mediated apoplastic barriers development involved in cadmium accumulation in root of hyperaccumulator *Sedum alfredii*. *Journal of Hazardous Materials, 403*, 123729. doi:10.1016/j.jhazmat.2020.123729.

Llugany, M., Martin, S. R., Barceló, J., & Poschenrieder, C. (2013a). Endogenous jasmonic and salicylic acids levels in the Cd-hyperaccumulator *Noccaea (Thlaspi) praecox* exposed to fungal infection and/or mechanical stress. *Plant Cell Reports, 32*(8), 1243–1249. doi:10.1007/s00299-013-1427-0.

Llugany, M., Tolrà, R., Martín, S. R., Poschenrieder, C., & Barceló, J. (2013b). Cadmium-induced changes in glutathione and phenolics of *Thlaspi* and *Noccaea* species differing in Cd accumulation. *Journal of Plant Nutrition and Soil Science, 176*(6), 851–858. doi:10.1002/jpln.201300096.

Llugany, M., Tolrà, R., Barceló, J., & Poschenrieder, C. (2019). Snails prefer it sweet: A multifactorial test of the metal defence hypothesis. *Physiologia Plantarum, 165*(2), 209–218. doi:10.1111/ppl.12821.

Lombi, E., Zhao, F. J., Dunham, S. J., & McGrath, S. P. (2000). Cadmium accumulation in populations of *Thlaspi caerulescens* and *Thlaspi goesingense*. *New Phytologist, 145*(1), 11–20. doi:10.1046/j.1469-8137.2000.00560.x.

Lombi, E., Zhao, F. J., McGrath, S. P., Young, S. D., & Sacchi, G. A. (2001). Physiological evidence for a high-affinity cadmium transporter highly expressed in a *Thlaspi caerulescens* ecotype. *New Phytologist, 149*(1), 53–60. doi:10.1046/j.1469-8137.2001.00003.x.

Lu, L., Tian, S. K, Yang, X. E., Wang, X. C., Brown, P., Li, T. Q., & He, Z. L. (2008). Enhanced root-to-shoot translocation of cadmium in the hyperaccumulating ecotype of *Sedum alfredii*. *Journal of Experimental Botany, 59*(11), 3203–3213. doi:10.1093/jxb/ern174.

Lux, A., Martinka, M., Vaculík, M., & White, P. J. (2011). Root responses to cadmium in the rhizosphere: A review. *Journal of Experimental Botany, 62*(1), 21–37. doi:10.1093/jxb/erq281.

Lux, A., Šottníková, A., Opatrná, J., & Greger, M. (2004). Differences in structure of adventitious roots in Salix clones with contrasting characteristics of cadmium accumulation and sensitivity. *Physiologia Plantarum, 120*(4), 537–545. doi:10.1111/j.0031-9317.2004.0275.x.

Macnair, M. R., Bert, V., Huitson, S. B., Saumitou-Laprade, P., & Petit, D. (1999). Zinc tolerance and hyperaccumulation are genetically independent characters. *Proceedings of the Royal Society of London. Series B: Biological Sciences, 266*(1434), 2175–2179. doi:10.1098/rspb.1999.0905.

Martens, S. N., & Boyd, R. S. (1994). The ecological significance of nickel hyperaccumulation: A plant chemical defense. *Oecologia, 98*(3–4), 379–384. doi:10.1007/BF00324227.

Mendoza-Cózatl, D. G., Xie, Q., Akmakjian, G. Z., Jobe, T. O., Patel, A., Stacey, M. G., Song, L., Demoin, D. W., Jurisson, S. S., Stacey, G., & Schroeder, J. I. (2014). OPT3 is a component of the iron-signaling network between leaves and roots and misregulation of OPT3 leads to an over-accumulation of cadmium in seeds. *Molecular Plant, 7*(9), 1455–1469. doi:10.1093/mp/ssu067.

Meyer, C. L., Juraniec, M., Huguet, S., Chaves-Rodriguez, E., Salis, P., Isaure, M. P., Goormaghtigh, E., & Verbruggen, N. (2015). Intraspecific variability of cadmium tolerance and accumulation, and cadmium-induced cell wall modifications in the metal hyperaccumulator *Arabidopsis halleri*. *Journal of Experimental Botany, 66*(11), 3215–3227. doi:10.1093/jxb/erv144.

Mishra, S., Mishra, A., & Küpper, H. (2017). Protein biochemistry and expression regulation of cadmium/zinc pumping ATPases in the hyperaccumulator plants *Arabidopsis halleri* and *Noccaea caerulescens*. *Frontiers in Plant Science, 8*. doi:10.3389/fpls.2017.00835.

Mittra, B., Ghosh, P., Henry, S. L., Mishra, J., Das, T. K., Ghosh, S., Babu, C. R., & Mohanty, P. (2004). Novel mode of resistance to *Fusarium* infection by a mild dose pre-exposure of cadmium in wheat. *Plant Physiology and Biochemistry, 42*(10), 781–787. doi:10.1016/j.plaphy.2004.09.005.

Mohiley, A. (2021). Between the devil and the deep blue sea: Herbivory induces foraging for and uptake of cadmium in a metal hyperaccumulating plant. *Proceedings of the Royal Society B, 288*. doi:10.1098/rspb.2021.1682.

Mohiley, A., Tielbörger, K., Seifan, M., Gruntman, M., & Barton, K. (2020). The role of biotic interactions in determining metal hyperaccumulation in plants. *Functional Ecology, 34*(3), 658–668. doi:10.1111/1365-2435.13502.

Morina, F., & Küpper, H. (2020). Direct inhibition of photosynthesis by Cd dominates over inhibition caused by micronutrient deficiency in the Cd/Zn hyperaccumulator *Arabidopsis halleri*. *Plant Physiology and Biochemistry, 155*, 252–261. doi:10.1016/j.plaphy.2020.07.018.

Morina, F., Küpper, H., & Dietz, K.-J. (2022). Trace metals at the frontline of pathogen defence responses in non-hyperaccumulating plants. *Journal of Experimental Botany, 73*(19), 6516–6524. doi:10.1093/jxb/erac316.

Morina, F., Mijovilovich, A., Mishra, A., Brückner, D., Vujić, B., Bokhari, S. N. H., Špak, J., Falkenberg, G., & Küpper, H. (2023). Cadmium and Zn hyperaccumulation provide efficient constitutive defense against *Turnip* yellow mosaic virus infection in *Noccaea caerulescens*. *Plant Science, 336*. doi:10.1016/j.plantsci.2023.111864.

Morina, F., Mishra, A., Mijovilovich, A., Matoušková, Šá, Brückner, D., Špak, J., & Küpper, H. (2020). Interaction between Zn deficiency, toxicity and turnip yellow mosaic virus infection in *Noccaea ochroleucum*. *Frontiers in Plant Science, 11*. doi:10.3389/fpls.2020.00739.

Naseer, S., Lee, Y., Lapierre, C., Franke, R., Nawrath, C., & Geldner, N. (2012). Casparian strip diffusion barrier in *Arabidopsis* is made of a lignin polymer without suberin. *Proceedings of the National Academy of Sciences, 109*(25), 10101–10106. doi:10.1073/pnas.1205726109.

Noctor, G., Reichheld, J. P., & Foyer, C. H. (2018). ROS-related redox regulation and signaling in plants. *Seminars in Cell and Developmental Biology, 80*, 3–12. doi:10.1016/j.semcdb.2017.07.013.

Noret, N., Meerts, P., Tolrà, R., Poschenrieder, C., Barceló, J., & Escarre, J. (2005). Palatability of *Thlaspi caerulescens* for snails: Influence of zinc and glucosinolates. *New Phytologist, 165*(3), 763–772. doi:10.1111/j.1469-8137.2004.01286.x.

Ó Lochlainn, S., Bowen, H. C., Fray, R. G., Hammond, J. P., King, G. J., White, P. J., Graham, N. S., & Broadley, M. R. (2011). Tandem quadruplication of HMA4 in the zinc (Zn) and cadmium (Cd) hyperaccumulator *Noccaea caerulescens*. *PLoS One, 6*(3). doi:10.1371/journal.pone.0017814.

Papoyan, A., & Kochian, L. V. (2004). Identification of *Thlaspi caerulescens* genes that may be involved in heavy metal hyperaccumulation and tolerance. Characterization of a novel heavy metal transporting ATPase. *Plant physiology, 136*(3), 3814–3823. doi:10.1104/pp.104.044503.

Pence, N. S., Larsen, P. B., Ebbs, S. D., Letham, D. L. D., Lasat, M. M., Garvin, D. F., Eide, D., & Kochian, L. V. (2000). The molecular physiology of heavy metal transport in the Zn/Cd hyperaccumulator *Thlaspi caerulescens*. *Proceedings of the National Academy of Sciences, 97*(9), 4956–4960. doi:10.1073/pnas.97.9.4956.

Peng, J. S., Wang, Y. J., Ding, G., Ma, H. L., Zhang, Y. J., & Gong, J. M. (2017). A pivotal role of cell wall in cadmium accumulation in the Crassulaceae hyperaccumulator *Sedum plumbizincicola*. *Molecular Plant, 10*(5), 771–774. doi:10.1016/j.molp.2016.12.007.

Piñeros, M. A., Shaff, J. E., & Kochian, L. V. (1998). Development, characterization, and application of a cadmium-selective microelectrode for the measurement of cadmium fluxes in roots of *Thlaspi* species and wheat. *Plant Physiology, 116*(4), 1393–1401. doi:10.1104/pp.116.4.1393.

Plaza, S., Weber, J., Pajonk, S., Thomas, J. érô, Talke, I. N., Schellenberg, M., Pradervand, S., Burla, B. o, Geisler, M., Martinoia, E., & Krämer, U. (2015). Wounding of *Arabidopsis halleri* leaves enhances cadmium accumulation that acts as a defense against herbivory. *BioMetals, 28*(3), 521–528. doi:10.1007/s10534-015-9829-9.

Pollard, A. J., & Baker, A. J. M. (1997). Deterrence of herbivory by zinc hyperaccumulation in *Thlaspi caerulescens* (Brassicaceae). *New Phytologist, 135*(4), 655–658. doi:10.1046/j.1469-8137.1997.00689.x.

Poschenrieder, C., Cabot, C., Martos, S., Gallego, B., & Barceló, J. (2013). Do toxic ions induce hormesis in plants? *Plant Science, 212*, 15–25. doi:10.1016/j.plantsci.2013.07.012.

Pu, L., Li, Z., Jia, M., Ke, X., Liu, H., Christie, P., & Wu, L. (2022). Effects of a soil collembolan on the growth and metal uptake of a hyperaccumulator: Modification of root morphology and the expression of plant defense genes. *Environmental Pollution, 303*, 119169. doi:10.1016/j.envpol.2022.119169.

Putra, R., Tölle, M., Krämer, U., & Müller, C. (2024). Effects of metal amendment and metalloid supplementation on foliar defences are plant accession-specific in the hyperaccumulator *Arabidopsis halleri*. *BioMetals, 37*(3), 649–669. doi:10.1007/s10534-023-00550-5.

Qi, X., Fung-yee Tam, N., Li, W. C, & Ye, Z. (2020). The role of root apoplastic barriers in cadmium translocation and accumulation in cultivars of rice (*Oryza sativa* L.) with different Cd-accumulating characteristics. *Environmental Pollution, 264*, 114736. doi:10.1016/j.envpol.2020.114736.

Reeves, R. D., Baker, A. J. M., Jaffré, T., Erskine, P. D., Echevarria, G., & van der Ent, A. (2018). A global database for plants that hyperaccumulate metal and metalloid trace elements. *New Phytologist, 218*(2), 407–411. doi:10.1111/nph.14907.

Romero-Puertas, M. C., Terrón-Camero, L. C., Peláez-Vico, M. Á., Molina-Moya, E., & Sandalio, L. M. (2021). An update on redox signals in plant responses to biotic and abiotic stress crosstalk: Insights from cadmium and fungal pathogen interactions. *Journal of Experimental Botany, 72*(16), 5857–5875. doi:10.1093/jxb/erab271.

Sachs, J. (1865). *Handbuch der Experimental-Physiologie der Pflanzen. §47. Handbuch der physiologischen Botanik* (pp. 153–154). Verlag von Wilhelm Engelmann. https://archive.org/details/handbuchderexper1865sach.

Sattelmacher, B. (2001). The apoplast and its significance for plant mineral nutrition. *New Phytologist, 149*(2), 167–192. doi:10.1046/j.1469-8137.2001.00034.x.

Schat, H., Llugany, M., Vooijs, R., Hartley-Whitaker, J., & Bleeker, P. M. (2002). The role of phytochelatins in constitutive and adaptive heavy metal tolerances in hyperaccumulator and non-hyperaccumulator metallophytes. *Journal of Experimental Botany, 53*(379), 2381–2392. doi:10.1093/jxb/erf107.

Schellingen, K., Van Der Straeten, D., Remans, T., Vangronsveld, J., Keunen, E., & Cuypers, A. (2015). Ethylene signalling is mediating the early cadmium-induced oxidative challenge in *Arabidopsis thaliana*. *Plant Science, 239*, 137–146. doi:10.1016/j.plantsci.2015.07.015.

Serra, O., & Geldner, N. (2022). The making of suberin. *New Phytologist, 235*(3), 848–866. doi:10.1111/nph.18202.

Shimizu-Inatsugi, R., Milosavljevic, S., Shimizu, K. K., Schaepman-Strub, G., Tanoi, K., & Sato, Y. (2021). Metal accumulation and its effect on leaf herbivory in an allopolyploid species *Arabidopsis kamchatica* inherited from a diploid hyperaccumulator *A. halleri*. *Plant Species Biology, 36*(2), 208–217. doi:10.1111/1442-1984.12304.

Shukla, V., & Barberon, M. (2021). Building and breaking of a barrier: Suberin plasticity and function in the endodermis. *Current Opinion in Plant Biology, 64*, 102153. doi:10.1016/j.pbi.2021.102153.

Sinclair, S. A., & Krämer, U. (2012). The zinc homeostasis network of land plants. *Biochimica et Biophysica Acta—Molecular Cell Research, 1823*(9), 1553–1567. doi:10.1016/j.bbamcr.2012.05.016.

Stein, R. J., Höreth, S., de Melo, J. R. F., Syllwasschy, L., Lee, G., Garbin, M. L., Clemens, S., & Krämer, U. (2017). Relationships between soil and leaf mineral composition are element-specific, environment-dependent and geographically structured in the emerging model *Arabidopsis halleri*. *New Phytologist, 213*(3), 1274–1286. doi:10.1111/nph.14219.

Stolpe, C., Giehren, F., Krämer, U., & Müller, C. (2017a). Both heavy metal-amendment of soil and aphid-infestation increase Cd and Zn concentrations in phloem exudates of a metal-hyperaccumulating plant. *Phytochemistry, 139*, 109–117. doi:10.1016/j.phytochem.2017.04.010.

Stolpe, C., Krämer, U., & Müller, C. (2017b). Heavy metal (hyper)accumulation in leaves of *Arabidopsis halleri* is accompanied by a reduced performance of herbivores and shifts in leaf glucosinolate and element concentrations. *Environmental and Experimental Botany, 133*, 78–86. doi:10.1016/j.envexpbot.2016.10.003.

Sun, Q., Ye, Z. H., Wang, X. R., & Wong, M. H. (2007). Cadmium hyperaccumulation leads to an increase of glutathione rather than phytochelatins in the cadmium hyperaccumulator *Sedum alfredii*. *Journal of Plant Physiology, 164*(11), 1489–1498. doi:10.1016/j.jplph.2006.10.001.

Talke, I. N., Hanikenne, M., & Krämer, U. (2006). Zinc-dependent global transcriptional control, transcriptional deregulation, and higher gene copy number for genes in metal homeostasis of the hyperaccumulator *Arabidopsis halleri*. *Plant Physiology, 142*(1), 148–167. doi:10.1104/pp.105.076232.

Tao, Q., Jupa, R., Luo, J., Lux, A., Ková, J., Wen, Y., Zhou, Y., Jan, J., Liang, Y., & Li, T. (2017). The apoplasmic pathway via the root apex and lateral roots contributes to Cd hyperaccumulation in the hyperaccumulator *Sedum alfredii*. *Journal of Experimental Botany, 68*(3), 739–751. doi:10.1093/jxb/erw453.

Tao, Q., Jupa, R., Liu, Y., Luo, J., Li, J., Kováč, J., Li, B., Li, Q., Wu, K., Liang, Y., Lux, A., Wang, C., & Li, T. (2019). Abscisic acid-mediated modifications of radial apoplastic transport pathway play a key role in cadmium uptake in hyperaccumulator *Sedum alfredii*. *Plant, Cell & Environment, 42*(5), 1425–1440. doi:10.1111/pce.13506.

Tewes, L. J., Stolpe, C., Kerim, A., Krämer, U., & Müller, C. (2018). Metal hyperaccumulation in the Brassicaceae species *Arabidopsis halleri* reduces camalexin induction after fungal pathogen attack. *Environmental and Experimental Botany, 153*, 120–126. doi:10.1016/j.envexpbot.2018.05.015.

Tian, S., Lu, L., Labavitch, J., Yang, X., He, Z., Hu, H., Sarangi, R., Newville, M., Commisso, J., & Brown, P. (2011). Cellular sequestration of cadmium in the hyperaccumulator plant species *Sedum alfredii*. *Plant Physiology, 157*(4), 1914–1925. doi:10.1104/pp.111.183947.

Uddin, N., Li, X., Ullah, M. W, Sethupathy, S., Ma, K., Zahoor, N. E, Khan, K. A, & Zhu, D. (2024). Lignin developmental patterns and Casparian strip as apoplastic barriers: A review. *International Journal of Biological Macromolecules, 260*(Pt 2), 129595. doi:10.1016/j.ijbiomac.2024.129595.

Ueki, S., & Citovsky, V. (2002). The systemic movement of a tobamovirus is inhibited by a cadmium-ion-induced glycine-rich protein. *Nature Cell Biology, 4*(7), 478–486. doi:10.1038/ncb806.

Ueno, D., Milner, M. J., Yamaji, N., Yokosho, K., Koyama, E., Clemencia Zambrano, M., Kaskie, M., Ebbs, S., Kochian, L. V., & Ma, J. F. (2011). Elevated expression of TcHMA3 plays a key role in the extreme Cd tolerance in a Cd-hyperaccumulating ecotype of *Thlaspi caerulescens*. *The Plant Journal, 66*(5), 852–862. doi:10.1111/j.1365-313x.2011.04548.x.

Vaculík, M., Lux, A., Luxová, M., Tanimoto, E., & Lichtscheidl, I. (2009). Silicon mitigates cadmium inhibitory effects in young maize plants. *Environmental and Experimental Botany, 67*(1), 52–58. doi:10.1016/j.envexpbot.2009.06.012.

Van De Mortel, J. E., Schat, H., Moerland, P. D., Van Themaat, E. V. L., Van Der Ent, S., Blankestijn, H., Ghandilyan, A., Tsiatsiani, S., & Aarts, M. G. M. (2008). Expression differences for genes involved in lignin, glutathione and sulphate metabolism in response to cadmium in *Arabidopsis thaliana* and the related Zn/Cd-hyperaccumulator *Thlaspi caerulescens*. *Plant, Cell and Environment, 31*(3), 301–324. doi:10.1111/j.1365-3040.2007.01764.x.

Van De Mortel, J. E., Villanueva, L. A., Schat, H., Kwekkeboom, J., Coughlan, S., Moerland, P. D., Van Themaat, E. V. L., Koornneef, M., & Aarts, M. G. M. (2006). Large expression differences in genes for iron and zinc homeostasis, stress response, and lignin biosynthesis distinguish roots of *Arabidopsis thaliana* and the related metal hyperaccumulator *Thlaspi caerulescens*. *Plant Physiology, 142*(3), 1127–1147. doi:10.1104/pp.106.082073.

van der Ent, A., Przybyłowicz, W. J., de Jonge, M. D., Harris, H. H., Ryan, C. G., Tylko, G., Paterson, D. J., Barnabas, A. D., Kopittke, P. M., & Mesjasz-Przybyłowicz, J. (2018). X-ray elemental mapping techniques for elucidating the ecophysiology of hyperaccumulator plants. *New Phytologist, 218*(2), 432–452. doi:10.1111/nph.14810.

Vázquez, M. D., Barceló, J., Poschenrieder, C. h., Mádico, J., Hatton, P., Baker, A. J. M., & Cope, G. H. (1992). Localization of zinc and cadmium in *Thlaspi caerulescens* (Brassicaceae), a metallophyte that can hyperaccumulate both metals. *Journal of Plant Physiology, 140*(3), 350–355. doi:10.1016/s0176-1617(11)81091-6.

Wang, B., Lin, L., Yuan, X., Zhu, Y., Wang, Y., Li, D., He, J., & Xiao, Y. (2023). Low-level cadmium exposure induced hormesis in peppermint young plant by constantly activating antioxidant activity based on physiological and transcriptomic analyses. *Frontiers in Plant Science, 14*. doi:10.3389/fpls.2023.1088285.

Wang, K., Sitsel, O., Meloni, G., Autzen, H. E., Andersson, M., Klymchuk, T., Nielsen, A. M., Rees, D. C., Nissen, P., & Gourdon, P. (2014). Structure and mechanism of Zn^{2+}-transporting P-type ATPases. *Nature, 514*(7253), 518–522. doi:10.1038/nature13618.

Waters, B. M., Chu, H. H., DiDonato, R. J., Roberts, L. A., Eisley, R. B., Lahner, B., Salt, D. E., & Walker, E. L. (2006). Mutations in *Arabidopsis* Yellow Stripe-Like1 and Yellow Stripe-Like3 reveal their roles in metal ion homeostasis and loading of metal ions in seeds. *Plant Physiology, 141*(4), 1446–1458. doi:10.1104/pp.106.082586.

Weber, M., Harada, E., Vess, C., Roepenack-Lahaye, E. v., & Clemens, S. (2004). Comparative microarray analysis of *Arabidopsis thaliana* and *Arabidopsis halleri* roots identifies nicotianamine synthase, a ZIP transporter and other genes as potential metal hyperaccumulation factors. *The Plant Journal, 37*(2), 269–281. doi:10.1046/j.1365-313x.2003.01960.x.

White, P. J., Whiting, S. N., Baker, A. J. M., & Broadley, M. R. (2002). Does zinc move apoplastically to the xylem in roots of *Thlaspi caerulescens*? *New Phytologist, 153*(2), 201–207. doi:10.1046/j.0028-646X.2001.00325.x.

Willems, G., Frérot, H., Gennen, J., Salis, P., Saumitou-Laprade, P., & Verbruggen, N. (2010). Quantitative trait loci analysis of mineral element concentrations in an *Arabidopsis halleri* × *Arabidopsis lyrata* petraea F 2 progeny grown on cadmium-contaminated soil. *New Phytologist, 187*(2), 368–379. doi:10.1111/j.1469-8137.2010.03294.x.

Williams, L. E., Pittman, J. K., & Hall, J. L. (2000). Emerging mechanisms for heavy metal transport in plants. *Biochimica et Biophysica Acta—Biomembranes, 1465*(1-2), 104–126. doi:10.1016/S0005-2736(00)00133-4.

Winter, T. R., Borkowski, L., Zeier, J., & Rostás, M. (2012). Heavy metal stress can prime for herbivore-induced plant volatile emission. *Plant, Cell & Environment, 35*(7), 1287–1298. doi:10.1111/j.1365-3040.2012.02489.x.

Wójcik, M., Vangronsveld, J., D'Haen, J., & Tukiendorf, A. (2005). Cadmium tolerance in *Thlaspi caerulescens*: II. Localization of cadmium in *Thlaspi caerulescens*. *Environmental and Experimental Botany, 53*(2), 163–171. doi:10.1016/j.envexpbot.2004.03.011.

Wu, L., Zhou, G., Bi, G., & Baker, Smith, L. (2013). *Sedum plumbizincicola* X.H. Guo et S.B. Zhou ex L.H. Wu (Crassulaceae): A new species from Zhejiang Province, China. *Plant Systematics and Evolution, 299*(3), 487–498. doi:10.1007/s00606-012-0738-x.

Xu, L., Lu, L., Lin, X., & Hu, Y. (2022). Cadmium armors the Cd hyperaccumulator *Sedum alfredii* against aphid attack. *Frontiers in Environmental Science, 10*. doi:10.3389/fenvs.2022.1036708.

Yan, J., Tang, Z., Fischel, M., Wang, P., Siebecker, M. G., Aarts, M. G. M., Sparks, D. L., & Zhao, F. J. (2022). Variation in cadmium accumulation and speciation within the same population of the hyperaccumulator *Noccaea caerulescens* grown in a moderately contaminated soil. *Plant and Soil, 475*(1–2), 379–394. doi:10.1007/s11104-022-05373-w.

Yang, X. E., Long, X. X., Ye, H. B., He, Z. L., Calvert, D. V., & Stoffella, P. J. (2004). Cadmium tolerance and hyperaccumulation in a new Zn-hyperaccumulating plant species (*Sedum alfredii* Hance). *Plant and Soil, 259*(1/2), 181–189. doi:10.1023/B:PLSO.0000020956.24027.f2.

Zhai, Z., Gayomba, S. R., Jung, H. I., Vimalakumari, N. K., Piñeros, M., Craft, E., Rutzke, M. A., Danku, J., Lahner, B., Punshon, T., Guerinot, M. L., Salt, D. E., Kochian, L. V., & Vatamaniuk, O. K. (2014). OPT3 is a phloem-specific iron transporter that is essential for systemic iron signaling and redistribution of iron and cadmium in *Arabidopsis*. *Plant Cell, 26*(5), 2249–2264. doi:10.1105/tpc.114.123737.

Zhang, P., Jin, C., Qiu, H., Peijnenburg, W. J. G. M., & He, E. (2023). Metals or metabolites? Leafy volatile metabolomics revealed a trade-off between elemental defense and organic defense in Zn/Cd hyperaccumulator *Sedum alfredii*. *Plant Ecology, 224*(2), 173–186. doi:10.1007/s11258-022-01287-w.

Zhang, P., Li, W., Qiu, H., Liu, M., Li, Y., & He, E. (2022). Metal resistant gut microbiota facilitates snails feeding on metal hyperaccumulator plant *Sedum alfredii* in the phytoremediation field. *Ecotoxicology and Environmental Safety, 236*, 113514. doi:10.1016/j.ecoenv.2022.113514.

Zhao, F. J., Hamon, R. E., Lombi, E., McLaughlin, M. J., & McGrath, S. P. (2002). Characteristics of cadmium uptake in two contrasting ecotypes of the hyperaccumulator *Thlaspi caerulescens*. *Journal of Experimental Botany, 53*(368), 535–543. doi:10.1093/jexbot/53.368.535.

Zhao, F. J., Jiang, R. F., Dunham, S. J., & McGrath, S. P. (2006). Cadmium uptake, translocation and tolerance in the hyperaccumulator *Arabidopsis halleri*. *New Phytologist, 172*(4), 646–654. doi:10.1111/j.1469-8137.2006.01867.x.

Zhao, F. J., Lombi, E., & McGrath, S. P. (2003). Assessing the potential for zinc and cadmium phytoremediation with the hyperaccumulator *Thlaspi caerulescens*. *Plant and Soil, 249*(1), 37–43. doi:10.1023/a:1022530217289.

Zhao, H., Wang, L., Zhao, F.-J., Wu, L., Liu, A., & Xu, W. (2019). SpHMA1 is a chloroplast cadmium exporter protecting photochemical reactions in the Cd hyperaccumulator *Sedum plumbizincicola*. *Plant, Cell & Environment, 42*(4), 1112–1124. doi:10.1111/pce.13456.

CHAPTER

13

Role of mycorrhiza in cadmium uptake and detoxification

Xin Zhang[a,b], *Baodong Chen*[a,b]

[a] State Key Laboratory of Regional and Urban Ecology, Research Centre for Eco-Environmental Sciences, Chinese Academy of Sciences, Beijing, China

[b] University of Chinese Academy of Sciences, Beijing, China

13.1 Introduction

Arbuscular mycorrhizal (AM) fungi are widely distributed in nature and colonize over 70% of terrestrial higher plants, forming an ecologically significant mutualistic symbiosis. In AM symbiosis, AM fungi promote the uptake of nutrients, particularly phosphorus (P), by the host plant. In return the plant supplies the fungi with carbohydrates and lipids for fungal growth and development (Jiang et al., 2017; Smith et al., 2003). There are also other mycorrhizal types, such as ectomycorrhizae (common in *Pinaceae* and *Fagaceae*), ericoid mycorrhizae (associated with *Ericales*), and orchid mycorrhizae (specific to *Orchidaceae*), whereas this chapter focuses specifically on AM fungi.

It has been well documented that AM fungi play an important role in the plant uptake and detoxification of cadmium (Cd). AM symbiosis can mitigate the impact of Cd pollution by enhancing mineral nutrient uptake and regulating the Cd uptake and accumulation by host plants (Riaz et al., 2021). The underlying mechanisms can be classified into indirect (nonspecific) and direct (specific) effects (Fig. 13.1). Indirect effects include activating and enhancing plant uptake of soil nutrients, promoting plant growth and increasing plant Cd tolerance by a growth dilution effect (Chen et al., 2019). Furthermore, AM symbiosis can also alleviate the Cd-induced damage to nonenzymatic and enzymatic antioxidant systems in plants by regulating the expression of relevant plant genes (Liang et al., 2023). Direct effects refer to the effects of AM fungi on Cd transport and transformation in the soil–plant continuum. First, AM fungi can directly take up and transport Cd (Guo et al., 1996). Second, AM fungi can mitigate the toxicity of Cd by retaining Cd through compartmentalization in the fungal structures, such as arbuscules, spores, and hyphae (Liu et al., 2022; Nayuki et al., 2014). Third, AM fungi can decrease Cd bioavailability by releasing glomalin-related soil protein (GRSP), a metal-chelating glycoprotein, which forms stable complexes with Cd in the rhizosphere (Jiang et al., 2021).

Cadmium Contamination in Soils and Plants: Mechanisms and Mitigation.
DOI: https://doi.org/10.1016/B978-0-443-34104-5.00007-4

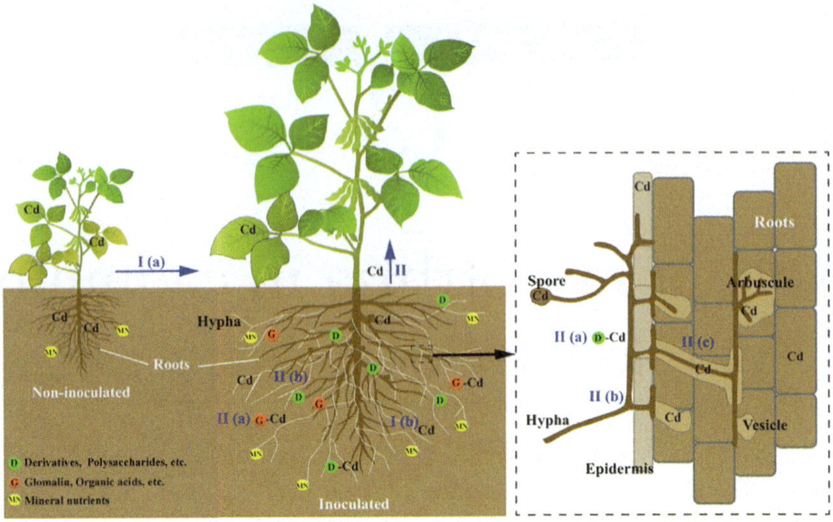

FIGURE 13.1 Arbuscular mycorrhizal (AM) fungi can indirectly (I) and directly (II) affect cadmium (Cd) uptake and distribution within plants. Indirect effects (I): AM fungi improve plant nutrient uptake and enhance plant Cd tolerance by growth dilution effect (I a); AM fungi influence root growth, altering Cd uptake characteristics (I b). Direct effects (II): AM secret polysaccharides, derivatives, glomalin, etc., may complex Cd or affect soil physicochemical properties, changing Cd bioavailability (II a); AM fungal extraradical mycelium takes up and sequesters Cd (II b), and Cd can be retained in symbiotic structures (compartmentalization) (II c), thus reducing Cd translocation to the aboveground parts of plants.

The uptake and detoxification of Cd by AM symbiosis are influenced by the AM fungal strains, host species, and soil Cd concentration. Different AM fungi may employ different strategies to detoxify Cd. This chapter focuses on the roles of AM fungi in facilitating the uptake, transformation, and detoxification of Cd in plants.

13.2 Uptake of cadmium by arbuscular mycorrhizal fungi

13.2.1 Mycorrhizal effects on cadmium uptake and accumulation by plants

AM fungi can enhance nutrient acquisition by host plants (Parmar et al., 2013); however, their role in Cd uptake remains inconsistent across studies. Most studies indicate that AM fungal symbiosis reduces Cd accumulation in plants (Rezvani et al., 2015; Xu et al., 2024; Zhao & Naeth, 2022). By contrast, some reports have shown increased Cd uptake under AM colonization. For instance, Han et al. (2021) observed that under treatment with Ca^{2+} channel blockers, nonmycorrhizal ryegrass (*Lolium perenne*) exhibited reduced Cd^{2+} influx, whereas mycorrhizal plants showed significantly higher Cd^{2+} uptake compared to noninoculated controls. It was hypothesized that AM fungi may employ nongated ion channels—unaffected by such blockers—to facilitate Cd^{2+} transport. Accordingly, colonized plants accumulated higher Cd concentrations in both roots and shoots than nonmycorrhizal plants. The difference in Cd uptake may be attributed to variations in plant species, AM fungal species, and environmental conditions, as discussed later.

Different genotypes of the same plant species may respond to AM symbiosis differently. Inoculation with AM fungi was found to increase Cd concentrations in the roots and shoots of some, but not all, cultivars of pea (*Pisum sativum* L.) (Rivera-Becerril et al., 2002). After being inoculated with three different AM fungal species the leguminous plant *Lotus japonicus* L. exhibited variations in Cd uptake, colonization rate, P uptake efficiency, and growth rate (Zhang et al., 2015). Inoculation with *Rhizophagus irregularis* promoted P uptake and plant growth and increased Cd uptake at high Cd levels; inoculation with *Acaulospora longula* promoted growth and P uptake only at low Cd levels; inoculation with *Gigaspora margarita* had no significant effects on growth or Cd uptake. Another study evaluated the effects of AM fungal mixtures from different provinces in Peru on Cd accumulation and growth of cocoa trees (*Theobroma cacao*) (Vallejos-Torres et al., 2022). Cocoa trees inoculated with different AM fungal mixtures exhibited variations in colonization rate, mycelium length, plant height, leaf area, and stem Cd concentration. In general a higher AM fungal diversity in the inoculant produced better performance in promoting plant growth and decreasing Cd uptake. The observed differences may stem from variations in AM fungal colonization rates, differential expression or activity of Cd transporters, and divergent capacities for synthesizing intracellular chelating compounds in plant cells. However, direct evidence supporting these hypotheses remains unavailable.

Some hyperaccumulator plants can accumulate a large amount of Cd (see Chapter 12), but AM fungi also have different effects on different hyperaccumulators. Vogel-Mikuš et al. (2006) showed that Cd concentrations in both roots and shoots of the hyperaccumulator *Noccaea praecox* Wulfen (formerly *Thlaspi praecox*) were lower after being inoculated with an indigenous AM fungi mixture. Because the root biomass was significantly reduced in inoculated plants while shoot biomass was not significantly affected, the total amount of Cd absorbed by plants also decreased. In contrast, Lu et al. (2020) reported that AM symbiosis promoted growth and increased Cd uptake by the Cd-tolerant species *Sphagneticola calendulacea* grown in soil amended with 50–100 mg/kg soluble Cd.

Environmental conditions affect plant growth and AM fungal colonization, consequently affecting plant Cd uptake. The formation and function of mycorrhizal associations are affected by many factors such as soil moisture, temperature, pH, and cation exchange capacity (Entry et al., 2002). Among them, soil P availability is one of the most important factors. Yazici et al. (2021) found that P fertilization caused a reduction of mycorrhizal colonization in wheat roots, thus reducing the effect of mycorrhizal immobilization of Cd. Lower mycorrhizal colonization could also reduce Zn uptake and, consequently, the competition on Cd uptake. Both effects may explain the increased Cd accumulation in wheat by P fertilization. You et al. (2022) found that under high Cd stress, P starvation reduced root Cd concentrations, whereas excess P increased Cd concentrations in roots of common reed (*Phragmites australis*). Liu et al. (2014) reported that Cd concentration and uptake in maize (*Zea mays* L.) increased with AM fungal colonization at a moderate Cd concentration (2.25 mg/kg) but decreased at an even higher Cd concentration (22.5 mg/kg) 6 weeks after planting. Other factors such as high temperature (Gao et al., 2022) and salinity (Alsadat Hashemi Nejad et al., 2021) synergize with AM fungal inoculation to increase plant Cd uptake. In contrast, the use of biochar (Zhao et al., 2021) or silicon (Khan et al., 2024) reduces Cd uptake of mycorrhizal plants, possibly by decreasing the availability of Cd in soil or inhibiting Cd uptake by plants.

13.2.2 Mycorrhizal fungi change the soil environment and affect cadmium bioavailability

Cd exists in various forms in soil. Cd in soil is often sequentially extracted with various chemical reagents and operationally defined as exchangeable, carbonate-bound, Fe–Mn oxide-bound, and residual fractions (Rauret et al., 1999) (see Chapter 9). AM fungi, as an important part of the soil microbial community, can affect Cd bioavailability by changing the soil conditions (Cheng et al., 2022). It was found that the AM fungus *Glomus mosseae* significantly increased soil pH, CEC, and organic matter in the rhizosphere soil of *Medicago truncatula* L., which decreased Cd availability (Jiang et al., 2021). They speculated that the increase in pH may be because of the secretion of glomalin by AM fungi, which may be protonated with H^+. For a long time, glomalin has been considered a glycoprotein (Wright & Upadhyaya, 1998). However, recent research suggests that glomalin is not a protein but is more likely a complex polysaccharide with a molecular weight ranging from 511 to 600 kDa (Alptekin et al., 2025). In soil, glomalin can form GRSP (González-Chávez et al., 2004). The structure of GRSP has not been precisely characterized yet. However, it is confirmed to be a glycoprotein complex. The protein part may be related to the adsorption by soil particles. The carbohydrate part is mainly composed of oligosaccharide chains formed of glucose and galactose. The hydroxyl and carboxyl groups of the sugar chains can form coordination bonds with metal ions in the soil while binding to soil organic matter through hydrogen bonds (Cornejo et al., 2008; González-Chávez et al., 2004). AM fungi can produce glomalin, and the GRSP content increases significantly under Cd stress. A quantity of 1 kg of GRSP could sorb about 1120 mg Cd, which reduces Cd availability in soil and Cd uptake by plants (González-Chávez et al., 2004). Zhang et al. (2021) showed that GRSP can promote the formation of soil macroaggregates, which may encapsulate heavy metals on their inner surfaces. Tan et al. (2023) conducted a meta-analysis of the influence of AM fungi on bioaccumulation and bioavailability of As and Cd. The results showed that the inoculation of AM fungi increased soil pH, organic carbon content, and easily-extractable and total GRSP and reduced As and Cd bioavailability in soil. However, the extent of the decrease was affected by many factors, including time, P fertilizer, AM fungal species, plant species, plant life cycle, and soil characteristics. There are also contradictory results. Wu et al. (2014) showed that the amount of Pb and Cd bound to GRSP only accounted for a very small portion of the total amount of Pb and Cd bound to soil organic matter, and the effect of GRSP on metal uptake by mycorrhizal plants was limited.

Citterio et al. (2005) showed that AM fungus *G. mosseae* has no effect on Cd bioavailability and the total amount of Cd absorbed by *Cannabis sativa*. Moreover, AM colonization suppressed the growth of *C. sativa*, with the decrease in biomass being proportional to the degree of mycorrhization. This may be related to the extensive root system of *C. sativa*, which can effectively absorb nutrients without AM symbiosis. In the study of Sun et al. (2023a) the AM fungus *Diversispora eburnea* increased exchangeable Cd in the soil, thereby increasing Cd uptake by *L. perenne* and *Amorpha fruticosa*. Transformed carrot roots, which have been introduced with root-inducing transfer DNA (Ri T-DNA) and possess the ability for autonomous growth, serve as a model system for studying the role of AM fungi in the transformation and plant accumulation of heavy metals (Barber-Zucker et al., 2021). Giasson et al. (2005b) studied the effect of AM fungi on the transformation of metal carbonates using in vitro culture of AM fungi with transformed carrot roots. The results showed that AM fungal hyphae secreted organic acids such as citric acid and oxalic acid to acidify the rhizosphere, resulting in the dissolution of $ZnCO_3$ and $CdCO_3$. As a result, AM fungi increased the amount of Zn and Cd absorbed by Ri T-DNA transformed carrot roots (Giasson et al., 2005a). Whether the results of these in vitro experiments are applicable to mycorrhizal plants growing in soil requires further investigation.

13.2.3 Direct evidence for cadmium uptake by mycorrhizal fungi: Isotope tracing and molecular evidence

There are numerous reports on the effects of AM fungi on the absorption of heavy metals by plants. However, there are only a few studies on the Cd uptake by AM fungi themselves. Compared with plant roots, AM fungal hyphae are very thin and can pass through a 60-μm mesh size of nylon nets (Haystead et al., 1988), which can be used to separate a hyphal compartment from the roots. Consequently, heavy metals or their radioisotopes added only to the hyphal compartment can only be accessed by AM fungi. Hutchinson et al. (2004) used nylon nets to separate *Plantago lanceolata* roots and AM fungal hyphae and exposed the hyphae to radioisotope ^{109}Cd. By calculating the proportion of ^{109}Cd in total Cd in the plant, they estimated that 10% of Cd in the plant was absorbed by the AM fungal hyphae. Chen et al. (2018) used a similar method and found a 90% decrease in the amount of Cd accumulated in plant shoots when only hyphae were able to access Cd in the hyphal compartment than when Cd was accessed by plant roots. In addition, they used the synchrotron radiation μX-ray fluorescence (SR-μXRF) method to localize Cd in mycorrhizal roots. SR-μXRF imaging showed that arbuscules and intercellular hyphae accumulated large amounts of Cd, followed by vesicles, whereas the Cd concentrations in root cells were below the detection limit. This distribution pattern indicates that after Cd is taken up by AM fungi, it is mainly retained in the fungal structure and not readily transferred to plant cells. This may be the key reason why mycorrhiza reduces Cd accumulation in plant shoots.

González-Guerrero et al. (2008) used energy-dispersive X-ray spectroscopy (EDS) to determine the metal concentrations in different subcellular locations in AM fungal hyphae and spores. Cd was mainly accumulated in fungal cell walls and vacuoles, with little accumulation in the cytoplasm, suggesting that Cd is compartmentalized to prevent toxicity in the fungal cytoplasm. However, the transporters for Cd uptake into AM fungi and for Cd sequestration into the vacuoles have not been elucidated yet (Barber-Zucker et al., 2021; Kaur & Garg, 2018).

13.3 Effects of arbuscular mycorrhizal fungi on cadmium tolerance and detoxification in plants

13.3.1 Improved plant nutrition and growth dilution effect

AM fungi can improve mineral nutrition of plants and alleviate the toxicity of Cd through the growth dilution effect. AM fungi have the ability to change the activity and quantity of enzymes in the soil, particularly the activity of acid phosphatases, resulting in enhanced nutrient availability for the absorption by the host plants (Bi et al., 2019). A phosphate transporter is present in AM fungi for the direct transport of inorganic P from soil to AM fungi. The inorganic P is eventually transported into the host plant, alleviating plant P deficiency (Johri et al., 2015). AM fungi can also dissolve, activate, and absorb other nutrients such as N and Zn (Parniske, 2008).

AM fungi extend the hyphae into the soil and increase the contact area, thereby enhancing nutrient uptake from the soil (Porras-Soriano et al., 2009). Moreover, AM symbiosis may promote plant growth by increasing stomatal conductance, transpiration rate, chlorophyll synthesis, and photosynthesis (Li et al., 2023). For instance, Shen et al. (2006) showed that inoculation of AM fungi increased maize growth through an improvement of P nutrition and enhanced plant tolerance to Cd possibly by a dilution effect. Kapoor and Bhatnagar (2007) demonstrated that AM fungal inoculation can significantly increase the concentrations of chlorophyll *a* and chlorophyll *b* in celery (*Apium graveolens* L.) leaves. The dilution effect caused by the increase of biomass and plant P uptake may have contributed to the decrease of Cd phytotoxicity. Gao et al. (2023) conducted a multilevel meta-analysis of crops under Cd stress and found that AM fungal colonization on average increased plant biomass by 65.7%, boosted the formation of photosynthetic pigments by 23.2%, and increased plant N and P uptake by 24.8% and 58.4%, respectively. The dilution effect in mycorrhizal plants decreased the Cd concentration by 25.2% in AM fungal plants compared to nonmycorrhizal ones.

13.3.2 Immobilization, sequestration, and translocation of cadmium

AM fungi can contribute to the detoxification of Cd by sequestering or converting Cd into less toxic forms. Using the radioisotope labeling method, Joner and Leyval (1997) showed that a large proportion of the increased Cd content after AM fungal inoculation was sequestered in the roots of *Trifolium subterraneum* L., concluding that the Cd transfer from AM fungi to plant is restricted because of fungal immobilization. Janoušková and Pavlíková (2010) found that the extraradical mycelium accumulated 10–20 times more Cd per unit of biomass than tobacco roots, thus contributing to Cd immobilization.

Inside AM fungi or in the host cell, Cd may be chelated with glutathione and phytochelatins and stored in the vacuole, thereby reducing its toxicity (Akhtar et al., 2024). Gao et al. (2021) reported that AM fungi increased pectin, hemicellulose, and lignin contents in the root cell wall of rice under Cd stress. Increased pectin content may promote Cd binding in the cell wall. Pan et al. (2023) found that AM fungal colonization increased cell wall polysaccharide content, which could enhance binding of Cd in the cell wall and reduce the transport of Cd to aboveground plant tissues. Kuang et al. (2023) studied the ultrastructure and Cd accumulation in the roots of *Eucalyptus grandis* by using transmission electron microscope with energy-dispersive X-ray spectroscopy. They found that AM inoculation increased Cd accumulation in roots but decreased Cd accumulation in shoots. In the roots a substantial quantity of Cd accumulates between the epidermis and the outermost layer of the cortex, whereas the stele accumulates a higher Cd concentration than the epidermis and cortex. Inoculation with AM fungi reduced the Cd signal in the stele, suggesting that AM fungal symbiosis may reduce Cd transport through the stele to the aboveground part.

13.3.3 Molecular and physiological mechanisms

Mycorrhizal symbiosis can induce upregulation of certain genes that are involved in Cd resistance and detoxification. Natural resistance-associated macrophage proteins (NRAMPs) are among the most important metal transporters in many plant species. AM fungal inoculation was found to reduce Cd accumulation and expression of *SlNRAMP3* but enhance the expression of *SlNRAMP1*, *SlNRAMP2*, and *SlNRMAP4* in tomato (*Solanum lycopersicum* L.) under Cd stress. However, whether these transporters are involved in Cd uptake remains unknown (Liu et al., 2024). In *Medicago sativa*, genes encoding phytochelatin synthase (*MsPCS1*), metallothionine (*MsMT2*), and Fe and Mn transporters (*MsIRT1* and *MsNRAMP1*) were downregulated following AM inoculation (Motaharpoor et al., 2019). In *C. sativa* L., inoculation with AM fungi downregulated the expression of some genes in the ABC transporter family but upregulated some genes regulating low molecular weight organic acids (Sun et al., 2023b). The concentrations of malic acid, citric acid, and lactic acid in the rhizosphere soil increased significantly, and they were significantly correlated with the concentrations of exchangeable Cd. A study with 2-D-based proteomics showed that AM fungal symbiosis with *M. truncatula* increased the accumulation of mycorrhiza-related proteins, including a cyclophilin, a guanine nucleotide-binding protein, an ubiquitin carboxylterminal hydrolase, a thiazole biosynthetic enzyme, an annexin, a glutathione *S*-transferase-like protein, and an *S*-adenosylmethionine (SAM) synthase (Aloui et al., 2009); whether these changes affect Cd uptake or detoxification remains unclear. Proteomic analysis in wheat showed that AM fungi significantly induced the expression of two enzymes involved in the chlorophyll biosynthetic pathway (magnesium chelatase and coproporphyrinogen oxidase) and improved the expression of two proteins related to CO_2 assimilation [Ribulose-1,5-bisphosphate carboxylase/oxygenase (RuBisCO) and malic enzyme], thereby enhancing carbohydrate accumulation (Gao et al., 2023). Moreover, AM fungi increased the expression of the SAM synthase gene, which may regulate Cd tolerance through DNA methylation.

13.4 Enhancing mycorrhizal effects through soil amendments and management

13.4.1 The impact of soil amendments on mycorrhizal effects

Soil amendments such as organic matter, lime, and biochar can improve soil fertility, potentially enhancing the growth and activity of AM fungi (Cheng et al., 2023; Garbowski et al., 2023). In addition, soil amendments can alter soil properties. For example, lime can increase soil pH, thereby reducing the bioavailability of heavy metals (Chen et al., 2019).

The presence of AM fungi themselves, however, is a key driver in enhancing plant metal resistance and reducing metal uptake. This effect can be strengthened when combined with soil amendments. For example, Wang et al. (2012) found that in Pb- and Cd-contaminated soil (1000 mg/kg of Pb and 100 mg/kg of Cd, respectively), the mixed use of AM fungi and organic amendment (cattle manure) reduced the Pb and Cd concentrations in the aboveground part of tobacco plants by 58% and 92%, respectively, compared with 8.2% and 10% reduction in the treatment with only organic amendment. In some studies, combined application of biochar and AM fungi was found to significantly strengthen the mycorrhizal effect and reduce the negative effects of Cd on plants. For example, in Cd-contaminated soil (6 mg/kg of Cd), biochar, AM fungi, and the combination of biochar and AM fungi increased the fresh weight of corn by 28%, 52%, and 70%, respectively. The combination of biochar and AM fungi reduced Cd concentration in various plant tissues by 50%–76%. The combined treatment also had a synergistic effect on soil alkalinization, resulting in decreased Cd bioavailability (Liu et al., 2018).

Vallejos-Torres et al. (2021) reported that in Cd-contaminated soil, the combined application of compost and AM fungi was more effective in increasing soil pH and organic matter content than composting or AM fungi alone. Urban waste compost applied at a dose of 2% in combination with the inoculation of a native AM fungus synergistically promoted the growth of *A. cytisoides* L. in a neutral mine tailing under field conditions (Kohler et al., 2015).

Combined treatments of humic/fulvic acids (HFA) and AM fungi can also enhance the growth of plants in heavy-metal-polluted soil. Ofori-Agyemang et al. (2024b) found that the combined application of HFA and AM fungi can reduce the concentrations of Cd and Zn in soil pore water. They found that using HFA or AM fungi alone had no significant effect on the aboveground biomass of sorghum (*Sorghum bicolor* L.) and barley (*Hordeum vulgare* L.). The combination of HFA and AM fungi significantly increased the stem dry weight of *Miscanthus* (*Miscanthus* × *giganteus* J.M. Greef, Deuter ex Hodk., Renvoize 2001) and hemp (*C. sativa* L.), thereby enhancing the plant's ability to accumulate Cd. Zhao and Naeth (2022) also found that the combined treatment of HFA and AM fungi can reduce the heavy metal contents in alfalfa (*Medicago ruthenica* L.) and barley, and combining nanohumus with AM fungi reduced Cd uptake by 18%.

13.4.2 The impact of soil management practices on mycorrhizal effects

Effective soil management can improve the physical, chemical, and biological properties of contaminated soils, which can facilitate plant establishment and decrease Cd availability (Tordoff et al., 2000). The application of pesticides is a commonly used method in soil management, whereas the impact of different pesticides on AM fungi can vary. Fungicides, such as fenpropimorph and fenhexamid, have a clear inhibitory effect on AM fungi (Zocco et al., 2011). Insecticides, such as aldicarb, fenamiphos, and dimethoate, may stimulate AM fungal colonization and P uptake (Spokes et al., 1981). Biological pesticides, such as spinosad, pyrethrum, and terpenes, do not significantly affect the colonization ability or diversity of AM fungi (Ipsilantis et al., 2012). The mechanisms by which pesticides affect AM fungi symbiosis remain unclear.

Studies have shown that the application of organic fertilizers can increase the biomass of AM fungi (Ma et al., 2022). For example, the application of biogas liquid fertilizer (an organic fertilizer) significantly increases the relative abundance of AM fungi in soil, especially in the dominant genus *Glomus* (Yu et al., 2022). Meanwhile the combined application of AM fungi and organic fertilizer can also reduce the Cd concentration in plants. Chen et al. (2023) found that compared with organic fertilizer treatment, the combined application of organic fertilizer and AM fungi significantly reduced the Cd concentrations in shoots and roots of ryegrass. The Cd concentrations in the aboveground and root parts decreased by 22% and 21%, respectively.

The mycelial density of AM fungi is one of the main factors affecting the remediation of Cd-contaminated soil. Soil tillage systems affect not only the mycelial density of AM fungi but also AM fungal activity, community structure, and glomalin production. Conventional tillage and zero tillage are two common soil management systems. The density of AM fungal hyphae in soil often decreases because of soil disturbance caused by conventional tillage. Soil plowing reduces the ability of AM fungi to colonize roots and disrupts their hyphal network (Alguacil et al., 2008). In contrast, zero tillage results in a higher density of AM fungal hyphae in soil compared with conventional tillage. Nonetheless, when applying zero tillage systems for a long time, the GRSP content, total hyphae, and water-stable aggregates still show a downward trend (Curaqueo et al., 2011). To ensure the remediation efficiency of the plant-AM fungus system on Cd-contaminated soil, both conventional tillage and appropriate zero tillage are needed; the former provides oxygen and water for plant roots, whereas the latter increases the density of AM fungal hyphae.

13.5 Applications of arbuscular mycorrhizal fungi in phytoremediation of cadmium-contaminated soils

13.5.1 Applications of arbuscular mycorrhizal fungi in phytostabilization of mining sites

Mine tailings are the residual materials produced during the extraction of minerals from ores. These residual materials contain high concentrations of heavy metals, which, if not properly managed, can pose a significant threat to the environment (Hassan et al., 2024). AM fungi can help restore soil fertility in harsh environments with low nutrients and moisture, providing a more favorable growth environment for plants and soil microorganisms (Chen et al., 2019). In substrates with elevated metal concentrations, AM fungi can protect host plants against metal toxicity (Leyval et al., 2002). For example, in Cu mine tailings, Chen et al. (2007) demonstrated that AM fungi decreased Cu and Cd concentrations in the aboveground tissues of *Pteris vittata, Coreopsis drummondii*, and *Trifolium repens* through the growth dilution effect. AM fungi exhibited a stronger dilution effect on Cd in the fronds of *P. vittata* but significantly increased Cd concentrations in the roots, suggesting that Cd may be more strongly bound in the mycorrhizal roots. However, other studies suggest that phytostabilization in areas contaminated with heavy metals is a complex process that cannot be achieved solely through the use of a single plant species. There are also many potential fungal strains to consider. Therefore, it is necessary to optimize AM fungi and plant species combinations for each specific phytoremediation practice (Owiny & Dusengemungu, 2024).

AM fungi can enhance the ability of plants to resist heavy metal pollution, thereby improving their capacity to establish in polluted soil. This has been demonstrated in a wide range of plant species exposed to Cd contamination in field conditions (see Table 13.1). For example, Ofori-Agyemang et al. (2024a) demonstrated that AM fungi can enhance the stress resistance of sorghum under abiotic stresses such as drought and Cd pollution. Szada-Borzyszkowska et al. (2022) demonstrated that mycorrhizal symbiosis increases metal uptake by the roots while decreasing metal uptake by the shoots. Although field studies conducted specifically on mining sites are relatively scarce, other studies have consistently demonstrated that AM fungi can effectively enhance plant metal tolerance in soils sourced from these contaminated areas (Table 13.1). A study conducted on mining subsidence sites in a semiarid region of northwest China used AM fungi inoculum for revegetation. The results indicated that AM inoculation improved soil quality and increased mycorrhizal colonization on *Artemisia annua, Astragalus membranaceus, Stipa grandiflorus*, and *Salix mongolica* (Qiu et al., 2019). Furthermore, local plants and AM fungi often coevolve to exhibit high tolerance to the stress induced by mining activities (Meharg & Cairney, 1999), offering a possible strategy for using AM fungi to stabilize contaminated soil in abandoned mines.

13.5.2 Application of arbuscular mycorrhizal fungi in farmland soil remediation

There are two ways for using plants and AM fungi to remediate farmland soil contaminated with Cd. One is phytoextraction, and the other is using AM fungi to enhance plant resistance

TABLE 13.1 Some applications of AM fungi for phytostabilization of Cd-contaminated soils.

AM fungal species	Host plants	Location	Soil condition	Results	References
Rhizophagus irregularis, Funneliformis geosporum BEG199, Funneliformis mosseae, Claroideoglomus lamentosum, Septoglomus deserticola	*Sorghum bicolor*	Evin-Malmaison (France)	Cd, Pb, Zn pollution	AM fungi can enhance the stress resistance of sorghum under abiotic stresses such as drought and Cd pollution	Ofori-Agyemang et al. (2024a)
Some indigenous Arbuscular mycorrhiza fungi (AMF)	*Miscanthus* hybrids (*M. sacchariflorus* × *M. sinensis*)	Bytom, Poland	Pb,Cd and Zn pollution	Mycorrhizal symbiosis increases metal uptake by the roots while decreasing uptake by the shoots	Szada-Borzyszkowska et al. (2022)
Glomus mosseae	Cattail	Sewage treatment center	Pb and Cd pollution	Mycorrhizal fungi boosted CO_2 release and mineralization rate in the soil, affecting soil pH and heavy metal solubility	Al-Maliki and Al-Shamary (2022)
R. irregularis	*Helianthus annuus*	Estarreja, North Portugal	Zn and Cd pollution	Compared to plants without AMF inoculation, the increase in Zn content in roots and stems of inoculated plants can reach up to approximately 6.5 times and 6.1 times, respectively, while the increase in Cd content in roots and stems is up to 15-folds	Paulo et al. (2023)
G. mosseae	*Miscanthus* (*M.* × *giganteus*)	Evin-Malmaison	Zn and Cd pollution	The addition of AMF inoculum increased metal (mainly Cd and Zn) accumulation in *Miscanthus* organs	Nsanganwimana et al. (2015)

AM, arbuscular mycorrhizal; *Cd*, cadmium.

to Cd. In the experiment by Ofori-Agyemang et al. (2024a), sorghum cultivation decreased the extractable soil Cd, Pb, and Zn concentrations while extracting 0.12 kg/Cd ha, 0.14 kg/Pb ha, and 1.47 kg/Zn ha in one growing season, and AM fungi enhanced the sorghum's resistance to abiotic stresses such as drought and Cd stress. In the study by Hu et al. (2022) in southern China, it was found that although *Glomus caledonium* can accelerate the extraction of Cd by stonecrop (*Sedum alfredii* Hance) to a certain extent, the process is very slow. Even under ideal conditions, it takes more than 5 years to reduce Cd in the experimental soil to an acceptable level. In summary, using phytoextraction to remediate Cd-contaminated farmland soil is difficult and requires a long time. Therefore using AM fungi to enhance plant resistance to Cd seems to be a more reasonable proposition. For heavily polluted farmland, extracting heavy metals is costly and offers limited benefits. In this case, restoration of vegetation is the priority, and inoculation of AM fungi can facilitate this process (Saia et al., 2020). As demonstrated by Aalipour et al. (2021), AM fungi helped reduce the transfer of Cd to the aboveground parts, allowing *Arizona cypress* trees to grow well in Cd-contaminated soil.

13.6 Conclusion and future prospects

The majority of studies show that AM fungi can be beneficial for plants in terms of enhancing tolerance to Cd and decrease its translocation to the aboveground parts. Enhanced sequestration and immobilization of Cd by AM fungi appears to be a common mechanism, although the effect varies between plant species and environmental conditions. The detoxification effect of AM fungi can be strengthened by soil amendments and management practices. Ongoing research should continue to explore these interactions and their potential applications in agriculture and environmental management.

The capacity of the AM fungi to take up and detoxify Cd is inconsistent among different studies. Current research lacks the determination and screening of functionally effective AM fungal strains. Future research should prioritize the isolation of natural and adapted AM fungal strains from Cd-contaminated soils for investigations on their detoxication function. Furthermore, development of straightforward evaluation criteria is essential for identifying strains with broad applicability, significant impact, and high field survival.

Contemporary molecular techniques can be used to elucidate the molecular basis and function of AM symbiosis under Cd stress and elucidate the molecular mechanisms associated with Cd tolerance. In addition, it is essential to ascertain the alterations in nutrient-related metabolic pathways regulated by AM fungi in response to Cd stress at the molecular level.

The uptake and detoxification of Cd by AM fungi are subject to various influencing factors. However, the current research primarily focuses on mycorrhizal fungi with insufficient considerations of other microorganisms in the mycorrhizosphere [the soil zone influenced by the mycorrhizal fungi, which encompasses both the rhizosphere and the hyphosphere (the soil zone around the fungal hyphae)]. Research on mycorrhizosphere microorganisms is still in its infancy. Future research should focus on mycorrhizosphere microorganisms, employing multiomics technology in conjunction with bioinformatics to elucidate the mechanisms by which AM fungi and other microorganisms in the mycorrhizosphere interact to mitigate Cd toxicity in soil and regulate plant Cd tolerance.

13.7 Summary

- AM symbiosis enhances the Cd tolerance of plants by regulating genes involved in Cd resistance and detoxification, strengthening cell wall binding of Cd, altering Cd distribution, improving plant nutrient uptake, and inducing growth dilution effects.
- The role of AM symbiosis in Cd uptake and detoxification is affected by biological factors such as AM fungal strains, host species, and genotypes, as well as environmental factors such as soil Cd concentration and P level.
- The role of AM symbiosis in Cd uptake and detoxification can be enhanced through soil amendments, such as biochar and compost, as well as management practices, including organic fertilizer application and zero tillage.
- The role of AM symbiosis in Cd uptake and detoxification has significant ecological implications. It can contribute to the phytostabilization of contaminated soils and also to the development of Cd-resistant crop varieties.

Acknowledgments

This study was financially supported by the National Natural Science Foundation of China (42377299, 42577326, and 42177109).

References

Aalipour, H., Nikbakht, A., & Etemadi, N. (2021). Physiological response of *Arizona cypress* to Cd-contaminated soil inoculated with arbuscular mycorrhizal fungi and plant growth promoting rhizobacteria. *Rhizosphere, 18*, 100354. doi:10.1016/j.rhisph.2021.100354.

Akhtar, O., Pandey, D., Zoomi, I., Singh, U., Chaudhary, K. L, Mishra, R., & Pandey, N. (2024). Role of arbuscular mycorrhizal fungi in heavy metals homoeostasis in plants. *Journal of Plant Growth Regulation, 43*(11), 3971–3985. doi:10.1007/s00344-024-11393-w.

Alguacil, M. M., Lumini, E., Roldán, A., Salinas-García, J. R., Bonfante, P., & Bianciotto, V. (2008). The impact of tillage practices on arbuscular mycorrhizal fungal diversity in subtropical crops. *Ecological Applications, 18*(2), 527–536. doi:10.1890/07-0521.1.

Al-Maliki, S., & Al-Shamary, A. (2022). Vital evidence for arbuscular mycorrhizal fungi, bacteria and cattail plant to remove Pb-Cd heavy metals from contaminated soils. *Acta Ecologica Sinica, 42*(4), 392–397. doi:10.1016/j.chnaes.2022.05.008.

Aloui, A., Recorbet, G., Gollotte, A., Robert, F., Valot, B. ît, Gianinazzi-Pearson, V., Aschi-Smiti, S., & Dumas-Gaudot, E. (2009). On the mechanisms of cadmium stress alleviation in *Medicago truncatula* by arbuscular mycorrhizal symbiosis: A root proteomic study. *Proteomics, 9*(2), 420–433. doi:10.1002/pmic.200800336.

Alptekin, B., Hirsch, H., Kleven, B., King, L., McLimans, C., Daniels, D., Irving, T., Floss, D., & Ané, J.-M. (2025). From glomalin to glomalose: Unraveling the molecular identity of the MAb32B11 antigen. *New Phytologist, 247*(5), 2328–2341. doi:10.1111/nph.70253.

Alsadat Hashemi Nejad, R., Kafi, M., Jari, S. K., Mozafari, H., & Motesharezadeh, B. (2021). Arbuscular mycorrhizal fungi improve growth, physiological status and nutrients accumulation of *Ailanthus altissima* seedlings under cadmium pollution and salinity. *Russian Journal of Plant Physiology, 68*(2), 266–273. https://rd.springer.com/journal/11183. doi:10.1134/S102144372102014X.

Barber-Zucker, S., Moran, A., & Zarivach, R. (2021). Metal transport mechanism of the cation diffusion facilitator (CDF) protein family—A structural perspective on human CDF (ZnT)-related diseases. *RSC Chemical Biology, 2*(2), 486–498. doi:10.1039/d0cb00181c.

Bi, Y., Xiao, L., & Liu, R. (2019). Response of arbuscular mycorrhizal fungi and phosphorus solubilizing bacteria to remediation abandoned solid waste of coal mine. *International Journal of Coal Science and Technology, 6*(4), 603–610. http://link.springer.com/journal/volumesAndIssues/40789. doi:10.1007/s40789-019-00270-7.

Chen, B., Nayuki, K., Kuga, Y., Zhang, X., Wu, S., & Ohtomo, R. (2018). Uptake and intraradical immobilization of cadmium by arbuscular mycorrhizal fungi as revealed by a stable isotope tracer and synchrotron radiation μx-ray fluorescence analysis. *Microbes and Environments, 33*(3), 257–263. https://www.jstage.jst.go.jp/article/jsme2/33/3/33_ME18010/_pdf. doi:10.1264/jsme2.ME18010.

Chen, B. D., Zhu, Y.-G., Duan, J., Xiao, X. Y., & Smith, S. E. (2007). Effects of the arbuscular mycorrhizal fungus *Glomus mosseae* on growth and metal uptake by four plant species in copper mine tailings. *Environmental Pollution, 147*(2), 374–380. doi:10.1016/j.envpol.2006.04.027.

Chen, B. D., Zhang, X., Wu, S. L., & Li, L. F. (2019). The role of arbuscular mycorrhizal fungi in heavy metal translocation, transformation and accumulation in the soil-plant continuum: Underlying mechanisms and ecological implications. *Yankuang Ceshi, 38*(1), 1–25. http://www.ykcs.ac.cn/en/article/2019/1. doi:10.15898/j.cnki.11-2131/td.201807110083.

Chen, J., Guo, J., Yang, Z., Yang, J., Dong, H., Wang, H., Wang, Y., & Zhan, F. (2023). The application of fertilizer and AMF promotes growth and reduces the cadmium and lead contents of ryegrass (*Lolium multiflorum L.*) in a copper mining area. *Phyton-International Journal of Experimental Botany, 92*(2), 471–485. https://file.techscience.com/ueditor/files/phyton/TSP_Phyton-92-2/TSP_Phyton_23660/TSP_Phyton_23660.pdf. doi:10.32604/phyton.2022.023660.

Cheng, H. Q., Giri, B., Wu, Q. S., Zou, Y. N., & Kuča, K. (2022). Arbuscular mycorrhizal fungi mitigate drought stress in citrus by modulating root microenvironment. *Archives of Agronomy and Soil Science, 68*(9), 1217–1228. http://www.tandf.co.uk/journals/titles/03650340.asp. doi:10.1080/03650340.2021.1878497.

Cheng, Z., Shi, J., He, Y., Chen, Y., Wang, Y., Yang, X., Wang, T., Wu, L., & Xu, J. (2023). Enhanced soil function and health by soybean root microbial communities during in situ remediation of Cd-contaminated soil with the application of soil amendments. *mSystems, 8*(3). https://journals.asm.org/doi/pdf/10.1128/msystems.01049-22. doi:10.1128/msystems.01049-22.

Citterio, S., Prato, N., Fumagalli, P., Aina, R., Massa, N., Santagostino, A., Sgorbati, S., & Berta, G. (2005). The arbuscular mycorrhizal fungus *Glomus mosseae* induces growth and metal accumulation changes in *Cannabis sativa* L. *Chemosphere, 59*(1), 21–29. doi:10.1016/j.chemosphere.2004.10.009.

Cornejo, P., Meier, S. án, Borie, G., Rillig, M. C., & Borie, F. (2008). Glomalin-related soil protein in a Mediterranean ecosystem affected by a copper smelter and its contribution to Cu and Zn sequestration. *Science of the Total Environment, 406*(1-2), 154–160. doi:10.1016/j.scitotenv.2008.07.045.

Curaqueo, G., Barea, J. M., Acevedo, E., Rubio, R., Cornejo, P., & Borie, F. (2011). Effects of different tillage system on arbuscular mycorrhizal fungal propagules and physical properties in a Mediterranean agroecosystem in central Chile. *Soil and Tillage Research, 113*(1), 11–18. doi:10.1016/j.still.2011.02.004.

Entry, J. A., Rygiewicz, P. T., Watrud, L. S., & Donnelly, P. K. (2002). Influence of adverse soil conditions on the formation and function of arbuscular mycorrhizas. *Advances in Environmental Research, 7*(1), 123–138. doi:10.1016/S1093-0191(01)00109-5.

Gao, M. Y., Chen, X. W., Huang, W. X., Wu, L., Yu, Z. S., Xiang, L., Mo, C. H., Li, Y. W., Cai, Q. Y., Wong, M. H., & Li, H. (2021). Cell wall modification induced by an arbuscular mycorrhizal fungus enhanced cadmium fixation in rice root. *Journal of Hazardous Materials, 416*. www.elsevier.com/locate/jhazmat. doi:10.1016/j.jhazmat.2021.125894.

Gao, Y., Jia, X., Zhao, Y., Zhao, J., Ding, X., Zhang, C., & Feng, X. (2022). Effect of arbuscular mycorrhizal fungi (*Glomus mosseae*) and elevated air temperature on Cd migration in the rhizosphere soil of alfalfa. *Ecotoxicology and Environmental Safety, 248*. http://www.elsevier.com/inca/publications/store/6/2/2/8/1/9/index.htt. doi:10.1016/j.ecoenv.2022.114342.

Gao, Y., An, T., Kuang, Q., Wu, Y., Liu, S., Liang, L., Yu, M., Macrae, A., & Chen, Y. (2023). The role of arbuscular mycorrhizal fungi in the alleviation of cadmium stress in cereals: A multilevel meta-analysis. *Science of the Total Environment, 902*. www.elsevier.com/locate/scitotenv. doi:10.1016/j.scitotenv.2023.166091.

Garbowski, T., Bar-Michalczyk, D., Charazińska, S., Grabowska-Polanowska, B., Kowalczyk, A., & Lochyński, P. (2023). An overview of natural soil amendments in agriculture. *Soil and Tillage Research, 225*, 105462. doi:10.1016/j.still.2022.105462.

Giasson, P., Jaouich, A., Gagné, S., & Moutoglis, P. (2005a). Arbuscular mycorrhizal fungi involvement in zinc and cadmium speciation change and phytoaccumulation. *Remediation, 15*(2), 75–81. doi:10.1002/rem.20044.

Giasson, P., Jaouich, A., Gagné, S., & Moutoglis, P. (2005b). Phytoremediation of zinc and cadmium: A study of arbuscular mycorrhizal hyphae. *Remediation, 15*(4), 113–122. doi:10.1002/rem.20064.

González-Chávez, M. C., Carrillo-González, R., Wright, S. F., & Nichols, K. A. (2004). The role of glomalin, a protein produced by arbuscular mycorrhizal fungi, in sequestering potentially toxic elements. *Environmental Pollution, 130*(3), 317–323. doi:10.1016/j.envpol.2004.01.004.

González-Guerrero, M., Melville, L. H., Ferrol, N., Lott, J. N. A., Azcón-Aguilar, C., & Peterson, R. L. (2008). Ultrastructural localization of heavy metals in the extraradical mycelium and spores of the arbuscular mycorrhizal fungus *Glomus intraradices*. *Canadian Journal of Microbiology, 54*(2), 103–110. http://article.pubs.nrc-cnrc.gc.ca/RPAS/RPViewDoc?_handler_=HandleInitialGet&calyLang=eng&journal=cjm&volume=54&articleFile=w07-119.pdf. doi:10.1139/W07-119.

Guo, Y., George, E., & Marschner, H. (1996). Contribution of an arbuscular mycorrhizal fungus to the uptake of cadmium and nickel in bean and maize plants. *Plant and Soil, 184*(2), 195–205. doi:10.1007/bf00010449.

Han, Y., Zveushe, O. K., Dong, F., Ling, Q., Chen, Y., Sajid, S., Zhou, L., & Resco de Dios, V. (2021). Unraveling the effects of arbuscular mycorrhizal fungi on cadmium uptake and detoxification mechanisms in perennial ryegrass (*Lolium perenne*). *Science of the Total Environment, 798*. www.elsevier.com/locate/scitotenv. doi:10.1016/j.scitotenv.2021.149222.

Hassan, S., Bhadwal, S. S., Khan, M., Khair-Ul Nissa, S., Shah, R. A., Bhat, H. M., Bhat, S. A., Lone, I. M., & Ganai, B. A. (2024). Revitalizing contaminated lands: A state-of-the-art review on the remediation of mine-tailings using phytoremediation and genomic approaches. *Chemosphere, 356*, 141889. doi:10.1016/j.chemosphere.2024.141889.

Haystead, A., Malajczuk, N., & Grove, T. S. (1988). Underground transfer of nitrogen between pasture plants infected with vesicular-arbuscular mycorrhizal fungi. *New Phytologist, 108*(4), 417–423. doi:10.1111/j.1469-8137.1988.tb04182.x.

Hu, B. o, Hu, S., Vymazal, J., & Chen, Z. (2022). Application of arbuscular mycorrhizal fungi for pharmaceuticals and personal care productions removal in constructed wetlands with different substrate. *Journal of Cleaner Production, 339*, 130760. doi:10.1016/j.jclepro.2022.130760.

Hutchinson, J. J., Young, S. D., Black, C. R., & West, H. M. (2004). Determining uptake of radio-labile soil cadmium by arbuscular mycorrhizal hyphae using isotopic dilution in a compartmented-pot system. *New Phytologist, 164*(3), 477–484. doi:10.1111/j.1469-8137.2004.01206.x.

Ipsilantis, I., Samourelis, C., & Karpouzas, D. G. (2012). The impact of biological pesticides on arbuscular mycorrhizal fungi. *Soil Biology and Biochemistry, 45*, 147–155. doi:10.1016/j.soilbio.2011.08.007.

Janoušková, M., & Pavlíková, D. (2010). Cadmium immobilization in the rhizosphere of arbuscular mycorrhizal plants by the fungal extraradical mycelium. *Plant and Soil, 332*(1-2), 511–520. doi:10.1007/s11104-010-0317-2.

Jiang, Y., Huang, R., Jiang, L., Chen, K., & Zhu, W. (2021). Alleviation of cadmium toxicity to *medicago truncatula* by AMF involves the changes of Cd speciation in rhizosphere soil and subcellular distribution. *Phyton, 90*(2), 403–415. https://www.techscience.com/phyton/v90n2/41477. doi:10.32604/phyton.2021.014376.

Jiang, Y., Wang, W., Xie, Q., Liu, N. a, Liu, L., Wang, D., Zhang, X., Yang, C., Chen, X., Tang, D., & Wang, E. (2017). Plants transfer lipids to sustain colonization by mutualistic mycorrhizal and parasitic fungi. *Science, 356*(6343), 1172–1175. doi:10.1126/science.aam9970.

Johri, A. K., Oelmüller, R., Dua, M., Yadav, V., Kumar, M., Tuteja, N., Varma, A., Bonfante, P., Persson, B. L., & Stroud, R. M. (2015). Fungal association and utilization of phosphate by plants: Success, limitations, and future prospects. *Frontiers in Microbiology, 6*. http://journal.frontiersin.org/article/10.3389/fmicb.2015.00984/full. doi:10.3389/fmicb.2015.00984.

Joner, E. J., & Leyval, C. (1997). Uptake of 109 Cd by roots and hyphae of a *Glomus mosseae/Trifolium subterraneum* mycorrhiza from soil amended with high and low concentrations of cadmium. *New Phytologist, 135*(2), 353–360. doi:10.1046/j.1469-8137.1997.00633.x.

Kapoor, R., & Bhatnagar, A. K. (2007). Attenuation of cadmium toxicity in mycorrhizal celery (*Apium graveolens L.*). *World Journal of Microbiology and Biotechnology, 23*(8), 1083–1089. doi:10.1007/s11274-006-9337-8.

Kaur, H., & Garg, N. (2018). Recent perspectives on cross talk between cadmium, zinc, and arbuscular mycorrhizal fungi in plants. *Journal of Plant Growth Regulation, 37*(2), 680–693. doi:10.1007/s00344-017-9750-2.

Khan, S. R., Ahmad, Z., Khan, Z., Khan, U., Asad, M., & Shah, T. (2024). Synergistic effect of silicon and arbuscular mycorrhizal fungi reduces cadmium accumulation by regulating hormonal transduction and lignin accumulation in maize. *Chemosphere, 346*, 140507. doi:10.1016/j.chemosphere.2023.140507.

Kohler, J., Caravaca, F., Azcón, R., Díaz, G., & Roldán, A. (2015). The combination of compost addition and arbuscular mycorrhizal inoculation produced positive and synergistic effects on the phytomanagement of a semiarid mine tailing. *Science of the Total Environment, 514*, 42–48. doi:10.1016/j.scitotenv.2015.01.085.

Kuang, Y., Li, X., Wang, Z., Wang, X., Wei, H., Chen, H., Hu, W., & Tang, M. (2023). Effects of arbuscular mycorrhizal fungi on the growth and root cell ultrastructure of *Eucalyptus grandis* under cadmium stress. *Journal of Fungi, 9*(2). www.mdpi.com/journal/jof. doi:10.3390/jof9020140.

Leyval, C., Joner, E. J., del Val, C., & Haselwandter, K. (2002). *Potential of arbuscular mycorrhizal fungi for bioremediation* (pp. 175–186). Springer Science and Business Media LLC. doi:10.1007/978-3-0348-8117-3_14.

Li, H., Zhang, L., Wu, B., Li, Y., Wang, H., Teng, H., Wei, D., Yuan, Z., & Yuan, Z. (2023). Physiological and proteomic analyses reveal the important role of arbuscular mycorrhizal fungi on enhancing photosynthesis in wheat under cadmium stress. *Ecotoxicology and Environmental Safety, 261*. http://www.elsevier.com/inca/publications/store/6/2/2/8/1/9/index.htt. doi:10.1016/j.ecoenv.2023.115105.

Liang, J., Wang, Z., Ren, Y., Jiang, Z., Chen, H., Hu, W., & Tang, M. (2023). The alleviation mechanisms of cadmium toxicity in *Broussonetia papyrifera* by arbuscular mycorrhizal symbiosis varied with different levels of cadmium stress. *Journal of Hazardous Materials, 459*. www.elsevier.com/locate/jhazmat. doi:10.1016/j.jhazmat.2023.132076.

Liu, D. i, Zheng, K., Wang, Y., Zhang, Y., Lao, R., Qin, Z., Li, T., & Zhao, Z. (2022). Harnessing an arbuscular mycorrhizal fungus to improve the adaptability of a facultative metallophytic poplar (*Populus yunnanensis*) to cadmium stress: Physiological and molecular responses. *Journal of Hazardous Materials, 424*(Pt B), 127430. doi:10.1016/j.jhazmat.2021.127430.

Liu, J., Bao, X., Qiu, G., Li, H., Wang, Y., Chen, X., Fu, Q., & Guo, B. (2024). Genome-wide identification and expression analysis of SlNRAMP genes in tomato under nutrient deficiency and cadmium stress during arbuscular mycorrhizal symbiosis. *International Journal of Molecular Sciences, 25*(15). http://www.mdpi.com/journal/ijms. doi:10.3390/ijms25158269.

Liu, L., Li, J., Yue, F., Yan, X., Wang, F., Bloszies, S., & Wang, Y. (2018). Effects of arbuscular mycorrhizal inoculation and biochar amendment on maize growth, cadmium uptake and soil cadmium speciation in Cd-contaminated soil. *Chemosphere, 194*, 495–503. www.elsevier.com/locate/chemosphere. doi:10.1016/j.chemosphere.2017.12.025.

Liu, L., Gong, Z., Zhang, Y., & Li, P. (2014). Growth, cadmium uptake and accumulation of maize (*Zea mays L.*) under the effects of arbuscular mycorrhizal fungi. *Ecotoxicology, 23*(10), 1979–1986. www.wkap.nl/journalhome.htm/0963-9292. doi:10.1007/s10646-014-1331-6.

Lu, R. R., Hu, Z. H., Zhang, Q. L., Li, Y. Q., Lin, M., Wang, X. L., Wu, X. N., Yang, J. T., Zhang, L. Q., Jing, Y. X., & Peng, C. L. (2020). The effect of *Funneliformis mosseae* on the plant growth, Cd translocation and accumulation in the new Cd-hyperaccumulator *Sphagneticola calendulacea. Ecotoxicology and Environmental Safety, 203*. http://www.elsevier.com/inca/publications/store/6/2/2/8/1/9/index.htt. doi:10.1016/j.ecoenv.2020.110988.

Ma, K., Wang, Y., Jin, X., Zhao, Y., Yan, H., Zhang, H., Zhou, X., Lu, G., & Deng, Y. e (2022). Application of organic fertilizer changes the rhizosphere microbial communities of a gramineous grass on Qinghai–Tibet Plateau. *Microorganisms, 10*(6), 1148. doi:10.3390/microorganisms10061148.

Meharg, A. A., & Cairney, J. W. G. (1999). Co-evolution of mycorrhizal symbionts and their hosts to metal-contaminated environments. *Advances in Ecological Research, 30*(C), 69–112. doi:10.1016/S0065-2504(08)60017-3.

Motaharpoor, Z., Taheri, H., & Nadian, H. (2019). *Rhizophagus irregularis* modulates cadmium uptake, metal transporter, and *chelator* gene expression in *Medicago sativa. Mycorrhiza, 29*(4), 389–395. doi:10.1007/s00572-019-00900-7.

Nayuki, K., Chen, B., Ohtomo, R., & Kuga, Y. (2014). Cellular imaging of cadmium in resin sections of arbuscular mycorrhizas using synchrotron micro X-ray fluorescence. *Microbes and Environments, 29*(1), 60–66. doi:10.1264/jsme2.me13093.

Nsanganwimana, F., Pourrut, B., Waterlot, C., Louvel, B., Bidar, G., Labidi, S., Fontaine, J., Muchembled, J., Lounès-Hadj Sahraoui, A., Fourrier, H., & Douay, F. (2015). Metal accumulation and shoot yield of *Miscanthus×giganteus* growing in contaminated agricultural soils: Insights into agronomic practices. *Agriculture, Ecosystems & Environment, 213*, 61–71. doi:10.1016/j.agee.2015.07.023.

Ofori-Agyemang, F., Burges, A., Waterlot, C., Sahraoui, A. Lès-H., Tisserant, B. ît, Mench, M., & Oustrière, N. è (2024a). Phytomanagement of a metal-contaminated agricultural soil with *Sorghum* bicolor, humic /fulvic acids and arbuscular mycorrhizal fungi near the former Pb/Zn metaleurop Nord smelter. *Chemosphere, 362*, 142624. doi:10.1016/j.chemosphere.2024.142624.

Ofori-Agyemang, F., Waterlot, C., Manu, J., Laloge, R., Francin, R., Papazoglou, E. G., Alexopoulou, E., Sahraoui, A. L.-H., Tisserant, B., Mench, M., Burges, A., & Oustrière, N. (2024b). Plant testing with hemp and miscanthus to assess phytomanagement options including biostimulants and mycorrhizae on a metal-contaminated soil to provide biomass for sustainable biofuel production. *Science of the Total Environment, 912*, 169527. doi:10.1016/j.scitotenv.2023.169527.

Owiny, A. A., & Dusengemungu, L. (2024). Mycorrhizae in mine wasteland reclamation. *Heliyon, 10*(13), e33141. doi:10.1016/j.heliyon.2024.e33141.

Pan, J., Cao, S., Xu, G., Rehman, M., Li, X., Luo, D., Wang, C., Fang, W., Xiao, H., Liao, C., & Chen, P. (2023). Comprehensive analysis reveals the underlying mechanism of arbuscular mycorrhizal fungi in kenaf cadmium stress alleviation. *Chemosphere, 314*, 137566. doi:10.1016/j.chemosphere.2022.137566.

Parmar, P., Kumari, N., & Sharma, V. (2013). Structural and functional alterations in photosynthetic apparatus of plants under cadmium stress. *Botanical Studies, 54*(1). doi:10.1186/1999-3110-54-45.

Parniske, M. (2008). Arbuscular mycorrhiza: The mother of plant root endosymbioses. *Nature Reviews Microbiology, 6*(10), 763–775. doi:10.1038/nrmicro1987.

Paulo, A. M. S., Caetano, N. S., Castro, P. M. L., & Marques, A. P. G. C. (2023). Phytomanagement of Zn- and Cd-contaminated soil: *Helianthus annuus* biomass production and metal remediation abilities with plant-growth-promoting microbiota assistance. *Soil Systems, 7*(3). www.mdpi.com/journal/soilsystems. doi:10.3390/soilsystems7030069.

Porras-Soriano, A., Soriano-Martín, M. L., Porras-Piedra, A., & Azcón, R. (2009). Arbuscular mycorrhizal fungi increased growth, nutrient uptake and tolerance to salinity in olive trees under nursery conditions. *Journal of Plant Physiology, 166*(13), 1350–1359. doi:10.1016/j.jplph.2009.02.010.

Qiu, L., Bi, Y., Jiang, B., Wang, Z., Zhang, Y., & Zhakypbek, Y. (2019). Arbuscular mycorrhizal fungi ameliorate the chemical properties and enzyme activities of rhizosphere soil in reclaimed mining subsidence in northwestern China. *Journal of Arid Land, 11*(1), 135–147. http://link.springer.com/journal/40333. doi:10.1007/s40333-018-0019-9.

Rauret, G., López-Sánchez, J. F., Sahuquillo, A., Rubio, R., Davidson, C., Ure, A., & Quevauviller, P. h. (1999). Improvement of the BCR three step sequential extraction procedure prior to the certification of new sediment and soil reference materials. *Journal of Environmental Monitoring, 1*(1), 57–61. doi:10.1039/a807854h.

Rezvani, M., Ardakani, M. R., Rejali, F., Zaefarian, F., Teimouri, S., Noormohammadi, G., & Miransari, M. (2015). Uptake of heavy metals by mycorrhizal barley (*Hordeum vulgare* L.). *Journal of Plant Nutrition, 38*(6), 904–919. www.tandf.co.uk/journals/titles/01904167.asp. doi:10.1080/01904167.2014.963114.

Riaz, M., Kamran, M., Fang, Y., Wang, Q., Cao, H., Yang, G., Deng, L., Wang, Y., Zhou, Y., Anastopoulos, I., & Wang, X. (2021). Arbuscular mycorrhizal fungi-induced mitigation of heavy metal phytotoxicity in metal contaminated soils: A critical review. *Journal of Hazardous Materials, 402*, 123919. doi:10.1016/j.jhazmat.2020.123919.

Rivera-Becerril, F., Calantzis, C., Turnau, K., Caussanel, J. P., Belimov, A. A., Gianinazzi, S., Strasser, R. J., & Gianinazzi-Pearson, V. (2002). Cadmium accumulation and buffering of cadmium-induced stress by arbuscular mycorrhiza in three *Pisum sativum* L. genotypes. *Journal of Experimental Botany, 53*(371), 1177–1185. http://jxb.oxfordjournals.org/. doi:10.1093/jexbot/53.371.1177.

Saia, S., Aissa, E., Luziatelli, F., Ruzzi, M., Colla, G., Ficca, A. G, Cardarelli, M., & Rouphael, Y. (2020). Growth-promoting bacteria and arbuscular mycorrhizal fungi differentially benefit tomato and corn depending upon the supplied form of phosphorus. *Mycorrhiza, 30*(1), 133–147. doi:10.1007/s00572-019-00927-w.

Shen, H., Christie, P., & Li, X. (2006). Uptake of zinc, cadmium and phosphorus by arbuscular mycorrhizal maize (*Zea mays* L.) from a low available phosphorus calcareous soil spiked with zinc and cadmium. *Environmental Geochemistry and Health, 28*(1–2), 111–119. doi:10.1007/s10653-005-9020-2.

Smith, S. E., Smith, F. A., & Jakobsen, I. (2003). Mycorrhizal fungi can dominate phosphate supply to plants irrespective of growth responses. *Plant Physiology, 133*(1), 16–20. http://www.plantphysiol.org/. doi:10.1104/pp.103.024380.

Spokes, J. R., David, R. M. M. D., & Hayman, S. (1981). Effects of plant protection chemicals on vesicular-arbuscular mycorrhizas. *Pesticide Science, 12*(3), 346–350. doi:10.1002/ps.2780120314.

Sun, J. H., Jia, Q., Li, Y. i, Dong, K., Xu, S., Ren, Y., Zhang, T., Chen, J., Shi, N., & Fu, S. (2023a). Effect of arbuscular mycorrhiza fungus *Diversispora eburnea* inoculation on *Lolium perenne* and *Amorpha fruticosa* growth, cadmium uptake, and soil cadmium speciation in cadmium-contaminated soil. *International Journal of Environmental Research and Public Health, 20*(1), 795. doi:10.3390/ijerph20010795.

Sun, S. M., Fan, X., Feng, Y., Wang, X., Gao, H., & Song, F. (2023b). Arbuscular mycorrhizal fungi influence the uptake of cadmium in industrial hemp (*Cannabis sativa* L.). *Chemosphere, 330*. http://www.elsevier.com/locate/chemosphere. doi:10.1016/j.chemosphere.2023.138728.

Szada-Borzyszkowska, A., Krzyżak, J., Rusinowski, S., Sitko, K., & Pogrzeba, M. (2022). Field Evaluation of arbuscular mycorrhizal fungal colonization in *Miscanthus × giganteus* and seed-based miscanthus hybrids grown in heavy-metal-polluted areas. *Plants, 11*(9), 1216. doi:10.3390/plants11091216.

Tan, Q., Guo, Q., Wei, R., Zhu, G., Du, C., & Hu, H. (2023). Influence of arbuscular mycorrhizal fungi on bioaccumulation and bioavailability of As and Cd: A meta-analysis. *Environmental Pollution, 316*. https://www.journals.elsevier.com/environmental-pollution. doi:10.1016/j.envpol.2022.120619.

Tordoff, G. M., Baker, A. J. M., & Willis, A. J. (2000). Current approaches to the revegetation and reclamation of metalliferous mine wastes. *Chemosphere, 41*(1-2), 219–228. doi:10.1016/s0045-6535(99)00414-2.

Vallejos-Torres, G., Torres, S. C., Gaona-Jimenez, N., Saavedra, J., Tuesta, J. C., Tuesta, O. A., Becerra, A. G., Marín, C., & del Mar Alguacil, M. (2022). The combined effect of arbuscular mycorrhizal fungi and compost improves growth and soil parameters and decreases cadmium absorption in cacao (*Theobroma cacao* L.) plants. *Journal of Soil Science and Plant Nutrition, 22*(4), 5174–5182. https://www.springer.com/journal/42729. doi:10.1007/s42729-022-00992-9.

Vallejos-Torres, G., Ruíz-Valles, R., Chappa-Santa María, C. E., Gaona-Jiménez, N., & Marín, C. (2021). Una alta diversidad de hongos micorrízicos arbusculares influye en la absorción de cadmio y crecimiento vegetal del cacao. *Bioagro, 34*(1), 75–84. doi:10.51372/bioagro341.7.

Vogel-Mikuš, K., Pongrac, P., Kump, P., Nečemer, M., & Regvar, M. (2006). Colonisation of a Zn, Cd and Pb hyperaccumulator *Thlaspi praecox* Wulfen with indigenous arbuscular mycorrhizal fungal mixture induces changes in heavy metal and nutrient uptake. *Environmental Pollution, 139*(2), 362–371. doi:10.1016/j.envpol.2005.05.005.

Wang, F. Y., Wang, L., Shi, Z. Y., Li, Y. J., & Song, Z. M. (2012). Effects of AM inoculation and organic amendment, alone or in combination, on growth, P nutrition, and heavy-metal uptake of tobacco in Pb-Cd-contaminated soil. *Journal of Plant Growth Regulation, 31*(4), 549–559. doi:10.1007/s00344-012-9265-9.

Wright, S. F., & Upadhyaya, A. (1998). A survey of soils for aggregate stability and glomalin, a glycoprotein produced by hyphae of arbuscular mycorrhizal fungi. *Plant and Soil, 198*(1), 97–107. doi:10.1023/a:1004347701584.

Wu, Z., McGrouther, K., Huang, J., Wu, P., Wu, W., & Wang, H. (2014). Decomposition and the contribution of glomalin-related soil protein (GRSP) in heavy metal sequestration: Field experiment. *Soil Biology and Biochemistry, 68*, 283–290. www.elsevier.com/inca/publications/store/3/3/2. doi:10.1016/j.soilbio.2013.10.010.

Xu, P. X., Li, R. J., Zhu, Q. Y., & Jing, Y. X. (2024). Transcriptome analysis shows that *Glomus versiforme* decrease the accumulation and toxicity of cadmium in *Ipomoea aquatic Forsk*. *Environmental Science and Pollution Research, 31*(30), 43023–43036. https://www.springer.com/journal/11356. doi:10.1007/s11356-024-34023-0.

Yazici, M. A., Asif, M., Tutus, Y., Ortas, I., Ozturk, L., Lambers, H., & Cakmak, I. (2021). Reduced root mycorrhizal colonization as affected by phosphorus fertilization is responsible for high cadmium accumulation in wheat. *Plant and Soil, 468*(1–2), 19–35. doi:10.1007/s11104-021-05041-5.

You, Y., Ju, C., Wang, L. i, Wang, X., Ma, F., Wang, G., & Wang, Y. (2022). The mechanism of arbuscular mycorrhizal enhancing cadmium uptake in *Phragmites australis* depends on the phosphorus concentration. *Journal of Hazardous Materials, 440*, 129800. doi:10.1016/j.jhazmat.2022.129800.

Yu, X. Y., Wang, B. T., Jin, L., Ruan, H. H., Lee, H. G., & Jin, F. J. (2022). Impacts of biogas slurry fertilization on arbuscular mycorrhizal fungal communities in the rhizospheric soil of poplar plantations. *Journal of Fungi, 8*(12). www.mdpi.com/journal/jof. doi:10.3390/jof8121253.

Zhang, J., Su, L., Yan, K., Li, M., He, Y., Zu, Y., Zhan, F., & Li, T. (2021). An arbuscular mycorrhizal fungus increased the macroaggregate proportion and reduced cadmium leaching from polluted soil. *International Journal of Phytoremediation, 23*(7), 684–692. https://www.tandfonline.com/loi/bijp20. doi:10.1080/15226514.2020.1849014.

Zhang, X., Chen, B., & Ohtomo, R. (2015). Mycorrhizal effects on growth, P uptake and Cd tolerance of the host plant vary among different AM fungal species. *Soil Science and Plant Nutrition, 61*(2), 359–368. doi:10.1080/00380768.2014.985578.

Zhao, Y., & Naeth, M. A. (2022). Soil amendment with a humic substance and arbuscular mycorrhizal fungi enhance coal mine reclamation. *Science of the Total Environment, 823*. www.elsevier.com/locate/scitotenv. doi:10.1016/j.scitotenv.2022.153696.

Zhao, Z., Chen, L. u, & Xiao, Y. (2021). The combined use of arbuscular mycorrhizal fungi, biochar and nitrogen fertilizer is most beneficial to cultivate *Cichorium intybus* L. in Cd-contaminated soil. *Ecotoxicology and Environmental Safety*, *217*, 112154. doi:10.1016/j.ecoenv.2021.112154.

Zocco, D., Van Aarle, I. M., Oger, E., Lanfranco, L., & Declerck, S. (2011). Fenpropimorph and fenhexamid impact phosphorus translocation by arbuscular mycorrhizal fungi. *Mycorrhiza*, *21*(5), 363–374. doi:10.1007/s00572-010-0344-0.

CHAPTER

14

Modeling cadmium transfer from soil to crops

Meie Wang[a], *Chi Peng*[b], *Rufus L. Chaney*[c], *Weiping Chen*[d]

[a]Research Center for Eco-Environmental Sciences, Chinese Academy of Sciences, Beijing, China [b]Institute of Environmental Engineering, School of Metallurgy and Environment, Central South University, Changsha, China [c]Chaney Environmental, Beltsville, MD, United States [d]Research Center for Eco-Environmental Sciences, Chinese Academy of Sciences, Beijing, China

14.1 Introduction

A detailed understanding of soil cadmium (Cd) dynamics in field is a prerequisite for predicting Cd transfer from soil to crops, but it is laborious and unrealistic to obtain long-term monitoring of such field data. Prediction by modeling Cd balance in agricultural systems, including inputs/output balance, is a primary step for prediction of Cd transfers. Moreover, to reduce the transfer of Cd to food chains, it is important to estimate the uptake of the metal by plants. The simple estimation of the biological enrichment factor [bioconcentration factor (BCF)] can serve as a rough guide of the likely range of metal uptake, but it does not reflect more detailed site-specific conditions of the soil. Mathematical approaches and regression models can be applied to predict specific dependent variables (e.g., Cd concentration within plant tissues) using soil properties (e.g., Cd concentration, pH, and soil organic matter content) as independent variables. Once generated, such models can be employed to predict Cd concentration in parts of crops and derive safety values for Cd-contaminated farmland. Both mechanistic and empirical models are used to predict Cd transfer, but the latter, such as multiple linear regression (MLR) models, are used more frequently.

In this chapter, models based on the whole process from Cd accumulation in soil and crop uptake, transportation, and accumulation will be summarized, and application scenarios will be shown by case studies.

Cadmium Contamination in Soils and Plants: Mechanisms and Mitigation.
DOI: https://doi.org/10.1016/B978-0-443-34104-5.00018-9

14.2 Environmental fate models based on mass balance for predicting cadmium accumulation in soils

The fundamental concept underlying all fate models is the integration of chemical transport and transfer processes into mass balance equations. In the 1970s, with growing concerns about the impacts of metals, including lead, Cd, mercury, and arsenic in multiple environmental media, rapid development of global and regional mass-balance models for these metals took place (Garrels et al., 1975; Nriagu, 1978, 1979). In the late 1970s regional or global mass-balance models for organic chemicals based on fugacity principles emerged (Mackay, 1979). By the late 1980s multimedia modeling had been further improved (Allen et al., 1989; Cowan et al., 1995; Mackay, 1991). However, most of the multimedia models have been developed for modeling the fate of organic pollutants in the environment.

The fate of heavy metals in the environment is predicted mostly by single-media models, such as in soil (Chen et al., 2007) and water (Zhou et al., 2023). In these models the net change of heavy metals in a specific environmental medium is the balance between the input and output fluxes that are either estimated by stochastic models (Peng et al., 2016) or dispersion models (Qiao et al., 2019; Zhou et al., 2023) or measured by monitoring (Hou et al., 2014; Tipping et al., 2010; Xia et al., 2014).

Although seldom used in predicting the fates of heavy metals in soil, multimedia models have the potential to be a useful tool for predicting the fates of heavy metals in the environment at a large scale because of advancements in concepts and computation techniques. Therefore a brief summary regarding the principle and application of spatial multimedia models (SMMs) is presented in this chapter. Moreover, models for predicting heavy metal accumulation in soil based on input and output to soil are reviewed, including their principles, features, and applicability.

14.2.1 Spatial multimedia models

SMMs are mass balance partitioning models, also called "box" or "compartmental models" or "Mackay type" models (Mackay, 1979). These types of models are usually composed of several connected environmental compartments (e.g., air, water, soil, sediments, and vegetation). Furthermore, over the last half century, the development and application of environmental fate models have been greatly expanded. For example, models with variation of spatiotemporal scales can be used in concert with field data to rationalize chemical concentrations and estimate emission rates of key fate processes with a range of spatial and temporal resolutions. The inclusion of climate variability and climate changes in different model scenarios provides the prediction of the influence of climate on the accumulation of environmental contaminants. When integrated with or coupled to multipathway exposure models describing the transfer of chemicals from the environment into the food ingested by humans, environmental fate models can be used to assess health risk of soil contaminants. Therefore environmental fate models play an important role in the chemical assessment process and, when coupled with monitoring data, can provide reliable information about exposing the environment to chemicals.

However, the inclusion of the real environment is more than the variability at large scale. For example, real environmental concentrations (in water, soil, air, biota, etc.) often depend not only on physical–chemical properties, partitioning, and emission strength but also on the variability of the natural environment [e.g., texture and organic carbon in soil, particulate organic carbon and dissolved organic carbon (Particulate Organic Carbon (POC) and Dissolved Organic Carbon (DOC), respectively), uptake rate for different plant species, and phytoavailability of chemicals in water and soil]. Although models are currently heavily relied upon to assess risks of environmental pollution, they are seldomly used as scientific and policy support tools in formulating chemical management strategies because of the manifold of challenges existing models face. One of the most important challenges is the inclusion of spatial variability of the real environment into the models.

From the emergence of the multimedia models in the late 1980s to the present, multimedia modeling approaches can be divided into four categories, depending on the level of complexity and treatment of spatial and temporal resolution (Cohen & Cooter, 2002; Hsieh & Ouimette, 1994):

1. *Integrated SMMs* in which transport and fate of the chemicals are described by spatial models considering intermedia boundary conditions and a set of medium-specific transport equations.
2. *Linked spatial single-medium models* in which a collection of single-media spatial models interacts with each other and the output of one model is used as the input to the others.
3. *Compartmental "well-mixed" media models* (both fugacity-based and concentration-based) in which the assumption of homogeneity is applied and all compartments are considered well-mixed.
4. *Integrated spatial-multimedia-compartmental models*, in which a combination of uniform and nonuniform compartments is considered to calculate the transport of the pollutant within the system.

The main advantage of SMMs is that they can provide sufficient information without a huge amount of input data and without the paradox of result validation.

14.2.2 Single-media models

In soil, detectable changes in heavy metal concentration may require years of accumulation. Therefore fate models are mostly used to predict temporal changes in heavy metal concentration in soil. The fate and transport of heavy metals in soils described in a single-media mass-balance mode follow the basic principle of mass balance, that is, the net change of an element in soil is the balance between the input at the soil surface and the output through leaching, crop harvest, and surface runoffs (see Chapter 5). As far as the input and output pathways for mass-balance equations are concerned, empirical models are the primary choice. Keller et al. (2001) described the regional heavy metal balance in agroecosystems with the following (Eq. 14.1):

$$\frac{\Delta M_{ij}}{\Delta t} = I_{\text{Atm}} + I_{\text{Agr},ij} - O_L \tag{14.1}$$

where ΔM_{ij} (g) is the amount of metal accumulated in the spatial unit ij during time Δt. I_{Atm}(g ha^{-1} year^{-1}) is the regional aggregated metal input flux by atmospheric deposition, $I_{\text{Agr},ij}$ (g ha^{-1} year^{-1}) is the net input flux of the spatial unit ij with farm type i and crop system j which considers both input and output flux specific to farm type i and crop system j, O_L (g ha^{-1} year^{-1}) is the output flux through leaching that is evaluated for the predominant soil types of the region and aggregated according to their fraction.

Peng et al. (2016) addressed the mass balance of input and output of soil heavy metals by a one-dimensional mass conservation equation:

$$\frac{\partial C_t}{\partial t} = \frac{I_t}{\partial t} - \frac{\partial U}{\partial t} - \frac{\partial L}{\partial t} \tag{14.2}$$

where C_t is the total metal in soil (mg dm^{-3}), t is the time step (an hour or a day), $\frac{I_t}{\partial t}$ is the total input flux of heavy metals to soil (mg dm^{-3} t^{-1}), $\frac{\partial U}{\partial t}$ denotes the metal outflux by plant uptake (mg dm^{-3} t^{-1}), and $\frac{\partial L}{\partial t}$ is the output flux by leaching.

Based on the mass-balance principle as described above, Chen et al. (2007) developed a Windows-based soil trace element model to estimate the accumulation of soil Cd in agricultural soil following long-term application of Cd-containing phosphatic fertilizer. It tracked 100 years of mass balance of Cd between external inputs applied to the surface of soil and outputs through leaching and plant uptake. de Vries and McLaughlin (2013) developed a mass-balance-based model to simulate soil and crop Cd in Australian agricultural systems over two centuries (1900–2100). In that model, input flux of Cd was calculated separately according to pathways, including fertilizer addition, atmospheric deposition, biosolids addition, and irrigation.

14.2.3 Equilibrium model composed of single media model

As widely acknowledged, each of the input or output terms in the mass balance equation is governed by a set of physical, chemical, and biological processes that are interactive and simultaneous. Therefore chemical equilibrium among metal species is usually coupled in the mass-balance-based single media model. For example, Tipping et al. (2006) simulated the changes of heavy metals Ni, Cu, Zn, Cd, and Pb in soil and water in three moorland subcatchments in Cumbria, United Kingdom, using CHemistry of the Uplands Model-Annual Metal (CHUM-AM). The model includes representations of speciation in soil solution, adsorption by Al and Fe oxide phases in the deeper soil, chemical weathering, dissolved and particulate transport, and speciation in surface waters besides hydrology and the uptake and storage of metals by plants. Chen et al. (2007) divided total trace element in soil into four fractions, and the total amount of the elements in soil matrix is defined by

$$C_T = \theta C + \mathrm{Ad} + \mathrm{MP} + \mathrm{Org} \tag{14.3}$$

where C_T is the total concentration of the element in soil (μmol kg soil^{-1}). θ is the volumetric water content (cm^3 cm soil^{-3}); C is the soil solution concentration of the element (μM); and Ad, MP, and Org refer to the element in the surface-adsorbed phase, immobilized mineral phase, and organic matter phase (μmol kg soil^{-1}), respectively.

Furthermore the soluble phase of trace elements in the soil solution is critical as it represents the readily mobile pool accessed by plant roots. The concentration of soluble trace elements in soil solution is dependent on the kinetics between those absorbed by solid phase and in solution. The surface adsorption and desorption reactions are rapid and reversible and usually characterized by a thermodynamic equilibrium distribution:

$$\mathrm{Ad} = k_a C \tag{14.4}$$

where the definition of Ad and C are as Eq. (14.3), k_a (soil L^{-1} soil solution) is the linear adsorption constant. The dissolution and immobilization because of precipitation and coprecipitation with other solid minerals and intro-particle sites are characterized by a first-order kinetic equilibrium:

$$\frac{\partial \mathrm{MP}}{\partial t} = k_f(k_a + \theta)C - k_b\mathrm{MP} \tag{14.5}$$

where ∂MP is the change of MP during time t, k_f, and k_b are the first-order forward (dissolution) and backward (immobilization) reaction rate constants (h^{-1}), respectively, and the definition of other parameters is as Eq. (14.3). In addition, a part of the trace element taken up by plants will return to the soil as crop residues. Mineralization of the crop residues in the soil may release the trace elements into the soil solution and surface-adsorbed phases, which may be described by a first-order kinetic as

$$\frac{\partial \mathrm{Org}}{\partial t} = -k_{\mathrm{im}}\mathrm{Org} \tag{14.6}$$

where ∂Org is the change of the element in organic phase during time t, k_{im} is the first-order organic mineralization rate constant (h^{-1}).

Meanwhile, to describe the spatial and temporal changes in partition coefficients driven by environmental factors such as soil type and acidification, changes of partition coefficients for heavy metals in soil are estimated by chemical equilibria models such as surface complexation models (Bonten et al., 2011) (see Chapter 6).

14.2.4 Single-media coupling process-based models

Single-media coupling process-based models are composed of equilibrium models and single-medium models that focus on the partitioning behavior of Cd within the soil matrix. Meanwhile, mass-balance-based single-media models are used to simulate fates of soil heavy metals in a homogenous soil matrix. Therefore the spatial heterogeneity of soil heavy metal at a regional scale may be demonstrated by composing spatial units that are considered homogenous within the unit (Keller et al., 2001). In this way, however, the interrelationship of heavy metal flux between adjacent units, for example, by the exchange of heavy metal flow, cannot be simulated.

Basically the exchange of heavy metal flow takes place when media such as air and water containing heavy metals disperse or transport over a large scale, for example, a regional or global scale. So far as those scenarios are considered, models have been developed by coupling hydrologic or atmospheric models. For example, watershed models have demonstrated their precision in simulating surface and subsurface hydrology and suspended solid loading, which are the major processes that transport heavy metals from contaminated sites. Therefore watershed models can provide a suitable platform to model the processes that determine the transport and distribution of heavy metals over the landscape. For example, Jacques et al. (2008) found, by coupling geochemical reactions with water flow and solute transport, that variations in water content and water fluxes can significantly influence the speciation, and thus the mobility and availability of elements such as Ca, P, and U. The incorporation of environmental models may increase the temporal and spatial resolution of single-media models. Bonten et al. (2011) constructed SMARTml by coupling hydrological model, the Soil Water Atmosphere Plant (SWAP) model that was used to calculate water fluxes, moisture contents, and soil temperatures for the Solling site, Germany. The outputs of hydrological model SWAP were the inputs for simulating vegetation growth, biogeochemical processes, and chemical equilibria. The soil and water assessment tool (SWAT) model is an independent flow-sediment model that has a Geographic Information System (GIS) compatible user interface and can be used to express physical processes associated with runoff and sediment movement, which makes it one of the most widely accepted models among hydrology and water quality management communities across the globe for evaluating the impacts of different management strategies on nutrient and suspended solid loading at watershed and basin scales. Furthermore, SWAT provides flexibility for accommodating the spatial details of both biophysical and human influence on the quantity and quality of surface flow, a common feature of industrial areas (Zhou et al., 2023).

Nair et al. (2022) elaborated an approach that couples SWAT and empirical modeling in simulating Hg flux at a watershed scale. In their approach, SWAT and empirical modeling were linked to represent the human-modified hydrology and multiple-dispersed, nonpoint Hg sources in an industrialized watershed. They estimated separately the Hg flux through water flow and suspended solid movement using empirical functions as shown in Eq. (14.7).

$$\widehat{Hg}_{\mathrm{OF}_i} = f(\mathrm{Flow}_{\mathrm{OF}_i}, S, \mathrm{FR}, \mathrm{FA}, C, \mathrm{HI}_{\mathrm{OF}_i}, \mathrm{Flow}_{\mathrm{OF}_i} \times S, \mathrm{Flow}_{\mathrm{OF}_i} \times \mathrm{FR}, \mathrm{Flow}_{\mathrm{OF}_i} \times \mathrm{FA}, \mathrm{Flow}_{\mathrm{OF}_i} \times C) \tag{14.7}$$

where i refers to aerosol outflow site (OF) numbers, $\widehat{Hg}_{\mathrm{OF}_i}$ (mg Ls^{-1}) and $\mathrm{Flow}_{\mathrm{OF}_i}$ (Ls^{-1}) represent the estimated Hg flux and simulated flow from these outflow sites. Season, S, flow regime, FR, storm drain cleanouts, C, and $\mathrm{HI}_{\mathrm{OF}_i}$ represent control variables introduced to isolate the influence of these variables on Hg remobilization. Here, they considered three major types of $\mathrm{HI}_{\mathrm{OF}_i}$ and three flow regimes. A similar approach may be applied to Cd.

Using SWAT, Qiao et al. (2019) integrated the quantity of heavy metal transported into runoff and sediment based on soil pH, soil particle size, and concentrations of heavy metals dissolved in runoff and adsorbed to particles. They assumed surface soil particles are carried into runoff. Water-soluble heavy metals dissolve in runoff, and insoluble heavy metals are absorbed to soil particles and transported by runoff. Therefore the ideal model for heavy metal transportation quantity ($Q_{\mathrm{tranHM-idel}}$, kg) simulation could be expressed as follows:

$$Q_{\mathrm{tranHM-idel}} = C_{\mathrm{Hins}} \times A_{\mathrm{sed}} + C_{\mathrm{Hw-s}} \times V_{\mathrm{flow}} \tag{14.8}$$

where C_{Hins} (mg kg^{-1}) is the average insoluble heavy metal concentrations in soils, $C_{\mathrm{Hw-s}}$ (mg L^{-1}) is the average water-soluble heavy metal concentrations, A_{sed}(ton) is the amount of soil erosion, and V_{flow} (m^3) is the runoff volume obtained from SWAT model. Moreover, the effects of soil pH and soil particle size on heavy metal migration were determined based on the soil pH value, soil particle size, runoff volume, sediment amount, insoluble heavy metal concentrations, and water-soluble heavy metal concentrations of soil samples that were collected from four runoff plots. These data were used to determine the form of the model first. The coefficients of the model were then adjusted by the observed data from whole watershed and subwatersheds. The model was generated as follows:

$$Q_{\mathrm{tranHM}} = \alpha \times \ln\mathrm{pH}(\alpha_1 \times 0.001 + \alpha_2 \times 0.026 + \alpha_3 \times 1.025)^{\beta} \times (C_{\mathrm{Hins}} \times A_{\mathrm{sed}} + C_{\mathrm{Hw-s}} \times V_{\mathrm{flow}}) \tag{14.9}$$

where Q_{tranHM} (kg) is the quantity of heavy metal transported in the watershed, pH is the average pH value of the study area, α_1, α_2, and α_3 represent the average percentages of clay, silt, and sand in catchment soils, respectively. The α values estimated based on the observed data for As, Pb, and Cd were 0.0015, 0.0017, and 0.0076, respectively. Similar approaches have been reported by Sun et al. (2025).

So far the SWAT-laden heavy metal module (SWAT-HM model) by Meng et al. (2018) is a relatively well-established model consisting of the following process models (modules): solid–liquid partitioning, complexation, slow transformation in soil and water, and transport in uplands and channels, among others. In their composite model the heavy metal transformation model takes three major types of reaction into account: sorption, complexation, and slow reaction, in which equilibrium constants such as K_d or K_L are used in modeling the local equilibrium of heavy metals. Moreover, in the heavy metal module used in the upland system, the SWAT model undertakes the hydrological and sediment processes that serve as external driving forces of the heavy metal module. The details of the climatic input, water balance, soil erosion, and channel routing, among others, are available in the SWAT documentation. In the heavy metal module used for water bodies a simple mass balance method is used to simulate the in-stream processes, including settling, resuspension, diffusion, and burial. For example, the settling and resuspension of heavy metal in the particulate phase (suspended sediment) and bed sediment phase accompany those processes for the sediment. Thus the sediment budget obtained in SWAT is harnessed to determine the settling or resuspension amount of heavy metal:

$$M_{\text{sed}} = \text{sed}_{\text{set}} \cdot \left(M_{1,p} + M_{n,p}\right) \cdot \text{UF} \tag{14.10}$$

$$M_{\text{res}} = \text{sed}_{\text{res}} \cdot \left(M_{1,b} + M_{n,b}\right) \cdot \text{UF} \tag{14.11}$$

In SWAT-HM the heavy metal module gains operability from the hydrological and sediment dynamics substantiated by SWAT. That is, the dynamic-related variables required for calculating metal behavior, such as w_{mobile}, w_{evap}, sed_{set}, and sed_{res}, can be provided by SWAT. Correspondingly the parameterization of the SWAT-HM consists of an exact parameterization of SWAT and that of the heavy metal module. Specifically, there are a total of eight parameters pertaining to the heavy metal module: the slow reaction kinetic rates (k_{-1}, k_1), the diffusion velocity V_d, the burial velocity V_b, the weathering rate r_w, and the partition coefficient for soil (K_d1), for channel sediment (K_d2), and for bed sediment (K_d3). Each of the eight parameters represents a certain physicochemical process and may vary in space. When applying the SWAT-HM the user may allow one parameter to take either a global value or different values across subbasins or hydrological response units, depending on the availability of information. If a parameter is known to be governed by certain crucial factors and the factor values are readily available, the parameter value can be estimated based on this knowledge. For instance the partition coefficient for soil (K_d1) is well known to be controlled by soil pH and soil organic carbon, as well as Fe and Mn oxides and clays, and the established formulae for Zn and Cd can be harnessed to obtain the K_d1 values.

14.3 Models for predicting cadmium transfer from soil to crops

The simple estimation of BCF can serve as a rough insight into the range of heavy metal accumulation in plants, but it does not reflect more detailed site-specific conditions. The relationship between metal concentrations in various soil extracts and plants is often described with a transfer function, which is usually determined by regression analysis. Linear functions are preferred in most cases. To improve the normality of the data, log-transformations of soil or plant metal concentrations or both are common. Moreover, crop uptake of Cd depends on its phytoavailability in soil, that is, on the flux of Cd^{2+} at the root surface, assuming that the free ion is the main species absorbed by root cells (see Chapter 9). In aerobic agricultural soils, Cd phytoavailability is mainly determined by sorption onto the solid organic and mineral phases (clay and oxides), by complexation with soluble ligands, generally organic compounds, and by transport to plant roots by diffusion and advection (see Chapter 6). Soil Cd, pH, organic matter, and, to a lesser extent, clay and Fe and Mn oxides have been found to be the major factors in statistical regression models to predict the soluble soil Cd and its accumulation by plants. An advantage of the Freundlich-type soil–plant transfer relationship is its simplicity and applicability. Except for Fe and Mn oxides, these variables are commonly measured in soil testing, and therefore, predicting plant Cd by using statistical models based on these variables is of great interest. However, it is challenging to identify the relationship between elements in soil and plant Cd concentration using linear regression methods because those relationships are often nonlinear and these equations should not be used for soils where concentrations of metals are outside the range from which the regressions were derived. In addition, one major drawback of such models is that they have not been tested on new data (cross-validation). Their real predictive value for new data is consequently unknown.

Machine learning (ML) approaches, such as artificial neural network (ANN), decision tree (DT), and random forest (RF), have been used to investigate nonlinear relationships between independent and dependent variables; they may better model a complex boundary between the two conformity datasets. For example, the RF approach consists of aggregating the predictions of a large number of trees (i.e., a forest) that are uncorrelated by bootstrapping the training dataset. In addition, many processes involved in plant uptake of Cd may be described with curvilinear functions. Such processes include the desorption of ions from soil solid phases, often described with Freundlich-type equations, the adsorption of ions to root surfaces, and the transport from root to shoot. However, process-based mechanistic models are data-intensive and computationally demanding, which hinders their practical use in the field. Because of the lack of field applications of the process-based mechanistic models, only empirical and ML-based approaches are reviewed and discussed in this section.

14.3.1 Regression-based empirical model

Regression-based models are typical empirical models predicting plant uptake of heavy metals from soil, in which concentrations of metals in plants are predicted by various soil parameters, mostly the total concentration of a metal in soil, soil pH, and soil organic carbon. Some regression-based models have included other parameters, for example, clay content, the dry weight of the plant, the concentrations of other metals in the soil, and redox and interaction terms between variables, which might reveal more than the usual measured soil properties. Cd is a metal that is typically monitored and modeled. Commonly investigated plant crops are cereals (rice, wheat, barley, and maize), root vegetables, leafy vegetables, and potatoes. Regressions were established for soil and plant paired sampling, and only those that measured concentrations above the detection limit. As widely acknowledged, the following measured parameters are considered to be of importance in explaining plant metal concentrations: pH value, organic carbon content, the concentrations of metal in soil, Eh, and silt and clay content. The general equation of the model was established as follows (Novotná et al., 2015):

$$M_{\text{plant}} = a + b \cdot \text{pH} + c \cdot M_c + d \cdot M_{\text{soil}} + e \cdot \text{silt} + f \cdot \text{clay} + g \cdot M_{\text{othermetals}} + h \cdot \text{Eh} + i \cdot \text{interactions} \quad (14.12)$$

where M_{plant} is the concentration of the given metal in the plant; M_{soil} is the total concentration of heavy metals in soil; pH is the pH measured in H_2O of the soil; Eh is the redox potential; M_c (%) is the content of organic carbon in soil; silt and clay are the content of silt and clay in soil;

$M_{othermetals}$ is the concentration of other metals such as Zn, Fe, and Mn; $a, b, c, d, e, f, g, h,$ and i are the regression coefficients; and interactions are the interactions of parameters in the regression equation.

Moreover, there is general agreement that the total metal concentration of a soil is usually not a good predictor of the metal concentrations in plants grown in that soil because only some fraction of the total metals is available for plant uptake. Therefore the potentially phytoavailable fraction of soil metal has been measured by extracting soil with relatively dilute solutions of salts, acids, or complexing agents (see Chapter 9). Further, experiments have indicated that the metal uptake by plant roots is not simply a function of the concentration of some particular chemical fraction in the soil. Soil acidity, organic content, cation exchange capacity, redox status, and temperature have all been shown to modify uptake, and the antagonistic or synergistic effects of other metals and phosphate must also be taken into account. Based on the knowledge of factors affecting the uptake of heavy metals by plants the early work by Davies (1992) pointed out that ideally it should be possible to formulate a quantitative relationship between the plant tissue concentration of a given metal and the concentration of that metal in soil which includes in the relationship appropriate weighting factors for other soil parameters. In his field plot experiment in old lead mining areas of Derbyshire in England and in Clwyd and Ceredigion in Wales, measurements of total and soluble metal contents (Pb, Zn, Cu, and Cd) in soil were considered predictors of metal uptake by an experimental plant (radish). Soil organic matter content, pH, cation exchange capacity, and soluble Mg, K, and P were also measured to investigate their possible influences on metal uptake. Results revealed that acetic acid-extractable Cd satisfactorily predicted Cd uptake by radish (see Eq. 14.13), which was not improved by including other soil properties.

$$\mathrm{Cd(leaf)} = 0.722\mathrm{Cd(acetic)} + 0.979 \tag{14.13}$$

Similar results were found by Zhang et al. (2016) for pot experiments with 25 soils (top 20 cm of the soil profile) from diverse contaminated agricultural soils in Guizhou Province, southwest China. In their work shoot Cd in Chinese cabbage could be predicted well by Cd concentration in situ sampled pore water or two different extractable fractions in soil (Eqs. 14.14–14.16).

$$[\mathrm{Cd}]_{\mathrm{sh}} = 10^{0.294}[\mathrm{Cd}]_{\mathrm{pw}}^{0.637}\ (R^2 = 0.828) \tag{14.14}$$

$$[\mathrm{Cd}]_{\mathrm{sh}} = 10^{4.264}[\mathrm{Cd}]_{\mathrm{HNO_3}}^{1.007}[\mathrm{H}^+]^{0.693}\ (R^2 = 0.859) \tag{14.15}$$

$$[\mathrm{Cd}]_{\mathrm{sh}} = 10^{3.891}[\mathrm{Cd}]_{\mathrm{DTPA}}^{0.785}[\mathrm{H}^+]^{0.826}\ (R^2 = 0.864) \tag{14.16}$$

Furthermore, metal solubility in soils may be modeled using an equation based on a pH-dependent Freundlich relation (Jopony & Young, 1994). This can be used to predict free metal ion activity in the soil pore water $[\mathrm{M}^{2+}]$ from total soil metal content $[M_{\mathrm{soil}}]$ and soil pH. As humus is often considered to be the primary adsorbent of metal ions in the soil (Sauvé et al., 2000), a further refinement is to assume that metals are exclusively adsorbed by humus to increase comparability between soils. Total metal concentrations in the soil $[M_{\mathrm{soil}}]$ were therefore reexpressed as if adsorbed on organic carbon $[M_{\mathrm{C}}]$. The activity of free metal ions $[\mathrm{M}^{2+}]$ was then predicted from $[M_{\mathrm{C}}]$ (mg kg^{-1}) and soil pH:

$$p(\mathrm{M}^{2+}) = \frac{p[M_{\mathrm{C}}] + k_1 = k_2\mathrm{pH}}{n_F} \tag{14.17}$$

where k_1 and k_2 are empirical, metal-specific constants, and n_F is the power term from the Freundlich equation.

As assumed by "free-ion activity model," plant uptake of metals may be controlled by metal ion activity in the soil pore water (Parker & Pedler, 1997). Therefore the transfer factor of plants could be expressed as the quotient of metal concentration in the plant [M_{plant}] to metal ion activity in soil pore water [M^{2+}] derived from Eq. (14.17):

$$TF = \log\frac{[M_{plant}]}{[M^{2+}]} \tag{14.18}$$

Combining Eqs. (14.17) and (14.18), an equation describes the relationship between [M_{plant}] and soil properties pH and [M_C] as follows:

$$\mathrm{Log}[M_{plant}] = C + \beta_1[\mathrm{pH}] + \beta_2\log[M_c] \tag{14.19}$$

where C, β_1, β_2 are empirical metal- and crop-specific constants. Values of C, β_1, and β_2 are derived by minimizing the error (residual standard error, RSD) in the prediction of [M_{plant}] from measured values of pH and [M_c] in Eq. (14.3).

The above log-transformed Freundlich equation is generally used for predicting the transfer of heavy metals from the soil to the plant system (Krauss et al., 2002), and soil properties such as pH, soil organic carbon (SOC), and cation exchange capacity (CEC) are added as covariates to describe the transfer of heavy metals in plants:

$$\mathrm{Log}[C_{plant}] = a + b\mathrm{Log}[C_{soil}] + c\mathrm{pH} + d\mathrm{Log}[\text{other soil properties}] \tag{14.20}$$

Based on 162 wheat and 215 barley grain samples collected from paired soil and crop surveys in Britain, and wheat and barley samples from two long-term sewage sludge experiments, Adams et al. (2004) reported that wheat grain Cd concentration could be predicted reasonably well using the following model: log (grain Cd) = a + b log(soil Cd) − c (soil pH), with 53% of variance accounted for, whereas the model prediction for barley grain Cd concentration was less satisfactory (22% of variance accounted for). Moreover, it was found in their work that similar controlling factors of Cd phytoavailability in sludge-amended and unamended soils.

To screen the most influencing factors for successful modeling, methods, such as stepwise regression analysis, principal component analysis (Zhou et al., 2019), redundancy analysis, and the structural equation model (Li et al., 2023), are widely coupled with multivariable regression models.

14.3.2 Machine learning approach

As addressed above, empirical models as well as corresponding algorithms such as multivariate linear regression and stepwise linear regression revealed unsatisfactory prediction results when applied to field system data because of the nonlinearity and nonstationary characteristics of complex environmental factors impacting the accumulation of Cd in crops, though they have good prediction for simple pot experiments. The ML approach has received remarkable attention in recent years in the environmental field and is an effective tool for solving nonlinear and nonstationary phenomena and uncovering hidden relationships in complex systems. In addition, ML models are unaffected by data statistics and can function with limited data availability. The supervised learning methods in ML with prescribed output labels normally include learning and prediction processes, with tuning of parameters during the learning process. The cross-validation approach in the ML training processes makes the model robust for generalization, which compensates for the potential overfitting in the validation of traditional multiple regression methods. In terms of crop Cd prediction modeling the majority of studies concentrated on the simulation of metal accumulation based on crop–soil paired field investigation data (Hu et al., 2020; Zhao et al., 2023). For example, Li et al. (2024) used RF models to successfully predict Cd in rice grains with soil pH, soil organic matter (SOM), total and extractable soil Cd, Zn, Mn, Fe, Mg, Ca, P, and Se as predictor variables based on 92 pairs of rice-rhizosphere soil samples. Ma et al. (2023) predicted rice Cd using 654 pairs of soil–rice data investigated in Guangxi province, China, by ML algorithms RF and ANN. Huang et al. (2024) found that a support vector machine model showed the best performance ($R^2 = 0.87$) in predicting grain Cd concentrations at a regional scale compared to RF and ANN based on a 3-year survey of 601 pairs of soil and rice samples from Xiangtan, Hunan Province, Southern China.

The usual process of ML includes three steps: (1) model performance, (2) hyperparameter optimization, and (3) model performance evaluation. Commonly used ML algorithms include RF, the ANN, gradient boosting regression trees, and extremely randomized trees. In the following sections, we take RF and ANN as examples to introduce modeling process briefly (Ma et al., 2023).

14.3.2.1 *Random forest*

RF is an improved DT model based on the bagging method; the trees of the RF are trained by multiple datasets obtained from replacement sampling, and the outcome is determined by a majority vote on the predictions of independent DT. The algorithm partitions the feature space through hierarchical rules, and each tree is built using a random subset of predictor variables. The final prediction is obtained by averaging the regression results from all the DTs in the ensemble. RF can overcome the overfitting problem in the face of a high number of environmental and crop-related factors to create better multiparameter prediction models.

14.3.2.2 *Artificial neural network*

ANN consists of layers of nodes containing input, hidden, and output layers. The neurons in the input layer receive input information, which is processed by the activation function and distributed to the neurons in the hidden layer. The information processed by the neurons in the hidden layer is sent to the output layer. Each node is also known as an artificial neuron and is connected to another node by the associated weights and thresholds. If the output of any individual node is above the specified threshold, then that node is activated, and the data are sent to the next layer of the network. Otherwise, no data are passed on to the nest layer of the network. Multiple factors will affect the stability of an ANN model, for example, the model chosen, modeling process, and training dataset. The optimal combination of neurons could not be determined in advance; thereafter the optimal structure of the network must be determined by testing multiple models by correlation coefficients and root mean square error (RMSE) guidelines. ANNs are trained using the resilient backpropagation algorithm, where the maximum step size of the neural network train can be set as is necessary. Moreover, the partial derivative of the error function is specified as a stopping criterion threshold. Resilient propagation is a variant of backpropagation and has advantages over backpropagation because it removes the effect of the partial derivatives on the step size of the weights that influence the weighting step.

14.3.2.3 *Hyperparameter optimization*

Hyperparameters are predefined settings that control the behavior of ML algorithms during the implementation process, rather than being learned from the training data. Selecting optimal hyperparameter combinations is crucial for improving the efficiency and generalization performance of ML algorithms. For RF the number of trees (*ntree*) and variables randomly sampled as candidates at each split (*mtry*) are two critical hyperparameters directly controlling learning behavior and significantly impacting model performance. Different combinations of *ntree* and *mtry* should be tested during the training process. For the ANN the number of hidden layers and neurons in each layer determines the overall architecture. However, there are no explicit rules for determining these hyperparameters. Typically, they are obtained through empirical attempts and trial-and-error. Although a single hidden layer is generally sufficient, adding a second hidden layer can improve prediction accuracy.

14.3.2.4 *Model performance evaluation*

Model performance is conducted by splitting crop–soil paired datasets into two subdatasets randomly. The two subdatasets usually consist of 85% and 15% of the whole and serve for model training and testing, respectively. Before the implementation of the models the input features are log-transformed for magnitude decrease and data normalization and thus accelerate the convergence of the ML algorithms. During the training process the robustness and prediction accuracy can be improved by cross-validation. Model performance is evaluated by the coefficient of determination (R^2), the RMSE, and the mean absolute error (MAE). The larger the R^2, the smaller the RMSE and MAE, and the better the model-fitting performance.

14.4 Case studies

14.4.1 Prediction of cadmium accumulation in paddy soil by mass-balance-based single media model: A case study in Youxian County, Hunan Province, China

14.4.1.1 *Description of the study area*

Youxian county (113.32°E, 27.01°N) consists primarily of townships of rice production. It has a population of ~800 thousand and occupies approximately 2600 km^2. The potential sources of Cd emissions include coal mining and combustion in rural scale industries such as metal smelters, cement manufacture, brick and roof tile firing, and in households for food preparation and space heating, through which the Cd fallout can spread across the entire region. There were approximately 387 km^2 of rice planting. The remainder is composed primarily of hilly mountainous topography.

In this study, we chose Taoshui and Wangling, two representative towns facing high Cd contamination in paddy soil because of mining and industrial activities, to demonstrate the model's ability to predict long-term changes in soil Cd. Taoshui is representative of soil Cd contamination from mining and metallurgical industries, including zinc oxide plants, and Wangling is the representative of multiple industrial point sources of Cd, including thermal power plants, cement plants, and metal processing plants. The availability of Cd in the paddy soils in the two towns is also enhanced because of soil acidity, with most of the soils lower

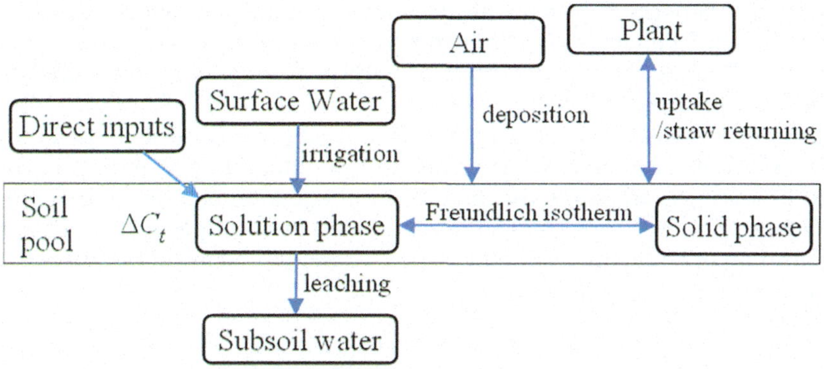

FIGURE 14.1 Framework of the pollutant accumulation model for heavy metals in soil. *From Peng et al. (2016) with permission.*

than pH 5.5. According to previous investigations the average Cd concentration in paddy soil in the two study towns ranged from 0.230 to 1.91 mg kg^{-1}, with 90% exceeding the allowable limit of 0.3 mg kg^{-1} stipulated by the China soil environmental quality standards (Wang et al., 2015).

14.4.1.2 Model establishment and parameters

In this study the accumulation of Cd in paddy soil is described as several mass-balanced processes involving the transportation of Cd among soil, water, and crop residues. The major input fluxes and output processes of Cd in the soil pool as well as their partitioning between soil solution and the solid phase. Fig. 14.1 shows the framework of the mass-balance-based model (so-called pollutant accumulation model-heavy metals). The governing mass-balance equation is as shown in Section 14.2.2, Eq. (14.2), namely, the total input flux, including irrigation water, fertilizers, and atmospheric deposition; and output flux through crop uptake and leaching. Each of those processes can be described by empirical or mechanism functions. For example, the equilibrium model describes the chemical equilibrium of Cd between soil water and particles as shown in Section 14.2.3. The uptake process of crops is described by the Michaelis–Menten kinetic equation as follows (Chen et al., 2007):

$$\frac{\partial U}{\partial t} = \frac{R_d \times J_{\max} \times C}{d \times (K_J + C)} \times \mathrm{UF} \tag{14.21}$$

where R_d denotes root density (cm root cm^{-2}), $J_{\max}$ denotes the maximum influx at the root surface (mg cm^{-1} $root^{-1}$), K_J denotes the root permeability coefficient (mg dm^{-3}), and d is the depth of the soil matrix (cm), UF is a unit conversion factor, 10^{-3}.

In terms of the normal field practice locally that straw residue is usually returned to the paddy soil, the amount of Cd in the straw returned periodically can be estimated as follows:

$$I_r = \frac{\partial U}{\partial t} \times \frac{T_r \times Rf}{d} \tag{14.22}$$

where I_r denotes the amount Cd in the straw returned to the soil matrix (mg dm^{-3} event^{-1}), T_r denotes the duration between straw returning events (180 or 365 days), and Rf is the straw return factor (%) indicating the amount returned to the soil as a percentage of the quantity taken up by the crop over the period T_r.

Though the process-based mechanism model describing crop uptake of heavy metals as shown in Eq. (14.23) is more general, the difficulty in obtaining parameters makes it hard to apply in field practice. As an alternative an empirical equation is proposed, in which the total amount of metal uptake by crops from soil is estimated by multiplying aboveground biomass by the concentration of heavy metals in the tissue. Therefore the metal output flux from soil by crop uptake can be described as follows:

$$\frac{\partial U}{\partial t} = \frac{C_t \times \text{PUF} \times G_{\text{Biomass}} \times R_p}{\rho \times d} \tag{14.23}$$

where G_{Biomass} denotes the rate of growth of aboveground biomass (kg cm^{-2} t^{-1}), R_p denotes the percentage of roots in the targe soil layer (%), and PUF is the plant uptake factor (unitless), which denotes the ratio of the concentration of metal in plant shoots to that in soil.

The output process by leaching can also be described by process-based models such as HYDRUS-1D (Chen et al., 2009) or empirical models as follows:

$$\frac{\partial L}{\partial t} = \frac{K_h \times \theta \times C}{d} \tag{14.24}$$

where K_h is the rate of percolation (cm t^{-1}). The volume of soil water leaching over time can be either estimated by hydraulic characteristics of the soil based on the van Genuchten equation (Chen et al., 2009) or by the balance of water among leaching, evapotranspiration, precipitation, irrigation, and surface runoff in the soil matrix.

As noted above the total input process is assumed to include irrigation, fertilizers, crop residues, and atmospheric deposition in this study. An empirical equation describes the total input flux as follows:

$$\frac{I_t}{\partial t} = \frac{I_a}{\partial t} + \frac{W_i}{\partial t} \times C_w + I_0 + I_r \tag{14.25}$$

where $\frac{I_a}{\partial t}$ denotes the metal input flux from atmospheric deposition (g dm^{-3} t^{-1}), $\frac{W_i}{\partial t}$ denotes the amount of irrigation water (cm t^{-1}), C_w is the concentration of the metal in irrigation water (mg dm^{-3}), and I_0 denotes direct inputs by other pathways such as application of fertilizers, crop residues, and soil amendments (mg dm^{-3} event^{-1}).

The data required by the model was collected by previous field investigation in the study area, for example, Cd concentration in paddy soils, rice grains, straw, irrigation water and sediment in irrigation canals, atmospheric dust, fertilizers, and lime (Table 14.1). The bulk atmospheric deposition was sampled by passive polypropylene tanks, which were 12.0 and 4.77 g ha^{-1} year^{-1} for Taoshui and Wangling, respectively, during 2013. The initial Cd concentration in the surface soil was set at 0.683 and 0.334 mg kg^{-1} based on investigated data in nondisturbed area. Site-specific partition coefficient of Cd was adopted from the literature by Zheng (2010). The crop uptake factors for rice grain and straw were calculated based on investigated Cd concentration in the study area.

TABLE 14.1 Default values of model parameters and mean Cd concentrations in the Youxian samples.

Model parameters	Values		Units	Sources
General settings				
Simulation period (T)	0–50		Year	Present study
Soil depth (d)	20		cm	
Simulation cycles of Monte Carlo	10.000			
Soil properties				
Volumetric moisture content (θ)	0.297		$dm^3 \cdot dm^{-3}$	Nie et al. (2010)
Bulk density (ρ)	1.05		$kg \cdot dm^{-3}$	
Percolation rate (K_h)	0.00533		$cm \cdot h^{-1}$	Ji (2007)
Isotherm parameters				
Freundlich partition coefficient (K_f)	108		$dm^3 \cdot kg^{-1}$	Zheng (2010)
Freundlich sorption parameter (n)	0.67		Unitless	
Plant information				
Plant uptake factor (PUF_{rice})	1.17		Unitless	Present study
Plant uptake factor ($PUF_{aboveground}$)	2.47		Unitless	
Biomass growth rate of rice grain ($BioG$)	14		$t \cdot year^{-1} \cdot ha^{-1}$	Mo et al. (2011)
Biomass growth rate of straw ($BioG$)	14		$t \cdot year^{-1} \cdot ha^{-1}$	Bi et al. (2009)
Root distribution (R_p)	90		%	Wu et al. (2001)
Input loads	Taoshui	Wangling		
Initial Cd conc. (C_0) in surface soil (0–20 cm)	0.683	0.334	$mg \cdot kg^{-1}$	Present study
Cd conc. in deep soil (20–40 cm)	0.311	0.121	$mg \cdot kg^{-1}$	
Cd input from atmospheric deposition (I_a)	12.02	4.77	$g \cdot ha^{-1} \cdot year^{-1}$	
Cd input from fertilizer (I_o)[a]	0.230	0.230	$g \cdot ha^{-1} \cdot season^{-1}$	
Cd input from lime (I'_o)	0.575	0.575	$g \cdot ha^{-1} \cdot season^{-1}$	
Cd conc. in irrigation water (C_w)	0.336	0.152	$\mu g \cdot kg^{-1}$	

(*continued on next page*)

TABLE 14.1 Default values of model parameters and mean Cd concentrations in the Youxian samples—cont'd

Model parameters	Values		Units	Sources
Irrigation water in early rice season (W_i)	12.7	12.7	cm•season^{-1}	Ji (2007)
Irrigation water in late rice season (W_i)	35.8	35.8	cm•season^{-1}	

[a] Compound fertilizer (composition NPK, etc.) of 750 kg•ha^{-1}•season^{-1}, lime 2250 kg•ha^{-1}•season^{-1}, and two growing seasons per year. *Cd*, cadmium. From Peng et al. (2016).

14.4.1.3 Model performance

To address the significant factors impacting the accumulation of Cd in paddy soil and offer strategies mitigating the risk of rice Cd, in this study, we assumed five representative scenarios to predict long-term changes in Cd concentration in paddy soil:

1. The default scenario: traditional cultivation and management practices; Cd input loads as measured by field monitoring. In this scenario, rice straw was assumed to be returned to field.
2. Increasing Cd removal from soil: rice straw not returned to field.
3. Decreasing industrial Cd emissions: Cd input flux through atmospheric deposition pathway is reduced to half of the original load as in the default scenario.
4. Decreasing Cd mobility by lime: based on previous study, liming with 2250 kg ha^{-1} decreased it approximately 20%–40%, and thus we assumed PUF and K_f would be 70% of those in the default scenario.
5. Integrated remediation measures: a combination of scenarios (B), (C), and (D).

The changes in the model parameters in terms of the assumptions made for the above scenarios are summarized in Table 14.2. The current concentrations of Cd in paddy soil in Taoshui and Wangling were used as the starting points for the simulation. Each scenario simulation was run for 50 years.

14.4.1.4 Results of simulation

As shown in Fig. 14.2, Scenario A demonstrated little change in Cd concentrations over the simulated 50 years, which indicates that the input and the output fluxes in the paddy soil were approaching a kinetic balance. However, monitored Cd concentrations in the deeper layer (20–40 cm) in the same field were 0.311 and 0.121 mg kg^{-1} in Taoshui and Wangling, respectively; both were approximately half of those in the upper layer (0–20 cm) (Table 14.1). The large difference in concentrations between the upper and deeper layers suggests higher Cd input loads in the past. Fig. 14.3 also demonstrates differences in Cd concentrations among the five scenarios in the future years: scenarios B, C, and E showed decrease of Cd in paddy soil in Taoshui and Wangling, whereas scenario D showed a steady increase. Table 14.2 summarizes the mass balance of Cd in the paddy soils under different scenarios. In scenario B the amount of Cd removed through crop uptake was three times that in scenario A, which suggests that removal of Cd-contaminated straw would be an effective measure to decrease paddy soil Cd. Scenario C showed a little reduction in paddy soil Cd as a result of decreasing Cd input flux through atmospheric deposition, which would take hundreds of years to reduce the current Cd

TABLE 14.2 Total Cd mass balance in the paddy soils after 50 years of continuous cultivation according to the scenario simulations (g ha^{-1}).

Scenario	Fertilization and lime (g ha^{-1})	Irrigation (g ha^{-1})	Atmospheric deposition (g ha^{-1})	Leaching (g ha^{-1})	Uptake removal (g ha^{-1})	Accumulation (g ha^{-1})
Taoshui						
A	23	81.3	601	110	518	76.3
B	23	81.3	601	58.9	1414	−768
C	23	81.3	300	96.7	472	−164
D	81	81.3	601	83.6	390	289
E	81	81.3	300	45.1	1081	−664
Wangling						
A	23	36.8	238	37.1	248	12.6
B	23	36.8	238	19.6	674	−395
C	23	36.8	119	33.1	230	−84.5
D	81	36.8	238	28.6	190	137
E	81	36.8	119	15.8	535	−314

Cd, cadmium.
From Peng et al. (2016).

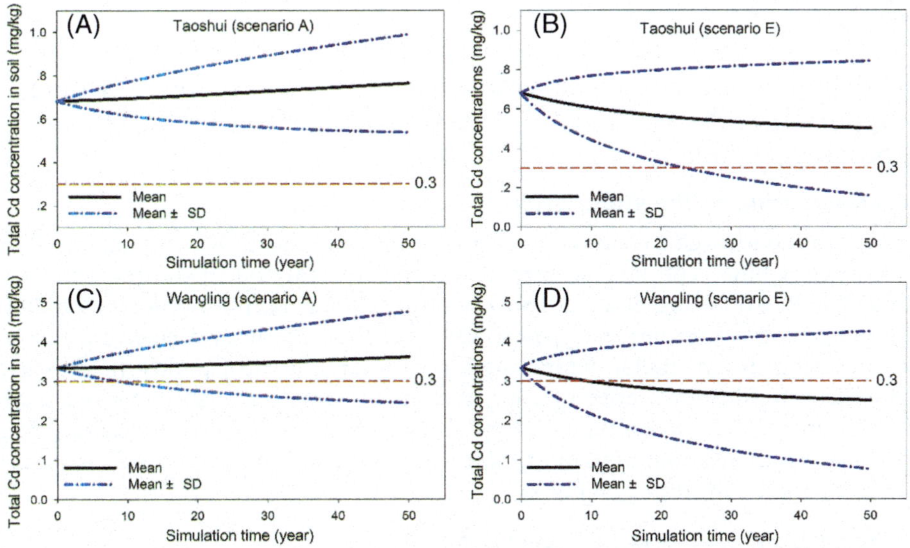

FIGURE 14.2 Changes in Cd concentrations among the five scenarios in the future years. *Cd*, Cadmium, A Taoshui (scenario A), B Taoshui (scenario E), C Wangling (scenario A), Wangling (scenario E). *From Peng et al. (2016) with permission.*

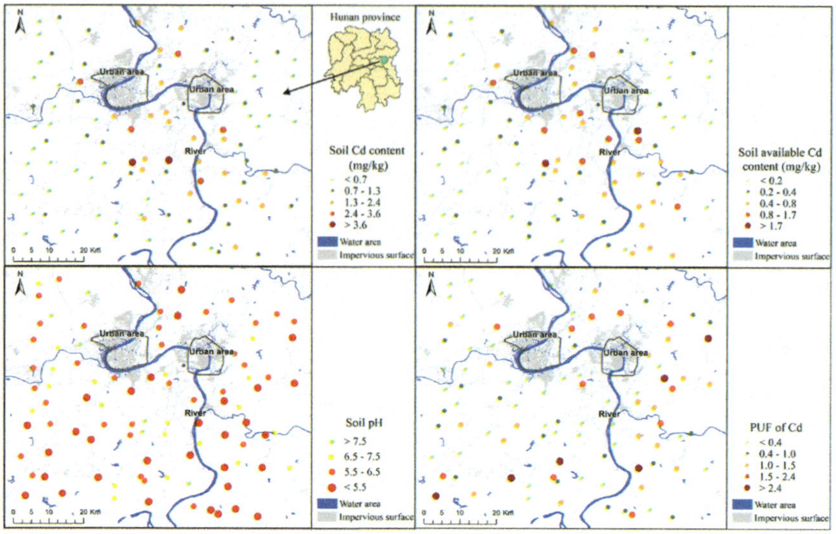

FIGURE 14.3 Spatial distribution of soil Cd, available Cd, pH, and rice plant uptake factor of Cd (PUF_{Cd}) in the study area. *Cd*, Cadmium. *From Peng et al. (2016) with permission.*

level, though it is effective in preventing further deterioration. Scenario D, long-term soil liming, however, increased paddy soil Cd (Table 14.2) but markedly reduced grain Cd. Scenario E, the combination of the above three measures, would decrease Cd concentration both in paddy soil and grains, suggesting the most effective way to mitigate Cd contamination in paddy soil on a large scale.

14.4.2 Prediction of cadmium accumulation in rice by machine learning approaches: A case study in mid-Xiangjiang River basin, Hunan province, China

14.4.2.1 Description of the study area

The study was conducted in a major rice-growing area in the middle Xiangjiang River basin, Hunan province, China. The largest part of the agricultural soil in the region is acidic and contaminated with Cd (Wang et al., 2016). The study area is rich in nonferrous metal mineral resources, particularly iron and manganese ore, and the mining and smelting industries are the main sources of heavy metal contamination in a large area of paddy fields surrounding these industries. In addition, rapid urbanization in the area has also increased heavy metal contamination in soil. The main type of agricultural soil is paddy soil, belonging to purplish red sand shale weathered material and Quaternary red clay. The paddy soil is weakly acidic and has organic matter around 24.6 mg kg^{-1}.

14.4.2.2 Machine learning model establishment

14.4.2.2.1 Support database

The study area was divided into grids, and altogether 92 pairs of rice–soil samples were collected. Rice grain Cd and 19 soil properties and components, including soil pH, SOM, total soil Cd, Zn, Mn, Fe, Mg, Ca, P, and Se, and acetic-extractable Cd, Zn, Mn, Fe, Mg, Ca, P, Si, and B, were analyzed to form the support database. The 19 soil variables were also selected as the input features to establish the grain Cd prediction model by ML approach. The overall soil properties and components are shown in Table 14.3. The spatial distributions of soil total and available Cd, soil pH, and rice plant uptake factor of Cd (PUF_{Cd}) in the study area are shown in Fig. 14.3.

14.4.2.2.2 Machine learning model performance

DT and RF were used in this study to establish prediction models. Software R 4.2.0 was used to build DT and RF models and plot DT diagrams and to validate models by three-time 10-fold cross-validation methods, obtaining model hyperparameters, including coefficient of determination (R^2), RMSE, and MAE.

14.4.2.3 Identification of primary factors affecting rice cadmium

The rice Cd uptake is influenced by multiple factors simultaneously, most of which, as well as their interactions, have nonlinear effects. DT and RF used in this work screened out the primary factors impacting rice Cd accumulation. DT demonstrated that Si_A, Zn_T, and Ca_T in soil were among factors with the greatest effects on Cd_g (Fig. 14.4), whereas soil total Cd (Cd_T) and acetic acid-extractable Cd (Cd_A) were not identified as the primary factors. The average Cd_g was 0.48 mg kg^{-1} when soil Si_A exceeded 88.5 mg kg^{-1}, significantly lower than that in soils with less Si_A. The increase in soil Zn_T ($\geq$162 mg kg^{-1}) with low Si_A would increase Cd_g, which may be because the increase in Cd always accompanies the increase of Zn in soil and the increase of soil Zn does not inhibit the uptake of Cd by rice under Zn-sufficient conditions. Contrary to the negative correlation between Cd_g and soil Ca_T in this study, Cd_g was high in soil with Ca_T higher than 2680 mg kg^{-1}. In all, soil with low Si_A and high Zn_T will lead to high Cd_g.

RF ranked the importance of factors by comprehensively considering linear and nonlinear relationships. The order of importance for factors is the following: $Si_A > Ca_T > Ca_A > Zn_T > Cd_g$ (Fig. 14.5). In the correlation analysis, soil pH was significantly correlated with rice grain Cd, while it became less significant by RF, which might be because soil pH is cocorrelated with soil Ca and Si_A. Feature analysis in RF also suggested that soil nutrient elements such as Si, Ca, and Fe rather than soil pH were the primary soil factors impacting PUF_{Cd}. It could be concluded in our study that the complicated nonlinear interaction among environmental factors under field conditions might result in divergent impacts on uptake and accumulation of Cd in rice, which might be missed in pot experiments and by MLR models.

14.4.2.4 Prediction of rice cadmium by established machine learning model

The prediction models for Cd_g constructed using DT, RF, and MLR based on the 19 soil variables were all significant ($P < .001$). The performance of the three models was compared by cross-validation (Table 14.4). The R^2s of the models ranged as follows: RF (0.543) > DT (0.464) > MLR (0.323). Moreover, RF had lower RMSE and MAE than DT and MLR, suggesting the priority of RF for the prediction in our study. The distribution of prediction errors can help to improve the model performance. As shown in Fig. 14.6 the deviations in rice grain Cd between the predicted and field measured were dependent on soil Cd to some degree. For example, when soil Cd was less than 1.08 mg kg^{-1}, the RF had much less deviations between the predicted and measured, and MLR greatly underestimated rice Cd. When soil Cd exceeded 1.08 mg kg^{-1}, the deviations by MLR and DT decreased, though RF was still better than the former two.

TABLE 14.3 Soil properties and concentrations of elements in paddy soil in the study area.

	SOM (g kg^{-1})	pH	Total concentration (mg kg^{-1})							Acetic acid-extractable concentration (mg kg^{-1})								
			Cd	Zn	Mg	Mn	Fe	Ca	P	Cd	Zn	Mg	Mn	Fe	Ca	P	Si	B
Mean	24.6	6.02	1.21	172	2665	305	33,527	5181	719	0.45	15.2	153	89.7	9.80	2370	19.9	142	0.40
SD.	1.71	0.82	0.98	71.2	1735	226	9688	6199	231	0.44	16.2	99.5	68.5	6.24	3647	17.5	100	0.26
Min.	18.5	4.83	0.31	70.4	831	86.9	12,663	135	80.3	0.03	2.15	18.6	13.2	1.44	150	3.64	37.0	0.09
Max.	28.1	8.03	6.19	473	10.0	1397	59,125	51,000	1490	3.07	96.5	684	359	37.9	31,168	115	636	1.39
CV	0.07	0.14	0.81	0.41	0.65	0.74	0.26	1.20	0.32	0.97	1.06	0.65	0.76	0.64	1.54	0.08	0.70	0.66

Cd, cadmium.
From Li et al. (2024).

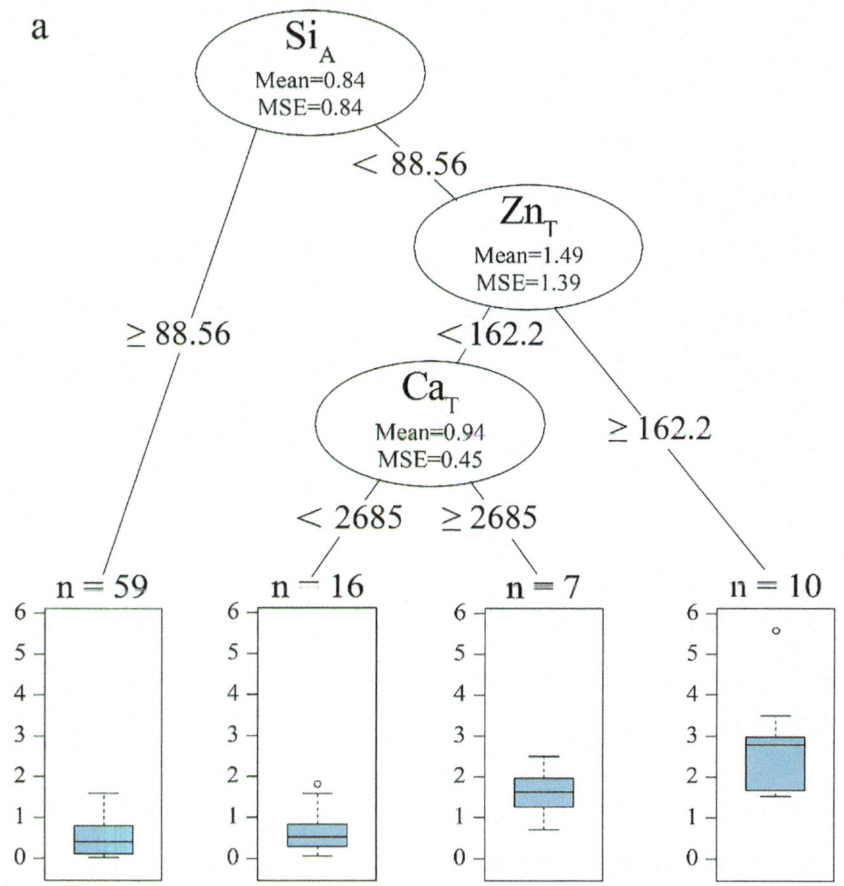

FIGURE 14.4 Decision trees for rice Cd. The concentrations of elements in soil are shown in the units of mg kg^{-1}. *Cd*, Cadmium. *From Li et al. (2024) with permission.*

14.5 Future prospects

Once pollutants are released into a medium, they can migrate and exchange with other adjacent media (air, water, soil, and sediment). To estimate the impact of pollutant release on human health a variety of multimedia models have been developed to simulate pollutants' fate and transport to perform risk analysis. Despite the advantage of multimedia environmental models in predicting the fate of pollutants at a regional scale, as far as heavy metals are concerned, mass-balance-based single medium models (e.g., soil models) are far more operable. The coupling of transport models such as HYDRUS-1D and SWAT improves the mechanistic explanation and expands from a plot scale to a regional scale.

MLR models are widely applied to predict crop Cd accumulation, particularly for pot experiments. However, large variation in environmental conditions and complicated interactions among factors, including soil properties, crop cultures, and local climate, limit the application of

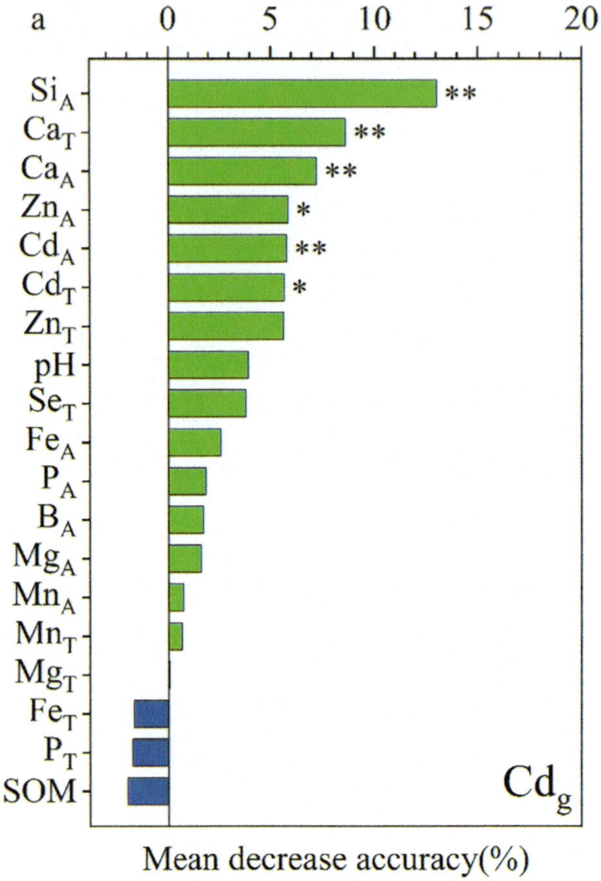

FIGURE 14.5 Feature importance of the environmental factors in the RF model for rice Cd. The subscripts T and A indicate the total and available concentration of the element in soil, respectively. "*" and "**" refer to significance at the 0.05 and 0.01 levels. *Cd*, Cadmium; *RF*, random forest. *From Li et al. (2024) with permission.*

TABLE 14.4 Performance of different prediction models for rice Cd (Cd_g) and plant uptake factor (PUF_{Cd}).

	Algorithm	R^2	RMSE	MAE
Cd_g	MLR	0.323	0.863	0.537
	DT	0.464	0.720	0.531
	RF	0.543	0.644	0.470
PUF_{Cd}	MLR	0.359	0.729	0.514
	DT	0.482	0.637	0.506
	RF	0.531	0.601	0.448

MLR, DT, and RF denote multiple linear regression, decision tree, and random forest, respectively. R^2, RMSE, and MAE dsenote the coefficient of determination, root mean square errors, and mean absolute error, respectively.

Cd, cadmium.
From Li et al. (2024).

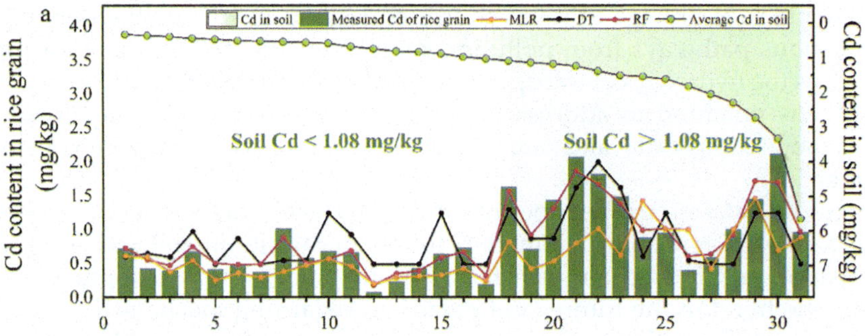

FIGURE 14.6 Comparison of predicted and measured mean values of rice Cd by different models. The unit for soil and rice Cd is mg kg^{-1}. *Cd*, Cadmium. *From Li et al. (2024) with permission.*

MLR in the field on a regional scale. With the advances of computer technology and the increasing availability of data, ML approaches are increasingly used to predict crop Cd accumulation based on environmental features.

However, shortcomings exist for current models. These limitations are as follows:

1. Steady-state empirical equations are still the primary model algorithms of current prediction models both for soil and crop Cd accumulation. Therefore the lumped parameters estimated through model calibration are site-specific. Achieving adequate quality assurance in modeling requires a battery of tests to establish the model's validity, trustworthiness, and relevance in performing a prediction.
2. The dark-box property of the mass-balance-based models for soil Cd prediction and regression-based empirical models for crop Cd prediction poses difficulty in analysis of uncertainty. The accuracy of predicted outcomes strongly relies on the accuracy of the data input. The accuracy of model prediction and the uncertainty analysis are highly dependent on the judgment and experience of the model user. Travis et al. (2001) found that even though all modelers modeled the same overall scenarios, there was considerable difference in model predictions. The most common reason for differences was differences in interpretation and choice of input parameters.
3. The complexity of model execution and the requirement that the user be able to implement the modeling system are ubiquitous challenges for the existing composed models, for example, coupling the process-based model SWAT with mass-balance-based soil model.

Thus more efforts are needed to improve the simulation capability of the current existing models in the following areas:

- Mechanism-based ML will be a promising approach for prediction of crop Cd in various soil environments. ML approaches have strong capability of data processing, but their feature importance is completely dependent on the input features. Therefore input feature screening based on scientific knowledge will be of significance for the subsequent explanation of feature importance. That is, model algorithms and parameterization to increase the fidelity of the model's descriptions of the real environment
- Process-based model coupling with mass-balanced models is the future trend of soil heavy metal accumulation models. The biggest limitation of single medium models based on mass

balance is the lack of linkage to point source emissions as well as the quantitative contribution of various pathways from pollution sources to soil. The coupling of process-based models such as the hydrological transport model SWAT and the atmospheric transport model CALPULFF has potential to address the whole source–pathway–sink process. However, balancing model complexity and simplified input parameters must be considered in the model design.

- Development of datasets to describe intermedia transfer coefficients in a variety of environmental scenarios as well as submodels for chemical fate in soil and sediments and for bioaccumulation and food chain transfer. As noted above the ML approach based on big data will be the trend in the future, and databases regarding mechanistic knowledge of Cd transport and transformation in environmental media are required. Nevertheless, with the ever-increasing scientific and technological advances in recent decades, substantial studies regarding the mechanisms of Cd transformation and transportation in the soil–plant system as well as source–pathway–sink relationships have been conducted, which greatly promotes the potential to establish comprehensive databases.

For future research on models predicting Cd transfer from soil to crops the limitations of modeling development are not avoidable because of the complex nature of chemical behavior and of environmental systems and also because of the current limited scientific knowledge. However, existing models or tools and current computing technology have allowed the development of models with an acceptable degree of complexity, reliability, and uncertainty to predict Cd transfer from soil to crops.

14.6 Summary

- Environmental fate models based on mass balance are the primary tools to predict Cd accumulation in soil both at a plot scale and a regional scale.
- Single-media coupled process-based models are widely applied to predict Cd accumulation in soil with the advantage of describing the vertical and spatial transport of Cd with hydrological processes.
- Multilinear regressions based on the Freundlich-type soil–plant transfer relationship have been applied despite their limitation in application scale.
- A comprehensive database regarding the mechanisms of Cd transformation and transportation in the soil–plant system as well as source–pathway–sink relationships is the prerequisite for the wide application of ML approaches.

References

Adams, M. L., Zhao, F. J., McGrath, S. P., Nicholson, F. A., & Chambers, B. J. (2004). Predicting cadmium concentrations in wheat and barley grain using soil properties. *Journal of Environmental Quality, 33*(2), 532–541. doi:10.2134/jeq2004.5320.

Allen, D. T., Cohen, Y., & Kaplan, I. R. (1989). *Intermedia pollutant transport: Modelling and field measurements*. Plenum Publisher.

Bi, Y., Gao, C., Wang, Y., & Li, B. (2009). Estimation of straw resources in China. *Transactions of the Chinese Society of Agricultural Engineering, 25*, 211–217. doi:10.3969/j.issn.1002-6819.2009.12.037.

Bonten, L. T. C., Groenenberg, J. E., Meesenburg, H., & de Vries, W. (2011). Using advanced surface complexation models for modelling soil chemistry under forests: Solling forest, Germany. *Environmental Pollution, 159*(10), 2831–2839. doi:10.1016/j.envpol.2011.05.002.

Chen, W., Chang, A. C., & Wu, L. (2007). Assessing long-term environmental risks of trace elements in phosphate fertilizers. *Ecotoxicology and Environmental Safety, 67*(1), 48–58. doi:10.1016/j.ecoenv.2006.12.013.

Chen, W., Wu, L., Chang, A. C., & Hou, Z. (2009). Assessing the effect of long-term crop cultivation on distribution of Cd in the root zone. *Ecological Modelling, 220*(15), 1836–1843. doi:10.1016/j.ecolmodel.2009.04.036.

Cohen, Y., & Cooter, E. J. (2002). Multimedia environmental distribution of toxics (Mend-Tox). I: Hybrid compartmental-spatial modeling framework. *Practice Periodical of Hazardous, Toxic, and Radioactive Waste Management, 6*(2), 70–86. doi:10.1061/(asce)1090-025x(2002)6:2(70).

Cowan, C. E., Mackay, D., Feijtel, T. C. J., Van De Meent, D., & Di Guardo, A. (1995). *Multi-media fate model: A vital tool for predicting the fate of chemicals.* Society of Environmental Toxicology and Chemistry.

Davies, B. E. (1992). Inter-relationships between soil properties and the uptake of cadmium, copper, lead and zinc from contaminated soils by radish (*Raphanus sativus* L.). *Water, Air, & Soil Pollution, 63*(3–4), 331–342. doi:10.1007/bf00475500.

de Vries, W., & McLaughlin, M. J. (2013). Modeling the cadmium balance in Australian agricultural systems in view of potential impacts on food and water quality. *Science of the Total Environment, 461–462*, 240–257. doi:10.1016/j.scitotenv.2013.04.069.

Garrels, R. M., Mackenzie, F. T., & Hunt, C. (1975). *Chemical cycles and the global environment: Assessing human influences.* Kaufmann.

Hou, Q., Yang, Z., Ji, J., Yu, T., Chen, G., Li, J., Xia, X., Zhang, M., & Yuan, X. (2014). Annual net input fluxes of heavy metals of the agro-ecosystem in the Yangtze River delta, China. *Journal of Geochemical Exploration, 139*, 68–84. doi:10.1016/j.gexplo.2013.08.007.

Hsieh, C. R., & Ouimette, J. R. (1994). Comparative study of multimedia modeling for dynamic partitioning of fossil fuels-related pollutants. *Journal of Hazardous Materials, 37*(3), 489–505. doi:10.1016/0304-3894(93)e0098-m.

Hu, Y., Zhang, P., Yang, M., Liu, Y., Zhang, X., Feng, S., Guo, D., & Dang, X. (2020). Biochar is an effective amendment to remediate Cd-contaminated soils—A meta-analysis. *Journal of Soils and Sediments, 20*(11), 3884–3895. doi:10.1007/s11368-020-02726-9.

Huang, B. Y., Lü, Q. X., Tang, Z. X., Tang, Z., Chen, H. P., Yang, X. P., Zhao, F. J., & Wang, P. (2024). Machine learning methods to predict cadmium (Cd) concentration in rice grain and support soil management at a regional scale. *Fundamental Research, 4*(5), 1196–1205. https://www.sciencedirect.com/science/journal/26673258. doi:10.1016/j.fmre.2023.02.016.

Jacques, D., Šimůnek, J., Mallants, D., & Genuchten, M. T. (2008). Modeling coupled hydrologic and chemical processes: Long-term Uranium transport following phosphorus fertilization. *Vadose Zone Journal, 7*(2), 698–711. doi:10.2136/vzj2007.0084.

Ji X., 2007. Studies on Regulation of Nutrient Loss in Typical Paddy Soils and Simulation of Watershed in Dongting Lake Area. Hunan Agricultural University. Doctoral Dissertation (in Chinese).

Jopony, M., & Young, S. D. (1994). The solid solution equilibria of lead and cadmium in polluted soils. *European Journal of Soil Science, 45*(1), 59–70. doi:10.1111/j.1365-2389.1994.tb00487.x.

Keller, A., Von Steiger, B., Van der Zee, S. E. A. T. M., & Schulin, R. (2001). A stochastic empirical model for regional heavy-metal balances in agroecosystems. *Journal of Environmental Quality, 30*(6), 1976–1989. doi:10.2134/jeq2001.1976.

Krauss, M., Wilcke, W., Kobza, J., & Zech, W. (2002). Predicting heavy metal transfer from soil to plant: Potential use of Freundlich-type functions. *Journal of Plant Nutrition and Soil Science, 165*(1), 3. doi:10.1002/1522-2624(200202)165:1⟨3::Aid-jpln3⟩3.0.co;2-b.

Li, A., Kong, L., Peng, C., Feng, W., Zhang, Y., & Guo, Z. (2024). Predicting Cd accumulation in rice and identifying nonlinear effects of soil nutrient elements based on machine learning methods. *Science of the Total Environment, 912*, 168721. doi:10.1016/j.scitotenv.2023.168721.

Li, R., Zhang, R., Yang, Y., & Li, Y. (2023). Accumulation characteristics, driving factors, and model prediction of cadmium in soil-highland barley system on the Tibetan Plateau. *Journal of Hazardous Materials, 453*, 131407. doi:10.1016/j.jhazmat.2023.131407.

Ma, X., Yu, T., Guan, D.-X., Li, C., Li, B., Liu, X., Lin, K., Li, X., Wang, L., & Yang, Z. (2023). Prediction of cadmium contents in rice grains from Quaternary sediment-distributed farmland using field investigations and machine learning. *Science of the Total Environment, 898*, 165482. doi:10.1016/j.scitotenv.2023.165482.

Mackay, D. (1979). Finding fugacity feasible. *Environmental Science & Technology, 13*(10), 1218–1223. doi:10.1021/es60158a003.

Mackay, D. (1991). *Multimedia environmental models: The fugacity approach.* Lewis Publishers.

Meng, Y., Zhou, L., He, S., Lu, C., Wu, G., Ye, W., & Ji, P. (2018). A heavy metal module coupled with the SWAT model and its preliminary application in a mine-impacted watershed in China. *Science of the Total Environment, 613–614*, 1207–1219. doi:10.1016/j.scitotenv.2017.09.179.

Mo, Z., Feng, L., Zou, H., Wang, J., Huang, W., & Yang, X. (2011). Validation and adaptability evaluation of rice growth model ORYZA2000 in double cropping rice area of Hunan province. *Shengtai Xuebao/Acta Ecologica Sinica, 31*(16), 4628–4637. http://www.ecologica.cn/stxb/ch/reader/create_pdf.aspx?file_no=stxb201008271237&year_id=2011&quarter_id=16&falg=1.

Nair, S. S., DeRolph, C., Peterson, M. J., McManamay, R. A., & Mathews, T. (2022). Integrated watershed process model for evaluating mercury sources, transport, and future remediation scenarios in an industrially contaminated site. *Journal of Hazardous Materials, 423*. doi:10.1016/j.jhazmat.2021.127049.

Nie, J., Zheng, S. X., Liao, Y. L., Xie, J., Wu, X. D., & Xiang, Y. W. (2010). Physical properties of paddy soils with different productivity in double-rice cropping region of Hunan Province. *Chinese Journal of Applied Ecology, 21*(11), 2777–2784.

Novotná, M., Mikeš, O., & Komprdová, K. (2015). Development and comparison of regression models for the uptake of metals into various field crops. *Environmental Pollution, 207*, 357–364. doi:10.1016/j.envpol.2015.09.043.

Nriagu, J. O. (1979). Global inventory of natural and anthropogenic emissions of trace metals to the atmosphere. *Nature, 279*(5712), 409–411. doi:10.1038/279409a0.

Nriagu, J. Q. (Ed.). (1978). Biogeochemistry of lead in the environment (pp. 137–184). Elsevier.

Parker, D. R., & Pedler, J. F. (1997). Reevaluating the free-ion activity model of trace metal availability to higher plants. *Plant and Soil, 196*(2), 223–228. doi:10.1023/A:1004249923989.

Peng, C., Wang, M., & Chen, W. (2016). Modelling cadmium contamination in paddy soils under long-term remediation measures: Model development and stochastic simulations. *Environmental Pollution, 216*, 146–155. doi:10.1016/j.envpol.2016.05.038.

Qiao, P., Lei, M., Yang, S., Yang, J., Zhou, X., Dong, N., & Guo, G. (2019). Development of a model to simulate soil heavy metals lateral migration quantity based on SWAT in Huanjiang watershed, China. *Journal of Environmental Sciences, 77*, 115–129. doi:10.1016/j.jes.2018.06.020.

Sauvé, S., Hendershot, W., & Allen, H. E. (2000). Solid-solution partitioning of metals in contaminated soils: Dependence on pH, total metal burden, and organic matter. *Environmental Science and Technology, 34*(7), 1125–1131. doi:10.1021/es9907764.

Sun, Y., Zhang, X., Peng, H., Zhou, W., Jiang, A., Zhou, F., Wang, H., & Zhang, W. (2025). Development of a coupled model to simulate and assess arsenic contamination and impact factors in the Jinsha River Basin, China. *Journal of Environmental Sciences (China), 147*, 50–61. doi:10.1016/j.jes.2023.09.038.

Tipping, E., Rothwell, J. J., Shotbolt, L., & Lawlor, A. J. (2010). Dynamic modelling of atmospherically-deposited Ni, Cu, Zn, Cd and Pb in Pennine catchments (northern England). *Environmental Pollution, 158*(5), 1521–1529. doi:10.1016/j.envpol.2009.12.026.

Tipping, E., Lawlor, A. J., Lofts, S., & Shotbolt, L. (2006). Simulating the long-term chemistry of an upland UK catchment: Heavy metals. *Environmental Pollution, 141*(1), 139–150. doi:10.1016/j.envpol.2005.08.019.

Travis, O., & Regens, W. (2001). Limitations of multimedia models for use in environmental decision making. *Environmental Monitoring and Assessment, 71*(1), 51–60. doi:10.1023/A:1011691628387.

Wang, M., Chen, W., & Peng, C. (2016). Risk assessment of Cd polluted paddy soils in the industrial and township areas in Hunan, Southern China. *Chemosphere, 144*, 346–351. doi:10.1016/j.chemosphere.2015.09.001.

Wang, M.-E., Peng, C., & Chen, W.-P. (2015). Effects of rice cultivar and typical soil improvement measures on the uptake of Cd in rice grains. *Huanjing Kexue/Environmental Science, 36*(11), 4283–4290. doi:10.13227/j.hjkx.2015.11.045.

Wu, W., Song, X., Yu, Y., & Zou, G. (2001). Comparison of root distribution between different type rice. *Chinese Journal of Rice Science, 15*, 276–280.

Xia, X., Yang, Z., Cui, Y., Li, Y., Hou, Q., & Yu, T. (2014). Soil heavy metal concentrations and their typical input and output fluxes on the southern Song-nen Plain, Heilongjiang Province, China. *Journal of Geochemical Exploration, 139*, 85–96. doi:10.1016/j.gexplo.2013.06.008.

Zhang, S., Song, J., Gao, H., Zhang, Q., Lv, M. C., Wang, S., Liu, G., Pan, Y. Y., Christie, P., & Sun, W. (2016). Improving prediction of metal uptake by Chinese cabbage (*Brassica pekinensis* L.) based on a soil-plant stepwise analysis. *Science of the Total Environment, 569–570,* 1595–1605. Www.elsevier.com/locate/scitotenv. doi:10.1016/j.scitotenv.2016.07.007.

Zhao, B., Zhu, W., Hao, S., Hua, M., Liao, Q., Jing, Y., Liu, L., & Gu, X. (2023). Prediction heavy metals accumulation risk in rice using machine learning and mapping pollution risk. *Journal of Hazardous Materials, 448,* 130879.

Zheng, S., 2010. Studies on the Transformation and Transport of Heavy Metals in Typical Chinese Agricultural Soils. Zhejiang University. Doctoral Dissertation (in Chinese).

Zhou, L., Zhao, X., Teng, M., Wu, F., Meng, Y., Wu, Y., Byrne, P., & Abbaspour, K. C. (2023). Model-based evaluation of reduction strategies for point and nonpoint source Cd pollution in a large river system. *Journal of Hydrology, 622,* 129701. doi:10.1016/j.jhydrol.2023.129701.

Zhou, Y., Jia, Z., Wang, J., Chen, L., Zou, M., Li, Y., & Zhou, S. (2019). Heavy metal distribution, relationship and prediction in a wheat-rice rotation system. *Geoderma, 354,* 113886. doi:10.1016/j.geoderma.2019.113886.

CHAPTER

15

Dietary intake of cadmium and impacts on human health

Di Zhao, Fang-Jie Zhao

College of Resources and Environmental Sciences, Nanjing Agricultural University, Nanjing, China

15.1 Introduction

Cadmium (Cd), a ubiquitous toxic heavy metal with no known biological function, poses significant long-term risks to human health because of its environmental persistence and propensity for bioaccumulation through trophic transfer. Its exceptionally long biological half-life in human bodies (10–30 years) facilitates preferential accumulation in critical organs, particularly the kidneys and liver, even at background environmental exposure levels (Satarug et al., 2023; Shimbo et al., 2000). Although inhalation and dermal contact contribute to the total body burden of Cd, dietary intake constitutes the dominant exposure pathway, accounting for approximately 90% of total exposure in nonsmokers (Clemens et al., 2013). For public health protection, it is, therefore, important to understand Cd transfer through the food chain.

The pervasive infiltration of Cd into agricultural systems and the food chain stems primarily from anthropogenic activities. Key sources include mining and smelting operations, industrial emissions, and the widespread application of phosphate fertilizers, which often contain Cd as an impurity (see Chapters 4 and 5). These activities contaminate irrigation water and soils, leading to increased Cd accumulation by crops. The extent of accumulation exhibits substantial variation, which is influenced by factors including soil chemistry, crop genetics, and agricultural management practices (Duan et al., 2017; Shi et al., 2020b). Beyond cereal grains and tubers, certain seafoods (e.g., shellfish and crustaceans) and organ meats (e.g., liver and kidney) also represent significant dietary Cd sources because of biomagnification in aquatic and terrestrial ecosystems.

Chronic dietary Cd exposure is associated with multisystem toxicity. Historical high-dose exposure, exemplified by the Itai-itai disease epidemic in Japan (Kobayashi, 1978), provides stark evidence of its devastating potential. Critically, epidemiological studies increasingly demonstrate that even subthreshold, low-level chronic exposure of Cd elevates risks for renal dysfunction, osteoporosis, cardiovascular disease, and hormone-sensitive cancers such as breast

Cadmium Contamination in Soils and Plants: Mechanisms and Mitigation.
DOI: https://doi.org/10.1016/B978-0-443-34104-5.00014-1

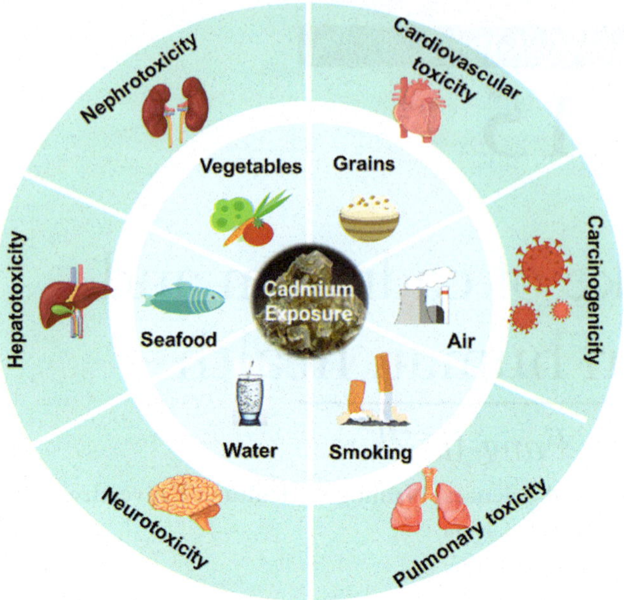

FIGURE 15.1 Cadmium exposure pathways (diet, smoking, and air) and associated health impacts.

and prostate cancer (Chen et al., 2016; Cirovic & Satarug, 2024; García-Esquinas et al., 2014). Individual susceptibility is modulated by factors including age, gender, and notably, nutritional status; for instance, iron deficiency enhances intestinal Cd absorption because of the upregulation of metal transporters (Reeves & Chaney, 2004).

This chapter provides a comprehensive synthesis of current knowledge regarding dietary Cd exposure. We systematically examine its environmental origins and pathways into the food system, evaluate factors governing Cd concentration, speciation, and bioavailability in major food commodities, and assess the consequent biological impacts on human health. Furthermore, we evaluate the efficacy of potential mitigation strategies, with specific focus on nutritional interventions (e.g., iron, zinc, and calcium supplementation) aimed at limiting Cd absorption and toxicity. Integrating evidence from environmental exposure science, nutrition, and epidemiology, this analysis highlights the need for robust, science-driven policies and targeted interventions to mitigate the global health burden associated with dietary Cd exposure.

15.2 Pathways of human exposure to cadmium

15.2.1 Dietary intake as the primary source

For the general population, dietary intake constitutes the predominant route of Cd exposure (Fig. 15.1), accounting for approximately 90% of the total exposure among nonsmokers (Clemens et al., 2013). This high contribution reflects the widespread contamination of food crops with Cd, which is readily absorbed by and accumulated in the edible parts of food crops (see Chapter 10). Cd enters the food chain mainly through crop uptake from soil, a process exacerbated by anthropogenic activities such as the application of Cd-contaminated phosphate and organic fertilizers, industrial discharges, mining and smelting operations, and the use of Cd-contaminated irrigation water (Shi et al., 2023).

Among staple crops, rice is of particular concern because of its relatively high capacity for Cd accumulation and bioavailability, especially in regions with Cd-contaminated soils (Zhao et al., 2017a). Wheat and various leafy vegetables and tubers also contribute significantly to dietary Cd intake (Shi et al., 2020a; Zhuang et al., 2023). In addition to plant-based foods, certain animal-derived products, especially seafoods (e.g., shellfish and crustaceans) and organ meats (e.g., liver and kidney), also serve as notable dietary Cd sources because of biomagnification through marine and terrestrial food chains (Schwarz et al., 2014). Collectively, these diverse dietary sources highlight the pervasive nature of Cd contamination in the human food supply.

15.2.2 Nondietary sources of cadmium exposure

Although dietary intake remains the primary source of Cd exposure for most individuals, tobacco smoking represents a major secondary pathway among smokers and those exposed to second-hand smoke (Domingo-Relloso et al., 2020) (Fig. 15.1). Tobacco plants have a pronounced capacity to accumulate Cd from soil, especially in areas with historical pollution or high fertilizer use (Siame et al., 2025). Following uptake, Cd is concentrated in tobacco leaves, where Cd concentration can reach 3–10 mg/kg even when grown in slightly contaminated soils (Clemens et al., 2013; Tang et al., 2017). Upon combustion, Cd volatilizes and is inhaled via smoke aerosols, leading to pulmonary deposition and systemic absorption. Epidemiological studies indicate that smokers have blood Cd levels approximately 4.3 times higher than nonsmokers and face a 1.5–2.0-fold increase in the risk of lung adenocarcinoma (International Agency for Research on Cancer, 2012).

Exposure through drinking water and ambient air generally contributes less than 10% of the total Cd intake in most populations (Olsson et al., 2002). The World Health Organization (WHO) has established a provisional guideline of 3 μg/L for Cd in drinking water, whereas China's national standard permits up to 5 μg/L. Assuming an average daily water intake of 2 L, consumption of water containing 3 μg/L Cd corresponds to an exposure of 6 μg/day—equivalent to the US Agency for Toxic Substances and Disease Registry's (ATSDR) provisional tolerable monthly intake of 3 μg/kg body weight (bw)/month for a 60 kg person. However, in areas with severely contaminated water sources, such as those near mining operations, industrial zones, or regions with inadequate environmental regulation, Cd exposure via drinking water may exceed safety thresholds and pose heightened risks of chronic toxicity.

Similarly, inhalation of airborne Cd can become a relevant exposure pathway in regions with poor air quality, especially near industrial facilities emitting Cd-containing particulates. A recent study in China reported that ambient Cd concentrations vary with particle size: fine particulate matter ($PM_{2.5}$) contains an average of 1.75 ng/m^3, whereas coarser particles (PM_{10} and total suspended particulate) contain 5.38 and 5.70 ng/m^3, respectively (Zhu et al., 2024). Given an average inhalation volume of 12 m^3/day the estimated Cd intake via inhalation for the general Chinese population is approximately 20.5 ng/day, which is minimal compared to dietary intake but potentially significant in industrialized areas with elevated atmospheric Cd levels.

Occupational exposure remains a major concern for workers in industries such as mining, smelting, battery manufacturing, and electronic waste recycling, where Cd is present in dust, fumes, and industrial byproducts (Byber et al., 2016). In recent decades, regulatory interventions, such as improved ventilation, use of personal protective equipment, and workplace exposure limits, have effectively reduced Cd exposure in regulated sectors. However, workers in informal or poorly regulated industries, particularly in developing countries, continue to face high exposure risks because of insufficient safety protocols (Zhang & Reynolds, 2019). These disparities underscore the urgent need for global occupational health standards and stronger enforcement measures to protect vulnerable worker populations.

15.3 Cadmium concentration, speciation, and bioavailability in food

15.3.1 Cadmium concentrations in food

Extensive global surveys have revealed substantial variation in Cd concentrations in rice samples (Table 15.1). A study covering 32 countries reported Cd concentrations in polished market rice ranging from <0.005 to 3.7 mg/kg, with a mean of 0.02 mg/kg (Shi et al., 2020b). Notably, they reported that 5% of samples exceeded 0.2 mg/kg, and 25% surpassed the more stringent 0.04 mg/kg threshold for processed baby foods (Shi et al., 2020b). International comparisons across 12 countries also revealed a broad variation, with Cd levels in polished rice ranging from <0.0005 to 1.31 mg/kg (Meharg et al., 2013). The variation is driven by factors such as soil contamination, agronomic practices, irrigation sources, and genetic differences between rice cultivars. For example, within a single region, Cd concentrations in rice may differ more than 5-fold between cultivars (Meharg et al., 2013), and a study of 471 rice cultivars grown under the same conditions reported a 10–32-fold variation (Duan et al., 2017), highlighting the potential for cultivar selection as a mitigation strategy.

The health risks associated with rice consumption are influenced not only by Cd concentrations but also by consumption patterns. In China, average daily rice intake is 219 g/day, varying from 116 g/day in northern China to 310 g/day in southern China. Consuming 219 g/day of rice at the regulatory limit of 0.2 mg/kg would result in a Cd intake of 43.8 μg/day, which is more than twice the (European Food Safety Authority, 2012) tolerable intake limit of 21.4 μg/day assuming a body weight of 60 kg (Fig. 15.2A). For typical rice consumers in northern and southern China eating rice at the regulatory Cd limit (0.2 mg/kg), estimated Cd exposures would amount to 23.2 and 62 μg/day, respectively, exceeding EFSA's limit by 1.1–2.8 times. Although 90%–97.8% of market basket rice samples in China complied with the 0.2 mg/kg Cd limit, surveys in contaminated areas of southern China revealed that 56%–87% of rice samples exceeded this threshold (Wang et al., 2019), highlighting a critical regional disparity that warrants urgent attention.

Wheat is another significant dietary Cd source, especially in regions where it is a staple. A meta-analysis of 138 studies in China (2000–22) reported an average wheat Cd concentration of 0.10 mg/kg (Wang et al., 2023), substantially higher than the 0.03 mg/kg reported in Swedish market basket surveys (Sand & Becker, 2012). Although wheat typically accumulates less Cd than rice, high consumption can still result in considerable exposure. For instance, with a wheat intake of 145 g/day and a mean Cd concentration of 0.02 mg/kg, dietary Cd intake in China is estimated at 2.9 μg/day (Song et al., 2017). However, in North China, where average wheat consumption is 248 g/day, consuming wheat with 0.1 mg/kg Cd would lead to 24.8 μg/day, exceeding the EFSA limit (Fig. 15.2B).

Potatoes are a major source of dietary Cd intake in Europe and America. For example, a large-scale survey in Serbia analyzed Cd concentrations in 11,227 food samples and found

TABLE 15.1 National surveys of Cd concentrations in rice grains across different countries.

			Rice Cd (mg/kg)			
Country	**Sampling site**	**Sample size *n***	**Range**	**Mean ± SD**	**Median**	**References**
Bangladesh	Market, field	260	<0.0005–1.31	0.099	0.057	Meharg et al. (2013)
	Market	144	0.001–0.18	0.044	–	Shahriar et al. (2020)
Brazil	Market	142	0.005–0.07	–	0.012	Shi et al. (2020b)
Cambodia	Market, field	14	0.001–0.03	0.006	0.001	Meharg et al. (2013)
China	Market	712	<0.001–0.74	0.05	–	Qian et al. (2010)
	Market, field	19,786	–	0.06 ± 0.13	–	Song et al. (2017)
	Market	160	<LOD–0.77	0.09	0.05	Chen et al. (2018)
	Market	495	0.005–3.71	–	0.069	Shi et al. (2020b)
France	Market, field	37	0.003–0.10	0.01	0.006	Meharg et al. (2013)
Ghana	Market, field	428	<0.005–0.27	0.02	0.013	Meharg et al. (2013)
	Market	426	0.005–0.27		0.013	Shi et al. (2020b)
India	Market, field	58	0.002–1.0	0.078	0.028	Meharg et al. (2013)
Indonesia	Market	80	0.005–0.60	–	0.021	Shi et al. (2020b)
Italy	Market, field	114	0.003–0.16	0.038	0.027	Meharg et al. (2013)
	Market	101	0–0.2	0.02–0.04	–	Sommella et al. (2013)
	Market	205	0.005–0.23	–	0.026	Shi et al. (2020b)
Ivory Coast	Market	88	0.011–0.09	–	0.022	Shi et al. (2020b)
Japan	Market	18	0.01–0.14	0.06	0.05	Meharg et al. (2013)
Nepal	Market, field	12	0.014–0.08	0.05	0.048	Meharg et al. (2013)
South Korea	Market	56	0.005–0.26	–	0.015	Shi et al. (2020b)
Spain	Market, field	92	0.0008–0.14	0.024	0.016	Meharg et al. (2013)
	Market	84	0.005–0.14	–	0.012	Shi et al. (2020b)
Sri Lanka	Market, field	75	<0.0005–0.8	0.081	0.024	Meharg et al. (2013)
	Market	137	0.005–0.37	–	0.028	Shi et al. (2020b)
Tanzania	Market	88	0.005–0.02	–	0.005	Shi et al. (2020b)
Thailand	Market, field	18	0.057–0.07	0.027	0.02	Meharg et al. (2013)
United States	Market, field	21	0.009–0.04	0.018	0.017	Meharg et al. (2013)
Vietnam	Market	107	0.005–0.23	–	0.019	Shi et al. (2020b)

Cd, cadmium.

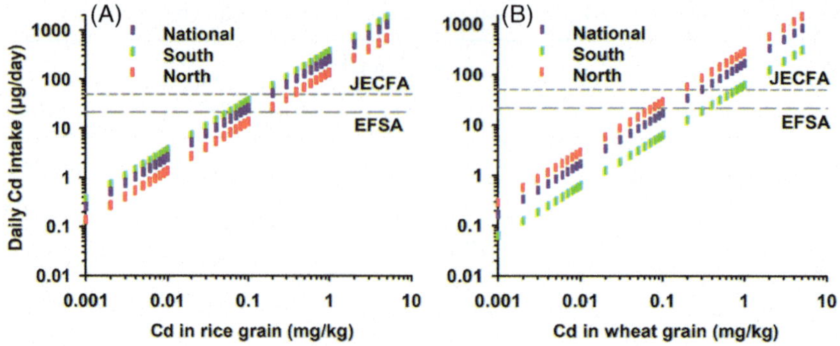

FIGURE 15.2 Daily dietary Cd intake for the Chinese population as functions of daily consumption rate and Cd concentrations in rice (A) and wheat (B) grains. Dashed *grey lines* are tolerable daily intake limits of 50 and 21.4 µg/day recommended by JECFA and EFSA assuming a body weight of 60 kg, respectively. *Cd*, Cadmium; *JECFA*, Joint FAO/WHO Expert Committee on Food Additives. *Food consumption data used in the calculation of daily Cd in take are from Song et al. (2017).*

that although the median Cd concentration in potatoes was 0.05 mg/kg, potato consumption contributed 41% of total Cd intake in adults, the highest among all food sources (Janković et al., 2023). Similarly, in Guizhou, China, Cd concentrations in 54 potato samples ranged from 0.023 to 0.18 mg/kg, with an average of 0.08 mg/kg (fresh weight) (Zhang et al., 2017). These findings highlight that despite relatively low Cd concentrations in potatoes, their high consumption rates make them a markable contributor to dietary Cd exposure in different regions.

Seafoods, particularly shellfish and mollusks, also contribute to Cd exposure. Concentrations of Cd vary widely by species and location, ranging from 0.01 to 4.8 mg/kg (fresh weight) (Joint FAO/WHO Expert Committee on Food Additives, 2011; Schwarz et al., 2014). Ocean acidification, driven by rising atmospheric CO_2, may further enhance Cd bioavailability in marine systems, potentially increasing Cd accumulation in seafood (Shi et al., 2016).

Offal products, especially liver and kidneys, are additional sources because of the tendency of Cd to accumulate in these organs. Following absorption, Cd binds to metallothionein and is transported to the kidneys, where it accumulates via glomerular filtration and tubular reabsorption (Zhao et al., 2024). A study of market pork kidney in China reported Cd concentrations ranging from 0.04 to 1.6 mg/kg (dry weight), with a mean of 0.39 mg/kg (Lin et al., 2024). In regions with high offal consumption, such products may significantly contribute to total Cd exposure, particularly in populations with low dietary diversity.

15.3.2 Cadmium speciation and distribution in rice

Beyond total Cd content, the chemical speciation and spatial distribution of Cd within rice grains are critical in determining its bioavailability and toxicity. Synchrotron-based K-edge X-ray absorption near-edge structure analysis of high Cd rice (3.51–11.4 mg/kg) has shown that most Cd in brown rice exists as Cd-thiolate complexes (66%–92%), with the remainder bound to carboxyl, histidine, or phytate ligands (Gu et al., 2020). Speciation of Cd differs between different grain tissues; Cd-carboxyl complexes (51%–68%) dominate in the husk, whereas Cd-thiolate species are predominant in the bran (53%–56%) and endosperm (80%–92%)

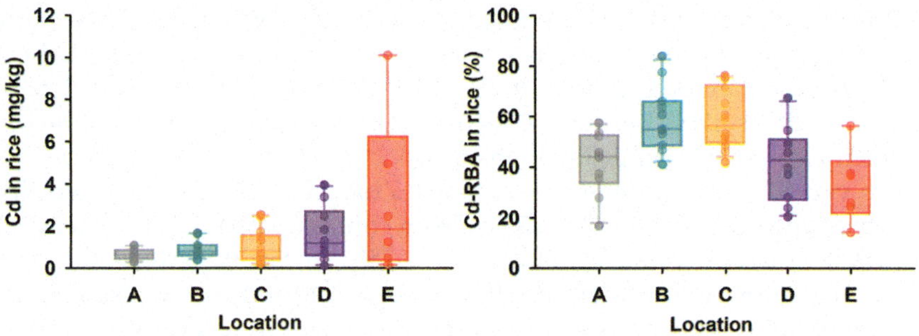

FIGURE 15.3 Cadmium concentration and relative bioavailability (RBA) in rice from different locations. *Data from Xiao et al. (2024); Yang et al. (2023); Zhao et al. (2017a, 2018); Zhuang et al. (2020).*

(Gu et al., 2020). However, accurate analysis of Cd speciation in rice with moderate levels of Cd remains analytically challenging because of insufficient analytical sensitivity.

Spatial distribution within rice grains also varies considerably. Micro-X-ray fluorescence imaging has identified two primary patterns: in some grains, Cd is evenly distributed; in others, it decreases from the aleurone layer and outer endosperm toward the inner endosperm (Gu et al., 2020). Conflicting reports exist, with some studies finding higher Cd in the endosperm than in the bran (Meharg et al., 2008), whereas others show the opposite (Wei et al., 2017).

The heterogeneity in Cd distribution has implications for rice milling, a common postharvest processing method. Although milling generally reduces Cd concentrations, on average by 23.5%, the effects are inconsistent, ranging from a 20.3% decrease to a 64.7% increase (Wang & Zhao, 2022). Other studies report more modest reductions of 12%–20% (Williams et al., 2009; Yao et al., 2020). The effectiveness of milling correlates with the Cd concentration in bran (Gu et al., 2020). In some cases, however, milling has been shown to increase Cd concentrations in polished rice by up to 20%, suggesting that assumptions of uniform Cd removal during milling may lead to underestimation of exposure risks (Gu et al., 2020; Meharg et al., 2013). Therefore regulatory limits for polished and brown rice should account for distributional variability to ensure accurate exposure assessments.

15.3.3 Bioavailability of cadmium in food

The toxicological impact of dietary Cd depends not only on total concentrations but also on its bioavailability. Only the fraction of ingested Cd that is absorbed into systemic circulation can contribute to toxic effects. Oral Cd bioavailability is typically expressed as relative bioavailability (RBA), defined as the proportion of Cd absorbed relative to a soluble reference compound (e.g., cadmium chloride) (Li et al., 2019; Zhao et al., 2025). Relative bioavailability is commonly assessed via in vivo animal models (e.g., mice or swine), which compare tissue Cd accumulation after consumption of contaminated food versus a reference standard (Xiao et al., 2024; Yang et al., 2023; Zhao et al., 2017a, 2018).

Substantial variability in rice Cd-RBA has been observed, depending on source, soil characteristics, and cultivar (Fig. 15.3). Zhao et al. (2017b) reported Cd-RBA values ranging from 17% to 57% in rice from an area impacted by Cd contamination from pottery, whereas higher values (41%–84%) were observed in rice from mining and smelting regions (Zhao et al., 2018).

Yang et al. (2023) similarly reported Cd-RBA values of 42%–76% in rice from mining-affected regions. Interestingly an inverse relationship has been noted between total Cd concentration and RBA in highly contaminated samples (Cd = 0.15–10.1 mg/kg), with RBA declining to 15%–56% (Zhuang et al., 2020), likely because of increased Cd binding to low-solubility ligands such as phytate, which reduces gastrointestinal absorption.

Cd-RBA also varies among different food types. Zhao et al. (2017b) reported RBA values of 37%–68% in wheat and 18%–78% in vegetables from contaminated regions. Wei et al. (2021) observed comparable variability in vegetables (24%–75%, mean = 52%) near mining areas in Guangxi, China. These findings underscore the necessity of food-specific bioavailability assessments, as relying on total Cd concentrations alone may result in risk misestimation.

Incorporating RBA into toxicokinetic modeling improves exposure estimates. A study in a Cd-contaminated area showed that using RBA-adjusted intake values in a one-compartment model accurately predicted urinary Cd concentrations (1.20 μg/g creatinine) compared to measured urinary Cd (1.07 μg/g creatinine) (Zhao et al., 2017a). These results suggest the need to include Cd bioavailability data in regulatory frameworks to improve exposure modeling and guide public health interventions.

15.4 Current status of dietary cadmium intake

15.4.1 Dietary cadmium intakes across countries

Given that dietary intake is the predominant route of Cd exposure for the general population, extensive studies have been conducted worldwide to assess intake levels and associated health risks across different regions. As shown in Fig. 15.4, dietary Cd intake exhibits considerable variability among different countries, ranging from 4.6 to 34.6 μg/day. These disparities reflect variation in dietary habits, agricultural practices, environmental contamination, and food safety regulations.

The United States reports the lowest average dietary Cd intake globally, at 4.6 μg/day, based on the Total Diet Studies (TDS) conducted by the US Food and Drug Administration between 2006 and 2013 (Kim et al., 2019). European countries, including Germany, Finland, Denmark, the United Kingdom, France, and Sweden, report moderate intake levels ranging from 12.9 to 19.1 μg/day, as estimated by the EFSA during the 2003–11 assessment period (European Food Safety Authority, 2012). These relatively low values likely reflect strict food safety regulations, diversified diets, and generally low concentrations of Cd in the staple food.

In contrast, populations in Asia typically experience higher dietary Cd exposure, with reported intake levels of between 14.5 and 34.6 μg/day. For example, South Korea's national nutrition survey (2001–03) estimated a mean intake of 14.5 μg/day (Kim & Wolt, 2011), whereas Chinese adult males consumed approximately 17.3 μg/day according to the sixth TDS (2016–19) (Zhao et al., 2022). Bangladesh reported the highest global intake at 34.6 μg/day partly because of high rice consumption (Al-Rmalli et al., 2012).

When compared to tolerable intake thresholds of 21.4 μg/day (EFSA) and 50 μg/day [Joint FAO/WHO Expert Committee on Food Additives (JECFA)], most countries remain below the limits. However, dietary Cd exposure in some Asian countries, particularly Japan and Bangladesh, exceeds EFSA's threshold. Geographic variation in Cd intake is related to dietary composition. In the United States, cereals and bread (34%), leafy vegetables (20%), and potatoes

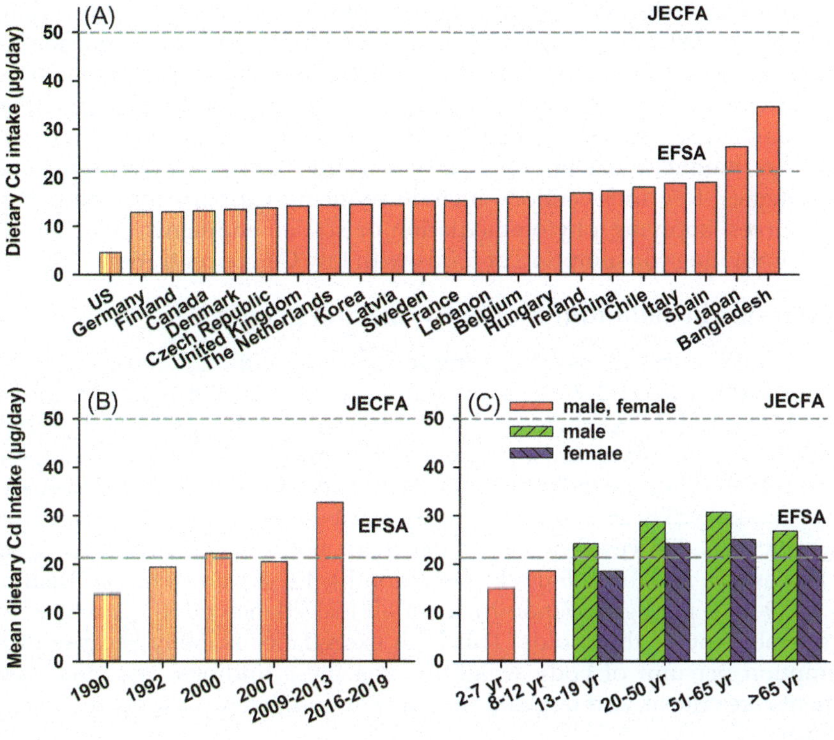

FIGURE 15.4 Dietary Cd intake levels in the general populations in different countries (A), temporal changes in dietary Cd intake for the general population in China (B), and dietary Cd intake by age groups in China (C). For countries with unspecified body weight, a default value of 60 kg was assumed. *Cd*, Cadmium. *Data from Zhao et al. (2022, 2023).*

(11%) are the main contributors (Kim et al., 2019); in Europe, cereals (27%), vegetables (17%), and tubers (12%) dominate (European Food Safety Authority, 2012); whereas in Asia, rice alone accounts for 49%–65% of the total Cd intake (Itoh et al., 2014; Zhao et al., 2022). The generally higher Cd intakes in Asian populations than in other populations are likely a result of higher per capita rice consumption.

15.4.2 Temporal trends in dietary cadmium intake

Temporal trends in dietary Cd intake reveal divergent patterns between developed and developing countries. In developed nations, regulatory interventions and agricultural improvements have significantly reduced dietary Cd exposure. For example, in the United States, women's mean Cd intake declined by 55% from 10.4 μg/day during 1991–96 to 4.7 μg/day during 2006–13 (Quraishi et al., 2016). This decline reflects state-level bans on Cd-containing fertilizers, which have helped limit Cd accumulation in crops (Quraishi et al., 2016).

Conversely, many developing countries have experienced rising Cd intake because of rapid industrialization, inadequate environmental regulation, and urban agriculture expansion (Moon et al., 2012; Song et al., 2017). In China, average dietary Cd intake increased from

13.8 μg/day in the early 1990s to 32.7 μg/day by the fifth TDS (2009–13), before declining by 47% to 17.3 μg/day in the sixth TDS (2016–19) (Zhao et al., 2022). This recent decline likely reflects China's strengthened pollution control measures, including tighter implementation of food Cd regulations and exclusion of highly contaminated soils from food production. Nevertheless, hotspots of contamination persist, particularly near mining and smelting regions.

These findings highlight the importance of sustained dietary monitoring and proactive mitigation strategies. Key interventions include reducing Cd emissions from industrial and agricultural sources, applying phytoremediation or soil amendments, selecting low-Cd crop cultivars, and promoting consumption of foods with low Cd concentrations.

15.4.3 Variations in cadmium intake by age and gender

Dietary Cd intake varies across demographic groups, with age and gender as significant determinants. Children aged 2–7 years have the lowest total Cd intake (mean: 15 μg/day) according to China's fifth TDS, whereas adolescents and adults (ages 8–65) show higher intakes (18.6–30.8 μg/day), primarily because of greater consumption of Cd-containing foods such as cereals and vegetables (European Food Safety Authority, 2012). Elderly individuals (65+) have slightly lower intakes (23.8–26.9 μg/day), likely attributable to reduced energy requirements or dietary restrictions. However, when adjusted for body weight, children aged 2–7 years exhibit the highest Cd intake (22.6 μg/kg bw/month), 1.6–2.1 times higher than that of other age groups (10.9–14.1 μg/kg bw/month), with an exceedance of 30.2% when comparing to the JECFA tolerable intake thresholds. This high exceedance is likely because of their higher food consumption per unit of body weight. Given these findings, age- and weight-specific risk assessments are critical, and targeted measures to mitigate Cd exposure in young children warrant priority.

Gender-related differences are also noteworthy. Males (11.2–13.4 μg/kg bw/month) generally ingest more Cd than females (10.9–12.7 μg/kg bw/month) based on China's fifth TDS, consistent with their higher food intake (Vahter et al., 2007). However, females, especially those with low iron stores, are more vulnerable to Cd toxicity. Iron deficiency upregulates intestinal divalent metal transporter 1 (DMT1), enhancing Cd absorption (Vahter et al., 2007). This mechanism contributed to the high incidence of Itai-itai disease among postmenopausal Japanese women, particularly those with multiple pregnancies and iron depletion, who developed severe Cd-induced osteomalacia and renal dysfunction (Kobayashi, 1978).

The above demographic patterns introduce uncertainty into risk assessments and underscore the need for targeted interventions. Recommended strategies include iron and calcium supplementation for at-risk groups, promotion of alternative low-Cd staple crops (e.g., oats and quinoa), and regular biomonitoring of vulnerable populations such as children and pregnant women.

15.5 Adverse health effects caused by dietary cadmium intake

15.5.1 Acute and chronic health effects of cadmium exposure

Cd exposure can pose serious health risks, with outcomes dependent on both the dose and duration of exposure. Acute Cd poisoning is rare in the general population but can occur in occupational settings. Inhalation of Cd-containing fumes or dust, common in industries such as battery manufacturing and metal smelting, can result in respiratory distress, including throat irritation, coughing, chemical pneumonitis, and pulmonary edema (Joint FAO/WHO Expert Committee on Food Additives, 2011). Acute ingestion of highly Cd-contaminated food or water causes gastrointestinal symptoms such as nausea, vomiting, abdominal pain, and diarrhea, typically within 15–30 minutes of exposure (Godt et al., 2006).

Chronic low-level Cd exposure is far more prevalent and poses a significant public health threat. Because of Cd's long biological half-life (10–30 years) in human bodies, even small daily intake can lead to cumulative accumulation, particularly in the kidneys and liver (Nordberg et al., 2007; Satarug et al., 2023). The kidneys are the principal target organs, where Cd induces proximal tubular dysfunction, as evidenced by increased urinary excretion of β2-microglobulin and *N*-acetyl-β-D-glucosaminidase. With continued exposure, glomerular damage may ensue, ultimately contributing to chronic kidney disease (Satarug et al., 2023). In addition, epidemiological studies link long-term Cd exposure to osteoporosis, hypertension, and neurodevelopmental disorders (Cirovic & Satarug, 2024; Li et al., 2021).

The most severe consequence of chronic Cd toxicity is Itai-itai disease, first documented in the Jinzu River basin, Toyama Prefecture, Japan, in the early 20th century. Between 1910 and 1950, irrigation of rice paddies with Cd-contaminated wastewater from upstream lead and zinc mining operations led to soil Cd concentrations reaching up to 8 mg/kg (Nogawa & Kido, 1993). A survey of 2446 rice samples collected from 88 hamlets in the Jinzu River basin from 1971 to 1976 showed an average Cd concentration of 0.38 mg/kg in rice grain, ranging from 0.02 to 0.95 mg/kg (Nogawa et al., 2017). Studies estimated that the lifetime Cd intake for the onset of mild and severe Itai-itai disease for half of the affected patients was 3.1 and 3.8 g, respectively (Inaba et al., 2005). Over time, those affected individuals developed debilitating symptoms, including severe bone and joint pain, pathological fractures, and kidney dysfunction. Between 1967 and 1993, 165 confirmed cases and 239 suspected cases were reported, predominantly in postmenopausal women, who are more vulnerable because of iron deficiency (Nogawa et al., 2017). The emergence of Itai-itai disease marked a pivotal moment in environmental toxicology and public health, highlighting the severe long-term consequences of dietary Cd exposure. This tragedy underscored the urgent need for stringent environmental regulations to prevent heavy metal contamination in agricultural systems.

15.5.2 Carcinogenic risks of low-dose environmental cadmium exposure

Cd is a well-established toxicant and a Group 1 human carcinogen, as classified by the International Agency for Research on Cancer (2012). Occupational studies have demonstrated a dose–response relationship between Cd exposure and cancer risk; individuals in the highest exposure categories exhibit up to a 61% increased incidence of lung cancer relative to those with the lowest exposure (Chen et al., 2016).

Although existing intake limits aim to prevent renal dysfunction, such as the JECFA (Joint FAO/WHO Expert Committee on Food Additives, 2011) provisional tolerable intake (50 μg/day) and the EFSA limit (25 μg/kg bw/month), growing evidence suggests that even low-dose Cd exposure may increase cancer risk. For instance, a large prospective cohort of 55,987 postmenopausal women found that dietary Cd intake of 16 μg/day was associated with a 27% higher risk of breast cancer compared to 13 μg/day [rate ratio (RR) = 1.27] (Julin et al., 2012). Similarly, Itoh et al. (2014) identified a statistically significant link between Cd intake and estrogen receptor-positive breast cancer, with women consuming >31.5 μg/day facing nearly double the risk (RR = 1.94) compared to those consuming 21.4 μg/day.

Emerging evidence also links Cd exposure to other cancers (Nawrot et al., 2015; Satarug et al., 2010). A population-based case–control study in Italy found that each 1 μg/day increase in Cd intake was associated with 11% higher odds of melanoma (Filippini et al., 2019). These findings challenge the assumption that current regulatory thresholds sufficiently protect against carcinogenic risks.

Given Cd's ubiquity in food, tobacco smoke, and industrial emissions, and its cumulative and persistent nature, there is an urgent need to reevaluate safety thresholds. Future research should aim to elucidate the molecular mechanisms underlying Cd-induced carcinogenesis, particularly at environmentally relevant exposure levels. This will enable the development of more protective, evidence-based regulatory standards.

15.6 Enhancing nutritional status to mitigate human exposure to cadmium

Since the outbreak of Itai-itai disease in Japan, considerable research has sought to identify the factors that influence Cd absorption and toxicity in humans. Epidemiological studies show that populations dependent on rice-based subsistence diets tend to exhibit higher Cd body burdens and greater toxicity risks compared to those with more diversified diets (Reeves & Chaney, 2004, 2008). This is largely because of the inherently low concentrations of essential minerals, particularly iron (Fe) and zinc (Zn), in rice compared to other staple crops (Scientific Panel on Responsible Plant Nutrition, 2023). For instance, Fe and Zn concentrations in rice typically range from 3 to 10 mg/kg and from 5 to 15 mg/kg, respectively, whereas millet contains 30–60 mg/kg Fe and 25–50 mg/kg Zn, and beans contain 30–70 mg/kg Fe and 20–50 mg/kg Zn. Moreover, the common practice of polishing rice to produce white rice further depletes its mineral content, exacerbating micronutrient deficiencies and inadvertently increasing Cd absorption (Chen et al., 2022; Zhao et al., 2023).

The low bioavailability of Fe and Zn in rice, primarily because of the presence of phytates and other dietary inhibitors, further contributes to enhanced Cd uptake (Chaney, 2015; Jin et al., 2004). Iron and zinc deficiencies upregulate the expression of intestinal metal transporters such as DMT1 (responsible for Fe^{2+} uptake) and ZIP8 (involved in Zn^{2+} uptake) in the duodenum (Kim et al., 2007; Reeves & Chaney, 2004). Because Cd^{2+} shares similar ionic characteristics with these essential metals, the upregulation of these transporters also facilitates Cd absorption (Peng et al., 2023).

Experimental studies using animal models support this mechanism. In a study using ^{109}Cd tracers, rats fed a diet marginally deficient in Fe, Zn, and calcium (Ca) exhibited Cd concentrations in enterocytes and kidneys that were approximately 10 times higher than those of rats receiving adequate mineral nutrition (Reeves & Chaney, 2004). Further interventional research has shown that mineral fortification can substantially reduce Cd bioavailability. For example, a mouse study found that supplementing Cd-contaminated rice with 150–5000 mg/kg Ca reduced Cd absorption by 31%–80%, highlighting the protective effect of Ca against intestinal Cd uptake (Zhao et al., 2017a).

Interestingly, populations consuming large amounts of high-Cd seafoods (e.g., shellfish) do not always exhibit correspondingly elevated Cd body burdens. This may be attributed to the coingestion of Fe and Zn, which competitively inhibit Cd absorption. For instance, Vahter et al. (1996) reported that young women who consumed shellfish had a threefold higher dietary Cd intake compared to nonconsumers yet showed no significant difference in blood Cd levels. This was likely because of higher serum ferritin levels resulting from the higher Fe content of shellfish, which reduced Cd uptake.

These findings emphasize the importance of dietary diversification and micronutrient adequacy as practical strategies to reduce Cd toxicity. Increasing the intake of nutrient-dense foods such as dairy products, lean meats, legumes, nuts, and leafy vegetables can improve mineral status and limit Cd bioavailability. Public health initiatives that promote mineral supplementation, biofortified crops, or balanced diets in Cd-exposed regions could provide dual benefits, alleviating micronutrient deficiencies while simultaneously reducing Cd-related health risks.

15.7 Concluding remarks and future perspectives

Cd is a highly toxic and persistent heavy metal that accumulates in agricultural soils and is readily taken up by food crops, posing widespread and long-term risks to human health through chronic dietary exposure. The global burden of dietary Cd intake varies significantly by region, largely because of differences in dietary composition and Cd concentration in staple food. Populations in rice-dependent regions, particularly across Asia, generally have higher Cd exposure compared to those reliant on wheat- or maize-based diets. This disparity likely results from higher Cd concentration in rice produced in some regions of Asia, higher bioavailability of Cd in rice grains, and inherently lower concentrations of essential minerals such as Fe, Zn, and Ca, which further increase intestinal Cd absorption.

To accurately assess the health risks of dietary Cd exposure, it is essential to consider more than total Cd concentrations in food. Critical factors include chemical speciation of Cd, its spatial distribution within edible plant tissues, RBA, and the consumption levels of contaminated foods. Emerging evidence supports the potential of nutritional interventions, particularly increased intake of Fe, Zn, and Ca, to competitively inhibit Cd absorption and reduce its accumulation in target organs such as the kidneys and liver. These strategies are practical, cost-effective, and especially valuable for populations with limited dietary diversity or access to safer food sources.

Although the adverse effects of high-level Cd exposure, such as nephrotoxicity, hypertension, and cancer, are well established, the health impacts of chronic low-level Cd intake remain an important and unresolved area of research. There is a critical need to refine our understanding of dose–response relationships at environmentally relevant exposure levels, particularly concerning latency, cumulative burden, and potential subclinical effects. To address this gap, large-scale prospective cohort studies with repeated biomarker measurements are required. These studies should investigate the long-term effects of low-dose Cd exposure on renal function, bone health (e.g., osteoporosis), cardiovascular outcomes, and the risk of hormone-sensitive cancers.

In addition, dietary Cd absorption is influenced by various physiological and nutritional factors that deserve systematic study. Dietary intervention trials examining how different food compositions or micronutrient profiles affect Cd bioavailability could inform targeted mitigation strategies. Moreover, individual susceptibility to Cd toxicity likely depends on genetic and epigenetic factors. Identifying these determinants and developing biomarkers of vulnerability will enable more personalized and effective risk management approaches. By integrating insights from environmental exposure science, nutrition, and epidemiology, more comprehensive strategies can be developed to minimize dietary Cd exposure and protect public health.

15.8 Summary

- Dietary intake is the primary source of cadmium (Cd) exposure in the general population.
- Chronic low-level Cd exposure is epidemiologically associated with multiple cancers.
- Current regulatory thresholds for dietary Cd intake may not sufficiently protect against chronic exposure and long-term health risks.
- Chemical speciation and relative bioavailability of Cd in food, as well as the concentrations and bioavailability of the essential micronutrients Fe, Zn, and Ca, are important factors influencing Cd absorption and toxicity.

Acknowledgments

This work was funded by the National Natural Science Foundation of China (42107430) and Jiangsu Provincial Double-Innovation Doctor Program (JSSCBS20210294).

AI disclosure

During the preparation of this work the author(s) used ChatGPT to improve readability. After using this tool/service the author(s) reviewed and edited the content as needed and take(s) full responsibility for the content of the publication.

References

Al-Rmalli, S. W., Jenkins, R. O., & Haris, P. I. (2012). Dietary intake of cadmium from Bangladeshi foods. *Journal of Food Science, 77*(1). doi:10.1111/j.1750-3841.2011.02467.x.

Byber, K., Lison, D., Verougstraete, V., Dressel, H., & Hotz, P. (2016). Cadmium or cadmium compounds and chronic kidney disease in workers and the general population: A systematic review. *Critical Reviews in Toxicology, 46*(3), 191–240. doi:10.3109/10408444.2015.1076375.

Chaney, R. L. (2015). How does contamination of rice soils with Cd and Zn cause high incidence of human Cd disease in subsistence rice farmers. *Current Pollution Reports, 1*(1), 13–22. springer.com/environment/pollution+and+remediation/journal/40726. doi:10.1007/s40726-015-0002-4.

Chen, C., Xun, P., Nishijo, M., & He, K. (2016). Cadmium exposure and risk of lung cancer: A meta-analysis of cohort and case–control studies among general and occupational populations. *Journal of Exposure Science & Environmental Epidemiology, 26*(5), 437–444. doi:10.1038/jes.2016.6.

Chen, H., Tang, Z., Wang, P., & Zhao, F. J. (2018). Geographical variations of cadmium and arsenic concentrations and arsenic speciation in Chinese rice. *Environmental Pollution, 238*, 482–490. www.elsevier.com/inca/publications/store/4/0/5/8/5/6. doi:10.1016/j.envpol.2018.03.048.

Chen, X., Chen, F., Sun, S., Li, Y., Li, Y., Mo, H., Li, Z., & Zhuang, P. (2022). Effect of polishing on lead and cadmium bioavailability in rice and its health implications. *Foods, 11*(17), 2718. doi:10.3390/foods11172718.

Cirovic, A., & Satarug, S. (2024). Toxicity tolerance in the carcinogenesis of environmental cadmium. *International Journal of Molecular Sciences, 25*(3), 1851. doi:10.3390/ijms25031851.

Clemens, S., Aarts, M. G. M., Thomine, S., & Verbruggen, N. (2013). Plant science: The key to preventing slow cadmium poisoning. *Trends in Plant Science, 18*(2), 92–99. doi:10.1016/j.tplants.2012.08.003.

Domingo-Relloso, A., Riffo-Campos, A. L., Haack, K., Rentero-Garrido, P., Ladd-Acosta, C., Fallin, D. M., Tang, W. Y., Herreros-Martinez, M., Gonzalez, J. R., Bozack, A. K., Cole, S. A., Navas-Acien, A., & Tellez-Plaza, M. (2020). Cadmium, smoking, and human blood DNA methylation profiles in adults from the strong heart study. *Environmental Health Perspectives, 128*(6). https://ehp.niehs.nih.gov/doi/pdf/10.1289/EHP6345. doi:10.1289/EHP6345.

Duan, G., Shao, G., Tang, Z., Chen, H., Wang, B., Tang, Z., Yang, Y., Liu, Y., & Zhao, F.-J. (2017). Genotypic and environmental variations in grain cadmium and arsenic concentrations among a panel of high yielding rice cultivars. *Rice, 10*(1). doi:10.1186/s12284-017-0149-2.

European Food Safety Authority. (2012). Cadmium dietary exposure in the European population. *EFSA Journal, 10*(1), 2551.

Filippini, T., Malagoli, C., Wise, L. A., Malavolti, M., Pellacani, G., & Vinceti, M. (2019). Dietary cadmium intake and risk of cutaneous melanoma: An Italian population-based case-control study. *Journal of Trace Elements in Medicine and Biology*, *56*, 100–106. www.urbanfischer.de/journals/jtraceelm/trace.htm. doi:10.1016/j.jtemb.2019.08.002.

García-Esquinas, E., Pollan, M., Tellez-Plaza, M., Francesconi, K. A., Goessler, W., Guallar, E., Umans, J. G., Yeh, J., Best, L. G., & Navas-Acien, A. (2014). Cadmium exposure and cancer mortality in a prospective cohort: The strong heart study. *Environmental Health Perspectives*, *122*(4), 363–370. http://ehp.niehs.nih.gov/wp-content/uploads/122/4/ehp.1306587.pdf. doi:10.1289/ehp.1306587.

Godt, J., Scheidig, F., Grosse-Siestrup, C., Esche, V., Brandenburg, P., Reich, A., & Groneberg, D. A (2006). The toxicity of cadmium and resulting hazards for human health. *Journal of Occupational Medicine and Toxicology*, *1*(1). doi:10.1186/1745-6673-1-22.

Gu, Y., Wang, P., Zhang, S., Dai, J., Chen, H. P., Lombi, E., Howard, D. L., Van Der Ent, A., Zhao, F. J., & Kopittke, P. M. (2020). Chemical speciation and distribution of cadmium in rice grain and implications for bioavailability to humans. *Environmental Science and Technology*, *54*(19), 12072–12080. http://pubs.acs.org/journal/esthag. doi:10.1021/acs.est.0c03001.

Inaba, T., Kobayashi, E., Suwazono, Y., Uetani, M., Oishi, M., Nakagawa, H., & Nogawa, K. (2005). Estimation of cumulative cadmium intake causing Itai–itai disease. *Toxicology Letters*, *159*(2), 192–201. doi:10.1016/j.toxlet.2005.05.011.

International Agency for Research on Cancer, Cadmium and cadmium compounds. IARC monographs on the evaluation of carcinogenic risks to humans no. 100C (2012), 121–145, IARC.

Itoh, H., Iwasaki, M., Sawada, N., Takachi, R., Kasuga, Y., Yokoyama, S., Onuma, H., Nishimura, H., Kusama, R., Yokoyama, K., & Tsugane, S. (2014). Dietary cadmium intake and breast cancer risk in Japanese women: A case–control study. *International Journal of Hygiene and Environmental Health*, *217*(1), 70–77. doi:10.1016/j.ijheh.2013.03.010.

Janković, S., Stošić, M., Miljaković, E. A., Ćurčić, M., Đukić Ćosić, D., Buha Đorđević, A., Bulat, Z., & Antonijević, B. (2023). Cadmium dietary exposure assessment in the adult population and pre-school children in the Republic of Serbia. *Food Additives and Contaminants—Part A*, *40*(1), 67–80. http://www.tandfonline.com/toc/tfac20/current. doi:10.1080/19440049.2022.2141467.

Jin, T., Nordberg, G., Ye, T., Bo, M., Wang, H., Zhu, G., Kong, Q., & Bernard, A. (2004). Osteoporosis and renal dysfunction in a general population exposed to cadmium in China. *Environmental Research*, *96*(3), 353–359. doi:10.1016/j.envres.2004.02.012.

Joint FAO/WHO Expert Committee on Food Additives. (2011). Evaluation of certain food additives and contaminants: Seventy-third report of the joint FAO/WHO expert committee on food additives. *WHO Technical Report Series, no. 960*. https://inchem.org/documents/jecfa/jecmono/v960je01.pdf.

Julin, B., Wolk, A., Bergkvist, L., Bottai, M., & Åkesson, A. (2012). Dietary cadmium exposure and risk of postmenopausal breast cancer: A population-based prospective cohort study. *Cancer Research*, *72*(6), 1459–1466. http://cancerres.aacrjournals.org/content/72/6/1459.full.pdf+html. doi:10.1158/0008-5472.CAN-11-0735Sweden.

Kim, D. W., Kim, K. Y., Choi, B. S., Youn, P., Ryu, D. Y., Klaassen, C. D., & Park, J. D. (2007). Regulation of metal transporters by dietary iron, and the relationship between body iron levels and cadmium uptake. *Archives of Toxicology*, *81*(5), 327–334. doi:10.1007/s00204-006-0160-7.

Kim, K., Melough, M. M., Vance, T. M., Noh, H., Koo, S. I., & Chun, O. K. (2019). Dietary cadmium intake and sources in the US. *Nutrients*, *11*(1). https://www.mdpi.com/2072-6643/11/1/2/pdf. doi:10.3390/nu11010002.

Kim, M., & Wolt, J. D. (2011). Probabilistic risk assessment of dietary cadmium in the South Korean population. *Food Additives and Contaminants—Part A*, *28*(1), 62–70. http://www.tandfonline.com/toc/tfac20/current. doi:10.1080/19440049.2010.529620.

Kobayashi, J. (1978). Toxicity of heavy metals in the environment. In F. W. Oehme (Ed.), *Pollution by cadmium and the Itai-itai disease in Japan*. Marcel Dekker.

Li, D., Lin, H. J., Zhang, M., Meng, J., Hu, L. Y., & Yu, B. (2021). Urine cadmium as a risk factor for osteoporosis and osteopenia: A meta-analysis. *Frontiers in Medicine*, *8*. doi:10.3389/fmed.2021.648902.

Li, H. B., Li, M. Y., Zhao, D., Li, J., Li, S. W., Juhasz, A. L., Basta, N. T., Luo, Y. M., & Ma, L. Q. (2019). Oral bioavailability of As, Pb, and Cd in contaminated soils, dust, and foods based on animal bioassays: A review. *Environmental Science and Technology*, *53*(18), 10545–10559. http://pubs.acs.org/journal/esthag. doi:10.1021/acs.est.9b03567.

Lin, X. Y., Ye, T. W., Duan, X., Wang, B. X., Zhou, D., & Li, H. B. (2024). Cadmium in market pork kidneys: A study on cadmium bioavailability and the health effects based on mouse models. *Environmental Science and Technology*, *58*(33), 14651–14661. http://pubs.acs.org/journal/esthag. doi:10.1021/acs.est.4c04801.

Meharg, A. A., Lombi, E., Williams, P. N., Scheckel, K. G., Feldmann, J., Raab, A., Zhu, Y., & Islam, R. (2008). Speciation and localization of arsenic in white and brown rice grains. *Environmental Science and Technology, 42*(4), 1051–1057. doi:10.1021/es702212p.

Meharg, A. A., Norton, G., Deacon, C., Williams, P., Adomako, E. E., Price, A., Zhu, Y., Li, G., Zhao, F. J., McGrath, S., Villada, A., Sommella, A., De Silva, P. M. C. S., Brammer, H., Dasgupta, T., & Islam, M. R. (2013). Variation in rice cadmium related to human exposure. *Environmental Science and Technology, 47*(11), 5613–5618. doi:10.1021/es400521h.

Moon, C. S., Lee, C. K., Lee, J. T., Kim, J. M., Paik, J. M., & Ikeda, M. (2012). Time trends in dietary cadmium intake of Korean women. *Toxicology Research, 1*(2), 145–150. doi:10.1039/c2tx00002d.

Nawrot, T. S., Martens, D. S., Hara, A., Plusquin, M., Vangronsveld, J., Roels, H. A., & Staessen, J. A. (2015). Association of total cancer and lung cancer with environmental exposure to cadmium: The meta-analytical evidence. *Cancer Causes & Control, 26*(9), 1281–1288. doi:10.1007/s10552-015-0621-5.

Nogawa, K., Sakurai, M., Ishizaki, M., Kido, T., Nakagawa, H., & Suwazono, Y. (2017). Threshold limit values of the cadmium concentration in rice in the development of Itai-itai disease using benchmark dose analysis. *Journal of Applied Toxicology, 37*(8), 962–966. doi:10.1002/jat.3444.

Nogawa, K., & Kido, T. (1993). Biological monitoring of cadmium exposure in Itai-itai disease epidemiology. *International Archives of Occupational and Environmental Health, 65*(S1), S43. doi:10.1007/bf00381306.

Nordberg, G. F., Fowler, B. A., Nordberg, M., & Friberg, L. T. (2007). *Handbook on the toxicology of metals handbook on the toxicology of metals*. Elsevier Inc http://www.sciencedirect.com/science/book/9780123694133.

Olsson, I. M., Bensryd, I., Lundh, T., Ottosson, H., Skerfving, S., & Oskarsson, A. (2002). Cadmium in blood and urine—Impact of sex, age, dietary intake, iron status, and former smoking—Association of renal effects. *Environmental Health Perspectives, 110*(12), 1185–1190. http://ehp.niehs.nih.gov/. doi:10.1289/ehp.021101185.

Peng, X., Li, C., Zhao, D., & Huang, L. (2023). Associations of micronutrients exposure with cadmium body burden among population: A systematic review. *Ecotoxicology and Environmental Safety, 256*, 114878. doi:10.1016/j.ecoenv.2023.114878.

Qian, Y., Chen, C., Zhang, Q., Li, Y., Chen, Z., & Li, M. (2010). Concentrations of cadmium, lead, mercury and arsenic in Chinese market milled rice and associated population health risk. *Food Control, 21*(12), 1757–1763. http://www.sciencedirect.com/. doi:10.1016/j.foodcont.2010.08.005.

Quraishi, S. M., Adams, S. V., Shafer, M., Meliker, J. R., Li, W., Luo, J., Neuhouser, M. L., & Newcomb, P. A. (2016). Urinary cadmium and estimated dietary cadmium in the Women's Health Initiative. *Journal of Exposure Science and Environmental Epidemiology, 26*(3), 303–308. http://www.nature.com/jes/index.html. doi:10.1038/jes.2015.40.

Reeves, P. G., & Chaney, R. L. (2008). Bioavailability as an issue in risk assessment and management of food cadmium: A review. *Science of the Total Environment, 398*(1–3), 13–19. doi:10.1016/j.scitotenv.2008.03.009.

Reeves, P. G., & Chaney, R. L. (2004). Marginal nutritional status of zinc, iron, and calcium increases cadmium retention in the duodenum and other organs of rats fed rice-based diets. *Environmental Research, 96*(3), 311–322. doi:10.1016/j.envres.2004.02.013.

Sand, S., & Becker, W. (2012). Assessment of dietary cadmium exposure in Sweden and population health concern including scenario analysis. *Food and Chemical Toxicology, 50*(3–4), 536–544. doi:10.1016/j.fct.2011.12.034.

Satarug, S., Garrett, S. H., Sens, M. A., & Sens, D. A. (2010). Cadmium, environmental exposure, and health outcomes. *Environmental Health Perspectives, 118*(2), 182–190. http://ehp03.niehs.nih.gov/article/fetchObjectAttachment.action?uri=info%3Adoi%2F10.1289%2Fehp.0901234&representation=PDF. doi:10.1289/ehp.0901234. United States.

Satarug, S., Vesey, D. A., & Gobe, G. C. (2023). Cadmium-induced proteinuria: Mechanistic insights from dose–effect analyses. *International Journal of Molecular Sciences, 24*(3). http://www.mdpi.com/journal/ijms. doi:10.3390/ijms24031893.

Schwarz, M. A., Lindtner, O., Blume, K., Heinemeyer, G., & Schneider, K. (2014). Cadmium exposure from food: The German LExUKon project. *Food Additives and Contaminants—Part A, 31*(6), 1038–1051. http://www.tandfonline.com/toc/tfac20/current. doi:10.1080/19440049.2014.905711.

Scientific Panel on Responsible Plant Nutrition. (2023). *Plant nutrients are essential for the alleviation of chronic and hidden hunger* (Issue Brief 5). http://www.sprpn.org.

Shahriar, S., Mahmudur Rahman, M., & Naidu, R. (2020). Geographical variation of cadmium in commercial rice brands in Bangladesh: Human health risk assessment. *Science of the Total Environment, 716*, 137049. doi:10.1016/j.scitotenv.2020.137049.

Shi, J., Zhao, D., Ren, F., & Huang, L. (2023). Spatiotemporal variation of soil heavy metals in China: The pollution status and risk assessment. *Science of the Total Environment, 871*, 161768. doi:10.1016/j.scitotenv.2023.161768.

Shi, R., Lv, J., & Liu, Y. (2020a). Spatial distribution and bio-accumulation of cadmium and lead in soil, rice and vegetables in typical pollution areas, China. *Bulletin of Environmental Contamination and Toxicology, 104*(2), 307–313. doi:10.1007/s00128-019-02765-0.

Shi, W., Zhao, X., Han, Y., Che, Z., Chai, X., & Liu, G. (2016). Ocean acidification increases cadmium accumulation in marine bivalves: A potential threat to seafood safety. *Scientific Reports, 6*(1). doi:10.1038/srep20197.

Shi, Z., Carey, M., Meharg, C., Williams, P. N., Signes-Pastor, A. J., Triwardhani, E. A., Pandiangan, F. I., Campbell, K., Elliott, C., Marwa, E. M., Jiujin, X., Farias, J. G., Nicoloso, F. T., De Silva, P. M. C. S., Lu, Y., Norton, G., Adomako, E., Green, A. J., Moreno-Jiménez, E., ... Meharg, A. A. (2020b). Rice grain cadmium concentrations in the global supply-chain. *Exposure and Health, 12*(4), 869–876. http://link.springer.com/journal/volumesAndIssues/12403. doi:10.1007/s12403-020-00349-6.

Shimbo, S., Zhang, Z. W., Moon, C. S., Watanabe, T., Nakatsuka, H., Matsuda-Inoguchi, N., Higashikawa, K., & Ikeda, M. (2000). Correlation between urine and blood concentrations, and dietary intake of cadmium and lead among women in the general population of Japan. *International Archives of Occupational and Environmental Health, 73*(3), 163–170. link.springer.de/link/service/journals/00420/index.htm. doi:10.1007/s004200050023.

Siame, T., Abolade, Y. A., Omotayo, F., Nyarko, A. J., Aminu, M. B., Ogwurumba, U. A., Akagbue, B. O., Abdulmalik, F., & Zabidi, H. (2025). Potentially toxic elements in local cigarettes and marijuana leaves of Bauchi State, Nigeria: Public health and environmental implications. *Pollutants, 5*(3). https://www.mdpi.com/journal/pollutants. doi:10.3390/pollutants5030026.

Sommella, A., Deacon, C., Norton, G., Pigna, M., Violante, A., & Meharg, A. A. (2013). Total arsenic, inorganic arsenic, and other elements concentrations in Italian rice grain varies with origin and type. *Environmental Pollution, 181*, 38–43. doi:10.1016/j.envpol.2013.05.045.

Song, Y., Wang, Y., Mao, W., Sui, H., Yong, L., Yang, D., Jiang, D., Zhang, L., Gong, Y., & Kunze, G. (2017). Dietary cadmium exposure assessment among the Chinese population. *PLOS ONE, 12*(5), e0177978. doi:10.1371/journal.pone.0177978.

Tang, Z., Cai, H., Li, J., Lv, Y., Zhang, W., & Zhao, F. J. (2017). Allelic variation of NtNramp5 associated with cultivar variation in cadmium accumulation in Tobacco. *Plant and Cell Physiology, 58*(9), 1583–1593. http://pcp.oxfordjournals.org/. doi:10.1093/pcp/pcx087.

Vahter, M., Berglund, M., Nermell, B., & Åkesson, A. (1996). Bioavailability of cadmium from shellfish and mixed diet in women. *Toxicology and Applied Pharmacology, 136*(2), 332–341. http://www.elsevier.com/inca/publications/store/6/2/2/9/5/1/index.htt. doi:10.1006/taap.1996.0040.

Vahter, M., Åkesson, A., Lidén, C., Ceccatelli, S., & Berglund, M. (2007). Gender differences in the disposition and toxicity of metals. *Environmental Research, 104*(1), 85–95. doi:10.1016/j.envres.2006.08.003.

Wang, C. C., Zhang, Q. C., Yan, C. A., Tang, G. Y., Zhang, M. Y., Ma, L. Q., Gu, R. H., & Xiang, P. (2023). Heavy metal(loid)s in agriculture soils, rice, and wheat across China: Status assessment and spatiotemporal analysis. *Science of the Total Environment, 882*. www.elsevier.com/locate/scitotenv. doi:10.1016/j.scitotenv.2023.163361.

Wang, P., Chen, H., Kopittke, P. M., & Zhao, F. J. (2019). Cadmium contamination in agricultural soils of China and the impact on food safety. *Environmental Pollution, 249*, 1038–1048. https://www.journals.elsevier.com/environmental-pollution. doi:10.1016/j.envpol.2019.03.063.

Wang, P., & Zhao, F. (2022). China national food safety standards of cadmium in staple foods: Issues and thinking. *Chinese Science Bulletin, 67*(27), 3252–3260. doi:10.1360/tb-2022-0138.

Wei, Y., Zheng, X., Zhang, Z., Liang, H., Gu, M., Shen, F., Shohag, M. J. I., & Li, X. (2021). In vivo-in vitro correlations for the assessment of cadmium bioavailability in vegetables. *Journal of Agricultural and Food Chemistry, 69*(41), 12295–12304. http://pubs.acs.org/journal/jafcau. doi:10.1021/acs.jafc.1c03284.

Wei, S., Guo, B., Feng, L., Jiang, T., Li, M., & Wei, Y. (2017). Cadmium distribution and characteristics of cadmium-binding proteins in rice (*Oryza sativa* L.) kernel. *Food Science and Technology Research, 23*(5), 661–668. https://doi.org/10.3136/fstr.23.661. doi:10.3136/fstr.23.661.

Williams, P. N., Lei, M., Sun, G., Huang, Q., Lu, Y., Deacon, C., Meharg, A. A., & Zhu, Y. G. (2009). Occurrence and partitioning of cadmium, arsenic and lead in mine impacted paddy rice: Hunan, China. *Environmental Science and Technology, 43*(3), 637–642. http://pubs.acs.org/doi/pdfplus/10.1021/es802412r. doi:10.1021/es802412r. China.

Xiao, W., Yang, Y., Tang, N., Huang, X., Zhang, Q., Zhao, S., De Chen, B. G., Zhao, Z., Jiang, Y., & Ye, X. (2024). Innovative accumulative risk assessment of co-exposure to Cd, As, and Pb in contaminated rice based on their in vivo bioavailability and in vitro bioaccessibility. *Science of the Total Environment, 912*, 168922. doi:10.1016/j.scitotenv.2023.168922.

Yang, L., Zhang, X., Zhao, D., Wang, P., & Zhao, F. (2023). Relative bioavailability of cadmium in rice: Assessment, modeling, and application for risk assessment. *Foods, 12*(5), 984. doi:10.3390/foods12050984.

Yao, B.-M., Chen, P., & Sun, G.-X. (2020). Distribution of elements and their correlation in bran, polished rice, and whole grain. *Food Science & Nutrition, 8*(2), 982–992. doi:10.1002/fsn3.1379.

Zhang, H., & Reynolds, M. (2019). Cadmium exposure in living organisms: A short review. *Science of the Total Environment, 678*, 761–767. doi:10.1016/j.scitotenv.2019.04.395.

Zhang, Y., Wu, Y., Yang, A., Wang, J., Zhang, K., & Hu, D. (2017). Assessment of cadmium content of potato grown in Weining County, Guizhou Province, China. *Environmental Monitoring and Assessment, 189*(5). doi:10.1007/s10661-017-5950-2.

Zhao, D., Liu, R. Y., Xiang, P., Juhasz, A. L., Huang, L., Luo, J., Li, H. B., & Ma, L. Q. (2017a). Applying cadmium relative bioavailability to assess dietary intake from rice to predict cadmium urinary excretion in nonsmokers. *Environmental Science and Technology, 51*(12), 6756–6764. http://pubs.acs.org/journal/esthag. doi:10.1021/acs.est.7b00940.

Zhao, D., Wang, J. Y., Tang, N., Yin, D. X., Luo, J., Xiang, P., Juhasz, A. L., Li, H. B., & Ma, L. Q. (2018). Coupling bioavailability and stable isotope ratio to discern dietary and non-dietary contribution of metal exposure to residents in mining-impacted areas. *Environment International, 120*, 563–571. www.elsevier.com/locate/envint. doi:10.1016/j.envint.2018.08.023.

Zhao, D., Lin, G. B., Liu, C., Juhasz, A. L., & Ma, L. Q. (2025). Health risk assessment of dietary cadmium exposure based on cadmium bioavailability in food: Opportunities and challenges. *Journal of Hazardous Materials, 488*. https://www.sciencedirect.com/science/journal/03043894. doi:10.1016/j.jhazmat.2025.137359.

Zhao, D., Juhasz, A. L., Luo, J., Huang, L., Luo, X. S., Li, H. B., & Ma, L. Q. (2017b). Mineral dietary supplement to decrease cadmium relative bioavailability in rice based on a mouse bioassay. *Environmental Science and Technology, 51*(21), 12123–12130. http://pubs.acs.org/journal/esthag. doi:10.1021/acs.est.7b02993.

Zhao, D., Wang, P., & Zhao, F. J. (2023). Dietary cadmium exposure, risks to human health and mitigation strategies. *Critical Reviews in Environmental Science and Technology, 53*(8), 939–963. www.tandf.co.uk/journals/titles/10643389.asp. doi:10.1080/10643389.2022.2099192.

Zhao, D., Wang, P., & Zhao, F. J. (2024). Toxic metals and metalloids in food: Current status, health risks, and mitigation strategies. *Current Environmental Health Reports, 11*(4), 468–483. https://www.springer.com/journal/40572. doi:10.1007/s40572-024-00462-7.

Zhao, X., Shao, Y., Ma, L., Shang, X., Zhao, Y., & Wu, Y. (2022). Exposure to lead and cadmium in the sixth total diet study—China, 2016–2019. *China CDC Weekly, 4*(9), 176–179. doi:10.46234/ccdcw2022.045.

Zhu, H., Tang, X., Gu, C., Chen, R., Liu, Y., Chu, H., & Zhang, Z. (2024). Assessment of human exposure to cadmium and its nephrotoxicity in the Chinese population. *Science of the Total Environment, 918*, 170488. doi:10.1016/j.scitotenv.2024.170488.

Zhuang, P., Sun, S., Zhou, X., Mao, P., McBride, M. B., Zhang, C., Li, Y., Xia, H., & Li, Z. (2020). Bioavailability and bioaccessibility of cadmium in contaminated rice by in vivo and in vitro bioassays. *Science of the Total Environment, 719*, 137453. doi:10.1016/j.scitotenv.2020.137453.

Zhuang, Z., Wang, Q., Huang, S., NiñoSavala, A. G, Wan, Y., Li, H., Schweiger, A. H., Fangmeier, A., & Franzaring, J. (2023). Source-specific risk assessment for cadmium in wheat and maize: Towards an enrichment model for China. *Journal of Environmental Sciences, 125*, 723–734. doi:10.1016/j.jes.2022.02.024.

CHAPTER

16

Ecological risk assessment of cadmium contamination in soils

Koen Oorts[a], *Steve McGrath*[b]

[a] **ARCHE Consulting, Leuven, Belgium** [b] **Sustainable Soils and Crops, Rothamsted Research, Harpenden, United Kingdom**

16.1 Introduction

A risk assessment of cadmium (Cd) contamination in soil compares the Cd concentration in soil with critical effect concentrations for different receptor pathways. The main pathways to be considered for ecotoxicological risk assessments are direct exposure and toxicity to soil-dwelling organisms (plants, invertebrates, and microorganisms) and secondary poisoning to wildlife via the terrestrial food chain. Critical soil concentrations for both groups of organisms are derived from dose–response studies where organisms are exposed to increasing Cd concentrations in soil or diet. From the resulting dose–response relationship, toxicity thresholds [no observed effect concentrations (NOEC), i.e., highest dose at which no significant inhibitory effect is observed; lowest observed effect concentrations (LOEC), i.e., lowest dose at which a significant inhibitory effect is observed; or EC_x value, i.e., effective concentration yielding x% inhibition in response] can be calculated for the effect of Cd on the endpoint studied (e.g., growth or reproduction).

This chapter presents an overview of the available data for Cd toxicity to soil organisms, birds, and mammals and explains how these data are used to derive critical effect concentrations according to the regulatory approaches in the European Union.

16.2 Effects of cadmium on soil-dwelling organisms

16.2.1 Toxicity data

Toxicity of Cd to terrestrial organisms (plants, invertebrates, and microorganisms) is typically studied by exposing these organisms to an uncontaminated soil recently amended with increasing doses of a soluble Cd salt [e.g., $Cd(NO_3)_2$, $CdCl_2$, or $CdSO_4$]. Numerous studies are published in the scientific literature that report such toxicity data of Cd to soil organisms. Data that can be used in a risk assessment must comply with the following criteria:

Cadmium Contamination in Soils and Plants: Mechanisms and Mitigation.
DOI: https://doi.org/10.1016/B978-0-443-34104-5.00008-6

- The substances tested must be soluble to present worst case bioavailability of Cd in soil.
- The toxicity of the accompanying anions (e.g., acetate, Cl^-, NO_3^-, SO_4^{2-}) must be negligible compared to the toxicity of the Cd^{2+} cation.
- Endpoints studied for plants and invertebrates are directly relevant on a population level, for example, survival, growth, reproduction, yield, or abundance. Data for molecular endpoints (e.g., biomarkers and enzyme activity) are not taken forward without information on their relevance at the organism or population level. Relevant endpoints for soil microorganisms must focus on functional parameters (such as respiration, nitrification, and mineralization) and microbial biomass or activity.
- Only data from observations in natural and artificial [Organisation for Economic Co-operation and Development (OECD)] soil media are considered. Tests performed in substrates that were judged as not representative of soils (e.g., nutrient solution, agar, pure quartz sand, and farmyard manure) are not included in the assessment.
- Only data for chronic exposure are considered. Because chronic exposure is a function of the life cycle of the test organisms, a priori fixed exposure durations are not relevant, and the duration should be related to the typical life cycle and should ideally encompass the entire life cycle or, for longer lived species, the most sensitive life stage. Exposure durations should also be related to recommendations from standard ecotoxicity (e.g., ISO, OECD, and ASTM) protocols. Typically, chronic test durations for the higher plants are within the range of 4 days [e.g., the root elongation test based on ISO 11269-1 (2012) and 28 days (e.g., the shoot yield test based on ISO 11269-2 (2012)]. The OECD 208 (plant seedling emergence and growth test, 2006) recommends a test duration of at least 14 days after the emergence of the seedlings. Testing with soil invertebrates has a typical chronic exposure duration of 4–8 weeks for the oligochaetes *Eisenia fetida*/*Eisenia andrei* (earthworms) (OECD 222, 2016), 3–6 weeks for the standard organism *Enchytraeus albidus* (white worm) (OECD 220, 2016), 2 weeks for predatory mite *Hypoaspis* (*Geolaelaps*) *aculeifer* reproduction test (OECD 226, 2016), and 4 weeks for the collembolan (e.g., *Folsomia candida*) survival and reproduction test (OECD 232, 2016). Reported test durations using soil microorganisms vary between 3 hours and 540 days. Typical exposure durations last up to 28 days for the carbon transformation test (OECD 217, 2000) and for the nitrogen transformation test (OECD 216, 2000).

The toxicological variables are estimated based on NOEC or EC_{10} values (Fig. 16.1). There is a growing preference to use regression-based EC_x values as calculated from the concentration-effect relationship for hazard and risk assessment over the hypothesis-based NOEC values (Beasley et al., 2015; Chapman et al., 1996). The EC_{10} value was selected here as the basis for the risk assessment, corresponding to the guidelines for European regulations on risk assessment of industrial chemicals [Regulation (EC) No. 1907/2006 concerning the Registration, Evaluation, Authorisation, and Restriction of Chemicals (REACH)], biocides [Regulation (EU) No. 528/2012 concerning the making available on the market and use of biocidal products], and plant protection products [Regulation (EC) No. 1107/2009 concerning the placing of plant protection products on the market], but it must be noted that other EC_x values are used in other jurisdictions, for example, EC_{20} in the United States.

When EC_x values are not reported in the original study, but dose–response data are available, allowing derivation of a concentration–effect relationship, the EC_{10} is calculated using a log-logistic (sigmoidal) dose–response curve by minimizing the unweighted squared residuals sum.

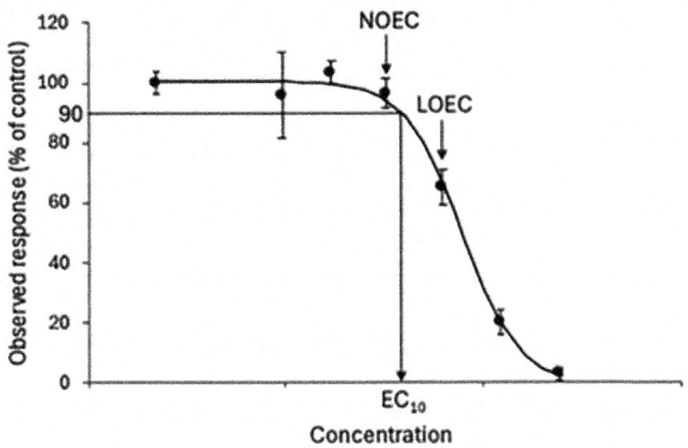

FIGURE 16.1 Example of a dose–response relationship from a toxicity test: The NOEC (no observed effect concentration, i.e., the highest tested dose at which no significant inhibitory effect is observed) and LOEC (lowest observed effect concentrations, i.e., lowest tested dose at which a significant inhibitory effect is observed) values are derived from statistical hypothesis testing. The line is the dose–response curve fitted by regression methods and is used to predict the EC_x values, that is, effective concentrations yielding x% inhibition in response.

A minimum of one control and three test concentrations are required for derivation of a dose–response curve. EC_{10} values, which lie outside the concentration range tested, that is, below the lowest (excluding control) or above the highest dose tested, respectively, are considered unreliable because of their large degree of uncertainty (OECD, 2006). If a reliable NOEC value can be derived from such results, this NOEC value can still be used. Unbounded NOEC (i.e., no significant effect at highest dose tested) are not selected. However, unbounded LOEC (i.e., significant effect at lowest dose tested) values can be used to derive an NOEC if the percentage effect at LOEC is <20% (ECHA, 2008). An NOEC can be calculated as LOEC/2 for LOEC values with >10% and <20% effect. If an effect of ≤10% is observed at the LOEC, the NOEC is set equal to this LOEC value. If the effect percentage of the LOEC is unknown or >20%, no reliable NOEC can be derived because of the large degree of uncertainty associated with such extrapolation. If no methodology is reported or if effect concentrations are "visually" derived, the data are not considered.

Reliable chronic toxicity data are available for the long-term effect of Cd on 68 terrestrial species or microbial endpoints covering the three trophic levels. Reliable EC_{10} and NOEC values vary between 0.5 and 4596.9 mg Cd/kg dry weight (dw) depending on the species and endpoints. There is no clear distinction in sensitivity across trophic levels or organism groups (Fig. 16.2). The data for invertebrates cover Cd toxicity to reproduction of eight different species of Annelida, five different arthropod species, one mollusk species, and one nematode species. The reliable NOEC and EC_{10} values for invertebrates vary between 2.5 mg Cd/kg dw for weight loss of *E. fetida* (Sinkakarimi et al., 2020) and 563.7 mg Cd/kg dw for mortality of *Enchytraeus crypticus* (Santos et al., 2022). Data for plants cover various endpoints (shoot or root biomass, length, grain or fruit yield, etc.) for 42 different plant species from 16 families. NOEC and EC_{10} values vary between 1.25 mg Cd/kg dw for root weight of *Physalis peruviana* (Çikili et al., 2016) and shoot biomass of *Spinacia oleracea* (Bingham et al., 1975) and 4000 mg Cd/kg dw for *Lolium perenne* shoot biomass (Moreno et al., 2006). Tests with microbial organisms are mainly performed with the native soil microbial community present in the soils, whereas three results were identified for specific microbial species (the fungi *Funneliformis mosseae* and *Piriformospora indica* and the bacterium *Rhizobium leguminosarum*). The selected toxicity data cover effects on nitrogen and carbon transformation processes, biomass, root colonization, and dehydrogenase activity. The reliable NOEC and EC_{10} values range between 2.6 mg Cd/kg dw

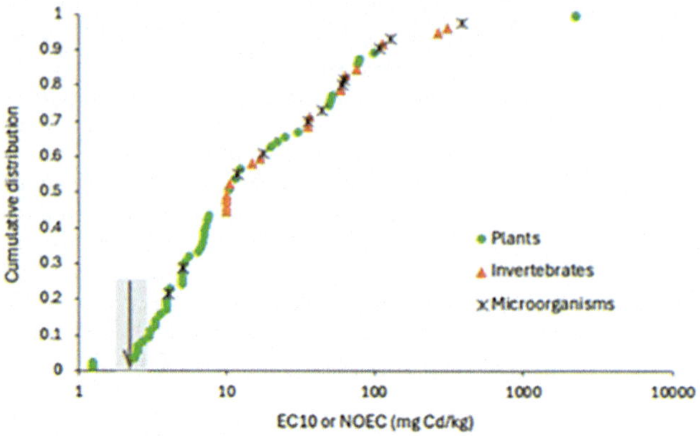

FIGURE 16.2 The cumulative distribution of species/process mean NOEC or EC_{10} values selected for derivation of a critical threshold concentration for direct exposure and toxicity of Cd to soil organisms. The arrow and shaded area show the HC_5 (hazard concentration protecting 95% of species) and its 5%–95% confidence interval derived from the fitted SSD (see text). *NOEC*, No observed effect concentration; *SSD*, species sensitivity distribution.

for glucose-induced respiration (Reber, 1989) and 4596.9 mg Cd/kg dw for basal respiration activity (Doelman & Haanstra, 1984).

The test soils represent a wide range in soil types, pH (2.8–8.4), organic carbon content (<0.1–46.4%), clay content (<1–60%) and ambient Cd background concentration (<0.1–2.3 mg Cd/kg).

16.2.2 Bioavailability of cadmium in soil

Metal toxicity to soil organisms not only depends on the total metal dose but can also be significantly affected by the time since contamination and by soil physicochemical properties (OECD, 2016; Smolders et al., 2009).

The ecological relevance of standardized laboratory tests immediately or shortly after freshly spiking with soluble metal salt can be considered limited. Standardized toxicity tests result in a sudden disturbance of the soil, whereas, in field conditions, Cd generally accumulates slowly with time because of, for example, industrial emissions or fertilizer applications. Spiking soils with soluble Cd salts not only increases the Cd content of a soil but also increases the ionic strength of the soil solution and decreases the soil pH by replacement of protons from the exchange complex with the metal cations (Speir et al., 1999). The higher ionic strength and lower soil solution pH in freshly amended soils compared to leached or field-contaminated soils (that are naturally leached) may affect the response of the organisms both directly and indirectly by increasing metal bioavailability. Testing soils immediately after adding Cd also ignores the "aging" reactions (a combination of diffusion and sorption processes, where Cd progressively binds to soil components with time) that further decrease bioavailability of added Cd beyond the duration of these tests. Standard toxicity experiments in soil therefore result in a potentially higher bioavailability compared to field conditions. Soil metal concentrations needed to produce toxic effects generally are lower for laboratory spiked than for field contaminated soils and therefore represent the worst case scenario. In addition, lower toxicity to microbial processes in field-contaminated or aged soils can also be caused by microbial acclimatization or adaptation of the microorganisms in the soil to gradually increased Cd concentrations (Mertens et al., 2010; Puglisi et al., 2012).

Only a few publications are available that study the effect of equilibration time on Cd toxicity in soil (Table 16.1). Doelman and Haanstra (1984) found that Cd was more toxic to soil respiration

TABLE 16.1 Effect of incubation time on Cd toxicity to soil organisms.

Test organism	Endpoint	Soil	Effect parameter	Freshly spiked (equilibration days) (mg Cd/kg)	Aged (equilibration days) (mg Cd/kg)	References
Native soil microorganisms	Basal respiration	Sand	EC_{50}	<1000 (14)	3469 (490)	Doelman and Haanstra (1984)
		Sandy loam	EC_{30}	732 (14)	814 (301)	
		Silty loam	EC_{50}	>8000 (14)	4542 (630)	
		Clay	EC_{50}	7239 (14)	>8000 (560)	
		Sandy peat	EC_{30}	1473 (14)	3704 (574)	
Native soil microorganisms	ATP content	Uncultivated soil	EC_{50}	3333 (0.125)	1232 (60)	Moreno et al. (2006)
		Cultivated soil	EC_{50}	6757 (0.125)	>8000 (60)	
	Dehydrogenase activity	Uncultivated soil	EC_{50}	1111 (0.125)	1429 (60)	
		Cultivated soil	EC_{50}	1190 (0.125)	4717 (60)	
Lolium perenne	Fresh shoot yield	Uncultivated soil	NOEC	1000 (0.125)	2000 (60)	
		Cultivated soil	NOEC	2000 (0.125)	4000 (60)	

when tested at 2 weeks of incubation after spiking soils with Cd chloride than after an extended period of up to 90 weeks of incubation. Moreno et al. (2006) also observed higher inhibition of soil adenosine triphosphate (ATP) content, soil dehydrogenase activity, and shoot yield of ryegrass (*L. perenne*) after incubating soil for 3 hours after spiking with Cd sulfate compared to after 60 days of incubation. The lower toxicity with longer equilibration times corresponded with the decrease in diethylenetriamine pentaacetic acid (DTPA) extractable Cd concentration with time in these soils. It must be noted that incubations took place in closed systems and that these soils were not leached to remove excess ions. The incubated treatments therefore still do not mimic field conditions. The available information does not allow a general correction factor for the extent of aging time and leaching on toxicity of Cd to soil organisms to be derived (Table 16.1).

Metal toxicity is well known to vary with soil properties (Smolders et al., 2009). Table 16.2 summarizes data from studies with ≥3 soils. Toxicity to plants and microorganisms shows more pronounced variation among soils (more than a factor of 10 differences) than toxicity to invertebrates, where the variation among soils is smaller than a factor of 7. Soil pH and organic matter (based on organic carbon content) are most often identified as the soil properties best correlated with Cd toxicity, with generally lower toxicity (higher EC_x) at higher pH or higher organic carbon content. However, negative correlations of toxicity with pH (lower toxicity at lower pH) were also observed for some plant species (Mahler et al., 1978; Zhang et al., 2020). The pH of the soil dominates the solid–liquid distribution of Cd in soil; higher pH leads to larger Cd sorption and lower Cd concentration in soil solution (see Chapter 6). It is often assumed that the metal concentration in soil solution represents the toxic dose for the ecosystem, and, therefore, a correlation between metal toxicity and pH can be expected. At higher pH, metal solubility is low, and the toxicity could be lower (higher EC_x) than at low pH where metals are more soluble. This, however, neglects the competition from protons in pore water for uptake of metal cations by the organisms, which can counteract the effect of pH on metal sorption. Toxicity studies in artificial solutions with the annelid *E. fetida* and the plant *Hordeum vulgare* indeed demonstrate that toxicity decreases with lower pH of the soil solution at a given Cd activity (Li et al., 2008; Wang et al., 2016). These competition effects also explain why soil solution–based toxicity values are no better a prediction of Cd toxicity to plants since intersoil variation did not decrease based on soluble Cd (Mahler et al., 1978; Zhang et al., 2020).

Next to pH and organic matter content, some significant correlations were also observed between Cd toxicity and clay content, effective cation exchange capacity, or Al oxides. Regression equations between Cd toxicity for single species and soil properties generally have R^2 values smaller than 0.6, indicating that soil properties only explain a limited proportion of the variation in toxicity. This is also confirmed by the lack of observation of strong overall correlations between soil toxicity data (EC_{10} or NOEC values) and soil properties for the combined dataset. The lowest toxicity values could not be related to specific combinations of soil properties. It must be noted that in these correlations, the effect of soil factors on the toxicity data is obscured by all other sources of variance, for example, the variance in sensitivity among species and endpoints. The available data nevertheless indicate that the impact of soil properties on toxicity is smaller than the observed differences in the Cd sensitivity among the tested species. For example, Bingham et al. (1975) observed NOEC or EC_{10} values between 1.25 and 78.3 mg Cd/kg for the yield of 12 plant species in the same soil, and the selected species mean values in this whole dataset (Fig. 16.2) vary by more than 1000-fold (1.25–2245 mg Cd/kg soil).

Because of the lack of strong comparative datasets of Cd toxicity for some major organism groups in different soils, some inconsistent effects of soil properties on Cd toxicity on soil

TABLE 16.2 Effect of soil properties on Cd toxicity to soil organisms.

Test organism	Endpoint	Number of soils	pH (0.01M $CaCl_2$)	Org. C (%)	Clay (%)	eCEC ($cmol_c$/kg)	EC_x (mg Cd/kg)	Soil property best related with toxicity[a]	References
Invertebrates									
Eisenia fetida	Reproduction	3 (OECD soils)	6	3.48, 5.80, 8.12	19–21	8.6–20.1	10.5–51 (EC_{50})	Org. C	Irizar et al. (2015)
Folsomia candida	Mortality	18	4.6–8.7	0.3–4.7	10–48	6.9–26.1	52.8–270.7 (LC_{50})	pH and org. C	Liu et al. (2019)
	Reproduction						41.4–146.8 (EC_{50})	pH and org. C	
Folsomia candida	Reproduction	3 (OECD soil)	4.5, 5.0, 6.0	5.8	20	9.9–14.3	224.4–348.1 (EC_{10})	pH	Sandifer and Hopkin (1996)
Paronchiurus kimi	Mortality	5 (OECD soil)	4.0, 5.2, 6.5	0, 2.9, 5.8	20	0.0–12.0	34.9–115.9 (LC_{50})	pH and org. C	Son et al. (2007)
	Reproduction						11.5–72.8 (EC_{50})	pH and org. C	
	Population growth						<12.3–49.2 (NOEC)	pH	
Paronchiurus kimi	Reproduction	9 (OECD soil)	4.0–6.5	0–5.8	20	0.0–12.5	6.4–38.4 (EC_{50})	pH and org. C	Son et al. (2009)
Helix aspera	Growth	3	5.5–6.9	1.9–9.6	5–9	10–20	66.8–131.5 (EC_{10})	pH, org. C, and eCEC	Sahraoui et al. (2021)

(*continued on next page*)

TABLE 16.2 Effect of soil properties on Cd toxicity to soil organisms—cont'd

Test organism	Endpoint	Number of soils	pH (0.01M $CaCl_2$)	Org. C (%)	Clay (%)	eCEC ($cmol_c$/kg)	EC_x (mg Cd/kg)	Soil property best related with toxicity[a]	References
Plants									
Triticum aestivum	Shoot yield	3	5.6–7.0	1.0–1.5	7–21	6.0–16.5	8.0–96.9 (EC_{20})	Not clear	Reber (1989)
Ligustrum japonicum	Shoot for fresh weight	13	4.7–7.9	0.7–7.3	9.2–39.2	4.8–19.5	8.83–44.79 ($EC_{50, added}$)	pH (–) and Al_{ox}	Zhang et al. (2020)
Theobroma cacao L.	Total biomass	4	4.0–5.2	0.5–11.6	12–50	15.3–28.6	<2–6.5 (EC_{10})	pH and org. C	Correa et al. (2021)
Beta vulgaris	Shoot yield	8	4.3–7.3	0.6–3.3	4–41	6.5–67.9	110.5–320 (EC_{50})	pH	Mahler et al. (1978)
Lactuca sativa	Shoot yield						20.0–277.5 (EC_{50})	pH (–) and org. C	
Glycine max	Shoot yield	9	4.0–7.4	Not reported	Not reported	2.3–30.3	1.1–86.8 (EC_{20})	pH and eCEC	Miller et al. (1976)

(continued on next page)

Microorganisms									
Basal respiration		5	3.9–7.2	0.9–7.4	2–60	3.7–41.9	187–6162 (EC_{20})	Clay and eCEC	Doelman and Haanstra (1984)
Basal respiration		5	5.4–7.1	2.81–5.47	52.7–85.4	19.8–41.6	2.79–6.64 (EC_{50})	Not clear	Prokop et al. (2003)
Glutamic acid–induced respiration		4	3.9–7.2	0.9–7.4	2–60	3.7–41.9	71.1–>1000 (EC_{10})	Not clear	Haanstra and Doelman (1984)
Glucose-induced respiration		3	5.6–7.0	1.0–1.5	7–21	6.0–16.5	8.7–32.5 (EC_{20})	Not clear	Reber (1989)
Fungal species	Growth	9	4.6–5.7	2.7–3.4	9.4–19.6	8.15–14.67	Not reported	Clay	Babich and Stotzky (1977)

[a] Positive relationship, unless noted as (–). Alox = Al oxide.

organisms and the rather limited variation in toxicity values among soils compared to other metals (e.g., Cu and Zn) and compared to differences in species sensitivity to Cd in soil, no representative bioavailability correction models were derived to correct for the impact of differences in soil properties on Cd toxicity to soil organisms.

16.2.3 Threshold for toxicity of cadmium to soil organisms

The available ecotoxicity database for the effect of Cd to soil organisms is large, and therefore, the use of the statistical extrapolation method is preferred for derivation of a threshold value for risk assessment of Cd in soil, rather than the use of an assessment factor on the lowest NOEC or EC_{10}. A species sensitivity distribution (SSD) is constructed through fitting of log-transformed toxicity data to different distributions. The Anderson–Darling best fitting statistic was used to select the best fitting distribution, and the 5th percentile value (HC_5, hazardous concentration at 5%, protecting 95% of the species) of this best fitting SSD is selected as the threshold value for risk assessment.

Species or process mean values are calculated as the geometric mean from all data for the most sensitive endpoint for each species or microbial process to avoid overrepresentation of widely tested species or processes in the SSD. A gamma distribution was defined as the optimal distribution according to the Anderson–Darling statistic, resulting in an HC_5 of 2.18 mg Cd/kg (5%–95% confidence limits: 1.74–2.84 mg Cd/kg) (Fig. 16.2).

Out of the 68 species mean values, only results for 2 species are below the HC_5 of 2.18 mg Cd/kg soil: an NOEC of 1.25 mg Cd/kg for root dw of *P. peruviana* and an NOEC of 1.25 mg Cd/kg for shoot yield of *S. oleracea*. Both values are within a factor of 2 of the HC_5 value.

Other species belonging to the same family (*Beta vulgaris* and *S. oleracea* for the plant family of Amaranthaceae and *Capsicum annuum, Nicotiana tabacum, P. peruviana, Solanum lycopersicum, Solanum melongena, Solanum nigrum*, and *Solanum tuberosum* for the plant family of Solanaceae) show species mean values that are consistently above the corresponding HC_5 value. Therefore there is no indication of organism groups that are not protected by the HC_5. Some individual NOEC or EC_{10} values for shoot yield of *Glycine max* and *Triticum aestivum* also fall below the HC_5 value, but the geometric values for these species (2.5 and 5.0 mg Cd/kg, respectively) are well above the HC_5 value.

16.2.4 Comparison with field data

Only a few field studies are available with only Cd contamination for comparison with the lab-derived toxicity data. A field trial with twofold replication on the effects of metal salts was started in Nagyhörcsök (Hungary) in 1991 (Kadar, 1995). Cd (as $CdSO_4$) was applied to the soil at 0, 90, 270, and 810 kg Cd/ha, resulting in measured total Cd concentrations in 1994 of 0.3, 18, 50, and 162 mg Cd/kg in the plough layer (0–20 cm). Maize, carrot, potato, pea, red beet, spinach, and wheat were grown successively. The soil was characterized by a high cation exchange capacity, high pH, and high base saturation, that is, not representing a worst case scenario for Cd. No toxic effects of Cd on yield of corn, carrot, potato, and pea were detected up to the highest rate during the first 4 years. However, toxic effects were very pronounced in the years when spinach and beetroot were grown, both plants belonging to the same family (Amaranthaceae) known to be Cd accumulators, and, to a lesser extent, in wheat grain. Pot trial studies confirm that spinach is far more sensitive to Cd than corn, wheat, or carrots (Bingham et al., 1975). The NOECs for spinach and beetroot in the field were both 18 mg Cd/kg, that is, well above the HC_5 of 2.2 mg Cd/kg soil as derived from the laboratory tests. The average plant yield was considerably reduced at this NOEC, but this reduction is not large enough to yield a statistically significant effect. The intrinsically higher variability in field data biases the comparison of field NOEC values with laboratory-derived NOEC values. The response of spinach and beetroot growth to Cd shows that yield is reduced by about 30% and 35%, respectively, at 18 mg Cd/kg. The EC_{25} of spinach grown in a pot trial was 4 mg Cd/kg, that is, more than fourfold lower than in the field (Bingham et al., 1975). The soil in that pot trial (pH 7.5) had a similar pH as in the field trial (7.3). Results from pot trials for beetroot in 8 different soils with varying soil properties showed EC_{10} values between 33.9 and 106.8 mg Cd/kg, with a geomean value of 52.6 mg Cd/kg (Mahler et al., 1978), comparable to the LOEC observed in the Hungarian field trial. The Bingham et al. study also reports an EC_{25} of Cd for wheat grain of 50 mg Cd/kg (same soil as with spinach), whereas even the highest Cd rate in the field did not result in a 25% reduction in wheat grain yield.

Another long-term field trial was started in 1988–90 in Bordeaux (France) in plots with two different pH treatments. Cd was more toxic to corn shoot yield in the most acid plots (pH 5.3–5.6) and had no significant effect on corn shoot yield up to 7–8 mg Cd/kg (ECB, 2007). The LOECs at 15 mg Cd/kg were associated with a 50% (high pH) and 61% (low pH) lower shoot yield than the control. A pot trial with corn was also reported in a similar soil as in the acid plots. Leaf and stem dw was reduced by 18% at 5.4 mg Cd/kg, the highest concentration tested (Mench et al., 1989). Interpolation of the field data between control and the first Cd rate suggests a similar inhibition of growth of corn by Cd as in the pot trials. In another field test, effect of Cd was measured in bush beans grown on a loamy sand located on the Savannah River Site, near Aiken in South Carolina, United States (Sajwani et al., 1996). No growth effects were noted at the highest tested concentration, that is, up to 6.7 mg Cd/kg soil dw (measured concentration), a value more than threefold above the HC_5 of 2.2 mg Cd/kg soil.

This comparison with available field studies shows that there is no indication of higher toxicity of Cd salts in the field than in the laboratory, and the HC_5 as derived from the laboratory tests is also protective in field conditions.

16.3 Effects of cadmium in soil on wildlife

16.3.1 Classic approach

16.3.1.1 Dietary toxicity of cadmium to birds and mammals

Standard risk assessment for secondary poisoning of Cd in mammals and birds through the terrestrial food chain is based on a threshold concentration for Cd in diet and information on the bioaccumulation factors from soil to earthworms and plants.

The threshold concentration in diet is derived from NOEC values from dietary toxicity studies with mammals and birds. For predators an additional assessment factor is often applied to the lowest NOEC value to compensate for both interspecies variation and lab-to-field extrapolation. The assessment factor should also compensate for differences in body weight to food intake ratios between wildlife species and laboratory animals, as well as metabolic stages (e.g., hibernation or migration) that might make predators more sensitive to contaminants in comparison with laboratory animals. In addition, the difference in calorific content of the diets of laboratory animals (e.g., grain) versus wildlife must be considered, meaning that wildlife must consume more food to obtain the same amount of energy, leading to a higher body burden of the pollutant (ECHA, 2008). An assessment factor of 30 is reported in European regulatory frameworks to be appropriate for chronic toxicity data for birds and mammals and a factor of 90–300 for subchronic (90 days) and subacute (28 days) data for mammals (ECHA, 2008).

Good quality chronic feeding studies using Cd-spiked diets were identified for three bird species (chicken, Japanese quail, and mallard duck) (Table 16.3). NOEC values for mortality, growth, and reproduction vary between 1.6 and ≥300 mg Cd/kg diet (fresh weight), depending on species and endpoint tested, whereas LOEC values vary between 10 and >300 mg Cd/kg diet. The lowest NOEC value was observed in a 90-day study with mallard ducks for spermatogenesis (White et al., 1978). Reliable chronic feeding studies for mammals cover four species (rat, rhesus monkey, lamb, and pig). NOEC and LOEC values for mortality, growth, and reproduction vary between 3.3 and ≥1350 mg Cd/kg diet (fresh weight) and between 10 and >1350 mg Cd/kg diet, respectively. The lowest NOEC value is observed for growth in a 9-year study with rhesus monkeys (Masaoka et al., 1994).

Applying the standard assessment factor of 30 in the EU to the lowest reliable $NOEC_{oral}$ data for birds and mammals results in the following threshold values: 0.053 and 0.11 mg/kg diet, respectively. It must be noted that these values are below most of the Cd concentrations in the control diets of the dietary studies (ranging from <0.03 to 0.32 mg/kg diet), stressing the overconservative nature of this approach. Lowering the default assessment factor by a factor of 5 to account for the availability of dietary toxicity data for several bird and mammal species yields dietary threshold concentrations of 0.27 mg Cd/kg diet and 0.55 mg Cd/kg diet for mammals and birds, respectively. These values are similar to or slightly above the upper range of the Cd concentrations in the control diets of the dietary studies (up to 0.32 mg/kg diet), indicating that the adapted assessment factors are more realistic but still conservative.

16.3.1.2 Bioaccumulation of cadmium in earthworms and plants

Bioaccumulation of Cd in earthworms is assessed based on data from field monitoring studies, measuring total Cd concentrations in both soil and earthworm samples (after depuration of their gut). This approach ensures that biota Cd burdens are in equilibrium with soil Cd concentrations. Data from laboratory studies where Cd was added to the soil as a Cd salt are hence excluded. Biota-to-soil accumulation factor (BSAF) values were identified for several earthworm species, belonging to different ecological groups of earthworms: anecic, endogeic, and epigeic earthworms.

BSAF values for Cd in earthworms vary between 0.1 and 24.2 $kg_{dw\ soil}/kg_{fw\ earthworm}$ (Andersen, 1979; Beyer et al., 1982; Dai et al., 2004; Ernst et al., 2008; Gish & Christensen, 1973; Hamers et al., 2006; Hobbelen et al., 2006; Huang et al., 2021; Nannoni et al., 2011; Spurgeon & Hopkin, 1996; van den Brink et al., 2010; Van Hook, 1974; Wang et al., 2018; Ma, 1982). The median BSAF for earthworms on a fresh weight basis is 2.3 kg_{dw}/kg_{fw} (10–90th percentiles are 0.7 and 8.5, respectively). No distinct differences in BSAF values across ecological groups could be identified. However, it must be noted that different species were sampled in different soils, which hampers a sound comparison of the data.

Cd is concentrated from the soil into earthworms (137 out of 171 BSAF values are higher than 1). The most important factor affecting the bioaccumulation of Cd by earthworms is the Cd concentration of the soil (Fig. 16.3). Soil pH, organic matter content, and cation exchange

TABLE 16.3 Dietary toxicity of cadmium to birds and mammals.

Test substance	Species	Exposure time (days)	Cd concentration in control diet	Most sensitive endpoint	NOEC (mg Cd/kg diet) (% effect)	LOEC (mg Cd/kg diet) (% effect)	References
Birds							
$CdCl_2$	Japanese quail (*Coturnix japonica*)	91	Not reported	Reproduction (egg number/bird)	<50	≤50 (10.4%)	Abou-kassem et al. (2020)
$CdSO_4$	Japanese quail (*Coturnix japonica*)	70	0.32 mg Cd/kg	Reproduction (egg mass g/day/bird)	5.3 (0.3%)	10.3 (7.1%)	Olgun (2015)
$CdCl_2$	Japanese quail (*Coturnix japonica*)	60	Not reported	Mortality	<150	≤150 (17.5%)	Saleemi et al. (2019)
$CdCl_2$	Mallard duck (*Anas platyrhynchos*)	84	0.1 mg Cd/kg	Kidney lesion	9.2	14.6	Cain et al. (1983)
$CdCl_2$	Mallard duck (*Anas platyrhynchos*)	90	0.08 mg Cd/kg	Spermatogenesis	1.6	15.2	White et al. (1978)
$CdSO_4$	Chicken (*Gallus domesticus*)	336	0.22 mg Cd/kg	Reproduction (egg production)	12.2 (−7.8%)	48.2 (39.2%)	Leach et al. (1979)

(*continued on next page*)

TABLE 16.3 Dietary toxicity of cadmium to birds and mammals—cont'd

Test substance	Species	Exposure time (days)	Cd concentration in control diet	Most sensitive endpoint	NOEC (mg Cd/kg diet) (% effect)	LOEC (mg Cd/kg diet) (% effect)	References
Mammals							
$CdCl_2$	Wistar rat (*Rattus norvegicus*)	30	<0.03 mg Cd/kg	Growth (body weight gain)	5 (5.0%)	25 (30.4%)	Callegaro et al. (2010)
$CdCl_2$	Wistar rat (*Rattus norvegicus*)	28	0.03 mg Cd/kg	Growth (body weight)	≥94.1 (6.5%)	>94.1	Hispard et al. (2008)
$CdCl_2$	Sprague-Dawley rats (*Rattus norvegicus*)	90	Not reported	Growth (body weight)	≥50 (3.6%)	>50	Huang et al. (2019)
$CdCl_2$	Wistar rats (*Rattus norvegicus*)	728	Not reported	Growth (body weight) males	10	50	Löser (1980)
$CdCl_2$	Rhesus monkey (*Macaca mulatta*)	3285	0.27 mg Cd/kg	Growth	3.3	10.3 (±25%)	Masaoka et al. (1994)
$CdCl_2$	Pigs (*Sus scrofa domesticus*)	83	0.11 mg Cd/kg	Growth (body weight gain)	5.1 (2.8%)	10.1 (18.6%)	Han et al. (2006a,b)
$CdCl_2$	Yorkshire pigs (*Sus scrofa domesticus*)	42	Not reported	Growth (weight gain/day)	50 (21.8%)	150 (60.0%)	Cousins et al. (1973)
$CdCl_2$	Lambs (*Ovis ammon aries*)	163	0.2 mg Cd/kg	Growth (body weight gain)	15.2 (16.7%)	30.2 (20%)	Doyle et al. (1974)

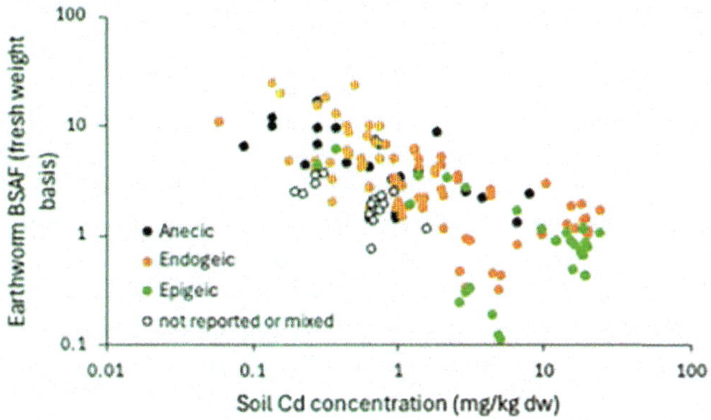

FIGURE 16.3 Field biota-to-soil accumulation factors (BSAF, fresh weight basis) for earthworms as a function of the Cd concentration in the soil (mg/kg dw).

capacity (CEC) all had a minor negative relationship with BSAF values for Cd in earthworms; that is, accumulation slightly decreased in soils with higher pH, organic matter content, and CEC. There was no significant distinction in accumulation of Cd between individual species or ecological groups of earthworms. The inverse relationship between the earthworm-to-soil accumulation factor for Cd and the Cd soil concentration is consistent with the general observation of inverse relationship between bioconcentration factor and exposure concentrations of metals. Uptake of metal ions by biota occurs via physiological mechanisms that exhibit saturable kinetics. In addition, physiological processes generally control elimination, sequestering, detoxification, and storage. As a result, biota often actively regulate metal bioaccumulation via dynamic feedback systems that respond to environmental loading and maintain homeostasis, and hence, the bioconcentration factor of metals is not independent of exposure (McGeer et al., 2003).

The influence of soil type, pH, CEC, and organic matter content of the soil was studied by Ma (1982). The body burden of Cd in earthworms was higher in sandy soils (3.4–5.5 mg/kg_{fw}) than in loamy soils (2.0–5.1 mg/kg_{fw}), although the soil Cd concentration was higher in the latter. The BSAF increased with decreasing pH, CEC, and soil organic carbon content. The soil pH was the most important soil factor for the BSAF. This is also reflected by the total dataset, where pH showed the strongest negative correlation with the BSAF values, followed by the organic carbon content of the soil (both $P < .001$). These observations for bioaccumulation of Cd in earthworms agree with the general findings for effect of soil properties on Cd toxicity of invertebrates (Table 16.2).

The median BSAF value for earthworms of 2.3 kg_{dw}/kg_{fw} was selected as the generic BSAF for Cd in earthworms for risk assessment. Omitting data for soils with Cd concentrations below 0.5 and above 10 mg/kg does not significantly affect this value.

Plants generally show a lower bioaccumulation of Cd from soils compared to earthworms. Soil–crop transfer models, predicting Cd concentrations in crops based on the pH, organic matter content, clay content, and total Cd concentration of a soil, have been developed for several crops. Predicted bioaccumulation factors vary between 0.03 and 1.03 (fresh weight based, average 0.28 kg_{dw}/kg_{fw}) for various crops at reasonable worst case soil conditions for Cd uptake (Cd concentration in soil of 1 mg/kg and reference soil properties of pH 6, 2% organic matter, and 10% clay) (Rietra et al., 2017).

16.3.1.3 Critical soil concentration for secondary poisoning of cadmium in soil for mammals and birds

A critical soil concentration for secondary poisoning of birds and mammals is calculated by dividing the lowest critical concentration in diet (0.27 mg/kg diet) by the BSAF values for Cd in earthworms (2.3) or plants (0.28). This yields soil Cd concentrations of 0.11 and 0.96 mg Cd/kg soil dw for carnivorous and herbivorous diet, respectively. The lowest value is more than 10-fold lower than the critical concentration for soil organisms directly exposed to Cd in soil (2.18 mg Cd/kg soil dw) and approximately the 25–35th percentile of background Cd concentrations in soils from Europe and conterminous United States, respectively (Reimann et al., 2014; Smith et al., 2013). According to this approach the risk of secondary poisoning in the soil-worm-bird/mammal is hence predicted to be far more critical than the risk of soil Cd affecting plants, invertebrates, or microorganisms.

16.3.2 Alternative approach based on critical cadmium concentration in kidneys

16.3.2.1 Introduction

The model for secondary poisoning of mammals and birds in the soil-worm-birds/mammals may, however, be questioned in the case of Cd: concentrations of Cd in earthworms sampled in uncontaminated areas typically range between 1 and 10 mg Cd/kg fresh weight and are generally higher than the critical concentration in the diet derived above (Spurgeon & Hopkin, 1996) have argued that the model is overly protective because of (1) a higher availability of Cd in metal salt spiked diets in the laboratory tests than Cd in worms and (2) problems of quantifying the BSAF: a mean BSAF would overestimate earthworm concentrations at more contaminated sites because the BSAF decreases with soil Cd (Fig. 16.3). Other authors, however, suggest that the model is not sufficiently protective and that additional safety factors should be used to account for differences in metabolic rate between laboratory and field animals, in calorific content between the laboratory diets and field prey, in food assimilation efficiency, and in bioavailability of the toxic compound (Smit et al., 2000).

A comparison of kidney Cd concentration in free-living common shrews (*Sorex araneus*), feeding on invertebrates, compiled by Shore and Douben (1994) with results from a laboratory feeding trial with shrews fed a $CdCl_2$-spiked diet (Dodds-Smith et al., 1992a, 1992b) clearly shows that the kidney Cd after only 11 weeks of exposure at the high Cd diets is significantly above that predicted by the field observations. This analysis therefore suggests that Cd is more available in laboratory spiked meals than in the diet of field animals and that the above approach for assessing risks of Cd to wildlife is overconservative.

An alternative strategy to estimate the risks of secondary poisoning by Cd using renal threshold was formulated by Spurgeon and Hopkin (1996) and presented in the EU Risk Assessment Report on Cd (ECB, 2007). The alternative method is based on kidney Cd concentrations of wildlife as an indicator of Cd exposure and risk and uses paired measurements of Cd concentration in kidney of wildlife and in the soil of their living area. The kidney is regarded as the critical organ in chronic Cd toxicity. With continued exposure, there is a continual increase in Cd concentration in the renal cortex until a critical value is reached and histopathological changes and renal dysfunction are found. This critical value should be regarded as a sublethal endpoint. The risk of secondary poisoning can hence be assessed by calculating the exposure at which this critical value is not exceeded for wildlife. The proposed approach overcomes the uncertainties in the traditional approach based on food-chain modeling because the proposed approach does not require assumptions about the diet (e.g., 100% earthworms) and about Cd bioavailability during the soil–food–wildlife transfer (i.e., the BSAF value and assumption of equal exposure at equal diet Cd between spiked meals and environmental Cd).

The ecological relevance of kidney damage as the critical endpoint in this assessment deviates from traditional endpoints such as growth or reproduction, which have obvious ecological relevance. Indeed, the relationship between ecological fitness and kidney damage is less known, which is the major difficulty in understanding effects of Cd in wildlife (Cooke, 2011). Seabirds showing signs of nephrotoxicity were "outwardly healthy" (Nicholson et al., 1983). A beaver population from the Mulde (a tributary of the Elbe in Germany) had average kidney Cd concentrations of 467 mg/kg dw, but the fertility of this population was higher than three other populations that had lower kidney Cd concentrations (Nolet et al., 1994). The authors conclude that the habitat quality seemed to determine the fertility and that Cd did not seem to interfere with fertility. Therefore the use of renal dysfunction as an endpoint with ecological relevance leads to a more conservative approach than traditional endpoints of growth or reproduction in most conditions.

There are several reviews of the critical kidney Cd concentrations in laboratory animals (ATSDR, 1999; International Programme on Chemical Safety (IPCS), 1992), small mammals (Cooke, 2011), and birds (Wayland & Scheuhammer, 2011). The World Health Organization (WHO) has reviewed several studies on test animals (mouse, rat, rabbit, pig, horse, monkey, and bird; International Programme on Chemical Safety (IPCS), 1992). Histological tubular lesions are usually observed at 200–300 mg Cd/kg fresh weight. In some studies on rats, monkeys, horses, and birds, certain effects were seen at lower levels. The WHO concludes that "if one wishes to establish a range of values for the critical concentration in individuals at which a small but significant part of the exposed population will show effects, animal studies indicate that a renal cortex level of about 100-200 mg/kg fresh weight is likely to coincide with such a range" (International Programme on Chemical Safety (IPCS), 1992). This range corresponds with whole kidney values of 400–800 mg/kg dw (fresh weight concentrations are about 0.2 times dw concentrations, and whole kidney concentrations are about 0.8 times kidney cortex values). Many experimental studies over 30 years using rats have supported the threshold of 100 mg/kg fresh weight (e.g., Mitsumori et al., 1998; Noël et al., 2004). The lower limit of the WHO-suggested critical range of Cd concentration, that is, 100 mg Cd/kg fresh weight in the renal cortex or 400 mg Cd/kg dw (International Programme on Chemical Safety (IPCS), 1992) is therefore taken forward for the risk assessment of Cd to wildlife. The WHO suggested range was based on the largest review of animal data and confirmed by more recent reviews (Cooke, 2011; Wayland & Scheuhammer, 2011).

16.3.2.2 Cadmium concentrations in soils and kidneys

Paired sets of measured soil Cd and kidney Cd concentrations were found in scientific literature for 23 mammal species and 3 bird species. Wildlife species are living and feeding in a limited area, whereas feed for domesticated animals originates from the local area. The mammal species cover 16 different rodent species (including hares, voles, beavers, shrews, rabbits, mice, moles, and gophers) and bats; larger wildlife species (fox, deer, and bison); and domesticated species (cattle, goat, and sheep). The bird species cover both wild species (crow and woodcock) and domesticated species (chicken). When Cd concentrations were reported on a fresh weight

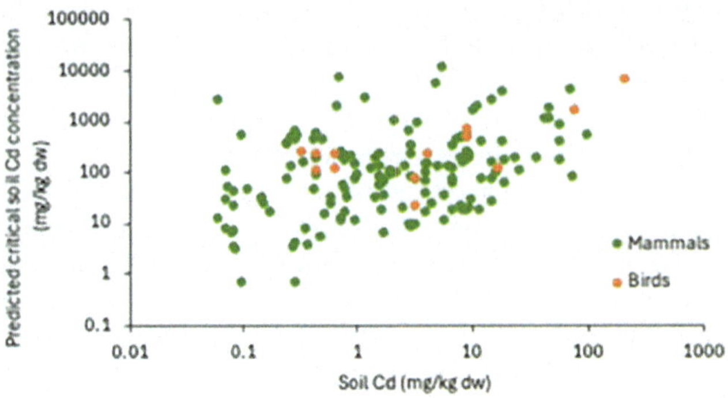

FIGURE 16.4 Predicted critical soil Cd concentrations at which critical renal concentration of 400 mg/kg dw is reached for mammals and birds as a function of the Cd concentration in the soil.

basis, they were recalculated to dw according to a typical moisture content in kidneys of 75%. Several of these studies have measured Cd concentrations at two or more locations differing in metal exposure (e.g., in the vicinity of smelters). The kidney concentrations expressed on a dw basis varied between 0.01 and 382 mg Cd/kg. Cd concentration in kidneys increases with increasing soil Cd concentrations in a nonproportional way, resulting in an inverse relationship of kidney to soil bioaccumulation factors for Cd and soil Cd concentrations. This is consistent with the bioaccumulation data of Cd in earthworms (Fig. 16.3) and general observations for metals (McGeer et al., 2003).

Several studies compared Cd accumulation in kidneys in different age groups of the same species (Carpenè et al., 2006; Janaydeh et al., 2018; Kar et al., 2018; Komarnicki, 2000; Samuel-Nakamura et al., 2017; Włostowski et al., 2006). All these results consistently showed higher Cd concentrations in kidneys of older animals compared to juveniles, confirming the continuous accumulation of Cd in kidneys with time.

Only a few studies reported soil properties of the soils studied. The pH was only reported for 36 out of the 164 observations but showed a general negative correlation with kidney-to-soil accumulation factors, indicating higher bioavailability of Cd at low pH, consistent with the results for bioaccumulation data for Cd in earthworms. Insufficient data are available to discuss the impact of other soil properties on the kidney-to-soil accumulation factors for Cd in mammals and birds.

16.3.2.3 Critical soil cadmium concentration for secondary poisoning

The critical soil Cd concentration is defined as the concentration at which a critical kidney Cd concentration of 400 mg/kg dw (whole kidney) is predicted using a proportional extrapolation from each paired soil Cd–kidney Cd concentration set. The lowest critical soil Cd concentration was selected for each species. This can be considered a conservative approach since kidney Cd only increases proportionally with the dose at low concentrations. The less than proportional increase of kidney Cd with soil Cd is observed in most studies and may be explained by, for example, the active uptake mechanisms of metal ions and avoidance of metal interactions for Cd absorption (most gradient studies are derived from sites with mixed metal pollution). So the lowest (i.e., most conservative) critical soil Cd concentrations are typically predicted at lower soil Cd and kidney Cd concentrations, whereas higher soil and kidney Cd concentrations result in higher predicted critical soil Cd concentrations (Fig. 16.4). Moreover, this approach selects the worst case conditions (soil properties, age of organisms) of all observations per species.

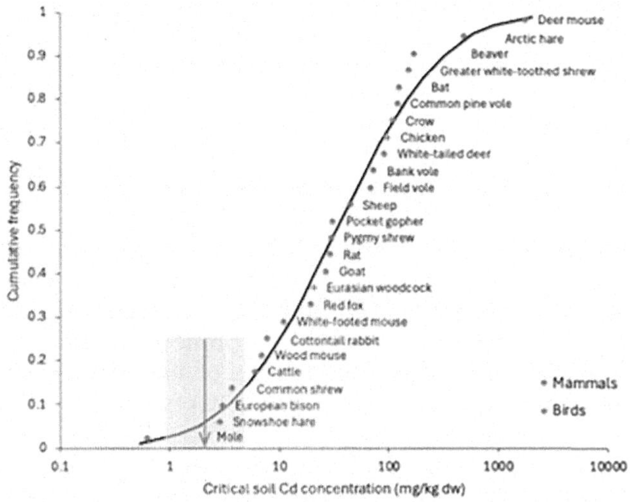

FIGURE 16.5 Cumulative frequency of the critical soil Cd concentration at which the critical kidney Cd concentration (400 mg/kg dw) may be exceeded in the average population of different species. The line is the normal distribution fitted on the log-transformed data. The arrow and shaded area show the HC_5 (hazard concentration protecting 95% of species) and its 5%–95% confidence interval derived from the fitted SSD (see text). *SSD*, Species sensitivity distribution.

The frequency distribution of the lowest predicted critical soil Cd concentrations for the 23 mammal and 3 bird species shows more than 3 orders of magnitude variation in predicted critical Cd concentrations among the species (Fig. 16.5). Moles have the lowest predicted critical Cd concentration in soil (0.6 mg/kg dw), whereas deer mice have the highest predicted critical Cd concentration in soil (1813 mg/kg dw).

The distribution of critical soil Cd concentrations reflects worst case species sensitivity if it is assumed that there are no differences in renal threshold between species. In this regard, this distribution is conceptually equal to an SSD of Cd thresholds from toxicity tests in soil. As for results for direct exposure and toxicity to soil organisms, the 5th percentile value (HC_5) of the best fitting SSD is selected as threshold value for risk assessment. The HC_5 associated with the best fitting distribution according to the Anderson–Darling statistic is 2.05 mg Cd/kg soil (5%–95% confidence limits: 0.88–4.91 mg Cd/kg).

Only two of the predicted critical soil Cd concentrations are below this HC_5 of 2.05 mg/kg soil. Both values (0.6 and 0.7 mg/kg soil) are derived for moles exposed to low Cd concentrations (≤0.3 mg/kg soil) in acidic soils (pH in 0.01 M $CaCl_2$ < 4.5) based on data from (Ma, 1987). Because of the low Cd concentration in these soils the predicted critical soil Cd concentrations derived from these data contain a safety factor because of the proportional extrapolation (see above). The predicted critical soil Cd concentration for moles in a soil with pH 5.4 and a Cd concentration of 1.8 mg/kg soil, as based on data from the same author, is 6.4 mg/kg soil, that is, well above the HC_5. No data were found on moles in soils that are acid and that contain high Cd concentrations. Other predicted critical soil Cd concentrations for moles vary between 3.2 and 16.7 mg/kg soil (Komarnicki, 2000; Ma, 1987). The overall geometric mean value of predicted critical soil Cd concentrations for moles is 4.1 mg/kg soil, a factor 2 above the HC_5. The next lowest predicted critical soil Cd concentration is 2.9 mg/kg for snowshoe hare (Amuno et al., 2018), but soil properties were not reported, and all other values are larger than 3.1 mg/kg soil. Therefore there is no indication of organisms or soil conditions that are not protected by the HC_5, and this value can be taken forward as a threshold value for assessment of secondary poisoning.

This HC_5 for protecting wildlife is slightly below, but very close to, the threshold value of 2.18 mg/kg soil derived for protecting plants, soil fauna, and microflora. Effectively, this assessment suggests that biotransfer of Cd from soil to higher trophic levels is a critical pathway for Cd and must be considered in the risk assessment of Cd in soil.

16.4 Concluding remarks and future perspectives

A wealth of data is available on Cd ecotoxicity in soil. Chronic NOEC and EC_{10} values for Cd toxicity to plants, soil invertebrates, and microbial processes vary between 0.5 and 4597 mg Cd/kg soil dw. Although pH and organic matter content of the soil affect the bioavailability and toxicity of Cd in soil, variation in toxicity across soil types is less pronounced than the variation in species sensitivities. A threshold for effects of direct exposure of soil organisms to Cd in soil can be predicted as 5th percentile of the species mean NOEC and EC_{10} values, yielding a concentration of 2.18 mg Cd/kg soil.

The assessment of secondary poisoning of Cd to wildlife via Cd exposure in food is complex, and we have shown that assumptions on food chain modeling and Cd availability in diet can result in unrealistic results. An alternative approach based on critical Cd concentrations in kidneys of mammals and birds in relation to soil Cd concentrations of their foraging area overcomes these assumptions and shows that secondary poisoning is indeed a critical pathway to consider for addressing the risks of Cd contamination in soil, with a 5th percentile of predicted critical soil concentrations of 2.05 mg Cd/kg soil, that is, slightly below the threshold for toxicity of soil Cd to soil-dwelling organisms.

The threshold values for Cd toxicity to soil organisms and wildlife are above the range in natural background concentrations in soils. For example, only 0.6%–1.2% of soils from Europe and conterminous United States have a measured Cd concentration larger than 2.05 mg/kg dw (Reimann et al., 2014; Smith et al., 2013).

Because of the large amount of toxicity data available the ecological risk assessments of Cd contamination in soil can be considered rather robust. A potential refinement for local assessments is a better quantitative assessment of the impact of soil properties, especially pH, on bioavailability and toxicity of Cd to soil organisms since the current information is rather scattered and not based on large comparative datasets as is available for some other metals (Smolders et al., 2009). There is currently only little field evidence that attributes toxicity to renal Cd concentrations above the critical kidney Cd concentration of 400 mg/kg dw and more field-based observations of Cd toxicity would therefore further strengthen the assessment of secondary poisoning.

16.5 Summary

- Using chronic EC_{10}/NOEC data for direct exposure and toxicity of Cd in soil to plants, invertebrates, and microbes gave a threshold of 2.18 mg Cd/kg dw soil.
- Soil pH and organic matter modulate Cd bioavailability, but the variation in Cd toxicity to soil organisms is mainly attributed to the interspecies differences in sensitivity to Cd. Field studies support the protectiveness of conclusions based on lab-based results.

- The classic assessment approach for secondary poisoning to mammals and birds based on dietary thresholds and accumulation in earthworms and plants can be overconservative for naturally occurring substances like Cd.
- An alternative renal-threshold approach results in a threshold of 2.05 mg Cd/kg dw soil, indicating biotransfer of Cd to wildlife is a critical risk pathway.
- Only ~0.6%–1.2% of European/US soils exceed both thresholds, so exceedance is uncommon under natural background conditions.

References

Abou-kassem, D. E., Abd El-Hack, M. E., Taha, A. E., Ajarem, J. S., Maodaa, S. N., & Allam, A. A. (2020). Detoxification impacts of ascorbic acid and clay on laying Japanese quail fed diets polluted by various levels of cadmium. *Animals, 10*(3). doi:10.3390/ani10030372.

Amuno, S., Jamwal, A., Grahn, B., & Niyogi, S. (2018). Chronic arsenicosis and cadmium exposure in wild snowshoe hares (*Lepus americanus*) breeding near Yellowknife, Northwest Territories (Canada), part 1: Evaluation of oxidative stress, antioxidant activities and hepatic damage. *Science of the Total Environment, 618*, 916–926. doi:10.1016/j.scitotenv.2017.08.278.

Andersen, C. (1979). Cadmium, lead and calcium content, number and biomass, in earth-worms (Lumbricidae) from sewage sludge treated soil. *Pedobiologia, 19*(5), 309–319. doi:10.1016/s0031-4056(23)02238-2.

ATSDR. (1999). *Toxicological profile for cadmium*. Agency for Toxic Substances and Disease Registry https://www.cadmium.org/wp-content/uploads/2021/11/UP_2010-06-27_11-11-38_Cd_ATSDR_99.pdf chrome-extension://efaidnbmnnnibpcajpcglclefindmkaj/.

Babich, H., & Stotzky, G. (1977). Effect of cadmium on fungi and on interactions between fungi and bacteria in soil: Influence of clay minerals and pH. *Applied and Environmental Microbiology, 33*(5), 1059–1066. doi:10.1128/aem.33.5.1059-1066.1977.

Beasley, A., Belanger, S. E., Brill, J. L., & Otter, R. R. (2015). Evaluation and comparison of the relationship between NOEC and EC10 or EC20 values in chronic *Daphnia* toxicity testing. *Environmental Toxicology and Chemistry, 34*(10), 2378–2384. doi:10.1002/etc.3086.

Beyer, W. N., Chaney, R. L., & Mulhern, B. M. (1982). Heavy metal concentrations in earthworms from soil amended with sewage sludge. *Journal of Environmental Quality, 11*(3), 381–385. doi:10.2134/jeq1982.00472425001100030012x.

Bingham, F. T., Page, A. L., Mahler, R. J., & Ganje, T. J. (1975). Growth and cadmium accumulation of plants grown on a soil treated with a cadmium-enriched sewage sludge. *Journal of Environmental Quality, 4*(2), 207–211. doi:10.2134/jeq1975.00472425000400020015x.

Cain, B. W., Sileo, L., Franson, J. C., & Moore, J. (1983). Effects of dietary cadmium on mallard ducklings. *Environmental Research, 32*(3), 286–297. doi:10.1016/0013-9351(83)90112-3.

Callegaro, M. G. K., Milbradt, B. G., Diettrich, T., Alves, E., Duarte, F. A., Flores, E. M. M., Dressler, V. L., Silva, L. P., & Emanuelli, T. (2010). Influence of cereal bran supplement on cadmium effects in growing rats. *Human and Experimental Toxicology, 29*(6), 467–476. doi:10.1177/0960327109357777.

Carpenè, E., Andreani, G., Monari, M., Castellani, G., & Isani, G. (2006). Distribution of Cd, Zn, Cu and Fe among selected tissues of the earthworm (*Allolobophora caliginosa*) and Eurasian woodcock (*Scolopax rusticola*). *Science of the Total Environment, 363*(1–3), 126–135. doi:10.1016/j.scitotenv.2005.06.023.

Chapman, P. M., Caldwell, R. S., & Chapman, P. F. (1996). A warning: NOECs are inappropriate for regulatory use. *Environmental Toxicology and Chemistry, 15*(2), 77–79. doi:10.1002/etc.5620150201.

Çikili, Y., Samet, H., & Dursun, S. (2016). Cadmium toxicity and its effects on growth and metal nutrient ion accumulation in Solanaceae plants. *Tarim Bilimleri Dergisi, 22*(4), 576–587. doi:10.1501/tarimbil_0000001416.

Cooke, J. A. (2011). Environmental contaminants in biota: Interpreting tissue concentrations. In *Cadmium in small mammals* (pp. 627–644). CRC Press. doi:10.1201/b10598-20.

Correa, J. E., Ramírez, R., Ruíz, O., & Leiva, E. I. (2021). Effect of soil characteristics on cadmium absorption and plant growth of *Theobroma cacao* L. seedlings. *Journal of the Science of Food and Agriculture, 101*(13), 5437–5445. doi:10.1002/jsfa.11192.

Cousins, R. J., Barber, A. K., & Trout, J. R. (1973). Cadmium toxicity in growing swine. *The Journal of Nutrition, 103*(7), 964–972. doi:10.1093/jn/103.7.964.

Dai, J., Becquer, T., Rouiller, J. H., Reversat, G., Bernhard-Reversat, F., Nahmani, J., & Lavelle, P. (2004). Heavy metal accumulation by two earthworm species and its relationship to total and DTPA-extractable metals in soils. *Soil Biology and Biochemistry, 36*(1), 91–98. doi:10.1016/j.soilbio.2003.09.001.

Dodds-Smith, M. E., Johnson, M. S., & Thompson, D. J. (1992a). Trace metal accumulation by the shrew *Sorex araneus*. *Ecotoxicology and Environmental Safety, 24*(1), 102–117. doi:10.1016/0147-6513(92)90039-6.

Dodds-Smith, M. E., Johnson, M. S., & Thompson, D. J. (1992b). Trace metal accumulation by the shrew *Sorex araneus*. II. Tissue distribution in kidney and liver. *Ecotoxicology and Environmental Safety, 24*(1), 118–130. doi:10.1016/0147-6513(92)90040-A.

Doelman, P., & Haanstra, L. (1984). Short-term and long-term effects of cadmium, chromium, copper, nickel, lead and zinc on soil microbial respiration in relation to abiotic soil factors. *Plant and Soil, 79*(3), 317–327. doi:10.1007/bf02184325.

Doyle, J. J., Peander, W. H., Grebing, S. E., & Pierce, J. O. (1974). Effect of dietary cadmium on growth, cadmium absorption and cadmium tissue levels in growing lambs. *Journal of Nutrition, 104*(2), 160–166. doi:10.1093/jn/104.2.160.

ECB. (2007). Cadmium oxide and cadmium metal. Part I - Environment. EUR 22919 EN. *European Union Risk Assessment report*: 72. Luxembourg: Office for Official Publications of the European Communities.

ECHA. (2008). Chapter R.10: Characterisation of dose [concentration]-response for environment. *Guidance on information requirements and chemical safety assessment*. European Chemicals Agency https://www.echa.europa.eu/documents/10162/17224/information_requirements_r10_en.pdf/bb902be7-a503-4ab7-9036-d866b8ddce69.

Ernst, G., Zimmermann, S., Christie, P., & Frey, B. (2008). Mercury, cadmium and lead concentrations in different ecophysiological groups of earthworms in forest soils. *Environmental Pollution, 156*(3), 1304–1313. doi:10.1016/j.envpol.2008.03.002.

Gish, C. D., & Christensen, R. E. (1973). Cadmium, nickel, lead, and zinc in earthworms from roadside soil. *Environmental Science and Technology, 7*(11), 1060–1062. doi:10.1021/es60083a011.

Haanstra, D. (1984). Glutamic acid decomposition as a sensitive measure of heavy metal pollution in soil. *Soil Biology and Biochemistry, 16*(6), 595–600. doi:10.1016/0038-0717(84)90078-6.

Hamers, T., van den Berg, J. H. J., van Gestel, C. A. M., van Schooten, F. J., & Murk, A. J. (2006). Risk assessment of metals and organic pollutants for herbivorous and carnivorous small mammal food chains in a polluted floodplain (Biesbosch, The Netherlands). *Environmental Pollution, 144*(2), 581–595. doi:10.1016/j.envpol.2006.01.020.

Han, X.-Y., Xu, Z.-R., Wang, Y.-Z., & Du, W.-L. (2006b). Effects of cadmium on serum sex hormone levels in pigs. *Journal of Animal Physiology and Animal Nutrition, 90*(9–10), 380–384. doi:10.1111/j.1439-0396.2006.00607.x.

Han, X. Y., Xu, Z. R., Wang, Y. Z., & Huang, Q. C. (2006a). Effect of cadmium on lipid peroxidation and activities of antioxidant enzymes in growing pigs. *Biological Trace Element Research, 110*(3), 251–263. doi:10.1385/BTER:110:3:251.

Hispard, F., de Vaufleury, A., Cosson, R. P., Devaux, S., Scheifler, R., Cœurdassier, M., Gimbert, F., Martin, H., Richert, L., Berthelot, A., & Badot, P.-M. (2008). Comparison of transfer and effects of Cd on rats exposed in a short experimental snail–rat food chain or to CdCl2 dosed food. *Environment International, 34*(3), 381–389. doi:10.1016/j.envint.2007.09.003.

Hobbelen, P. H. F., Koolhaas, J. E., & van Gestel, C. A. M. (2006). Bioaccumulation of heavy metals in the earthworms *Lumbricus rubellus* and *Aporrectodea caliginosa* in relation to total and available metal concentrations in field soils. *Environmental Pollution, 144*(2), 639–646. doi:10.1016/j.envpol.2006.01.019.

Huang, C., Ge, Y., Yue, S., Qiao, Y., & Liu, L. (2021). Impact of soil metals on earthworm communities from the perspectives of earthworm ecotypes and metal bioaccumulation. *Journal of Hazardous Materials, 406*, 124738. doi:10.1016/j.jhazmat.2020.124738.

Huang, X., Liu, T., Zhao, M., Fu, H., Wang, J., & Xu, Q. (2019). Protective effects of moderate Ca supplementation against Cd-induced bone damage under different population-relevant doses in young female rats. *Nutrients, 11*(4), 849. doi:10.3390/nu11040849.

International Programme on Chemical Safety (IPCS). (1992). *Cadmium. Environmental health criteria*: 134. World Health Organization, Geneva.

Irizar, A., Rodríguez, M. P., Izquierdo, A., Cancio, I., Marigómez, I., & Soto, M. (2015). Effects of soil organic matter content on cadmium toxicity in *Eisenia fetida*: Implications for the use of biomarkers and standard toxicity tests. *Archives of Environmental Contamination and Toxicology, 68*(1), 181–192. doi:10.1007/s00244-014-0060-4.

Janaydeh, M., Ismail, A., Omar, H., Zulkifli, S. Z., Bejo, M. H., & Aziz, N. A. A. (2018). Relationship between Pb and Cd accumulations in house crow, their habitat, and food content from Klang area, Peninsular Malaysia. *Environmental Monitoring and Assessment., 190*(1). doi:10.1007/s10661-017-6416-2.

Kadar, I. (1995). Effect of heavy metal load on soil and crop. *Acta Agronomica Hungarica, 43*(1-2), 3–9.

Kar, I., Mukhopadhayay, S. K., Patra, A. K., & Pradhan, S. (2018). Bioaccumulation of selected heavy metals and histopathological and hematobiochemical alterations in backyard chickens reared in an industrial area, India. *Environmental Science and Pollution Research., 25*(4), 3905–3912. doi:10.1007/s11356-017-0799-z.

Komarnicki, G. J. K. (2000). Tissue, sex and age specific accumulation of heavy metals (Zn, Cu, Pb, Cd) by populations of the mole (*Talpa europaea* L.) in a central urban area. *Chemosphere, 41*(10), 1593–1602. doi:10.1016/S0045-6535(00)00018-7.

Leach, R. M., Wang, K. W-L. i, & Baker, D. E. (1979). Cadmium and the food chain: The effect of dietary cadmium on tissue composition in chicks and laying hens. *The Journal of Nutrition, 109*(3), 437–443. doi:10.1093/jn/109.3.437.

Li, L. Z., Zhou, D. M., Luo, X. S., Wang, P., & Wang, Q. Y. (2008). Effect of major cations and pH on the acute toxicity of cadmium to the earthworm *Eisenia fetida*: Implications for the biotic ligand model approach. *Archives of Environmental Contamination and Toxicology, 55*(1), 70–77. doi:10.1007/s00244-007-9100-7.

Liu, H., Xuan, L., Zhou, J., Zhou, D., & Wang, Y. (2019). Effects of soil properties on cadmium toxicity to *Folsomia candida* (Collembola). *Bulletin of Environmental Contamination and Toxicology, 103*(1), 90–97. doi:10.1007/s00128-018-2514-2.

Löser, E. (1980). A 2 year oral carcinogenicity study with cadmium on rats. *Cancer Letters, 9*(3), 191–198. doi:10.1016/0304-3835(80)90086-5.

Ma, W.-c. (1982). The influence of soil properties and worm-related factors on the concentration of heavy metals in earthworms. *Pedobiologia, 24*(1), 109–119. doi:10.1016/s0031-4056(23)05872-9.

Ma, W.-c. (1987). Heavy metal accumulation in the mole, *Talpa europea*, and earthworms as an indicator of metal bioavailability in terrestrial environments. *Bulletin of Environmental Contamination and Toxicology, 39*(6), 933–938. doi:10.1007/BF01689581.

Mahler, R. J., Bingham, F. T., & Page, A. L. (1978). Cadmium-enriched sewage sludge application to acid and calcareous soils: Effect on yield and cadmium uptake by lettuce and chard. *Journal of Environmental Quality, 7*(2), 274–281. doi:10.2134/jeq1978.00472425000700020025x.

Masaoka, T., Akahori, F., Arai, S., Nomiyama, K., Nomiyama, H., Kobayashi, K., Nomura, Y., & Suzuki, T. (1994). A nine-year chronic toxicity study of cadmium ingestion in monkeys. I. Effects of dietary cadmium on the general health of monkeys. *Veterinary and Human Toxicology, 36*(3), 189–194.

McGeer, J. C., Brix, K. V., Skeaff, J. M., Deforest, D. K., Brigham, S. I., Adams, W. J., & Green, A. (2003). Inverse relationship between bioconcentration factor and exposure concentration for metals: Implications for hazard assessment of metals in the aquatic environment. *Environmental Toxicology and Chemistry, 22*(5), 1017–1037. doi:10.1897/1551-5028(2003)022⟨1017:IRBBFA⟩2.0.CO;2.

Mench, M., Tancogne, J., Gomez, A., & Juste, C. (1989). Cadmium bioavailability to *Nicotiana tabacum* L., *Nicotiana rustica* L., and *Zea mays* L. grown in soil amended or not amended with cadmium nitrate. *Biology and Fertility of Soils, 8*(1), 48–53. doi:10.1007/BF00260515.

Mertens, J., Wakelin, S. A., Broos, K., Mclaughlin, M. J., & Smolders, E. (2010). Extent of copper tolerance and consequences for functional stability of the ammonia-oxidizing community in long-term copper-contaminated Soils. *Environmental Toxicology and Chemistry, 29*(1), 27–37. doi:10.1002/etc.16.

Miller, J. E., Hassett, J. J., & Koeppe, D. E. (1976). Uptake of cadmium by soybeans as influenced by soil cation exchange capacity, pH, and available phosphorus. *Journal of Environmental Quality, 5*(2), 157–160. doi:10.2134/jeq1976.00472425000500020009x.

Mitsumori, K., Shibutani, M., Sato, S., Onodera, H., Nakagawa, J., Hayashi, Y., & Ando, M. (1998). Relationship between the development of hepato-renal toxicity and cadmium accumulation in rats given minimum to large amounts of cadmium chloride in the long-term: Preliminary study. *Archives of Toxicology, 72*(9), 545–552. doi:10.1007/s002040050541.

Moreno, J. L., Sanchez-Marín, A., Hernández, T., & García, C. (2006). Effect of cadmium on microbial activity and a ryegrass crop in two semiarid soils. *Environmental Management, 37*(5), 626–633. doi:10.1007/s00267-004-5006-6.

Nannoni, F., Protano, G., & Riccobono, F. (2011). Fractionation and geochemical mobility of heavy elements in soils of a mining area in northern Kosovo. *Geoderma, 161*(1-2), 63–73. doi:10.1016/j.geoderma.2010.12.008.

Nicholson, J. K., Kendall, M. D., & Osborn, D. (1983). Cadmium and mercury nephrotoxicity. *Nature, 304*(5927), 633–635. doi:10.1038/304633a0.

Noël, L., Guérin, T., & Kolf-Clauw, M. (2004). Subchronic dietary exposure of rats to cadmium alters the metabolism of metals essential to bone health. *Food and Chemical Toxicology, 42*(8), 1203–1210. doi:10.1016/j.fct.2004.02.017.

Nolet, B. A., Dijkstra, V. A. A., & Heidecke, D. (1994). Cadmium in beavers translocated from the Elbe river to the Rhine/Meuse estuary, and the possible effect on population growth rate. *Archives of Environmental Contamination and Toxicology, 27*(2), 154–161. doi:10.1007/BF00214257.

OECD (2006). Current approaches in the statistical analysis of ecotoxicity data: A guidance to application. In *OECD Environment Health and Safety Publications, Series on Testing and Assessment: No 54* (pp. 1–147). Paris: OECD Publishing. https://doi.org/10.1787/9789264085275-en.

OECD (2016). Guidance on the incorporation of bioavailability concepts for assessing the chemical ecological risk and/or environmental threshold values of metals and inorganic metal compounds. In *OECD Environment, Health and Safety Publications, Series on Testing and Assessment: No 259* (pp. 1–85). Paris: OECD Publishing. https://doi.org/10.1787/9789264274839-en.

Olgun, O. (2015). The effect of dietary cadmium supplementation on performance, egg quality, tibia biomechanical properties and eggshell and bone mineralisation in laying quails. *Animal, 9*(8), 1298–1303. doi:10.1017/s1751731115000579.

Prokop, Z., Cupr, P., Zlevorova-Zlamalikova, V., Komarek, J., Dusek, L., & Holoubek, I. (2003). Mobility, bioavailability, and toxic effects of cadmium in soil samples. *Environmental Research, 91*(2), 119–126. doi:10.1016/s0013-9351(02)00012-9.

Puglisi, E., Hamon, R., Vasileiadis, S., Coppolecchia, D., & Trevisan, M. (2012). Adaptation of soil microorganisms to trace element contamination: A review of mechanisms, methodologies, and consequences for risk assessment and remediation. *Critical Reviews in Environmental Science and Technology, 42*(22), 2435–2470. doi:10.1080/10643389.2011.592735.

Reber, H. H. (1989). Threshold levels of cadmium for soil respiration and growth of spring wheat (*Triticum aestivum* L.), and difficulties with their determination. *Biology and Fertility of Soils, 7*(2), 152–157. doi:10.1007/bf00292574.

Reimann, C., Birke, M., Demetriades, A., Filzmoser, P., & O'Connor, P. (2014). *Chemistry of Europe's agricultural soils – Part A: Methodology and interpretation of the GEMAS data set*. Schweizerbart Science Publisher https://www.schweizerbart.de/publications/detail/isbn/9783510968466/Geologisches_Jahrbuch_Reihe_B_Heft_B102.

Rietra, R. P. J. J., Mol, G., Rietjens, I. M. C. M., & Römkens, P. F. A. M. (2017). *Cadmium in soil, crops and resultant dietary exposure*. Wageningen Environmental Research. doi:10.18174/403611.

Sahraoui, A. S., Verweij, R. A., Belhiouani, H., Cheriti, O., van Gestel, C. A. M., & Sahli, L. (2021). Dose-dependent effects of lead and cadmium and the influence of soil properties on their uptake by *Helix aspersa*: An ecotoxicity test approach. *Ecotoxicology, 30*(2), 331–342. doi:10.1007/s10646-020-02331-z.

Sajwani, K. S., Ornes, W. H., Youngblood, T. V., & Alva, A. K. (1996). Uptake of soil applied cadmium, nickel and selenium by bush beans. *Water, Air, & Soil Pollution, 91*(3-4), 209–217. doi:10.1007/bf00666258.

Saleemi, M. K., Tahir, M. W., Abbas, R. Z., Akhtar, M., Ali, A., Javed, M. T., … Hassan, Z. U. (2019). Amelioration of toxicopathological effects of cadmium with silymarin and milk thistle in male Japanese quail (*Coturnix japonica*). *Environmental Science and Pollution Research, 26*(21), 21371–21380. doi:10.1007/s11356-019-05385-7.

Samuel-Nakamura, C., Robbins, W. A., & Hodge, F. S. (2017). Uranium and associated heavy metals in *Ovis aries* in a mining impacted area in northwestern New Mexico. *International Journal of Environmental Research and Public Health, 14*(8). doi:10.3390/ijerph14080848.

Sandifer, R. D., & Hopkin, S. P. (1996). Effects of pH on the toxicity of cadmium, copper, lead and zinc to *Folsomia candida* Willem, 1902 (Collembola) in a standard laboratory test system. *Chemosphere, 33*(12), 2475–2486. doi:10.1016/S0045-6535(96)00348-7.

Santos, F. C. F., Verweij, R. A., van Gestel, C. A. M., & Amorim, M. J. B. (2022). Toxicokinetics and toxicodynamics of copper and cadmium in the soil invertebrate *Enchytraeus crypticus* (Oligochaeta). *Ecotoxicology and Environmental Safety, 236*. doi:10.1016/j.ecoenv.2022.113485.

Shore, R. F., & Douben, P. E. T. (1994). The ecotoxicological significance of cadmium intake and residues in terrestrial small mammals. *Ecotoxicology and Environmental Safety, 29*(1), 101–112. doi:10.1016/0147-6513(94)90035-3.

Sinkakarimi, M. H., Solgi, E., & Hosseinzadeh Colagar, A. (2020). Subcellular partitioning of cadmium and lead in *Eisenia fetida* and their effects to sperm count, morphology and apoptosis. *Ecotoxicology and Environmental Safety, 187*. doi:10.1016/j.ecoenv.2019.109827.

Smit, C. E., van Wezel, A. P., Jager, T., & Traas, T. P. (2000). *Secondary poisoning of cadmium, copper and mercury: Implications for the maximum permissible concentrations and negligible concentrations in water and soil. RIVM report 601501 009*. RIVM.

Smith, D. B., Cannon, W. F., Woodruff, L. G., Solano, F., Kilburn, J. E., & Fey, D. L. (2013). *Geochemical and mineralogical data for soils of the conterminous United States: U.S. Geological Survey Data Series 801* https://pubs.usgs.gov/ds/801/.

Smolders, E., Oorts, K., Van Sprang, P., Schoeters, I., Janssen, C. R., McGrath, S. P., & McLaughlin, M. J. (2009). Toxicity of trace metals in soil as affected by soil type and aging after contamination: Using calibrated bioavailability models to set ecological soil standards. *Environmental Toxicology and Chemistry, 28*(8), 1633–1642. doi:10.1897/08-592.1.

Son, J., Mo, H. h., Kim, J. h., Ryoo, M. I., & Cho, K. (2007). Effect of soil organic matter content and pH on toxicity of cadmium to *Paronychiurus kimi* (Lee) (Collembola). *Journal of Asia-Pacific Entomology, 10*(1), 55–61. doi:10.1016/S1226-8615(08)60331-3.

Son, J., Shin, K.-, & Cho, K. (2009). Response surface model for predicting chronic toxicity of cadmium to *Paronychiurus kimi* (Collembola), with a special emphasis on the importance of soil characteristics in the reproduction test. *Chemosphere, 77*(7), 889–894. doi:10.1016/j.chemosphere.2009.08.047.

Speir, T. W., Kettles, H. A., Percival, H. J., & Parshotam, A. (1999). Is soil acidification the cause of biochemical responses when soils are amended with heavy metal salts? *Soil Biology and Biochemistry, 31*(14), 1953–1961. doi:10.1016/s0038-0717(99)00115-7.

Spurgeon, D. J., & Hopkin, S. P. (1996). Risk assessment of the threat of secondary poisoning by metals to predators of earthworms in the vicinity of a primary smelting works. *Science of the Total Environment, 187*(3), 167–183. doi:10.1016/0048-9697(96)05132-7.

van den Brink, N., Lammertsma, D., Dimmers, W., Boerwinkel, M. C., & van der Hout, A. (2010). Effects of soil properties on food web accumulation of heavy metals to the wood mouse (*Apodemus sylvaticus*). *Environmental Pollution, 158*(1), 245–251. doi:10.1016/j.envpol.2009.07.013.

Van Hook, R. I. (1974). Cadmium, lead, and zinc distributions between earthworms and soils: Potentials for biological accumulation. *Bulletin of Environmental Contamination and Toxicology, 12*(4), 509–512. doi:10.1007/bf01684990.

Wang, K., Qiao, Y., Zhang, H., Yue, S., Li, H., Ji, X., & Liu, L. (2018). Bioaccumulation of heavy metals in earthworms from field contaminated soil in a subtropical area of China. *Ecotoxicology and Environmental Safety, 148*, 876–883. doi:10.1016/j.ecoenv.2017.11.058.

Wang, X., Wu, M., Ma, J., Chen, X., & Hua, L. (2016). Modeling of acute cadmium toxicity in solution to barley root elongation using biotic ligand model theory. *Journal of Environmental Sciences (China), 42*, 112–118. doi:10.1016/j.jes.2015.06.019.

Wayland, M., & Scheuhammer, A. M. (2011). *Cadmium in birds environmental contaminants in biota: Interpreting tissue concentrations* (2nd ed., pp. 645–668). doi:10.1201/b10598-21.

White, D. H., Finley, M. T., & Ferrell, J. F. (1978). Histopathologic effects of dietary cadmium on kidneys and testes of mallard ducks. *Journal of Toxicology and Environmental Health, 4*(4), 551–558. doi:10.1080/15287397809529678.

Włostowski, T., Bonda, E. lż, & Krasowska, A. (2006). Free-ranging European bisons accumulate more cadmium in the liver and kidneys than domestic cattle in north-eastern Poland. *Science of the Total Environment, 364*(1-3), 295–300. doi:10.1016/j.scitotenv.2005.12.009.

Zhang, X., Wu, H., Ma, Y., Meng, Y. u, Ren, D., & Zhang, S. (2020). Intrinsic soil property effects on Cd phytotoxicity to *Ligustrum japonicum* 'Howardii' expressed as different fractions of Cd in forest soils. *Ecotoxicology and Environmental Safety, 206*, 110949. doi:10.1016/j.ecoenv.2020.110949.

CHAPTER

17

Soil cadmium standards for agricultural land

Sha Zhang[a,b], *Jing Song*[c,d], *Yongming Luo*[c], *Roger Brewer*[e]

[a]State Key Laboratory of Regional and Urban Ecology, Ningbo Observation and Research Station, Institute of Urban Environment, Chinese Academy of Sciences, Xiamen, China [b]Zhejiang Key Laboratory of Pollution Control for Port-Petrochemical Industry, CAS Haixi Industrial Technology Innovation Center in Beilun, Ningbo, China [c]State Key Laboratory of Soil and Sustainable Agriculture, Institute of Soil Science, Chinese Academy of Sciences, Nanjing, China [d]University of Chinese Academy of Sciences, Nanjing, China [e]Hawai'i Department of Health, Honolulu, HI, United States

17.1 Introduction

Cadmium (Cd) contamination in agricultural soils represents a significant environmental challenge to the sustainable development of global agriculture (Faroon et al., 2012; McLaughlin et al., 2021; Zhao et al., 2023) (see Chapter 3). As a highly toxic heavy metal, Cd not only harms crop growth but can also enter the human body through the food chain, posing serious health risks (Zhang et al., 2012; Zhao et al., 2023). Dietary intake is the primary source of Cd exposure for the general population, particularly nonsmokers (World Health Organization, 2014; Zhao et al., 2023) (see Chapter 15). Consequently, establishing scientifically sound and practically feasible soil Cd standards for agricultural land is a critical component of environmental protection and food safety management in many countries. However, because of variations in soil type and other natural environmental conditions, socioeconomic development levels, and policy objectives, significant differences exist in how nations set their soil Cd standards. These differences not only reflect each country's priorities in addressing soil pollution but also highlight their distinct strategies in balancing agricultural production, food safety, and human health protection.

Cadmium Contamination in Soils and Plants: Mechanisms and Mitigation.
DOI: https://doi.org/10.1016/B978-0-443-34104-5.00012-8

17.2 Cadmium behaviors in a soil environment

Cd is widely present in agricultural environments because of both geological processes and anthropogenic influences. Low levels of Cd are common in natural soils, whereas high levels can cooccur with other heavy metals in mineral deposits such as in lead–zinc ores (Zhang et al., 2012), phosphorites (Page et al., 1987), or pyrite (Cao et al., 2021). Mining activities often result in the widespread release of Cd into the environment. Industrial uses of Cd include manufacturing of products such as nickel–Cd batteries, dyes, electroplating, and plastic additives (Faroon et al., 2012). Amendment of agricultural soils with sludges from such industries or use of wastewater for irrigation can result in contaminated land as well as crops. In addition, the application of phosphate fertilizers can lead to a significant increase in soil Cd levels (Chapter 5).

Cd exhibits relatively high mobility in soils and can be readily absorbed by plants. The bioavailability of Cd in soils is highly sensitive to pH. As pH decreases the availability of Cd increases sharply (see Chapter 6). Continuously flooded field conditions can lead to the formation of insoluble Cd sulfide. Redox fluctuations can subsequently lead to the rerelease of stable Cd into soil solution, making it available for uptake into plants (see Chapter 7). As a result, Cd can exhibit high mobility in both lowland paddy soils and upland soils.

17.3 Basis of soil and food cadmium standards

17.3.1 Development of standards

Cd standards for agricultural soils are developed based on specific protection goals, which can be broadly classified into three categories (Table 17.1):

1. Direct-exposure standards based on health risk assessments that consider the protection of farm workers and their families from exposure to Cd via incidental ingestion, dermal absorption, and dust inhalation.
2. Food safety standards, with standards based on targeted limitations of Cd in food crops and statistical relationships between soil and crop Cd concentrations.
3. Ecological protection standards, which assess risks to soil biota and ecosystem functions because of adverse uptake of Cd (see Chapter 16). Standards might also be based on other factors, such as natural background. Different countries emphasize different approaches depending on regulatory priorities. For example, Canada focuses on health and ecological risks, whereas Germany emphasizes crop quality and food safety.

This chapter focuses on standards derived from the perspective of food safety and the uptake of Cd into edible crops. To derive soil Cd standards (as demonstrated in Fig. 17.1), two key relationships are typically established: (1) The dose–response relationship between dietary Cd exposure and adverse health effects, used to develop standards for food (Faroon et al., 2012; Zhao et al., 2023), and (2) The quantitative relationship between Cd concentrations in soil and the corresponding uptake of Cd into crops, used to develop corresponding standards for Cd in soil (Environment Agency, 2009). The European Food Safety Authority (EFSA) Panel on Contaminants in the Food Chain has established a tolerable weekly intake (TWI) for Cd at 2.5 $\mu g\ kg^{-1}$ body weight per week (European Food Safety Authority (EFSA), 2009). This is more stringent than the provisional TWI (PTWI) of 7 $\mu g\ kg^{-1}$ body weight per week previously

TABLE 17.1 Soil Cd standards for agricultural land worldwide.

Country/region	Protection goal	Basis of standard	Example soil Cd standard (mg kg^{-1})	Key features	References
China	Food safety	Risk-based (GB 15618-2018)	0.3–0.8[a]	Hierarchical: priority protection, safe utilization, strict control	Ministry of Ecology and Environment of the People's Republic of China and Ministry of Agriculture and Rural Affairs of the People's Republic of China (2018)
European Union	Food safety and ecological protection	National standards vary	0.5–3.0[b]	Not harmonized EU-wide; precautionary, often crop-focused	Carlon and D'Alessandro (2007), Carlon et al. (2008)
Germany	Food safety	Statistical (soil–crop data correlation)	1.0–3.0	Emphasizes soil–crop thresholds to limit crop Cd uptake	Carlon and D'Alessandro (2007), Carlon et al. (2008)
Canada	Direct exposure and ecological protection	Health/ecological risk assessment	0.6–1.5[c]	Multipathway exposure considered; includes background levels	Canadian Council of Ministers of the Environment (1999)
United States	Food safety and direct exposure	Risk-based biosolids regulations	No uniform soil Cd limit[d]	Limits Cd in biosolids applied to land (e.g., 39 mg/kg in biosolids)	United States Environmental Protection Agency (1993, 2024), Brown et al. (2023), Fitzpatrick et al. (2017)
Australia	Food safety and direct exposure	Risk-based (NEPM guidelines)	1.0–3.0	Based on exposure scenarios and crop uptake modeling	Sneath (2020), Gray and Cavanagh (2023)
Japan	Food safety	Risk-based, soil content standard (soil leaching standard, mg L^{-1})	45 (0.003/0.009[e])	Focused on rice safety; low thresholds to prevent crop exceedance	Ministry of the Environment, Government of Japan (2020)

(continued on next page)

TABLE 17.1 Soil Cd standards for agricultural land worldwide—cont'd

Country/region	Protection goal	Basis of standard	Example soil Cd standard (mg kg^{-1})	Key features	References
Netherlands	Direct exposure, natural background	Target and intervention values	0.8 (target), 12 (intervention)	Differentiates background from remediation trigger levels	Carlon and D'Alessandro (2007), Carlon et al. (2008)
Czech Republic	Natural background	Screening values for negligible risk	0.5	Estimated and compared or added to (average) background concentrations	Vácha et al. (2014)

[a] China's soil Cd standards for food safety and direct exposure vary by land use class and region; for example, 0.2 mg kg^{-1} for priority protection to up to 1.0 mg kg^{-1} under strict control.

[b] EU values vary by member states, for example, 0.5 mg kg^{-1} in some countries, up to 3 mg kg^{-1} in others.

[c] Canada's soil guidelines depend on land use and exposure; 0.6 mg kg^{-1} often cited for agricultural land.

[d] United States does not have a national agricultural soil Cd standard but regulates Cd content in sludges applied to agricultural land based on potential uptake into food crops and food safety concerns (USEPA 1993).

[e] The threshold of 0.009 mg L^{-1} (i.e., threefold the soil leaching standard) is applied when the contaminated soil is separated from the groundwater table and groundwater Cd concentrations do not exceed 0.003 mg L^{-1}.*Cd*, cadmium.

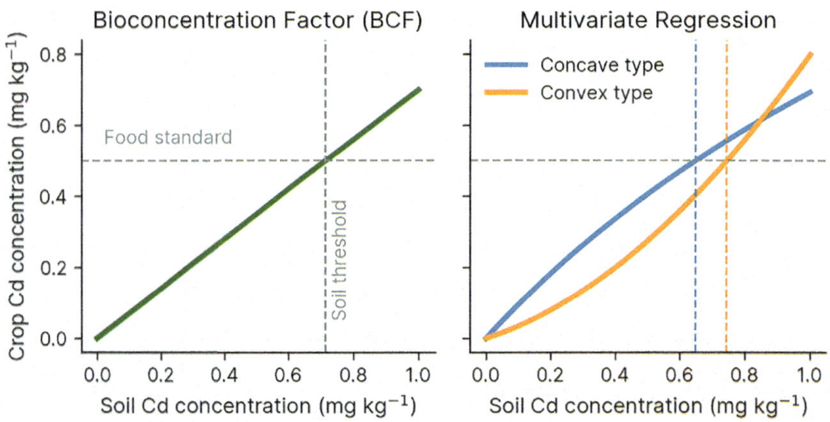

FIGURE 17.1 Conceptual comparison of bioconcentration factor and multivariate regression approaches for estimating Cd transfer from soil to crops. Left: The BCF model assumes a constant linear relationship between soil and crop Cd concentrations. Right: Nonlinear relationships (e.g., concave or convex) assume variable Cd uptake into a crop as a function of the concentration of Cd in the soil and/or other environmental or physiological factors (single variable Cd concentration shown for simplicity). When a food safety threshold for crop Cd is specified, for example, 0.5 mg kg^{-1}, a corresponding soil Cd threshold is derived as indicated by vertical lines. *BCF*, Bioconcentration factor; *Cd*, cadmium.

set by the Joint FAO/WHO Expert Committee on Food Additives (Schrenk et al., 2020). This health-based guidance value is derived from toxicological studies investigating the relationship between dietary Cd intake, kidney Cd accumulation, and the dose–effect relationship between urinary Cd levels and kidney damage. A more comprehensive risk assessment framework follows the general principles outlined in recent EFSA chemical risk assessments such as the 2020 update for nickel, which incorporates benchmark dose modeling, dietary exposure assessment, and margin of exposure considerations to ensure a comprehensive evaluation (Schrenk et al., 2020). However, it is widely recognized that dietary exposure is the primary pathway of Cd exposure for the general population (Zhao et al., 2023). Based on dietary Cd risk assessments, food safety standards can be derived. For example, the maximum allowable Cd concentration in rice is set at 0.2 mg kg^{-1} by the China National Food Safety Standard (GB2762-2022), 0.4 mg kg^{-1} by the Codex Alimentarius Commission and Japan, and 0.15 mg kg^{-1} by the EFSA (see Chapter 15 for more details). Similarly, Cd standards for other agricultural products can be developed.

Food safety standards for Cd are primarily designed to protect human health over a lifetime of low-level exposure, reflecting the cumulative nature of Cd-induced diseases. Although lifetime exposure forms the basis of risk assessment and regulatory limits such as PTWIs, special attention is given to protecting sensitive subpopulations, notably young children from birth to approximately 5 or 6 years of age, who are particularly vulnerable during critical developmental periods (United States Environmental Protection Agency, 1989a, 1989b). Minimizing exposure to ensure the protection of developing children is anticipated to be protective of potential adverse health effects for adults over longer periods of exposure. Toxicity factors, including PTWIs and food standards, are thus intended to safeguard against the mean daily Cd exposure averaged over a lifetime but incorporate additional safety considerations for early childhood exposure. Although daily Cd concentrations in foods like rice may vary, assessment of risk and the derivation of standards are based on the mean exposure concentrations over the specified exposure duration, as this sufficiently captures the chronic exposure relevant to health outcomes.

17.3.2 Collection of sample data for comparison to standards

Soil and/or crop samples must correspondingly be collected in a manner that directly estimates the mean Cd concentration in a targeted area and volume of crop or soil. "Grab" or "discrete" samples collected from single points or inadequately prepared "composite" samples recommended by some regulatory agencies in the past have since been demonstrated to yield unreliable data (Brewer et al., 2017). The collection of a representative sample for assessment of health risk begins with the designation of a well-thought-out "decision unit (DU)" area and volume of crop or soil of concern. A single "multiple increment" or "Multi Increment" sample is prepared by collecting and combining small portions (increments) of the media from many points within the designated DU (Association of American Feed Control Officials, 2015, 2018; CRC for Contamination Assessment and Remediation of the Environment, 2024b; Follett, 2010; Gourley & Weaver, 2019; National Standardization Administration and State Administration for Market Regulation, 2023) (Multi Increment is a registered trademark of Envirostat, Inc.). Field studies indicate that a representative, 1–3 kg sample can normally be prepared by the collection and combination of 40–60 increments of soil from within a designated DU in a systematic, random manner. The sample is then air-dried and sieved to the targeted particle size, and a representative subsample collected for testing by collecting a minimum 10 g, Multi Increment subsample. This method of sample collection and processing ensures that the data ultimately provided by the laboratory is representative of the sample provided and the sample provided is representative of the targeted area and volume of soil in the field. A similar approach is used to collect a sample of a targeted area and volume of a crop (refer to CRC for Contamination Assessment and Remediation of the Environment, 2024a). The resulting data can then be confidently assumed to represent the mean concentration of Cd for the targeted DU area and volume of crop or soil.

The careful collection and testing of food crop and soil samples using DU and "Multi Increment Sample" (DU–MIS) approaches is especially important for reliable derivation of Cd uptake factors. Independent, replicate MISs (minimum two) can be collected and tested to verify the precision of the overall sampling method (CRC for Contamination Assessment and Remediation of the Environment, 2024b). Data for soil and crops grown under similar soil and related environmental conditions can be used to develop reliable bioconcentration factors for different crops and predict a threshold of Cd in soil at which the food standard for the target crop will not be exceeded (Römkens et al., 2011). Establishing robust relationships between Cd concentrations in soil and crops for fields that have similar environmental conditions allows the development of reliable and soil Cd thresholds specific to those conditions and crop type (Environment Agency, 2009). The use of multivariate regression analysis of soil and crop data for different environmental conditions could, in theory, allow for development of an all-encompassing model that is able to reliably predict Cd uptake into a specified crop under different soil and environmental conditions. Refer to National Standardization Administration and State Administration for Market Regulation (2023) and CRC for Contamination Assessment and Remediation of the Environment (2024b) for additional discussion of sample collection methods.

Nationwide-scale mapping of Cd distribution in soil can occur at a resolution of several tens of kilometers. Low-cost and time-intensive samples are often collected from relatively small areas of land and assumed to be representative of immediately surrounding soil. Trends between isolated data points are assumed to be linear. Both assumptions are required for the use of geostatistical models to generate isoconcentration maps. Such sampling methods allow for large-scale patterns of Cd distribution in soil to be identified but can be unreliable at a regional or local level.

More advanced, DU–MIS sampling methods discussed in Section 17.3 are required for reliable characterization of Cd distribution in soil at a regional and local scale (CRC for Contamination Assessment and Remediation of the Environment, 2024b; National Standardization Administration and State Administration for Market Regulation, 2023). Preparation of a sample by collection of increments across an entire, targeted DU will not be practical in some cases. Collaboration with local soil experts and agencies familiar with historical contamination risks will be required to group land areas into DUs for the collection of soil samples. Nonagricultural areas should be excluded from DUs and sample collection areas. DU designation and sample collection could be further restricted to lands specifically proposed for food crop cultivation. Geostatistical modeling and knowledge of the historical industrial and agricultural use of the area can be used to fill in data gaps between individually tested and discontinuous DUs.

Based on the associated risk levels and functions, threshold values can be roughly divided into three categories: Target values, screening values, and intervention values (Fig. 17.2). Target values, often close to natural background levels of Cd in soil, are based on negligible-to-low risk levels and represent long-term goals for soil environmental protection. Screening values are derived from moderate risk levels for different land use types. If soil contamination is below the

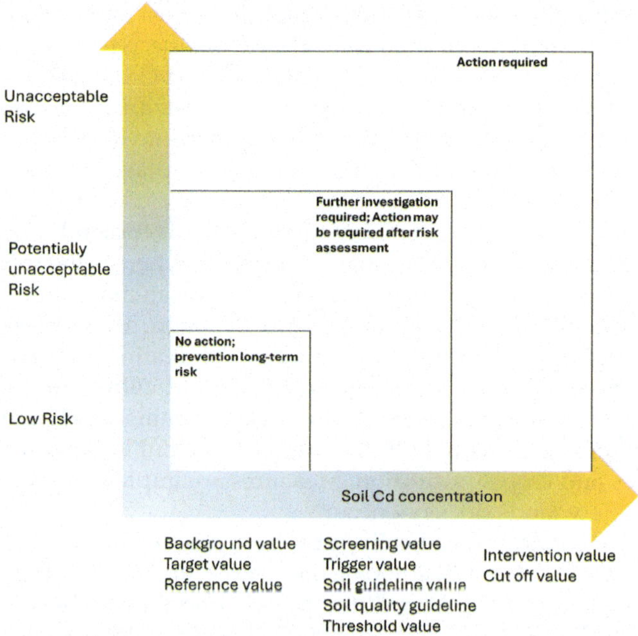

FIGURE 17.2 Overview of soil Cd standards and risk management. Threshold values are classified into target, screening, and intervention levels according to associated risk levels and management objectives. Target values represent long-term soil protection goals near background concentrations; screening values indicate acceptable risk ranges guiding monitoring and land-use management; and intervention values denote unacceptable risk levels requiring remediation or restrictions on food crop cultivation based on site-specific risk assessment. *Cd*, Cadmium.

corresponding screening level, the risks are considered acceptable. More frequent monitoring and investigation are recommended if the screening level is approached or exceeded. Risks might still be mitigated through adjustments in land use without necessarily requiring remediation. Intervention values are based on potentially unacceptable risk levels for agricultural land uses. Remedial action or a restriction on the cultivation of food crops is typically required if soil contamination exceeds the intervention level. Soil cleanup targets are determined based on site-specific risk assessments that consider such parameters as the physiochemical nature of the soil in question, contaminant mobility and potential for leaching to groundwater or uptake into crops, and the specific type of crop proposed for cultivation.

17.4 Current status and differences in soil cadmium standards for agricultural land

Establishment of a universally applicable, fixed soil standard protective of adverse uptake of Cd into food crops is not practicable because of the inherent variability of uptake between different crops, location-specific environmental and cultivation conditions that affect uptake, variability of dietary consumption rates for different crops, perceived acceptable risk, and population sensitivity across countries and regions. As a result, countries and regions have developed soil Cd management systems specifically tailored to their agriculture cultivation characteristics, soil properties, dietary habits, and risk management needs. Several examples are given in the following sections.

China's Cd standard system for agricultural soils is more developed than most other countries' and structured to be hierarchical and regionally adaptive. Based on the paired soil–crop data the standard *Soil Environmental Quality—Risk Control Standard for Soil Contamination of Agricultural Land* (GB 15618-2018) classifies farmland into three categories with respect to potential contamination risk (Table 17.2): (1) minimal risk of adverse Cd uptake into food crops, (2) potential risk of adverse uptake into food crops, and (3) high risk of adverse uptake into food crops.

Different Cd screening values are established based on soil type (paddy fields/upland) and pH levels. For instance the risk screening values range from 0.3 to 0.8 mg kg^{-1}, whereas the risk control values range from 1.5 to 4.0 mg kg^{-1}. A mean concentration of Cd for an agricultural field below the Risk Screening Level normally does not require additional action. These lands are placed under high priority for protection against future contamination. Mitigation methods are limited for agricultural fields that exceed the Cd Risk Control Level, and cultivation of food crops is generally banned. A concentration of Cd between the Risk Screening Level and the Risk Control Level requires additional monitoring of cultivated food crops but is generally manageable through agronomic practices and crop substitution. Measures are implemented to ensure that additional contamination of these lands does not occur.

In addition, China introduced stricter standards specifically for wheat production in 2022, that is, *Safety Thresholds of Cd, Pb, Cr, Hg, As in Soil for Wheat Production*. These standards specify soil Cd thresholds ranging from 0.2 to 0.36 mg kg^{-1} under different pH conditions to ensure that wheat grain Cd content does not exceed 0.1 mg kg^{-1} (GB2762-2022). Local regulations such as DB36/T 1819.2-2023 further refine management measures by conducting site-specific risk assessments for areas exceeding the risk screening values, combining soil–plant sampling and analysis. However, practical challenges remain. For example, highly acidic soils have been noted to significantly increase Cd bioavailability in priority protection areas with naturally low levels of Cd. Even if soil Cd concentrations do not exceed Tier 1 standards (0.3 mg kg^{-1} under pH <5.5), crops might still accumulate Cd at levels that exceed food safety limits (Yang et al., 2021). Conversely, certain crops have been shown to still be safely grown in neutral to alkaline soils, even when soil Cd concentrations exceed risk control levels (Feng et al., 2020; Xiao et al., 2022; Xu et al., 2022; Yang et al., 2021). This indicates the need for soil Cd standards to be dynamically adjustable in accordance with case-specific soil properties and crop types.

New Zealand adopted a related approach that combines ecotoxicology and human health risk assessment (HHRA) (Cavanagh, 2013) and developed a national *Cadmium Management Strategy* (CMS) (Gray & Cavanagh, 2023; Sneath, 2020). The country also offers an online tool *Cadmium Balance Model Decision Support Tool (CadBal)* for fertilizer management in agricultural soils (https://www.fertiliser.org.nz/cadbal/Calculator). The CMS recommends that soil Cd concentrations in agricultural land for food crops not exceed 1.0 mg kg^{-1} (based on safe application of biosolids). Adjustments are allowed based on specific site conditions. This system introduces a five-tier management framework. Tier 0 represents natural background levels of Cd. Tier 1's trigger value (0.6 mg Cd kg^{-1} soil) is defined as the 99th percentile of background concentrations. More proactive management measures are required when the Tier 1 level is exceeded, such as a reduction of phosphate fertilizer application or monitoring of Cd accumulation trends. Tiers 2–4 correspond to 1.0, 1.4, and 1.8 mg kg^{-1} Cd in soil, respectively, serving as progressively increasing Cd management thresholds and testing of food crops cultivated in the field. Notably, New Zealand's agriculture is predominantly livestock-based. Soil Cd standards subsequently focus more on the safety of forage and feed crops and the uptake of Cd into livestock to be used

TABLE 17.2 Soil Cd standards for agricultural land in China (after GB 15618-2018).

			Soil pH			
Standard type	**Action**	**Cultivation method**	**pH: ≤5.5**	**5.5 < pH ≤ 6.5**	**6.5 < pH ≤ 7.5**	**pH > 7.5**
Risk screening level	• Minimal risk of exceedance of food safety standard • No action required	Paddy	≤0 3	≤0.4	≤0.6	≤0.8
		Upland	≤0.3	≤0.3	≤0.3	≤0.6
Risk screening level <Cd< risk control level	• Potential risk of exceedance of food safety standard • Generally manageable through agronomic measures or crop substitution • Expanded monitoring of soil and crops required	Paddy	>0.3 ≤1.5	>0.4 ≤2.0	>0.6 ≤3.0	>0.8 ≤4.0
		Upland	≥0.3 ≤1.5	>0.3 ≤2.0	>0.3 ≤3.0	>0.6 ≤4.0
Risk control level	• High risk of exceedance of food safety standard • Limited mitigation options • Strict control measures against the cultivation of food crops are generally required	All crop cultivation methods	>1.5	>2.0	>3.0	>4.0

Cd, cadmium.

for human consumption or adverse health effects to the livestock in general rather than directly targeting grain or other crops to be directly used for human consumption. This flexibility allows the standards to better adapt to the natural and socioeconomic conditions of different regions and is reflective of a risk-based management philosophy.

The European Union (EU) does not have a unified soil Cd limit for agricultural soils. Individual member states are instead permitted to develop region-specific management policies based on soil characteristics, agricultural practices, and local food standards (Carlon & D'Alessandro, 2007; Carlon et al., 2008). For example, the Netherlands has set relatively strict soil Cd limits, stipulating that the concentration of Cd in agricultural soil should not exceed 0.8 mg kg^{-1} (dry weight, based on the mean concentration of Cd in 75.5% of investigated sites). Risk target values and intervention values are used to implement tiered management. By strictly controlling Cd content in food in EU countries the producers have the responsibilities to manage soil to minimize Cd accumulation in crops through the promotion of low-Cd phosphate fertilizers, control of fertilizer application rates, and regular monitoring of high-risk crop-growing areas.

Japan's Cd standard system is based on the Food Sanitation Law, with a primary focus on Cd content in rice. Japan stipulates that the Cd content of rice should not exceed 0.4 mg kg^{-1} (Japan Ministry of the Environment, 2002). This threshold in theory leads to an exceedance of the JECFA recommended exposure limit of 25 μg (kg body weight)$^{-1}$ month^{-1}, assuming an average rice intake for 1- to 6-year-old children in Japan of approximately 1.35 kg week^{-1} (Shinsugi & Takimoto, 2025) and an average body weight for 1- to 6-year-old children of 15 kg (United States Environmental Protection Agency, 2024). The Japan standard for farmland soil Cd concentration is 45 mg kg^{-1} (Ministry of the Environment, Government of Japan, 2020). However, the measured concentration of Cd in rice takes precedence over the soil Cd standard. This is because bioavailable Cd in soil is influenced by numerous factors, such as water management during rice cultivation, making it difficult to establish a uniform, maximum allowable concentration of Cd in soil for all fields in general. Japan's approach involves incentivizing farmers to reduce Cd content in rice through agricultural management practices (e.g., water management), thereby achieving food safety goals (Vannan et al., 2020). This method avoids limitations on agricultural production caused by overly stringent soil standards while promoting innovation and application of agricultural technologies.

The United States publishes standards for Cd in biosolids applied to agricultural land but not to soil (United States Environmental Protection Agency, 1993). Considerations for biosolids include ceiling concentrations for heavy metals in biosolids and limits on monthly average concentrations of heavy metals in applied biosolids, annual loading rates to soil, and cumulative loading rates to soil. The United States Environmental Protection Agency also published risk-based "Regional Screening Levels" for contaminants in soil based on direct exposure of children and adults to contaminants in soil and current or anticipated future land use (Canadian Council of Ministers of the Environment, 2012; United States Environmental Protection Agency, 2024). Site-specific screening levels can be developed as needed. Tools such as the integrated exposure uptake biokinetic model for children's lead exposure (also applicable to Cd with adjustment) (Brown et al., 2023) and the HHRA framework (United States Environmental Protection Agency, 1989a, 1989b) can be considered to evaluate potential dietary and nondietary exposure pathways (Fitzpatrick et al., 2017). Canada adopts a similar approach under the guidance of the Canadian Council of Ministers of the Environment. Cd soil guidelines are established using quantitative risk assessment models that integrate exposure from multiple sources (soil, water, and food) and consider both human health and ecological endpoints (Canadian Council of Ministers of the Environment, 1999). Restrictions are also placed on the concentration of Cd in biosolids to be placed on agricultural lands in a manner similar to that utilized in the United States (Canadian Council of Ministers of the Environment, 2012).

17.5 Main causes for the differences in soil cadmium standards

Natural geographic conditions significantly influence the establishment of soil Cd standards. In complex geological conditions with high Cd, establishment of a standard requires comprehensive consideration of background values and avoidance of a one-size-fits-all approach. Cd naturally occurs in soil because of processes such as rock weathering. Accurate estimation of natural Cd backgrounds helps identify additional accumulation caused by human activities. In certain regions, especially those with metal-rich parent materials (e.g., southwestern China and parts of Europe), natural Cd concentrations in soil exceed regulatory limits, posing challenges for implementation of standards (Li et al., 2021; Xiao et al., 2022; Xu et al., 2022; Yang et al., 2021). However, whether high natural Cd levels in soil routinely translate into elevated risks to food safety or human health remains a subject of active debate (Adler et al., 2023; Li et al., 2021; Xu et al., 2022; Yang et al., 2021).

Similar problems have been identified in acidic agricultural soils of New Zealand, where volcanic activity has led to naturally elevated levels of Cd (Gray & Cavanagh, 2023). When formulating soil Cd standards, New Zealand pays particular attention to the impact of geological background levels and adopts context-specific management strategies. In contrast, northern Europe, with generally lower soil pH (Ap horizon: 5.2) and moderate Cd background levels, has stricter standards (Ballabio et al., 2024; Faroon et al., 2012).

Dietary habits also largely determine differences in standards used. Rice is a staple crop in East Asia, and rice plants have a relatively high capacity to absorb Cd. Therefore countries such as Japan and China pay particular attention to soil Cd levels in rice-growing areas (Wang & Zhao, 2022). Differences in rice Cd content and dietary structures among countries or regions, especially concerning sensitive groups like children, are not always thoroughly considered. This leads to variations in food Cd standards ranging from 0.04 to 0.4 mg kg^{-1} (China Food and Drug Administration and National Health and Family Planning Commission, 2017; Vannan et al., 2020). Based on health guidance values from the Joint FAO/WHO Expert Committee on Food Additives and considering dietary structures and regional differences, it is estimated that for the general population in southern China, where rice is the primary grain, the rice Cd limit should not exceed 0.10 mg kg^{-1} (estimated), and for northern populations, where wheat is also consumed, it should not exceed 0.15 mg kg^{-1} (Wang & Zhao, 2022).

Wheat and corn, in contrast, are the main staple crops of European and American countries and have relatively low Cd absorption capacities (Chen et al., 2014; Feng et al., 2020). Soil standards for cultivation of these commodities are therefore somewhat higher. The EU distinguishes between agricultural land and pastures, setting stricter limits for soils used for vegetable cultivation compared to pastures.

Technological capabilities and economic costs are also significant factors. The EU and the United States have more detailed baseline data on soil Cd levels, with comprehensive Cd maps (Ballabio et al., 2024; Smith et al., 2019). Regular soil monitoring and agricultural product surveys help further identify high-risk areas and implement soil management and crop substitution plans (Adler et al., 2023). This allows these regions to set higher, area- and crop-specific soil standards and implement them effectively. In contrast, some countries or regions, because of technical and financial constraints, might not have local standards and tend to use national uniform standards.

Differences in policy objectives and risk management strategies also lead to diversification of standards. Some countries emphasize additional safety factors for the establishment of food standards, with the EU tending to set stricter standards to minimize risks. In contrast, countries like the United States more often focus on safety factors already incorporated into toxicity factors used to establish food standards and established risk assessment-based approaches, without the need for additional and often subjective downward adjustment of standards.

17.6 Proposed framework for optimizing soil cadmium standards

In vast regions with significant differences in soil types, land use, and agricultural practices, there are substantial variations in soil environmental background levels, crop types and varieties, sensitivity to pollutants, accumulation levels, water and fertilizer management, and consumption habits (Adler et al., 2023; Li et al., 2024; Wang & Zhao, 2022). Uniform, conservative soil Cd standards, typically represented by fixed values, can serve as useful tools to rapidly identify and eliminate low-risk soils from further investigation but can be unnecessarily conservative for areas where the screening level is exceeded. An easily implemented approach that allows region-specific assessment of crop uptake risk is, therefore, warranted.

Building upon the discussions in the preceding sections, we propose the following framework for optimizing soil Cd standards (Fig. 17.3): (1) improvement of the resolution and reliability of soil sample data to more accurately identify and optimize land suitable for cultivation of food crops and (2) adjustment of soil and/or food safety standards to reflect local (regional) Cd uptake risk and dietary considerations. Data resolution is improved by decreasing the area of land represented by a data point. Grid cells on the left side of Fig. 17.3 represent "DU" areas of land targeted for collection of a single, Multi Increment soil or crop sample (CRC for Contamination Assessment and Remediation of the Environment, 2024a) (refer to Section 17.3). A resolution of "5 km" noted on the figure reflects DU–MIS data considered to be representative of individual, 25 km^2 areas. Similar to pixels in a digital photo, decreasing the size of the DU increases the resolution of the data.

The framework begins with a scientifically sound soil–crop Cd monitoring strategy, informed by historical data (as available). In the example depicted in Fig. 17.3 the lower left quadrant of the targeted investigation area is quickly cleared for unrestricted cultivation of crops based on sample data and the known history of the area, without the need for additional sampling (Category C). Data resolution is progressively improved in the remaining quadrants to cost-effectively isolate land that is unsuitable for the cultivation of food crops (Categories A and B) and allow the area of land acceptable for crop cultivation to be optimized (Categories C and D).

Soil–crop Cd data for individual DUs would ideally fall either in Quadrant C (<crop standard and <soil standard) or Quadrant B (>crop standard and >soil standard). Cases where the soil standard is met but the crop standard is exceeded (Quadrant A) or the soil standard is exceeded but the crop standard is met (Quadrant D) are inevitable, however. These mismatches highlight the limitations of uniform thresholds and underscore the need for more region-specific assessment. In such cases, intensified monitoring is warranted, incorporating additional

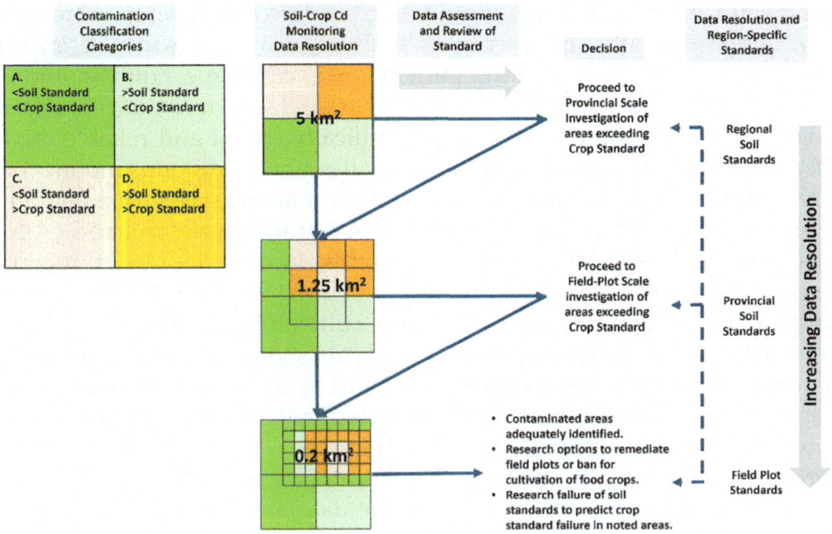

FIGURE 17.3 Conceptual framework illustrating enhanced data resolution and region-specific adjustment of soil Cd standards to isolate areas where crops exceed food standards and improve management and utilization of agricultural lands by reducing soil–crop Cd model uncertainties. For example, both soil standards for the cultivation of crops and crop (food) standards are exceeded in the dark orange areas. Crop standards are exceeded in the light orange areas, but soil standards are not exceeded. The data maps reflect sampling methods discussed in National Standardization Administration and State Administration for Market Regulation (2023) and CRC for Contamination Assessment and Remediation of the Environment (2024b). The noted resolutions are for example only. Categorization of agricultural land suitable for food crop cultivation can be optimized by (A) increasing the resolution and reliability of soil data through the transition from traditional grab sampling to decision unit and Multi Increment Sample collection methods (DU–MIS), and (B) using the soil and crop data to develop reliable soil standards for different cultivation scenarios and improve identification of high-risk land prior to cultivation of a food crop. *Cd*, Cadmium.

variables such as crop genotype, specific chemical nature and bioavailability of Cd in the soil, source of irrigation water, application of fertilizers, and atmospheric deposition.

Establishment of soil–crop Cd relationships at medium spatial resolutions (e.g., district or county level >10 km) and progressively improved resolutions as cost-beneficial can help identify such areas of mismatch. The resulting refined models can be used to support the development of region-specific soil Cd thresholds. Although high-resolution predictions (e.g., field or plot level) may seem ideal, they are often constrained by legal, administrative, and resource limitations. A practical and policy-relevant definition of "region" could align with agroecological zones, counties, or small watersheds, offering a balance between local specificity and administrative feasibility. For example, the EU LUCAS survey, with a data node resolution of approximately 14 km, reflecting 196 km^2 grid cells, exemplifies a medium-resolution national soil monitoring system suitable for regional-scale spatial modeling and threshold development (Ballabio et al., 2024).

In more high-risk cases, for example, in areas of known industrial pollution, more intensive soil–crop monitoring might be necessary at a scale of 1 km or individual fields. Additional local parameters, including knowledge of the site history, crop characteristics, atmospheric deposition, and soil Cd bioavailability, should be considered to develop more precise models and reduce misjudgment rates. Interventions such as increasing soil pH, improving water management, and changing crop planting patterns (e.g., crop type, cultivar, and crop season) should be considered to reduce food safety risks associated with soil Cd contamination.

The approach described can lead to more scientifically robust and reliable regional soil Cd standards. Regularly updating soil–crop Cd models through continuous monitoring will allow for iterative refinement of regional and national soil Cd standards, ensuring better protection of agricultural products and human health. Reviews of region-specific soil and crop data might, in some cases, allow for a relaxation of soil standards. In other cases a reduction of the soil standard might be necessary because of unanticipated uptake of Cd into a food crop. This could be because of the physiochemical nature of the soil, cultivation methods, or the specific species of crop being cultivated. Modification of soil standards should be carried out in coordination with regional agriculture and food safety experts and in accordance with applicable national guidelines, as available. Modification of food safety standards should include input from a toxicologist experienced in health risks posed by Cd.

It is important to ensure that both soil and crop samples are collected in a manner demonstrated to be scientifically supportable and that the data are reproducible (CRC for Contamination Assessment and Remediation of the Environment, 2024b; National Standardization Administration and State Administration for Market Regulation, 2023) (refer to Section 17.3). Preparation of a sample by collection of increments across an entire, targeted DU will not be practical in some cases. Collaboration with local soil experts and agencies familiar with historical contamination risks will be required to group land areas into DUs for the collection of soil samples. Nonagricultural areas should be excluded from DUs and sample collection areas. DU designation and sample collection could be further restricted to lands specifically proposed for food crop cultivation. Geostatistical modeling and knowledge of the historical industrial and agricultural use of the area can be used to fill in data gaps between individually tested and discontinuous DUs.

17.7 Uncertainty in soil–crop cadmium transfer models for deriving cadmium thresholds

An important uncertainty in formulating soil Cd standards (Table 17.1; see also Fig. 17.3) lies in the inherent, unknown error associated with soil–crop Cd transfer models. These models are used to back-calculate threshold levels in soil that ensure food safety limits will be met in edible crop parts. One widely adopted approach incorporates soil Cd concentration and pH into empirical or semimechanistic equations, such as the extended Freundlich-type model (Zhang et al., 2018):

$$\text{Log}_{10}(\text{Crop Cd}) = a \cdot \text{Log}_{10}(\text{Soil Cd}) + b \cdot \text{pH} + c \tag{17.1}$$

Although such models capture first-order effects of pH on Cd bioavailability (Zhu et al., 2016), their predictive accuracy is limited. Model coefficients vary markedly across datasets because of differences in crop species, soil properties, management practices, and climatic conditions (Fig. 17.4). Furthermore, these linear or log-linear models often fail at low and high Cd concentrations—tending to overpredict crop Cd when soil Cd is low and underpredict when it is high (Huang et al., 2024b; Li et al., 2025). These limitations suggest that Cd uptake is

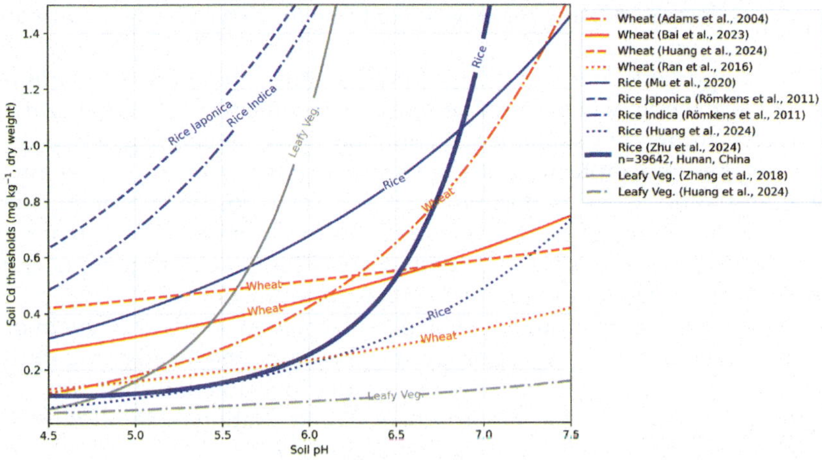

FIGURE 17.4 Relationship between soil Cd thresholds (mg kg^{-1}, dry weight) and soil pH at a given crop Cd standard (0.1 mg kg^{-1}, dry/fresh weight) for various crops. Individual lines highlight conditions where crop Cd uptake equals 0.1 mg kg^{-1} fresh/dry weight. Soil–plant transfer models (Eq. 17.1) for wheat (n = 204–272) (Adams et al., 2004; Bai et al., 2023; Huang et al., 2024a; Ran et al., 2016), rice (n = 66–39,642) (Huang et al., 2024a; Mu et al., 2023; Römkens et al., 2011; Zhu et al., 2016), and leafy vegetables (n = 70–204) (Huang et al., 2024a; Mu et al., 2023; Zhang et al., 2018) are adopted from literature. *Cd*, Cadmium.

influenced by nonlinear and interacting factors not captured in traditional two-variable models (Ballabio et al., 2024; Environment Agency, 2009; Li et al., 2021, 2025; Xia et al., 2023).

A particularly instructive case is rice, which presents substantial challenges for the construction of reliable soil–plant Cd transfer models. Two major sources of uncertainty make rice fundamentally more complex than most upland crops. The first is dynamic soil redox conditions during grain filling. Soil redox potential (E_h) exerts a dominant control on Cd speciation and solubility in flooded paddy systems. During the reproductive stage and grain filling, even small topographic differences can cause sharp gradients in E_h and, consequently, in increased availability of Cd for uptake. Field studies have demonstrated that spatial variation in E_h and pH alone can explain large intrafield differences in rice grain Cd (Chen et al., 2020). However, E_h is rarely monitored in large-scale surveys and is not routinely included in model structures because of its temporal variability and measurement difficulty. The second is the large genotypic variability in grain Cd accumulation. Among rice cultivars, natural variation in grain Cd content is exceptionally high—up to 20-fold or more under identical soil conditions (Duan et al., 2017).

In contrast the prediction of Cd uptake into upland crops such as wheat is more straightforward. Redox variations are largely absent, and cultivar differences in grain Cd accumulation tend to be moderate—typically within a 2- to 4-fold range (Chen et al., 2020). As a result, soil–crop models for wheat are generally more robust and transferable across regions. Nevertheless, crop type remains one of the largest sources of uncertainty in soil Cd standard derivation. As shown in Fig. 17.4, even within neutral pH soils, Cd threshold values estimated for different crops can differ by more than an order of magnitude. This variability underscores the need for crop-specific models and emphasizes the danger of adopting uniform soil Cd limits across diverse cropping systems and using them as a fixed tool for determining if cultivation of a food crop is acceptable.

This raises an important question for threshold setting: Should soil Cd limits be derived based on the average, median, maximum, or upper percentile (e.g., 95th) of cultivar responses? Each choice reflects a different risk management philosophy—balancing between inclusivity of high-yielding varieties and protection of food safety—and can lead to substantially different recommended thresholds. Risk-based standards for direct exposure to contaminants, including food safety standards, are based on the average concentration of a contaminant in the subject media over the duration of exposure—normally 6 years for developing young children and 24 years for adults (United States Environmental Protection Agency, 1989a, 1989b). This requires consideration of the mean concentration of Cd in a subject food crop. Assessment of crop uptake risk posed by Cd in soil is, thus, most appropriately based on the mean concentration of Cd for the subject area and volume of soil (i.e., the DU) under investigation (CRC for Contamination Assessment and Remediation of the Environment, 2024b; National Standardization Administration and State Administration for Market Regulation, 2023).

17.8 High cadmium risks and management in low pH soils

In regions such as Hunan Province in central-southern China the combination of mining and smelting, high industrial activity, and soil acidification has led to widespread Cd contamination of soil. A comprehensive survey of rice paddies in key areas revealed that the average Cd concentration in brown rice was 0.43 mg kg^{-1} (n = 39,642). Notably, 86.9% of surface soil samples had pH values between 4.4 and 6.0, and 74.8% of soil samples exceeded the national standard of 0.30 mg Cd kg^{-1} dry soil (Zhu et al., 2016). Utilizing established soil–crop transfer models (Fig. 17.4), it was observed that under widespread acidic conditions, even low soil Cd concentrations (<0.2 mg kg^{-1}) can result in rice Cd concentrations exceeding 0.2 mg kg^{-1}. However, by increasing soil pH to neutral levels, the standard of 0.2 mg kg^{-1} can be achieved without altering soil Cd concentrations. Government-led initiatives to raise soil pH and promote low-Cd rice varieties have shown promising results in addressing Cd contamination (Li et al., 2024). For instance, applying lime to acidic soils can increase pH, thereby reducing the bioavailability of Cd and its uptake by rice plants (Li et al., 2024; Zhang et al., 2023). In addition, selecting rice varieties with low Cd accumulation traits can further mitigate Cd levels in rice grains (Li et al., 2024; Sun et al., 2022; Xue et al., 2023; Yang et al., 2025; Zhang et al., 2023). Thus management of soil pH through liming and cultivation of low-Cd accumulating rice varieties have proven to be effective strategies to reduce Cd risks in rice production.

17.9 Risks and management of high soil cadmium in pH-neutral-to-alkaline regions

In regions with naturally elevated soil Cd levels, such as the karst areas in southwestern China, soil Cd concentrations often exceed standard thresholds for typical agricultural lands (Xiao et al., 2022; Yang et al., 2021). However, this does not necessarily indicate contamination or potential health risks. Application of uniform regulatory limits could otherwise prematurely declare extensive, safe agricultural lands unsuitable for farming.

For instance, in Guangxi province of China, 48% of the area has a soil pH greater than 6.5, with 59% of soil samples exceeding national screening values. Yet, only 4% of the rice grown in this area exceeds the food safety standard (Yang et al., 2021). In areas with a pH < 6.5 and total soil Cd < 0.6 mg kg^{-1} (covering 37% of the area), 32% of soil samples exceed screening values, but only 17% of rice samples exceed the standard (Yang et al., 2021). In another extensive investigation, 65% of soil samples exceed risk screening values, yet only 28% of rice samples exceed food safety standards (Xiao et al., 2022). In a more typical example in Yunnan Province, paddy soil Cd content ranges from 0.47 to 1.87 mg kg^{-1}, averaging 0.78 mg kg^{-1}—several times higher than national screening values (0.3–0.8 mg kg^{-1}). However, brown rice Cd content is only 0.007–0.062 mg kg^{-1}, which is below the national food standard of 0.2 mg kg^{-1} (Liu et al., 2024). This is often associated with the high pH resulting from carbonate development in karst regions and with flooded rice cultivation that decreases Cd mobility.

Although pH-dependent soil Cd screening values have been introduced in China's regulatory framework for paddy soils, current implementations fall short in capturing the full range of conditions found in high-Cd background regions. As shown in Fig. 17.4, most pH-dependent models suggest an exponential increase in allowable Cd levels under neutral to alkaline conditions. However, the lack of site-specific data hampers accurate threshold setting at the regional level.

One potential solution is to adopt models based on bioavailable or labile forms of soil Cd, rather than total concentrations. This could allow a more realistic estimation of the fraction of Cd accessible to crops. Nevertheless, as discussed earlier, the bioavailability of Cd is highly dynamic and influenced by fluctuating soil redox status, plant root activity, and irrigation regimes—especially in rice paddies. Consequently a single-time-point measurement of bioavailable Cd may not reliably reflect the risk throughout the growing season.

Despite these challenges, greater spatial and temporal resolution in soil monitoring as well as the use of improved sample collection methods can improve model calibration and allow for the development of integrated risk assessment frameworks. These should account for soil pH, redox dynamics, cultivar-specific uptake patterns, and regionally specific biogeochemical processes, enabling more nuanced and scientifically grounded land-use decisions.

17.10 Challenges in establishing soil cadmium standards for agricultural land

Despite progress in formulating soil Cd standards for agricultural land, several challenges persist, encompassing technical issues, policy coordination, economic costs, and public awareness.

Development of scientifically sound soil Cd standards relies on extensive foundational data, including soil background levels, factors affecting the Cd availability, and potential impacts on crops and human health. However, many regions lack comprehensive soil surveys and long-term monitoring systems, hindering data acquisition. For instance, parts of Africa and South Asia have yet to establish extensive soil databases. Even in areas with existing databases, such as the EU (Ballabio et al., 2024) or the United States (Adhikari et al., 2024), the spatial variability of soil Cd remains inadequately addressed (Adler et al., 2023), complicating the development of region-specific standards.

Furthermore, current risk assessment models often rely on laboratory simulations and short-term field trials, which may not fully capture the complexities of natural environments (Li et al., 2025). A key process is to establish soil–plant Cd transfer models. However, Cd bioavailability in soil is influenced by various factors, such as pH, organic matter content, and soil redox potential, and existing models may not accurately predict these dynamic changes. In addition, soil–plant systems are under complex feedback, and plant physiological factors are often overlooked in assessing Cd accumulation in edible parts (Li et al., 2025). Therefore the development of more precise and universally applicable assessment tools remains a challenge. In recent years, machine learning (ML) has shown great promise for improving soil Cd assessment by integrating complex datasets and enhancing prediction accuracy of Cd bioavailability and distribution (Huang et al., 2024a). ML models, such as eXtreme Gradient Boosting, can efficiently analyze diverse soil and environmental variables to identify key factors affecting Cd accumulation (Huang et al., 2024a). However, their effectiveness depends on the availability of high-quality data, and challenges remain regarding model interpretability and generalizability across varied agroecosystems. Thus, although ML offers valuable supplementary tools for soil Cd risk assessment, it should complement—rather than replace—traditional methods and expert judgment.

Formulating and implementing soil Cd standards involves collaboration among multiple government departments, including environmental protection, agriculture, health, and food safety (Holdaway & Wang, 2018; Leaman et al., 2025; Li et al., 2024). In practice, differing responsibilities and interests among these departments can lead to conflicts, complicating policy coordination. For example, environmental protection agencies might advocate stricter Cd limits to reduce environmental risks, whereas agricultural departments might prioritize production efficiency and economic benefits, favoring less conservative standards. Such disagreements can delay standard development and weaken policy enforcement. Moreover, inconsistencies in standards pose significant challenges. Because of variations in natural environmental conditions, socioeconomic development levels, and policy objectives, soil Cd standards often differ regionally. This inconsistency not only increases technical barriers in international trade but also complicates cross-border pollution issues. For example, agricultural products that pass standards in the country where they were produced might be rejected by importing nations if they do not meet local Cd standards or even standards for soil in which the crops are grown, affecting farmers' incomes and regional economic development (Steven & Spencer, 2004). Therefore balancing regional differences while promoting international standard harmonization is a crucial future policy consideration.

Soil Cd remediation is a high-investment, long-term endeavor (Hou et al., 2020; Li et al., 2024). In severely contaminated areas, remediation costs may exceed the financial capacities of local governments and farmers. For example, physical–chemical remediation technologies, such as soil leaching and stabilization, are effective but often prohibitively expensive for many developing countries. Even low-cost bioremediation techniques, like phytoremediation, require extended periods to show results, making it challenging to meet urgent remediation needs. In addition, the adoption and application of new technologies face resistance. Farmers' limited awareness and acceptance of new technologies, especially in regions dominated by traditional agriculture, lead them to prefer familiar cultivation methods. Moreover, technology dissemination requires accompanying financial support and training systems, which remain lacking in many resource-constrained areas. Therefore a reduction of remediation costs and enhancement of technology accessibility through policy incentives and technological innovation are pressing issues to address.

Public understanding of soil Cd pollution directly influences the effectiveness of standards implementation (Li et al., 2024). In many regions, public concern about soil pollution is low, and there is a lack of basic awareness of its potential hazards. This deficiency may lead to farmers' low compliance with soil Cd standards and consumers' insufficient vigilance regarding Cd-contaminated agricultural products, thereby weakening the practical effectiveness of standards. Furthermore the absence of public participation mechanisms is a significant constraint on standard implementation. In some countries the formulation of soil Cd standards is primarily led by government departments and research institutions, with insufficient reflection of public opinions and needs. This top-down decision-making model can trigger public resistance and even impede policy enforcement. Therefore, enhancing public awareness through science popularization and community participation, and encouraging active involvement in governance, should be emphasized in the future.

17.11 Global perspective and future directions for soil cadmium standards

In summary, soil Cd standards for agricultural land across countries reflect their respective agricultural characteristics, dietary structures, and risk management needs. Overall, China and Japan focus more on strict control of Cd content in food, whereas Australia, New Zealand, and Western countries tend to combine ecotoxicology and health risk assessments based on local variations in diet and other factors to develop flexible but still scientifically sound standard systems. In the future, as global understanding of soil Cd pollution deepens, these standard systems are likely to become more refined and dynamic, better adapting to the natural and socioeconomic conditions of different regions. Emerging technologies, such as ML, offer promising tools to enhance risk assessment by integrating complex environmental and soil datasets, improving prediction accuracy, and supporting adaptive management strategies. However, these approaches require high-quality data and expert interpretation to ensure reliable application.

Establishing Cd standards in agricultural soils is a complex interplay of scientific risk assessment, environmental variability, and socioeconomic considerations. Although global regulatory thresholds differ, the ultimate goal remains unchanged: ensuring food safety and protecting human health. Case studies emphasize the importance of localized and adaptive approaches, especially in high-background and historically contaminated areas. Looking ahead the integration of risk-based assessments with innovative mitigation strategies will enhance the effectiveness of global soil Cd management policies.

17.12 Summary

- Global soil Cd standards for agriculture reflect diverse natural environmental conditions, dietary habits, and policy goals.
- Consideration of Pierre Gy's theory of sampling and use of "DU" and "MIS" soil and crop characterization methods is critical for ensuring the representativeness of sample data.
- Soil pH and redox conditions strongly influence Cd bioavailability and crop uptake, especially in paddy systems.
- Soil Cd thresholds should be derived based on health risk, food safety, and ecological protection objectives.

- Uniform standards often fail to account for regional variability, calling for adaptive, localized management strategies.
- Future development of Cd standards requires improved modeling, dynamic monitoring, and multistakeholder collaboration.

Acknowledgments

The authors gratefully acknowledge the financial support from the National Key R&D Program of China (Grant No. 2023YFC3708705) and the Key R&D Program of Guizhou Province, China (Grant No. 2025041).

AI disclosure

During the preparation of this work the author(s) used ChatGPT to improve readability. After using this tool/service the author(s) reviewed and edited the content as needed and take(s) full responsibility for the content of the publication.

References

Adams, M. L., Zhao, F. J., McGrath, S. P., Nicholson, F. A., & Chambers, B. J. (2004). Predicting cadmium concentrations in wheat and barley grain using soil properties. *Journal of Environmental Quality, 33*(2), 532–541. doi:10.2134/jeq2004.5320.

Adhikari, K., Mancini, M., Libohova, Z., Blackstock, J., Winzeler, E., Smith, D. R., Owens, P. R., Silva, S. H. G., & Curi, N. (2024). Heavy metals concentration in soils across the conterminous USA: Spatial prediction, model uncertainty, and influencing factors. *Science of the Total Environment, 919*. https://www.sciencedirect.com/science/journal/00489697. doi:10.1016/j.scitotenv.2024.170972.

Adler, K., Persson, K., Söderström, M., Eriksson, J., & Pettersson, C.-G. (2023). Digital soil mapping of cadmium: Identifying arable land for producing winter wheat with low concentrations of cadmium. *Agronomy, 13*(2), 317. doi:10.3390/agronomy13020317.

Association of American Feed Control Officials. (2015). *GOODSamples: Guidance on obtaining defensible samples*. Association of American Feed Control Officials http://www.aafco.org/Publications/GOODTestPortions.

Association of American Feed Control Officials. (2018). *Good test portions: Guidance on obtaining defensible test portions*. Association of American Feed Control Officials http://www.aafco.org/Publications/GOODTestPortions.

Bai, L., Ding, S., Huang, X., Chen, X., Chen, Y., Cao, X., Wang, X., Yu, X., & Dai, J. (2023). Prediction of the cadmium content in grains of low-accumulating wheat cultivars and soil cadmium threshold for safe production. *Journal of Cleaner Production, 417*, 138081. doi:10.1016/j.jclepro.2023.138081.

Ballabio, C., Jones, A., & Panagos, P. (2024). Cadmium in topsoils of the European Union—An analysis based on LUCAS topsoil database. *Science of the Total Environment, 912*, 168710. doi:10.1016/j.scitotenv.2023.168710.

Brewer, R., Peard, J., & Heskett, M. (2017). A critical review of discrete soil sample data reliability: Part 2—Implications. *Soil and Sediment Contamination: An International Journal, 26*(1), 23–44. doi:10.1080/15320383.2017.1244172.

Brown, J. S., Spalinger, S. M., Weppner, S. G., Hicks, K. J. W., Thorhaug, M., Thayer, W. C., Follansbee, M. H., & Diamond, G. L. (2023). Evaluation of the integrated exposure uptake biokinetic (IEUBK) model for lead in children. *Journal of Exposure Science and Environmental Epidemiology, 33*(2), 187–197. https://www.nature.com/jes/. doi:10.1038/s41370-022-00473-2.

Canadian Council of Ministers of the Environment. (1999). *Canadian soil quality guidelines for the protection of environmental and human health: Cadmium*. Canadian Council of Ministers of the Environment https://ccme.ca/en/res/cadmium-canadian-soil-quality-guidelines-for-the-protection-of-environmental-and-human-health-en.pdf.

Canadian Council of Ministers of the Environment (2012). *Guidance document for the beneficial use of municipal biosolids, municipal sludge and treated septage (PN 1473)* (pp. 1–122). Canadian Council of Ministers of the Environment.

Cao, J., Ren, S., Wang, C., She, J., Jiang, Y., Liu, Y., Zhou, Y., Wang, L., Wang, J., Wang, Y., Liu, J., & Chen, Y. (2021). Cadmium and lead distribution in pyrite ores: Environmental concerns over geochemically mobile fractions. *Elementa, 9*(1). https://www.elementascience.org/articles/. doi:10.1525/elementa.2021.00093.

Carlon, C., & D'Alessandro, M. (2007). *Derivation methods of soil screening values in Europe. A review and evaluation of national procedures towards harmonisation* (pp. 1–306). European Commission, Joint Research Centre.

Carlon, C., Pizzol, L., Critto, A., & Marcomini, A. (2008). A spatial risk assessment methodology to support the remediation of contaminated land. *Environment International, 34*(3), 397–411. www.elsevier.com/locate/envint. doi:10.1016/j.envint.2007.09.009.

Cavanagh, J. (2013). *Methodologies for deriving cadmium soil guideline values for the protection of ecological receptors and food standards (Landcare Research Report LC1713)*. Landcare Research, Lincoln https://www.envirolink.govt.nz/assets/Envirolink/R10-420Development20of20soil20guideline20values20for20the20protection20of20ecological20receptors20-20Technical20document.pdf.

Chen, H., Wang, P., Gu, Y., Kretzschmar, R., Kopittke, P. M., & Zhao, F. J. (2020). The within-field spatial variation in rice grain Cd concentration is determined by soil redox status and pH during grain filling. *Environmental Pollution, 261*. https://www.journals.elsevier.com/environmental-pollution. doi:10.1016/j.envpol.2020.114151.

Chen, J., Yu, W., Zu, Y., & Yu, Y. (2014). Variety diference of Cd accumulation and translocation in *Zea mays*. *Ecology and Environmental Sciences, 23*(10), 1671–1676. https://www.jeesci.com/CN/Y2014/V23/I10/1671.

National Health and Family Planning Commission. (2017). *National food safety standard—Maximum levels of contaminants in foods (GB 2762-2017)*. China Standards Press https://www.chinesestandard.net/PDF-EN/GB2762-2017EN-P06P-H5002H-258298.pdf?srsltid=AfmBOoqqLeGcL6pN6HaB5MAm2VZpr8nBjzAv4RfBMeZNAF2_o2o9yQNc.

CRC for Contamination Assessment and Remediation of the Environment. (2024a). *Guidance for the risk-based investigation of per- and polyfluoroalkyl substances (PFAS) impacts to agricultural fields and food crops (in press) Technical Report series*. CRC Care Pty Ltd.

CRC for Contamination Assessment and Remediation of the Environment. (2024b). *Risk-based investigation of environmental contamination: Implementation (Part 2) CRC CARE technical report*. CRC Care Pty Ltd https://www.crccare.com/publications/technical-reports/.

Duan, G., Shao, G., Tang, Z., Chen, H., Wang, B., Tang, Z., Yang, Y., Liu, Y., & Zhao, F. J. (2017). Genotypic and environmental variations in grain cadmium and arsenic concentrations among a panel of high yielding rice cultivars. *Rice, 10*(1). http://www.springerlink.com/content/1939-8425. doi:10.1186/s12284-017-0149-2.

Environment Agency. (2009). *Supplementary information for the derivation of Soil Guideline Values for cadmium (Science Report SC050021—Technical Review: Cadmium)*. Environment Agency http://assets.publishing.service.gov.uk/media/5a7c94b8ed915d6969f45e46/scho0309bpqm-e-e.pdf.

European Food Safety Authority (EFSA). (2009). Cadmium in food—Scientific opinion of the Panel on Contaminants in the Food Chain. *EFSA Journal, 7*(3). doi:10.2903/j.efsa.2009.980.

Faroon, O., Ashizawa, A., Wright, S., Tucker, P., Jenkins, K., Ingerman, L., & Rudisill, C. (2012). *Toxicological profile for cadmium ATSDR toxicological profiles*. Agency for Toxic Substances and Disease Registry https://www.ncbi.nlm.nih.gov/books/NBK158838/.

Feng, L., Yan, H., Dai, C., Xu, W., Gu, F., Zhang, F., Li, T., Xian, J., He, X., Yu, Y., Ma, M., Wang, F., & He, Z. (2020). The systematic exploration of cadmium-accumulation characteristics of maize kernel in acidic soil with different pollution levels in China. *Science of the Total Environment, 729*, 138972. doi:10.1016/j.scitotenv.2020.138972.

Fitzpatrick, J., Schoeny, R., Gallagher, K., Deener, K., Dockins, C., Firestone, M., Jordan, W., McDonough, M., Murphy, D., Olsen, M., & Raffaele, K. (2017). US Environmental Protection Agency's framework for human health risk assessment to inform decision making. *International Journal of Risk Assessment and Management, 20*(1), 3–20. http://www.inderscience.com/ijram. doi:10.1504/IJRAM.2017.082558.

Follett, R. F. (2010). *Sampling protocols introduction*. USDA Agricultural Research Service (ARS).

Gourley, C. J. P., & Weaver, D. (2019). *A guide for fit for purpose soil sampling*. Fertilizer Australia https://library.dpird.wa.gov.au/lr_consultrpts/27.

Gray, C. W., & Cavanagh, J. A. E. (2023). The state of knowledge of cadmium in New Zealand agricultural systems: 2021. *New Zealand Journal of Agricultural Research, 66*(4), 285–335. https://www.tandfonline.com/loi/tnza20. doi:10.1080/00288233.2022.2069130.

Holdaway, J., & Wang, W. (2018). From soil pollution to cadmium rice to public health impacts: An interdisciplinary analysis of influencing factors and possible responses. *Journal of Resources and Ecology, 9*(1), 1–12. doi:10.5814/j.issn.1674-764x.2018.01.001.

Hou, D., O'Connor, D., Igalavithana, A. D., Alessi, D. S., Luo, J., Tsang, D. C. W., Sparks, D. L., Yamauchi, Y., Rinklebe, J., & Ok, Y. S. (2020). Metal contamination and bioremediation of agricultural soils for food safety and sustainability. *Nature Reviews Earth and Environment, 1*(7), 366–381. nature.com/natrevearthenviron/. doi:10.1038/s43017-020-0061-y.

Huang, B. Y., Lü, Q. X., Tang, Z. X., Tang, Z., Chen, H. P., Yang, X. P., Zhao, F. J., & Wang, P. (2024a). Machine learning methods to predict cadmium (Cd) concentration in rice grain and support soil management at a regional scale. *Fundamental Research, 4*(5), 1196–1205. https://www.sciencedirect.com/science/journal/26673258. doi:10.1016/j.fmre.2023.02.016.

Huang, X., Li, X., Zheng, L., Zhang, Y., Sun, L., Feng, Y., Du, J., Lu, X., & Wang, G. (2024b). Comprehensive assessment of health and ecological risk of cadmium in agricultural soils across China: A tiered framework. *Journal of Hazardous Materials, 465*, 133111. doi:10.1016/j.jhazmat.2023.133111.

Japan Ministry of the Environment. (2002). *Soil Contamination Countermeasures Act (Act No. 53 of 2002)*. Ministry of Justice https://www.env.go.jp/content/900452902.pdf.

Leaman, S. M., McEntire, J. C., Abley, M., Choiniere, C., Draper, A., Krout-Greenberg, N., Lopez, T., & Davis, D. A. (2025). Collaborating to address heavy metals in fresh produce supply—A case study of cadmium in spinach and carrots grown in Arizona and California. *Food Protection Trends, 45*(1), 19–26. https://www.foodprotection.org/members/fpt-archive-articles/2025-01-collaborating-to-address-heavy-metals-in-fresh-produce-supply-a-case-study-of-cadmium-in-spi/. doi:10.4315/FPT-24-026.

Li, C., Yang, Z., Yu, T., Hou, Q., Liu, X., Wang, J., Zhang, Q., & Wu, T. (2021). Study on safe usage of agricultural land in karst and non-karst areas based on soil Cd and prediction of Cd in rice: A case study of Heng County, Guangxi. *Ecotoxicology and Environmental Safety, 208*, 111505. doi:10.1016/j.ecoenv.2020.111505.

Li, Y., Fan, G., Gao, Y., Chen, W., Shi, G., Tong, F., Liu, L., & Zhou, D. (2025). Wheat tends to accumulate higher levels of cadmium in the grains than rice under a wide range of soil pH and Cd concentrations: A field study on rice-wheat rotation farmland. *Environmental Pollution, 367*, 125574. doi:10.1016/j.envpol.2024.125574.

Li, Y., Tao, H., Cao, H., Wan, X., & Liao, X. (2024). Achieving synergistic benefits through integrated governance of cultivated cadmium contamination via multistakeholder collaboration. *Nature Communications, 15*(1). https://www.nature.com/ncomms/. doi:10.1038/s41467-024-53626-2.

Liu, C., Chen, M., Lei, F., Huang, Y., Wang, X., Wang, J., Deng, G., & Zhang, Y. (2024). Cd accumulation and human health risk assessment of rice in high background areas of heavy metals: A case study of Nagu Town, Huize County, Yunnan Province. *Sedimentary Geology and Tethyan Geology, 44*(1), 194–204. https://www.cjyttsdz.com.cn/en/article/2024/1. doi:10.19826/j.cnki.1009-3850.2022.11002.

McLaughlin, M. J., Smolders, E., Zhao, F. J., Grant, C., & Montalvo, D. (2021). Managing cadmium in agricultural systems. *Advances in Agronomy, 166*, 1–129. http://www.elsevier.com/wps/find/bookdescription.cws_home/675816/description#description. doi:10.1016/bs.agron.2020.10.004.

Ministry of Agriculture and Rural Affairs of the People's Republic of China. (2018). *Soil environmental quality—Risk control standard for soil contamination of agricultural land (GB 15618–2018) National Environmental Quality Standard of the People's Republic of China* https://www.mee.gov.cn/ywgz/fgbz/bz/bzwb/trhj/201807/t20180703_446029.shtml.

Ministry of the Environment, Government of Japan. (2020). *Review of environmental quality standards for soil contamination and specifications under the Soil Contamination Countermeasures Act (Fourth Report) Environmental Quality Standards Review under the Soil Contamination Countermeasures Act* https://www.env.go.jp/en/water/soil/contami_law.html.

Mu, D., Zheng, S., Lin, D., Xu, Y., Dong, R., Pei, P., & Sun, Y. (2023). Derivation and validation of soil cadmium thresholds for the safe farmland production of vegetables in high geological background area. *Science of the Total Environment, 873*, 162171. doi:10.1016/j.scitotenv.2023.162171.

State Administration for Market Regulation. (2023). *Soil quality—Decision unit and Multi Increment Sampling Methods GB/T 42489-2023*. China National Standardization Administration Committee https://openstd.samr.gov.cn/bzgk/gb/newGbInfo?hcno=B5E8F45C1E990CE2E3BDDE7158DAA835 2023.

Page, A. L., Chang, A. C., & El-Amamy, M. (1987). Cadmium levels in soils and crops in the United States. *Lead, mercury, cadmium and arsenic in the environment: SCOPE 31*: 8 https://scope.dge.carnegiescience.edu/SCOPE_31/SCOPE_31.html.

Ran, J., Wang, D., Wang, C., Zhang, G., & Zhang, H. (2016). Heavy metal contents, distribution, and prediction in a regional soil-wheat system. *Science of the Total Environment, 544*, 422–431. www.elsevier.com/locate/scitotenv. doi:10.1016/j.scitotenv.2015.11.105.

Römkens, P. F. A. M., Brus, D. J., Guo, H. Y., Chu, C. L., Chiang, C. M., & Koopmans, G. F. (2011). Impact of model uncertainty on soil quality standards for cadmium in rice paddy fields. *Science of the Total Environment, 409*(17), 3098–3105. doi:10.1016/j.scitotenv.2011.04.045.

Schrenk, D., Bignami, M., Bodin, L., Chipman, J. K., del Mazo, J., Grasl-Kraupp, B., Hogstrand, C., Hoogenboom, L., Leblanc, J. C., Nebbia, C. S., Ntzani, E., Petersen, A., Sand, S., Schwerdtle, T., Vleminckx, C., Wallace, H., Guérin, T., Massanyi, P., Van Loveren, H., Baert, K., Gergelova, P., & Nielsen, E. (2020). Update of the risk assessment of nickel in food and drinking water. *EFSA Journal, 18*(11). efsa.onlinelibrary.wiley.com/journal/18314732/. doi:10.2903/j.efsa.2020.6268.

Shinsugi, C., & Takimoto, H. (2025). Food consumption trends in Japanese children and adolescents: The National Health and Nutrition Survey, 2001–2019. *Foods, 14*(8), 1392. doi:10.3390/foods14081392.

Smith, D. B., Solano, F., Woodruff, L. G., Cannon, W. F., & Ellefsen, K. J. (2019). Geochemical and mineralogical maps, with interpretation, for soils of the conterminous United States. *USGS Scientific Investigations Report 2017-5118* http://pubsdata.usgs.gov/pubs/sir/2017/5118/index.html. doi:10.3133/sir20175118.

Sneath, G. (2020). Nutrient management in farmed landscapes revision of the tiered fertiliser management system for soil cadmium. *Occasional Report (Farmed Landscapes Research Centre). Occasional Report No. 33* https://flrc.massey.ac.nz/workshops/20/Manuscripts/Paper_Sneath_2020.pdf.

Steven, J., & Spencer, H. (2004). *Standards and agro-food exports from developing countries: Rebalancing the debate, Policy Research Working Paper Series*. World Bank Publications https://hdl.handle.net/10986/14061. doi:10.1596/1813-9450-3348.

Sun, L., Wang, R., Tang, W., Chen, Y., Zhou, J., Ma, H., Li, S., Deng, H., Han, L., Chen, Y., Tan, Y., Zhu, Y., Lin, D., Zhu, Q., Wang, J., Huang, D., & Chen, C. (2022). Robust identification of low-Cd rice varieties by boosting the genotypic effect of grain Cd accumulation in combination with marker-assisted selection. *Journal of Hazardous Materials, 424*(Pt D), 127703. doi:10.1016/j.jhazmat.2021.127703.

United States Environmental Protection Agency. (1989a). *Risk assessment guidance for Superfund. Volume I: Human Health Evaluation Manual (Part A), (EPA/540/1-89/002)*. United States Environmental Protection Agency, Office of Emergency and Remedial Response https://semspub.epa.gov/work/11/191.pdf.

United States Environmental Protection Agency. (1989b). *Risk assessment guidance for Superfund. Volume II: Environmental Evaluation Manual (EPA/540/1-89/001)*. U.S. EPA Office of Emergency and Remedial Response https://nepis.epa.gov/Exe/ZyPURL.cgi?Dockey=10001FQY.TXT.

United States Environmental Protection Agency. (1993). Code of Federal Regulations, Title 40 standards for the use or disposal of sewage sludge, 40 CFR Part 503. *Federal Register, 58*. https://www.ecfr.gov/current/title-40/chapter-I/subchapter-O/part-503.

United States Environmental Protection Agency. (2024). *Regional Screening Levels (RSLs) for chemical contaminants Regional Screening Level (RSL) Tables*. United States Environmental Protection Agency https://www.epa.gov/risk/regional-screening-levels-rsls.

Vácha, R., Sáňka, M., Hauptman, I., Zimová, M., & Čechmánková, J. (2014). Assessment of limit values of risk elements and persistent organic pollutants in soil for Czech legislation. *Plant, Soil and Environment, 60*(5), 191–197. doi:10.17221/909/2013-pse.

Vannan, R. R. M., Moorthy, T. V., Hariharan, P., & Gnanavel, B. K. (2020). Effect of physical vapour deposition coatings on high speed steel single point cutting tool. In *Lecture notes in mechanical engineering: 27* (pp. 1–5). doi:10.1007/978-981-13-8297-0_1.

Wang, P., & Zhao, F. (2022). China national food safety standards of cadmium in staple foods: Issues and thinking. *Chinese Science Bulletin, 67*(27), 3252–3260. doi:10.1360/tb-2022-0138.

World Health Organization. (2014). *Evaluation of certain food additives and contaminants: seventy-seventh report of the Joint FAO/WHO Expert Committee on Food Additives (WHO Technical Report Series No. 983) WHO Technical Report Series*. World Health Organization https://iris.who.int/server/api/core/bitstreams/b8ed427c-9a83-4476-b9f6-928e4e57968c/content.

Xia, R., Zhou, J., Sun, Y., Zeng, Z., Liu, H., Cui, H., Yan, J., Kou, L., Hu, K., Zhang, H., & Zhou, J. (2023). Stable isotope ratios trace the rice uptake of cadmium from atmospheric deposition via leaves and roots. *Environmental Science and Technology, 57*(44), 16873–16883. http://pubs.acs.org/journal/esthag. doi:10.1021/acs.est.3c04820.

Xiao, N., Wang, F., Tang, L., Zhu, L., Song, B., & Chen, T. (2022). Recommended risk screening values for Cd in high geological background area of Guangxi, China. *Environmental Monitoring and Assessment, 194*(3). https://link.springer.com/journal/10661. doi:10.1007/s10661-022-09802-2.

Xu, M., Yang, L., Chen, Y., Jing, H., Wu, P., & Yang, W. (2022). Selection of rice and maize varieties with low cadmium accumulation and derivation of soil environmental thresholds in karst. *Ecotoxicology and Environmental Safety, 247,* 114244. doi:10.1016/j.ecoenv.2022.114244.

Xue, T., Liao, X., Li, H., Xie, Y., Wei, W., Chen, J., Liu, Z., & Ji, X. (2023). Remediation of Cd contaminated paddy fields by intercropping of the high- and low- Cd-accumulating rice cultivars. *Science of the Total Environment, 878.* www.elsevier.com/locate/scitotenv. doi:10.1016/j.scitotenv.2023.163133.

Yang, J. M., Chen, H. L., Wang, X., Guan, D. X., Ji, X. H., Xie, Y. H., Huang, R., Peng, B., & Qin, Q. B. (2025). Toward safe rice production in As-Cd co-contaminated paddy soils: Biogeochemical mechanisms and remediation strategies. *Critical Reviews in Environmental Science and Technology, 55*(1), 1–24. www.tandf.co.uk/journals/titles/10643389.asp. doi:10.1080/10643389.2024.2373949.

Yang, Q., Yang, Z., Zhang, Q., Liu, X., Zhuo, X., Wu, T., Wang, L., Wei, X., & Ji, J. (2021). Ecological risk assessment of Cd and other heavy metals in soil-rice system in the karst areas with high geochemical background of Guangxi, China. *Science China Earth Sciences, 64*(7), 1126–1139. http://www.springer.com/earth+sciences/journal/11430. doi:10.1007/s11430-020-9763-0.

Zhang, L., Sheng, H., Yin, Z., Xue, Y., Huang, D., Zhang, L., Zhou, Q., Zhou, P., & Thompson, M. L. (2023). Climate and soil pH control the reduction of cadmium in rice during 7 years of biannual liming. *Plant and Soil, 492*(1–2), 471–484. https://www.springer.com/journal/11104. doi:10.1007/s11104-023-06192-3.

Zhang, S., Song, J., Cheng, Y., & McBride, M. B. (2018). Derivation of regional risk screening values and intervention values for cadmium-contaminated agricultural land in the Guizhou Plateau. *Land Degradation and Development, 29*(8), 2366–2377. http://onlinelibrary.wiley.com/journal/10.1002/(ISSN)1099-145X. doi:10.1002/ldr.3034.

Zhang, X., Yang, L., Li, Y., Li, H., Wang, W., & Ye, B. (2012). Impacts of lead/zinc mining and smelting on the environment and human health in China. *Environmental Monitoring and Assessment, 184*(4), 2261–2273. doi:10.1007/s10661-011-2115-6.

Zhao, D., Wang, P., & Zhao, F. J. (2023). Dietary cadmium exposure, risks to human health and mitigation strategies. *Critical Reviews in Environmental Science and Technology, 53*(8), 939–963. www.tandf.co.uk/journals/titles/10643389.asp. doi:10.1080/10643389.2022.2099192.

Zhu, H., Chen, C., Xu, C., Zhu, Q., & Huang, D. (2016). Effects of soil acidification and liming on the phytoavailability of cadmium in paddy soils of central subtropical China. *Environmental Pollution, 219,* 99–106. www.elsevier.com/inca/publications/store/4/0/5/8/5/6. doi:10.1016/j.envpol.2016.10.043.

CHAPTER

18

Amendments and agronomic practices to reduce cadmium uptake by crops

Zhong-Rui Xu[a], *Fang-Jie Zhao*[a], *Peng Wang*[a,b]

[a]College of Resources and Environmental Sciences, Nanjing Agricultural University, Nanjing, China [b]College of Sciences, Nanjing Agricultural University, Nanjing, China

18.1 Introduction

Cadmium (Cd) is a highly toxic heavy metal that poses serious threats to human health. Chronic Cd exposure has been associated with renal dysfunction, skeletal damage, and carcinogenic effects (Nawrot et al., 2006; Satarug et al., 2010). In agricultural systems, Cd can easily penetrate the soil–plant barrier and accumulate in edible plant tissues, making it one of the most concerning toxic metals affecting food safety through the soil–food chain (Clemens et al., 2013; Zhao & Wang, 2020). For nonsmoking populations, dietary intake is the primary source of Cd exposure (Clemens et al., 2013) (see Chapter 15). Globally, approximately 9% of agricultural soils are estimated to be contaminated with Cd, posing an escalating risk to public health (Hou et al., 2025). Therefore the development of effective management practices to reduce Cd accumulation in crops while maintaining crop productivity is of paramount importance.

Cd accumulation in plants is governed by both the concentration of bioavailable Cd in soil and the plant's capacity for Cd uptake and translocation. A wide range of soil amendments and agronomic practices have been developed to lower Cd bioavailability and reduce plant uptake (Grant et al., 1999; McLaughlin et al., 2021). Soil Cd bioavailability is strongly influenced by pH, redox potential (*Eh*), organic matter content, cation exchange capacity, ionic strength of the soil solution, and the presence of competing or complexing ions (Alloway, 2013; Bolan et al., 2014; de Livera et al., 2011; Naidu et al., 1994; Sauvé et al., 2000) (see Chapter 9). Soil amendments, such as lime, biochar, organic matter, clay minerals, inorganic sorbents, and mineral element fertilizers, can alter these physicochemical properties to decrease the mobility of Cd (Bolan et al., 2014; McLaughlin et al., 2021). In addition, agronomic practices—most notably

Cadmium Contamination in Soils and Plants: Mechanisms and Mitigation.
DOI: https://doi.org/10.1016/B978-0-443-34104-5.00013-X

water management in paddy fields—also play a crucial role in reducing Cd mobility and uptake (Chen et al., 2020; Honma et al., 2016).

This chapter examines the mechanisms through which soil amendments and agronomic practices mitigate Cd availability in contaminated soils. It summarizes recent research findings and evaluates the effectiveness, advantages, and limitations of different approaches. By integrating current knowledge the chapter provides insights into the practical implementation of these strategies in diverse agricultural systems. Ultimately, it aims to establish a scientific foundation for the development of sustainable and effective management practices to minimize Cd contamination in agricultural soils and protect food safety and human health.

18.2 Paddy water management

Rice is a major dietary source of Cd exposure, particularly in Asian populations (Song et al., 2017) (see Chapter 15). Rice cultivation involves alternating cycles of submergence and drainage, which strongly influence Cd solubility and bioavailability in soil. These water management cycles alter soil *Eh* and pH, thereby controlling Cd speciation and mobility (Chen et al., 2020; Honma et al., 2016). Flooding lowers soil *Eh* and stimulates sulfate reduction mediated by sulfate-reducing bacteria, leading to the formation of insoluble Cd sulfide (CdS) (Cornu et al., 2007; de Livera et al., 2011). Under submerged conditions the pH of initially acidic soils tends to rise toward neutral values because of proton consumption during reductive reactions (Kirk, 2004), further reducing Cd solubility. In contrast, soil drainage introduces oxidizing conditions that promote Cd solubilization (Fulda et al., 2013). This process is primarily driven by the rapid oxidation of CdS and the accompanying decline in pH, which diminishes Cd sorption by iron–manganese (oxyhydr)oxides and organic matter (Wang et al., 2019).

The influence of paddy water status on Cd bioavailability and subsequent grain Cd accumulation was demonstrated in a field experiment using a gentle gradient—an agricultural practice commonly employed to facilitate water flow from inlet to outlet (Chen et al., 2020). Variations in redox potential and pH across the slope during soil drainage led to large spatial differences—by three to four orders of magnitude—in soluble Cd concentrations, resulting in six- to sevenfold spatial variation in the concentration of Cd in rice grain.

In conventional rice cultivation, paddy soils remain flooded during the vegetative stage, followed by a short drainage period at late tillering to suppress unproductive tillers. Water is also drained during the late grain-filling stage to facilitate harvest. Most Cd accumulation in rice grain occurs during this grain-filling period, coinciding with Cd solubilization upon soil drainage (Huang et al., 2022a). Consequently, continuous flooding or prolonged submergence during grain filling has been shown to effectively reduce Cd concentrations in rice grain (Fig. 18.1) (Honma et al., 2016; Huang et al., 2022a). However, these practices may simultaneously increase arsenic bioavailability in soil and arsenic accumulation in rice grain (Arao et al., 2009; Honma et al., 2016; Huang et al., 2022a; Xu et al., 2008). Thus careful management is required in fields cocontaminated with both Cd and arsenic. Moreover, maintaining flooding until harvest is often impractical, as it complicates machine harvesting operations.

18.3 Liming

Soil pH is one of the most critical factors influencing Cd availability in soils (Helmke, 1999) (see Chapters 6 and 7). In acidic soils, increasing soil pH markedly reduces Cd availability

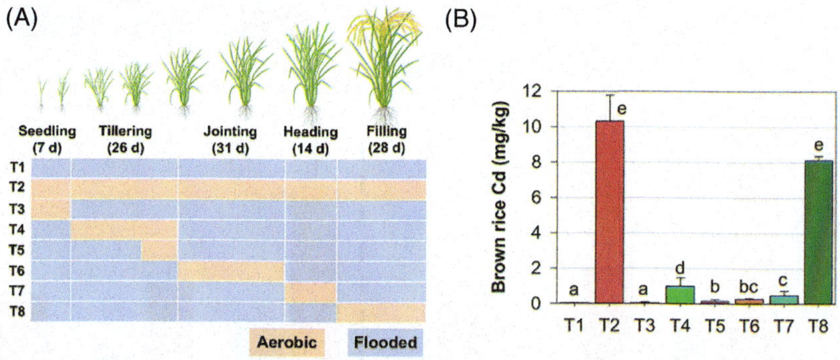

FIGURE 18.1 The effect of water management on rice grain Cd concentrations in pot experiments: (A) schematic of the different water management treatments and (B) grain Cd concentrations. *Cd*, Cadmium. *From Huang et al. (2022a) with permission.*

by enhancing Cd adsorption onto soil particles. Numerous studies have reported an inverse relationship between soil pH and Cd extractability (Blake & Goulding, 2002; Singh et al., 1995; Zhu et al., 2016). A meta-analysis of 70 studies demonstrated that soil pH alone accounts for about 50% of the variation in Cd distribution coefficients (K_d, defined as the ratio of solid- to solution-phase Cd concentrations) (Sauvé et al., 2000). Similarly, Cd accumulation in plants shows a strong negative correlation with soil pH (He & Singh, 1993; Liu et al., 2015; Singh et al., 1995; Zhu et al., 2016). For example, a field survey in southern Norway revealed a clear negative relationship between soil pH and Cd concentration in grass (He & Singh, 1993). Comparable trends have been reported for rice and wheat grains (Liu et al., 2015; Yu et al., 2016). In southern China, soil acidification is a major cause of high Cd levels in rice grain (Zhao et al., 2015; Zhu et al., 2016). A large-scale survey (n = 39,642) found that the median soil pH in this region decreased by nearly one unit over the past three decades (Zhu et al., 2016). When soil pH decreased from 7.0 to 5.0, the median transfer factor (i.e., the ratio of grain Cd concentration to total soil Cd concentration) increased by approximately 10 folds (Zhu et al., 2016).

In acidic paddy soils, liming is a widely adopted and effective practice for controlling Cd accumulation in rice grain (Bolan et al., 2013; Chen et al., 2018; Zhao et al., 2015; Zhu et al., 2016). Common agricultural liming materials include limestone ($CaCO_3$), burnt lime (CaO), and hydrated lime [$Ca(OH)_2$]. Although other alkaline materials can be used, lime is generally more effective and economical. For example, the application of dolomite or agricultural lime to a paddy field increased soil pH from 6.4 to approximately 8.2, reducing Cd concentrations in rice grain six folds (Kim et al., 2016). Similarly a field trial conducted in Taoshui, Hunan Province, in southern China, revealed that elevating soil pH from 5.2 to 5.8 via liming (CaO) decreased Cd concentration in brown rice by 40% (Wang et al., 2018). The efficiency of liming depends strongly on both the application rate and method. Insufficient lime application is often ineffective (Chen et al., 2018; Wang et al., 2018). In China the usual practice involves applying reactive CaO before planting or during early growth, which can damage seedlings if overapplied. A 3-year field trial in Xiangtan, Hunan Province, China, demonstrated that a single application of ground limestone ($CaCO_3$, 7.5 t ha^{-1}) raised soil pH from approximately 5.5 to 6.5 and reduced rice grain Cd concentrations by 70%–80% over three consecutive seasons (Fig. 18.2) (Chen et al., 2018; Wang et al., 2019).

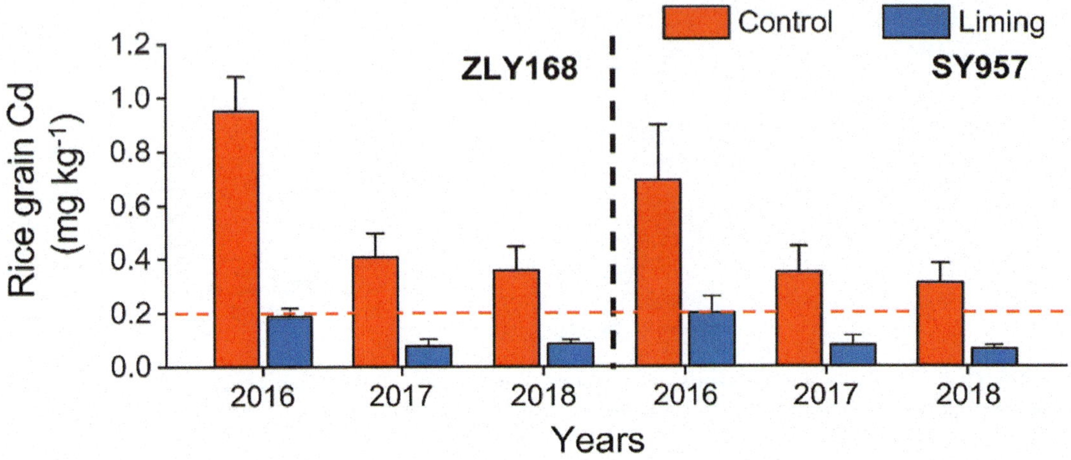

FIGURE 18.2 Effect of a one-time $CaCO_3$ application at 7.5 t ha^{-1} on rice grain Cd concentrations in continuous 3-year field trials. The field experiment was carried out in Xiangtan, Hunan province, in southern China, with two widely grown rice cultivars (SY957 and ZLY168). *Cd*, Cadmium. Dotted line represents the Cd limit for rice in China. *Data from Chen et al. (2018) and Wang et al. (2019).*

Liming is also effective in acid dryland soils. For example, increasing soil pH from 5.7 to 7.3 through liming reduced Cd concentrations in wheat grain from 2.24 to 1.36 mg kg^{-1} (Tlustos et al., 2006). In a Cd-contaminated soil from Dayu, Jiangxi Province, China, a high-dose application of hydrated lime (11.15 t ha^{-1}) decreased wheat grain Cd concentrations by 54%, whereas a low dose (2.2 t ha^{-1}) had negligible effects (Guo et al., 2018). Similar reductions were observed in oat, ryegrass, carrot, and spinach with increased soil pH in both sandy and loam soils (He & Singh, 1995).

However, the effects of liming on Cd uptake are not always consistent and can vary with soil type and crop species. Although increasing soil pH generally reduces Cd bioavailability, several studies have reported little or no effect (Li et al., 1996; Sims & Kline, 1991; Singh et al., 1995). It is expected that liming would not be effective on neutral to alkaline soils (Li et al., 1996; MacLean, 1976). However, increasing the pH of an acid clay loam from 5.5 to 7.5 significantly reduced Cd accumulation in carrot and lettuce but not in wheat (Singh et al., 1995). In some cases, liming may even increase Cd accumulation (Bolan et al., 2003; Chaney et al., 2006, 2009; He & Singh, 1995; Xu et al., 2024). One possible explanation is that Ca supplied through lime can displace adsorbed Cd from the solid phase, increasing Cd concentration in the soil solution (Bolan et al., 2003; He & Singh, 1995). In soils with high Cd/Zn ratios, liming may induce Zn deficiency, which upregulates Zn/Cd transporter genes and increases Cd uptake in plants such as lettuce (Chaney et al., 2006, 2009).

A recent systematic study examined the effect of soil pH on Cd accumulation in different crop types (Xu et al., 2024). Gradual pH increases from 5.0 to 7.9 markedly decreased $CaCl_2$-extractable Cd concentrations in soil and Cd accumulation in rice, a monocot species, but increased Cd accumulation in dicotyledonous plants such as lettuce, pakchoi, Chinese cabbage, and *Arabidopsis thaliana* when pH was raised from 5.0 to 5.8–7.0 (Fig. 18.3). This increase in dicots was attributed to Fe deficiency induced by liming, which upregulated the Fe transporter gene

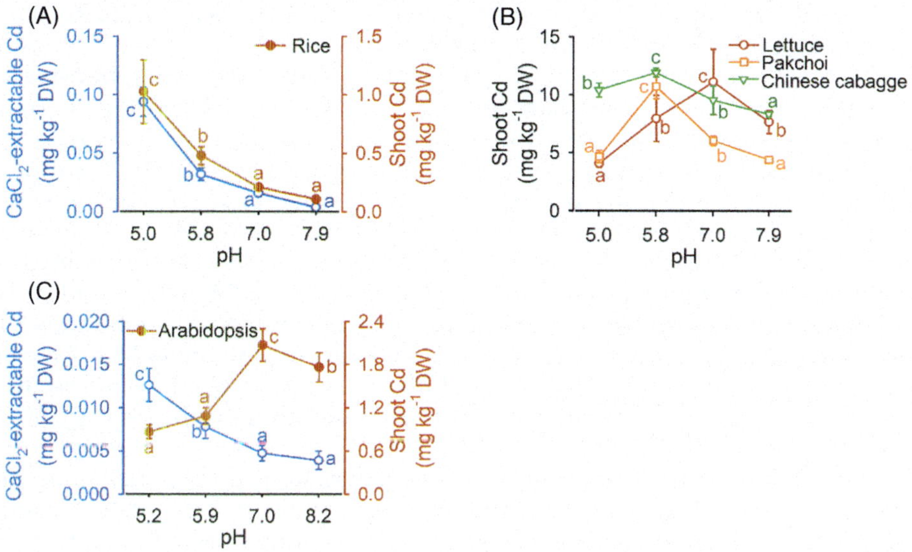

FIGURE 18.3 Effects of liming-induced soil pH increase on $CaCl_2$-extractable Cd concentrations in soil and Cd accumulation in various crop species. *Cd*, Cadmium. *From Xu et al. (2024) with permission.*

IRT1—responsible for the uptake of Fe^{2+} and Cd^{2+} in *A. thaliana* (Vert et al., 2002). This effect can counteract or even override the pH-induced decrease in Cd availability. In contrast, Cd uptake in rice is mainly mediated by the Mn transporter *OsNRAMP5*, which is unaffected by Fe deficiency (Ishikawa et al., 2012; Sasaki et al., 2012) (see Chapter 10).

Because the effectiveness of liming depends on soil type, crop species, and environmental conditions, site-specific evaluation is essential before applying this strategy for Cd mitigation.

18.4 Applications of manganese fertilizers

Manganese (Mn) is an essential micronutrient for both plants and humans. It is relatively abundant in soils and primarily exists as Mn oxides under aerobic conditions (Gilkes & McKenzie, 1988). In flooded paddy soils, Mn solubility increases because Mn oxides are reduced to Mn^{2+}. The application of Mn fertilizers or Mn-rich materials has been shown to reduce Cd accumulation in crops through several mechanisms.

First the formation of Mn oxides in soil can enhance Cd adsorption (Fang et al., 2021; Suda & Makino, 2016). Both Mn and Fe oxides are natural soil constituents with highly reactive surfaces and low solubility, allowing them to adsorb trace elements such as Cd at high concentrations (Kumpiene, 2010; Suda & Makino, 2016). Among them, Mn oxides, such as hausmannite and cryptomelane, exhibit stronger Cd adsorption capacities than Fe oxides such as ferrihydrite and goethite (Backes et al., 1995), although Fe oxides are generally more abundant in soils.

Second, in flooded paddy soils, MnS formed under anoxic conditions can influence Cd mobility through a voltaic effect during drainage. When soil redox potential increases upon oxidation, MnS, which has a lower electrochemical potential than CdS, acts as an anode and preferentially oxidizes, thereby suppressing the oxidation and remobilization of CdS, which serves as a cathode (Huang et al., 2021) (see Chapter 7). This electrochemical interaction helps maintain Cd in its insoluble sulfide form during drainage.

Third, Mn and Cd share common uptake pathways in plant roots. In rice, transporters, such as OsNRAMP5 and OsNRAMP1, mediate both Mn^{2+} and Cd^{2+} uptake, leading to competitive inhibition of Cd absorption when Mn availability is high (Chang et al., 2020; Ishikawa et al., 2012; Sasaki et al., 2012; Yang et al., 2014). The antagonistic effect of Mn on Cd uptake has been confirmed in several rice studies (Hu et al., 2024; Wang et al., 2022; Yang et al., 2014). In a hydroponic experiment with a very low Cd concentration (2 nM Cd^{2+}), increasing Mn concentration from 0.08 to 800 μM progressively reduced Cd concentrations in shoots and roots of wild-type rice but not in *osnramp5* mutants (Yang et al., 2014). Other studies, however, reported that Mn supplementation reduced Cd uptake only at high Mn/Cd ratios (Hu et al., 2024; Wang et al., 2022), suggesting that competition occurs mainly under Mn-rich conditions. In contrast, this antagonistic relationship is not observed in dicotyledonous species such as spinach, where Cd uptake is primarily mediated by Fe and Zn transporters (e.g., *IRT1*) rather than Mn transporters (Xu et al., 2024).

Applications of Mn oxides effectively decreased Cd accumulation in rice grain in several studies (Fang et al., 2021; Huang et al., 2022a). In 2 years, multisite field trials, the application of 200 mg Mn kg^{-1} soil (approximately 500 kg ha^{-1}) in the form of MnO_2 (mainly ramsdellite and akhtenskite) or Mn-loaded biochar to acidic paddy soils with low native Mn reduced Cd concentrations in rice grain by 44%–53% and 78%–82%, respectively (Fang et al., 2021). However, MnO_2 amendments were ineffective in soils already rich in Mn (Fang et al., 2021).

In neutral to alkaline dryland soils, large Mn fertilizer doses (0.1%–1.0% of soil mass in the top 20 cm) have also been reported to reduce Cd availability and Cd accumulation in wheat and *Brassica chinensis*, mainly by increasing the proportion of Cd bound to Mn oxides (Huang et al., 2022b; Lu et al., 2022; Yong et al., 2022). A pot experiment comparing addition of Mn oxides, Fe/Mn mixed oxides, and Fe oxides showed that Mn oxides were most effective in decreasing Cd concentrations in wheat grain (Lu et al., 2022). The application of Mn oxides at 1.0% of soil mass (approximately 26 t ha^{-1}) reduced soil-available Cd and wheat grain Cd concentrations by 50% and 70%, respectively (Lu et al., 2022). The high doses required in these soils, likely because of the low availability of Mn under neutral/alkaline conditions, substantially increase costs, reducing the economic feasibility of this approach.

18.5 Applications of zinc fertilizers

Zinc (Zn) is an essential micronutrient for both plants and humans but can become phytotoxic at excessive concentrations. In contrast to Cd, Zn deficiency is common in many agricultural soils and poses a threat to crop productivity and human nutrition (Alloway, 2009). Owing to their close chemical similarity, Zn and Cd interact strongly in soil–plant systems, influencing Cd mobility, uptake, and translocation. Depending on soil and plant conditions, these interactions can be either antagonistic or synergistic. Consequently the application of Zn fertilizers has produced inconsistent effects on Cd accumulation in plants, largely depending on the relative concentrations of Cd and Zn, soil properties, and plant species (McLaughlin et al., 2021).

Zinc can influence Cd availability in soil by competing for adsorption sites and forming stable complexes (de Livera et al., 2011; Helmke, 1999). In most contaminated soils, Zn concentrations are typically more than 200 times higher than those of Cd (Chaney, 2010). At high concentrations, Zn may compete with Cd for sorption sites, thereby decreasing Cd adsorption and increasing its solubility, which can enhance plant Cd uptake (Helmke, 1999; McLaughlin et al., 2021). However, in paddy soils, Zn application can reduce Cd availability by inhibiting the oxidative dissolution of CdS during drainage through the voltaic cell mechanism (Huang et al., 2021) (see Chapter 7). Under flooded conditions, Zn facilitates the formation of ZnS, which has a lower electrochemical potential than CdS, thereby acting as an anode and suppressing CdS oxidation upon reoxidation (Huang et al., 2021). In pot experiments, the application of $ZnSO_4$ reduced $CaCl_2$-extractable Cd concentrations in paddy soils after drainage by 32%–64%, whereas field trials showed reductions of 16%–30% (Huang et al., 2022c).

Zinc also influences Cd accumulation through plant physiological processes affecting uptake and translocation, but the results are variable. In hydroponic studies, Zn supplementation reduced Cd uptake in wheat, lettuce, and spinach (Green et al., 2003; Hart et al., 2002, 2005; McKenna et al., 1993), likely because of competition for shared transporters at the root plasma membrane. This explanation, however, may not apply to rice, where Cd uptake occurs mainly via Mn transporters such as OsNRAMP5 (Ishikawa et al., 2012; Sasaki et al., 2012). In rice, some studies reported that increasing Zn supply from adequate to subphytotoxic levels did not inhibit Cd uptake (Green et al., 2017), whereas others found that Zn application suppressed Cd uptake by roots (Fontanili et al., 2016).

Zn and Cd are translocated into the xylem through the same P-type ATPase transporters, HMA2 and HMA4 (Takahashi et al., 2012; Wong & Cobbett, 2009; Yamaji et al., 2013). Thus elevated Zn supply is expected to inhibit Cd translocation from roots to shoots. However, several studies found the opposite—high Zn levels enhanced Cd translocation in rice (Cai et al., 2019; Chang et al., 2023; Green et al., 2017; Liu et al., 2007). This effect was attributed to Zn-induced upregulation of the *OsHMA2* gene; knockout of *OsHMA2* abolished the Zn-promoted Cd translocation effect (Chang et al., 2023).

The absolute and relative concentrations of Zn and Cd are therefore key determinants of their interactions (Zare et al., 2018). At very low Zn levels, Zn fertilization typically decreases Cd accumulation in plants, primarily by restoring plasma membrane integrity impaired by Zn deficiency (Oliver et al., 1994), suppressing Zn transporter upregulation that enhances Cd influx (Chaney et al., 2009; Talke et al., 2006; Zare et al., 2018), and by biomass dilution effects associated with improved plant growth. Conversely, when Zn availability is sufficient, additional Zn application often has little or no effect on Cd accumulation (Rojas-Cifuentes et al., 2012). Therefore, the effect of Zn fertilization on Cd behavior in crops is highly context-dependent and varies with soil properties, Zn nutritional status, and crop species.

18.6 Biochar

Biochar, a carbon-rich material produced through the pyrolysis of biomass under limited oxygen conditions, has been widely investigated for its potential to immobilize heavy metals in soil (He et al., 2019). It is commonly derived from agricultural residues such as crop straw, wood chips, and animal manures. Owing to its high specific surface area, porous structure, large particle size, high cation exchange capacity, abundant functional groups, and alkaline nature, biochar can significantly modify soil physicochemical properties and influence Cd bioavailability (Atkinson et al., 2010; Beesley et al., 2010; Lehmann & Joseph, 2015; Major, 2010; Puga et al., 2015).

The mechanisms by which biochar immobilizes Cd include surface adsorption, cation exchange, and complexation or precipitation with functional groups (e.g., –COOH and –OH)

(He et al., 2019; Lei et al., 2019; Li et al., 2015; Qin et al., 2018). In addition, biochar application can increase soil organic carbon content, cation exchange capacity, and soil pH, further contributing to Cd immobilization (He et al., 2019). Applications of large amounts of biochar (>20 t ha^{-1}) to Cd-contaminated soils were found to increase soil pH and total organic carbon and significantly decrease soil extractable Cd and Cd concentration in rice grain (Bian et al., 2013, 2014).

The effects of biochar on Cd availability vary depending on biochar's physicochemical properties, such as surface area and functional groups. Biochar with smaller particle sizes typically has higher surface areas and greater Cd adsorption capacity (Fahmi et al., 2018; Ramezanzadeh et al., 2021). The pyrolysis temperature used to produce biochar affects the physiochemical properties of biochar, such as the functional groups on the surface (Tan et al., 2020). Biochar produced at lower temperatures contains more O-containing functional groups, which can enhance Cd immobilization (Ahmad et al., 2014). Modified biochars, such as those treated with metals, metal oxides, or other functional groups, often exhibit higher Cd adsorption capacities and have a high remediation potential (Rahim et al., 2022).

Despite its recognized benefits for soil quality enhancement and crop productivity, the application of biochar may introduce contaminants such as polycyclic aromatic hydrocarbons, heavy metals, and polychlorinated dibenzodioxins into the soil, thereby posing potential ecological and health risks (He et al., 2019). Another limitation is that effective reduction of Cd availability often requires the application of biochar in large quantities, resulting in high costs at current market prices. Moreover, the long-term efficacy of biochar may be compromised over time because of surface aging, oxidation, and alterations in functional groups, which collectively diminish its sorption capacity (O'Connor et al., 2018).

In summary, biochar application can be an effective strategy for reducing Cd bioavailability and improving soil health, especially in moderately contaminated soils. Its long-term performance depends on biochar type, application rate, and site-specific soil and crop characteristics. Combining biochar with other amendments, such as lime or Mn oxides, may provide more stable and synergistic effects on Cd immobilization.

18.7 Organic fertilizers

Organic fertilizers—including manures, composts, peats, or other organic waste products—are widely applied to improve soil fertility and structure. They also influence the behavior of heavy metals such as Cd in soil through both chemical and biological mechanisms. The addition of organic matters can alter soil pH, redox potential, and cation exchange capacity and provides functional groups capable of binding Cd, thereby affecting its mobility and bioavailability (Grant et al., 1999; McLaughlin et al., 2021). The potential of organic fertilizers to reduce Cd availability in soil is appealing, especially given their additional soil quality benefits. However, the effect of organic fertilizers on Cd availability in soil and Cd accumulation in crops varies with the type and origin of organic fertilizer as well as soil properties.

Animal manures are commonly used as fertilizers in agricultural soil and can either decrease or increase Cd availability in the soil. Application of farmyard manure was reported to reduce Cd uptake of maize through forming Cd–organic complexes and increasing soil cation exchange capacity (Ram & Verloo, 1985). Additions of farmyard manure, pig manure, and cow manure in paddy soils were found to decrease available Cd and reduce Cd accumulation in rice grain (Huang et al., 2020; Li et al., 2019; Wang et al., 2021a). Manures from different sources may produce different effects. A study found that application of cow manure in a naturally metal-rich soil was more effective in reducing wheat Cd accumulation than either pig manure or peat soil because cow manure increased the soil pH slightly (Narwal & Singh, 1998). However, manure may also increase Cd accumulation in crops because of the direct introduction of Cd by manure application or the solubilization of native Cd in the soil. Applications of animal manures containing Cd could increase Cd concentration in the soil and Cd uptake by crops (see Chapter 5).

Composts from green wastes (tree leaves, straw, grass, and mushrooms) are ideal soil amendments because of their high carbon and nitrogen levels but low heavy metal contents. The hydroxyl and phenolic groups in humic compounds of compost are the dominant binding sites for heavy metals and could immobilize Cd in soil (Yang et al., 2021). Applications of mushroom compost and other organic wastes (tobacco dust and grape marc) in a 6-month soil incubation experiment increased soil organic matter, but only mushroom compost and grape marc decreased DTPA-extractable Cd in soil. In contrast, the application of tobacco dust significantly increased the concentration of DTPA-extractable Cd in the soil, likely because of a decrease in soil pH and its own high Cd concentration (Karaca, 2004). The application of clinoptilolite- or bentonite-amended compost derived from green waste in contaminated soils reduced the leaching of Cd in laboratory leaching tests, with a maximum reduction of 48%, resulting in an efficient reduction in plant Cd accumulation (van Herwijnen et al., 2007).

Other organic materials, such as peats, humic substances, and lignite, can also immobilize Cd to some extent (Eriksson, 1988; Meng et al., 2017; Simmler et al., 2013). The addition of peat was reported to reduce NH_4AcO-extractable Cd in soils, especially those with low organic matter content, in field studies (Eriksson, 1988). In contrast, applications of peat were reported to increase Cd concentration in soil pore water (Egene et al., 2018). Leonardite-derived alkaline humic substances show a high capacity for Cd adsorption and could significantly reduce extractable Cd in contaminated soils (Meng et al., 2017). Lignite could effectively sorb Cd mainly via the sulfur-containing functional groups, thus reducing soluble Cd in soils and plant Cd uptake (Simmler et al., 2013).

In general, the effectiveness of organic fertilizers in reducing crop Cd accumulation is limited (Narwal & Singh, 1998) and may even increase Cd accumulation. Applications of organic fertilizers alone are unlikely to be sufficient for remediation of Cd-contaminated soils but can be a component of the remediation strategy.

18.8 Other amendments

A wide variety of materials, including inorganic sorbents, clay minerals, and elemental fertilizers, have been reported to reduce Cd accumulation in crops grown in contaminated soils. These materials may reduce Cd availability through direct reactions (e.g., adsorption, precipitation, complexation) or indirect effects (e.g., increasing soil pH) or suppress Cd uptake by plant roots.

Zeolites are natural microporous silicates with a three-dimensional cage-like structure and a high cation-exchange capacity, which facilitates Cd adsorption (Cataldo et al., 2021). Although the structure and exchange capacity of zeolites vary with their origin, these materials show potential in reducing Cd availability for plants. Pot experiments using Cd-spiked soils have shown that zeolite amendments reduce Cd accumulation in barley, maize, and *Chamomilla recutita* (Chlopecka & Adriano, 1997; Grejtovsky et al., 1998). Zeolites have also been used as carriers for other immobilizing agents. For instance, zeolite-supported manganese oxides effectively decreased Cd accumulation in wheat grain in a weakly alkaline Cd-contaminated soil (Wang et al., 2021). In addition, zeolites can improve soil properties and enhance crop yields in acidic soils (Cataldo et al., 2021).

Sepiolite, a fibrous clay mineral with the chemical formula [$Mg_8Si_{12}O_{30}(OH)_4(H_2O)_4 \bullet 8H_2O$], possesses a porous structure consisting of two tetrahedral silica sheets sandwiching a central octahedral magnesium oxide/hydroxide layer (Suárez & García-Romero, 2011). Owing to its distinctive structure and high ion-exchange capacity, sepiolite exhibits a strong potential for adsorption of heavy metals (Brigatti et al., 2000; Sheikhhosseini et al., 2013). It can also raise soil pH, as natural sepiolite often contains substantial amounts of $CaCO_3$ (Chen et al., 2020). These properties have led to its use in the remediation of Cd-contaminated soils (Xu et al., 2017). In a 2-year field study in southern China, a single application of sepiolite at 0.1%, 0.5%, and 1% (w/w) to soil effectively reduced Cd concentrations in rice grain by 47%–49% in the first year and 44%–50% in the second year (Chen et al., 2020). Consecutive applications over two years resulted in a 75% reduction in grain Cd concentration. Applications of sepiolite have also been shown to lower Cd accumulation in various vegetables, including spinach, lettuce, rape, and radish (Xu et al., 2017). Owing to its low cost, universal applicability, ease of application, and broad compatibility with other soil amendments—such as limestone, phosphate fertilizers, and biochar—sepiolite represents a practical option for the remediation of Cd-polluted soils (Xu et al., 2017). However, care needs to be taken to avoid inhalation of dust because of the presence of silica crystals, and some sepiolite may contain asbestos, a class one carcinogen.

Palygorskite, also referred to as attapulgite, is a magnesium aluminum phyllosilicate with the formula $(Mg,Al)_2Si_4O_{10}(OH) \bullet 4(H_2O)$ (Murray et al., 2011). Its high Cd sorption capacity and ability to increase soil pH make it suitable for Cd remediation (Álvarez-Ayuso & García-Sánchez, 2007; Liang et al., 2014). In a field experiment, the application of palygorskite at 10–20 t ha^{-1} increased soil pH slightly and significantly reduced both the exchangeable Cd in soil and Cd concentration in rice grain by 23%–56% (Liang et al., 2014). However, at the same application rate, palygorskite was less effective than sepiolite in reducing rice grain Cd. The Cd^{2+} sorption capacity of palygorskite can be substantially improved through mercapto-based modification (Liang et al., 2013). Consequently, mercapto-modified palygorskite has emerged as a high-performance amendment, with efficacy confirmed in several field trials (Li et al., 2022; Liang et al., 2019; Ma et al., 2024; Wu et al., 2022). Notably, this modified material is highly effective even in alkaline soils, where immobilization of Cd is more challenging (Liang et al., 2019; Ma et al., 2024). Nevertheless, the improved performance comes with higher costs, which may compromise its cost-effectiveness and limit its large-scale agricultural adoption.

Layered double hydroxides (LDHs) are a class of two-dimensional anionic clays consisting of positively charged brucite-like host layers and charge-balancing interlayer anions. Their general formula is $[M(II)_{1-x}M(III)_x(OH)_2]^{x+}(A^{n-})_{x/n} \bullet mH_2O$, where M(II) and M(III) are divalent and trivalent metal cations, respectively, for example, MgAl, MgFe, CaAl, and CaFe, and A^{n-} represents interlayer anions such as NO_3^-, CO_3^{2-}, SO_4^{2-}, MoS_4^{2-}, and L-cysteine (Fan et al., 2014). This layered structure enables LDHs to adsorb heavy metals through multiple mechanisms: (1) isomorphic substitution of divalent cations (e.g., Mg^{2+} and Ca^{2+}) in the host layers by Cd^{2+}, leading to stable mineralization; (2) surface adsorption via abundant –OH groups on the layers; and (3) intercalation of chelating agents that replace original anions and provide additional metal-binding sites (He et al., 2023). Field studies have shown that various LDHs, including MgAl-, CaAl-, and Fe-modified types, can reduce Cd bioavailability and plant accumulation, though their effectiveness varies (Kong et al., 2021; Li et al., 2025; Sun et al., 2023;

Zhang et al., 2022). For instance an application of CaAl-LDH at 3 t ha^{-1} reduced Cd in wheat grain from 0.22 to 0.08 mg kg^{-1} in the first year, with the effect lasting for 3 years (Kong et al., 2021). The compositional tunability of LDHs, both in host layers and interlayer anions, offers promising opportunities to enhance their performance. For instance, mercapto-modified MgAl/ZnAl-LDHs exhibited a 1.6–1.8-fold increase in Cd^{2+} sorption capacity and more effectively reduced Cd accumulation in pakchoi compared to unmodified LDHs (Gao et al., 2025).

Red mud is a by-product of the aluminum industries produced in enormous quantities (Hua et al., 2017). Red mud is rich in iron oxides and is strongly alkaline (pH ranging from 9.0 to 13.1); both characteristics are useful for reducing soil Cd availability (Hua et al., 2017; Lombi et al., 2002). In pot experiments with heavily polluted soils, red mud application increased plant yields and lowered Cd concentrations in oilseed rape, pea, wheat, and lettuce (Lombi et al., 2002). However, red mud may contain elevated levels of heavy metals, raising concerns about its use in agriculture. Other alkaline industrial by-products, such as beringite and wood ash, have also been shown to immobilize Cd primarily through pH elevation (Lombi et al., 2002; Mortensen et al., 2018; Oste et al., 2001).

Silicon (Si) fertilization has been observed to reduce Cd accumulation in rice under hydroponic conditions, likely by downregulating the expression of Cd transporter genes *OsNRAMP5* and *OsHMA2* (Kim et al., 2014; Shao et al., 2017). Some pot and field studies also attribute the effect on reducing Cd accumulation in plants to Si-induced increases in soil pH (Liang et al., 2005; Wang et al., 2016). However, the effects of Si fertilizers are not consistent across studies. A multisite, 2-year field trial series revealed that Si application did not reliably decrease Cd in rice grain (Fang et al., 2021). In that study, Si fertilization raised soil pH at certain sites but had little impact on soil porewater pH and did not elevate porewater Si concentrations to levels comparable with those used in hydroponic experiments.

18.9 Conclusion and future perspectives

Cd contamination in agricultural soils continues to pose a major threat to food safety and human health. A range of soil amendments and agronomic practices, including water management, liming, and the application of micronutrient fertilizers, biochar, organic matter, and inorganic sorbents, have been developed to reduce Cd bioavailability and plant uptake. However, the effectiveness of these measures varies widely depending on soil type, crop species, climatic conditions, and management practices. No single strategy can universally mitigate Cd accumulation across all agroecosystems.

Water management in paddy fields is an effective means of reducing Cd uptake by rice. Continuous flooding or prolonged submergence during the grain-filling period markedly decreases Cd solubility and accumulation in rice grain, although these practices may increase arsenic uptake and are often impractical for mechanized harvesting. Liming is highly effective in acidic soils, where pH adjustment substantially reduces Cd mobility and plant uptake, but its benefits diminish in neutral or alkaline soils and may even increase Cd uptake in certain dicotyledonous vegetables.

The application of Mn and Zn fertilizers can suppress Cd uptake through competitive inhibition, redox-mediated immobilization, and transporter regulation. Yet, their effects are strongly influenced by the Mn/Cd and Zn/Cd ratios, soil properties, and plant species. Biochar and organic amendments improve soil health and contribute to Cd immobilization by enhancing adsorption and complexation processes and increasing soil pH in the case of biochar application.

Nonetheless, their Cd immobilization capacity is generally limited, and risks of introducing additional contaminants cannot be overlooked. Clay minerals and inorganic sorbents, such as zeolites, hydroxyapatite, and LDHs, exhibit relatively high Cd-binding capacities and provide promising options for in situ immobilization, although cost and field applicability remain challenges.

Effective Cd mitigation requires an integrated, site-specific approach that considers soil chemistry, plant physiology, economic and environmental trade-offs, and feasibility of management. Combining multiple strategies—such as liming with biochar or optimized water management—may produce synergistic effects and improve long-term stability of Cd immobilization.

Future research should focus on the following aspects:

Mechanistic understanding—Elucidating molecular and physicochemical mechanisms governing Cd interactions with soil components and plant transporters under varying environmental conditions.

Targeted amendment development—Designing multifunctional, cost-effective materials capable of simultaneously immobilizing Cd and improving soil fertility.

Field-scale validation—Conducting long-term and large-scale trials to evaluate the sustainability and practicality of integrated Cd mitigation strategies.

Precision management—Leveraging soil and crop diagnostics, remote sensing, and data-driven modeling to guide selection of amendments and optimize application timing and rates for specific spatially varying conditions.

Through such targeted, interdisciplinary efforts, it will be possible to develop sustainable and economically viable strategies to minimize Cd accumulation in crops, safeguard food safety, and promote the long-term health of agricultural ecosystems.

18.10 Summary

- **No single strategy** provides universal effectiveness for Cd immobilization; management practices must be tailored to specific soil–plant systems.
- **Liming in acidic soils** effectively reduces Cd uptake in rice and other monocot species but may increase Cd accumulation in some dicotyledonous plants because of nutrient imbalances such as Fe deficiency.
- **Manganese (Mn) and zinc (Zn) fertilizers** can suppress Cd uptake through competitive inhibition and redox-mediated immobilization, although their effectiveness varies with soil properties and crop species.
- **Biochar and other organic amendments** improve soil quality and contribute to Cd immobilization through adsorption and complexation, but their overall effectiveness is often limited and site-dependent.

Acknowledgments

This work was supported by National Key Research and Development Program of China (2024YFD1701800).

AI disclosure

During the preparation of this work the author(s) used ChatGPT to polish English. After using this tool/service the author(s) review(s) and edit(s) the content as needed and take(s) full responsibility for the content of the publication.

References

Ahmad, M., Rajapaksha, A. U., Lim, J. E., Zhang, M., Bolan, N., Mohan, D., Vithanage, M., Lee, S. S., & Ok, Y. S. (2014). Biochar as a sorbent for contaminant management in soil and water: A review. *Chemosphere, 99*, 19–33. doi:10.1016/j.chemosphere.2013.10.071.

Alloway, B. J. (2009). Soil factors associated with zinc deficiency in crops and humans. *Environmental Geochemistry and Health, 31*(5), 537–548. doi:10.1007/s10653-009-9255-4.

Alloway, B. J. (2013). *Heavy metals in soils: Trace metals and metalloids in soils and their bioavailability*. Springer Science & Business Media.

Álvarez-Ayuso, E., & García-Sánchez, A. (2007). Removal of cadmium from aqueous solutions by palygorskite. *Journal of Hazardous Materials, 147*(1-2), 594–600. doi:10.1016/j.jhazmat.2007.01.055.

Arao, T., Kawasaki, A., Baba, K., Mori, S., & Matsumoto, S. (2009). Effects of water management on cadmium and arsenic accumulation and dimethylarsinic acid concentrations in Japanese rice. *Environmental Science and Technology, 43*(24), 9361–9367. doi:10.1021/es9022738.

Atkinson, C. J., Fitzgerald, J. D., & Hipps, N. A. (2010). Potential mechanisms for achieving agricultural benefits from biochar application to temperate soils: A review. *Plant and Soil, 337*(1), 1–18. doi:10.1007/s11104-010-0464-5.

Backes, C. A., McLaren, R. G., Rate, A. W., & Swift, R. S. (1995). Kinetics of cadmium and cobalt desorption from iron and manganese oxides. *Soil Science Society of America Journal, 59*(3), 778–785. doi:10.2136/sssaj1995.03615995005900030021x.

Beesley, L., Moreno-Jiménez, E., & Gomez-Eyles, J. L. (2010). Effects of biochar and greenwaste compost amendments on mobility, bioavailability and toxicity of inorganic and organic contaminants in a multi-element polluted soil. *Environmental Pollution, 158*(6), 2282–2287. doi:10.1016/j.envpol.2010.02.003.

Bian, R. J., Joseph, S., Cui, L. Q., Pan, G. X., Li, L. Q., Liu, X. Y., Zhang, A. F., Rutlidge, H., Wong, S., Chia, C., Marjo, C., Gong, B., Munroe, P., & Donne, S. (2014). A three-year experiment confirms continuous immobilization of cadmium and lead in contaminated paddy field with biochar amendment. *Journal of Hazardous Materials, 272*, 121–128. doi:10.1016/j.jhazmat.2014.03.017.

Bian, R. J., Chen, D., Liu, X. Y., Cui, L., Li, L. Q., Pan, G. X., Xie, D., Zheng, J. W., Zhang, X. H., Zheng, J. F., & Chang, A. (2013). Biochar soil amendment as a solution to prevent Cd-tainted rice from China: Results from a cross-site field experiment. *Ecological Engineering, 58*, 378–383. doi:10.1016/j.ecoleng.2013.07.031.

Blake, L., & Goulding, K. W. T. (2002). Effects of atmospheric deposition, soil pH and acidification on heavy metal contents in soils and vegetation of semi-natural ecosystems at Rothamsted Experimental Station, UK. *Plant and Soil, 240*(2), 235–251. doi:10.1023/a:1015731530498.

Bolan, N., Kunhikrishnan, A., Thangarajan, R., Kumpiene, J., Park, J., Makino, T., Kirkham, M. B., & Scheckel, K. (2014). Remediation of heavy metal(loid)s contaminated soils—To mobilize or to immobilize? *Journal of Hazardous Materials, 266*, 141–166. doi:10.1016/j.jhazmat.2013.12.018.

Bolan, N. S., Adriano, D. C., Mani, P. A., & Duraisamy, A. (2003). Immobilization and phytoavailability of cadmium in variable charge soils. II. Effect of lime addition. *Plant and Soil, 251*(2), 187–198. doi:10.1023/a:1023037706905.

Bolan, N. S., Makino, T., Kunhikrishnan, A., Kim, P. J., Ishikawa, S., Murakami, M., Naidu, R., & Kirkham, M. B. (2013). Cadmium contamination and its risk management in rice ecosystems. *Advances in Agronomy, 119*, 183–273. doi:10.1016/B978-0-12-407247-3.00004-4.

Brigatti, M. F., Lugli, C., & Poppi, L. (2000). Kinetics of heavy-metal removal and recovery in sepiolite. In *Applied Clay Science: 16* (pp. 45–57). doi:10.1016/S0169-1317(99)00046-0.

Cai, Y., Xu, W., Wang, M., Chen, W., Li, X., Li, Y., & Cai, Y. (2019). Mechanisms and uncertainties of Zn supply on regulating rice Cd uptake. *Environmental Pollution, 253*, 959–965. doi:10.1016/j.envpol.2019.07.077.

Cataldo, E., Salvi, L., Paoli, F., Fucile, M., Masciandaro, G., Manzi, D., Masini, C. M., & Mattii, G. B. (2021). Application of zeolites in agriculture and other potential uses: A review. *Agronomy, 11*(8). doi:10.3390/agronomy11081547.

Chaney, R. L., Filcheva, E., Green, C. E., & Brown, S. L. (2006). Zn deficiency promotes Cd accumulation by lettuce from biosolids amended soils with high Cd:ZN ratio. *Journal of Residuals Science and Technology, 3*(2), 79–85.

Chaney, R. L. (2010). Cadmium and zinc. In *Trace elements in soils* (pp. 409–439). John Wiley and Sons. doi:10.1002/9781444319477.ch17.

Chaney, R. L., Green, C. E., & Ajwa, H. A. (2009). Zinc fertilization plus liming to reduce cadmium uptake by Romaine lettuce on Cd-mineralized Lockwood soil. *The Proceedings of the International Plant Nutrition Colloquium, 16*(7), 1252.

Chang, J. D., Huang, S., Wiseno, I., Sui, F. Q., Feng, F., Zheng, L., Ma, J. F., & Zhao, F. J. (2023). Dissecting the promotional effect of zinc on cadmium translocation from roots to shoots in rice. *Journal of Experimental Botany, 74*(21), 6790–6803. doi:10.1093/jxb/erad330.

Chang, J. D., Huang, S., Yamaji, N., Zhang, W., Ma, J. F., & Zhao, F. J. (2020). OsNRAMP1 transporter contributes to cadmium and manganese uptake in rice. *Plant Cell and Environment, 43*(10), 2476–2491. doi:10.1111/pce.13843.

Chen, H., Zhang, W., Yang, X., Wang, P., McGrath, S. P., & Zhao, F. J. (2018). Effective methods to reduce cadmium accumulation in rice grain. *Chemosphere, 207*, 699–707. doi:10.1016/j.chemosphere.2018.05.143.

Chen, H., Wang, P., Gu, Y., Kretzschmar, R., Kopittke, P. M., & Zhao, F. J. (2020). The within-field spatial variation in rice grain Cd concentration is determined by soil redox status and pH during grain filling. *Environmental Pollution, 261*. doi:10.1016/j.envpol.2020.114151.

Chlopecka, A., & Adriano, D. C. (1997). Influence of zeolite, apatite and Fe-oxide on Cd and Pb uptake by crops. *Science of the Total Environment, 207*(2–3), 195–206. doi:10.1016/s0048-9697(97)00268-4.

Clemens, S., Aarts, M. G. M., Thomine, S., & Verbruggen, N. (2013). Plant science: The key to preventing slow cadmium poisoning. *Trends in Plant Science, 18*(2), 92–99. doi:10.1016/j.tplants.2012.08.003.

Cornu, J. Y., Denaix, L., Schneider, A., & Pellerin, S. (2007). Temporal evolution of redox processes and free Cd dynamics in a metal-contaminated soil after rewetting. *Chemosphere, 70*(2), 306–314. doi:10.1016/j.chemosphere.2007.06.003.

de Livera, J., McLaughlin, M. J., Hettiarachchi, G. M., Kirby, J. K., & Beak, D. G. (2011). Cadmium solubility in paddy soils: Effects of soil oxidation, metal sulfides and competitive ions. *Science of the Total Environment, 409*(8), 1489–1497. doi:10.1016/j.scitotenv.2010.12.028.

Egene, C. E., Van Poucke, R., Ok, Y. S., Meers, E., & Tack, F. M. G. (2018). Impact of organic amendments (biochar, compost and peat) on Cd and Zn mobility and solubility in contaminated soil of the Campine region after three years. *Science of the Total Environment, 626*, 195–202. doi:10.1016/j.scitotenv.2018.01.054.

Eriksson, J. E. (1988). The effects of clay, organic matter and time on adsorption and plant uptake of cadmium added to the soil. *Water, Air, and Soil Pollution, 40*(3–4), 359–373. doi:10.1007/bf00163740.

Fahmi, A. H., Samsuri, A. W., Jol, H., & Singh, D. (2018). Bioavailability and leaching of Cd and Pb from contaminated soil amended with different sizes of biochar. *Royal Society Open Science, 5*(11). doi:10.1098/rsos.181328.

Fan, G., Li, F., Evans, D. G., & Duan, X. (2014). Catalytic applications of layered double hydroxides: Recent advances and perspectives. *Chemical Society Reviews, 43*(20), 7040–7066. doi:10.1039/c4cs00160e.

Fang, X., Wang, J., Chen, H., Christl, I., Wang, P., Kretzschmar, R., & Zhao, F. J. (2021). Two-year and multi-site field trials to evaluate soil amendments for controlling cadmium accumulation in rice grain. *Environmental Pollution, 289*. doi:10.1016/j.envpol.2021.117918.

Fontanili, L., Lancilli, C., Suzui, N., Dendena, B., Yin, Y. G., Ferri, A., Ishii, S., Kawachi, N., Lucchini, G., Fujimaki, S., Sacchi, G. A., & Nocito, F. F. (2016). Kinetic analysis of zinc/cadmium reciprocal competitions suggests a possible Zn-insensitive pathway for root-to-shoot cadmium translocation in rice. *Rice, 9*(1). doi:10.1186/s12284-016-0088-3.

Fulda, B., Voegelin, A., & Kretzschmar, R. (2013). Redox-controlled changes in cadmium solubility and solid-phase speciation in a paddy soil as affected by reducible sulfate and copper. *Environmental Science and Technology, 47*(22), 12775–12783. doi:10.1021/es401997d.

Gao, X., Chen, X., Shu, Y., Chen, K., Gao, G., Liu, X., Huang, Q., Wang, L., Sun, Y., Zhao, Y., & Liang, X. (2025). Mercapto-modified layered double hydroxide: High-efficiency emerging amendment for cadmium immobilization in cadmium-contaminated acidic and alkaline soil. *Journal of Hazardous Materials, 498*, 139788. doi:10.1016/j.jhazmat.2025.139788.

Gilkes, R. J., & McKenzie, R. M. (1988). Geochemistry and mineralogy of manganese in soils. In *Manganese in soils and plants: Proceedings of the International Symposium on 'Manganese in Soils and Plants'* (pp. 23–35). Glen Osmond, South Australia: Springer.

Grant, C. A., Bailey, L. D., McLaughlin, M. J., & Singh, B. R. (1999). Management factors which influence cadmium concentrations in crops. In M. J. McLaughlin, & B. R. Singh (Eds.), *Cadmium in soils and plants*. Springer.

Green, C. E., Chaney, R. L., & Bouwkamp, J. (2017). Increased zinc supply does not inhibit cadmium accumulation by rice (*Oryza sativa* L.). *Journal of Plant Nutrition, 40*(6), 869–877. doi:10.1080/01904167.2016.1262407.

Green, C. E., Chaney, R. L., & Bouwkamp, J. (2003). Interactions between cadmium uptake and phytotoxic levels of zinc in hard red spring wheat. *Journal of Plant Nutrition, 26*(2), 417–430. doi:10.1081/PLN-120017144.

Grejtovsky, A., Repcak, M., & Gianits, L. (1998). The influence of soil cadmium eliminating sorbents on *Chamomilla recutita*. *Journal of Environmental Science and Health, B, 33*(3), 307–316. doi:10.1080/03601239809373146.

Guo, F., Ding, C., Zhou, Z., Huang, G., & Wang, X. (2018). Effects of combined amendments on crop yield and cadmium uptake in two cadmium contaminated soils under rice-wheat rotation. *Ecotoxicology and Environmental Safety, 148*, 303–310. doi:10.1016/j.ecoenv.2017.10.043.

Hart, J. J., Welch, R. M., Norvell, W. A., & Kochian, L. V. (2002). Transport interactions between cadmium and zinc in roots of bread and durum wheat seedlings. *Physiologia Plantarum, 116*(1), 73–78. doi:10.1034/j.1399-3054.2002.1160109.x.

Hart, J. J., Welch, R. M., Norvell, W. A., Clarke, J. M., & Kochian, L. V. (2005). Zinc effects on cadmium accumulation and partitioning in near-isogenic lines of durum wheat that differ in grain cadmium concentration. *New Phytologist, 167*(2), 391–401. doi:10.1111/j.1469-8137.2005.01416.x.

He, L., Zhong, H., Liu, G., Dai, Z., Brookes, P. C., & Xu, J. (2019). Remediation of heavy metal contaminated soils by biochar: Mechanisms, potential risks and applications in China. *Environmental Pollution, 252*, 846–855. doi:10.1016/j.envpol.2019.05.151.

He, T., Li, Q., Lin, T., Li, J., Bai, S., An, S., Kong, X., & Song, Y. F. (2023). Recent progress on highly efficient removal of heavy metals by layered double hydroxides. *Chemical Engineering Journal, 462*. doi:10.1016/j.cej.2023.142041.

He, Q. B., & Singh, B. R. (1995). Cadmium availability to plants as affected by repeated applications of phosphorus fertilizers. *Acta Agriculturae Scandinavica, Section B – Soil & Plant Science, 45*(1), 22–31. doi:10.1080/09064719509410930.

He, Q. B., & Singh, B. R. (1993). Plant availability of cadmium in soils: II. factors related to the extractability and plant uptake of cadmium in cultivated soils. *Acta Agriculturae Scandinavica Section B: Soil and Plant Science, 43*(3), 142–150. doi:10.1080/09064719309411232.

Helmke, P. A. (1999). Chemistry of cadmium in soil solution. In M. J. McLaughlin, & B. R. Singh (Eds.), *Cadmium in soils, plants and the food chain*. Springer.

Honma, T., Ohba, H., Kaneko-Kadokura, A., Makino, T., Nakamura, K., & Katou, H. (2016). Optimal soil Eh, pH, and water management for simultaneously minimizing arsenic and cadmium concentrations in rice grains. *Environmental Science and Technology, 50*(8), 4178–4185. doi:10.1021/acs.est.5b05424.

Hou, D., Jia, X., Wang, L., McGrath, S. P., Zhu, Y. G., Hu, Q., Zhao, F. J., Bank, M. S., O'Connor, D., & Nriagu, J. (2025). Global soil pollution by toxic metals threatens agriculture and human health. *Science, 388*(6744), 316–321. doi:10.1126/science.adr5214.

Hu, R., Limmer, M. A., & Seyfferth, A. L. (2024). How manganese affects rice cadmium uptake and translocation in vegetative and mature plants. *Plant and Soil, 504*(1), 941–954. doi:10.1007/s11104-024-06669-9.

Hua, Y., Heal, K. V., & Friesl-Hanl, W. (2017). The use of red mud as an immobiliser for metal/metalloid-contaminated soil: A review. *Journal of Hazardous Materials, 325*, 17–30. doi:10.1016/j.jhazmat.2016.11.073.

Huang, B. Y., Zhao, F. J., & Wang, P. (2022a). The relative contributions of root uptake and remobilization to the loading of Cd and As into rice grains: Implications in simultaneously controlling grain Cd and As accumulation using a segmented water management strategy. *Environmental Pollution, 293*. doi:10.1016/j.envpol.2021.118497.

Huang, Q., Wan, Y., Luo, Z., Qiao, Y., Su, D., & Li, H. (2020). The effects of chicken manure on the immobilization and bioavailability of cadmium in the soil-rice system. *Archives of Agronomy and Soil Science, 66*(13), 1753–1764. doi:10.1080/03650340.2019.1694146.

Huang, Q., Wang, Y., Qin, X., Zhao, L., Liang, X., Sun, Y., & Xu, Y. (2022b). Soil application of manganese sulfate effectively reduces Cd bioavailability in Cd-contaminated soil and Cd translocation and accumulation in wheat. *Science of the Total Environment, 814*, 152765. doi:10.1016/j.scitotenv.2021.152765.

Huang, H., Tang, Z. X., Qi, H. Y., Ren, X. T., Zhao, F. J., & Wang, P. (2022c). Soil amendments with $ZnSO_4$ or $MnSO_4$ are effective at reducing Cd accumulation in rice grain: An application of the voltaic cell principle. *Environmental Pollution, 294*. doi:10.1016/j.envpol.2021.118650.

Huang, H., Chen, H. P., Kopittke, P. M., Kretzschmar, R., Zhao, F. J., & Wang, P. (2021). The voltaic effect as a novel mechanism controlling the remobilization of cadmium in paddy soils during drainage. *Environmental Science and Technology, 55*(3), 1750–1758. doi:10.1021/acs.est.0c06561.

Ishikawa, S., Ishimaru, Y., Igura, M., Kuramata, M., Abe, T., Senoura, T., Hase, Y., Arao, T., Nishizawa, N. K., & Nakanishi, H. (2012). Ion-beam irradiation, gene identification, and marker-assisted breeding in the development of low-cadmium rice. *Proceedings of the National Academy of Sciences, 109*(47), 19166–19171. doi:10.1073/pnas.1211132109.

Karaca, A. (2004). Effect of organic wastes on the extractability of cadmium, copper, nickel, and zinc in soil. *Geoderma, 122*(2–4), 297–303. doi:10.1016/j.geoderma.2004.01.016.

Kim, S. C., Kim, H. S., Seo, B. H., Owens, G., & Kim, K. R. (2016). Phytoavailability control based management for paddy soil contaminated with Cd and Pb: Implications for safer rice production. *Geoderma, 270*, 83–88. doi:10.1016/j.geoderma.2015.11.031.

Kim, Y. H., Khan, A. L., Kim, D. H., Lee, S. Y., Kim, K. M., Waqas, M., Jung, H. Y., Shin, J. H., Kim, J. G., & Lee, I. J. (2014). Silicon mitigates heavy metal stress by regulating P-type heavy metal ATPases, *Oryza sativa* low silicon genes, and endogenous phytohormones. *BMC Plant Biology, 14*(1). doi:10.1186/1471-2229-14-13.

Kirk, G. (2004). Reduction and oxidation. In *The biogeochemistry of submerged soils* (pp. 93–134). Wiley. doi:10.1002/047086303X.ch4.

Kong, X., Ge, R., Liu, T., Xu, S., Hao, P., Zhao, X., Li, Z., Lei, X., & Duan, H. (2021). Super-stable mineralization of cadmium by calcium-aluminum layered double hydroxide and its large-scale application in agriculture soil remediation. *Chemical Engineering Journal, 407*, 127178. doi:10.1016/j.cej.2020.127178.

Kumpiene, J. (2010). Trace element immobilization in soil using amendments. In P. S. Hooda (Ed.), *Trace elements in soils*. Wiley. doi:10.1002/9781444319477.ch15.

Lehmann, J., & Joseph, S. (2015). *Biochar for environmental management: An introduction* (pp. 1–13). Routledge.

Lei, S., Shi, Y., Qiu, Y., Che, L., & Xue, C. (2019). Performance and mechanisms of emerging animal-derived biochars for immobilization of heavy metals. *Science of the Total Environment, 646*, 1281–1289. doi:10.1016/j.scitotenv.2018.07.374.

Li, F., Li, Z., Mao, P., Li, Y., Li, Y., McBride, M. B., Wu, J., & Zhuang, P. (2019). Heavy metal availability, bioaccessibility, and leachability in contaminated soil: Effects of pig manure and earthworms. *Environmental Science and Pollution Research, 26*(20). doi:10.1007/s11356-018-2080-5.

Li, J., Wu, B., Luo, Z., Lei, N., Kuang, H., & Li, Z. (2022). Immobilization of cadmium by mercapto-functionalized palygorskite under stimulated acid rain: Stability performance and micro-ecological response. *Environmental Pollution, 306*, 119400. doi:10.1016/j.envpol.2022.119400.

Li, M., Lou, Z., Wang, Y., Liu, Q., Zhang, Y., Zhou, J., & Qian, G. (2015). Alkali and alkaline earth metallic (AAEM) species leaching and Cu(II) sorption by biochar. *Chemosphere, 119*, 778–785. doi:10.1016/j.chemosphere.2014.08.033.

Li, X., Wang, L., Huang, C., Hou, R., & Hou, D. (2025). Long-term soil remediation using layered double hydroxides: Field evidence for simultaneous immobilization of both cations and oxyanions. *Environmental Pollution, 366*, 125417. doi:10.1016/j.envpol.2024.125417.

Li, Y. M., Chaney, R. L., Schneiter, A. A., & Johnson, B. L. (1996). Effect of field limestone applications on cadmium content of sunflower (*Helianthus annum* L.) leaves and kernels. *Plant and Soil, 180*(2), 297–302. doi:10.1007/BF00015313.

Liang, X., Xu, Y., Tan, X., Wang, L., Sun, Y., Lin, D., Sun, Y., Qin, X., & Wang, Q. (2013). Heavy metal adsorbents mercapto and amino functionalized palygorskite: Preparation and characterization. *Colloids and Surfaces A: Physicochemical and Engineering Aspects, 426*, 98–105. doi:10.1016/j.colsurfa.2013.03.014.

Liang, X., Han, J., Xu, Y., Sun, Y., Wang, L., & Tan, X. (2014). In situ field-scale remediation of Cd polluted paddy soil using sepiolite and palygorskite. *Geoderma, 235–236*, 9–18. doi:10.1016/j.geoderma.2014.06.029.

Liang, X., Li, N., He, L., Xu, Y., Huang, Q., Xie, Z., & Yang, F. (2019). Inhibition of Cd accumulation in winter wheat (*Triticum aestivum* L.) grown in alkaline soil using mercapto-modified attapulgite. *Science of the Total Environment, 688*, 818–826. doi:10.1016/j.scitotenv.2019.06.335.

Liang, Y., Wong, J. W. C., & Wei, L. (2005). Silicon-mediated enhancement of cadmium tolerance in maize (*Zea mays* L.) grown in cadmium contaminated soil. *Chemosphere, 58*(4), 475–483. doi:10.1016/j.chemosphere.2004.09.034.

Liu, H. J., Zhang, J. L., Christie, P., & Zhang, F. S. (2007). Influence of external zinc and phosphorus supply on Cd uptake by rice (*Oryza sativa* L.) seedlings with root surface iron plaque. *Plant and Soil, 300*(1–2), 105–115. doi:10.1007/s11104-007-9393-3.

Liu, K., Lv, J., He, W., Zhang, H., Cao, Y., & Dai, Y. (2015). Major factors influencing cadmium uptake from the soil into wheat plants. *Ecotoxicology and Environmental Safety, 113*, 207–213. doi:10.1016/j.ecoenv.2014.12.005.

Lombi, E., Zhao, F. J., Wieshammer, G., Zhang, G., & McGrath, S. P. (2002). In situ fixation of metals in soils using bauxite residue: Biological effects. *Environmental Pollution, 118*(3), 445–452. doi:10.1016/S0269-7491(01)00295-0.

Lu, T., Wang, W., Liu, L., Wang, L., Hu, J., Li, X., & Qiu, G. (2022). Remediation of cadmium-polluted weakly alkaline dryland soils using iron and manganese oxides for immobilized wheat uptake. *Journal of Cleaner Production, 365*, 132794. doi:10.1016/j.jclepro.2022.132794.

Ma, W., Ge, C., Sun, Y., Wang, M., & Zhou, D. (2024). Cadmium immobilization by mercapto-palygorskite in alkaline soil: Impacts on soil microbial communities and wheat rhizosphere metabolism. *Science of the Total Environment, 955*, 176734. doi:10.1016/j.scitotenv.2024.176734.

MacLean, A. J. (1976). Cadmium in different plant species and its availability in soils as influenced by organic matter and additions of lime, P, Cd and Zn. *Canadian Journal of Soil Science, 56*(3), 129–138. doi:10.4141/cjss76-021.

Major, J. (2010). Guidelines on practical aspects of biochar application to field soil in various soil management systems. *International Biochar Initiative, 8*(1), 5–7.

McKenna, I. M., Chaney, R. L., & Williams, F. M. (1993). The effects of cadmium and zinc interactions on the accumulation and tissue distribution of zinc and cadmium in lettuce and spinach. *Environmental Pollution, 79*(2), 113–120. doi:10.1016/0269-7491(93)90060-2.

McLaughlin, M. J., Smolders, E., Zhao, F. J., Grant, C., & Montalvo, D. (2021). Managing cadmium in agricultural systems. *Advances in Agronomy, 166*, 1–129. doi:10.1016/bs.agron.2020.10.004.

Meng, F., Yuan, G., Wei, J., Bi, D., & Wang, H. (2017). Leonardite-derived humic substances are great adsorbents for cadmium. *Environmental Science and Pollution Research, 24*(29), 23006–23014. doi:10.1007/s11356-017-9947-8.

Mortensen, L. H., Rønn, R., & Vestergård, M. (2018). Bioaccumulation of cadmium in soil organisms – With focus on wood ash application. *Ecotoxicology and Environmental Safety, 156*, 452–462. doi:10.1016/j.ecoenv.2018.03.018.

Murray, H. H., Pozo, M., & Galán, E. (2011). An introduction to palygorskite and sepiolite deposits-location, geology and uses. *Developments in Clay Science, 3*, 85–99. doi:10.1016/B978-0-444-53607-5.00004-9.

Naidu, R., Bolan, N. S., Kookana, R. S., & Tiller, K. G. (1994). Ionic-strength and pH effects on the sorption of cadmium and the surface charge of soils. *European Journal of Soil Science, 45*(4), 419–429. doi:10.1111/j.1365-2389.1994.tb00527.x.

Narwal, R. P., & Singh, B. R. (1998). Effect of organic materials on partitioning, extractability and plant uptake of metals in an alum shale soil. *Water, Air, and Soil Pollution, 103*(1–4), 405–421. doi:10.1023/a:1004912724284.

Nawrot, T., Plusquin, M., Hogervorst, J., Roels, H. A., Celis, H., Thijs, L., Vangronsveld, J., Van Hecke, E., & Staessen, J. A. (2006). Environmental exposure to cadmium and risk of cancer: A prospective population-based study. *Lancet Oncology, 7*(2), 119–126. doi:10.1016/S1470-2045(06)70545-9.

O'Connor, D., Peng, T., Zhang, J., Tsang, D. C. W., Alessi, D. S., Shen, Z., Bolan, N. S., & Hou, D. (2018). Biochar application for the remediation of heavy metal polluted land: A review of in situ field trials. *Science of the Total Environment, 619–620*, 815–826. doi:10.1016/j.scitotenv.2017.11.132.

Oliver, D. P., Hannam, R., Tiller, K. G., Wilhelm, N. S., Merry, R. H., & Cozens, G. D. (1994). The effects of zinc fertilization on cadmium concentration in wheat grain. *Journal of Environmental Quality, 23*(4), 705–711. doi:10.2134/jeq1994.00472425002300040013x.

Oste, L. A., Dolfing, J., Ma, W. C., & Lexmond, T. M. (2001). Effect of beringite on cadmium and zinc uptake by plants and earthworms: More than a liming effect? *Environmental Toxicology and Chemistry, 20*(6), 1339–1345. doi:10.1002/etc.5620200624.

Puga, A. P., Abreu, C. A., Melo, L. C. A., & Beesley, L. (2015). Biochar application to a contaminated soil reduces the availability and plant uptake of zinc, lead and cadmium. *Journal of Environmental Management, 159*, 86–93. doi:10.1016/j.jenvman.2015.05.036.

Qin, P., Wang, H., Yang, X., He, L., Müller, K., Shaheen, S. M., Xu, S., Rinklebe, J., Tsang, D. C. W., Ok, Y. S., Bolan, N., Song, Z., Che, L., & Xu, X. (2018). Bamboo- and pig-derived biochars reduce leaching losses of dibutyl phthalate, cadmium, and lead from co-contaminated soils. *Chemosphere, 198*, 450–459. doi:10.1016/j.chemosphere.2018.01.162.

Rahim, H. U., Akbar, W. A., & Alatalo, J. M. (2022). A comprehensive literature review on cadmium (Cd) status in the soil environment and its immobilization by biochar-based materials. *Agronomy, 12*(4). doi:10.3390/agronomy12040877.

Ram, N., & Verloo, M. (1985). Influence of organic materials on the uptake of heavy metals by corn in a polluted Belgian soil. *Pedologie, 35*(2), 147–153.

Ramezanzadeh, H., Reyhanitabar, A., Oustan, S., Mohammadi, M. H., & Zee, S. E. A. T. M. (2021). Enhanced sorption of cadmium by using biochar nanoparticles from ball milling in a sandy soil. *Eurasian Soil Science, 54*(2), 201–211. doi:10.1134/s1064229321020125.

Rojas-Cifuentes, G. A., Johnson, B. L., Berti, M. T., & Norvell, W. A. (2012). Zinc fertilization effects on seed cadmium accumulation in oilseed and grain crops grown on North Dakota soils. *Chilean journal of agricultural research, 72*(1), 117–124. doi:10.4067/S0718-58392012000100019.

Sasaki, A., Yamaji, N., Yokosho, K., & Ma, J. F. (2012). Nramp5 is a major transporter responsible for manganese and cadmium uptake in rice. *Plant Cell, 24*(5), 2155–2167. doi:10.1105/tpc.112.096925.

Satarug, S., Garrett, S. H., Sens, M. A., & Sens, D. A. (2010). Cadmium, environmental exposure, and health outcomes. *Environmental Health Perspectives, 118*(2), 182–190. doi:10.1289/ehp.0901234.

Sauvé, S., Hendershot, W., & Allen, H. E. (2000). Solid-solution partitioning of metals in contaminated soils: Dependence on pH, total metal burden, and organic matter. *Environmental Science and Technology, 34*(7), 1125–1131. doi:10.1021/es9907764.

Shao, J. F., Che, J., Yamaji, N., Shen, R. F., & Ma, J. F. (2017). Silicon reduces cadmium accumulation by suppressing expression of transporter genes involved in cadmium uptake and translocation in rice. *Journal of Experimental Botany, 68*(20), 5641–5651. doi:10.1093/jxb/erx364.

Sheikhhosseini, A., Shirvani, M., & Shariatmadari, H. (2013). Competitive sorption of nickel, cadmium, zinc and copper on palygorskite and sepiolite silicate clay minerals. *Geoderma, 192*(1), 249–253. doi:10.1016/j.geoderma.2012.07.013.

Simmler, M., Ciadamidaro, L., Schulin, R., Madejón, P., Reiser, R., Clucas, L., Weber, P., & Robinson, B. (2013). Lignite reduces the solubility and plant uptake of cadmium in pasturelands. *Environmental Science and Technology, 47*(9), 4497–4504. doi:10.1021/es303118a.

Sims, J. T., & Kline, J. S. (1991). Chemical fractionation and plant uptake of heavy metals in soils amended with co-composted sewage sludge. *Journal of Environmental Quality, 20*(2), 387–395. doi:10.2134/jeq1991.00472425002000020009x.

Singh, B. R., Narwal, R. P., Jeng, A. S., & Almas, Å. (1995). Crop uptake and extractability of cadmium in soils naturally high in metals at different pH levels. *Communications in Soil Science and Plant Analysis, 26*(13–14), 2123–2142. doi:10.1080/00103629509369434.

Song, Y., Wang, Y., Mao, W., Sui, H., Yong, L., Yang, D., Jiang, D., Zhang, L., Gong, Y., & Kunze, G. (2017). Dietary cadmium exposure assessment among the Chinese population. *PLoS One, 12*(5), e0177978. doi:10.1371/journal.pone.0177978.

Suárez, M., & García-Romero, E. (2011). Advances in the crystal chemistry of sepiolite and palygorskite. *Developments in Clay Science, 3*, 33–65. doi:10.1016/B978-0-444-53607-5.00002-5.

Suda, A., & Makino, T. (2016). Functional effects of manganese and iron oxides on the dynamics of trace elements in soils with a special focus on arsenic and cadmium: A review. *Geoderma, 270*, 68–75. doi:10.1016/j.geoderma.2015.12.017.

Sun, Z., Wang, Y., Liu, T., Kong, X., Pan, T., Zhang, F., Lei, X., & Duan, X. (2023). Super-stable mineralization of Cu, Cd, Zn and Pb by CaAl-layered double hydroxide: Performance, mechanism, and large-scale application in agriculture soil remediation. *Journal of Hazardous Materials, 447*, 130723. doi:10.1016/j.jhazmat.2023.130723.

Takahashi, R., Ishimaru, Y., Shimo, H., Ogo, Y., Senoura, T., Nishizawa, N. K., & Nakanishi, H. (2012). The OsHMA2 transporter is involved in root-to-shoot translocation of Zn and Cd in rice. *Plant, Cell and Environment, 35*(11), 1948–1957. doi:10.1111/j.1365-3040.2012.02527.x.

Talke, I. N., Hanikenne, M., & Krämer, U. (2006). Zinc-dependent global transcriptional control, transcriptional deregulation, and higher gene copy number for genes in metal homeostasis of the hyperaccumulator Arabidopsis halleri. *Plant Physiology, 142*(1), 148–167. doi:10.1104/pp.105.076232.

Tan, Z., Yuan, S., Hong, M., Zhang, L., & Huang, Q. (2020). Mechanism of negative surface charge formation on biochar and its effect on the fixation of soil Cd. *Journal of Hazardous Materials, 384*, 121370. doi:10.1016/j.jhazmat.2019.121370.

Tlustos, P., Száková, J., Korinek, K., Pavlíková, D., Hanc, A., & Balík, J. (2006). The effect of liming on cadmium, lead, and zinc uptake reduction by spring wheat grown in contaminated soil. *Plant Soil and Environment, 31*(52).

van Herwijnen, R., Hutchings, T. R., Al-Tabbaa, A., Moffat, A. J., Johns, M. L., & Ouki, S. K. (2007). Remediation of metal contaminated soil with mineral-amended composts. *Environmental Pollution, 150*(3), 347–354. doi:10.1016/j.envpol.2007.01.023.

Vert, G., Grotz, N., Dédaldéchamp, F., Gaymard, F., Guerinot, M. L., Briat, J. F., & Curie, C. (2002). IRT1, an Arabidopsis transporter essential for iron uptake from the soil and for plant growth. *Plant Cell, 14*(6), 1223–1233. doi:10.1105/tpc.001388.

Wang, F., Peng, L., Zhou, X., Zeng, Q., & Luo, S. (2021a). Typical sources of Cd to paddy fields in different contaminated areas and their impacts on Cd accumulation in topsoil and rice in Changzhutan, China. *Environmental Research, 193*, 110523. doi:10.1016/j.envres.2020.110523.

Wang, H. Y., Wen, S. L., Chen, P., Zhang, L., Cen, K., & Sun, G. X. (2016). Mitigation of cadmium and arsenic in rice grain by applying different silicon fertilizers in contaminated fields. *Environmental Science and Pollution Research, 23*(4), 3781–3788. doi:10.1007/s11356-015-5638-5.

Wang, M., Ma, W., Chaney, R. L., Green, C. E., & Chen, W. (2022). Effects of Mn^{2+} on Cd accumulation and ionome in rice and spinach. *Journal of Environmental Quality, 51*(5), 890–898. doi:10.1002/jeq2.20358.

Wang, M., Yang, Y., & Chen, W. (2018). Manganese, zinc, and pH affect cadmium accumulation in rice grain under field conditions in Southern China. *Journal of Environmental Quality, 47*(2), 306–311. doi:10.2134/jeq2017.06.0237.

Wang, W., Lu, T., Liu, L., Yang, X., Sun, X., Qiu, G., Hua, D., & Zhou, D. (2021). Zeolite-supported manganese oxides decrease the Cd uptake of wheat plants in Cd-contaminated weakly alkaline arable soils. *Journal of Hazardous Materials, 419*, 126464. doi:10.1016/j.jhazmat.2021.126464.

Wang, J., Wang, P. M., Gu, Y., Kopittke, P. M., Zhao, F. J., & Wang, P. (2019b). Iron-manganese (oxyhydro)oxides, rather than oxidation of sulfides, determine mobilization of cd during soil drainage in paddy soil systems. *Environmental Science and Technology, 53*(5), 2500–2508. doi:10.1021/acs.est.8b06863.

Wong, C. K. E., & Cobbett, C. S. (2009). HMA P-type ATPases are the major mechanism for root-to-shoot Cd translocation in *Arabidopsis thaliana*. *New Phytologist, 181*(1), 71–78. doi:10.1111/j.1469-8137.2008.02638.x.

Wu, Y., Yang, H., Wang, M., Sun, L., Xu, Y., Sun, G., Huang, Q., & Liang, X. (2022). Immobilization of soil Cd by sulfhydryl grafted palygorskite in wheat-rice rotation mode: A field-scale investigation. *Science of the Total Environment, 826*, 154156. doi:10.1016/j.scitotenv.2022.154156.

Xu, Z. R., You, T. T., Liu, W. Y., Ye, K., Zhao, F. J., & Wang, P. (2024). Mitigating cadmium accumulation in dicotyledonous vegetables by iron fertilizer through inhibiting Fe transporter IRT1-mediated Cd uptake. *Chemosphere, 346*. doi:10.1016/j.chemosphere.2023.140559.

Xu, X. Y., McGrath, S. P., Meharg, A. A., & Zhao, F. J. (2008). Growing rice aerobically markedly decreases arsenic accumulation. *Environmental Science & Technology, 42*(15), 5574–5579. doi:10.1021/es800324u.

Xu, Y., Liang, X., Xu, Y., Qin, X., Huang, Q., Wang, L., & Sun, Y. (2017). Remediation of heavy metal-polluted agricultural soils using clay minerals: A review. *Pedosphere, 27*(2), 193–204. doi:10.1016/S1002-0160(17)60310-2.

Yamaji, N., Xia, J., Mitani-Ueno, N., Yokosho, K., & Ma, J. F. (2013). Preferential delivery of zinc to developing tissues in rice is mediated by P-type heavy metal ATPase OsHMA2. *Plant Physiology, 162*(2), 927–939. doi:10.1104/pp.113.216564.

Yang, F., Tang, C., & Antonietti, M. (2021). Natural and artificial humic substances to manage minerals, ions, water, and soil microorganisms. *Chemical Society Reviews, 50*(10), 6221–6239. doi:10.1039/d0cs01363c.

Yang, M., Zhang, Y., Zhang, L., Hu, J., Zhang, X., Lu, K., Dong, H., Wang, D., Zhao, F. J., Huang, C. F., & Lian, X. (2014). OsNRAMP5 contributes to manganese translocation and distribution in rice shoots. *Journal of Experimental Botany, 65*(17), 4849–4861. doi:10.1093/jxb/eru259.

Yong, Y., Xu, Y., Huang, Q., Sun, Y., Wang, L., Liang, X., Qin, X. u, & Zhao, L. (2022). Remediation effect of mercapto-palygorskite combined with manganese sulfate on cadmium contaminated alkaline soil and cadmium accumulation in pak choi (*Brassica chinensis* L.). *Science of the Total Environment, 813*, 152636. doi:10.1016/j.scitotenv.2021.152636.

Yu, H. Y., Liu, C., Zhu, J., Li, F., Deng, D. M., Wang, Q., & Liu, C. (2016). Cadmium availability in rice paddy fields from a mining area: The effects of soil properties highlighting iron fractions and pH value. *Environmental Pollution, 209*, 38–45. doi:10.1016/j.envpol.2015.11.021.

Zare, A. A., Khoshgoftarmanesh, A. H., Malakouti, M. J., Bahrami, H. A., & Chaney, R. L. (2018). Root uptake and shoot accumulation of cadmium by lettuce at various Cd:ZN ratios in nutrient solution. *Ecotoxicology and Environmental Safety, 148*, 441–446. doi:10.1016/j.ecoenv.2017.10.045.

Zhang, S., Chen, J., Yu, J., Yu, Q., & Qiu, X. (2022). Remediation of Cd-contaminated soil through different layered double hydroxides: The weakness of delamination and mechanism. *Journal of Environmental Chemical Engineering, 10*(3), 107815. doi:10.1016/j.jece.2022.107815.

Zhao, F. J., Ma, Y., Zhu, Y. G., Tang, Z., & McGrath, S. P. (2015). Soil contamination in China: Current status and mitigation strategies. *Environmental Science and Technology, 49*(2), 750–759. doi:10.1021/es5047099.

Zhao, F. J., & Wang, P. (2020). Arsenic and cadmium accumulation in rice and mitigation strategies. *Plant and Soil, 446*(1–2). doi:10.1007/s11104-019-04374-6.

Zhu, H., Chen, C., Xu, C., Zhu, Q., & Huang, D. (2016). Effects of soil acidification and liming on the phytoavailability of cadmium in paddy soils of central subtropical China. *Environmental Pollution, 219*, 99–106. doi:10.1016/j.envpol.2016.10.043.

CHAPTER

19

Breeding and biotechnology to limit cadmium accumulation in crops

Xin-Yuan Huang, Fang-Jie Zhao

College of Resources and Environmental Sciences, Nanjing Agricultural University, Nanjing, China

19.1 Introduction

Cadmium (Cd) contamination in agricultural soils poses a significant threat to food safety and human health because of its relatively high bioavailability. Cd is readily absorbed by crops, particularly staple foods such as rice, wheat, and vegetables, leading to its accumulation in edible parts. Cd accumulation in crops varies significantly among plant species and among different varieties within the same species. For example, rice is known to accumulate relatively high levels of Cd in its grains. The concentration of Cd in rice grains can range from undetectable levels to several milligrams per kilogram, depending on soil Cd content, rice variety, and growing conditions (Zhao et al., 2015). As dietary intake, especially from staple foods (see Chapter 15), is the primary route of Cd exposure in humans, reducing Cd accumulation in food crops is essential to ensure food safety and protect public health. Breeding and biotechnology provide promising solutions to address this challenge by developing crop varieties with reduced Cd accumulation while maintaining high yields and nutritional quality.

This chapter reviews breeding and biotechnological approaches to limit Cd accumulation in crops. It presents traditional breeding strategies, including the screening of low-Cd varieties, genetic mapping of Cd-related loci, marker-assisted selection (MAS), and mutagenesis for identifying low-Cd mutants. Advanced biotechnological methods, such as genome editing, gene overexpression, and targeted mutagenesis, are also discussed. The chapter highlights successful case studies and discusses prospects and challenges.

Cadmium Contamination in Soils and Plants: Mechanisms and Mitigation.
DOI: https://doi.org/10.1016/B978-0-443-34104-5.00004-9

19.2 Traditional breeding strategies to limit cadmium accumulation

Traditional breeding has long been used in agriculture to improve traits such as yield, quality, and disease resistance. These strategies rely on genetic principles and natural variation within crop populations. In this section, we examine key traditional breeding methods used to develop crops with reduced Cd accumulation. Each method offers unique advantages and challenges but contributes to the shared goal of producing high-yielding, safer crop varieties. We begin with the screening of crop varieties for low Cd accumulation, outlining its processes, significance, and factors affecting its effectiveness.

19.2.1 Screening for low-cadmium crop varieties

Screening for low-Cd crop varieties is one of the most economical and effective strategies in traditional breeding to reduce Cd accumulation in crops. This strategy takes advantage of the natural variation of Cd accumulation among different varieties or accessions under specific conditions to identify those with low-Cd-accumulating traits. The systematic screening of crop varieties for low Cd accumulation involves a multistep process, including germplasm collection, crop planting, precise phenotyping, and data analysis.

Screening of low-Cd crop varieties begins with a comprehensive collection of a diverse panel of germplasm, including commercial cultivars and landraces. The geographic and ecological diversity must be taken into account when collecting the crop germplasm. A well-curated germplasm pool maximizes genetic diversity, enhancing the likelihood of identification of rare alleles associated with low Cd accumulation. For practical breeding purposes, it is advantageous to include locally adapted cultivars, especially those that dominate regional agriculture.

Three primary screening methodologies are commonly employed to identify cultivars with low Cd accumulation: hydroponic systems, pot experiments, and field trials. Each method offers distinct advantages and addresses specific research objectives, enabling effective screening for low-Cd varieties as well as comprehensive characterization of Cd uptake, translocation, and sequestration mechanisms.

Hydroponic screening provides a high level of precision in isolating genetic factors by eliminating soil heterogeneity. Plants are cultivated in nutrient solutions with controlled Cd concentrations, allowing direct quantification of root uptake and root-to-shoot translocation. However, because of the labor-intensive and time-consuming nature of nutrient solution replacement, hydroponic screening faces considerable challenges in scaling up for large-scale screening, although automated systems for nutrient solution renewal can be designed and implemented. Moreover, growing plants to maturity in hydroponic conditions for seed harvest is particularly difficult. In addition, hydroponic systems largely neglect rhizosphere interactions, such as Cd mobilization mediated by root exudates in the rhizosphere soil. Consequently, crop varieties identified through hydroponic screening as having low Cd accumulation in roots or shoots do not necessarily exhibit low Cd accumulation in grains.

Pot experiments with soil serve as an intermediate approach between the highly controlled hydroponic systems and the variable conditions encountered in the field. This method involves growing plants in pots filled with Cd-spiked soil or naturally contaminated soil collected from the field, thereby minimizing soil heterogeneity while retaining soil–root interactions. Pot experiments are particularly useful for crops such as soybean and vegetables, where root architecture and rhizosphere processes significantly influence Cd uptake (Li et al., 2012). They enable repeated sampling of roots and shoots to monitor Cd dynamics over time. Furthermore, pot experiments facilitate evaluation of soil amendments in combination with genetic traits, providing insights into integrated mitigation strategies. However, limitations include restricted rooting volume, altered microbial activity, and agronomic traits that may not fully represent field conditions.

Field-based screening remains the gold standard for assessing crop performance in Cd-contaminated environments, as it captures interactions among genotypes, soil properties, and agronomic practices. By exposing crops to natural soil conditions, field trials encompass the complex interplay between soil characteristics (e.g., pH, organic matter content, and microbial activity), Cd bioavailability, and environmental factors (e.g., rainfall, irrigation, and temperature fluctuations), all of which critically influence Cd uptake dynamics. This ecological realism ensures that observed Cd accumulation patterns reflect real-world scenarios, including rhizosphere interactions and long-term soil-Cd partitioning, thereby enhancing the practical relevance of selected varieties for contaminated farmland. In addition, field trials capture genotype × environment (G × E) interactions, enabling identification of varieties with stable low-Cd traits across diverse agronomic conditions. Field trials are typically conducted in soils with elevated Cd levels; for example, paddy fields with historical Cd contamination in southern China have been widely utilized to identify low-Cd rice cultivars (Duan et al., 2017). In field-based screening, Cd accumulation and other agronomic traits, including yield, disease resistance, and environmental adaptability, can be evaluated simultaneously. This integrated approach facilitates a holistic assessment of cultivation potential alongside Cd accumulation capacity during varietal promotion. However, field trials face challenges such as controlling confounding factors, including spatial heterogeneity in soil Cd distribution and bioavailability, and unexpected weather events, which may obscure treatment effects and increase data variability. High operational costs, extended timelines, and dependence on seasonal cycles further constrain scalability for rapid screening.

To overcome the inherent challenges of field trials in screening low-Cd crop varieties, integrated approaches can enhance screening precision and efficiency. First, selecting field sites with relatively uniform soil properties and Cd concentrations is critical to minimize spatial variation within the field. Second, standardized field management practices should be applied consistently, particularly maintaining uniform irrigation frequency and controlled drainage duration. Third, planting appropriate control varieties with known Cd accumulation profiles throughout the field can help assess spatial variation in Cd bioavailability at the field scale. Planting identical control varieties adjacent to all test varieties is the most effective way to compare differences in Cd accumulation among cultivars; however, this approach significantly increases screening workload. Fourth, multiyear and multilocation trials mitigate environmental unpredictability by evaluating varieties across diverse agroclimatic zones and growing seasons, enabling robust analysis of G × E interactions. Multiyear trials account for seasonal weather variability to confirm trait stability over time, whereas multilocation trials expose cultivars to diverse soils, climates, and farming practices, ensuring adaptability across environments. Such trials improve statistical robustness, reduce site-specific biases, and facilitate identification of genotypes exhibiting consistent low Cd accumulation under real-world conditions. For example, through 2-year and three-location trials, several rice cultivars with stable low grain Cd accumulation have been identified (Duan et al., 2017; Table 19.1).

Conducting crop variety screening for low Cd accumulation usually involves measurements of Cd and other mineral elements in a large number of samples. Therefore high-throughput

TABLE 19.1 Low-Cd-accumulating rice cultivars selected based on multiyear and multilocation trials.

Low-Cd cultivars	Heading days	Youxian (2014)	Youxian (2015)	Fuyang (2014)	Fuyang (2015)	Xiangtan (2015)
		Grain Cd (mg kg^{-1})	Grain Cd (mg kg^{-1})	Grain Cd (mg kg^{-1})	Grain Cd (mg kg^{-1})	Grain Cd (mg kg^{-1})
ShenYou957	91	0.07	0.03	0.09	0.09	0.06
LongPing602	98	0.09	0.03	0.15	0.13	0.10
T-You535	84	0.15	0.03	0.08	0.19	0.06
JieFengYou1	87	0.12	0.03	0.11	0.19	0.11
I-You899	81	0.07	0.07	0.06	0.15	0.17
WeiYou402	88	n.g.	0.03	n.g.	0.15	0.06
WeiYou463	83	n.g.	0.05	n.g.	0.15	0.06
ZhuLiangYou168	81	0.04	0.05	0.06	0.16	0.13
Median	100	0.32	0.12	0.18	0.22	0.40

Data from Duan et al. (2017).
Youxian: field site at Hunan province, China; Fuyang: field site at Zhejiang province, China; Xiangtan: field site at Hunan province, China; *Cd*, cadmium; *n.g.*: not grown.

methods for the determination of Cd and other elements are very important. Plant tissues, including roots, shoots, and grains, are harvested and analyzed using techniques such as inductively coupled plasma mass spectrometry (ICP-MS), which offers high-throughput and simultaneous measurements of multiple elements and has been widely used in variety screening. This method requires pretreatment of plant samples through acid digestion to solubilize Cd and other mineral elements. In contrast, X-ray fluorescence (XRF) is a nondestructive analytical technique that identifies elemental composition by measuring characteristic secondary X-rays emitted from a sample. Compared to ICP-MS, XRF offers advantages, including minimal or no sample preparation, rapid in situ analysis, and lower operational costs, making it ideal for field applications and bulk material screening. However, XRF is less sensitive than ICP-MS and lacks the capability for multielement measurement. More information about sample preparation and Cd determination can be found in Chapter 2.

19.2.2 Genetic mapping of cadmium accumulation in crops

Genetic mapping of Cd accumulation in crops aims to identify the quantitative trait loci (QTLs) or genes associated with Cd accumulation traits through the construction of genetic maps and analysis of the relationship between molecular markers and Cd accumulation traits. Insights gained from QTL mapping are crucial for MAS (Box 19.1) and for identifying candidate genes for subsequent functional characterization and biotechnological manipulation.

QTL mapping of Cd accumulation can be performed through linkage analysis using bi-parental mapping populations or via genome-wide association studies (GWAS) of natural populations. The typical process for genetic mapping of Cd accumulation in crops involves several key steps. First, a mapping population is established. For linkage analysis a segregating population is generated by crossing two parental lines. Common mapping populations include F_2 populations, recombinant inbred lines, doubled haploid lines, and chromosome segment substitution lines. In contrast, GWAS requires a natural population with diverse germplasm. Selecting parental lines or natural populations with significant differences or wide variation in Cd accumulation traits increases the likelihood of detecting relevant QTLs. Second, the mapping population is phenotyped for Cd accumulation traits. Plants are cultivated in Cd-contaminated soils or hydroponic solutions, and Cd concentrations in various tissues are quantified. Multiple replications and multienvironment trials are often conducted to enhance the accuracy and reliability of phenotypic data. Third, molecular markers are selected and genotyped. A wide range of molecular markers, such as simple sequence repeats, restriction fragment length polymorphisms, and amplified fragment length polymorphisms, have been used historically, whereas single nucleotide polymorphisms (SNPs) derived from whole genome resequencing have become the most commonly employed markers for QTL mapping. These markers are distributed genome-wide to construct high-density genetic maps.

Once both the phenotypic and genotypic data are obtained, statistical methods are employed to identify linkages or associations between marker loci and Cd accumulation. Linkage analysis or association mapping approaches are commonly used. Linkage analysis involves constructing a genetic linkage map based on the segregation of markers and then performing interval mapping or composite interval mapping to detect QTLs. Association mapping, on the other hand, utilizes linkage disequilibrium within diverse germplasm collections to identify marker-trait associations. A significant association between a marker and the Cd accumulation phenotype indicates the presence of a QTL in that genomic region. The output of genetic mapping is

BOX 19.1

Key terms with definitions

Gene overexpression: Gene overexpression is a genetic engineering approach that increases the expression level of a specific gene beyond its natural baseline, thereby amplifying the production of its encoded protein. This is commonly achieved by introducing additional copies of the gene or placing the gene under the control of strong, constitutive, or tissue-specific promoters. The elevated gene expression can enhance or modify physiological or biochemical pathways, leading to improved traits such as stress tolerance, nutrient uptake, or detoxification capacity. Gene overexpression is widely applied in research to elucidate gene function and in biotechnology to develop plants with enhanced performance or novel characteristics. However, overexpression may sometimes lead to metabolic imbalances or unintended effects, so careful design and regulation of expression levels are important considerations in its application.

Genome editing technologies: Genome editing technologies encompass a set of molecular tools that enable precise, targeted modifications to an organism's DNA sequence. These technologies use engineered nucleases, such as CRISPR–Cas systems, transcription activator-like effector nucleases (TALENs), and zinc finger nucleases (ZFNs), to introduce double-strand breaks at specific genomic loci. The natural DNA repair pathways in cells then resolve these breaks, resulting in insertions, deletions, or precise sequence changes. CRISPR–Cas9, the most widely adopted system, utilizes a guide RNA to direct the Cas9 nuclease to the target DNA sequence, allowing efficient and versatile editing. Recent advancements include base editors and prime editors, which enable single-nucleotide changes without inducing double-strand breaks, thereby reducing off-target effects and improving precision. Genome editing is extensively used in functional genomics, crop improvement, and therapeutic development, allowing the creation of plants with improved traits such as disease resistance, abiotic stress tolerance, and reduced accumulation of harmful substances, often without introducing foreign DNA sequences.

Marker-assisted selection (MAS): Marker-assisted selection is a breeding strategy that utilizes molecular markers linked to specific genes or QTLs to facilitate the selection of plants carrying desirable traits. The process involves identifying DNA markers, such as single nucleotide polymorphisms (SNPs) or simple sequence repeats (SSRs), which are closely associated with traits of interest. Once these markers are validated, breeders can screen large populations at the seed or seedling stage for the presence of favorable alleles without waiting for phenotypic expression. MAS accelerates the breeding cycle by improving selection accuracy, especially for complex traits controlled by multiple genes or those difficult to measure directly. It is extensively used in crop improvement programs to enhance yield, disease resistance, abiotic stress tolerance, and quality traits, enabling efficient incorporation of beneficial genes from diverse germplasm into elite cultivars.

Targeted mutagenesis: Targeted mutagenesis is a biotechnological approach that combines traditional random mutagenesis with modern high-throughput sequencing techniques to efficiently identify mutations in specific target genes. The process begins by inducing mutations in crop seeds, callus tissues, or other plant materials using physical (e.g., ion beam irradiation) or chemical mutagens (e.g., EMS). This generates a large and diverse population of mutated individuals. Subsequently, DNA from pooled samples of these mutants is subjected to high-throughput sequencing focused on the target gene regions. Bioinformatics analysis is then employed to detect and characterize mutations within the targeted genes. This strategy significantly enhances the efficiency and precision of mutant screening compared to conventional methods, enabling the rapid isolation of plants carrying desirable allelic variants without introducing foreign DNA. Targeted mutagenesis has emerged as a powerful tool in crop breeding programs, particularly for developing varieties with reduced Cd accumulation, by enabling precise modification of genes involved in Cd uptake and transport.

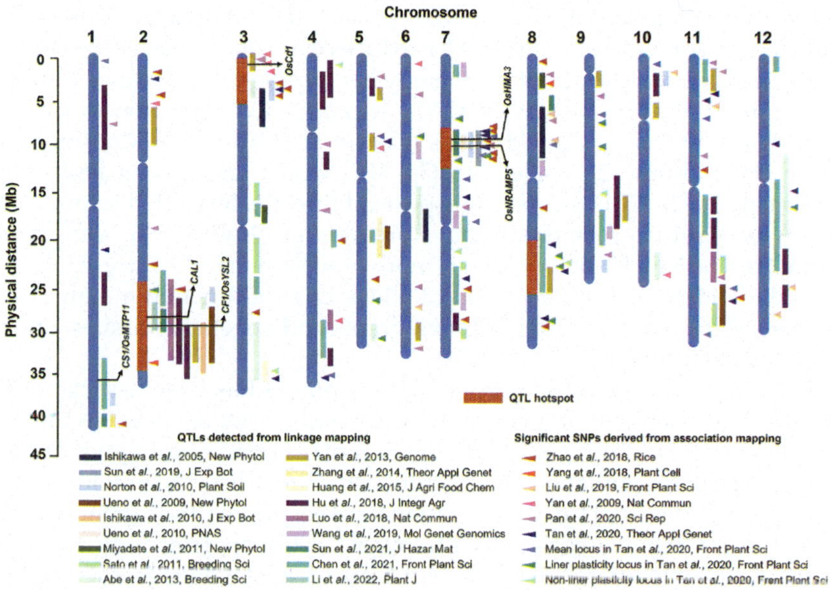

FIGURE 19.1 Summary of QTLs that control Cd accumulation in rice. QTLs were identified through linkage mapping or association mapping studies according to the studies labeled in the reference list. The vertical blue bars with arc-shaped ends represent the 12 rice chromosomes. The chromosome regions highlighted in red represent the QTL hotspots. The bars with right-angled ends on the right side of chromosomes represent the Cd-associated QTL identified by linkage analysis from various studies, and the arrows represent SNPs significantly associated with Cd accumulation identified by GWAS. Representative cloned genes controlling Cd accumulation were shown. *Cd*, Cadmium; *GWAS*, genome-wide association studies; *SNPs*, single nucleotide polymorphisms. *Modified from with detailed information listed in Sun et al. (2023).*

typically a set of chromosomal regions (QTLs) associated with Cd accumulation, along with estimates of their phenotypic variance explained and the additive/dominance effects.

Considerable progress has been made in the genetic mapping of Cd accumulation in crops, particularly in rice. Numerous QTLs associated with Cd uptake, translocation, and sequestration have been identified across the rice genome. Among them, several QTL hotspots have emerged repeatedly among independent studies, notably those located on chromosomes 2, 3, 7, and 8 (Fig. 19.1), indicating that these regions harbor key genes influencing Cd accumulation traits. The QTL hotspot on chromosome 7 has been intensively studied, and two major genes have been identified in this region: *OsHMA3* and *OsNRAMP5*. *OsHMA3* encodes a P1B-type ATPase transporter localized to the vacuolar membrane in root cells. It functions by sequestering Cd into root vacuoles, thus limiting its translocation to the shoot and grains. This gene accounts for a large proportion of phenotypic variation in Cd accumulation in rice shoots or grains, and natural variation in OsHMA3 alleles significantly contributes to differences among cultivars (Sui et al., 2019; Ueno et al., 2010). *OsNRAMP5*, another key gene in this region, encodes a transporter involved in Cd and Mn uptake from the soil and also plays a central role in the natural variation of grain Cd accumulation (see Chapter 10).

Another important QTL hotspot resides on chromosome 2, where two genes with distinct mechanisms have been cloned and functionally characterized: *CAL1* and *CF1*. *CAL1* (*Cadmium*

Accumulation in Leaf 1) encodes a defensin-like protein predominantly expressed in the root exodermis and xylem parenchyma cells. CAL1 reduces cytosolic Cd levels by chelating Cd and facilitating its secretion into the apoplast, thereby promoting long-distance transport through xylem vessels (Luo et al., 2018). Notably, CAL1 increases Cd allocation to leaves and straw without significantly altering grain Cd levels (Luo et al., 2018). *CF1* (Cadmium and Iron 1), also located on chromosome 2, is allelic to *OsYSL2*, which encodes an Fe(II)-nicotianamine transporter (Li et al., 2022). CF1 does not transport Cd directly. Instead, it improves Fe nutrition in the shoots, which in turn suppresses the expression of *OsNRAMP5* in the roots and thus reduces Cd uptake. The $CF1^{TQ}$ allele, which is more prevalent in indica rice, confers lower grain Cd levels than the widely distributed japonica allele. Loss-of-function mutants of *CF1* exhibit significantly increased Cd accumulation in both grains and vegetative tissues, underscoring its regulatory role (Li et al., 2022).

The QTL hotspot on the upper region of chromosome 3 has been associated with *OsCd1*, which encodes a major facilitator superfamily (MFS) transporter localized to the plasma membrane of root cells (Yan et al., 2019). OsCd1 facilitates root Cd uptake and significantly contributes to variation in grain Cd accumulation. A naturally occurring missense mutation (Val449Asp) in OsCd1 underlies the differential Cd accumulation between indica and japonica subspecies. The japonica allele ($OsCd1^{V449}$) is associated with lower Cd transport activity and reduced grain Cd levels. CRISPR/Cas9-mediated disruption of *OsCd1* leads to decreased Cd accumulation in roots, shoots, and grains without major impacts on essential micronutrients (Yan et al., 2019).

Several causal genes for Cd QTLs outside these hotspot regions have also been uncovered. For example, the *CS1* (*Cd-safe 1*)/*OsMTP11* gene, associated with a QTL on chromosome 1, encodes a tonoplast-localized transporter that promotes Cd and Mn sequestration in the vacuoles of leaf vascular cells. This mechanism restricts Cd distribution from leaves to grains (Liu et al., 2024). To date, only a limited number of causal genes, such as those described above, have been successfully cloned from QTLs and functionally characterized, whereas many others, including the major hotspot on chromosome 8, remain to be cloned and elucidated.

Genetic mapping of Cd accumulation in crops not only helps elucidate the genetic basis of Cd accumulation traits but also provides molecular markers for MAS breeding. Once QTLs or causal genes associated with Cd accumulation are identified, molecular markers linked to low-Cd loci can be developed. These markers can be used in MAS breeding to rapidly and accurately select individuals with low Cd accumulation traits, improving breeding efficiency.

19.2.3 Marker-assisted selection breeding for low-cadmium crop varieties

MAS is a powerful tool that enhances the efficiency and precision of traditional breeding by using molecular markers linked to traits of interest (Box 19.1). In the context of breeding low-Cd crop varieties, MAS enables the identification and selection of individuals carrying favorable alleles at loci associated with reduced Cd accumulation. This technique is particularly useful for traits with complex inheritance patterns, low heritability, or those that are difficult, time-consuming, or expensive to phenotype directly.

The standard process of MAS breeding for low-Cd crop varieties typically includes several key steps. First, molecular markers linked with Cd accumulation traits are identified (see Section 19.2.2). These markers serve as selection tools throughout the MAS breeding process. Second, a breeding population is developed. Varieties with low Cd accumulation are used as the donor parents, whereas elite varieties with high Cd accumulation are used as the recipient parents. Hybrids (F_1) are produced through crossing, followed by the generation of a segregating population (F_2) via self-pollination. Third, MAS selection is conducted. The segregated population is genotyped using molecular markers to identify individuals carrying the target alleles linked with low Cd accumulation. These individuals are retained for further breeding through backcrossing with the elite variety (BC_1F_1). This backcrossing can be repeated for several generations, using individuals from each backcross generation (e.g., BC_1F_2) that carry the desired low-Cd loci. Finally, comprehensive selection is conducted. In addition to selecting for low Cd accumulation traits, comprehensive selection for other agronomic traits such as yield, quality, and disease resistance is performed to develop low-Cd crop varieties with excellent comprehensive performance. To ensure that the genetic background of the selected lines closely resembles that of the recipient parent, multiple rounds of backcrossing can be conducted, with MAS applied in each generation. This approach minimizes the genetic differences between the selected lines and the recipient parent while preserving the desired low Cd accumulation trait.

MAS breeding for low-Cd crop varieties has several advantages. First, it significantly improves breeding efficiency. Molecular markers can detect the presence of low-Cd gene(s) at the seedling stage or even before planting, enabling early selection and reducing the breeding cycle. Second, it enhances breeding accuracy. Molecular markers can directly target the genes associated with Cd accumulation traits, avoiding the influence of environmental factors and improving the accuracy of selection. Third, it enables gene pyramiding. By combining multiple molecular markers linked to different Cd-related QTLs/genes, MAS breeding can accumulate multiple favorable loci or genes in a single variety, achieving a more substantial reduction in Cd accumulation. MAS can also be integrated with genomic selection (GS), which uses genome-wide marker data to predict the breeding value of individuals. Although MAS targets a few major QTLs, GS captures both major and minor loci contributing to Cd accumulation. This integration is particularly valuable for complex traits influenced by multiple genes and strong genotype-by-environment interactions.

The effectiveness of MAS depends heavily on the accuracy and reliability of the markers employed. The linkage relationship between molecular markers and target genes may be affected by genetic background and environmental conditions. As a result the performance of molecular markers may vary across different breeding populations or environments. Therefore it is necessary to verify and evaluate the reliability of molecular markers in different breeding populations and environments. Functional markers, which are derived from polymorphisms within the causal gene itself, offer the highest selection precision. For example, a functional marker for the Val449Asp polymorphism in the coding region of *OsCd1* has been developed to distinguish the low-Cd japonica allele from high Cd indica alleles, facilitating marker-based pyramiding of favorable alleles (Yan et al., 2019).

19.2.4 Mutagenesis and screening for low-cadmium mutants

Mutagenesis is a breeding method that induces genetic variations in crops through exposure to physical or chemical mutagens (Box 19.1). Screening for low-Cd mutants among these variations can provide new germplasm resources for breeding low-Cd crop varieties. By using elite cultivars as the starting material, mutagenesis can be employed to induce genetic variation, followed by the selection of individuals exhibiting reduced Cd accumulation. Through successive backcrossing with the wild type parent to purify the mutation background, it is possible to rapidly develop new low-Cd varieties that retain high yield and superior quality. Ultimately, this approach can lead to the development of low-Cd crop varieties suitable for direct use in agricultural production.

Various physical and chemical mutagens are used in mutagenesis, including ethyl methanesulfonate (EMS), fast neutrons, ion beams, and ultraviolet radiation. These mutagens induce various types of mutations, such as base substitutions, insertions, deletions, and chromosomal structural variations, leading to changes in gene functions. For example, EMS primarily causes guanine-to-adenine base substitutions, inducing point mutations in genes and altering their functions. Fast neutrons and ion beams can cause chromosomal breakage and rearrangements, generating large-scale genetic variations.

The process of screening for low-Cd mutants involves inducing mutations in crop seeds or callus tissues using mutagens, cultivating the mutated materials in Cd-contaminated soils or hydroponic solutions, and then determining the Cd concentrations in different plant tissues to identify low-Cd mutants. The frequency of induced mutations depends on the mutagen type, dosage, and plant species, with typical mutation rates ranging from approximately 1 mutation per 25–1000 kilobases (kb) in major crops. For example, in diploid species such as rice, EMS mutagenesis can achieve mutation frequencies of about 1 per 135–1000 kb, whereas polyploid crops like wheat have higher mutation rates with up to 1 per 24–40 kb because of genetic redundancy (Sikora et al., 2011). To obtain at least one desirable mutation in a specific gene associated with Cd accumulation, it is often necessary to screen several thousand to tens of thousands of individuals to ensure coverage of all possible mutations in target genes. The required population size is positively correlated with the genome size of the species and the functional redundancy of the target gene family. For instance, rice, with a genome size of ~400 Mb, generally requires a screening population of at least 5000–20,000 individuals to ensure sufficient mutation coverage (Sikora et al., 2011). Therefore, the construction and management of large mutant populations, along with the implementation of high-throughput screening and precise analytical methods, are essential for the efficient identification and subsequent breeding of low-Cd crop varieties. With the development of modern analytical technologies, such as ICP-MS and atomic absorption spectrometry, the Cd concentrations in large numbers of plant samples can be accurately determined, providing technical support for the screening of low-Cd mutants. In addition, high throughput elemental profiling techniques, such as synchrotron radiation-based XRF imaging, facilitate rapid evaluation of Cd accumulation and distribution, improving the efficiency of mutant screening (Ren et al., 2023).

Several successful cases of mutagenesis-based identification of low-Cd mutants in rice have been reported. For example, rice seeds of the elite japonica cultivar "Koshihikari" were subjected to carbon ion-beam mutagenesis, and three low-Cd mutants (*lcd-kmt1*, *lcd-kmt2*, and *lcd-kmt3*) were identified by screening for grain Cd concentrations of 2592 M2 plants grown in Cd-polluted soil using ICP-MS (Ishikawa et al., 2012). These mutants exhibited Cd concentrations of <0.05 mg kg^{-1} in their grains, compared to an average of 1.73 mg kg^{-1} in the parent variety grown on the same soil. Linkage analysis on the F_2 population derived from the crossing of *lcd-kmt1* with the wild type identified as *OsNRAMP5* as the causal gene responsible for reduced Cd uptake. Specifically, *lcd-kmt1* harbored a 433-bp insertion in exon 5, *lcd-kmt2* had a single-nucleotide deletion in exon 4, and *lcd-kmt3* exhibited an ~227-kb deletion encompassing the entire *OsNRAMP5* genomic region, all resulting in complete loss of gene function. Similarly, EMS mutagenesis of the elite indica rice cultivar 9311 led to the discovery of a low Cd accumulation mutant, *lcd1* (*low cadmium 1*), which exhibited markedly reduced Cd levels in both rice straws and grains. Molecular analysis revealed that the causal mutation was an SNP resulting in a Pro236Leu amino acid substitution in a highly conserved region of OsNRAMP5, significantly reducing its Cd transport activity (Cao et al., 2019). In another study, Kuramata et al. (2022) mutagenized the japonica rice cultivar "Koshihikari" using the alkylating agent diepoxybutane (DEB) and identified a mutant carrying the Gln337Lys substitution in OsNRAMP5. This single amino acid change (glutamine to lysine) produced an intermediate phenotype in Cd and Mn accumulation, positioned between the wild-type and knockout lines (Kuramata et al., 2022).

Although mutagenesis breeding can accelerate the development of low-Cd rice varieties, it also poses several challenges. One major limitation is the randomness and unpredictability of induced mutations. Although mutagenesis can generate a wide range of genetic variations, most of these mutations are harmful to plant growth and development, and only a small fraction may confer beneficial traits such as reduced Cd accumulation. As a result, large populations of mutated lines must be screened to identify promising low-Cd mutants. The mutants identified most likely also contain multiple mutations in the genome. The unwanted mutations introduced during mutagenesis can be removed through successive backcrossing with the original wild type variety to recover the desirable genetic background while retaining the low-Cd trait. Another challenge lies in the genetic stability of these mutants. Some low-Cd lines may exhibit unstable traits, with Cd accumulation levels varying across environments or generations. This instability often arises because the low-Cd phenotype may be influenced by environmental conditions such as soil pH, Cd availability, or micronutrient levels. Some mutants may only exhibit the low-Cd phenotype under specific levels of Cd contamination, failing to maintain the low-Cd trait across a broader range of field conditions. In addition, interactions among unintended mutations introduced during mutagenesis may influence trait expression in variable environments or over time. Therefore multilocation and multiyear trials are necessary to confirm the stability of the low-Cd trait.

19.2.5 Successful breeding of a low-cadmium rice variety: A case study of Koshihikari Kan No. 1 (Kan1)

Koshihikari Kan No. 1 (Kan1) is a prominent example of successful breeding for low-Cd rice varieties through the integration of mutagenesis and screening. Kan1 is a derivative of the widely cultivated japonica rice variety "Koshihikari," which is known for its superior grain quality and market acceptance. To reduce Cd accumulation while preserving its agronomic traits, Koshihikari seeds were irradiated with 320 MeV carbon ions ($^{12}C^{6+}$) at a dose of 40 Gy and obtained approximately 4000 M1 (mutant generation 1) plants. About 3000 M2 progenies were grown in Cd-contaminated paddy soil (1.8 mg Cd kg^{-1}), managed under controlled water conditions to increase Cd bioavailability, and evaluated for grain Cd content using ICP-MS (Ishikawa et al., 2012). Three mutants (*lcd-kmt1*, *lcd-kmt2*, and *lcd-kmt3*) were identified in the initial screen, all showing grain Cd concentrations <0.05 mg kg^{-1} compared to 1.73 mg kg^{-1} in wild-type Koshihikari. Among them, *lcd-kmt2* exhibited stable low Cd accumulation across multiple environments and was registered as low-Cd rice variety Kan1. Hydroponic assays confirmed that *Kan1* had reduced Cd uptake by roots because of a mutation in the *OsNRAMP5* gene.

When Kan1 and the parent Koshihikari were cultivated in three geographically distinct Cd-polluted paddy fields (soil Cd concentrations ranging from 0.35 to 1.35 mg kg^{-1}, Koshihikari grains consistently exceeded the Codex maximum limit of 0.4 mg kg^{-1} Cd in all three sites, whereas Kan1 grains contained Cd levels below the limit of quantification (<0.01 mg kg^{-1}). Straw Cd concentrations were also significantly lower in Kan1 compared to Koshihikari, confirming the reduced Cd accumulation in both edible and vegetative tissues of Kan1.

Rice grain contamination with arsenic (As) is another critical food safety concern. Typically, flooded paddy conditions increase grain As levels, whereas aerobic conditions increase Cd bioavailability in soil and accumulation in rice grain, creating a trade-off in water management strategies. Kan1 maintained nearly undetectable Cd levels in grains under three water regimes, including continuous flooding (FLD), alternate wetting and drying (AWD), and water-saving (WAS) (Ishikawa et al., 2016). AWD and WAS treatments significantly reduced grain As concentrations in Kan1 by approximately 27%–48% compared to FLD. This demonstrates that aerobic cultivation of Kan1 offers a practical solution to simultaneously minimize both Cd and As accumulation in rice grains, addressing two major toxic metal risks in rice-based diets.

Most of the agronomic traits of Kan1 were indistinguishable from Koshihikari when grown in the paddy field, including plant height, panicle length, grain shape, culm length, panicle number, grain yield, and 1000-grain weight (Ishikawa et al., 2019). A sensory evaluation to assess eating quality based on glossiness, smell, taste, stickiness, and hardness of cooked rice revealed that Kan1 performs similarly to Koshihikari, indicating that the mutation did not compromise palatability or consumer acceptance.

OsNRAMP5 functions as a transporter for the uptake of both Cd and Mn (see Chapter 10). The mutation of *OsNRAMP5* in Kan1 impairs Mn uptake and may exhibit manganese deficiency symptoms, particularly when grown in Mn-deficient soils. Although Kan1 shows no significant differences in growth compared to Koshihikari under normal paddy conditions, it is important to monitor Mn nutrition, as Kan1 exhibits substantially lower Mn concentrations in grains and straw, depending on soil Mn availability and water management. Kan1 is more susceptible to brown spot disease under low-Mn and low-Si conditions, but this can be alleviated by Mn supplementation, which improves both disease resistance and yield (Suda et al., 2021).

19.3 Biotechnological approaches to limit cadmium accumulation

Although traditional breeding strategies have made substantial progress in reducing Cd accumulation in crops, they often depend on naturally occurring or induced genetic variation, and the breeding process can be time-consuming and labor-intensive. Recent advances in biotechnology have provided more precise and efficient alternatives by enabling direct manipulation of the genes responsible for Cd uptake, transport, and sequestration. These biotechnological approaches complement conventional breeding and offer new opportunities to develop crop varieties with lower Cd accumulation while maintaining desirable agronomic traits. This section highlights three major strategies currently used in this context: (1) gene knockdown and knockout techniques, including RNA interference (RNAi) and genome editing technologies, to suppress or eliminate the function of Cd-related genes, (2) gene overexpression to enhance the expression of genes that reduce Cd accumulation, and (3) targeted mutagenesis (Box 19.1) combined with high-throughput sequencing to identify and utilize novel alleles associated with reduced Cd uptake and translocation.

19.3.1 Gene knockdown and knockout

Disruption of genes involved in Cd uptake, translocation, and remobilization is an effective strategy to limit Cd accumulation in crops. RNAi and genome editing are two technologies that are commonly used to knockdown (reduce gene expression) and knockout (disrupt gene function) of a target gene, respectively. RNAi is a powerful biotechnology tool used to achieve gene knockdown by selectively reducing the expression of target genes at the posttranscriptional level. Unlike genome editing described below, which induces permanent and precise modifications in the DNA sequence, RNAi transiently suppresses gene function by degrading messenger RNA (mRNA) or inhibiting its translation, thereby decreasing the production of the corresponding protein (Borges & Martienssen, 2015). The RNAi pathway is naturally present in eukaryotic cells and is triggered by double-stranded RNA molecules that are processed into small interfering RNAs or microRNAs (miRNAs). These small RNAs guide the RNA-induced silencing complex to complementary mRNA targets, leading to their cleavage or translational repression (Borges & Martienssen, 2015). In plants, RNAi can be engineered by introducing constructs expressing hairpin RNAs or artificial miRNAs specific to the gene of interest. RNAi has been widely explored for its potential applications in disease management, plant development, and crop improvement by reducing the expression of genes involved in undesirable traits (Das & Sherif, 2020). Although RNAi does not alter the genomic DNA and its effects may vary in efficiency and duration, it remains a valuable approach for functional gene analysis and trait modification, especially when permanent gene knockout is not feasible or desirable.

Genome editing is a revolutionary biotechnology that enables precise, targeted modifications of DNA sequences of an organism. Unlike traditional mutagenesis, which induces random mutations, genome editing allows for specific gene knockout, knock-in, base substitution, or gene regulation with high accuracy. The most widely used genome editing tools include zinc finger nucleases, transcription activator-like effector nucleases, and the clustered regularly interspaced palindromic repeats (CRISPR)/Cas9 systems, with CRISPR/Cas9 being the most powerful tool because of its simplicity, efficiency, and versatility (Gao, 2021). The CRISPR–Cas9 system utilizes a guide RNA to direct the Cas9 nuclease to a complementary DNA target sequence, where Cas9 induces a double-strand break (DSB). The endogenous DNA repair mechanisms in the cell, including the nonhomologous end joining or homology-directed repair, then repair the break and result in insertions or deletions (indels). These modifications can disrupt gene function (knockout). Genome editing technology can also precisely change DNA sequence to introduce desired alleles (knock-in) if a repair template is provided. More recently, base editors and prime editors have emerged, enabling single nucleotide changes without DSBs, further enhancing precision and reducing off-target effects.

Numerous studies have demonstrated the effectiveness of genome editing technology and RNAi in limiting Cd accumulation in crops by editing or suppressing genes involved in Cd uptake, sequestration, or translocation. A prominent example is *OsNRAMP5*, a major Cd and Mn transporter gene in rice. Knockout mutants of *OsNRAMP5* created by using CRISPR/Cas9 technology exhibit drastically reduced Cd uptake and accumulation in both roots and grains without severely compromising plant growth under Mn-sufficient conditions. For example, CRISPR/Cas9-mediated knockout of *OsNRAMP5* in Huazhan and Longke638S, two extensively used indica parental lines in the current two-line hybrid rice breeding, resulted in a complete loss of function of *OsNRAMP5*. Field trials in Cd-contaminated sites demonstrated that Cd concentration in the grains of *osnramp5* knockout mutants was consistently lower than

0.05 mg kg^{-1}, in contrast to high Cd concentrations from 0.33 to 2.90 mg kg^{-1} in the grains of wild-type indica rice (Tang et al., 2017). Furthermore the hybrid produced by crossing two knockout parental lines contains extremely low-Cd content in grains compared to the hybrid derived from the crossing of wild-type parental lines without yield penalty when grown in Mn-sufficient paddy fields. Several other genes have also been successfully edited to lower Cd accumulation. OsZIP2, a plasma membrane-localized transporter, is involved in root-to-shoot Cd translocation and Cd transfer within nodes. Disruption of *OsZIP2* via CRISPR/Cas9 significantly decreased Cd concentration in grains by 41.2%–55.9% compared to wild-type plants (Li et al., 2024). The *oszip2* knockout mutants show no difference in agronomic traits from the wild type. Knockout of a putative cation/calcium exchanger gene *OsCCX2* or a cation/H^+ exchanger gene *OsCAX2* led to a significant reduction in grain Cd concentration by 50% and 38%–63%, respectively (Hao et al., 2018; Liu et al., 2025). Most agronomic traits of *oscax2* knockout mutant, including grain yield per plant, seed setting rate, grain number per panicle, panicle number per plant, thousand grain weight, and plant height, were generally comparable to the wild type. However, knockout of *OsCCX2* significantly reduces the thousand grain weight. Knockout of the MFS protein OsCd1 also leads to significant reduction of Cd accumulation in rice grains but has a negative impact on plant growth and reduces yield at harvest (Yan et al., 2019), suggesting that it is not an optimal strategy to create low-Cd rice by simply knocking out *OsCd1*.

RNAi technology is another effective strategy to limit Cd accumulation in crops. Knockdown of *OsLCT1* by RNAi did not affect xylem-mediated Cd transport but reduced phloem-mediated Cd transport and lowered grain Cd concentration by half compared with wild type (Uraguchi et al., 2011). Similarly, RNAi-mediated knockdown of rice P-type heavy metal ATPase gene *OsHMA2*, which is involved in loading Cd into the xylem and participates in root-to-shoot translocation of Cd, significantly reduces the accumulation of Cd in leaves and grains (Takahashi et al., 2012). However, the plant height and grain yield of *OsHMA2* RNAi lines were impaired, possibly because of Zn deficiency in the aboveground tissues, which limits its application in breeding low-Cd rice variety. In addition to rice, RNAi has been applied to limit Cd accumulation in other crops. In barley, RNAi-mediated knockdown of *HvNRAMP5* gene resulted in a significant reduction of Cd accumulation in roots and shoots (Wu et al., 2016).

Although both RNAi and genome editing have proven effective in limiting Cd accumulation by targeting key transport-related genes, they differ substantially in mechanisms, precision, and regulatory implications. RNAi is relatively simple to implement and allows for partial or inducible gene suppression, which can be advantageous when complete knockout leads to severe phenotypes. However, RNAi constructs are transgene-based and often raise regulatory and public acceptance concerns because of their genetically modified organism (GMO) status. Furthermore, RNAi efficiency can vary depending on expression levels, tissue specificity, and environmental factors, potentially leading to incomplete or unstable silencing effects. In contrast, genome editing, particularly CRISPR/Cas-based systems, enables precise, stable, and heritable changes to the target gene, whereas the foreign Cas9 gene can be removed by segregation in the offspring. In many cases, genome-edited plants that do not retain transgenic components after segregation may be exempt from GMO regulations in certain countries. Therefore genome editing is generally considered a more robust and publicly acceptable tool for developing low-Cd crop varieties, especially for traits requiring complete gene knockout.

19.3.2 Gene overexpression

Gene overexpression is another important biotechnological approach used to reduce Cd accumulation in crops. Gene overexpression involves enhancing the transcriptional or translational activity of a target gene to increase its protein product, thereby amplifying its biological function. The basic principle of gene overexpression is to construct a gene overexpression vector containing the target gene and a strong promoter. The vector is then introduced into crop cells using methods such as *Agrobacterium*-mediated transformation or biolistic transformation. After plant regeneration and screening, transgenic plants overexpressing the target gene are obtained. Unlike gene knockdown and knockout strategies that disrupt endogenous genes involved in Cd uptake or translocation to grains, overexpression aims to increase the expression of endogenous genes or exogenous genes from other organisms that are involved in vacuolar sequestration, chelation, or efflux of Cd, enhancing a plant's ability to retain Cd in nonedible tissues or vacuoles, thereby preventing its accumulation in grains or edible parts.

One of the most extensively studied genes is *OsHMA3*, which encodes a tonoplast-localized P1B-type ATPase that sequesters Cd into vacuoles of rice root cells (see Chapter 10). Overexpression of *OsHMA3* by using a constitutive CaMV 35S promoter reduced the translocation of Cd from roots to shoots in hydroponic growth conditions and Cd accumulation in grains under soil-grown conditions (Ueno et al., 2010). Overexpression of *OsHMA3* in an indica rice reduced Cd concentrations in brown rice by 94%–98% in two Cd-contaminated paddy soils, reaching levels near the detection limit, and had no significant effect on the concentrations of essential micronutrients in brown rice (Lu et al., 2019). Overexpression of a rice functional *OsHMA3* gene in wheat driven by a strong constitutive ubiquitin promoter decreased root-to-shoot Cd translocation by nearly 10-fold and Cd accumulation in wheat grain by 96% (Zhang et al., 2020). HvHMA3 is also responsible for transporting Cd into the vacuoles in barley. Overexpression of *HvHMA3* driven by different promoters dramatically decreases grain Cd concentration in barley (Lei et al., 2020). Apart from OsHMA3, OsABCC9 also contributes to vacuolar sequestration of Cd in rice roots (Yang et al., 2021). OsABCC9 is located at the tonoplast in the parenchyma cells inside the stele, and its expression is strongly induced by Cd exposure. Overexpression of *OsABCC9* likely limits Cd accumulation in rice grains as its knockout leads to increased Cd accumulation.

Overexpression of genes involved in Cd uptake and xylem loading may lead to reduction of Cd accumulation in grains. For instance, overexpression of *OsNRAMP5* driven by the promoters of *OsActin1* or maize *ubiquitin* decreased Cd concentration in the grain by 49%–94% compared with the wild type, which may result from the ectopic expression of *OsNRAMP5* disrupting the radial transport of Cd into the stele for xylem loading (Chang et al., 2020). Similarly, overexpression of *OsHMA2* by using a constitutive CaMV 35S promoter decreases the Cd content in leaves and grains (Takahashi et al., 2012). This is likely because ectopic expression of *OsHMA2* in the roots enhanced exclusion of Cd from root cells, resulting in a reduced flux of Cd toward the xylem.

Although gene overexpression is effective in conferring low-Cd phenotypes, it often involves the insertion of foreign DNA or the use of constitutive promoters, thereby classifying the resulting plants as GMOs under most regulatory frameworks. This may limit the public acceptance and commercial deployment of such varieties. In addition, overexpression strategies may lead to unintended effects on other physiological processes because of ectopic expression or gene dosage imbalances. Tissue-specific, in situ overexpression or inducible expression systems can

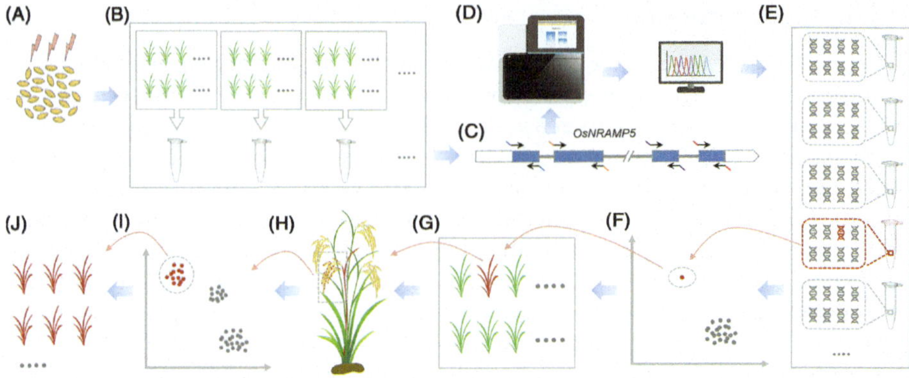

FIGURE 19.2 The workflow of M1TDSA for identification of low-Cd rice mutants of loss-of-function of *OsNRAMP5*. Rice seeds are irradiated with heavy-ion beams (A) to induce genome-wide mutations. M1 plants are cultivated, and leaf samples are collected and pooled for genomic DNA extraction (B). The coding region of *OsNRAMP5* is PCR-amplified using primers tagged with UMTs (C) and subjected for NGS (D). Candidate pools harboring mutations are identified (E) and are further analyzed by KASP to genotype individual plants (F). The M1 mutants with loss-of-function alleles of *OsNRAMP5* (G and H) are advanced to the M2 generation for isolation of homozygous lines (I and J). *Cd*, Cadmium; *KASP*, Kompetitive allele-specific PCR; *NGS*, next-generation sequencing; *UMTs*, unique molecular tags. *Based on the concept from Shao et al. (2022).*

help mitigate these concerns by restricting gene activity to relevant organs or developmental stages. For instance, expression of *OsHMA3* under the control of the *OsHMA2* promoter can effectively reduce Cd accumulation in rice grains through sequestering more Cd into the vacuoles of various tissues (Shao et al., 2018).

19.3.3 Targeted mutagenesis

Targeted mutagenesis is a biotechnological approach that combines random mutagenesis with high-throughput sequencing to specifically screen for mutants with alterations in target genes. It has become a new strategy for breeding crops with limited Cd accumulation. The principle of targeted mutagenesis involves inducing mutations in crop seeds or callus tissues using physical (e.g., ion beam irradiation) or chemical (e.g., EMS) mutagens, generating a large number of mutated materials. High-throughput sequencing technologies are then used to sequence the target genes related to Cd accumulation in the mutated materials, and bioinformatics analysis is conducted to identify mutants with mutations in the target genes. Unlike conventional mutant screening, which relies on laborious phenotyping of mutants, the targeted mutagenesis strategy uses reverse genetics platforms such as Targeting Induced Local Lesions in Genomes or high-throughput sequencing to identify mutations in Cd-related genes and link mutants to the low-Cd trait.

A notable example in targeted mutagenesis is the M1 generation targeted deep sequencing (M1TDS) technology (Fig. 19.2), which enables high-throughput detection of low-frequency chimeric mutations of *OsNRAMP5* in the M1 generation resulting from physical mutagenesis (Shao et al., 2022). The M1TDS strategy combines heavy-ion beam irradiation with targeted deep sequencing, allowing for early identification of mutated alleles in the *OsNRAMP5* gene responsible for Cd uptake. The M1TDS process begins with the irradiation of rice seeds using heavy-ion beams (e.g., $^{12}C^{6+}$ at 120 Gy) to induce genome-wide mutations. M1 plants are grown, and at the tillering stage, leaf samples are collected and pooled into high-redundancy sample pools (e.g., 30–500 individuals per pool). Genomic DNA is extracted from these pools, and the coding sequences of *OsNRAMP5* are polymerase chain reaction (PCR)-amplified using specific primers with unique molecular tags. The PCR products were then subjected to library preparation for ultradeep next-generation sequencing, achieving average coverage depths exceeding 6000× to 30,000×, which is critical for detecting low-frequency mutations present in chimeric M1 tissues (as low as 0.1%–2%). Once candidate pools containing low-frequency mutations at *OsNRAMP5* are identified, Kompetitive allele-specific PCR is used to genotype individual plants within the positive pools. Sanger sequencing is then applied to confirm the exact mutation at the genomic sequence of *OsNRAMP5*. Plants with mutations at the exons of *OsNRAMP5* or at exon–intron splice junctions, which can lead to aberrant mRNA splicing and altered gene function, are selected as candidate mutants. Putative mutants at M1 generation are advanced to the M2 generation, where homozygous lines can be isolated. The Cd concentrations in the grains of homozygous putative mutants are evaluated in Cd-contaminated paddy fields to confirm the low grain Cd of putative mutants.

The M1TDS technology has been successfully used to develop two low-Cd, two-line hybrid rice varieties, Zhenliangyou 8612 and Zhuoliangyou 1126, both harboring mutations in *OsNRAMP5* (Shao et al., 2022, 2025). The original two-line hybrid rice Zhenliangyou 8612 is derived from the crossing of the sterile line Longzhen 36S and the restorer line Huahui 8612. Through heavy-ion irradiation mutagenesis followed by M1TDS screening, a mutant with a 3-base pair deletion at the splice junction between intron 10 and exon 11 of *OsNRAMP5*, which leads to abnormal mRNA splicing and significantly reduced gene expression, was identified in the restorer line Huahui 8612. The mutated restorer line was named low-Cd restorer line R001. By using the same strategy a low-Cd sterile line named Lian 1S was also identified from mutation population of Longzhen 36S, which harbors a deletion of 18 base pairs at the splice site between intron 8 and exon 9 of *OsNRAMP5*, resulting in disruption of its normal transcription. The low-Cd Zhenliangyou 8612, also known in national trials as Lianliangyou 1, is developed by crossing the low-Cd sterile line Lian 1S and the low-Cd restorer line R001 (Shao et al., 2022). In a field contaminated with an $CaCl_2$-extractable Cd of 0.443 mg kg^{-1} (soil pH 4.93), the grain Cd content of the hybrid rice Lianliangyou 1 was stable below 0.03 mg kg^{-1}, much lower than that of original hybrid rice Zhenliangyou 8612 (0.38 mg kg^{-1}) and the national limit standard of 0.20 mg kg^{-1} (Shao et al., 2022).

Similarly the low-Cd two-line hybrid rice Zhuoliangyou 1126 was developed using the sterile line Longke 638S and the restorer line Huazhan, both of which had mutations in *OsNRAMP5* and were identified by M1TDS (Shao et al., 2025). When grown in Cd-contaminated soil with a $CaCl_2$-extractable Cd of 0.677 mg kg^{-1} and pH 5.6, grain Cd content of Zhuoliangyou 1126 was stably below 0.05 mg kg^{-1}, whereas that of the original hybrid was 0.91 mg kg^{-1}, demonstrating the broad applicability of M1TDS for generating low-Cd germplasm in diverse genetic backgrounds.

The targeted mutagenesis approaches, such as M1TDS technology, enable targeted, high-throughput screening of induced mutations directly in the M1 generation, significantly shortening breeding cycles and reducing costs compared to traditional M2 generation-based screening. It breaks the technical bottleneck of detecting rare chimeric mutations in large mutagenized populations and facilitates the rapid development of low-Cd-crop germplasm. One of the major advantages of this approach is its nontransgenic nature. Unlike overexpression or RNAi, targeted mutagenesis does not involve the insertion of foreign DNA, which may facilitate regulatory approval and consumer acceptance in certain regions. Furthermore, once useful mutations are discovered, they can be introgressed into elite lines via conventional or marker-assisted backcrossing, allowing precise transfer of traits without linkage drag.

One of the challenges of targeted mutagenesis is that it requires advanced high-throughput sequencing technologies capable of detecting low-frequency mutations within pooled populations. In addition, the interpretation of ultradeep sequencing data demands sophisticated bioinformatics pipelines and substantial computational resources to accurately identify, annotate, and prioritize candidate mutations.

19.4 Future prospects and challenges

Despite remarkable advances in both traditional breeding and biotechnological approaches to limit Cd accumulation in crops, there are still limitations and gaps in these methods. Traditional breeding methods, such as screening for low-Cd crop varieties and genetic mapping of Cd accumulation, while foundational, are time-consuming and labor-intensive. These methods are often constrained by the limited availability of natural low-Cd alleles within existing germplasm pools. Furthermore the genetic basis of Cd accumulation traits is complex, involving multiple genes and prone to influence by environmental factors. The QTLs detected through genetic mapping often have small effects and are generally environment-dependent. MAS and QTL mapping have accelerated the identification and introgression of low-Cd alleles, but their effectiveness depends on the precision of phenotyping, the stability of QTLs across environments, and the availability of robust molecular markers. Biotechnological strategies, including genome editing and gene overexpression, provide powerful tools for precise manipulation of Cd-related genes. However, these approaches face regulatory and societal hurdles, particularly regarding the acceptance and commercialization of genetically modified or gene-edited crops in many countries.

Another critical gap lies in the complex interactions between plant genotype, soil properties, and agronomic practices. Environmental factors, such as soil pH, redox status, organic matter content, and irrigation regime, can significantly influence Cd bioavailability and plant uptake, sometimes overriding genetic improvements. In addition, current breeding programs often rely on a limited number of well-characterized genes such as *OsNRAMP5* and *OsHMA3*, whereas the broader network of genes involved in Cd uptake, detoxification, and distribution remains underexplored. There is a lack of comprehensive understanding of the mechanisms of Cd accumulation and transport in crops, and the overall regulatory network of Cd accumulation traits remains unclear, limiting the precision and effectiveness of breeding strategies. Furthermore, many genes that control Cd uptake and translocation also play essential roles in the homeostasis of vital micronutrients such as Mn, Fe, and Zn. A typical example is OsNRAMP5, which functions as a transporter for both Cd and Mn. Genome editing or targeted mutagenesis that disrupts or knocks out *OsNRAMP5* effectively reduces Cd absorption but simultaneously decreases Mn uptake. This reduction in Mn absorption can lead to Mn deficiency in crops, particularly under Mn-limited soil conditions, which may subsequently impair plant disease resistance and reduce tolerance to abiotic stress. Such functional overlap creates a trade-off between reducing Cd accumulation and maintaining plant nutrition and yield stability, especially under variable field conditions.

Future research will benefit from emerging precision breeding tools such as base editing, prime editing, and epigenome editing, which allow single-nucleotide modifications or transcriptional regulation without inducing DSBs or introducing foreign DNA. These approaches may circumvent regulatory barriers while offering higher specificity and fewer off-target effects. Advances in high-throughput sequencing and multiomics integration (transcriptomics, proteomics, metabolomics, and ionomics) are revealing new candidate genes and regulatory networks involved in Cd uptake, transport, and detoxification. Advancements in synthetic biology and inducible gene expression systems may also enable dynamic, environment-responsive control of Cd transporters, reducing Cd accumulation only under high-exposure scenarios or the growth stage of Cd absorption.

Recent progress in structural biology also offers exciting prospects for precision breeding. By resolving the three-dimensional structures of Cd transporters such as OsNRAMP5, researchers aim to identify the specific amino acid residues involved in the transport of Cd versus essential micronutrients like Mn. This structural insight enables the rational design of amino acid substitutions at critical substrate-binding or gating sites. Using base editing or prime editing technologies, these residues can be precisely modified to disrupt Cd transport while preserving or enhancing the uptake of beneficial elements. Such targeted modifications hold great promise for developing crop varieties that maintain nutrient efficiency without accumulating toxic metals.

19.5 Conclusion

This chapter has comprehensively reviewed both traditional and modern biotechnological strategies for limiting Cd accumulation in crops, with a focus on the underlying genetic and physiological mechanisms, practical breeding approaches, and successful case studies. Traditional breeding methods, including low-Cd variety screening, genetic mapping, MAS, and mutagenesis, have yielded valuable low-Cd germplasms, as exemplified by varieties such as Koshihikari Kan No. 1. Biotechnological innovations, particularly genome editing and gene overexpression, have enabled precise manipulation of key genes such as *OsNRAMP5* and *OsHMA3*, resulting in substantial reductions in grain Cd concentration without compromising yield or quality. The integration of advanced mutation detection technologies, such as M1TDS (Fig. 19.2), and next-generation genome editing tools is revolutionizing the speed and precision of breeding for low-Cd crops. However, the successful deployment of these technologies requires careful consideration of nutrient trade-offs, environmental variability, and regulatory and societal acceptance. Sustained progress will require interdisciplinary efforts integrating molecular biology, soil science, breeding, agronomy, and regulatory policy. Challenges, such as nutrient trade-offs, mutation detection efficiency, and public acceptance, must be addressed through collaborative research and innovation. Overall the integration of traditional and modern approaches holds great promise for delivering safe, high-quality crops suitable for cultivation in Cd-contaminated environments.

19.6 Summary

- Genetic analysis and breeding have identified QTLs and developed low-Cd rice varieties through phenotypic screening and marker-assisted selection.

- CRISPR/Cas9-based genome editing allows precise modification of key genes involved in Cd uptake and transport, such as *OsNRAMP5, OsCCX2*, and *OsZIP2, facilitating the development of rice materials with low grain Cd.*
- Gene overexpression of *OsHMA3* and *OsABCC9* enhances vacuolar sequestration in roots and thus reduces the Cd accumulation in grains but is limited by regulatory and transgenic concerns.
- Targeted mutagenesis such as M1TDS technology enables rapid identification of loss-of-function of *OsNRAMP5* alleles in early mutant populations and speeds up the breeding of low-Cd rice varieties.
- Future innovations, including base/prime editing, structural biology-guided transporter engineering, and inducible gene systems, hold promise for further reducing Cd accumulation without compromising micronutrient uptake.

Acknowledgment

This work was supported by the National Key Research and Development Program of China (2023YFC3708104).

AI disclosure

During the preparation of this work the author(s) used ChatGPT to improve readability. After using this tool/service the author(s) reviewed and edited the content as needed and took full responsibility for the content of the publication.

References

Borges, F., & Martienssen, R. A. (2015). The expanding world of small RNAs in plants. *Nature Reviews Molecular Cell Biology, 16*(12), 727–741. doi:10.1038/nrm4085.

Cao, Z. Z., Lin, X. Y., Yang, Y. J., Guan, M. Y., Xu, P., & Chen, M. X. (2019). Gene identification and transcriptome analysis of low cadmium accumulation rice mutant (lcd1) in response to cadmium stress using MutMap and RNA-seq. *BMC Plant Biology, 19*(1). doi:10.1186/s12870-019-1867-y.

Chang, J. D., Huang, S., Konishi, N., Wang, P., Chen, J., Huang, X. Y., Ma, J. F., & Zhao, F. J. (2020). Overexpression of the manganese/cadmium transporter OsNRAMP5 reduces cadmium accumulation in rice grain. *Journal of Experimental Botany, 71*(18), 5705–5715. doi:10.1093/jxb/eraa287.

Das, P. R., & Sherif, S. M. (2020). Application of exogenous dsRNAs-induced RNAi in agriculture: Challenges and triumphs. *Frontiers in Plant Science, 11*. doi:10.3389/fpls.2020.00946.

Duan, G., Shao, G., Tang, Z., Chen, H., Wang, B., Tang, Z., Yang, Y., Liu, Y., & Zhao, F. J. (2017). Genotypic and environmental variations in grain cadmium and arsenic concentrations among a panel of high yielding rice cultivars. *Rice, 10*(1). doi:10.1186/s12284-017-0149-2.

Gao, C. (2021). Genome engineering for crop improvement and future agriculture. *Cell, 184*(6), 1621–1635. doi:10.1016/j.cell.2021.01.005.

Hao, X., Zeng, M., Wang, J., Zeng, Z., Dai, J., Xie, Z., Yang, Y., Tian, L., Chen, L., & Li, D. (2018). A node-expressed transporter OsCCX2 is involved in grain cadmium accumulation of rice. *Frontiers in Plant Science, 9*. doi:10.3389/fpls.2018.00476.

Ishikawa, S., Ishimaru, Y., Igura, M., Kuramata, M., Abe, T., Senoura, T., Hase, Y., Arao, T., Nishizawa, N. K., & Nakanishi, H. (2012). Ion-beam irradiation, gene identification, and marker-assisted breeding in the development of low-cadmium rice. *Proceedings of the National Academy of Sciences of the United States of America, 109*(47), 19166–19171. doi:10.1073/pnas.1211132109.

Ishikawa, S., Makino, T., Ito, M., Harada, K., Nakada, H., Nishida, I., Nishimura, M., Tokunaga, T., Shirao, K., Yoshizawa, C., Matsuyama, M., Abe, T., & Arao, T. (2016). Low-cadmium rice (*Oryza sativa* L.) cultivar can simultaneously reduce arsenic and cadmium concentrations in rice grains. *Soil Science and Plant Nutrition, 62*(4), 327–339. doi:10.1080/00380768.2016.1144452.

Ishikawa, S., Abe, T., Kuramata, M., & Hayashi, S. (2019). Cadmium toxicity. In S. Himeno, & K. Aoshima (Eds.), *Current topics in environmental health and preventive medicine*. Springer Nature Singapore Pte Ltd.

Kuramata, M., Abe, T., Tanikawa, H., Sugimoto, K., & Ishikawa, S. (2022). A weak allele of OsNRAMP5 confers moderate cadmium uptake while avoiding manganese deficiency in rice. *Journal of Experimental Botany, 73*(18), 6475–6489. doi:10.1093/jxb/erac302.

Lei, G. J., Fujii-Kashino, M., Wu, D. Z., Hisano, H., Saisho, D., Deng, F., Yamaji, N., Sato, K., Zhao, F. J., & Ma, J. F. (2020). Breeding for low cadmium barley by introgression of a Sukkula-like transposable element. *Nature Food, 1*(8), 489–499. doi:10.1038/s43016-020-0130-x.

Li, J. T., Baker, A. J. M., Ye, Z. H., Wang, H. B., & Shu, W. S. (2012). Phytoextraction of Cd-contaminated soils: Current status and future challenges. *Critical Reviews in Environmental Science and Technology, 42*(20), 2113–2152. doi:10.1080/10643389.2011.574105.

Li, L., Mao, D., Sun, L., Wang, R., Tan, L., Zhu, Y., Huang, H., Peng, C., Zhao, Y., Wang, J., Huang, D., & Chen, C. (2022). CF1 reduces grain-cadmium levels in rice (*Oryza sativa*). *Plant Journal, 110*(5), 1305–1318. doi:10.1111/tpj.15736.

Li, M. Z., Hu, D. W., Liu, X. Q., Zhang, R., Liu, H., Tang, Z., Zhao, F. J., & Huang, X. Y. (2024). The OsZIP2 transporter is involved in root-to-shoot translocation and intervascular transfer of cadmium in rice. *Plant Cell and Environment, 47*(10), 3865–3881. doi:10.1111/pce.14993.

Liu, P., Sun, L., Zhang, Y., Tan, Y., Zhu, Y., Peng, C., Wang, J., Yan, H., Mao, D., Liang, G., Liang, G., Li, X., Liang, Y., Wang, F., He, Z., Tang, W., Huang, D., & Chen, C. (2024). The metal tolerance protein OsMTP11 facilitates cadmium sequestration in the vacuoles of leaf vascular cells for restricting its translocation into rice grains. *Molecular Plant, 17*(11), 1733–1752. doi:10.1016/j.molp.2024.09.012.

Liu, X. Q., Liu, H., Fu, M. J., Zhang, L. W., Yin, S. F., Tang, Z., Zhao, F. J., & Huang, X. Y. (2025). The cation/H+ exchanger OsCAX2 is involved in cadmium uptake and contributes to differential grain cadmium accumulation between Indica and Japonica rice. *Journal of Hazardous Materials, 487*. doi:10.1016/j.jhazmat.2025.137252.

Lu, C., Zhang, L., Tang, Z., Huang, X. Y., Ma, J. F., & Zhao, F. J. (2019). Producing cadmium-free indica rice by overexpressing OsHMA3. *Environment International, 126*, 619–626. doi:10.1016/j.envint.2019.03.004.

Luo, J. S., Huang, J., Zeng, D. L., Peng, J. S., Zhang, G. B., Ma, H. L., Guan, Y., Yi, H. Y., Fu, Y. L., Han, B., Lin, H. X., Qian, Q., & Gong, J. M. (2018). A defensin-like protein drives cadmium efflux and allocation in rice. *Nature Communications, 9*(1). doi:10.1038/s41467-018-03088-0.

Ren, Z. W., Yang, M., McKenna, B. A., Lian, X. M., Zhao, F. J., Kopittke, P. M., Lombi, E., & Wang, P. (2023). Fast X-ray fluorescence microscopy provides high-throughput phenotyping of element distribution in seeds. *Plant Physiology, 191*(3), 1520–1534. doi:10.1093/plphys/kiac534.

Shao, J. F., Xia, J., Yamaji, N., Shen, R. F., & Ma, J. F. (2018). Effective reduction of cadmium accumulation in rice grain by expressing OsHMA3 under the control of the OsHMA2 promoter. *Journal of Experimental Botany, 69*(10), 2743–2752. doi:10.1093/jxb/ery107.

Shao, Y., Hu, Y., Peng, Y., Mao, B., Liu, H., Tang, C., Lei, B., Tang, L., Yu, L., Li, W., Luo, W., Luo, Z., Yuan, Y., Li, Y., Zhang, D., Zhou, L., Bai, L., Tang, W., & Zhao, B. (2025). Directed improvement of hybrid rice Zhuoliangyou 1126 by heavy ion beam mutagenesis based on M1TDS targeted screening technology. *Chinese Journal of Rice Science, 39*(5), 624–634. doi:10.16819/j.1001-7216.2025.240106.

Shao, Y., Peng, Y., Mao, B., Yu, L., Tang, L., Li, Y., Hu, Y., Zhang, D., Yuan, Z., Luo, W., Peng, X., Li, W., Zhou, L., Bai, L., & Zhao, B. (2022). M1TDS technology and creation of low-cadmium accumulation parents for hybrid rice breeding. *Hybrid Rice, 37*(1), 1–11. doi:10.16267/j.cnki.1005-3956.20211020.361.

Sikora, P., Chawade, A., Larsson, M., Olsson, J., & Olsson, O. (2011). Mutagenesis as a tool in plant genetics, functional genomics, and breeding. *International Journal of Plant Genomics, 2011*. doi:10.1155/2011/314829.

Sun, L., Tan, Y., & Chen, C. (2023). The road toward Cd-safe rice: From mass selection to markerassisted selection and genetic manipulation. The Crop Journal, 11(4), 1059–1071. doi:10.1016/j.cj.2023.01.007.

Suda, A., Abe, T., & Ishikawa, S. (2021). Amelioration in manganese uptake by a low-cadmium rice cultivar with application of several manganese fertilizers. *Soil Science and Plant Nutrition, 67*(5), 585–593. doi:10.1080/00380768.2021.1955602.

Sui, F., Zhao, D., Zhu, H., Gong, Y., Tang, Z., Huang, X. Y., Zhang, G., & Zhao, F. J. (2019). Map-based cloning of a new total loss-of-function allele of OsHMA3 causes high cadmium accumulation in rice grain. *Journal of Experimental Botany, 70*(10), 2857–2871. doi:10.1093/jxb/erz093.

Takahashi, R., Ishimaru, Y., Shimo, H., Ogo, Y., Senoura, T., Nishizawa, N. K., & Nakanishi, H. (2012). The OsHMA2 transporter is involved in root-to-shoot translocation of Zn and Cd in rice. *Plant, Cell and Environment, 35*(11), 1948–1957. doi:10.1111/j.1365-3040.2012.02527.x.

Tang, L., Mao, B., Li, Y., Lv, Q., Zhang, L., Chen, C., He, H., Wang, W., Zeng, X., Shao, Y., Pan, Y., Hu, Y., Peng, Y., Fu, X., Li, H., Xia, S., & Zhao, B. (2017). Knockout of OsNramp5 using the CRISPR/Cas9 system produces low Cd-accumulating indica rice without compromising yield. *Scientific Reports, 7*(1). doi:10.1038/s41598-017-14832-9.

Ueno, D., Yamaji, N., Kono, I., Huang, C. F., Ando, T., Yano, M., & Ma, J. F. (2010). Gene limiting cadmium accumulation in rice. *Proceedings of the National Academy of Sciences of the United States of America, 107*(38), 16500–16505. doi:10.1073/pnas.1005396107.

Uraguchi, S., Kamiya, T., Sakamoto, T., Kasai, K., Sato, Y., Nagamura, Y., Yoshida, A., Kyozuka, J., Ishikawa, S., & Fujiwara, T. (2011). Low-affinity cation transporter (OsLCT1) regulates cadmium transport into rice grains. *Proceedings of the National Academy of Sciences of the United States of America, 108*(52), 20959–20964. doi:10.1073/pnas.1116531109Japan.

Wu, D., Yamaji, N., Yamane, M., Kashino-Fujii, M., Sato, K., & Ma, J. F. (2016). The HvNramp5 transporter mediates uptake of cadmium and manganese, but not iron. *Plant Physiology, 172*(3), 1899–1910. doi:10.1104/pp.16.01189.

Yan, H., Xu, W., Xie, J., Gao, Y., Wu, L., Sun, L., Feng, L., Chen, X., Zhang, T., Dai, C., Li, T., Lin, X., Zhang, Z., Wang, X., Li, F., Zhu, X., Li, J., Li, Z., Chen, C., … He, Z. (2019). Variation of a major facilitator superfamily gene contributes to differential cadmium accumulation between rice subspecies. *Nature Communications, 10*(1). doi:10.1038/s41467-019-10544-y.

Yang, G., Fu, S., Huang, J., Li, L., Long, Y., Wei, Q., Wang, Z., Chen, Z., & Xia, J. (2021). The tonoplast-localized transporter OsABCC9 is involved in cadmium tolerance and accumulation in rice. *Plant Science, 307*, 110894. doi:10.1016/j.plantsci.2021.110894.

Zhang, L., Gao, C., Chen, C., Zhang, W., Huang, X. Y., & Zhao, F. J. (2020). Overexpression of Rice OsHMA3 in wheat greatly decreases cadmium accumulation in wheat grains. *Environmental Science and Technology, 54*(16), 10100–10108. doi:10.1021/acs.est.0c02877.

Zhao, F. J., Ma, Y., Zhu, Y. G., Tang, Z., & McGrath, S. P. (2015). Soil contamination in China: Current status and mitigation strategies. *Environmental Science and Technology, 49*(2), 750–759. doi:10.1021/es5047099.

CHAPTER

20

Phytoremediation of cadmium-contaminated soils

Longhua Wu, Pengjie Hu, Tong Zhou, Jiawen Zhou, Zhu Li, Peter Christie

State Key Laboratory of Soil and Sustainable Agriculture, Institute of Soil Science, Chinese Academy of Sciences, Nanjing, China

20.1 Introduction

Phytoremediation is a technique that uses plants to remove, transfer, or stabilize contaminants in soils (McGrath et al., 2002). Among the various phytoremediation techniques available, phytoextraction—the use of accumulators or hyperaccumulators to remove potentially toxic metals or metalloids from soils—is considered to be a cost-efficient and environment-friendly method, although it still faces many challenges. This chapter presents case studies on the phytoremediation of cadmium (Cd)–contaminated soils using a Cd and zinc (Zn) hyperaccumulator, *Sedum plumbizincicola*. This plant species was identified in eastern China and has been used to remove Cd from contaminated soils. Factors influencing the Cd phytoremediation efficiency of *S. plumbizincicola* are examined. Field applications at different sites with varying soil properties, pollution levels, and climatic zones are reviewed. Biomass posttreatment techniques following phytoremediation are discussed. Future prospects, including the development of new hyperaccumulator materials, intercropping with cash crops, and cost-efficient environmentally sensitive disposal of biomass, are presented.

20.2 Brief description of *Sedum plumbizincicola*

The success of metal phytoextraction depends on several factors. First, an ideal hyperaccumulator plant would produce a large biomass, grow rapidly, and be easily cultivated. Second, suitable agronomic measures such as optimum moisture conditions, plant nutrition, plant density, and harvest strategies are required. Intercropping with other crop species, for example, might be a feasible method for remediating contaminated soils without interrupting routine agricultural production. Third, the harvested biomass must be safely treated.

Cadmium Contamination in Soils and Plants: Mechanisms and Mitigation.
DOI: https://doi.org/10.1016/B978-0-443-34104-5.00002-5

S. plumbizincicola (Crassulaceae) was first discovered in 2005 in a Zn–lead (Pb) mine area in Zhejiang Province in East China and was identified as a Cd–Zn hyperaccumulator in 2007 (Wu et al., 2006, 2013b). It is a perennial, light green or yellowish-green herb that flowers from early June to August and fruits from July to September. This species superficially resembles *Sedum alfredii* Hance and three other *Sedum* species (*S. emarginatum* Migo, *S. hangzhouense* K.T. Fu and G.Y. Rao and *S. bailey* Praeger) found growing in the same area but differs from these other taxa in bearing tetramerous flowers. The specific epithet "*plumbizincicola*" refers to the distribution of the plant in Pb and Zn mining areas.

This species has a strong ability to hyperaccumulate and tolerate Zn and Cd. The soil in mining areas where it is native is naturally enriched with Cd (36–157 mg kg^{-1}), Zn (1930–7250 mg kg^{-1}), copper (Cu) (530–8340 mg kg^{-1}), and Pb (71–6940 mg kg^{-1}) because of pollution from mine tailings and wastewater processing (Hu et al., 2015). The shoot Cd and Zn concentrations of this species range from 574 to 1470 and 9020 to 14,600 mg kg^{-1}, respectively. Moreover, Cd concentrations in the leaves are approximately twice as high as in the stems (Hu et al., 2015). In the stems, 22.1%, 46.2%, and 31.8% of total Zn and 17.1%, 71.6%, and 11.3% of total Cd were present in the epidermis, cortex, and central cylinder, respectively. In the leaves, 81.3% of total Cd and 55.2% of total Zn were distributed in the mesophyll. The parenchyma cells, for example, the cortex in the stems and the mesophyll in the leaves, play more important roles in Cd storage and detoxification than in those of Zn in *S. plumbizincicola* (Hu et al., 2015). Genes encoding Heavy Metal ATPase 1 (*SpHMA1*) and 3 (*SpHMA3*) have been reported to play important roles in Cd ion transport and detoxification in *S. plumbizincicola* (Liu et al., 2017; Zhao et al., 2019).

20.3 Phytoremediation efficiency, influencing factors, and regulatory measures

20.3.1 Phytoextraction performance in different soils

Metal phytoextraction efficiency varies with soil properties and duration of phytoextraction. It is therefore necessary to investigate the phytoextraction performance of the hyperaccumulator *S. plumbizincicola* across a wide range of soil types during repeated phytoextraction.

20.3.1.1 Phytoextraction in four metal-contaminated soils over successive crops

The dynamic changes in metal uptake by *S. plumbizincicola* during an extended phytoextraction period were investigated by conducting a glasshouse pot experiment in which the seedlings of *S. plumbizincicola* were replanted over seven successive crops during a 3-year period (Li et al., 2014). Four metal-contaminated soils (two acids and two calcareous) were collected from agricultural fields in different parts of China. *S. plumbizincicola* grew well in all four soils. Shoot Zn and Cd concentrations decreased markedly in the two acid soils as phytoextraction proceeded but showed no apparent decline in the two calcareous soils. This indicates that plant metal uptake under low available metal conditions depended on both soil and plant metal processes, including rhizosphere activity. After the seven consecutive crops were harvested, the hyperaccumulating plant had removed 0.40 and 11.6 mg Cd pot^{-1} and 162 and 371 mg Zn pot^{-1} from the two acid soils, which led to decreases in soil Cd and Zn concentrations of 64%–89% and 37%–52%, respectively. The decreases in concentration were smaller in the two calcareous soils (Cd, 37%–38%; Zn, 12%–19%; equal to amounts of metal removed of 3.04 and 0.65 mg Cd pot^{-1}

and 726 and 685 mg Zn pot^{-1}, respectively). Moreover, the decrease in soil available metal concentration assessed by extraction with 0.01 mol L^{-1} $CaCl_2$ was >84% in the acid soils but only 25% (Zn) and 8%–12% (Cd) in the calcareous soils. The metal resupply capacity from the soil solid phase was further evaluated using the DGT-induced fluxes in soils model (Li et al., 2016). Before phytoextraction the calcareous soils showed a faster metal supply from the solid phase than the acid soils. During repeated phytoextraction, large decreases in metal desorption rate constants were found, and different metal pools regulated metal release to the soil solution in the calcareous soils. Sequential extraction indicates that the acid-soluble metal fraction was easily removed by *S. plumbizincicola* from the acid soils but not from the calcareous soils. Reducible and oxidizable metal fractions showed discernible decreases in both acid and calcareous soils, suggesting that *S. plumbizincicola* can mobilize less labile metal fractions, particularly in the later stages of long-term phytoextraction. The results of this pot study suggest that phytoextraction with *S. plumbizincicola* is suitable for the removal of both total and available Cd and Zn from acid soils.

20.3.1.2 Phytoextraction from a wide range of soils

Predicting metal uptake by hyperaccumulators and estimating the duration of phytoextraction from various soil types is very important in the large-scale field application of phytoextraction. *S. plumbizincicola* was grown in 108 agricultural soils collected from different regions of China to assess how soil properties affect phytoextraction efficiency (Wu et al., 2018; Zhou et al., 2019). The soils varied widely in pH (3.4–8.5), soil organic carbon content (5.95–56.5 g kg^{-1}), clay content (7.12%–61.5%), cation exchange capacity [3.67–31.3 cmol (+) kg^{-1}], total Zn concentration (26.6–8803 mg kg^{-1}), and total Cd concentration (0.06–95.4 mg kg^{-1}). The bioconcentration factors for Cd (BCF_{Cd}) and Zn (BCF_{Zn}), defined as the ratio of shoot to soil metal concentrations, ranged from 2.00 to 213 [mean $\pm$ SD (standard deviation): 41.2 $\pm$ 36.1] and from 0.55 to 37.1 (mean $\pm$ SD: 8.31 $\pm$ 7.69), respectively. Soil pH was the most influential factor for the variation in BCF_{Cd} and BCF_{Zn} among the 108 soils, explaining 45.6% and 30.1% of the variation, respectively (Fig. 20.1). BCF_{Cd} and BCF_{Zn} increased significantly with decreasing pH from 7.5 to 5.5 and changed little at pH < 5.5 or pH > 7.5 (Fig. 20.2), suggesting an optimum pH of around 5.5 for phytoextraction. In addition, the soil total manganese (Mn) concentration was the second major factor influencing BCF_{Cd}, accounting for 14.1% of the variation in BCF_{Cd}, whereas soil texture played an important role in affecting BCF_{Zn}, explaining 38.2% of the variation. *S. plumbizincicola* had larger shoot Zn concentrations growing in loamy or clay loam soils (clay and sand contents of about 20%–25% and 30%–50%, respectively) than in clay or sandy soils. The empirical piecewise equations in different ranges of soil pH were further derived to predict metal concentrations in the shoots, and this significantly improved the prediction performance (Fig. 20.3). Estimation of the potential of *S. plumbizincicola* for metal removal in the field using the piecewise function indicates that it is feasible to decontaminate polluted acid soils (pH < 5.5) with total Cd concentrations of $\leq$5 mg kg^{-1} and total Zn concentrations of $\leq$330 mg kg^{-1} to below the risk screening level of soil environmental quality standard of China [GB15618-2018: Cd 0.30 mg kg^{-1}, Zn 200 mg kg^{-1} for soils with pH < 5.5] using *S. plumbizincicola* with nine cycles of consecutive cropping.

The empirical regression equation based on the linear relationship between metal concentrations in plants and soils may not be applicable to a wide range of soils because of multiple factors affecting metal uptake. A mechanical model based on the multisurface model (MSM) framework was therefore developed to predict Cd and Zn uptake by *S. plumbizincicola*, which

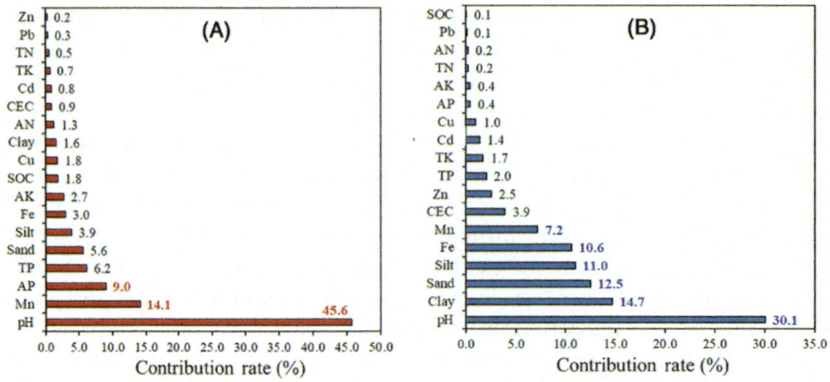

FIGURE 20.1 Contributions of soil properties to (A) BCF_{Cd} and (B) BCF_{Zn}. *AK*, Available potassium content; *AN*, available nitrogen content; *AP*, available phosphorus content; *CEC*, soil cation exchange capacity; *Silt, Sand, Clay*: soil silt, sand, clay relative contents; *SOC*, soil organic carbon content; *TK*, total potassium content; *TN*, total nitrogen content; *TP*, total phosphorus content; *Zn, Pb, Cd, Cu, Fe, Mn*: soil total Zn, Pb, Cd, Cu, Fe, and Mn concentrations. *Modified from Wu et al. (2018) and Zhou et al. (2019) with permission.*

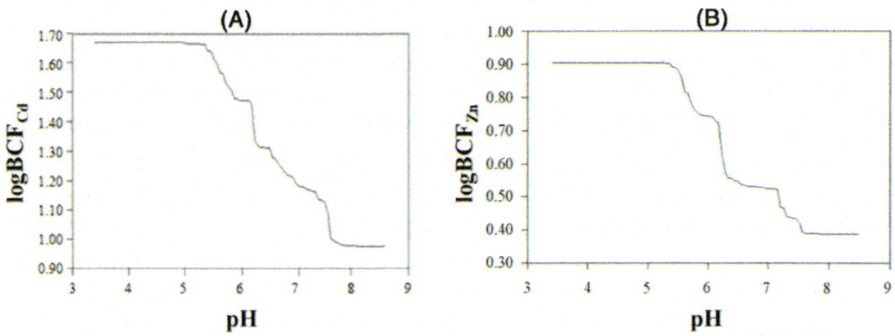

FIGURE 20.2 Partial dependency plots of (A) BCF_{Cd} and soil pH and (B) BCF_{Zn} and soil pH. *Modified from Wu et al. (2018) and Zhou et al. (2019) with permission.*

comprehensively considers the effects of enhanced root exudation and rhizosphere acidification of hyperaccumulators relative to nonhyperaccumulators (Li et al., 2024). In the MSMs, soil is regarded as a system with different active surfaces for metal adsorption comprising soil organic matter, dissolved organic matter, clay minerals, crystalline ferric oxides, and amorphous hydrous ferric oxides, and only the dissolved metal fraction is thought to be available to plants. The results show that the dissolved Cd/Zn predicted by MSMs were strongly correlated with shoot metal uptake (determination coefficients, $R^2 = 0.825$ and 0.802 for Cd and Zn, respectively, $n = 95$) and outperformed chemical extraction methods. In addition, MSMs further integrated with phytoextraction time and Cd and Zn interactions to form dynamic-MSM combined (D-MSM-C) models that predicted well metal concentrations remaining in soils during repeated phytoextraction over six consecutive crops with mean absolute percentage errors of 20.4% (Cd) and 2.46% (Zn) ($n = 66$). MSM is a useful tool for predicting phytoextraction efficiency and duration across diverse soil properties.

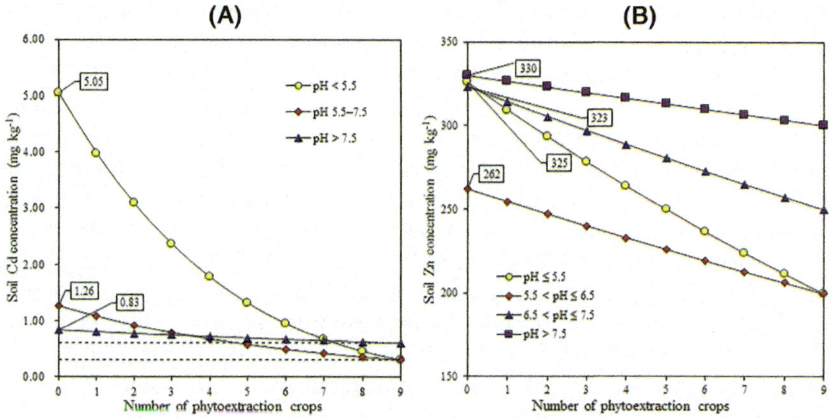

FIGURE 20.3 Soil (A) Cd and (B) Zn concentrations predicted with the piecewise equations derived at different pH ranges after successive crops of *Sedum plumbizincicola. Cd*, cadmium; *Zn*, zinc. *Modified from Wu et al. (2018) and Zhou et al. (2019) with permission.*

20.3.1.3 *Phytoextraction of soils with high geological background levels of cadmium*

Soils in some areas of southwest China are often naturally enriched with Cd. Crops grown on these soils can still accumulate Cd to levels exceeding the Chinese food safety standard (GB 2762-2022). The feasibility of cleaning up high background Cd soils with *S. plumbizincicola* was investigated in a pot experiment over three cropping cycles (Wu et al., 2024). Three soils were collected from Yunnan Province with total Cd concentrations ranging from 0.72 to 0.83 mg kg^{-1} and soil pH values from 5.60 to 6.72. The average BCF_{Cd} and BCF_{Zn} were 41.7 ± 22.8 and 3.71 ± 1.51, respectively. BCF_{Cd} values were comparable to those found in 108 agricultural soils that were contaminated mainly from anthropogenic sources, whereas BCF_{Zn} was smaller. After three cropping cycles of *S. plumbizincicola*, total soil Cd and Zn concentrations decreased by 24.8%–58.8% and 3.7%–26.3%, respectively. Average Cd removal efficiency (21.8%) was slightly lower than that in the anthropogenically polluted soils with similar total Cd levels (32.5%), reflecting a lower Cd bioavailability in soils with geologically high Cd backgrounds. The European Community Bureau of Reference (BCR) sequential extraction method shows that metals in the soils with high geological background values before phytoextraction occurred mainly in the residual fraction (Cd, 29%–55%; Zn, 87%–90%). Successive phytoextraction largely decreased the percentage of acid-soluble and reducible Cd, whereas Zn chemical fractions changed only slightly.

20.3.2 Copper and aluminum toxicities inhibit the phytoextraction potential of *Sedum plumbizincicola*

S. plumbizincicola is highly tolerant to Cd and Zn but shows limited tolerance to other metals such as Cu and aluminum (Al). In soils with high available concentrations of Cu or Al the toxicities of these metals may decrease Cd and Zn phytoextraction. In hydroponic experiments, 5–50 μM Cu showed no significant effect on plant Cd and Zn accumulation, but Cu at 200 μM induced root cell plasmolysis and chloroplast structural disorder (Li et al., 2013). The plants primarily detoxified Cu by retaining it in the roots and cell walls and by complexing it in insoluble forms. Four days of exposure to 200 μM Cu inhibited Cd uptake and translocation but did not affect Zn concentrations in roots and stems. Translocation of Cd and Zn from stems to leaves was inhibited compared to the control (0.31 μM Cu). Cu toxicity suppressed plant growth and interfered with Cd uptake and Cd/Zn translocation to leaves in *S. plumbizincicola*.

In soils with pH < 5.0, Al dissolution from aluminosilicate minerals may cause Al toxicity and inhibit plant growth. Excessive Al in growth media interferes with various physiological and cellular processes, including nutrient uptake, enzyme activities, DNA replication, and cell division (Singh et al., 2017). Both long-term (7 weeks) and short-term (72 hours) hydroponic experiments, together with a soil-based pot experiment, were conducted to investigate Al toxicity in *S. plumbizincicola* (Zhou et al., 2020). In the long-term hydroponic experiments, 100 μM Al at pH 5.00 (soluble Al activity 8.74 μM predicted by Visual MINTEQ) decreased shoot Cd accumulation by 39.3% compared to Al-free controls. A scanning ion-selective electrode technique shows significant inhibition of root Cd^{2+} influx after 48 hours of exposure to 200 μM Al at pH 4.0. The pot experiment confirmed that Al toxicity inhibited plant growth and metal uptake in acid soil containing 0.33 cmol (+) kg^{-1} of exchangeable Al. Decreased root surface Cd adsorption under Al stress may contribute significantly to decreased shoot Cd accumulation at low soil pH. Aluminum toxicity also impaired Cd and Zn translocation. Al toxicity commonly occurs in highly acid soils or those undergoing pH decline during phytoextraction, necessitating mitigation strategies such as lime or biochar application.

20.3.3 Agronomic measures to increase cadmium and zinc phytoextraction

20.3.3.1 Cultivation measures to increase phytoextraction efficiency

S. plumbizincicola is well-suited to clonal propagation for which shoot cuttings are typically selected. Tissue culture methods have been established for mass propagation. For field applications, optimum plant density, soil moisture, light conditions, and harvest strategies have been established to maximize biomass production and Cd/Zn uptake while minimizing remediation time. A field plot experiment investigated planting density effects on *S. plumbizincicola* growth and metal uptake (Liu et al., 2009). The optimum planting density was 4.4×10^5 plants ha^{-1} (15 cm × 15 cm spacing), yielding maximum biomass and Cd/Zn removal. Maintaining soil moisture at around 70% of soil water-holding capacity optimizes plant growth and metal accumulation (Cui et al., 2009). *S. plumbizincicola* is a perennial species, and this permits 2–3 annual harvests. Harvesting should retain 3–5 cm stubble to facilitate shoot regrowth (Li et al., 2009).

S. plumbizincicola is shade-tolerant and does not require high light intensity for growth. Metal accumulation remained stable at 50% natural light intensity (Li et al., 2010). This shade tolerance enables intercropping with tall crops such as maize, sorghum, sugarcane, and wheat. An 8-year-long intercropping experiment with maize decreased soil Cd from 3.50 to 0.54 mg kg^{-1} and Zn from 1799 to 1465 mg kg^{-1} (Deng et al., 2016). However, *S. plumbizincicola* root activity may mobilize soil Cd, potentially increasing uptake by the intercropped species. For example, Zhao et al. (2011) report that intercropping increased $NaNO_3$-extractable Zn/Cd in soil and wheat uptake compared to monoculture. Shen et al. (2010) also observed elevated Zn/Cd concentrations in rice following *S. plumbizincicola* cultivation compared to the control without the hyperaccumulator. The effects of *S. plumbizincicola* root activity should therefore be considered during field application.

20.3.3.2 Chemical measures to increase phytoextraction efficiency

In calcareous or neutral soils, application of elemental sulfur (S) can acidify soils to enhance Cd and Zn availability for phytoextraction. A pot experiment investigated S application rates combined with wet–dry cycles in an *S. plumbizincicola*-rice rotation system in Cd-contaminated neutral and calcareous soils (Wu et al., 2019). Results demonstrate that S oxidation under aerobic conditions significantly lowered soil solution pH while increasing soluble SO_4^{2-}, Cd, and Fe concentrations in both soils. Compared to controls, S addition increased shoot Cd concentrations in *S. plumbizincicola* 1.7–5.5 times in calcareous soil and 1.7–2.3 times in neutral soil. During rice cultivation the soil solution redox (E_h) fell below −200 mV, and the solution pH normalized to neutral within the initial days of flooding. Soluble SO_4^{2-} and Cd declined significantly with prolonged flooding. Moreover, rice yields increased, but Cd concentrations decreased at S addition rates of 0.5 to 2.0 g kg^{-1} (equal to 0.975–3.9 t ha^{-1}).

S. plumbizincicola thrives in soils with pH 5.5–7.5. In strongly acid soils, lime application promotes both plant growth and metal uptake (Han et al., 2013). Organic amendments with diverse functional groups enhance metal bioavailability. Rice straw or ground clover amendments increased shoot Cd concentrations 1.92 and 1.71 times compared to the control (Wu et al., 2012a).

20.3.3.3 Microbial measures for increasing phytoextraction efficiency

Soil microorganisms play important roles in hyperaccumulator-based remediation, as microbial communities significantly influence plant growth and metal uptake. Monitoring of *S. plumbizincicola* growth and Cd uptake revealed that the rhizosphere microbial community stabilized during peak growth stages. The stable bacterial community and enriched biomarker taxa (*Paenibacillus* and *Streptacidiphilus*) were positively correlated with Cd uptake (Hou et al., 2021). Both microbial inoculation and soil sterilization influenced hyperaccumulator remediation efficiency. Inoculation with plant growth-promoting *Rhodococcus* sp. NSX2 increased its rhizosphere abundance and enhanced cooperation with indigenous microorganisms, thereby increasing remediation efficiency (Hou et al., 2019). Conversely, vancomycin amendment altered soil microbial composition, decreasing Cd (38%–46%) and Zn (35%–53%) uptake by *S. plumbizincicola* and was correlated with changes in bacterial gene function (upregulated biofilm formation and downregulated siderophore synthesis) (Hou et al., 2017). Soil fauna also influence phytoextraction. The common soil collembolan *Folsomia candida* enhanced *S. plumbizincicola* shoot Cd concentrations by 27.9%–66.3% across a range of metal-contaminated soils (Li et al., 2021). This enhancement resulted primarily from collembolan-root interactions.

20.4 Field applications of phytoextraction of cadmium-contaminated soils

S. plumbizincicola has been employed for field phytoextraction of metal-contaminated soils across diverse climatic, edaphic, and topographic conditions. Generally, *S. plumbizincicola* demonstrates significantly higher Cd removal efficiency in slightly to moderately polluted acid soils than in neutral or calcareous soils.

FIGURE 20.4 Demonstration of Cd phytoextraction with *Sedum plumbizincicola* in a Cd-polluted acid soil at Xiangtan, Hunan Province, Southern China. *Cd*, cadmium. *From Wu et al. (2021).*

20.4.1 Phytoextraction of cadmium from slightly to moderately contaminated soils

20.4.1.1 Acid soils

A 1-ha field experiment in Xiangtan County, Hunan Province, Southern China, evaluated Cd removal efficiency from a contaminated acid paddy soil (pH 5.0, Cd 0.6 mg kg^{-1}) through repeated *S. plumbizincicola* phytoextraction from October to June annually (Fig. 20.4) (Hu et al., 2019; Wu et al., 2021). *S. plumbizincicola* exhibited robust growth, producing 1.8–5.9 t shoot dry biomass ha^{-1} containing a concentration of 51.4–94.9 mg Cd kg^{-1} and accumulating 169–353 g Cd ha^{-1}. Cd removal rates from the rooting zone (top 15 cm of the soil profile) ranged from 13.8% to 34.7% (mean 21.8%) in different seasons. Total soil Cd content decreased from 0.64 to 0.29 mg kg^{-1} after two phytoextraction seasons. Declining Cd occurred primarily at 2–10 cm depth, corresponding to the root distribution of *S. plumbizincicola*. Subsequent Cd risk assessment in five rice cultivars shows that grain Cd concentrations decreased by 37%–67% in phytoextracted versus nonphytoextracted soils. Low-accumulating cultivars MY12085 and MY12086 showed brown rice Cd < 0.2 mg kg^{-1} (meeting Chinese standards), whereas high-accumulating cultivars required additional immobilization. This demonstrates that *S. plumbizincicola* can remediate slightly contaminated acid soils to regulatory standards within 2–5 years. Strategic cultivar selection and soil amendments can further ensure the safety of agricultural produce.

20.4.1.2 Slightly acid soils

A field phytoremediation experiment using *S. plumbizincicola* was conducted in Cd-contaminated paddy soils in the south of Jiangsu Province, East China (Fig. 20.5) (Hu et al., 2024). The soil pH ranged from 6.0 to 6.5 with a total Cd concentration averaging 1 mg kg^{-1}. *S. plumbizincicola* was cultivated annually from November to June between 2016 and 2020. Elemental S was applied (0.75–1.5 t ha^{-1}) to acidify the soil, increase Cd availability, and increase the Cd removal rate by *S. plumbizincicola*. Shoot dry biomass ranged from 2.25 to 4.62 t ha^{-1} and shoot Cd concentrations ranged from 72.1 to 139 mg kg^{-1}. The cumulative Cd removal by the plants was 1256 g ha^{-1} in plot 1 over four seasons of remediation and 1053 g ha^{-1} in plot 2 over three seasons of remediation. The soil total Cd concentrations determined in plots 1 and 2 decreased from 1.16 to 0.52 mg kg^{-1} and 0.93 to 0.34 mg kg^{-1}, respectively, with a cumulative decrease of 55.2%–63.4% (average decline 59.3%). The Cd concentration in the remediated soil

FIGURE 20.5 Demonstration of Cd phytoextraction with *Sedum plumbizincicola* in a Cd-polluted slightly acid soil in the south of Jiangsu Province, East China. *Cd*, cadmium.

TABLE 20.1 Shoot biomass, Zn and Cd concentrations, and Zn and Cd uptakes of *Sedum plumbizincicola* over three successive crops in a field experiment on a calcareous soil.

		Concentration (mg kg^{-1})		Uptake (g ha^{-1})	
Crop	Shoot biomass (t ha^{-1})	Cd	Zn	Cd	Zn
2012–13	2.50 ± 0.72b	101 ± 9b	1007 ± 86a	252 ± 77c	2523 ± 762b
2013–14	3.84 ± 0.40a	153 ± 15a	1019 ± 146a	582 ± 20a	3959 ± 940a
2014–15	4.03 ± 0.78a	123 ± 23ab	978 ± 266a	496 ± 18b	3941 ± 207ab

Data from Zhou et al. (2018).

Different lowercase characters in the same column indicate significant differences among the three successive crops by one-way analysis of variance at $P < .05$. *Cd*, cadmium; *Zn*, zinc.

reached or fell below the Chinese agricultural land risk screening values, significantly mitigating the pollution risks.

20.4.1.3 Calcareous soils

A field experiment was conducted in the northwest of Henan Province, Southern China, in a Cd-contaminated calcareous soil (pH 7.91) near a metal smelter (Zhou et al., 2018). The soil contained 3.55 mg total Cd kg^{-1} and 108 mg total Zn kg^{-1}. From October 2012 to July 2015, three successive crops of *S. plumbizincicola* were cultivated, and the results are shown in Table 20.1. After three successive phytoextraction crops the total uptakes of Cd and Zn by *S. plumbizincicola* were 1.33 ± 0.17 and 10.4 ± 0.8 kg ha^{-1}, representing 14.4% and 4.64%, respectively, of the total Cd and Zn at 0–20 cm depth in the soil. The repeated phytoextraction did not significantly alter the percentage distribution of Cd and Zn among the soil fractions (acid-soluble, reducible, oxidizable, and residual). The experiment also showed that the enhancement of soil fertility and structure resulting from organic fertilizer application significantly increased the shoot biomass and Cd and Zn uptake by *S. plumbizincicola*.

20.4.2 Phytoextraction of cadmium from highly polluted soils near a mining area

An 8-ha field experiment was conducted in Yunnan Province, Southwest China (Wang et al., 2023), at a site highly contaminated with Cd from long-term mining activities at the Jinding Pb–Zn mine. The soil pH ranged from 4.82 to 6.86 (mean 5.61) and total Cd concentrations from 1.50 to 5.45 mg kg^{-1} (mean 2.58 mg kg^{-1}). Phytoextraction using *S. plumbizincicola* was conducted for two growing seasons at this polluted field from 2018 to 2019. Across the field, 144 paired soil-plant samples were collected at each harvest. *S. plumbizincicola* shoot dry biomass production was 1.95 ± 1.02 t ha^{-1} in the first crop and decreased by 53.3% to 0.91 ± 0.83 t ha^{-1} in the second crop (Table 20.2). Shoot Cd concentrations ranged from 59.1 to 373 mg kg^{-1} with an average of 170 ± 49 mg kg^{-1}, and the amounts of Cd taken up ranged from 61.3 g to 1390 g ha^{-1} with an average of 339 ± 22 g ha^{-1} in the first crop. In the second crop the shoot Cd concentration and uptake were 71.9–386 mg kg^{-1} and 8.40–840 g ha^{-1}, with mean values of 172 ± 57 mg kg^{-1} and 142 ± 121 g ha^{-1}, respectively. The average total amount of shoot Cd taken up by the two crops was 478 ± 259 g ha^{-1}.

Repeated phytoextraction did not change the soil pH significantly after the first (5.67 ± 0.27) or second (5.66 ± 0.38) crops (Table 20.3). The soil 0.01 mol L^{-1} $CaCl_2$-extractable Cd concentration was 0.22 ± 0.12 mg kg^{-1} before phytoextraction. Repeated phytoextraction significantly decreased soil $CaCl_2$-extractable Cd concentrations by 44.2% and 39.3% to 0.12 ± 0.06 and 0.14 ± 0.07 mg kg^{-1}, respectively, after the first and second crops. However, no significant difference in soil extractable Cd concentration was observed between the two phytoextraction seasons. Average total Cd concentrations in the soil decreased from 2.58 ± 0.69 mg kg^{-1} before phytoextraction to 1.82 ± 0.45 and 1.53 ± 0.43 mg kg^{-1} after the first and second seasons of phytoextraction, respectively. After two seasons of phytoextraction the Cd removal rate in the polluted soil (0–15 cm depth) calculated by plant Cd uptake was 11.8% ± 6.8%, a smaller value than the measured decrease in soil Cd concentration of 39.6% ± 13.8%. This discrepancy may be because of leaching and surface runoff, the cropping pattern, and the sampling error.

According to the Chinese Soil Environmental Quality Risk Control Standard for Agricultural Land (GB 15618-2018), soil Cd risk control values vary with soil pH as follows: 1.5 mg kg^{-1} ($pH \leq 5.5$), 2.0 mg kg^{-1} ($5.5 < pH \leq 6.5$), and 3.0 mg kg^{-1} ($6.5 < pH \leq 7.5$). The risk screening value is 0.3 mg kg^{-1} at $pH \leq 7.5$ in upland fields. As shown in Fig. 20.6, initial soil Cd concentrations at all 144 sampling points exceeded the risk screening value >5 times, with 91.7% surpassing the risk control value, classifying the site in the "strict management and control" category, thereby prohibiting food crop cultivation. After two seasons of phytoextraction with *S. plumbizincicola*, soil pH values ranged from 4.67 to 6.98, and soil total Cd concentrations decreased to 0.65–3.60 mg kg^{-1}, reducing the exceedance of the risk control values to 20.9% of samples. Consequently, 70.8% of the area transitioned to the "safe utilization" category, permitting agricultural production when implementing appropriate Cd-limiting practices.

In a field experiment, six low-Cd-accumulation maize varieties were also intercropped with *S. plumbizincicola*. The Cd concentrations in maize grains were 0.013–0.026 mg kg^{-1} in 2018 and 0.023–0.058 mg kg^{-1} in 2019, lower values than the maximum allowable Cd concentration

TABLE 20.2 Descriptive statistics of the shoot biomass production (dry $biomass_{Shoot}$), Cd concentration (Cd_{Shoot}) and Cd uptake (Cd_{Uptake}) of *Sedum plumbizincicola.*

	Dry $biomass_{Shoot}$ ($t\ ha^{-1}$)			Cd_{Shoot} ($mg\ kg^{-1}$)		Cd_{Uptake} ($g\ ha^{-1}$)		
Variable	**First**	**Second**	**Total**	**First**	**Second**	**First**	**Second**	**Total**
Numbers	144	141	144	144	141	144	141	144
Max	5.77	4.34	8.77	373	386	1390	840	1522
Min	0.40	0.048	0.85	59.1	71.9	61.3	8.40	78.8
Mean	1.95	0.91	2.85	170	172	339	142	478
SD	1.02	0.83	1.43	48.6	56.5	222	121	259

Data from Wang et al. (2023).

TABLE 20.3 Changes in pH and extractable and total Cd contents of contaminated soils before and after phytoextraction with *Sedum plumbizincicola* in a highly contaminated soil.

		Cd concentrations (mg kg^{-1})	
Season	**pH**	**Extractable**	**Total**
Before	5.61 ± 0.35a	0.22 ± 0.12a	2.58 ± 0.69a
First	5.67 ± 0.27a	0.12 ± 0.06b	1.82 ± 0.45b
Second	5.66 ± 0.38a	0.14 ± 0.07b	1.53 ± 0.43c

Data from Wang et al. (2023).
Different lowercase letters in the same column indicate significant differences by one-way analysis of variance at $P < .05$.

(0.1 mg kg^{-1}) recommended by the Chinese National Food Safety Standard (GB 2762-2022). These results indicate that intercropping *S. plumbizincicola* with low-Cd accumulation maize varieties is a feasible strategy for Cd-polluted soils.

20.4.3 Integration of phytoextraction and photovoltaic solar energy production on highly contaminated soils

Phytoextraction using hyperaccumulator plant species is relatively cost-effective and environmentally friendly. However, maintaining economic outputs during the remediation process remains challenging. Combining photovoltaic (PV) solar energy production systems with soil remediation in polluted arable areas represents a new land-use option. Renewable electricity generation can sustain the economic output of the land while phytoextraction restores its long-term agricultural value and provides environmental benefits by reducing CO_2 emissions and removing soil pollutants.

A 3-year continuous field experiment was conducted from 2018 to 2020 at a PV power station in Renhua County, Guangdong Province, Southern China (Li et al., 2023). The agricultural soil at this site was contaminated with Cd, Pb, and Zn from a Zn–Pb mine. The construction of the power station began in November 2016 and was completed in December 2017. At the initiation of the study in early 2018 the surface soil (0–15 cm depth) had a pH of 6.13 (H_2O extract, 1:2.5 w/v) and a bulk density of 1.10 g cm^{-3} and contained 28.2 g organic matter kg^{-1}. The soil was highly polluted with Cd (5.95 mg kg^{-1}) and Zn (1996 mg kg^{-1}). Because of meteorological heterogeneity beneath the solar panels, three beds were treated as distinct experimental treatments (Fig. 20.7).

Shoot dry biomass of *S. plumbizincicola* did not differ significantly among the three beds during the 2018 and 2020 seasons, as this species is adapted to partial shading. Beds 1 and 3, which received less solar radiation, exhibited greater plant height, single-leaf area, and chlorophyll content than the other two beds. These observations suggest a typical shade-avoidance response of *S. plumbizincicola* under the PV panels, similar to that of most higher plant species. At harvest in 2018 the average shoot concentrations of Cd and Zn were up to 509 and 13,200 mg kg^{-1}, respectively. No significant differences in Cd or Zn concentrations were observed among beds except in 2019 when beds 1 and 3 had significantly lower concentrations than bed 2. Cd uptake per unit area did not vary significantly across beds in any season. In 2019, Zn uptake was higher in beds 1 and 3 (higher biomass) than in bed 2, but Zn uptake declined in 2020 compared to previous seasons. PV panel shading reduced annual solar irradiance by 58.4%, yet this appeared to have no inhibitory effect on *S. plumbizincicola* growth or metal uptake. Over

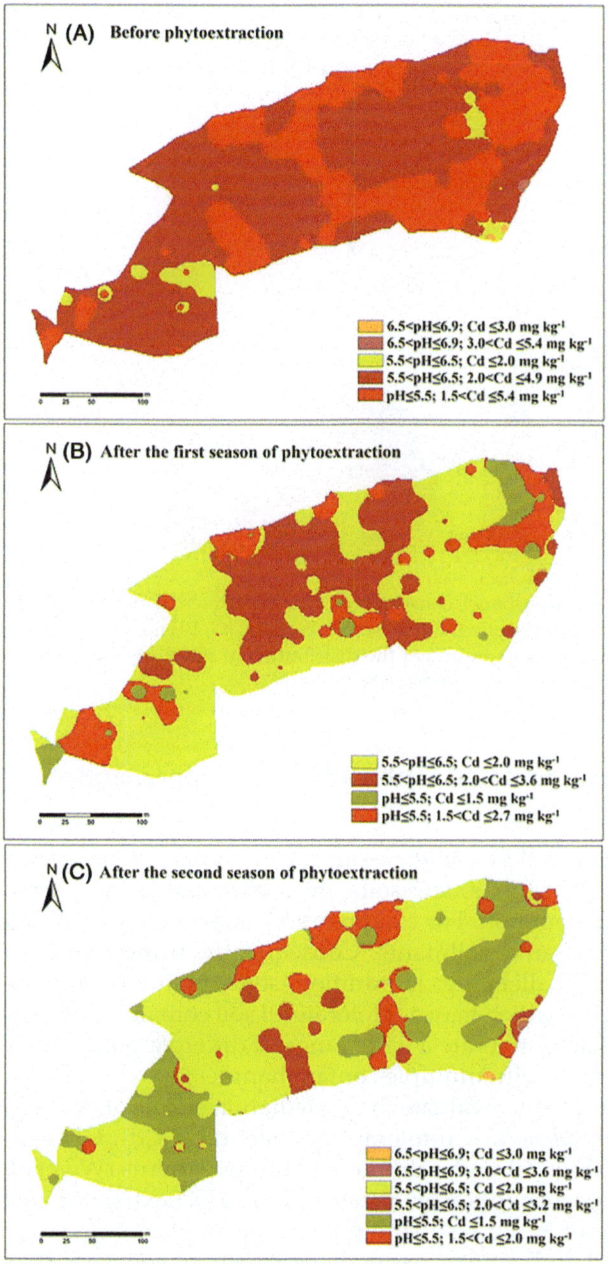

FIGURE 20.6 Distribution of soil cadmium concentrations (A) before and (B) after the first and (C) after the second seasons of phytoextraction. Different colors on the map refer to the different classifications of Cd pollution risk according to Chinese standard GB 15618-2018. *Cd*, cadmium. *Modified from Wang et al. (2023) with permission.*

three seasons, phytoextraction under PV panels achieved high removal efficiencies, with Cd and Zn uptake rates of 3.40 and 96.2 kg ha^{-1}, respectively. This study demonstrates that combining soil phytoextraction with PV electricity generation can simultaneously achieve soil remediation and energy production.

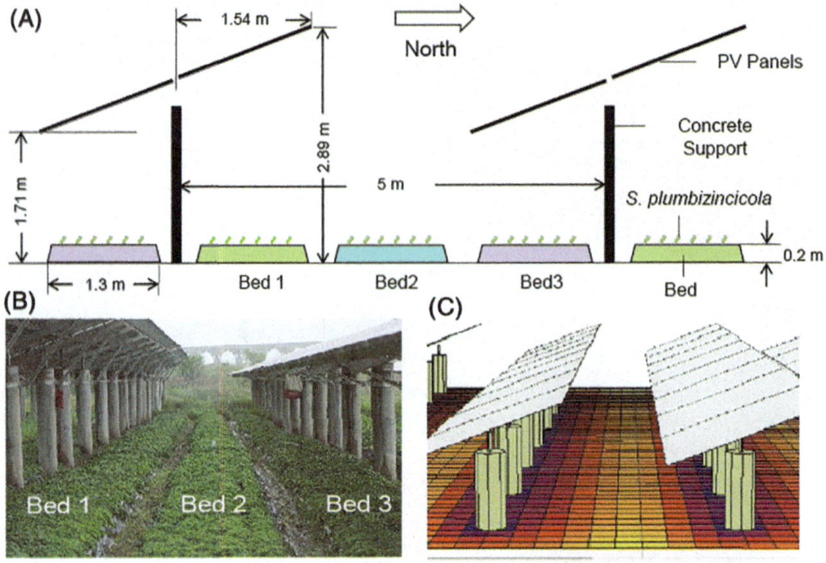

FIGURE 20.7 Schematic diagram of photovoltaic (PV) panel construction and experimental design. (A) PV modules and planting pattern; (B) picture of the experimental site (with *S. plumbizincicola*); (C) PV module modeling and radiation simulation by means of Ecotect software; colors represent the distribution of main annual radiation, with yellow to blue representing high to low. *Modified from Li et al. (2023) with permission.*

20.4.4 Phytoextraction of soils cocontaminated with cadmium, copper, and polychlorinated biphenyls

One major drawback of phytoremediation is that a single plant species typically remediates only one or two pollutants. In practical applications, few soils are contaminated by a single pollutant, and multiple contaminations are common. For example, soils may contain multiple potentially toxic elements and persistent organic pollutants. Consequently, using one plant species to remove all harmful pollutants is challenging. In addition, some pollutants may be toxic to the selected remediation plant species, which may not tolerate all soil contaminants. Use of multiple plant species that can accumulate or tolerate a wide range of different pollutants is therefore necessary for the remediation of soils with multiple contaminants.

A field microcosm experiment was conducted to evaluate the remediation effects of *Medicago sativa*, *S. plumbizincicola*, and *Elsholtzia splendens* (a Cu-tolerant species) on a Cd-, Cu-, and polychlorinated biphenyl (PCB)-contaminated soil under different planting patterns (Wu et al., 2012b). The field soil exhibited acidic properties (pH 4.56, measured via H_2O extract at 1:2.5 w/v ratio), with an organic carbon content of 36.5 g kg^{-1}, and total nitrogen, phosphorus, and potassium concentrations of 2.10, 0.60, and 19.7 g kg^{-1}, respectively. The total concentrations of Zn, Cd, and Cu were 369 $\pm$ 1, 7.67 $\pm$ 0.51, and 179 $\pm$ 5 mg kg^{-1}, respectively, and the average PCB concentration was 323 $\pm$ 12 μg kg^{-1}, primarily consisting of trichlorinated (41.8%) and tetrachlorinated (44.6%) congeners. After 6 months, intercropping of *S. plumbizincicola* with *E. splendens* achieved the highest Cd uptake (18.5 $\pm$ 1.8 mg per microcosm), followed by *S. plumbizincicola* intercropped with *M. sativa* (15.9 $\pm$ 1.9 mg per microcosm). In addition,

S. plumbizincicola intercropped with *M. sativa* significantly enhanced PCB degradation. Thus, under limed soil conditions (pH 4.56), the *S. plumbizincicola* and *M. sativa* intercropping system effectively removed Cd, Cu, and PCBs from the soil.

20.5 Disposal of plant biomass after phytoremediation

20.5.1 Dewatering and drying of fresh biomass and the treatment of waste liquor

Fresh *S. plumbizincicola* biomass typically has a high-water content (>90%) after harvest, making solar drying a lengthy process requiring substantial space. Efficient dehydration and rapid drying are necessary before thermal disposal (e.g., incineration). A rapid disposal technology for fresh *S. plumbizincicola* biomass was therefore developed (Wang et al., 2017).

Fresh biomass was cut and compressed to separate residue and waste liquor, significantly reducing the drying time. Field tests show that compressing lowered the drying time from 15 to 2–6 days by rupturing plant tissue structures, particularly the thick leaf cuticle. In addition, increasing the squeezing rate (percentage of extracted tissue water) progressively reduced the volume of residue. For example, at 60% water extraction, the volume decreased by over 70%, substantially saving drying space in practice. However, the resulting Cd-containing waste liquor required treatment before disposal. Higher squeezing rates increased Cd concentrations in the liquor from 0.60 mg L^{-1} at 20% to 2.00 mg L^{-1} at 80% (Wang et al., 2017). Thus the optimum dehydration rate must balance the drying efficiency and waste treatment costs.

Several cost-effective treatment techniques were developed to efficiently remove Cd from the waste liquor, including flocculation-precipitation and an electro-Fenton (EF) process. The effects of different flocculants, pH values, and metal capture agents on Cd removal efficiency were investigated to determine the optimum flocculation–precipitation method. Addition of 0.3% polyaluminum chloride, 0.1% NaOH, and 0.05% trimercaptotriazine reduced the Cd concentration in the effluent from 1.98 to 0.066 mg L^{-1}, a value below the 0.1 mg L^{-1} limit set by the Chinese National Wastewater Comprehensive Discharge Standard (GB 8978-1996; Wang et al., 2017).

An EF process was also used to remove Cd and chemical oxygen demand (COD) from wastewater generated during biomass dewatering of the hyperaccumulator *S. plumbizincicola*, following flocculation–precipitation treatment (Hu et al., 2021). The relative influence of operational parameters on COD and Cd removal efficiency followed the sequence: pH > electrical current density > H_2O_2 dosage. This increased the pH-enhanced Cd removal but negatively affected COD removal. Both Cd and COD removal rates exhibited an initial increase followed by a decrease with increasing current density and H_2O_2 dosage. Compared with conventional electrocoagulation, H_2O_2 addition in the EF process significantly increased Cd and Zn removal efficiencies. Speciation analysis shows that the majority of Cd and Zn in the initial wastewater occurred as complexes with organic and/or inorganic ligands. Under optimized conditions (pH 5, current density 15 mA cm^{-2}, and H_2O_2 dosage 9 g L^{-1}), removal efficiencies after 80 min of EF treatment were up to 99.4% (Cd), 99.9% (Zn), and 55.5% (COD). These findings demonstrate that the EF process represents an effective method for rapid trace metal removal from hyperaccumulator-derived waste liquors.

20.5.2 Incineration and pyrolysis of *Sedum plumbizincicola* dry biomass

Dry plant biomass collected from various contaminated sites was subjected to various incineration processes. Thermogravimetric analysis shows that the mass percentage of bottom ash relative to the original biomass decreased with increasing temperature in a horizontal tube furnace. Weight losses of *S. plumbizincicola* were 6.17%–10.8%, 49.0%–50.4%, and 82.7%–89.4% at three temperature steps of <185°C, 185°C–400°C, and 400°C–697°C, respectively (Zhong et al., 2015).

The distribution of metals in flue gas constitutes a critical consideration for gas cleaning system design during *S. plumbizincicola* incineration. With total Cd and Zn concentrations of 152 and 7480 mg kg^{-1} in *S. plumbizincicola*, the corresponding concentrations in control flue gas were 0.101 and 46.4 mg m^{-3} during incineration (Wu et al., 2013a). The addition of aluminum oxide (Al_2O_3), CaO, kaolin, and activated carbon decreased Cd concentrations in flue gas by 33.0%, 45.2%, 91.2%, and 97.6%, respectively, meeting Chinese emission standards. These adsorbents simultaneously removed 21.6%, 50.8%, 88.1%, and 99.1% of Zn from the gas phase. Metal recovery rates from fly ash exceeded those from bottom ash with increasing airflow rates, indicating that both oxidizing and reducing conditions at elevated temperatures promoted the transfer of Cd, Zn, and Pb to the fly ash phase (Zhong et al., 2015). In summary, optimum thermal disposal conditions for *S. plumbizincicola* were established as temperatures exceeded 700°C with an excess air supply ratio of 1.2–1.6.

Pyrolysis represents an alternative thermal treatment for *S. plumbizincicola* biomass. During pyrolysis the tar yield decreased from 6.34% at 650°C to 3.19% at 750°C. The biooil yield increased from 22.0% at 450°C to a maximum of 31.7% at 650°C and declined to 15.9% at 750°C. Char yield showed a consistent decrease from 32.4% at 450°C to 22.5% at 750°C, whereas gas yield remained relatively constant up to 650°C before increasing significantly at 750°C. The chemical composition of pyrolysis oils transitioned from predominantly acidic compounds at 450°C to higher yields of low-oxygen alkanes at 650°C. Kinetic analysis of the pyrolysis process shows activation energies ranging from 146.4 to 232.4 kJ mol^{-1} and frequency factors varying between 1.34×10^{11} and 8.99×10^{15} s^{-1} (Zhong et al., 2016).

Metal concentrations in pyrolysis gas remained below detection limits. At temperatures of 450°C–750°C, >43.6% of Zn was retained in the char, and 54.4% was recovered in biooil at 750°C. More than 87.6% of Cd was found in the biooil. Zn contents in flue gas showed a temperature-dependent increase, but Cu and Cd concentrations showed fluctuating patterns in the entrained flow-tube furnace. Comparative analysis between pyrolysis and incineration in a horizontal tube furnace demonstrates that bottom ash from pyrolysis contained 3.5, 2.7, and 2.3 times the concentrations of Cd, Pb, and Zn, respectively, confirming that oxidizing conditions preferentially promoted the transfer of these metals to the gaseous phase.

20.5.3 Recovery of valuable metals from incineration ash

The high concentrations of Cd and Zn in incineration ash, especially fly ash, present significant potential for valuable metal recovery. Extracting these metals from fly ash not only mitigates environmental risks associated with secondary metal leaching but also enhances the economic viability of phytoextraction.

Fluidized bed incineration cyclone ash was used to evaluate the feasibility of metal recovery from *S. plumbizincicola* incineration ash (Zhao et al., 2022). This study examined factors affecting metal leaching characteristics and assessed the safety of leaching residues. The results reveal that the extractant type and concentration and liquid–solid ratio significantly influenced metal leaching efficiency, whereas leaching time exhibited minimum impact. Acidic extractants (HCl or HNO_3) achieved higher leaching efficiencies than ammonium chloride. Leaching efficiencies of these metals increased proportionally with extractant concentration and liquid/solid ratio, achieving maximum recovery rates of 94.9%, 90.2%, and 92.4%, respectively. The environmental risk from treated residues was effectively decreased by optimizing the liquid/solid ratio. The leaching behavior of Cd, Zn, and Pb in fly ash followed a cationic leaching pattern as a function of pH. The recommended optimum leaching conditions for routine *S. plumbizincicola* fly ash treatment were 1 mol L^{-1} HCl solution, a liquid/solid ratio of 16:1, and a processing time of 30 min. This treatment protocol met landfill requirements while producing leachates suitable for subsequent high-value metal recovery.

20.6 Future prospects for phytoremediation

Further studies are required to explore novel Cd hyperaccumulators and their integration with complementary technologies to enhance phytoremediation efficiency. Although phytoremediation offers environmental, ecological, and economic advantages, significant challenges remain in large-scale application. Few hyperaccumulators or accumulators currently combine both high remediation efficiency and economic value. Genetic engineering and gene editing technologies show some potential for developing novel germplasm resources.

Intercropping and crop rotation systems incorporating *S. plumbizincicola* with cash crops may simultaneously provide soil remediation and safe food production in metal-contaminated areas. Despite the widespread use of *S. plumbizincicola* in phytoextraction, efficient processing of harvested biomass remains challenging. The extraction of valuable metals from biomass may increase the economic viability of phytoremediation while decreasing the environmental risks associated with residual materials.

20.7 Summary

- *Sedum plumbizincicola*, a Cd and Zn hyperaccumulator demonstrating strong adaptability across various climatic zones in China, has potential as an important plant species for the remediation of Cd-contaminated soils.
- Optimized agronomic practices, including cultivation management, intercropping systems, and chemical/microbial amendments, can significantly increase Cd and Zn phytoextraction efficiency using *S. plumbizincicola.*
- Field validation studies confirm that *S. plumbizincicola*-based phytoextraction represents a viable remediation strategy for slightly and moderately Cd-contaminated soils while allowing continuing agricultural productivity.
- The integration of phytoremediation with photovoltaic systems in highly contaminated areas offers a novel land-use approach combining environmental restoration with renewable energy production.
- Thermal treatment (incineration and pyrolysis) and subsequent acid extraction provide effective solutions for both biomass disposal and energy and metal recovery from *S. plumbizincicola.*

Acknowledgment

This work was supported by the National Natural Science Foundation of China (U2002210).

References

Cui, L., Wu, L., Li, N., Li, S., Li, L., Pan, G., & Luo, Y. (2009). Effects of soil moisture on growth and uptake of heavy metals of *Sedum plumbizincicola*. *Soils, 41*(4), 572–576.

Deng, L., Li, Z., Wang, J., Liu, H., Li, N., Wu, L., Hu, P., Luo, Y., & Christie, P. (2016). Long-term field phytoextraction of zinc/cadmium contaminated soil by *Sedum plumbizincicola* under different agronomic strategies. *International Journal of Phytoremediation, 18*(2), 134–140. doi:10.1080/15226514.2015.1058328.

Han, C., Wu, L., Tan, W., & Luo, Y. (2013). Bioavailability and accumulation of cadmium and zinc by *Sedum plumbizincicola* after liming of an agricultural soil subjected to acid mine drainage. *Communications in Soil Science and Plant Analysis, 44*(6), 1097–1105. doi:10.1080/00103624.2012.750333.

Hou, J., Liu, W., Wu, L., Hu, P., Ma, T., Luo, Y., & Christie, P. (2017). Modulation of the efficiency of trace metal phytoremediation by *Sedum plumbizincicola* by microbial community structure and function. *Plant and Soil, 421*(1–2), 285–299. doi:10.1007/s11104-017-3466-8.

Hou, J., Liu, W., Li, Y., Huang, Y., Christie, P., Wu, L., & Luo, Y. (2021). Rhizosphere bacterial community dynamics of the cadmium hyperaccumulator *Sedum plumbizincicola* under a cadmium concentration gradient during phytoextraction. *Plant and Soil, 468*(1–2), 375–388. doi:10.1007/s11104-021-05123-4.

Hou, J., Liu, W., Wu, L., Ge, Y., Hu, P., Li, Z., & Christie, P. (2019). *Rhodococcus* sp. NSX2 modulates the phytoremediation efficiency of a trace metal-contaminated soil by reshaping the rhizosphere microbiome. *Applied Soil Ecology, 133*, 62–69. doi:10.1016/j.apsoil.2018.09.009.

Hu, P., Zhang, Q., Yang, Y., Dong, B., Ying, R., Wu, L., Liu, H., Luo, Y., & Christie, P. (2021). An electro-Fenton process to treat waste liquor of a hyperaccumulator that contains potentially toxic elements and the COD. *International Journal of Phytoremediation, 23*(7), 715–725. doi:10.1080/15226514.2020.1852390.

Hu, P., Wang, Y., Przybyłowicz, W. J., Li, Z., Barnabas, A., Wu, L., Luo, Y., & Mesjasz-Przybyłowicz, J. (2015). Elemental distribution by cryo-micro-PIXE in the zinc and cadmium hyperaccumulator *Sedum plumbizincicola* grown naturally. *Plant and Soil, 388*(1–2), 267–282. doi:10.1007/s11104-014-2321-4.

Hu, P., Zhang, Y., Dong, B., Gao, W., Cheng, C., Luo, Y., Christie, P., & Wu, L. (2019). Assessment of phytoextraction using *Sedum plumbizincicola* and rice production in Cd-polluted acid paddy soils of south China: A field study. *Agriculture, Ecosystems & Environment, 286*, 106651. doi:10.1016/j.agee.2019.106651.

Hu, P., Liang, Y., Wu, L., Jiang, H., & Qiu, D. (2024). *Theory of safe utilization and remediation of heavy metal polluted cultivated land and practice in Jiangsu Province*. Science Press.

Li, Z., Jia, M., Wu, L., Christie, P., & Luo, Y. (2016). Changes in metal availability, desorption kinetics and speciation in contaminated soils during repeated phytoextraction with the Zn/Cd hyperaccumulator *Sedum plumbizincicola*. *Environmental Pollution, 209*, 123–131. doi:10.1016/j.envpol.2015.11.015.

Li, Z., Ding, Y., Ke, X., Wu, L., & Christie, P. (2021). Enhancement by soil micro-arthropods of phytoextraction of metal-contaminated soils using a hyperaccumulator plant species. *Plant and Soil, 464*(1–2), 335–346. doi:10.1007/s11104-021-04958-1.

Li, Z., Wu, L., Hu, P., Luo, Y., Zhang, H., & Christie, P. (2014). Repeated phytoextraction of four metal-contaminated soils using the cadmium/zinc hyperaccumulator *Sedum plumbizincicola*. *Environmental Pollution, 189*, 176–183. doi:10.1016/j.envpol.2014.02.034.

Li, Z., Sun, X., Zhou, J., Wu, L., Bi, D., Zhao, J., Zhu, R., & Christie, P. (2023). Sustainable phytoextraction of metal-polluted agricultural land used for commercial photovoltaic power generation. *Journal of Cleaner Production, 391*, 136093. doi:10.1016/j.jclepro.2023.136093.

Li, N., Wu, L., Luo, Y., Tang, M., Tan, C., Jiang, Y., He, X., & Teng, C. (2009). Effects of harvesting way of *Sedum plumbizincicola* on its zinc and cadmium uptake in a mixed heavy metal contaminated soil. *Acta Pedologica Sinica, 46*(4), 725–728.

Li, N., Tang, M., Cui, L., Wu, L., & Luo, Y. (2010). Effects of light intensity on plant growth and zinc and cadmium uptake by *Sedum plumbizincicola*. *Acta Pedologica Sinica, 47*(2), 370–373.

Li, S., Gu, X., Zhou, J., Wu, L., & Christie, P. (2024). Prediction of cadmium and zinc phytoextraction by the hyperaccumulator *Sedum plumbizincicola* using a dynamic geochemical mechanical combination model. *Science of the Total Environment, 906*, 167627. doi:10.1016/j.scitotenv.2023.167627.

Li, Z., Wu, L., Hu, P., Luo, Y., & Christie, P. (2013). Copper changes the yield and cadmium/zinc accumulation and cellular distribution in the cadmium/zinc hyperaccumulator *Sedum plumbizincicola*. *Journal of Hazardous Materials, 261*, 332–341. doi:10.1016/j.jhazmat.2013.07.050.

Liu, H., Zhao, H., Wu, L., Liu, A., Zhao, F. J., & Xu, W. (2017). Heavy metal ATPase 3 (HMA3) confers cadmium hypertolerance on the cadmium/zinc hyperaccumulator *Sedum plumbizincicola*. *New Phytologist, 215*(2), 687–698. doi:10.1111/nph.14622.

Liu, L., Wu, L. H., Li, N., Cui, L. Q., Li, Z., Jiang, J. P., Jiang, Y. G., Qiu, X. Y., & Luo, Y. M. (2009). Effect of planting densities on yields and zinc and cadmium uptake by *Sedum plumbizincicola*. *Huanjing Kexue/Environmental Science, 30*(11), 3422–3426.

McGrath, S. P., Zhao, J., & Lombi, E. (2002). Phytoremediation of metals, metalloids, and radionuclides. *Advances in Agronomy, 75*, 1–56. doi:10.1016/s0065-2113(02)75002-5.

Shen, L. B., Wu, L. H., Tan, W. N., Han, X. R., Luo, Y. M., Ouyang, Y. N., Jin, Q. Y., & Jiang, Y. G. (2010). Effects of *Sedum plumbizincicola—Oryza sativa* rotation and phosphate amendment on Cd and Zn uptake by *O.* sativa. *Chinese Journal of Applied Ecology, 21*(11), 2952–2958.

Singh, S., Tripathi, D. K., Singh, S., Sharma, S., Dubey, N. K., Chauhan, D. K., & Vaculík, M. (2017). Toxicity of aluminium on various levels of plant cells and organism: A review. *Environmental and Experimental Botany, 137*, 177–193. doi:10.1016/j.envexpbot.2017.01.005.

Wang, P., Hu, P., Zhong, D., Chen, Y., & Wu, L. (2017). Rapid disposal technology for fresh biomass of harvested Cd/Zn hyperaccumulator *Sedum plumbizincicola*. *Chinese Journal of Environmental Engineering, 11*(9), 5307–5312. doi:10.12030/j.cjee.201610116.

Wang, Y., Zhou, T., Wang, W., Zhao, J., Li, Z., Ge, Y., Wang, Z., Wu, L., & Christie, P. (2023). Phytoextraction of highly cadmium-polluted agricultural soil by *Sedum plumbizincicola*: An eight-hectare field study. *Science of the Total Environment, 905*, 167216. doi:10.1016/j.scitotenv.2023.167216.

Wu, G., Hu, P., Zhou, J., Dong, B., Wu, L., Luo, Y., & Christie, P. (2019). Sulfur application combined with water management enhances phytoextraction rate and decreases rice cadmium uptake in a *Sedum plumbizincicola—Oryza sativa* rotation. *Plant and Soil, 440*(1–2), 539–549. doi:10.1007/s11104-019-04095-w.

Wu, L., Li, Z., Akahane, I., Liu, L., Han, C., Makino, T., Luo, Y., & Christie, P. (2012a). Effects of organic amendments on Cd, Zn and Cu bioavailability in soil with repeated phytoremediation by *Sedum plumbizincicola*. *International Journal of Phytoremediation, 14*(10), 1024–1038. doi:10.1080/15226514.2011.649436.

Wu, L., Zhong, D., Du, Y., Lu, S., Fu, D., Li, Z., Li, X., Chi, Y., Luo, Y., & Yan, J. (2013a). Emission and control characteristics for incineration of *Sedum Plumbizincicola* biomass in a laboratory-scale entrained flow tube furnace. *International Journal of Phytoremediation, 15*(3), 219–231. doi:10.1080/15226514.2012.687021.

Wu, L., Zhou, J., Zhou, T., Li, Z., Jiang, J., Zhu, D., Hou, J., Wang, Z., Luo, Y., & Christie, P. (2018). Estimating cadmium availability to the hyperaccumulator *Sedum plumbizincicola* in a wide range of soil types using a piecewise function. *Science of the Total Environment, 637–638*, 1342–1350. doi:10.1016/j.scitotenv.2018.04.386.

Wu, L., Li, Z., Han, C., Liu, L., Teng, Y., Sun, X., Pan, C., Huang, Y., Luo, Y., & Christie, P. (2012b). Phytoremediation of soil contaminated with cadmium, copper and polychlorinated biphenyls. *International Journal of Phytoremediation, 14*(6), 570–584. doi:10.1080/15226514.2011.619227.

Wu, L. H., Liu, Y. J., Zhou, S. B., Guo, F. G., Bi, D., Guo, X. H., Baker, A. J. M., Smith, J. A. C., & Luo, Y. M. (2013b). *Sedum plumbizincicola* X.H. Guo et S.B. Zhou ex L.H. Wu (*Crassulaceae*): A new species from Zhejiang Province, China. *Plant Systematics and Evolution, 299*(3), 487–498. doi:10.1007/s00606-012-0738-x.

Wu, T., Zhan, J., Zhou, J., Zhou, T., Li, Z., Hu, P., Huang, H., Li, T., Zhang, X., & Luo, Y. (2024). Phytoextraction of the soils from the east of Yunnan Province with a high cadmium geological background and its effect on rice cadmium uptake. *Acta Pedologica Sinica, 61*(5), 1349–1359. doi:10.11766/trxb202303060092.

Wu, L., Zhou, S., Bi, D. e, Guo, X., Qin, W., Wang, H., Wang, C., & Luo, Y. (2006). *Sedum plumbizincicola*, a new species of the *Crassulaceae* from Zhejiang. *Soils, 38*(5), 632–633.

Wu, L., Zhou, T., Hu, P., Zhao, J., Jiang, Y., & Liu, D. (2021). *Cultivation and management of* Sedum plumbizincicola *– A cadmium and zinc hyperaccumulator*. Science Press.

Zhao, B., Shen, L. B., Cheng, M. M., Wang, S. F., Wu, L. H., Zhou, S. B., & Luo, Y. M. (2011). Effects of intercropping *Sedum plumbizincicola* in wheat growth season under wheat-rice rotation on the crops growth and their heavy metals uptake from different soil types. *Chinese Journal of Applied Ecology, 22*(10), 2725–2731.

Zhao, B., Zhong, D. X., Zhang, Y. P., & Wu, L. H. (2022). Leaching of potentially toxic elements and safe treatment of fly ash following incineration of the cadmium and zinc hyperaccumulator *Sedum plumbizincicola*. *Chinese Journal of Ecology, 41*(1), 116–123. doi:10.13292/j.1000-4890.202201.007.

Zhao, H., Wang, L., Zhao, F. J., Wu, L., Liu, A., & Xu, W. (2019). SpHMA1 is a chloroplast cadmium exporter protecting photochemical reactions in the Cd hyperaccumulator *Sedum plumbizincicola*. *Plant Cell and Environment, 42*(4), 1112–1124. doi:10.1111/pce.13456.

Zhong, D., Zhong, Z., Wu, L., Ding, K., Luo, Y., & Christie, P. (2016). Pyrolysis of *Sedum plumbizincicola*, a zinc and cadmium hyperaccumulator: Pyrolysis kinetics, heavy metal behaviour and bio-oil production. *Clean Technologies and Environmental Policy, 18*(7), 2315–2323. doi:10.1007/s10098-016-1150-y.

Zhong, D. X., Zhong, Z. P., Wu, L. H., Xue, H., Song, Z. W., & Luo, Y. M. (2015). Thermal characteristics and fate of heavy metals during thermal treatment of *Sedum plumbizincicola*, a zinc and cadmium hyperaccumulator. *Fuel Processing Technology, 131*, 125–132. doi:10.1016/j.fuproc.2014.11.022.

Zhou, J., Li, Z., Zhou, T., Xin, Z., Wu, L., Luo, Y., & Christie, P. (2020). Aluminum toxicity decreases the phytoextraction capability by cadmium/zinc hyperaccumulator *Sedum plumbizincicola* in acid soils. *Science of the Total Environment, 711*. doi:10.1016/j.scitotenv.2019.134591.

Zhou, J. W., Wu, L. H., Zhou, T., Li, Z., Sun, X. Y., Luo, Y. M., & Christie, P. (2019). Comparing chemical extraction and a piecewise function with diffusive gradients in thin films for accurate estimation of soil zinc bioavailability to *Sedum plumbizincicola*. *European Journal of Soil Science, 70*(6), 1141–1152. doi:10.1111/ejss.12810.

Zhou, T., Zhu, D., Wu, L., Xing, W., Luo, Y., & Christie, P. (2018). Repeated phytoextraction of metal contaminated calcareous soil by hyperaccumulator *Sedum plumbizincicola*. *International Journal of Phytoremediation, 20*(12), 1243–1249. doi:10.1080/15226514.2016.1156641.

Index

T

V

W

X

Z

Printed and bound by CPI Group (UK) Ltd, Croydon, CR0 4YY
14/09/2026
02216408-0014